EUROPEAN MINERALOGICAL UNION
NOTES IN MINERALOGY

Commissioning Editor: R. Oberti

Volume 17

REDOX-REACTIVE MINERALS: PROPERTIES, REACTIONS AND APPLICATIONS IN CLEAN TECHNOLOGIES

UNIVERSITY TEXTBOOK

Edited by
I.A.M. AHMED AND K.A. HUDSON-EDWARDS

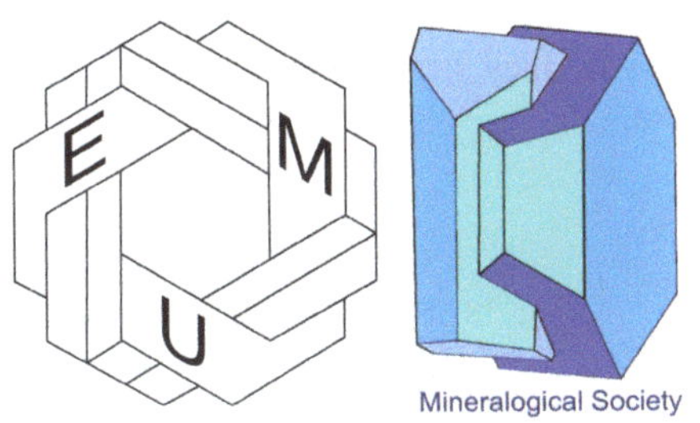

Published by the European Mineralogical Union and the
Mineralogical Society of Great Britain & Ireland,
London, 2017

The publication of this textbook is supported by the European Mineralogical Union

Editors of this Volume
I.A.M. Ahmed, Department of Earth Sciences, University of Oxford, UK and
K.A. Hudson-Edwards, Department of Earth and Planetary Sciences,
Birkbeck University of London, UK

Managing Editor and Indexer: Kevin Murphy, London

Front cover design: Michel H. Guay

On the front cover: SEM image of hexagonal Fe(II,III) hydroxysulfate green rust particles (500–800 nm) as they grow from poorly ordered crystallized phases in an oxygen-free atmosphere.

ISSN: 1417 2917
ISBN: 978-0903056-57-1

Published by the European Mineralogical Union and the Mineralogical Society of Great Britain & Ireland (12, Baylis Mews, Amyand Park Road, Twickenham TW1 3HQ, UK)
Printed by Lightning Source, USA

The EMU Notes in Mineralogy Series

Published volumes

Volume	*Year*	*Editors*	*Title*
1	1997	S. Merlino	*Modular aspects of minerals*
2	2000	D.J. Vaughan R. Wogelius	*Environmental mineralogy*
3	2001	C.A. Geiger	*Solid solutions in silicate and oxide systems*
4	2002	C. Gramaccioli	*Energy modelling in minerals*
5	2003	D.A. Carswell R. Compagnoni	*Ultrahigh pressure metamorphism*
6	2004	A. Beran E. Libowitzky	*Spectroscopic methods in mineralogy*
7	2005	R. Miletich	*Mineral behaviour at extreme conditions*
8	2010	F.E. Brenker G. Jordan	*Nanoscopic approaches in earth and planetary sciences*
9	2010	G. Christidis	*Advances in the characterization of industrial minerals*
10	2010	M. Prieto	*Ion-partitioning in ambient-temperature aqueous systems*
11	2011	M.F. Brigatti A. Mottana	*Layered mineral structures and their application in advanced technologies*
12	2012	J. Dubessy M.-C. Caumon F. Rull	*Applications of Raman Spectroscopy to earth sciences and cultural heritage*
13	2013	D.J. Vaughan R.A. Wogelius	*Environmental mineralogy II*
14	2013	F. Nieto K.J.T. Livi	*Minerals at the nanoscale*
15	2015	M.R. Lee H. Leroux	*Planetary mineralogy*
16	2017	W. Heinrich R. Abart	*Mineral reaction kinetics: microstructures, textures, chemical and isotopic signatures*
17	2017	I.A.M. Ahmed K.A. Hudson-Edwards	*Redox-reactive minerals: properties, reactions and applications in clean technologies*

Copies of the EMU Notes (volumes 1–7) are distributed in Europe by the larger member societies of the European Mineralogical Union:

Società Italiana di Mineralogia e Petrologia:
www.socminpet.it

Mineralogical Society of Great Britain & Ireland:
www.minersoc.org

Société Française de Minéralogie et de Cristallographie:
www.sfmc-fr.org

and by the Secretary of the EMU, Prof. Juraj Majzlan:
juraj.majzlan@uni-jena.de

in America by the Mineralogical Society of America:
www.minsocam.org

Institutional orders as well as individual requests from outside Europe and America should be sent to the Secretary of EMU, Prof. Juraj Majzlan:
juraj.majzlan@uni-jena.de

For volumes 8 onwards, the Mineralogical Society of Great Britain acts as co-publisher and copies may be ordered from www.minersoc.org

or from:

Mineralogical Society
12 Baylis Mews,
Amyand Park Road,
Twickenham TW1 3HQ
UK

E-mail: admin@minersoc.org
Tel. + 44 (0)20 8891 6600
Fax: + 44 (0)20 8891 6599

Contents

Chapter 1. Introduction to redox-reactive minerals in natural systems and clean technologies . . . 1
by I. A. M. Ahmed and K. A. Hudson-Edwards

Recommended reading . . . 3

Chapter 2. Influence of iron oxide structure and size on redox reactivity . . . 5
by C. Chanéac and J.-P. Jolivet

1. Introduction . . . 5

2. Main size effects on oxide nanoparticle properties . . . 6
   2.1. Thermodynamic stability . . . 6
   2.2. Crystal structure and structural defects . . . 8
   2.3. Electronic structure . . . 10

3. Crystal structure and formation of iron oxides with the control of nanoparticle size . . . 11

4. Examples of size effect on the reactivity of iron oxide nanoparticles . . . 15
   4.1. Reactivity towards proton adsorption . . . 15
   4.2. Redox transformations induced by Fe(II) adsorption . . . 18
   4.3. Redox catalysis . . . 22
   4.4. Size-dependent reduction reactions induced by adsorption onto magnetite. 24
   4.5. Bioreduction . . . 25

5. Conclusions . . . 26

Acknowledgements . . . 26

References . . . 27

Chapter 3. Mössbauer and Raman spectroscopic *in situ* characterization of iron-bearing minerals in Mars: Exploration and cultural heritage . . . 33
by G. Klingelhöfer, F. Rull, G. Venegas, F. Gázquez and J. Medina

1. Introduction . . . 33

2. Instrument descriptions . . . 38
   2.1. The Mössbauer instrument MIMOS II . . . 38
   2.2. The Raman spectrometer for the Exomars 2018 mission . . . 39

3. Rio Tinto and Jaroso Ravine Mars analogue sites . . . 41

4. Results and discussion . . . 42

5. Historical and cultural heritage . . . 46
   5.1. Archaeological artefacts analysed by Mössbauer spectroscopy . . . 47
   5.2. Rock paintings in the field . . . 48

6. Concluding remarks. 49
Acknowledgements 51
References 51

Chapter 4. The timescales of mineral redox reactions. 55
by B. Gilbert and G. A. Waychunas

1. Introduction 55
   1.1. Timescales relevant to geochemical reaction kinetics 58
   1.2. Thermodynamic basis for mineral redox reactions 60
   1.3. Mechanistic descriptions of mineral redox reactions 61
      1.3.1. Reductive dissolution of iron(III) oxides 63
      1.3.2. Additional features of mineral redox reactions 64
2. Chemical kinetics analysis of mineral redox reactions 65
   2.1. Experimental determination of reaction rates 65
      2.1.1. Initiating reactions by reagent mixing 65
      2.1.2. Measuring reaction progress 65
      2.1.3. Electrochemical experiments 66
   2.2. Activation barriers in mineral redox reactions 67
      2.2.1. Arrhenius equation 69
      2.2.2. Marcus equation for electron transfer. 69
      2.2.3. Interfacial electron transfer. 73
   2.3. Free energy relationships 73
      2.3.1. Linear free energy relationships 74
      2.3.2. Marcus theory free energy relationships. 75
   2.4. Additional constraints on the rates of multistep reactions. 75
      2.4.1. Multi-electron redox reactions 76
      2.4.2. Formation of radical species 76
      2.4.3. Redox reactions coupled to precipitation 77
3. Chemical dynamics methods for studying mineral-reaction intermediates 80
   3.1. The pump-probe method 80
      3.1.1. Optical pump-probe methods 81
      3.1.2. X-ray pump-probe methods 82
   3.2. Other time-resolved methods 82
      3.2.1. Relaxation methods 82
      3.2.2. Time-resolved electrochemical methods. 83
   3.3. Pump-probe studies of photochemical reactions 84
   3.4. Initiating chemical reactions with light 84
      3.4.1. Acid-base reactions 84
      3.4.2. Electron-transfer reactions 84
4. Outlook 87
Acknowledgements 87
References 87

Chapter 5. Biogeochemical redox processes of sulfide minerals 95
by D. J. Vaughan and V. S. Coker

1. Introduction 95

2. Structures and compositions 96

3. Formation of sulfide minerals through biological processes 99
   3.1. Microbial sulfate reduction 100
   3.2. Microbial metal reduction 101
   3.3. Iron sulfide formation in marine sediments 102

4. Dissolution of sulfide minerals through biological processes 104
   4.1. Pyrite oxidation 105
   4.2. Galena oxidation 107
   4.3. Arsenopyrite oxidation 108
   4.4. Chalcopyrite oxidation 109

5. Toxic metals bound to sulfide mineral surfaces 109
   5.1. Uptake by pyrite 110
   5.2. Uptake by mackinawite 111

6. Applications in clean technologies 113

References 114

Chapter 6. Iron minerals as archives of Earth's redox and biogeochemical evolution 121
by C. L. Peacock, S. V. Lalonde and K. Konhauser

1. Introduction 121

2. Iron minerals as archives of chemical information 122
   2.1. Iron mineralogy as a function of redox 122
   2.2. Uptake of trace elements by iron minerals 124
      2.2.1. Iron oxyhydroxides 124
      2.2.2. Iron sulfides 126
      2.2.3. Exemplar molybdenum 128
   2.3. Iron mineral archives 129
      2.3.1. Banded iron formations 130
      2.3.2. Ferromanganese crusts 131
      2.3.3. Black shales 133

3. Stable isotope fractionation of trace elements 133
   3.1. Stable isotope nomenclature and conventions 134
   3.2. Stable isotope fractionation theory 135
      3.2.1. Mass-dependent stable isotope fractionation 135
      3.2.2. Fractionation factors 136
      3.2.3. Processes producing isotopically distinct reservoirs 137
   3.3. Predicted and measured fractionations of trace elements in geological systems 139
   3.4. Emerging non-traditional stable isotope systems 140
      3.4.1. Iron 140

3.4.2. Molybdenum. 142
3.4.3. Chromium 144

4. Iron-based records of Earth's redox and biogeochemical evolution 145
4.1. Banded iron formations (BIF). 145
4.1.1. Rare earth elements 145
4.1.2. Trace element isotope compositions 147
4.1.3. Trace-element concentrations 150
4.2. Ferromanganese crusts. 155
4.3. Black shales 156
4.3.1. Trace element concentrations 156
4.3.2. Molybdenum isotopes in shales. 158

5. Conclusions and future directions. 159

Acknowledgements 160

References 160

Chapter 7. Formation of manganese oxide minerals by bacteria. 173
by S.-W. Lee, M. Jones, C. Romano and B. M. Tebo

1. Introduction 173

2. Mn(II) oxidation 175
2.1. Mn(II)-oxidizing bacteria. 175
2.2. Mn(II) oxidases 177
2.2.1. MCOs of Bacillus species. 178
2.2.2. MCOs in other Mn(II)-oxidizing bacteria 180
2.2.3. Mn(II) oxidation mediated by Ca^{2+}-binding animal heme peroxidases 182
2.2.4. Combined enzymatic and non-enzymatic Mn(II) oxidation. 184
2.2.5. Bacteriogenic Mn oxides 184

3. Environmental applications of Mn-oxidizing enzymes and bacteriogenic Mn oxides 188

4. Conclusions 189

Acknowledgements 190

References 190

Chapter 8. Bioconversion of Fe(III) oxides into magnetic nanoparticles: Processes and applications. 197
by V. S. Coker, M. P. Watts and J. R. Lloyd

1. Introduction 197

2. Redox cycling 198
2.1. Fe(III) reduction 198
2.1.1. Mechanisms of bacterial Fe(III) reduction 200
2.2. Fe(II) oxidation. 201

3. Primary and secondary mineral phases 203

3.1. Primary Fe(III) phases 203
3.2. Magnetite 204
3.3. Greigite 206

4. Tailoring magnetic minerals 206

5. Applications of magnetic biominerals 207
5.1. Magnetite–contaminant interactions 208
5.1.1. Chromium(VI) 208
5.1.2. Mercury(II) 210
5.1.3. Arsenic(V) 210
5.1.4. Uranium(VI) 211
5.1.5. Technetium(VII) 213
5.1.6. Neptunium(V) 214
5.1.7. Plutonium(V) and plutonium(VI) 214
5.1.8. Organic compounds 215
5.2. Biomedical applications 217

6. Conclusions 218

Acknowledgements 218

References 218

Chapter 9. Impact of iron redox chemistry on nuclear waste disposal 229
by C. I. Pearce, K. M. Rosso, R. A. D. Pattrick and A. R. Felmy

1. Introduction and importance 229

2. Post-closure redox conditions in a GDF 234

3. Redox-active Fe components in the near field 236
3.1. Corrosion behaviour of Fe-bearing waste canisters 236
3.2. Corrosion of Fe-bearing phases at the interface between the waste package and the buffer material 238
3.3. Effect of microbial activity on Fe-bearing phases in the near field 241
3.4. Effect of radiation damage on Fe-bearing phases in the near field 241

4. Fe-bearing minerals in candidate far-field lithologies 242

5. Radionuclide interaction with redox-active Fe components: natural samples and synthetic analogues 251

6. Radionuclide interaction with redox-active Fe components: Thermodynamic and molecular modelling 258

7. Future research needs 264

Acknowledgements 265

References 265

Chapter 10. Biogeochemical modelling of redox processes in low-temperature natural systems **273**
by I. A. M. Ahmed, P. Acero and L. F. Auqué

1. Introduction 273

2. Theoretical background 276
   2.1. Equilibrium redox potentials 276
   2.2. Eh-pH stability diagrams and the range of redox potentials in natural systems 278
   2.3. Modelling redox reactions 281

3. Redox equilibrium and disequilibrium in natural systems. 282
   3.1. Estimation of the redox potential of natural systems and its use in equilibrium calculations 284
   3.2. Main redox reactions in natural systems. 287
   3.3. The role of biological processes in redox reactions 292
   3.4. Redox zonations 294
   3.5. Geochemical modelling of redox transformations 295
      3.5.1. Redox equilibrium *vs.* kinetic assumption. 295
      3.5.2. Some applications of geochemical modelling of redox processes . . 298

4. General limitations of the geochemical modelling of redox processes and final remarks 306

References 306

Chapter 11. Breakdown of organic contaminants in soil by manganese oxides: a short review **313**
by K. L. Johnson, C. M. McCann and C. E. Clarke

1. Introduction 314
   1.1. Overview 314
   1.2. Soil contamination 314
   1.3. Organic contaminants in soils. 317
      1.3.1. Phenols 318
      1.3.2. Polychlorinated biphenyls (PCBs) 319
      1.3.3. N-containing aromatic compounds 320
      1.3.4. Polycyclic aromatic hydrocarbons (PAHs) 320

2. Manganese oxides and natural organic matter 321
   2.1. Manganese oxide geochemistry 321
   2.2. Natural Organic Matter (NOM). 324
      2.2.1. Background 324
      2.2.2. Dissolved organic carbon (DOC). 325
      2.2.3. Interactions between Mn oxides and DOC 325

3. Interaction mechanisms between Mn oxides and synthetic organic contaminants 329
   3.1. Background 329
      3.1.1. Sorption 332
      3.1.2. Oxidation 334

3.1.3. Polymerization 335
3.2. Phenols 335
3.2.1. Halogenated phenols 338
3.2.2. Polychlorinated biphenyls 339
3.3. N-containing compounds 340
3.3.1. Azo compounds. 341
3.4. Polycyclic aromatic hydrocarbons. 342
4. Summary 346
5. Future outlook 347
Acknowledgements 348
References 348

Chapter 12. Role of redox-reactive minerals in the reuse and remediation of mine wastes 357
by K. A. Hudson-Edwards and D. Kossoff

1. Introduction 357
2. Reuse of mine wastes containing redox-reactive minerals. 359
2.1. Definition of, and potential problems associated with, reuse of mine wastes 359
2.2. Industrial reuse of mine waste 360
2.3. Mine wastes as sources of plant nutrients. 362
2.4. Water purification 363
3. Phytostabilization of mine waste 364
4. Remediation of mine waste waters. 365
4.1. Abiotic systems: lime addition and limestone drains 366
4.2. Biological systems 368
4.2.1. Off-line sulfidic bioreactors 368
4.2.2. Permeable reactive barriers. 368
4.2.3. Aerobic wetlands. 370
4.2.4. Anaerobic composter reactors and wetlands 370
4.2.5. Packed bed iron-oxidation bioreactors 371
5. Conclusions 371
References 372

Chapter 13. Functionality and applications of redox-active porous metal–organic framework structures 379
by T. Devic and C. Serre

1. Introduction 379
2. Nature of the redox-active species 380
3. Synthesis issues and electrochemical characterization 382
3.1. Conventional synthesis 382
3.2. Electrochemically assisted synthesis 383

3.3. Electrochemical characterization 384

4. Redox-driven insertion.. 384
4.1. Insertion of cationic species through the reduction of the framework 385
4.2. Insertion of neutral species into pores through the oxidation of the framework 386
4.3. Insertion of anionic species through the oxidation of the framework.... 386

5. Applications 387
5.1. Enhanced gas storage or capture 387
5.2. MOFs as electrodes for Li-ion batteries 390
5.3. MOFs as supercapacitors 392
5.4. Electronic conductivity 392
5.5. Catalysis 393
5.6. Photo-induced redox activity 395

6. Conclusion 396

Acknowledgements 396

References 396

Chapter 14. Removal of arsenic, nitrate, persistent organic pollutants and pathogenic microbes from water using redox-reactive minerals 405
by A. A. Bogush, J. K. Kim and L. C. Campos

1. Introduction 405
1.1 Properties of some redox-reactive minerals......................... 407

2. Removal of arsenic.. 414
2.1. Chemical precipitation 415
2.2 Oxidation and adsorption 418
2.3. *In situ* treatment technologies................................. 421

3. Removal of nitrate 423
3.1. Application of iron-bearing minerals........................... 423
3.2. Electrochemical methods 425

4. Removal of persistent organic pollutants............................ 427
4.1. Conventional treatment technologies for POPs 427
4.2. Reduction and oxidation process for POPs........................ 428

5. Removal of bacteria and other microorganisms........................ 431
5.1. Effects of redox compounds on health and DBP production 432

6. Conclusions 432

Acknowledgements 433

References 433

Index 443

EMU Notes in Mineralogy, Vol. 17 (2017), Chapter 1, 1–4

Introduction to redox-reactive minerals in natural systems and clean technologies

IMAD A. M. AHMED[1] and KAREN A. HUDSON-EDWARDS[2]

[1] *Department of Earth Sciences, University of Oxford, South Parks Road, Oxford OX1 3AN, UK, e-mail: imad.ahmed@earth.ox.ac.uk*
[2] *Department of Earth and Planetary Sciences, Birkbeck, University of London, Malet St., London WC1E 7HX, UK, e-mail: k.hudson-edwards@bbk.ac.uk*

Minerals are naturally occurring inorganic solids that make up the solid part of most solar terrestrial planets. Redox-active elements such as iron, manganese, titanium and sulfur, in these minerals allow them to engage in a wide range of electron-transfer reactions including those mediated by biota or processes involved in palaeo-weathering and biogeochemical cycling. The importance of redox-reactive minerals in many natural and industrial processes has been demonstrated by a plethora of scientific publications and industrial applications in recent decades. In this book, the influence of redox-reactive minerals on key biogeochemical processes and opportunities for their application in environmental technologies are outlined and illustrated in 14 comprehensive chapters. The book will be a key reference for Earth science students, geologists, geochemists and engineers and other researchers and practitioners in this rapidly growing interdisciplinary field.

For several centuries, mineralogy has been dealing with the occurrence, classification and description of materials that appear in nature as rocks. Recent advances in analytical techniques, such as X-ray and neutron diffraction, and computational power have enabled accurate identification of crystal structure and atomic-scale simulation of mineral behaviour. These advances have led to greater understanding of the relationships between the atomic-scale structure of minerals and their function under different environmental conditions. Modern mineralogy has opened frontiers and given inspiration to material chemists in the synthesis of new materials for clean-energy generation, environmental remediation and energy-efficiency technologies. Redox-reactive minerals are among the most attractive materials in many industries including waste minimization and recycling, reduction of atmospheric pollution, carbon sequestration and novel energy storage using materials such as electronic ceramics. Redox-reactive minerals are also abundant in nature, occurring in environments such as aquatic sediments, hydromorphic soils, sewage sludge, waterlogged peat soils, hypolimnia of stratified lakes, sediments of eutrophic rivers and seafloor hydrothermal vents, to name just a few. A molecular-level understanding of the electron-transfer reactions is the key to innovate many society-formative and clean technologies

DOI: 10.1180/EMU-notes.17.1

including ore processing, waste recycling and environment protection which are based on stringent control of interfacial processes.

The objectives of this book are: (1) to outline the importance and complexity of interfacial electron-transfer reactions in geoenvironments; (2) to critically review the recent developments and future trends in the identification, characterization and study of the formation and reactivity of naturally occurring, chemically or biologically engineered redox-reactive minerals; and (3) to highlight key practical innovations and potential applications of redox-reactive minerals in environmental clean technologies and designing engineered remediation systems. The book also aims to serve the needs of undergraduate and postgraduate students who have groundings in the principles of geochemistry and materials science, to allow environmental scientists, chemists, engineers and material scientists to access and and utilize fields related to clean materials technology or environmental materials and allow policy makers and more specialist groups (*e.g.* inorganic chemists and materials R&D departments) to gain information on this topic, especially from an environmental and mineralogical perspective.

The chapters of the book have been grouped under two themes. The first focuses on redox-reactive minerals in natural systems. The structure and size of iron oxides (C. Chanéac and J.P. Jolivet), spectroscopic characteristics of iron-bearing minerals (G. Klingelhöfer, F. Rull, G. Venegas, F. Gázquez and J. Medina), timescales of mineral redox reactions (B. Gilbert and G.A. Waychunas), biogeochemical redox processes of sulfide minerals (D.J. Vaughan and V.S. Coker), geochemical proxies for biogeochemical cycling and ocean anoxia (C.L. Peacock, A. Lalonde and K. Konhauser) and formation of manganese oxide minerals by bacteria (S.-W. Lee, M. Jones, J. Romano and B.M. Tebo). The succeeding seven chapters deal with clean technologies and engineered systems, namely nano-sized magnetic materials (V.S. Coker, M. Watts and J.R. Lloyd), nuclear waste disposal (C.I. Pearce, K.M. Rosso, R.A.D. Pattrick and A.R. Felmy), modelling of low-temperature redox systems (I.A.M. Ahmed, P. Acero and L.F. Auqué), breakdown of organic contaminants in soil (K.L. Johnson, C.M. McCann and C.E. Clarke), the recycling and remediation of mine wastes (K.A. Hudson-Edwards and D. Kossoff), porous metal-organic framework structures (T. Devic and C. Serre) and the removal of arsenic, iron, manganese and other pollutants from waters (A.A. Bogush, J.K. Kim and L.C. Campos). While this is not a comprehensive list, we hope that the examples used illustrate the complex properties and widely varying applications of redox-reactive minerals.

In closing this introductory chapter, we note the significant implications of mineral crystal structure and particle size on their electrochemical properties and in determining their overall chemical and biological reactivity and function. Mineral redox reactions are complex, and studying them can be challenging, especially for those occurring at small timescales. Gilbert and Waychunas (Chapter 4) provide an excellent example of the insights gained by comparing molecular-scale and mineral redox reactions and the potential of employing ultrafast time-resolved methods in characterizing the steps and mechanisms involved in complex geochemical processes. While this book has, unintentionally, focused on iron and manganese minerals, other natural redox-reactive minerals should be considered. Synthetic redox-reactive solids form an extensive class of materials used industrially to overcome diverse challenges in the realms of energy, health and pollution. One important example of this class of materials is the metal–organic frameworks whose structures and applications are reviewed in Chapter 13 by Devic and Serre.

Redox-reactive materials have undoubtedly attracted enormous interest over the past several decades and we anticipate substantial growth in applied and basic research in this field in the future, especially in areas related to molecular-scale understanding of biogeochemical cycling and palaeoenvironmental change, mechanisms of biochemical weathering and environmental toxicology of nanoparticles. Further industrial innovations that utilize redox-reactive materials are also anticipated, particularly in addressing waste (*e.g.* nuclear, mining and chemical), energy and health challenges.

It is hoped that this volume may provide a useful reference source and a picture of the present status of the chemistry, geochemistry and mineralogy of redox-reactive materials. Although in this volume some progress has been made in this direction, this aim is as yet by no means achieved, especially with this extremely broad, diverse and vibrant area of research. We are indebted to the thirty-one expert reviewers for their time and effort reviewing the individual chapters, to all authors in this volume, to Roberta Oerti for support from the European Mineralogical Union in getting the project started and to Kevin Murphy for much devoted work in compiling this volume.

Recommended reading

Arumugam, K. and Becker, U. (2014) Computational redox potential predictions: applications to inorganic and organic aqueous complexes, and complexes adsorbed to mineral surfaces. *Minerals*, **4**, 345–387.

Ahmed, I.A.M., Benning, L.G., Kakonyi, G., Sumoondur, A.D., Terrill, N.J. and Shaw, S. (2010) Formation of green rust sulfate: A combined *in situ* time-resolved X-ray scattering and electrochemical study. *Langmuir*, **26**, 6593–6603.

Birkner, N. and Navrotsky, A. (2014) Rapidly reversible redox transformation in nanophase manganese oxides at room temperature triggered by changes in hydration. *Proceedings of the National Academy of Science USA*, **111**, 6209–6214.

Charlet, L., Scheinost, A.C., Tournassat, C., Greneche, J.M., Géhin, A., Fernández-Martínez, A., Coudert, S., Tisserand, D. and Grendle, J. (2007) Electron transfer at the mineral/water interface: Selenium reduction by ferrous iron sorbed on clay. *Geochimica et Cosmochimica Acta*, **71**, 5731–5749.

Cornell, R.M. and Schwertmann, U. (2003) *The Iron Oxides: Structure, Properties, Reactions, Occurrences and Uses*. Wiley-VCH, Weinheim, Germany.

Dehouck, E., Gaudin, A., Chevrier, V. and Mangold, N. (2016) Mineralogical record of the redox conditions on early Mars. *Icarus*, **271**, 1–9.

Faivre, D. and Frankel, R.B. (2016) *Iron Oxides: From Nature to Applications*. Wiley, Chichester, UK.

Fernández-García, M., Martínez-Arias, A., Hanson, J.C. and Rodriguez, J.A. (2004) Nanostructured oxides in chemistry: Characterization and properties. *Chemical Reviews*, **104**, 4063–4104.

Hudson-Edwards, K.A. (2016) Tackling mine wastes. *Science*, **352(6283)**, 288–290.

Jolivet, J.P., Chanéac, C. and Tronc, É. (2004) Iron oxide chemistry. From molecular clusters to extended solid networks. *Chemical Communications*, **5**, 477–483.

Kille, P., Morgan, A.J., Powell, K., Mosselmans, J.F.W., Hart, D., Gunning, P., Hayes, A., Scarborough, D., McDonald, I. and Charnock, J.M. (2016) 'Venus trapped, Mars transits': Cu and Fe redox chemistry,

cellular topography and in situ ligand binding in terrestrial isopod hepatopancreas. *Open Biology*, **6**, 150270.
Maher, B.A. and Taylor, R.M. (1988) Formation of ultrafine-grained magnetite in soils. *Nature*, **336(6197)**, 368–370.
Navrotsky, A., Mazeina, L. and Majzlan, J. (2008) Size-driven structural and thermodynamic complexity in iron oxides. *Science*, **319(5870)**, 1635–1638.
Nieto, F. and Livi, K.J. (editors) (2013) *Minerals at the Nanoscale.* EMU Notes in Mineralogy, **14**, European Mineralogical Union and the Mineralogical Society of Great Britain & Ireland, London
Sander, M., Hofstetter, T.B. and Gorski, C.A. (2015) Electrochemical analyses of redox-active iron minerals: A review of non-mediated and mediated approaches. *Environmental Science & Technology*, **49**, 5862–5878.
Sander, S.G. and Koschinsky, A. (2011) Metal flux from hydrothermal vents increased by organic complexation. *Nature Geoscience*, **4**, 145–150.
Schüring, J., Schulz, H.D., Fischer, W.R., Böttcher, J. and Duijnisveld, W.H. (editors) (2013) *Redox: Fundamentals, Processes and Applications.* Springer-Verlag, Berlin.
Stucki, J.W., Goodman, B.A. and Schwertmann, U. (1985) *Iron in Soils and Clay Minerals.* D. Reidel Publishing Company, Doredrecht, Holland.
Sun, C., Li, H. and Chen, L. (2012) Nanostructured ceria-based materials: synthesis, properties, and applications. *Energy & Environmental Science*, **5**, 8475–8505.
Taylor, K.G., Hudson-Edwards, K.A., Bennett, A.J. and Vishyakov, V. (2008) Early diagenetic vivianite [$Fe_3(PO_4)_2 \cdot 8H_2O$] in a contaminated freshwater sediment and insights into zinc uptake: A μ-EXAFS, μ-XANES and Raman study. *Applied Geochemistry*, **23**, 1623–1633.
Vaughan, D.J. (2006) Sulfide mineralogy and geochemistry: Introduction and overview. Pp. 1–5 in: *Sulfide Mineralogy and Geochemistry* (D.J. Vaughan, editor). Reviews in Mineralogy and Geochemistry, **61**. Mineralogical Society of America and Geochemical Socoety, Washington D.C.
Widdel, F., Schnell, S., Heising, S., Ehrenreich, A., Assmus, B. and Schink, B. (1993) Ferrous iron oxidation by anoxygenic phototrophic bacteria. *Nature*, **362(6423)**, 834–836.
Yakushev, E.V. (editor) (2012) *Chemical Structure of Pelagic Redox Interfaces: Observation and Modeling.* The Handbook of Environmental Chemistry, **22**. Springer, Heidelberg.
Zhang, H., Waychunas, G.A. and Banfield, J.F. (2015) Molecular dynamics simulation study of the early stages of nucleation of iron oxyhydroxide nanoparticles in aqueous solutions. *Journal of Physical Chemistry B*, **119**, 10630–10642.

EMU Notes in Mineralogy, Vol. 17 (2017), Chapter 2, 5–31

Influence of iron oxide structure and size on redox reactivity

CORINNE CHANÉAC and JEAN-PIERRE JOLIVET

Laboratoire de Chimie de la Matière Condensée, UMR 7574, Université Paris 6 - Sorbonne Universités, CNRS, Collège de France, 11 place Marcelin Berthelot, 75005 Paris, France, e-mail: corinne.chaneac@upmc.fr

Decreasing the particle size of a solid to the nanoscale has consequences for its crystalline and electronic structures. It also involves a large extension of the surface area, thus modifying its reactivity. This chapter focuses on the effect of particle size on thermodynamic, structural and electronic characteristics and on redox properties. It is commonly accepted that a decrease in particle size increases the reactivity of materials. This is discussed here through the description of three main iron oxide structural groups, namely spinel (magnetite, maghemite), hematite and goethite, which exhibit specific behaviour *vs.* size effect. The role of structural defects and nanoparticle morphology in terms of reactivity is highlighted.

1. Introduction

For several decades, ultrafine matter has been of considerable scientific and technological interest leading to the development of the 'nanomaterial sciences'. Specific properties appear when at least one dimension of the object is of the order of magnitude of fundamental physical length, *e.g.* the mean free path of electrons in a solid, the wavelength of light or of phonons in a solid, as well as the size of magnetic domain. These properties related to nanoscale solids have motivated the development of various scientific and technical fields such as plasmonics, photonics, spintronics and nanomagnetism, for which physical laws are, nowadays, relatively well established (Scholes and Rumbles, 2006). The high degree of chemical reactivity of the ubiquitous nanoscale matter demands that attention be paid to tackling environmental issues also.

Nanoscale effects are not only related to a volume effect but also to a surface effect because the surface/volume ratio increases significantly when the size of the object decreases. For example, as this ratio varies as $3/r$ (where r is the radius), ~5% of atoms are located at the surface for 20 nm diameter spheres considering a compact packing, while the proportion rises to 15% and 60% for 15 nm and 2 nm spheres, respectively. This significant increase of the surface atom number leading to a large concentration of surface active sites is already used in heterogeneous catalysis (Li and Somorjai, 2010). Size effect is the source of superparamagnetism of ferrimagnetic materials due to relaxation of magnetization in very small volumes (Dormann, 1997) such as in

DOI: 10.1180/EMU-notes.17.2

magnetite and maghemite nanoparticles. For nano-goethite, the surface magnetic canting gives a weak ferromagnetic component (Lemaire *et al.*, 2002) compared to an antiferromagnetic coupling in the particle core. In densified nanomaterials, huge amounts of interfaces (grain boundaries) have different behaviours to those of bulk solids. For instance, the grain boundary allows mechanical deformation of the solid (superplastic ceramics) and limits the sliding of atomic planes (ultrahard metals), or controls the wall displacement of magnetic domains (solids having high magnetic coercivity) (Hadjipanayis and Siegel, 1994).

If the size effects relating to physical (optics, electronics, magnetism) and physicochemical (aggregation, dispersion) properties are mostly well understood, the influence of nanoparticle size on (electro)chemical reactivity is much less understood and rationalized. It is, however, a very important challenge since many nanoscale systems are used for their reactivity, especially in heterogeneous catalysis and in the development of renewable energy (Li and Somorjai, 2010; Grätzel, 1983). Some nanomaterials such as iron and manganese oxides are also important in the environment because of their structural and chemical diversity (Hochella *et al.*, 2008). These oxides give rise to redox phenomena implicated in the regulation of the cycle of many elements in the environment (Hem, 1977; Stumm and Sulzberger, 1992; Borch *et al.*, 2010) and also play a crucial role in immobilization and transformation of heavy elements such as arsenic and lead (Dong *et al.*, 2000; Charlet *et al.*, 2011). Understanding these phenomena requires that we understand the influence of size on the reactivity of nanoparticles. The issue is complex and despite the large number of experimental studies, there is still uncertainty in this area.

It is currently assumed that solid reactivity increases with the decrease in crystal size as a consequence of the large specific surface area, the presence of defects (steps, terraces,...) and structural disorder giving a kinetic advantage for many reactions (Fernandez-Garcia *et al.*, 2004). Nevertheless, in some cases, nanoparticle reactivity decreases surprisingly as size decreases. This chapter aims to show how the size effect influences the redox reactivity of iron oxides. Manganese oxides can also give rise to rich redox chemistry, but as only a few examples of size effect are reported in the literature, they will not be examined in this chapter.

2. Main size effects on oxide nanoparticle properties

The decrease in size of metal oxide particles induces strong modifications particularly in three of their basic characteristics: thermodynamic stability and crystal and electronic structures. These changes obviously have a deep influence on particle reactivity.

2.1. Thermodynamic stability

Increasing the surface area with the size reduction of mineral nanoparticles from the micron to nanometer range means that the surface-energy contribution (ΔG_{surf}, always positive) becomes competitive with the bulk component (ΔG_{bulk}, negative) in the free enthalpy of formation's balance ($\Delta G = \Delta G_{bulk} + \Delta G_{surf}$). This shows that the ultradivided material is intrinsically metastable compared to bulk material: if some changes are kinetically possible (by Ostwald ripening or aggregation in aqueous dispersions), the system evolves spontaneously to reduce the amount of interfaces. In the case of many polymorphic systems (iron, manganese, titanium, aluminium oxides...), some changes in the relative stabilities of different phases have been reported. Because the bulk enthalpies (ΔH) of different polymorphs are not substantially different and their entropies are similar, the metastability ($\Delta G = \Delta H + T\Delta S$) of the least stable phases is quite relative. It follows that a metastable phase in the solid state may be the most stable polymorph at the nanometric scale (Fig. 1) (Fritsch *et al.* 1997; Laberty and Navrotsky, 1998; Zhang and Banfield, 1998; Navrotsky, 2003; Navrotsky *et al.* 2008; Birkner and Navrotsky, 2012). It should be noted that the phase stability is also related to aggregation and sintering phenomena at high temperature. This is evidenced by the dispersion of metastable nanoparticles in a rigid inert matrix such as silica glass or polymer. For instance, the transformation of anatase (titanium dioxide) nanoparticles into rutile shifts from 300°C for powdered nanoparticles to 800°C for confined nanoparticles (Pottier *et al.*, 2003). The same results are observed with maghemite nanoparticles dispersed in silica glass, the hematite crystallization is delayed by more than 1000°C (Chanéac *et al.*, 1996). Interestingly, a rare phase of iron oxide, ε-Fe_2O_3, has been stabilized at 1200°C due to the very limited aggregation of maghemite nanoparticles in silica (Tronc *et al.*, 1998) at this temperature.

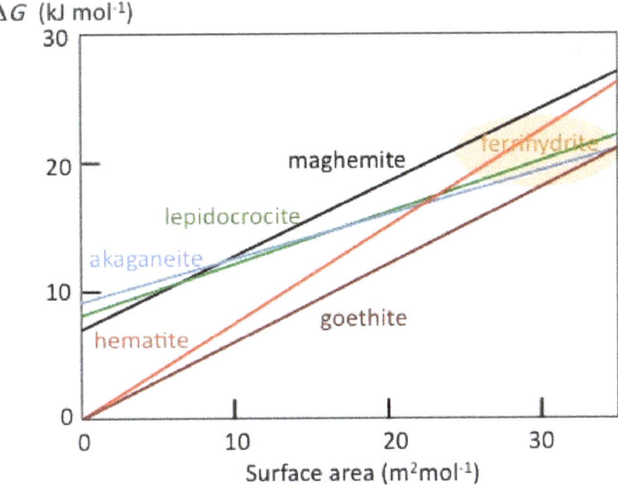

Figure 1. Free enthalpy of transformation of iron oxides and oxyhydroxides (with hydrated surfaces) to hematite and liquid water at 298 K, as a function of surface area (expressed per mole of Fe, for thermodynamic consistency when comparing different compositions). Values for ferrihydrite are approximate because of sample variability and are represented as an elliptical area. The slopes of the lines represent the hydrated surface energy; the crossovers correspond to changes in relative stability of the different phases (reproduced from Navrotsky *et al.*, 2008 with the permission of the American Association for the Advancement of Science).

Surface hydration also plays a very important role in thermodynamic stabilization of nanoparticles because it significantly reduces the interfacial tension. In addition, particle solubility that usually increases with decreasing size (Stumm and Morgan, 1996) is one of the relevant parameters to be taken into account in the chemical reactivity especially when biogeochemical phenomena are involved. Therefore, complex phase transformations can occur in the aqueous environment. For example, goethite becomes thermodynamically more stable than hematite for a size smaller than 60 nm and lepidocrocite more stable than maghemite for particle size smaller than 12 nm (Navrotsky *et al.*, 2008). Similar complex energetic crossovers occur with manganese oxides (Birkner and Navrotsky, 2012). However, if thermodynamics helps to explain the relative stability or the formation of some crystalline phases in relation to particle size, kinetic effects are sometimes such that no transformation is observed.

2.2. Crystal structure and structural defects

Nanocrystals are often associated with crystalline disorder. Indeed, some internal distortions are induced by strong surface reconstructions whose the effects of which can spread to the cores of particles a few nm in diameter (Gilbert *et al.*, 2004). These surface reconstructions may result from the small particle size, and therefore require a strong surface curvature and under-coordinated surface atoms (Waychunas *et al.*, 2005). Due to this surface disorder, the structure and chemical composition near the surface can differ from the bulk. For example, maghemite particles smaller than 11 nm exhibit an altered crystal structure with the creation of cation vacancies in the surface tetrahedral sites (Brice-Profeta *et al.*, 2005).

Due to such surface defects, maghemite nanoparticles exhibit an unusual behaviour, which can be demonstrated by the adsorption of As(III) (Auffan *et al.*, 2008). Arsenic adsorption is an important environmental topic because its presence in large quantities in natural groundwaters causes serious public health problems in countries around the world. Thus, arsenic adsorption by oxides could be an interesting way to avoid dissemination of the element. For mean particle diameters of 300 nm and 20 nm, the amounts of As(III) adsorbed normalized per unit area were ~3.6 atom nm^{-2} (Yean *et al.*, 2005); for smaller particles, the adsorption capacity increases and can reach 11 atom nm^{-2} for particles with a mean diameter of 11 nm. The decrease in occupancy of surface tetrahedral Fe(III) sites with decreasing maghemite particle size to <20 nm (Brice-Profeta *et al.*, 2005) creates highly reactive sites for adsorption of various ions (Jolivet *et al.*, 2004; Navrotsky *et al.*, 2008) and especially arsenic (III) which is usually tetrahedrally coordinated (Auffan *et al.*, 2008). Accordingly, two mechanisms are involved in As adsorption by maghemite nanoparticles: at low coverage rates, As is coordinated into the tetrahedral Fe(III) vacancies at the centre of hexagons made by edge-sharing FeO_6 octahedra which are the most reactive sites (Fig. 2); after filling these sites, subsequent As adsorption occurs as tridentate complexes bound to the classical sites available only with larger particles (Morin *et al.*, 2009).

Similarly, an EXAFS structural study of goethite nanoparticles showed that a decrease in particle size generates a distortion in the geometry of Fe(III) surface sites

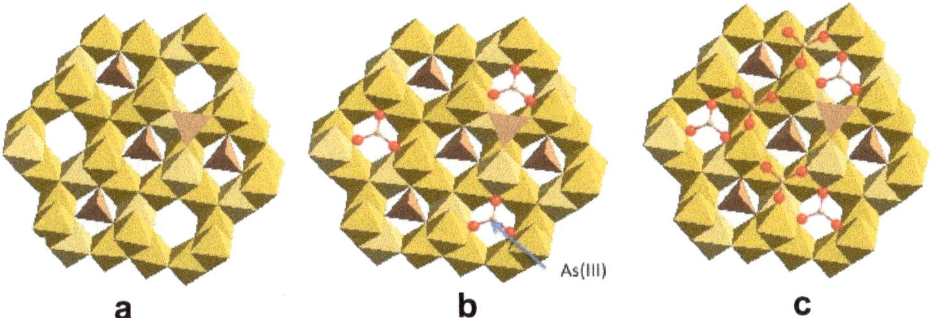

Figure 2. Schematic adsorption mechanisms of As(III) on 6 nm sized maghemite nanoparticles based on the structural information derived from the XRD and As K-edge XAS analyses (a) Initial nanoparticle (octahedra and tetrahedra are FeO_6 and FeO_4 respectively) (b) Filling of the more reactive sites at low surface coverage ; (c) Adsorption on lattice positions at higher surface coverage allowing a decrease of surface energy. (From Auffan *et al.*, 2008)

(Waychunas *et al.*, 2005). This change disadvantages cation adsorption: the adsorption rate of Hg(II) on goethite particles 5 nm in diameter is much smaller than for 25 or 75 nm (Waychunas *et al.*, 2005). In other cases, distortions of the surface geometry can promote adsorption. Thus, the presence of more numerous defects on hematite particle surfaces of decreasing size (88, 25, 7 nm) promotes the coordination of Cu^{2+} ions which are themselves highly distorted by the Jahn-Teller effect (Madden *et al.*, 2006). Defects are here favourable to sites matching with the low symmetry of the cation environment.

A somewhat surprising effect has been reported for hematite nanoparticles with varyious sizes in the range 7–120 nm. It was suggested that the concentration of tetrahedral Fe(III) ions increased with decreasing particle size (Chernyshova *et al.*, 2007) while bulk material contains only octahedral Fe(III) ions. The existence of tetrahedral Fe(III) in 2–3 nm sized ferrihydrite also gives rise to an intense debate (Gilbert *et al.*, 2013; Manceau *et al.*, 2014). To our knowledge, tetrahedrally coordinated Fe(III) never appears in soluble ferric species (except in a very alkaline medium) or in solid structures other than the spinel structure. A decrease in occupancy of tetrahedral sites with reduction in size of spinel nanoparticles has been observed. In addition, ferrihydrite never evolves to spinel or other structures containing tetrahedral Fe(III), at least without the presence of divalent ions. Consequently, despite the quality and variety of experimental techniques of investigation, a spontaneous reduction in iron coordination number and the presence of tetrahedral iron ions in nanohematite and ferrihydrite seem very unlikely.

Note that it is difficult to compare samples of the 'same' material produced by different procedures. For instance, nano-hematite particles (12–150 nm), synthesized by decomposition in air of iron pentacarbonyl at 1000–1200°C using an aerosol technique, exhibit more irregular surfaces than aqueous-synthesized particles of similar sizes. Surface disorder could be the cause of their larger affinity and site density towards U(VI) adsorption, and of the increase in reactivity with decreasing particle size (Zeng *et al.*, 2009).

2.3. Electronic structure

Size effects result in significant changes to the electronic structure of ultradivided solids compared to bulk materials (Scholes and Rumbles, 2006). The key factors of the reactivity of transition metal complexes or molecules are the energy gap between the highest-occupied molecular orbital (HOMO) and the lowest-unoccupied molecular orbital (LUMO), as well as electron population of the HOMO. We can roughly deduce the electronic structure of the solid from coordination complexes by extrapolation (Sherman, 1984; Burdett, 1995). In solid matter, the multiplication of orbital levels due to the high quantity of orbitals leads to the concept of bands and, by analogy to HOMO and LUMO, are named valence band (VB) and the conduction band (CB), respectively.

For insulators and semiconductors, the VB is fully occupied and the CB is empty. These two bands are separated by a band gap (BG) or energy gap which is generally not as wide as the energy difference between HOMO and LUMO in the corresponding molecular unit (Fig. 3). For iron oxides and oxyhydroxides (hematite and goethite, for example), the BG is ~2 eV. For an insulator such as SiO_2, the BG is near 9 eV. Magnetite (Fe_3O_4), with a BG of ~0.1 eV, is a semi-metal. The BG value represents the energy needed to excite an electron from the VB to the CB where it may move freely outside the electrostatic interaction with a hole created in the VB (electron-hole pair). At room temperature, for iron oxides and hydroxides, having a band gap of ~2 eV, the thermal or light energy allows the excitation of a small number of electrons to the CB giving a very small electrical conductivity. The size of the interaction domain (bound state) of the lowest energy of the electron-hole pair (exciton) is characterized by the Bohr radius. At the nanoscale, the particle size becomes of the order of the Bohr radius so that an electron and a hole can never be sufficiently far away from each other for their interaction to be negligible. Consequently, the lowest excitation energy of an electron is greater than the bulk BG. This is the effect of confinement which involves the BG opening (Gilbert and Banfield, 2005). The phenomenon is well illustrated by the blue shift occurring in optical absorption of CdSe semiconductors of decreasing size from ~10 to 2 nm (Murray *et al.*, 1993). The BG width is usually determined experimentally

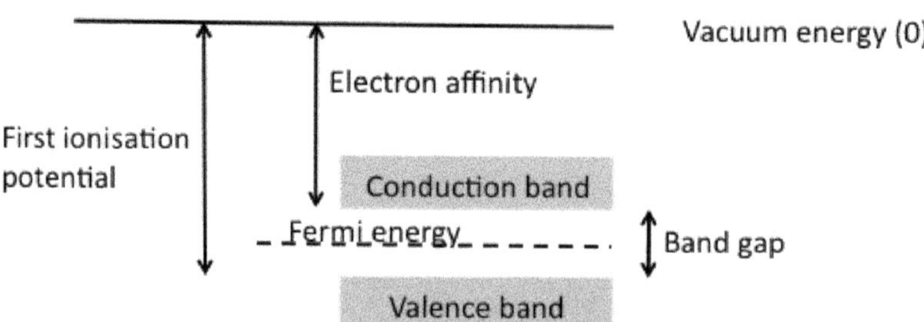

Figure 3. Scheme of the electronic structure of a solid showing valence band (VB) and conduction band (CB) separated by the band gap. For insulators and semi-conductors, the Fermi energy, that represents the electronic chemical potential, is located inside the gap, between the top of the valence band and the bottom of the conduction band

by optical UV-visible spectroscopy and the positions in energy of VB and CB are obtained from combined X-ray absorption and emission spectroscopies (Gilbert *et al.*, 2009; Chernyshova *et al.*, 2010).

For bulk solids, experimental data found in the literature are generally consistent with each other while in the nanometer range, there are more conflicting data. For instance, in the case of TiO_2 in which there are no 3d electron exchange effects, conduction band states should be delocalized, and any true quantum confinement should be properly modelled by the electron-in-a-box model (Gilbert and Banfield, 2005). However, there has been a lot of debate about whether there is quantum confinement in TiO_2 (*e.g.* Monticone *et al.*, 2000; Satoh *et al.*, 2008). The conclusion seems to be that there is a confinement effect for only diameters of <1 nm for TiO_2. Similar discrepancies related to electronic structure also exist for hematite due to the two competing contributions in energy transfer, the ligand-to-metal charge transfer (located at the top of valence band) and the d–d ligand field intraband electronic transition (located at the bottom of conduction band). It has been shown from UV-Vis absorption and XPS spectra that the defected octahedral coordination of iron ions in small particles induces an increase in band gap (BG) from 2.18 eV to 2.55 and 2.95 eV with decreasing size from 120 nm to 38 nm and 7 nm, respectively (Chernyshova *et al.*, 2010). However, an *ab initio* cluster-model strategy consisting of the addition of an extra electron into hematite indicates that the wavefunctions appear to extend out to only one or two neighbours (Rosso *et al.*, 2003b). Therefore, it is difficult to imagine how such a localized wavefunction could experience any confinement in a 7 nm sized 'box'. Accordingly, no change in the BG width is observed by soft X-ray spectroscopy for bulk hematite and 30–8 nm-sized particles (Gilbert *et al.*, 2009). Theoretical and experimental work is obviously required to fully understand such d-electron semiconductors. Great care must also be taken regarding the purity of the phases studied. As an example, hematite nanoparticles synthesized by thermolysis of acid Fe(III) solutions are very often contaminated by ferrihydrite or goethite undetectable by XRD but to which spectroscopies are very sensitive. The presence of such impurities could be an important cause of divergence in experimental data.

3. Crystal structure and formation of iron oxides with the control of nanoparticle size

Because of the kinetic instability of ultradispersed systems, nanoparticle elaboration needs methods of 'soft chemistry' such as precipitation from solution at room or moderate temperatures. In such conditions, the acidity of the medium is the main parameter for the control of the structure and size of particles, especially with iron oxides for which the structural chemistry appears to be very complex (Cornell and Schwertmann, 2003; Jolivet *et al.*, 2006).

Briefly, hydroxylation of ferric ions in solution by addition of base at room temperature ($pH > 3$) leads quasi instantaneously to 2-line ferrihydrite, a poorly defined and highly hydrated phase exhibiting only two broad bands on X-ray diffraction patterns (Jambor and Dutrizac, 1998; Schwertmann *et al.*, 1999). Due to its poor structural organization, ferrihydrite is thermodynamically unstable. It transforms into different crystalline phases *via* different pathways depending on the acidity of the medium. At $5 \leqslant \text{pH} \leqslant 8$, ferrihydrite transforms into very small particles of hematite, $\alpha\text{-}Fe_2O_3$ (Combes *et al.*, 1989). Because of the very low solubility of the solid ($\sim 10^{-10}$ mol L^{-1}), the transformation can proceed only by *in situ* dehydration and local rearrangement. This explains the formation of an oxide with very small size of crystalline domains. When the solubility of ferrihydrite is greater ($\text{pH} < 4$ or $\text{pH} > 8$), the transformation can proceed more easily *via* a dissolution-crystallization process, leading to goethite, α-FeOOH, the more thermodynamically stable phase in these conditions. Unlike aluminium hydroxide ($Al(OH)_3$), ferric oxide ($Fe(OH)_3$) does not exist and has never been identified. Complexing ligands such as silicate and especially phosphate delay or hinder the transformation of ferrihydrite into crystalline phases (Parfitt *et al.*, 1992; Rose *et al.*, 1996, 1997; Doelsch *et al.*, 2000). This is the case, for instance, in very old soils in New Zealand (200,000 years!) or in living organisms, in ferritin nanoparticles in which iron(III) is stored.

Goethite is an attractive material because of its strongly anisotropic morphology, appearing in more or less long rod-shaped particles. The origin of this morphology is related to the crystal structure consisting of chains of edge-shared $FeO(OH)_6$ octahedra bounded by corner sharing (Fig. 4) that provides a preferred direction of crystal growth (Jolivet *et al.*, 2006). Such anisotropic morphology gives rise to a differentiated reactivity of the crystal faces.

Hematite also can be obtained by thermolysis of acidic ferric solutions at 90–100°C ($\text{pH} \leqslant 3$) (Matijevic 1988; Cornell and Schwertmann, 2003). Very small particles of 6-line ferrihydrite are first formed at low chloride concentration ($C < 10^{-3}$ mol L^{-1}, ~3 nm in diameter) (Jambor and Dutrizac, 1998; Schwertmann *et al.*, 1999; Cornell and Schwertmann, 2003). This phase, better organized than 2-line ferrihydrite, is then converted to hematite during thermolysis. It is interesting to observe that the size of hematite particles increases strongly with the acidity of the medium (Jolivet *et al.*, 2006). The role of the acidity is to facilitate oxide crystallization and to favour the dissolution-crystallization process leading to particle growth. Hematite, exhibiting the same crystal structure type as corundum, is the most compact and dense of all iron oxides, which justifies its highest thermodynamic stability. The crystal structure consists of hexagonal close packed stacking of oxide ions in which 2/3 of the octahedral sites are occupied by ferric cations (Fig. 4). There are strong distortions in the lattice because pairs of FeO_6 octahedra share common faces, inducing intense repulsions between Fe^{3+} ions. Typical morphologies of particles at nanoscale are rhombohedra and platelets.

Quasi-stoichiometric magnetite nanoparticles may be synthesized by precipitation of the Fe(III)-Fe(II) mixture with the composition 2Fe(III) + Fe(II) in alkaline aqueous medium in anaerobic conditions. The fine control of acidity allows the particle size to be tuned, from 2 nm at pH 9 to 15 nm at pH 12 (Vayssières *et al.*, 1998; Jolivet *et al.*, 2004). The inverse spinel structure of iron oxide (Fig. 4) is characterized by face centred cubic stacking of oxide ions in which 1/2 of octahedral sites (O_h) and 1/8 of

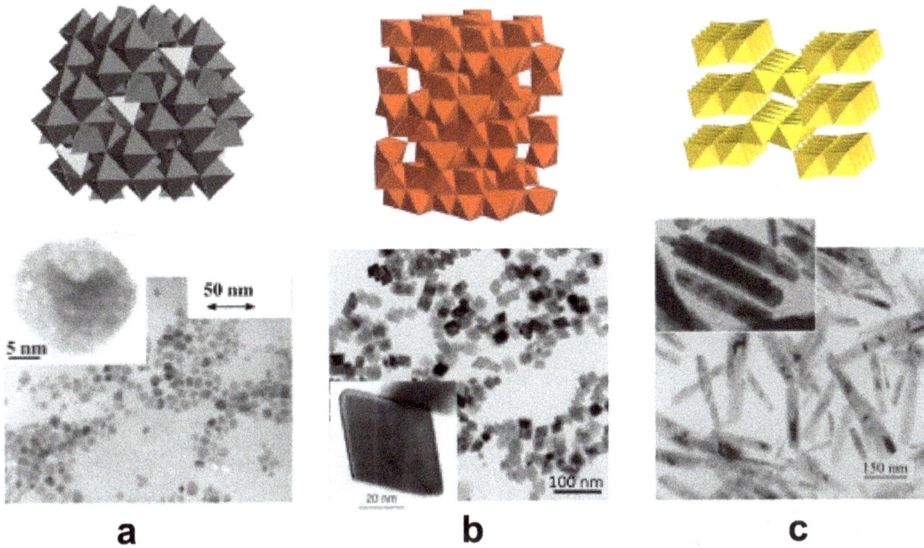

Figure 4. Crystal structures and typical morphologies of some iron oxides synthesized from aqueous solutions: (a) magnetite, (b) hematite and (c) goethite. In the models, polyedra are FeO_6 and FeO_4 coordination polyhedra.

tetrahedral sites (T_d) are occupied by cations. Ferrous cations and half of ferric cations occupy the O_h sites because the other half of the ferric cations occupies the T_d sites following the formula $[Fe^{3+}]_{Td}[Fe^{3+}Fe^{2+}]_{Oh}O_4$ (Cornell and Schwertmann, 2003). Magnetite is a semi-metal with a narrow band gap (Eg $\approx$ 0.1 eV). It permits fast electron hopping at room temperature between Fe(II) and Fe(III) octahedral cations, as evidenced by the $Fe^{2.5+}$ component of the Mössbauer spectrum (Cornell and Schwertmann, 2003). This intervalence transfer is the origin of the black colour of this oxide. Due to the cubic structure, the nanoparticles are typically spheroidal shape.

The presence of Fe(II) ions during precipitation of Fe(III) ions combined with electron mobility plays a fundamental role in crystallization of the spinel framework. Indeed, the fully oxidized form of magnetite, namely maghemite, $Fe_{2.67}O_4$ or γ-Fe_2O_3, ($[Fe^{3+}]_{Td}[Fe^{3+}_{5/3}\square_{1/3}]_{Oh}O_4$, where $\square$ represents the cationic vacancies) is never obtained only by the precipitation of Fe(III) ions in neutral to alkaline media nor by thermolysis in acidic medium (Jolivet *et al.*, 2006). Fe(III) ion precipitation immediately forms a very poorly organized solid (2- or 6-line ferrihydrite). Addition of Fe(II) ions to this early amorphous solid, even in small proportion (~10% of atoms), induces a very fast crystallization of spinel oxide (Jolivet *et al.*, 1992; Tronc *et al.*, 1992; Yang *et al.*, 2010). Whatever the amount between 10 and 50%, Fe(II) ions are incorporated completely in the precipitate leading to a mixed-valence material where fast electron hopping occurs. The result of electron delocalization is the presence of $Fe^{2.5+}$ ions as shown by Mössbauer spectroscopy above 100 K (Fig. 5). For $T < 100$ K, the spectrum exhibits two components, one for ferrous and another for ferric species.

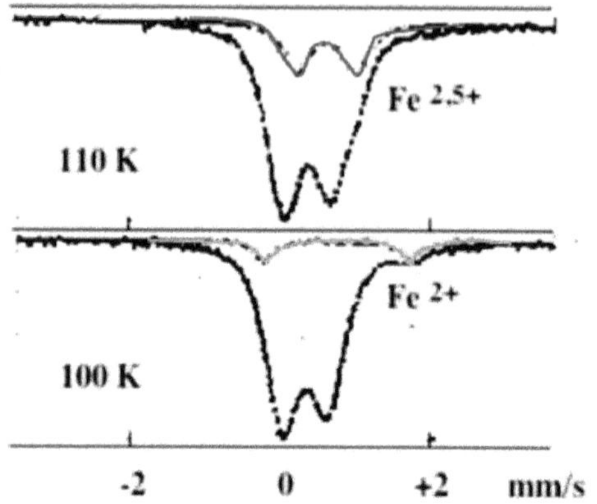

Figure 5. Observed (black lines) and calculated Fe^{+2} and $Fe^{+2.5}$ components (red and green lines) Mössbauer spectra of the early non-crystalline suspended mixed-valence hydroxide (FeII/FeIII = 0.15) (From Tronc *et al.*, 1992).

Interestingly, ferrihydrite containing other divalent cations such as Co(II), Mn(II), Ni(II) or Zn(II) is much more difficult to crystallize in spinel structure without any heating step at 100°C over several days. Easier crystallization of magnetite was attributed to Fe(II)-Fe(III) electron transfer which is not observed for other divalent cations. Such electron transfer has been shown by Mössbauer spectroscopy following the adsorption of $^{56}Fe(II)$ on ^{57}Fe ferrihydrite and the reverse (adsorption of $^{57}Fe^{(II)}$ on $^{56}Fe^{(II)}$ ferrihydrite) (Williams and Scherer, 2004). As only the ^{57}Fe isotope is sensitive in Mössbauer spectroscopy, the presence of ferric ions in the adsorbed layer and the presence of ferrous ions in the core of the particles, respectively, clearly demonstrate electron transfer without ion diffusion into the solid.

Magnetite nanoparticles are powerful reducers and consequently are very sensitive to oxidation in forming maghemite, γ-Fe_2O_3. The magnetite-reduction process involves the formation of continuous solid solution from magnetite (Fe^{II}/Fe^{III} 0.5) to maghemite (Fe^{II}/Fe^{III} 0). Great care must be taken for the synthesis of truly stoichiometric magnetite nanoparticles in order to avoid any oxidation reaction. The oxidation process is quicker because the particles are small as the diffusion of iron ions from the particle core to the surface is the limiting step of this oxidation reaction. The effective composition of the particles is a very sensitive factor because material properties such as adsorption ability, magnetism and especially reactivity (Gorski and Scherer, 2009; Gorski *et al.*, 2010a) are heavily dependent on stoichiometry. For instance, the reduction rate of nitrobenzene $ArNO_2$ derivatives by 20 nm-sized spinel iron oxide nanoparticles increases by five orders of magnitude when their initial composition, given by the Fe^{II}/Fe^{III} ratio, is changed from 0.31 to 0.50 (Gorski *et al.*, 2010a). This parameter is neglected or ignored in many studies, though it can be determined easily using three techniques: chemical titration, Mössbauer spectroscopy and X-ray diffraction (Jolivet *et al.*, 1992; Tronc *et al.*, 1992; Gorski and Scherer, 2010b).

Some other phases of iron oxyhydroxides may be formed in various conditions (Cornell and Schwertmann, 2003), but they are not considered here.

4. Examples of size effect on the reactivity of iron oxide nanoparticles

Among the most common iron oxides, one may distinguish three structural types which have different behaviours. In the first, the spinel structure (magnetite, maghemite), which presents high electron mobility through the three-dimensional framework, gives rise to a very interesting redox surface chemistry strongly affected by size effects. Next, hematite nanoparticles exhibiting somewhat contrasting behaviours and contradictory size effects have been reported. Contrary to what might be expected, the reactivity of the smallest particles is sometimes found to be less than that of larger ones. Finally, the oxyhydroxide phases, especially goethite, do not seem to exhibit size effects on reactivity, but are influenced heavily by morphological anisotropy of the crystalline structure (Fig. 4). Indeed, significant shape anisotropy involves different crystalline faces with local structure, energy and consequently reactivity which, themselves, vary significantly.

Three reasons are invoked to account for the influence of particle size on reactivity towards various reactants: the evolution of the electronic structure, changes in the organization of crystalline solids and/or the aggregation of particles. In many cases, it is somewhat difficult to obtain reliable experimental data. Different sized particles of one type of material are sometimes synthesized following different protocols leading to different surface states, different morphologies or to the presence of impurities or parasitic phases. Aggregation is also an important drawback in modifying, in an uncontrolled manner, the surface area of the system studied. Some typical reactions of the main structural groups of iron oxides are examined by highlighting their redox reactivity.

4.1. Reactivity towards proton adsorption

Magnetite nanoparticles can be converted to maghemite, as previously described, involving different ionic and/or electronic transfers through the solid–solution interface. The main mechanism involves, simultaneously, the oxidation of Fe(II) and the creation of a cationic vacancy in the octahedral network to maintain electric neutrality. Thanks to the electron mobility, the particle composition spans the whole range from magnetite, Fe_3O_4, to maghemite, Fe_2O_3.

In a neutral or alkaline medium, oxidation of magnetite by oxygen proceeds by reduction of dissolved oxygen at the particle surface with the coordination of oxide ions O^{2-} by surface Fe(III) ions (Fig. 6). Transformation induces the increase in particle size by the formation of oxide layers.

In an acidic medium, the oxidation process of nanomagnetite is completely different. Around pH 2 and under anaerobic conditions, oxidation proceeds in a different way involving surface Fe^{2+} ion desorption as soluble hexa-aquo complexes, following the chemical equation:

$$[Fe^{3+}]_{Td}[Fe_2^{2.5+}]_{Oh}O_4 + 2\,H^+ \rightarrow 0.75\,[Fe^{3+}]_{Td}[Fe_{5/3}^{3+}\square_{1/3}]_{Oh}O_4 + [Fe(OH_2)_6]^{2+} + H_2O$$

which corresponds formally to a net exchange $Fe^{2+}/2H^+$ at solid/solution interface (Jolivet and Tronc, 1988). The process is illustrated in Fig. 7:

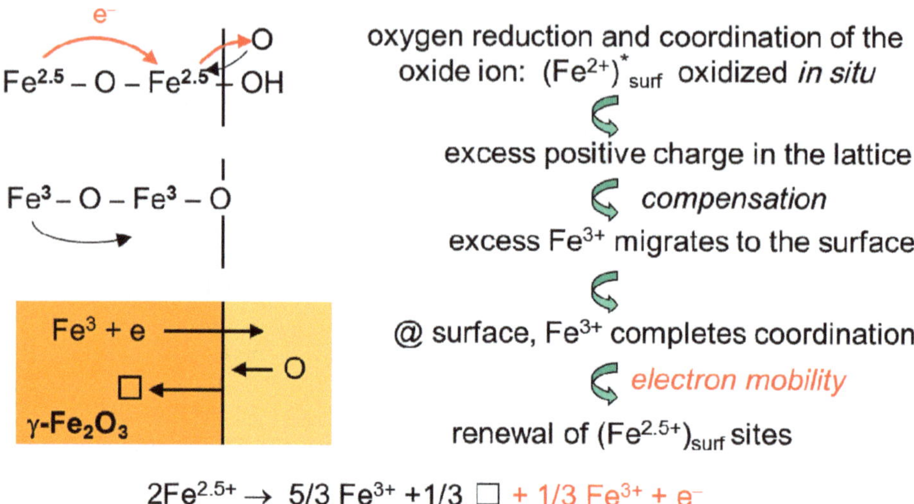

Figure 6. Different steps of the oxidation of magnetite to maghemite nanoparticles in neutral or alkaline medium.

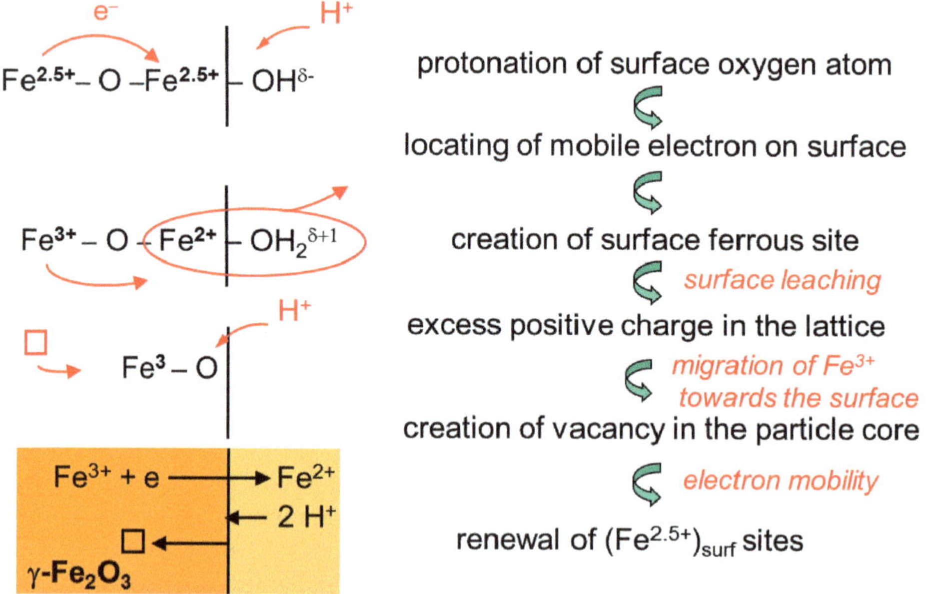

Figure 7. Different steps of the oxidation process of magnetite to maghemite nanoparticles in acidic suspension. The oxidation reaction corresponds to the elimination of Fe(II) ions, the consumption of two protons/Fe^{2+} and diffusion of cation vacancies in the particle core (Jolivet and Tronc, 1988).

Because of nanoscale particles, oxidation of magnetite does not lead to a noticeable change in size. From the kinetic point of view, ~50% of the reaction is almost instantaneous and complete oxidation of 8 nm sized particles proceeds in ~1 month at room temperature. Of course the smaller the particles, the quicker is the reaction. The contribution of electron mobility to magnetite oxidation in acid medium was demonstrated by comparison with cobalt ferrite, $CoFe_2O_4$, in which no intervalence transfer occurs. Under similar conditions, only ~10% of Co^{2+} ions are desorbed, which corresponds to surface cations only (Jolivet and Tronc, 1988).

Goethite, in which there is no electron mobility, has a very different behaviour in acidic medium, where it dissolves. The dissolution rate, which depends on particle size, depends also on particle morphology. In nitric acid (pH 2) and in the dark, nanorods (75 nm × 6 nm) dissolve much faster than microrods (1006 nm × 25 nm) (Rubasinghege *et al.*, 2010). The initial specific surface area-normalized dissolution rate is multiplied by a factor close to 3 and soluble species are Fe(III) ions. Under sunlight, the dissolution produces more Fe(II) for nanorods than for microrods (Fig. 6). This suggests that reactivity under both dark and light conditions in goethite dissolution is size-dependent. However, electron microscopy shows that the length of the nanorods decreased by ~15% while the thickness did not change significantly after 46 hours of treatment. In addition, the (021) faces located on the particles' extremities, initially well defined, become rounded after acid treatment. It thus appears that the end faces (021) are more reactive than the lateral faces (110). In fact, as the reactivity of goethite essentially comes from the ends of particles, kinetic data should be normalized by the surface area of the end facets in order to prove a size-driven reactivity linked to crystalline disorder, to specific structural or electronic characteristics. In addition, aggregation of nanoparticles in some physicochemical conditions used for their study could be a source of misinterpretation of data.

It is interesting to note that while acid-base reactions confined on the surface lead to dissolution in the case of goethite only they give rise to a structural transformation induced by a redox transfer in the core of particles for spinel iron oxides. The kinetics of

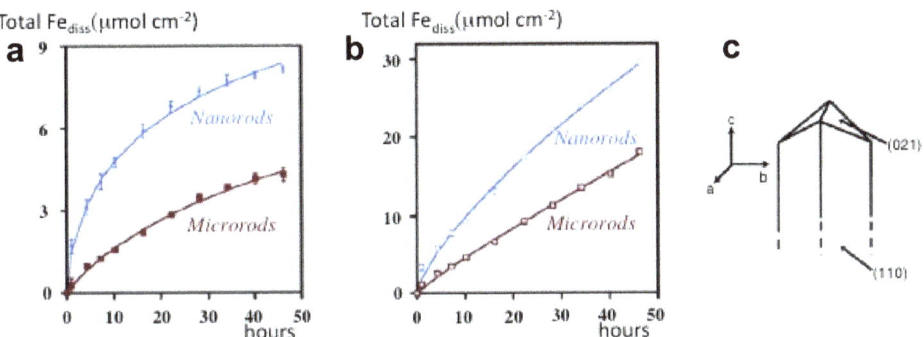

Figure 8. Dissolution kinetics of goethite nanorods and microrods in acidic medium (HNO_3, pH2) at RT (a) under dark and (b) light conditions (concentrations are given on surface area basis). (c) Scheme of goethite rod morphology (From Rubasinghege *et al.*, 2010)

the reaction, apparently size-dependent in the two cases, seems, however, to be more sensitive to the surface structure than to the specific surface area for anisotropic nanogoethite.

4.2. Redox transformations induced by Fe(II) adsorption

Interaction of ferric oxides with Fe(II) ions in suspension is probably the most studied iron redox system because of its specificity (Gorski and Scherer, 2011). Adsorption of ferrous ions onto a ferric surface enhances the reducing power of Fe(II) ions and allows the reduction of various species in solution (Mulvaney *et al.*, 1988; Wehrli *et al.*, 1989; Silvester *et al.*, 2005) due to the ability of the Fe(III) ions produced to be integrated structurally into the solid and to be strongly stabilized. An another feature is the ability of solid/solution atomic exchange due to Fe(II)-Fe(III) electron transfer. This has been shown nicely by Scherrer's group by using ^{57}Fe and ^{56}Fe isotopes in Mössbauer spectroscopy studies. ^{57}Fe(II) adsorption onto ^{56}Fe(III) oxides leads to ^{57}Fe(III) oxide with leaching of ^{56}Fe(II) ions in solution (Williams and Scherer, 2004). The kinetics and the rate of exchange depend on the solid structure: ferrihydrite, goethite, lepidocrocite, hematite, magnetite (Jeon *et al.*, 2003; Gorski and Scherer, 2011). So, Fe(II) adsorption may induce specific redox effects because electron injection in the solid is permitted. The crystalline structure and its electronic properties are consequently key elements of the reactivity of iron oxide nanoparticles.

Fe(II) adsorption from solution by metastable phases such as ferrihydrite induces the transformation to spinel (Jolivet *et al.*, 1992; Pedersen *et al.*, 2005). In a similar way, adsorption of Fe(II) ions on maghemite nanoparticles (lacunar spinel) at pH $\approx$ 6 (out of Fe(II) precipitation range), occurs until the magnetite composition (Fe(II)/Fe(III) = 0.5) is reached. This would suggest that the oxidation of magnetite in acidic medium is reversible. However, even if all iron ions appear in Mössbauer spectroscopy as "spinel-type iron" (Tronc *et al.*, 1984, 1989), the solid remains lacking in iron indicating that there is no cation diffusion from the surface to the cores of particles. In fact, adsorption of Fe(II) ions on surface hydroxo ligands induces the formation of a ferrous hydroxide surface layer and the diffusion of electrons (and protons to keep the charge balance) inside the particle cores. The coexistence of Fe(III) and Fe(II) ions in this surface layer induces its crystallization into spinel. The reaction proceeds until equipopulation of Fe(III) and Fe(II) in the octahedral network and the end of the reaction, corresponding to the beginning of Fe(II) precipitation in solution, shown well by titration (Fig. 9). $^{57}Fe^{2+}$ adsorption on ^{56}Fe spinel oxide (and $^{56}Fe^{2+}$ adsorption on ^{57}Fe spinel oxide) in various ratios of non-stoichiometry (Gorski and Scherer, 2009) has shown clearly the growth of a spinel layer on the surfaces of particles and the presence of Fe(II) ions in the cores of particles. In addition, the uptake of Fe(II) increases when the initial deviation of spinel particles from stoichiometry is significant. The adsorption of Co(II) or Ni(II), which do not give rise to intervalence transfers with Fe(III) ions in corresponding ferrites, is very limited. This further demonstrates the fundamental role of electron mobility in the interfacial chemistry of spinel iron oxide nanoparticles.

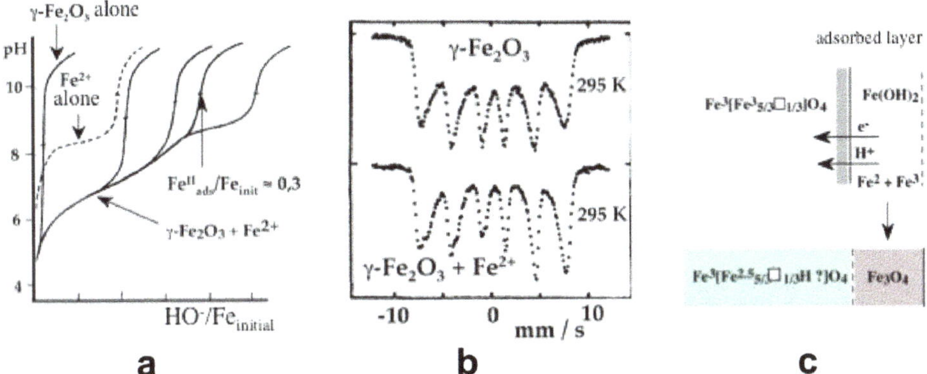

Figure 9. (a) pH-titration curves of maghemite suspensions with various quantities of added Fe(II) ions. (b) Mössbauer spectra of maghemite before and after Fe(II) adsorption. (c) Reaction balance corresponding to adsorption of Fe(II) ions on maghemite (From Tronc *et al.*, 1984).

Without any oxidizing species in suspension other than ferrous solid phase, such exchanges of iron species between solid and solution suggest that the reactivity of the Fe(II) ion-Fe(III) oxide system involves the reconstruction of the solid. This remarkable phenomenon was shown on tabular micrometric hematite particles in acidic medium in the presence of oxalate (Yanina and Rosso, 2008; Rosso *et al.*, 2010). The reaction leads to a morphological change which manifests itself by corrosion of the side faces and growth of hematite pyramids in perfect epitaxy on the basal surfaces (Fig. 10).

The phenomenon is due to the different physicochemical properties of basal and lateral surfaces. Because of their topologies and the coordination of oxygen atoms, the basal faces (001 type) are practically inert against protonation and do not bear any electrostatic charge, while all the positive charges gained in an acidic medium are carried by the lateral faces. Thus, soluble Fe(II) ions are preferentially sorbed on the basal faces and are able to inject electrons into the solid through these faces (Mulvaney *et al.*, 1988). The potential difference between the basal and lateral faces (a few tenths

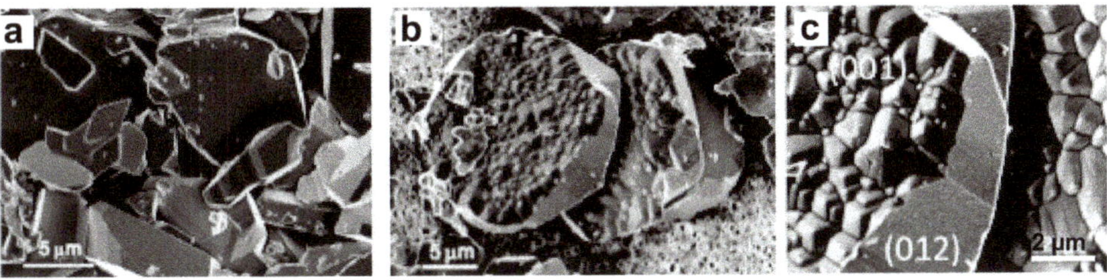

Figure 10. Scanning electron micrographs of synthetic tabular hematite platelets (a) before and (b) after reaction at 75°C for 12 hours in pH 2–3 in 1 mM Fe(II) and 10 mM oxalate solution free of oxygen. (c) Magnification of (b) image showing that pyramidal (001) growth coupled to (012) edge surface dissolution is clearly displayed on these hematite samples (From Yanina and Rosso, 2008).

of a volt at pH 2) drives the free charges towards the positively charged faces from which reductive desorption takes place, producing a redox cycle.

Two coupled interfacial processes are at work (Fig. 11):

- the growth on basal faces by oxidative sorption: $Fe^{2+}_{aq} \rightarrow Fe^{2+}_{001} \rightarrow Fe^{3+}_{001} + e^-$
- the reductive dissolution on lateral faces: $Fe^{3+}_{hk0} + e^- \rightarrow Fe^{2+}_{hk0} \rightarrow Fe^{2+}_{aq}$.

The coupling of these two reactions is mediated by the charge transport through the crystal bulk which is, at RT, sufficiently easy to create a small current (Fig. 10).

This mechanism, which involves interfacial electron transfer and solid/solution ion exchange, was confirmed by the study of sorption of $^{57}Fe(II)$ onto $^{56}Fe_2O_3$. It induces the growth of $^{57}Fe_2O_3$ on basal faces and enriches the solution with $^{56}Fe(II)$ (Rosso *et al.*, 2010). In terms of implementation, layers of ~100 nm can be deposited in 12 hours, without any limitation occurring. The same mechanism can occur in a neutral medium at pH 7 (Catalano *et al.*, 2010).

Much smaller hematite nanoparticles (30 m^2g^{-1}), synthesized by precipitation in solution, behave differently and give rise to an obvious size effect. The sorption of Fe(II) ions also induces an interfacial electron transfer and Fe(II) ions exchange with the solution, but a saturation phenomenon occurs after the formation of a monolayer. Subsequent Fe(II) ion sorption forms ferrous hydroxide (Larese-Casanova and Scherer, 2007). It seems that the electron transfer cannot occur beyond the first sorbed and

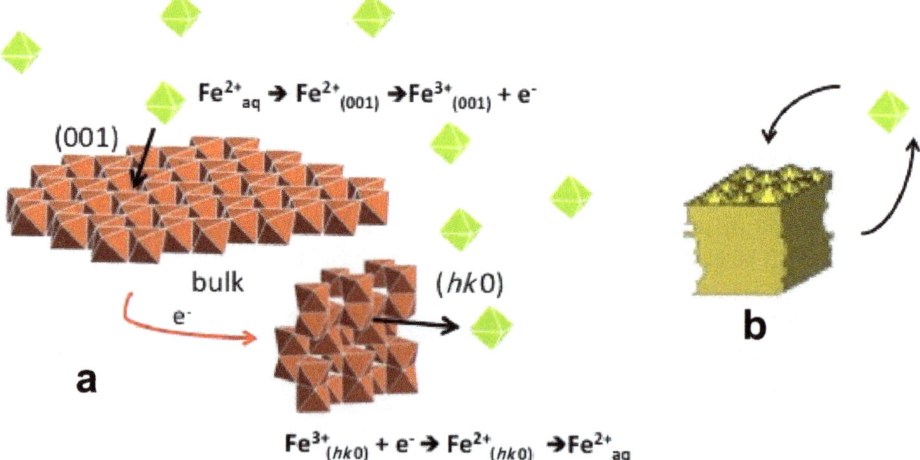

Figure 11. (a) Scheme depicting the coupled interfacial electron transfer process operative for the hematite single crystals in the presence of Fe(II) ions. The chemically self-induced surface potential gradient across the crystal directs current flow through the bulk. The low electrical resistivity is fed by net injection of electrons at (001) surfaces and net release of electrons at (*hk*0) surfaces. (b) Scheme of a crystal with pyramidal growth on (001) face and corrosion of lateral (*hk*0) faces. Green octahedra stand for soluble ferrous complexes and orange octahedra for structural $Fe(III)O_6$ polyhedra (From Yanina and Rosso, 2008).

coordinated Fe(II) ions. This limitation may be due to a surface disorder inhibiting the epitaxial crystallization of a hematite layer. Geometric constraints can distort the orbital overlaps allowing the Fe(II)–Fe(III) charge transfer (Sherman, 1987). The incorporation of structural defects during the adsorption of Fe(II) ions beyond the first adsorbed layer can also limit hematite crystallization (Larese-Casanova, 2007; Rosso *et al.*, 2010). It is, nevertheless, possible that the crystal faces differentiation in nanoscale particles is lower, the crystalline faces could be less well defined and the electrical field gradient through the bulk could be too low to bring the driving force to the process.

The sorption of Fe(II) ions may occur with different outcomes for the electrons injected in the solid, the process being heavily dependent on the electronic properties of the solid. With maghemite, Fe(II) ions are sorbed in building an epitaxial surface layer with delocalization of the injected electrons in the conduction band. With a semi-conducting oxide such as hematite, the electrons injected take part in reductive dissolution of the solid, as discussed above. These electrons may also be localized in a trapping site as surface state (Gorski and Scherer, 2011). The reducing power of such a system is obviously dependent on its band gap and probably on the particle size, but to our knowledge no data have been published. The surface disorder, which is even more important for small particle sizes, may be due in part to the under-coordination of sub-surface atoms. It may be reduced by coordinating strongly chelating ligands, such as enediol (Fig. 12), which can complete the coordination number of sub-surface cations (Rajh *et al.*, 2002). Thus, ligands can not only restructure the surface but also modify the electronic structure of the solid in forming surface charge-transfer complexes leading to a strong electronic coupling of ligand to surface. That introduces energy states inside the band gap and the optical absorption of coated particle suspensions is thus shifted to the lower energy, creating the phenomenon of 'sensitization'. Using enediols as chelating ligands with 6 nm-sized hematite nanoparticles, the width of the band gap decreases from 2 eV to 1.40, 1.12 and 0.89 eV with alizarin, ascorbic acid and dopamine, respectively (Rajh *et al.*, 2002).

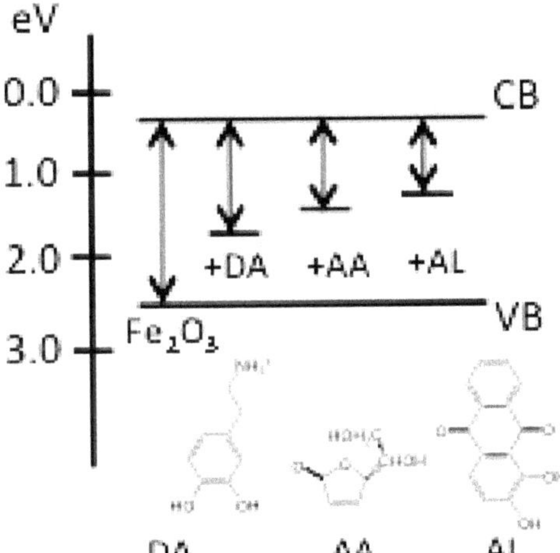

Figure 12. Scheme of the modification of hematite electronic structure by coordination of enediol ligands (dopamine (DA), ascorbic acid (AA) and alizarin (AL)). Additional electronic states are introduced inside the band gap of the oxide and less energy is required to excite electrons from the valence band (VB) into the conduction band (CB) (From Rajh *et al.*, 2002).

Adsorbed Fe(II) on an insulator such as goethite may reduce oxidizing species in solution. Thus, goethite gives rise to enhanced redox reactivity by adsorption of Fe(II) from solution (Silvester *et al.*, 2005; Handler *et al.*, 2009), *e.g.* in reduction of Cr(VI) (Deng and Stone, 1996) or in transformation of arsenic (Amstaetter *et al.*, 2010). The reaction of Fe(II) ions with differently sized goethite nanorods (80 nm × 7 nm, 330 nm × 14 nm, 670 nm × 25 nm) was followed in pH ranging from 6.1 to 7.5 (Cwiertny *et al.*, 2009a). In these conditions, the particles are unfortunately aggregated strongly and exhibit very similar reactivities for both nanorods and microrods as regard to Fe(II) sorption, Fe(II)–Fe(III) interfacial electron transfer and nitrobenzene reduction. Aggregation results also in a comparable behaviour of goethite particles of different sizes in reductive dissolution or not in the presence of oxalate (Cwiertny *et al.*, 2009b). It is of course very difficult to assess the active surface in suspension, which is certainly less than that measured by BET. However, in a more acidic medium (pH 3.75) where aggregation and flocculation of particles are at a minimum, the reactivity of goethite nanorods (64 nm × 5 nm) and microrods (367 nm × 22 nm) are clearly differentiated from the reductive dissolution by hydroquinone (Anschutz and Penn, 2005). The specific surface-normalized dissolution rate constant is nearly twice as high for nano- than for micro-rods, but the change in reactivity seems to be the result of structural differences between end and lateral faces rather than from an electronic origin.

4.3. Redox catalysis

Similarly to the interaction of Fe(II) ions with iron oxide, adsorption of other divalent cations on iron oxides may also give rise to enhanced redox reactivity. For instance, the oxidation of Mn^{2+} ions by oxygen leads to manganite γ-MnOOH according to the reaction:

$$Mn^{2+} + 1/4\ O_2 + 3/2\ H_2O \rightarrow \gamma\text{-}MnOOH + 2\ H^+$$

This reaction proceeds extremely slowly in a homogeneous phase at pH <8.5. In the presence of hematite with decreasing particle size the reaction is found to be accelerated (Madden and Hochella, 2005). For particles of 37 or 7 nm in diameter, with normalized surface area, the reaction rate was found to be ~1 and 1½ orders of magnitude greater. The acceleration of the reaction appears to be due to three main effects:

- the adsorption of the Mn cation which modifies its redox potential relative to that of the ion in solution (Werhli, 1989; Silvester, 2005) and reduces the free energy of the reaction;
- the presence of more surface defects on small hematite crystals introducing low-symmetry surface coordination sites and allowing stabilization of Mn^{3+} ions strongly distorted by the Jahn-Teller effect (Junta and Hochella, 1994; Madden *et al.*, 2006);
- the increase in the Lewis base character of the surface oxygen atoms as the particle size decreases (Noguera *et al.*, 2002) which enhances interaction with adsorbed cations.

Thus, the catalytic effect results from greater adsorption on the particles and reduction of the activation energy of the electron transfer (Madden and Hochella, 2005).

Surprisingly, the same oxidation reaction of Mn^{2+} in the presence of hematite, carried out under similar conditions, was found to be much slower when the size of the hematite nanoparticles was small (Chernyshova *et al.*, 2011a). The study was conducted with particles 7, 9 and 38 nm in diameter obtained by thermal hydrolysis, and with commercial particles 150 nm in size. The rate constants of pseudo-first order normalized by a specific surface are nearly identical for particles of 7 and 9 nm in both studies, but for the larger particles, the results are divergent. The reason seems to be a more or less abrupt change in pH during the introduction of Mn^{2+} ions in the non-buffered suspension leading to differences of acidity in the different study areas, as well as different approximations of processing kinetic data (Chernyshova *et al.*, 2011a). However, it seems more likely that commercial particles and homemade particles must have somewhat different surface and/or structural characteristics. Other than this distinction, it is interesting to have another interpretation of the catalytic effect, based on an electrochemical model instead of the chemical paradigm based on the activated complex model, which assumes the proximity of the redox partners to the surface (Fig. 13). Similar to the previous scheme for the interaction of Fe(II) oxides with Fe(III), the model is based on the fact that the two redox half reactions are separated spatially and are coupled electrically by charge transfer through the bulk or the surface of the solid catalyst (Chernyshova *et al.*, 2011a,b). Thus the electronic structure of the particle and its change with size determine its (electro)chemical reactivity.

In such a model, the lowering of reactivity with the decrease in particle size results essentially from the band gap opening with the decrease in particle size (see

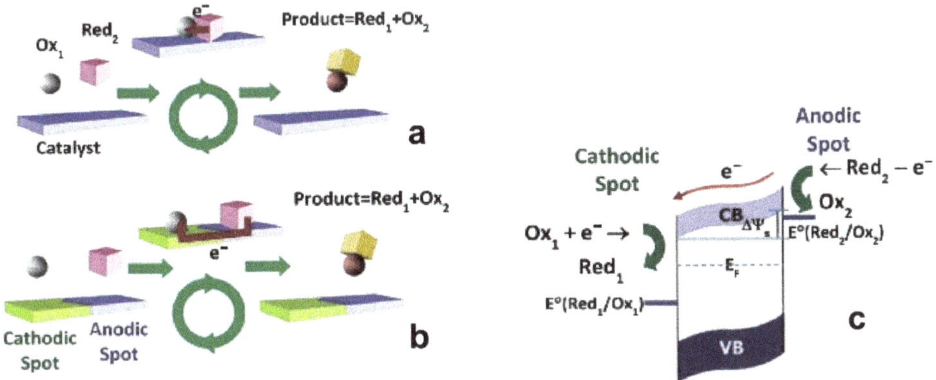

Figure 13. Schematic representations of (a) chemical and (b) electrochemical mechanisms involved in heterogeneous catalysis of redox reaction Ox_1+Red_2 ∞ Red_1+Ox_2. A chemical mechanism is shown in the case of a charge transfer between two proximate adsorbates Red_2 and Ox_1. (c) Electrochemical mechanism is shown for an n-type semiconductor with a heterogeneous surface having anodic and cathodic spots. $\Delta\Psi s$ = gradient of potential, VB and CB are valence band and conduction band respectively. (From Chernyshova *et al.*, 2011b)

Section 2.3. Electronic structure). The origin of band-gap opening is attributed to weakening of the Fe(3d)−O(2p) orbital hydridization due to lattice disorder and relaxation of trigonal distortion of the octahedral coordination of ferric ions. So, the upward shift of the conduction band hampers the electron injection from the reducing species, reducing the redox reaction.

These considerations raise an open question concerning the role of size effect in the electronic structure of semi-conducting solids and especially hematite. Indeed, experimental data related to the opening of band gap with decreasing size are sometimes contradictory: the decrease in size leads to a hypochromic shift of the optical absorption edge related to confinement effect in most studies (Chernyshova *et al.*, 2011b) or less distortion of the crystal lattice (Sivula *et al.*, 2010). A bathochromic shift has also been observed (Pailhé *et al.*, 2008) while other experiments report that there is no change in the band gap for hematite nanoparticles up to 8 nm in size (Gilbert *et al.*, 2009). Based on theoretical consideration, a change in electronic structure with size is expected but not always observed, probably because for very small particles many defects, *e.g.* inclusions of impurities such as ferrihydrite or protons in the structure, affect the electronic structure giving non-reproducible results depending on the synthesis method and on purification techniques of materials. It is likely that for hematite, an effective confinement effect exists, maybe for smaller sizes, and efforts to achieve a clear understanding of such phenomena are still needed.

4.4. Size-dependent reduction reactions induced by adsorption onto magnetite

Due to fast electron hopping in the spinel iron oxide network, magnetite nanoparticles in aqueous suspension may be used as a strong reductant. The oxidation of nanometric magnetite may occur by Fe(III) ion adsorption at pH $\approx$ 8 (Belleville *et al.*, 1992). Sorbed Fe(III) ions immediately form a layer of ferrihydrite which is partially reduced by electron transfer from the core of the particle, allowing layer crystallization in epitaxy on oxidized particles. Such oxidation of magnetite proceeds with an increase in size corresponding to a thickness of around two spinel crystal cells. The reaction stops at this stage because the diffusion of cations from the core of the particle and through the sorbed layer is the limiting rate of the reaction. A similar process occurs with easily reducible metal cations such as silver, platinum or palladium (Jolivet *et al.*, 1990). The reaction proceeds quasi immediately without Fe(III) desorption but induces surface hydroxylation allowing the coordination of Fe(III) ions that are diffused from the core, with an abrupt acidification of the solution according to Fig. 14. The reaction forms a mixture of maghemite and silver nanoparticles. All these reactions are enhanced by a decrease in particle size, but quantitative data on kinetics are needed.

Another example concerns the reduction by nanomagnetite of carbon tetrachloride, CCl_4, mainly to chloroform, $CHCl_3$. The reaction rate is influenced heavily by the particle size (Vikesland *et al.*, 2007). In kinetic conditions corresponding to the pseudo-first-order law of the CCl_4 degradation reaction, the rate constant, normalized by specific surface area, is greater by one order of magnitude for 9 nm particles than for

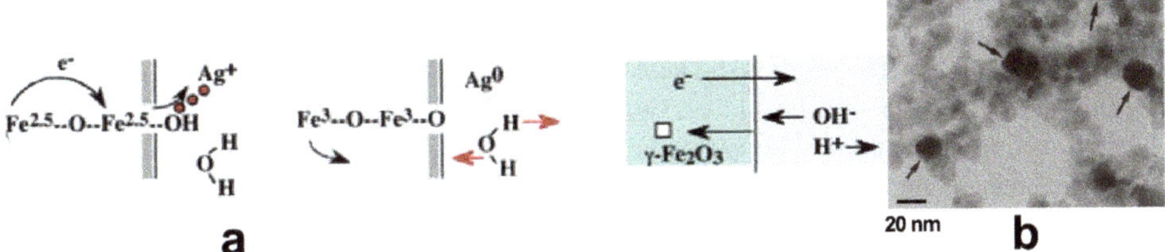

Figure 14. (a) Adsorption of Ag^+ ions onto magnetite induces, similarly to the protonation mechanism (Fig. 7), a mobile electron to be localized on a surface Fe ion and causing the reduction of silver ions. Charge balance in spinel particles proceeds by diffusion of Fe^{3+} towards the surface. Surface solvation and deprotonation of a water molecule allow the local charge balance of Fe(III) ions and explains the rapid acidification from pH 8 to 4 of the medium. The balance reaction corresponds to the elimination of an electron from magnetite and migration of vacancies towards the core of a particle, to the hydroxylation of the surface and the release of one proton in the solution. (b) Silver nanoparticules (marked by arrows) resulting from the reduction of silver nitrate by magnetite nanoparticles in suspension at pH 8 (final pH 4) (Jolivet *et al.*, 1990).

80 nm particles. Even if it is difficult to assess quantitatively the size effect because other factors, such as particle aggregation, could modify their behaviour and the available surface area, these results clearly demonstrated increased reactivity with the reduction in size of the nanoparticles of magnetite. As the redox transfer involved in the reduction implies likely adsorption of the CCl_4 molecule by hydroxylated surface groups, the kinetic efficiency of the smallest particles could be due to greater surface energy leading to stronger adsorption energy. Such an increase in reactivity could also be due to surface structural defects, as illustrated by As adsorption on maghemite nanoparticles (Auffan *et al.*, 2008).

Many other studies of reductive adsorption of metal ions on magnetite have been carried out, *e.g.* Cr(VI) reduced to Cr(III) and coordinated to the surface as inner-sphere complexes (Peterson *et al.*, 1996). Unfortunately, particle-size influence is not described. However, as in the examples above, an increase in reactivity with a decrease in particle size may be anticipated.

4.5. Bioreduction

Bioreduction of iron oxides is probably quantitatively the most important contribution to mobilization of iron in the environment. Of course the reactivity of nanoparticles is one of the main characteristics of the process and one may expect a significant particle-size effect, but that seems to be less evident.

Reduction of hematite nanoparticles 10, 20 and 50 nm in mean diameter (synthesized by an aerosol technique) by *Geobacter sulfurreducens* has been performed at pH 7 (Yan *et al.*, 2008). The mass-normalized reduction rates, similar for the 10 and 30 nm-sized particles are significantly larger than for 50 nm particles. However, standardization of the specific surface area shows that 30 nm particles are the most reactive. This suggests that particle aggregation must play a significant role in their reactivity by decreasing the available surface. Bioreduction at pH 7 by *Shewanella oneidensis* MR-1 of hematite nanoparticles with a large size range (11–99 nm) (Bose *et al.*, 2009) shows that the surface area-normalized reduction rate is one order of magnitude greater for the 99 nm particles than for the smaller (11–43 nm) particles which have very similar reactivity. In this case, these last particles were synthesized by thermohydrolysis of acidic solution of ferric nitrate and are significantly aggregated, while the larger (99 nm) are a commercial product used as received. Bioreduction by *Shewanella alga* of goethite nanoparticles with various surface areas between 30 and 150 m^2 g^{-1} showed no size effect on the initial reduction rate, but the long-term reduction progress was correlated linearly with the specific surface area of the samples (Roden and Zachara, 1996).

In these examples, it is difficult to identify a significant effect of size, mainly because of aggregation. However, it has been shown that the reduction rate of hematite by *Shewanella putrefaciens* is linearly correlated to the surface of the bacteria covered by nanoparticles (Bonneville *et al.*, 2006). As the reduction of iron by *Geobacter sulfurreducens* occurs at the outer surface of the cell, one might expect the reduction rate to be correlated with the bacteria-area contact rather than the total surface area of the particles, and aggregation does not appear to be a serious drawback. In addition, bioreduction may involve different mechanisms depending on the nature of bacteria and reaction conditions. Electron transfer may proceed from contact between the oxide and the outer membrane of bacteria or *via* electron shuttles synthesized by bacteria (*e.g.* quinones, riboflavin, riboflavin 5′-phosphate and flavin mono nucleotide) allowing the reduction of ferric ions (Bose *et al.*, 2009). It is also possible that the mechanism is controlled by a combination of these possibilities (Rosso *et al.*, 2003a). It seems, therefore, that the crystalline state of the material is a very important factor in its (bio)reactivity (Cutting *et al.*, 2009); the better the crystallinity, the lower the reactivity. It also seems difficult to compare samples issued from different synthesis procedures, giving in general, particles with different surface states or different crystalline faces.

5. Conclusions

This short review shows the difficulty in providing a correlation between the change in size of the oxide particles, especially for hematite, and their (electro)chemical reactivity. It is clear that at the nanoscale, a very precise description of the particle itself becomes a challenge. Due to internal or superficial structural distortions, the surface and/or the core are subjected to various reconstructions and to structural as well as chemical defects, making the electronic characterization very difficult. Nevertheless, a large majority of experiments reports that the reactivity of a solid increases with decreasing particle size, as that seems intuitive from structural and thermodynamic considerations. Differences in reactivity are, of course, related to the crystal and electronic structure and also to the morphology, as illustrated by iron oxides.

Acknowledgements

Benjamin Gilbert and Alain Manceau are warmly acknowledged for fruitful discussions.

References

Amstaetter, K., Borsch, T., Larese-Casanova, Ph. and Kappler, A. (2010) Redox transformation of arsenic by Fe(II)-activated goethite (α-FeOOH). *Environmental Science & Technology*, **44**, 102–108.

Anschutz, A.J. and Penn, R.L. (2005) Reduction of crystalline iron(III) oxyhydroxides using hydroquinone: influence of phase and particle size. *Geochemical Transactions*, **6**, 60–66.

Auffan, M., Rose, J., Proux, O., Borschneck, D., Masion, A., Chaurand, P., Hazemann, J.L., Chanéac, C., Jolivet, J.P., Wiesner, M.R., Van Geen, A. and Bottero, J.Y. (2008) Enhanced adsorption of arsenic onto maghemite nanoparticles: As(III) as a probe of the surface structure and heterogeneity. *Langmuir*, **24**, 3215–3222.

Belleville, Ph., Jolivet, J.P., Tronc, E. and Livage J. (1992) Crystallization of ferric hydroxide into spinel by adsorption on colloidal magnetite. *Journal of Colloid and Interface Science*, **150**, 453–460.

Birkner, N. and Navrotsky, A. (2012) Thermodynamics of manganese oxides: effects of particle size and hydration on oxidation-reduction equilibria among hausmannite, bixbyite, and pyrolusite. *American Mineralogist*, **97**, 1291–1298.

Bonneville, S., Behrends, T., Van Cappellen, P., Hyacinthe, C. and Roling, W.F.M. (2006) Reduction of Fe(III) colloids by *Shewanella putrefaciens*: a kinetic model. *Geochimica et Cosmochimica Acta*, **70**, 5842–5854.

Borch, T., Kretzschmar, R., Kappler, A., Van Cappellen, Ph., Ginder-Vegel, M., Voegelin, A. and Campbell, K. (2010) Biogeochemical redox processes and their impact on contaminant dynamics. *Environmental Science & Technology*, **44**, 15–23.

Bose, S., Hochella, M.F. Jr., Gorby, Y.A., Kennedy, D.W., McCready, D.E., Madden, A.S. and Lower, B.H. (2009) Bioreduction of hematite nanoparticles by the dissimilatory iron reducing bacterium *Shewanella oneidensis* MR-1. *Geochimica et Cosmochimica Acta*, **73**, 962–976.

Brice-Profeta, S., Arrio, M.-A., Tronc, E., Menguy, N., Letard, I., Cartier dit Moulin, C., Noguès, M., Chanéac, C., Jolivet, J.-P. and Sainctavit, Ph. (2005) Magnetic order in γ-Fe_2O_3 nanoparticles: a XMCD study. *Journal of Magnetism and Magnetic Materials*, **288**, 354–365.

Burdett, J.K. (1995) *Chemical Bonding in Solids*. Oxford University Press, New York, 319 pp.

Catalano, J.G., Fenter, P., Park, C., Zhang, Z. and Rosso, K.M. (2010) Structure and oxidation state of hematite surfaces reacted with aqueous Fe(II) at acidic and neutral pH. *Geochimica et Cosmochimica Acta*, **74**, 1498–1512.

Chanéac, C., Tronc, E. and Jolivet, J.P. (1996) Magnetic iron oxide-silica nanocomposites. Synthesis and characterization. *Journal of Materials Chemistry*, **6**, 1905–1911.

Charlet, L., Morin, G., Rose, J., Wang, Y., Auffan, M., Burnol, A. and Fernandez-Martinez, A. (2011) Reactivity at (nano)particle-water interfaces, redox processes, and arsenic transport in the environment. *Comptes Rendus Geoscience*, **343**, 123–139.

Chernyshova, I.V., Hochella, M.F., Jr. and Madden, A.S. (2007) Size-dependent structural transformations of hematite nanoparticles. 1. Phase transition. *Physical Chemistry Chemical Physics*, **9**, 1736–1750.

Chernyshova, I.V., Ponnurangam, S. and Somasundaran, P. (2010) On the origin of an unusual dependence of (bio)chemical reactivity of ferric hydroxides on nanoparticle size. *Physical Chemistry Chemical Physics*, **12**, 14045–14056.

Chernyshova, I.V., Ponnurangam, S. and Somasundaran, P. (2011a) Effect of nanosize on catalytic properties of ferric (hydr)oxides in water: Mechanistic insights. *Journal of Catalysis*, **282**, 25–34.

Chernyshova, I.V., Ponnurangam, S. and Somasundaran, P. (2011b) Tailoring (bio)chemical activity of semiconducting nanoparticles: critical role of deposition and aggregation. *Journal of the American Chemical Society*, **133**, 9536–9544.

Combes, J.M., Manceau, A., Calas, G. and Bottero, J.-Y. (1989) Formation of ferric oxides from aqueous solutions: A polyhedral approach by X-ray absorption spectroscopy: I. Hydrolysis and formation of ferric gels. *Geochimica et Cosmochimica Acta*, **53**, 583–594.

Cornell, R.M. and Schwertmann, U. (2003) *The Iron Oxides: Structure, Properties, Reactions, Occurrence, and Uses*. VCH, New York, 664 pp.

Cutting, R.S., Coker, V.S., Fellowes, J.W., Lloyd, J.R. and Vaughan, D.J. (2009) Mineralogical and morphological constraints on the reduction of Fe(III) minerals by *Geobacter sulfurreducens*. *Geochimica et Cosmochimica Acta*, **73**, 4004–4022.

Cwiertny, D.M., Handler, R.M., Schaefer, M.V., Grassian, V.H. and Scherer, M.M. (2009a) Interpreting nanoscale size-effects in aggregated Fe-oxide suspensions: reaction of Fe(II) with goethite. *Geochimica et Cosmochimica* Acta, **72**, 1365–1380.

Cwiertny, D.M., Hunter, G.J., Pettibone, J.M., Scherer, M.M. and Grassian, V.H. (2009b) Surface chemistry and dissolution of α-FeOOH nanorods and microrods: environmental implications of size-dependent interactions with oxalate. *Journal of Physical Chemistry C*, **113**, 2175–2186.

Deng, B. and Stone, A.T. (1996) Surface-catalyzed chromium(VI) reduction: reactivity comparisons of different organic reductants and different oxide surfaces. *Environmental Science & Technology*, **30**, 2484–2494.

Doelsch, E., Rose, J., Masion, A., Bottero, J.-Y., Nahon, D. and Bertsch, P.M. (2000) Speciation and crystal chemistry of iron(III) chloride hydrolyzed in the presence of SiO_4 ligands. 1. An Fe K-edge EXAFS study. *Langmuir*, **16**, 4726–4731.

Dong, D.M., Nelson, Y.M., Lion, L.W., Shuler, M.L. and Ghiorse, W.C. (2000) Adsorption of Pb and Cd onto metal oxides and organic material in natural surface coatings as determined by selective extractions: new evidence for the importance of Mn and Fe oxides. *Water Research*, **34**, 427–436.

Dormann, J.L., Fiorani, D. and Tronc E. (1997) Magnetic relaxation in fine-particle systems. *Advances in Chemical Physics*, **98**, 283–494.

Fernandez-Garcia, M., Martinez-Arias, A., Hanson, J.C. and Rodriguez J.A. (2004) Nanostructured oxides in chemistry: characterization and properties. *Chemical Reviews*, **104**, 4063–4104.

Fritsch, S., Post, J.E. and Navrotsky A. (1997) Energetics of low-temperature polymorphs of manganese dioxide and oxyhydroxide. *Geochimica et Cosmochimica Acta*, **61**, 2613–2616.

Gilbert, B. and Banfield, J.F. (2005) Molecular-scale processes involving nanoparticulate minerals. Pp. 109–155 in: *Molecular Geomicrobiology* (J.F. Banfield, J. Cervini-Silva and K.H. Nealson, editors). Reviews in Mineralogy and Geochemistry, **59**, Mineralogical Society of America and Geochemical Society, Chantilly, Virginia, USA.

Gilbert, B., Huang, F., Zhang, H., Waychunas, G.A. and Banfield J.F. (2004) Nanoparticles: strained and stiff. *Science*, **305**, 651–654.

Gilbert, B., Frandsen, C., Maxey, E.R. and Sherman, D.M. (2009) Band-gap measurements of bulk and nanoscale hematite by soft X-ray spectroscopy. *Physical Review B*, **79**, 035108.

Gilbert, B., Erbs, J.J., Penn, R.L., Petkov, V., Spagnoli, D. and Waychunas, G.A. (2013) A disordered nanoparticle model for 6-line ferrihydrite. *American Mineralogist*, **98**, 1465–1476.

Gorski, C.A. and Scherer, M.M. (2009) Influence of magnetite stoichiometry on Fe^{II} uptake and nitrobenzene reduction. *Environmental Science & Technology*, **43**, 3675–3680.

Gorski, C.A., Nurmi, J.T., Tratnyek, P.G., Hofstetter, T.B. and Scherer, M.M. (2010a) Redox behavior of magnetite: implications for contaminant reduction. *Environmental Science & Technology*, **44**, 55–60.

Gorski, C.A. and Scherer, M.M. (2010b) Determination of nanoparticulate magnetite stoichiometry by Mössbauer spectroscopy, acidic dissolution, and powder X-ray diffraction: a critical review. *American Mineralogist*, **95**, 1017–1026.

Gorski, C.A. and Scherer, M.M. (2011) Fe sorption at the Fe oxide-water interface: a revised conceptual framework. Pp. 315–343 in: *Aquatic Redox Chemistry* (P.G. Tratnyek, T.J. Grundl and S.B. Haderlein, editors). ACS Symposium Series, Washington, D.C..

Grätzel, M. (1983) *Energy Resources Through Photochemistry and Catalysis*. Academic Press, New York, 573 pp.

Hadjipanayis, G.C. and Siegel, R.W. (1994) Nanophase materials, synthesis-properties-applications. *NATO ASI Series E: Applied Sciences*, Vol. **260**.

Handler, R.M., Beard, B.L., Johnson, C.M. and Scherer, M.M. (2009) Atom exchange between aqueous Fe(II) and goethite: an Fe isotope tracer study. *Environmental Science & Technology*, **43**, 1102–1107.

Hem, D.J. (1977) Reactions of metal ions at surfaces of hydrous iron oxide. *Geochimica et Cosmochimica Acta*, **41**, 527–538.

Hochella, M.F. Jr., Lower, S.K., Maurice, P.A., Penn, R.L., Sahai, N., Sparks, D.L. and Twining, B.S. (2008) Nanominerals, mineral nanoparticles and Earth systems. *Science*, **319**, 1631–1635.

Jambor, L. and Dutrizac, J.E. (1998) Occurrence and constitution of natural and synthetic ferrihydrite, a widespread iron oxyhydroxide. *Chemical Review*, **98**, 2549–2585.

Jeon, B.H., Dempsey, B.A. and Burgos, W.D. (2003). Kinetics and mechanisms for reactions of Fe(II) with iron(III) oxides. *Environmental Science & Technology*, **37**, 3309–3315.

Jolivet, J.P. and Tronc, E. (1988) Interfacial electron transfer in colloidal spinel iron oxide. Conversion of Fe_3O_4 into γ-Fe_2O_3 in aqueous medium. *Journal of Colloid and Interface Science*, **125**, 688–701.

Jolivet, J.P., Tronc, E., Barbé, C. and Livage, J. (1990) Interfacial electron transfer in colloidal spinel iron oxide. Silver ion reduction in aqueous medium. *Journal of Colloid and Interface Science*, **138**, 465–472.

Jolivet, J.P., Belleville, P., Tronc, E. and Livage, J. (1992) Influence of Fe(II) ions on the formation of spinel iron oxide in aqueous medium. *Clays and Clay Minerals*, **40**, 531–540.

Jolivet, J.P., Froidefond, C., Pottier, A., Chanéac, C., Cassaignon, S., Tronc, E. and Euzen, P. (2004) Size tailoring of oxide nanoparticles by precipitation in aqueous medium. A semi-quantitative modelling. *Journal of Materials Chemistry*, **14**, 3281–3288.

Jolivet, J.P., Tronc, E. and Chanéac, C. (2006) Iron oxides: from molecular clusters to solid. A nice example of chemical versatility. *Comptes Rendus Geoscience*, **338**, 488–497.

Junta, J.L. and Hochella, M.F. Jr. (1994) Manganese (II) oxidation at mineral surfaces: a microscopic and spectroscopic study. *Geochimica et Cosmochimica Acta*, **58**, 4985–4999.

Laberty, C. and Navrotsky, A. (1998) Energetics of stable and metastable low-temperature iron oxides and oxyhydroxides. *Geochimica et Cosmochimica Acta*, **62**, 2905–2913.

Larese-Casanova, Ph. and Scherer, M.M. (2007) Fe(II) sorption on hematite: new insights based on spectroscopic measurements. *Environmental Science & Technology*, **41**, 471–477.

Lemaire, B.J., Davidson, P., Ferré, J., Jamet, J.P., Paninev, D.I. and Jolivet, J.P. (2002) Outstanding magnetic properties of nematic suspensions of goethite (α-FeOOH) nanorods. *Physical Review Letters*, **88**, 125507.

Li, Y. and Somorjai, G.A. (2010) Nanoscale advances in catalysis and energy applications. *Nano Letters*, **10**, 2289–2295.

Madden, A.S. and Hochella, M.F. Jr. (2005) A test of geochemical reactivity as a function of mineral size: manganese oxidation promoted by hematite nanoparticles. *Geochimica et Cosmochimica Acta*, **69**, 389–398.

Madden, A.S., Hochella, M.F. Jr. and Luxton, T.P. (2006) Insights for size-dependent reactivity of hematite nanomineral surfaces through Cu^{2+} sorption. *Geochimica et Cosmochimica Acta*, **70**, 4095–4104.

Manceau, A., Skanthakumar, S. and Soderholm, L. (2014) PDF analysis of ferrihydrite: Critical assessment of the under-constrained akdalaite model. *American Mineralogist*, **99**, 102–108.

Matijevic, E. (1988) Colloid science of ceramic powders. *Pure and Applied Chemistry*, **60**, 1479–1491.

Monticone, S., Tufeu, R., Kanaev, A.V., Scolan, E. and Sanchez, C. (2000) Quantum size effect in TiO_2 nanoparticles: does it exist? *Applied Surface Science*, **162–163**, 565–570.

Morin, G., Wang, Y., Ona-Nguema, G., Juillot, F., Calas, G., Menguy, N., Aubry, N., Bargar, J.R. and Brown, G.E. Jr. (2009) EXAFS and HRTEM evidence for As(III)-containing surface precipitates on nanocrystalline magnetite: implications for As sequestration. *Langmuir*, **25**, 9119–9128.

Mulvaney, P., Cooper, R., Grieser, F. and Meisel, D. (1988) Charge trapping in the reductive dissolution of colloidal suspensions of iron(III) oxides. *Langmuir*, **4**, 1206–1211.

Murray, C.B., Norris, D.J. and Bawendi, M.G. (1993) Synthesis and characterization of nearly monodisperse CdE (E = S, Se, Te) semiconductor nanocrystallites. *Journal of the American Chemical Society*, **115**, 8706–8715.

Navrotsky, A. (2003) Energetics of nanoparticle oxides: interplay between surface energy and polymorphism. *Geochemical Transactions*, **4**, 34–37.

Navrotsky, A., Mazeina, L. and Majzlan, J. (2008) Size-driven structural and thermodynamic complexity in iron oxides. *Science*, **319**, 1635–1638.

Noguera, C., Pojani, A., Casek, P. and Finocchi, F. (2002) Electron redistribution in low-dimensional oxide structures. *Surface Science*, **507–510**, 245–255.

Pailhé, N., Wattiaux, A., Gaudon, M. and Demourgues, A. (2008) Impact of structural features on pigment properties of α-Fe_2O_3 haematite. *Journal of Solid State Chemistry*, **181**, 2697– 2704.

Parfitt, R.L., van der Gast, S.J. and Childs, C.W. (1992). A structural model for natural siliceous ferrihydrite. *Clays and Clay Minerals*, **40**, 675–681.

Pedersen, H.D., Postma, D., Jakobsen, R. and Larsen, O. (2005) Fast transformation of iron oxyhydroxides by the catalytic action of aqueous Fe(II). *Geochimica et Cosmochimica Acta*, **69**, 3967–3977.

Peterson, M.L., Brown, G.E. Jr. and Parks, G.A. (1996) Direct XAFS evidence for heterogeneous redox reaction at the aqueous chromium/magnetite interface. *Colloids and Surfaces*, **107**, 77–88.

Pottier, A., Cassaignon, S., Chanéac, C., Villain, F., Tronc, E. and Jolivet, J.P. (2003) Size tailoring of TiO_2 anatase nanoparticles in aqueous medium and synthesis of nanocomposites. Characterization by Raman spectroscopy. *Journal of Materials Chemistry*, **13**, 877–882.

Rajh, T., Chen, L.X., Lukas, K., Liu, T., Thurnauer, M.C. and Tiede, D.M. (2002) Surface restructuring of nanoparticles: an efficient route for ligand-metal oxide crosstalk. *Journal of Physical Chemistry B*, **106**, 10543–10552.

Roden, E.E. and Zachara, J.M. (1996) Microbial reduction of crystalline iron(III) oxides: influence of oxide surface area and potential for cell growth. *Environmental Science & Technology*, **30**, 1618–1626.

Rose, J., Manceau, A., Masion, A., Bottero, J.-Y. and Garcia, F. (1996) Nucleation and growth mechanisms of Fe oxyhydroxide in the presence of PO_4 ions. 1. Fe K-edge EXAFS study. *Langmuir*, **12**, 6701–6707.

Rose, J., Flanck, A.M., Masion, A., Bottero, J.Y. and Elmerich, P. (1997) Nucleation and growth mechanisms of Fe oxyhydroxide in the presence of PO_4 ions. 2. P K-Edge EXAFS study. *Langmuir*, **13**, 1827–1834.

Rosso, K.M., Zachara, J.M., Fredrickson, J.K., Gorby, Y.A. and Smith, S.C. (2003a) Nonlocal bacterial electron transfer to hematite surfaces. *Geochimica et Cosmochimica Acta*, **67**, 1081–1087.

Rosso, K.M., Smith, D.M.A. and Dupuis, M. (2003b) An ab initio model of electron transport in hematite (α-Fe_2O_3) basal planes. *Journal of Chemical Physics*, **118**, 6455–6466.

Rosso, K.M., Yanina, S.V., Gorski, C.A., Larese-Casanova, Ph. and Scherer, M.M. (2010) Connecting observations of hematite (α-Fe_2O_3) growth catalysed by Fe(II). *Environmental Science & Technology*, **44**, 61–67.

Rubasinghege, G., Lentz, R.W., Scherer, M.M. and Grassian, V.H. (2010) Simulated atmospheric processing of iron oxyhydroxide minerals at low pH: roles of particle size and acid anion in iron dissolution. *Proceedings of the National Academy of Science*, **107**, 6628–6633.

Satoh, N., Nakashima, T., Kamikura, K. and Yamamoto K. (2008) Quantum size effect in TiO_2 nanoparticles prepared by finely controlled metal assembly on dendrimer templates. *Nature Nanotechnology*, **3**, 106–111.

Scholes, G.D. and Rumbles, G. (2006) Excitons in nanoscale systems. *Nature Materials*, **5**, 683–696.

Schwertmann, U., Friedl, J. and Stanjek, H. (1999). From Fe(III) ions to ferrihydrite and then to haematite. *Journal of Colloid and Interface Science*, **209**, 215–223.

Sherman, D.M. (1984) The electronic structures of manganese oxide minerals. *American Mineralogist*, **69**, 788–799.

Sherman, D.M. (1987) Molecular orbital (SCF-Xa-SW) theory of metal-metal charge transfer processes in minerals II. Application to $Fe^{2+} \rightarrow Ti^{4+}$ charge transfer transitions in oxides and silicates. *Physics and Chemistry of Minerals*, **14**, 364–367.

Silvester, E., Charlet, L., Tournassat, C., Géhin, A., Grenèche, J.M. and Liger, E. (2005) Redox potential measurements and Mössbauer spectrometry of FeII adsorbed onto FeIII (oxyhydr)oxides. *Geochimica et Cosmochimica Acta*, **69**, 4801–4815.

Sivula, K., Zboril, R., Le Formal, F., Robert, R., Weidenkaff, A., Tucek, J., Frydrych, J. and Grätzel, M. (2010) Photoelectrochemical water splitting with mesoporous hematite prepared by a solution-based colloidal approach. *Journal of the American Chemical Society*, **132**, 7436–7444.

Stumm, W. and Morgan, J.J. (1996) *Aquatic Chemistry: Chemical Equilibria and Rates in Natural Waters.* Wiley, New York, 583 pp.

Stumm, W. and Sulzberger, B. (1992) The cycling of iron in natural environments: considerations based on laboratory studies of heterogeneous redox processes. *Geochimica et Cosmochimica Acta*, **56**, 3233–3257.

Tronc, E., Jolivet, J.P., Lefebvre, J. and Massart, R. (1984) Ion adsorption and electron transfer in spinel-like iron oxide colloids. *Journal of the Chemical Society Faraday Transactions I*, **80**, 2619–2629.

Tronc, E., Jolivet, J.P., Belleville, P. and Livage, J. (1989) Redox phenomena in spinel iron-oxide colloids induced by adsorption. *Hyperfine Interactions*, **46**, 637–643.

Tronc, E., Belleville, P., Jolivet, J.P. and Livage, J. (1992) Transformation of ferric hydroxide into spinel by Fe(II) adsorption. *Langmuir*, **8**, 313–319.

Tronc, E., Chanéac, C. and Jolivet, J.P. (1998) Structural and magnetic characterization of ε-Fe_2O_3. *Journal of*

Solid State Chemistry, **139**, 93–104.

Vayssières, L., Chanéac, C., Tronc, E. and Jolivet J.P. (1998) Size tailoring of magnetite particles formed by aqueous precipitation: an example of thermodynamic stability of nanometric oxide particles. *Journal of Colloid and Interface Science*, **205**, 205–212.

Vikesland, P.J., Heathcock, A.M., Rebodos, R.L. and Makus, K.E. (2007) Particle size and aggregation effects on magnetite reactivity toward carbon tetrachloride. *Environmental Science & Technology*, **41**, 5277–5283.

Waychunas, G.A., Kim, C.S. and Banfield, J.F. (2005) Nanoparticulate iron oxide minerals in soils and sediments: unique properties and contaminant scavenging mechanisms. *Journal of Nanoparticle Research*, **7**, 409–433.

Wehrli, B., Sulzberger, B. and Stumm, W. (1989) Redox processes catalysed by hydrous oxide surfaces. *Chemical Geology*, **78**, 167–179.

Williams, A.G.B. and Scherer, M.M. (2004) Spectroscopic evidence for Fe(II)-Fe(III) electron transfer at the iron oxide-water interface. *Environmental Science & Technology*, **38**, 4782–4790.

Yan, B., Wrenn, B.A., Basak, S., Biswas, P. and Giammar, D.E. (2008) Microbial reduction of Fe(III) in hematite nanoparticles by *Geobacter sulfurreducens*. *Environmental Science & Technology*, **42**, 6526–6531.

Yang, L., Steefel, C.I., Marcus, M.A. and Bargar, J.R. (2010) Kinetics of Fe(II)-catalyzed transformation of 6-line ferrihydrite under anaerobic flow conditions. *Environmental Science & Technology*, **44**, 5469–5475.

Yanina, S.V. and Rosso, K.M., 2008. Linked reactivity at mineral-water interfaces through bulk crystal conduction. *Science*, **320**, 218–222.

Yean, S., Cong, L., Yavuz, C.T., Mayo, J.T., Yu, W.W., Kan, A.T., Colvin, V.L. and Tomson, M.B. (2005) Effect of magnetite particle size on adsorption and desorption of arsenite and arsenate. *Journal of Materials Research*, **20**, 3255–3264.

Zeng, H., Singh, A., Basak, S., Ulrich, K.U., Sahu, M., Biswas, P., Catalano, J.G. and Giammar, D.E. (2009) Nanoscale size effects on uranium(VI) adsorption to hematite. *Environmental Science & Technology*, **43**, 1373–1378.

Zhang, H. and Banfield, J.F. (1998) Thermodynamic analysis of phase stability of nanocrystalline titania. *Journal of Materials Chemistry*, **8**, 2073–2076.

EMU Notes in Mineralogy, Vol. 17 (2017), Chapter 3, 33–54

Mössbauer and Raman spectroscopic *in situ* characterization of iron-bearing minerals in Mars: Exploration and cultural heritage

GÖSTAR KLINGELHÖFER[1,*], FERNANDO RULL[2], GLORIA VENEGAS[2], FERNANDO GÁZQUEZ[3] and JESÚS MEDINA[2]

[1]*Institut für Anorganische und Analytische Chemie, Universität Mainz, Germany, email: klingel@mail.uni-mainz.de*
[2]*Unidad Asociada UVa-CSIC al Centro de Astobiología, University of Valladolid, Spain*
[3]*Department of Earth Sciences, University of Cambridge, UK*
** Corresponding author*

An increasing number of minerals have been described from the Martian surface in recent times. The Mössbauer spectrometers onboard the rovers *Opportunity* and *Spirit* have revealed the presence of several hydrous minerals. The development of these miniaturized analytical instruments represented a milestone for planetary research. Future missions to Mars such as EXOMARS 2018 and Mars 2020 will be equipped with miniaturized Raman spectrometers, intending to carry out an exhaustive mineralogical study of the Mars surface. Furthermore, detection of the presence of organic compounds related to present or ancient life on Mars will be addressed. Investigation of mineral formation, biogenesis and other water-related processes on Earth in similar conditions to those experienced in the past on Mars, is of considerable interest in relation to present and future Mars missions. Accordingly, the present work reviews some recent results obtained in the framework of several field expeditions to potential Mars analogue sites, mainly Rio Tinto and Jaroso Ravine in Spain by combining Mössbauer and Raman spectroscopies.

The prototypes of the instruments used for Mars exploration can also be used in several other interesting terrestrial applications such as the investigation of cultural heritage. This is relevant because in many cases sampling and even physical contact is not possible and only techniques performing *in situ* and non-destructive analysis are allowed. Some examples are addressed in this work, including Roman and Greek artefacts and prehistoric rock art in caves of great importance, such as Altamira cave (northern Spain). The outcomes of these analyses demonstrate the advantages of these portable techniques in detecting, *in situ*, and in a totally non-destructive way, mineral phases, and in the case of Mössbauer, Fe-bearing minerals in particular.

1. Introduction

In the Meridiani Planum of Mars (Fig. 1), hydrated iron-bearing sulfates and iron oxyhydroxides were detected along a 25 km traverse by the miniaturized Mössbauer spectrometer MIMOS II onboard the rover *Opportunity* (Klingelhöfer *et al.*, 2004;

DOI: 10.1180/EMU-notes.17.3

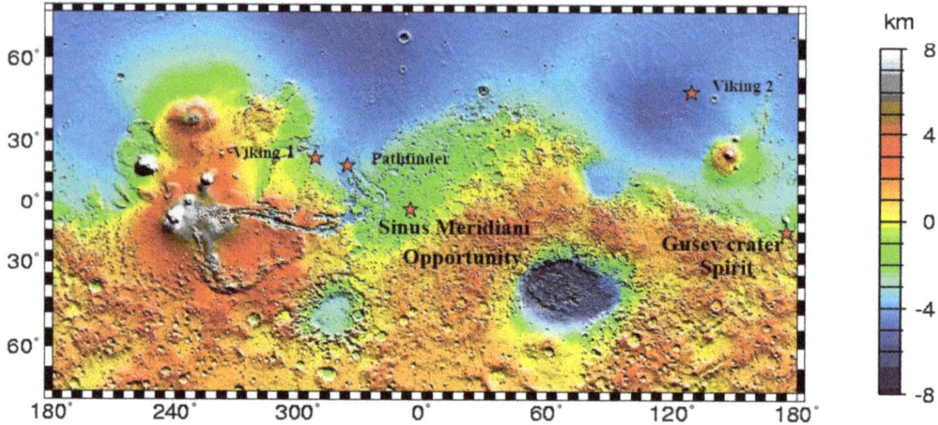

Figure 1. Geological map of the surface of Mars, showing some of the landing sites of successful missions, including the Martian rovers *Opportunity* and *Spirit*.

Squyres *et al.*, 2004) (Fig. 2). In particular, the presence of jarosite ($KFe_3^{3+}(SO_4)_2(OH)_6$) has been regarded as strong mineralogical evidence of wet and probably highly acidic conditions that prevailed during the Hesperian (3.7 to 3.0 Gy ago) age of Mars (Clark *et al.*, 2005; Ming *et al.*, 2006; Cockell, 2015).

Upon this outstanding discovery, potential Mars analogue sites on Earth in which jarosite had been detected have become the subject of a variety of studies (Frost *et al.*, 2005; Edwards *et al.*, 2007; Sobrón *et al.*, 2009; Rull *et al.*, 2005, 2008, 2009, 2014a;

Figure 2. NASA Mars-Exploration-Rover *Opportunity*, artist's view (courtesy of NASA, JPL, Cornell University, USA). On the front of the Rover the robotic arm carries the Mössbauer spectrometer and other instruments.

Guerrero, 2012). Among them, the Jaroso Ravine (Almería, SE Spain), the world type locality of jarosite, and the Rio Tinto site (Huelva, SW Spain), where a wide range of hydrated sulfates can be found (Frost *et al.*, 2005; Sobrón *et al.*, 2009; Rull *et al.*, 2014a), are rated as potential analogue sites of the mineralogenetic processes which occurred on Mars. Remarkably, in addition to jarosite, other hydrated sulfates such as kieserite ($MgSO_4{\cdot}H_2O$), epsomite ($MgSO_4{\cdot}7H_2O$), hexahydrite ($MgSO_4{\cdot}6H_2O$), gypsum ($CaSO_4{\cdot}2H_2O$), pentahydrite ($MgSO_4{\cdot}5H_2O$) and starkyite ($MgSO_4{\cdot}4H_2O$) (Zhu *et al.*, 2006) have been detected on the Mars surface, as well as the possible presence of szomolnokite ($Fe^{2+}SO_4{\cdot}H_2O$) and ferricopiapite ($Fe^{3+}_{2/3}Fe^{3+}_4(SO_4)_6(OH)_2{\cdot}20H_2O$) (Lane *et al.*, 2004). Most of these minerals have been described both in the Jaroso Ravine (Frost *et al.*, 2005; Venegas *et al.*, 2012) and in Rio Tinto (Sobrón *et al.*, 2009; Rull *et al.*, 2014a).

The ExoMars mission of the European Space Agency (ESA), scheduled for launch in 2018, will be equipped with a Raman spectrometer (Fig. 3). Raman spectroscopy was chosen as a technique for planetary exploration due to its versatility for structural

Figure 3. ExoMars mission rover, artist's view (courtesy of ESA). On the front of the rover, the drill that will take samples from Martian subsoil at depths of up to 2 m can be seen. The powdered samples prepared from the subsurface cores will be analysed by several instruments onboard the rover vehicle, including Raman spectroscopy.

identification of solids, liquids and gases, and a high potential for mineral identification (Rull and Martínez-Frías, 2006) and also organic detection. The so-called Raman effect is an inelastic interaction process between the exciting laser radiation and the illuminated matter. The observed spectrum is based on the measurement of the difference in energy between the incident light and the Raman scattered photons, which corresponds to the energy of the vibrational transitions. The analytical capabilities of this technique, both in macro- and micro-modes and in contact or at stand-off distances give significant advantages over other techniques for the analysis of minerals. The main objectives of the Raman instrument, part of the Pasteur payload of the ExoMars Rover, are to obtain information on the mineralogical structure and composition of surfaces by allowing the identification of minerals (silicates, carbonates, phosphates, sulfates, nitrates) and organic molecules *in situ* at the micro scale and hence identification of mineral components and their chemical structures (Haskin *et al.*, 1997; Rull and Martínez-Frías, 2006).

In the Exomars mission, Raman spectra will be collected from micro-scale powdered samples. These samples will be prepared by the crushing and dosing system inside the rover from cores obtained at different depths in the subsurface of Mars. For further details, see http://exploration.esa.int/mars/46048-programme-overview.

Mössbauer spectroscopy on the other hand, is a nuclear technique based on the recoil-free emission and absorption of γ-rays after the decay of certain radioactive isotopes. The Mössbauer effect, the recoilless emission and resonant absorption of γ-rays by certain nuclei in solid materials (in our application, ^{57}Fe), has been reviewed extensively in the literature (*e.g.* Greenwood and Gibb, 1971; Burns and Solberg, 1990; Burns, 1993). We give a brief overview here of the use of Mössbauer spectroscopy from the perspective of its application to Martian surface materials.

Resonant emission and absorption of γ-rays means that absorbed and emitted γ-rays have approximately the same energy. The requirement for this to occur with non-zero probability is that the emitting and absorbing nuclei must be in solid lattices having sufficient mass so that there is negligible nuclear recoil energy (*i.e.* 'recoil-free' nuclear transitions). Because kinetic energy ($= p^2/2m$, where p and m are momentum of the absorbed or emitted photon and mass of the absorbing or emitting body, respectively) is inversely proportional to mass at constant momentum, the recoil energy is virtually zero when the mass of the solid lattice is $\sim 10^{18}$ times the nuclear mass, *i.e.* a small particle weighing ~0.1 mg. Even in a solid, however, the emission (or absorption) of the γ-rays can involve energy transfer to lattice vibrations (phonons) in the crystal so that the Mössbauer effect is not observed. The probability of emissions occurring without inducing any phonons is referred to as the Debye-Waller factor, f, as in X-ray diffraction. In general, this factor depends on temperature, the Debye temperature of the crystal and the γ-ray energy (Gonser, 1986). The probability of recoil-free emission or absorption is 0.76 for ^{57}Fe in α-iron metal (α-Fe^0) at room temperature (Gonser, 1986). Normally, ^{57}Co is used as the emission source for Fe Mössbauer spectroscopy. ^{57}Co decays by electron capture to the second excited state of ^{57}Fe at 136.4 keV (~99.8% probability) which subsequently undergoes decay to its first excited state at

122.0 keV (~91% probability) and ground state (~9% probability). The first excited state undergoes recoil-free decay to the ground state by emission of the 14.4 keV Mössbauer γ-ray (10% probability), internal conversion electrons (~90% probability), and X-rays (at ~6.4 keV).

The energies of γ-rays from recoilless emissions are very precise (14.401 keV for ^{57}Fe), and their energy distribution ideally has a Lorentzian line shape with a width of ~4.7 neV that approaches the theoretical width imposed by the Heisenberg uncertainty principle. This width is, in general, less than the difference in energy between nuclear energy levels that result from different nuclear environments caused by, for example, different oxidation states and/or mineralogical environments. Therefore, recoil-free emission of 14.4 keV γ-rays by ^{57}Fe in one lattice will not undergo recoilless absorption in another lattice if the nuclear environments are too different, unless the energy of the emitted γ-ray or the energy of the nuclear energy levels in the absorber is changed slightly. Usually, the energy of emitted recoilless γ-rays is changed by exploiting the Doppler effect, *i.e.* by moving the source. For ^{57}Fe, energy 'scanning' through limits of ±10 mm/s is sufficient; positive velocities refer to motion of the source toward the absorber. The Mössbauer spectrum is then counts passing through (or emitted by) an absorbing sample as a function of the relative velocity of the source and sample.

The nuclear energy levels of ^{57}Fe (ground and first excited states with nuclear spins of 1/2 and 3/2, respectively) can be changed (*i.e.* shifted or split) by electric and/or magnetic fields acting at the nucleus. The electrical quadrupole interaction leads to both the isomer shift (IS) and the electric quadrupole splitting/shift (QS). The nuclear Zeeman hyperfine interaction leads to the internal magnetic field (B_{hf}) (Gonser *et al.*, 1991). The hyperfine interactions can be translated into three parameters that are measured from Mössbauer spectra by curve fitting, and it is these parameters that characterize the oxidation state and mineralogical environment of the ^{57}Fe nucleus, which has a natural abundance of 2.2%.

The combination of Mössbauer and Raman spectroscopy will probably be considered for future missions to Mars. As a result, the identification of sulfate-bearing minerals by Raman and Mössbauer spectroscopy using samples from potential Martian analogues represents an important step towards the development of the techniques and the methodology for these missions. In the present work, both techniques have been applied *in situ* on mineral samples at Rio Tinto and Jaroso Ravine and in the laboratory of the Associated Unit UVa-CSIC of the Astrobiology Centre (Valladolid, Spain). In addition, some examples of Mössbauer analyses performed by the *MER* rovers on the Mars surface have been reported (Klingelhöfer *et al.*, 2004).

The use of miniaturized and portable prototypes of the instruments developed for Mars exploration is not limited to field mineralogical and geochemical analysis. Other applications are also interesting; among these, the *in situ* analysis of cultural heritage sites is of great relevance. Key in investigating precious and rare objects from cultural and historical heritage is the need to avoid disturbing physically or chemically the original materials. These portable techniques are able to identify minerals *in situ* and non-destructively (Fe-bearing minerals in the case of Mössbauer). These capabilities are unique and in the present work some examples of application are presented. Several paintings of the archaeological site at Santana do Riacho, located near Belo Horizonte, in the 'Serra do Cipó' region (Brazil) (Klingelhöfer *et al.*, 2001, and references therein), and a Lekytos vase of the Greek Epoch (~500 years B.C.) have been analysed by Mössbauer spectroscopy. In addition, we performed preliminary analyses of paintings in the archaeological site at the world-famous Altamira Cave, located in Cantabria (northern Spain) and referred to as the 'Sistine Chapel' of the Palaeolithic.

2. Instrument descriptions

2.1. The Mössbauer instrument MIMOS II

The MIMOS II instrument (Fig. 4) is a miniaturized (with dimensions of 5 cm × 5.5 cm × 9.5 cm and a weight of ~400 g) Mössbauer spectrometer compared to standard laboratory Mössbauer spectrometers and is optimized for low power consumption and high detection efficiency (Klingelhöfer *et al.*, 1996, 2003, 2004; Squyres *et al.*, 2003).

MIMOS II operates in backscattering geometry, irradiating the samples through a collimator and detecting the resonantly scattered 14.4 keV Mössbauer radiation, and simultaneously the 6.4 keV Fe-X-ray's (accompanying the resonant absorption internal conversion process) with its 4 Si-PIN diodes (each with a sensitive area of 1 cm^2) (Klingelhöfer *et al.*, 1995, 1996, 2003).

In the flight version for Mars exploration all components were selected to withstand high acceleration forces and shocks, temperature variations over the Martian diurnal cycle, and cosmic-ray irradiation. Mössbauer measurements can be done during both day and night covering the whole diurnal temperature variation between ~−100°C (min) and ~+10°C (max) of a Martian day (Bell, 2008). To minimize experiment time, a very intense ^{57}Co/Rh source was used, with an initial strength of ~350 mCi at launch. Internal calibration of the instrument is accomplished by a second, less intense radioactive source mounted on the end of the velocity transducer opposite the main source and in transmission measurement geometry with a reference sample (Klingelhöfer *et al.*, 2003).

The MIMOS II Mössbauer spectrometer sensor head is located at the end of the Instrument Deployment Device IDD on the *Spirit* and *Opportunity* rovers (Fig. 4d). On *Mars-Express Beagle-2*, a European Space Agency (ESA) mission in 2003, the sensor head was also mounted on a robotic arm integrated into the Position Adjustable Workbench (PAW) instrument assembly (Gibson *et al.*, 2004). The sensor head shown in Fig. 4a carries the electromechanical transducer with the main and reference ^{57}Co/Rh sources and detectors, a contact plate and a sensor. The contact plate and sensor are used in conjunction with the IDD to apply a small preload when it places the sensor head, holding it firmly against the target. No sample preparation is required. Both the 14.41 keV γ-rays and 6.4 keV Fe X-rays are detected simultaneously.

MIMOS II has three temperature sensors, one on the electronics board and two on the sensor head. One temperature sensor in the sensor head is mounted near the internal reference absorber. The other sensor is mounted outside the sensor head at the contact

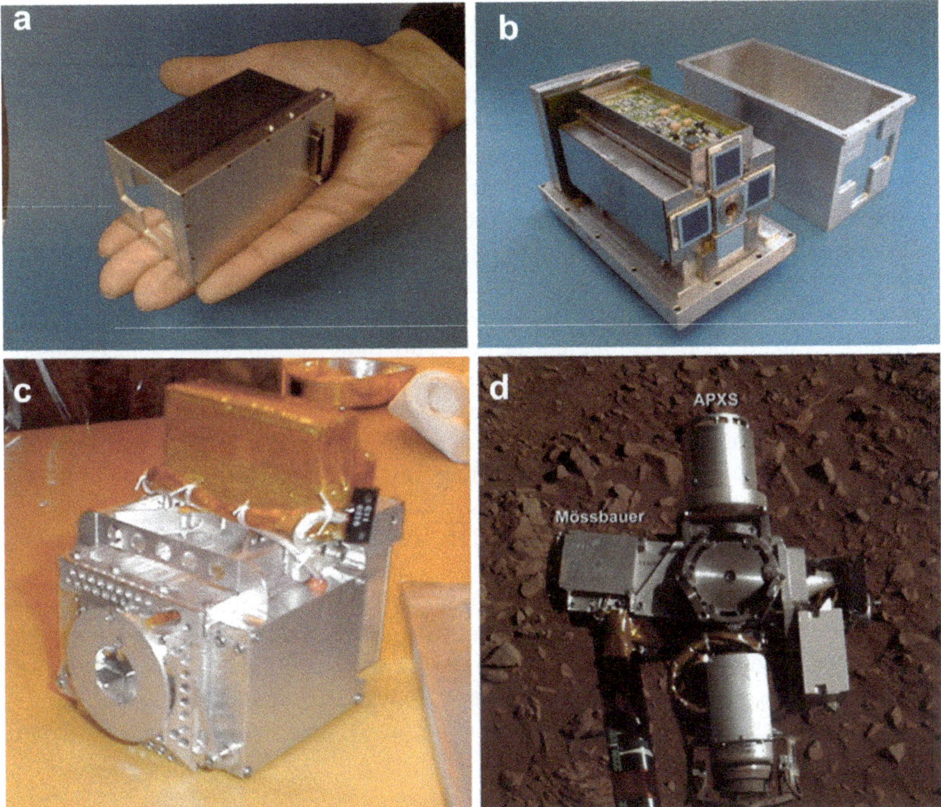

Figure 4. (a) External view of the MIMOS II sensor head without contact plate assembly; (b) internal view of the sensor head (cover removed), showing the 4 Si-PIN-diodes (dark squares; 1 cm × 1 cm) and the front-end electronics. In the centre the collimator for the MB source can be seen; (c) flight unit of the MIMOS II Mössbauer spectrometer sensor head, with the circular contact plate assembly (front). The circular opening in the contact plate has a diameter of 15 mm, defining the field of view of the instrument; (d) contact instruments on the robotic arm (IDD) of the *Spirit* rover at Gusev Crater, Mars. The MIMOS II instrument can be seen to the left of the instrument turret (marked as Mössbauer).

ring assembly to determine the temperature of the sample at the Martian surface. This temperature is used to route the Mössbauer data to the different temperature intervals assigned in memory. Materials at Rio Tinto were analysed using a twin prototype of the device used on the MER rovers (Figs 4a and 4b).

2.2. The Raman spectrometer for the Exomars 2018 mission

A Raman laser spectrometer (RLS) will be part of the payload in the 2018 rover vehicle of ESA's ExoMars mission (Rull and Martínez-Frías, 2006). The ExoMars rover constitutes the next logical step in Mars exploration. It will be the first mission combining mobility and access to subsurface locations, thus allowing investigation of

Mars' third dimension: depth. For that, the ExoMars rover will be equipped with a drill to obtain samples from the Martian sub-surface up to a depth of 2 m where material is not affected chemically, in order to have the opportunity to detect and analyse intrinsic organic matter. The samples collected are delivered to the Raman instrument in the form of powder deposited in a small container.

The RLS main characteristics (Fig. 5) are: the illumination is made with a 532 nm continuous wave laser with irradiance ~0.6 kW/cm^2 on the sample under a 50 μm spot size. The spectral range is between 100 and 4000 cm^{-1} with a spectral resolution of 6 cm^{-1}, on average (López-Reyes *et al.*, 2013). The detection of the Raman signal is made with a cooled (−10°C) CCD (2048 × 256 pixels).

The RLS instrument will operate in both automatic scanning and smart scanning modes on the surface of the powdered samples. During automatic scanning, the rover will follow a preconfigured sequence of movements for the RLS optical head, including performing 20 Raman analyses per sample. Subsequently, during the smart scanning, the rover will decide autonomously whether any particular part of the sample is of interest for further analyses (López-Reyes *et al.*, 2013).

To support instrument development and the associated science, a detailed simulator of the Exomars operation science cycle has been developed at the laboratory in Valladolid. This simulator consists of a miniaturized spectrometer similar in mass, dimensions and performance to that under development for flight (Fig. 5a) and a system designed to flatten the powdered surface in the same way as for the Exomars rover. This device also has an XYZ positioning system under Raman scanning (Fig. 6). The simulator works in autonomous mode, performing positioning along the sample

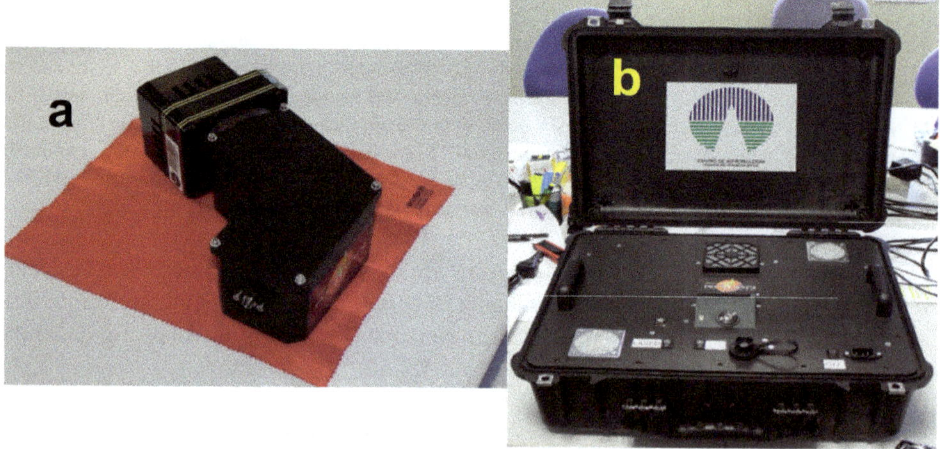

Figure 5. (a) The science prototype of the Raman spectrometer for Exomars with similar characteristics (geometry, dimensions, mass, optics, *etc.*) to those of the flight model. (b) The prototype integrated in a specific carrying case for fieldwork. The system includes the laser, electronics, batteries, a heating system and filters for working under dust and difficult ambient conditions.

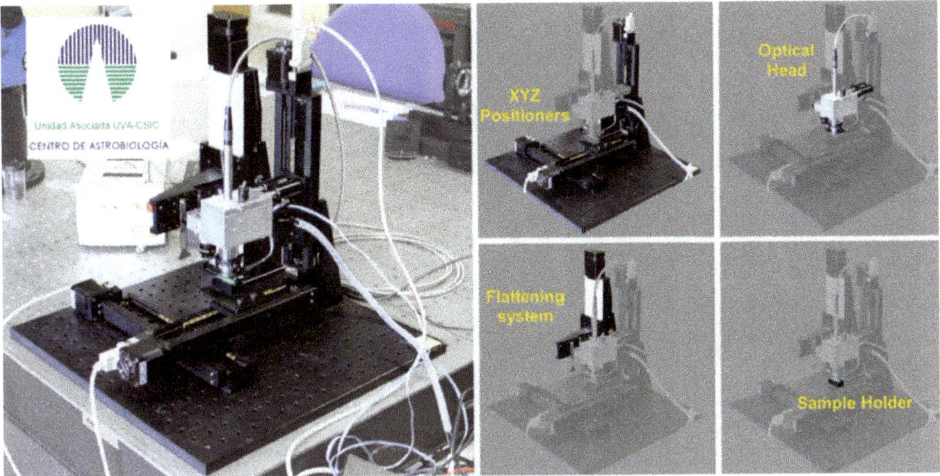

Figure 6. ExoMars RLS Simulator prototype and schematics (after López-Reyes *et al.*, 2013).

automatically and with autofocus on the mineral grains and automatic spectral acquisition.

The spectrometer prototype can also be adapted for field applications (Fig. 5b). In the current work, samples from both Martian analogue sites, Rio Tinto and Jaroso Ravine, were analysed using the prototype of the RLS adapted for field analysis, as well as the simulator of the full scientific operation mode that will be carried by the rover's analytical laboratory. In some cases, field analyses were also carried out using a commercial portable Raman instrument from B&W TEC Inc (Lübeck, Germany) specifically adapted to work outside.

3. Rio Tinto and Jaroso Ravine Mars analogue sites

Rio Tinto and the Jaroso Ravine (Spain) are two extremely interesting fluvial stream systems, with and without water, respectively, which have been proposed as potential Martian analogues (Fernández Remolar *et al.*, 2004; Martinez-Frias *et al.*, 2004) from geological and astrobiological points of view.

Rio Tinto (Huelva, SW Spain) is considered to be a potential Martian analogue of modern sulfate minerals precipitating in wet conditions, linked to significant acidophilic biogenic activity (Fernández-Remolar *et al.*, 2004, 2005). Sulfates derive from aqueous alteration of iron-rich sulfide minerals of the Iberian Pyrite Belt (Boulter, 1996; Leistel *et al.*, 1997). The average pH of the river is ~2.2 along its 100 km length. Its acid mine drainage is associated with low pH and very high heavy-metal concentrations leading to the well known reddish colour of its water (Fig. 7), which is the consequence of the chemical reaction of Fe(II) oxidation with water superposed by catalytic action of chemolitotrophic bacteria, especially *Thiobacillus ferrooxidans* (López-Archilla *et al.*, 1993, 2001). A great variety of iron-bearing minerals, mostly

Figure 7. Field analyses by (miniaturized) Mössbauer spectroscopy using a MIMOS II instrument (a, b) and the Raman spectroscopy prototype developed for the ExoMars mission (c, d) at the Rio Tinto site.

hydrated sulfates, have been described from the riverbanks of the Rio Tinto (Sobrón *et al.*, 2009; Rull *et al.*, 2014a).

In contrast, the Jaroso Ravine, located in Sierra Almagrera (Almería, Spain) and the world type locality of jarosite, is part of the Jaroso Hydrothermal System (JHS). This system is an ancient model of formation of supergene sulfates associated with polymetallic (Fe, Pb, Ag) sulfides and sulfosalts which are genetically linked to the calc-alkaline shoshonitic volcanism (Upper Miocene) of the SE Mediterranean margin of Spain (Martínez-Frías *et al.*, 2004).

The deposits are polymetallic veins and hydrothermal breccias hosted in the Permian–Triassic basement. The JHS includes oxyhydroxides, gold, silver, Hg Sb and base metal sulfides and sulfosalts (Martínez-Frías *et al.*, 1989; Martínez-Frías, 1991). Hydrothermal acidic solutions weathering the ores have generated large quantities of oxides and sulfate minerals of which jarosite is the most abundant. At present, some small ore veins are visible only at the surface, varying between 25 and 70 cm wide (Fig. 9b).

4. Results and discussion

Primary materials, secondary alteration minerals, as well as evaporite minerals and efflorescences were analysed at both the Rio Tinto and Jaroso Ravine analogue sites. In the case of Rio Tinto (Fig. 7), Mössbauer spectroscopy was applied to the outcrops of primary volcanic rocks in the vicinity of this area, where minor signs of weathering on the surface were found. This technique revealed the presence of chlorite, goethite and hematite. Similarly, Raman spectroscopic analysis identified goethite and hematite as well as quartz and muscovite. These originally porous volcano-sedimentary materials were altered by hydrothermal acidic fluids, resulting in 'gossan rocks' comprising hard black materials interlayered between softer red materials (Klingelhöfer *et al.*, 2011). Mössbauer spectroscopic analysis of these gossan rocks identified crystalline and superparamagnetic goethite (α-FeOOH) and hematite (α-Fe_2O_3) with different degrees of crystallinity and ferrous silicates. Raman spectroscopy was used to detect muscovite, hematite, goethite and quartz (Fig. 7c).

The Rio Tinto riverbed comprises conglomeratic cemented materials. Rock surfaces are visibly affected by the acidic water and exhibit a dark crust. Goethite, ferrous silicate and hematite were identified by Mössbauer spectroscopy (Fig. 8a), and the presence of jarosite ($KFe^{3+}_3(SO_4)_2(OH)_6$), hematite (Fe_2O_3), goethite ($FeO(OH)$), quartz (SiO_2) and rozenite ($Fe^{2+}SO_4.4H_2O$) was confirmed by Raman spectroscopy (Fig. 8b). On the riverbanks, efflorescences, layered crusts and popcorn-textured evaporites can be found. Copiapite ($Fe^{2+}Fe^{3+}_4(SO_4)_6(OH)_2.20H_2O$), jarosite (Fig. 8c,d) and rozenite ($Fe^{2+}SO_4.4H_2O$) were detected using both techniques, whereas the presence of ferricopiapite ($Fe^{3+}_{2/3}Fe^{3+}_4(SO_4)_6(OH)_2.20H_2O$), coquimbite ($Fe^{3+}_2(SO_4)_3.9H_2O$), hexahydrite ($MgSO_4.6H_2O$) and epsomite ($MgSO_4.7H_2O$) was revealed by *in situ* Raman spectroscopy. Baryte ($BaSO_4$), gypsum ($CaSO_4.2H_2O$), halotrichite ($FeAl_2(SO_4)_4.22H_2O$), melanterite ($Fe^{2+}SO_4.7H_2O$), rhomboclase ($H_5Fe^{3+}O_2(SO_4).2H_2O$) and szomolnokite ($Fe^{2+}SO_4.H2O$) were detected in the laboratory by Raman spectroscopy (Sobrón *et al.*, 2009). Additionally, Mössbauer spectroscopy was used to identify schwertmannite ($Fe^{3+}_{16}O_{16}(OH)_{9.6}(SO_4)_{3.2}.10H_2O$) as part of the ochre sediments in the riverbed. This unstable mineral has been suggested as a precursor of other iron-bearing minerals in Rio Tinto (Fernández-Remolar *et al.*, 2005) and has been observed in laboratory experiments and natural acid water systems in which this mineral has been found (Bigham *et al.*, 1996).

In summary, up to 13 minerals were identified in Rio Tinto by applying the *in situ* combination of Mössbauer and Raman spectroscopy. This number rises to 19 minerals with the additional Raman analyses using laboratory instruments. In order to avoid ambiguity about whether the analysed minerals are primary or secondary phases, the samples were analysed immediately after arrival in the laboratory. In general, however, only small mineralogical changes were observed in a proportion of samples several months after their collection. In the case of samples collected from caves with very long periods of obscurity, low temperatures and high humidity, new secondary phases were observed in the laboratory.

Raman spectroscopy can supply useful information about the biogenic processes related to mineral formation under acidic conditions. One example is shown in Fig. 9 of a sample collected from the bank of the Rio Tinto displaying the effect of bacterial acidophilic activity on the primary minerals (quartz and pyrite). In addition to these phases, the Raman results show a small amount of jarosite and a strong and broad spectral feature corresponding to the disordered carbon D (~1350 cm^{-1}) and G (1500–1600 cm^{-1}) bands. The D and G bands are regarded as key evidence of bacterial activity in the Rio Tinto environment.

The Mössbauer and Raman spectrometers used during field work at Jaroso Ravine are shown in Fig. 10. As mentioned above, the deposits are polymetallic veins and

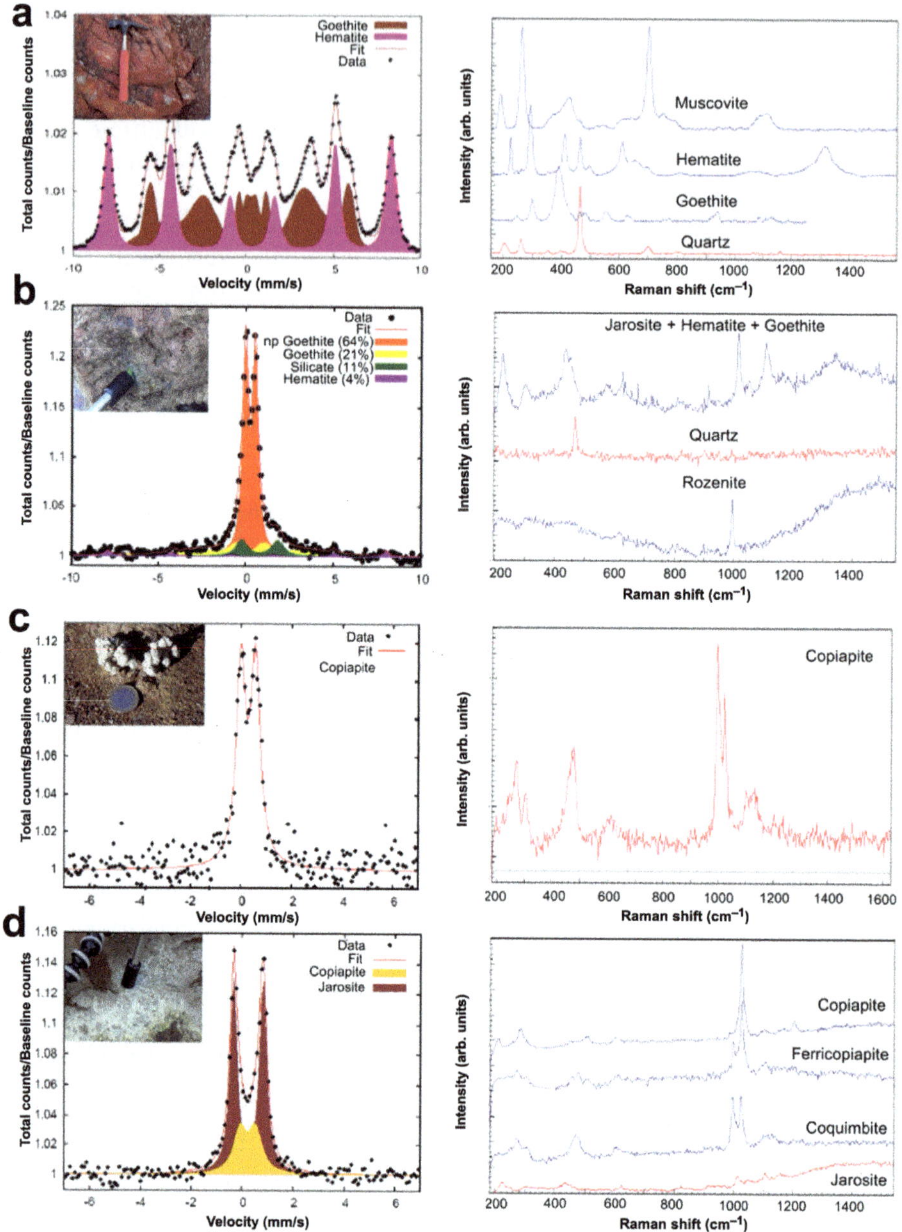

Figure 8. Results of *in situ* Mössbauer (left) and Raman (right) spectroscopy applied on different substrates and minerals at the Rio Tinto site: (a) hydrothermally altered rocks ('gossan rocks'); (b) river bedrock; (c) popcorn-texture efflorescences; (d) white and yellow crust on red substrate.

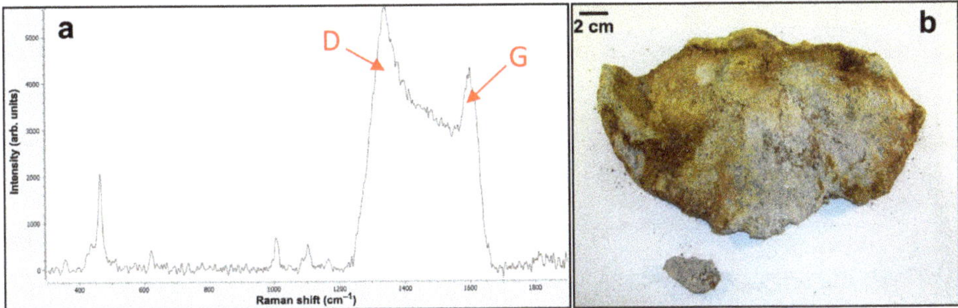

Figure 9. Raman spectrum (a) of a sample (b) collected from the Rio Tinto source banks. The broad spectral features correspond to disordered carbon D (~1350 cm^{-1}) and G (1500–1600 cm^{-1}) bands originating from bacterial activities. Besides the primary minerals (quartz and pyrite), small amounts of jarosite can be identified.

hydrothermal breccias hosted in the Permian–Triassic basement. Raman results of minerals from the host rocks of Jaroso reveal the dominance of muscovite and graphite, in addition to oxy-hydroxides (anatase, rutile, hematite, magnetite, goethite and lepidocrocite) and silicates (zircon, quartz, orthoclase and anorthoclase). Raman results also indicate the presence of pyrite and sphalerite along with gangue minerals (baryte, siderite and quartz) from the analysis of the ore minerals of the veins.

Although jarosite is the best known sulfate mineral described from the Jaroso Ravine (Fig. 11), a wide range of other sulfates has been identified by the Raman analysis of hydrothermal breccias, including gypsum, anglesite, natrojarosite, alunite, botryogen,

Figure 10. Field analyses by Raman spectroscopy (a,b) and Mössbauer spectroscopy (c) at the Jaroso Ravine.

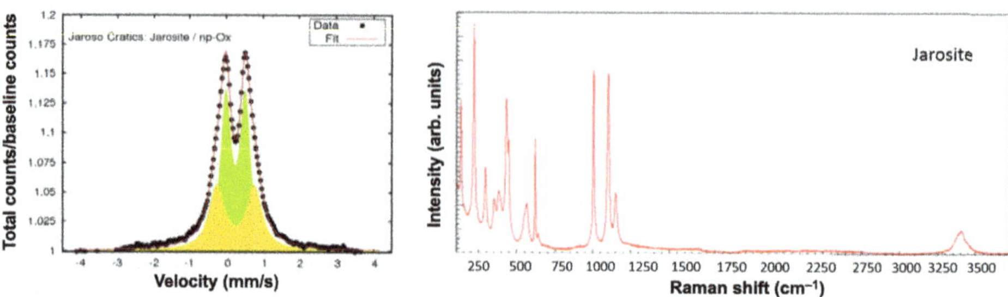

Figure 11. The *in situ* Mössbauer analysis at Jaroso ravine shows the presence of jarosite (yellow) and nanophase Fe oxides (light green) (left), and the *in situ* Raman spectrum of jarosite at an outcrop from Jaroso ravine indicates the presence of jarosite (right).

copiapite, ferricopiapite, halotrichite, epsomite, hexahydrite, rozenite and szomolnokite (Fig. 12) (Frost *et al.*, 2005; Venegas *et al.*, 2012; Venegas, 2014).

Most sulfate minerals in the Jaroso Ravine derive from supergene alteration of pre-existing minerals (especially jarosite) forming mineral veins inside the metamorphic host rock. Nevertheless, contemporary ephemeral hydrated sulfate efflorescences also appear on the ravine walls. They appear to be controlled seasonally by climate variability and landscape features. Most hydrated sulfates found in the Jaroso Ravine have also been detected on Mars, whilst others such as szomolnokite, copiapite and rozenite have been suggested to be present on the basis of simulation models (Marion *et al.*, 2006).

5. Historical and cultural heritage

The instrument prototypes developed for Mars exploration can also be used for several other applications on Earth, and in particular in the study of the historical and cultural

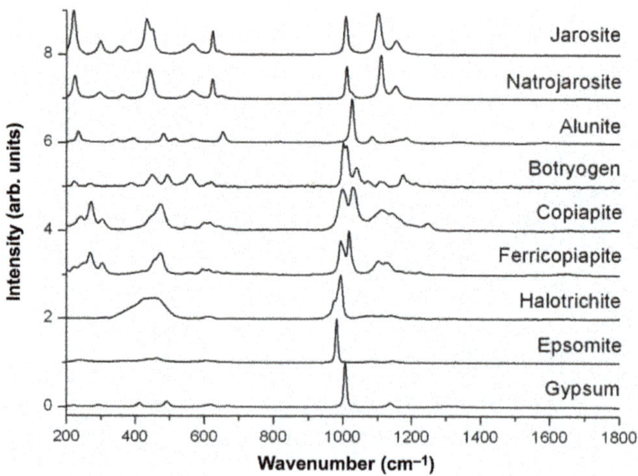

Figure 12. Mineral phases collected from the Jaroso Ravine analysed by laboratory-based Raman spectroscopy.

heritage. The non-destructive nature of both the Mössbauer and Raman techniques and the possibility of analysing these materials *in situ* provide unique capabilities for the analysis and characterization of materials.

5.1. Archaeological artefacts analysed by Mössbauer spectroscopy

The Lekythos vase (500 y B.C.) (Fig. 13) was analysed by backscattering Mössbauer spectroscopy. This Lekythos vase has three black figures with red details painted on yellow fired clay. The Lekythos was built as an oil or perfume jar, used chiefly for ointments and religious (funerary) ceremonies. This is substantiated by the fact that Lekythoi (plural of Lekythos) have been found in and around tombs and excavated from ancient homes.

Using a MIMOS II instrument, Mössbauer backscattering spectra were recorded at room temperature from two different parts of the Lekythos vase. One was on non-painted surfaces and the other was on the painted surfaces (black only, and black with red detail). The unpainted surface shows a broad spectrum that can be associated with poorly crystallized iron oxides produced during the firing clay process. The painted surfaces show, in addition to the characteristic unpainted area, a well defined sextet. The Mössbauer parameters of this sextet correspond to well crystallized hematite. The Mössbauer spectrum taken from the red-painted details shows no significant differences from the unpainted surface. Therefore, the red details are presumably iron free.

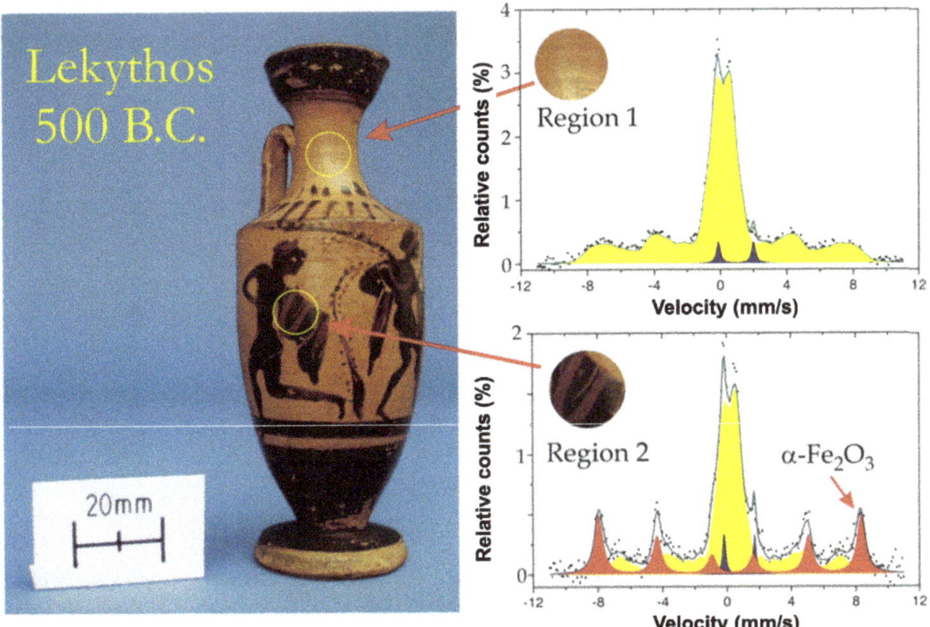

Figure 13. Photo of the Lekythos vase (500 y B.C.) and two examples of Mössbauer spectra obtained from two different spots using the portable MIMOS II spectrometer.

5.2. Rock paintings in the field

Non-destructive analysis with MIMOS II was performed on an ancient rock painting in Brazil (near Belo Horizonte, Minas Gerais). The archaeological site of Santana do Riacho is located ~90 km north-east of Belo Horizonte, in the Serra do Cipó region. Excavations of all sorts of materials (bones, stones, pigmented materials, *etc.*) have been performed over the years and a vast body of information has been collected. Several prehistoric paintings are found in this locality, but so far no conclusive proof exists about the nature of the materials (*e.g.* paints) that were used in these drawings. Such information is valuable for increasing our understanding of the technical capacity and expertise in obtaining and processing minerals and plants in prehistoric times. The vast majority of the paintings are red (70%) and yellow (20%), but there are others in orange, black, brown and ochre. There is increasing evidence that iron oxides are probably among the principal pigments used in prehistoric times (*e.g.* Walton *et al.*, 2010). The impressive paintings in Santana do Riacho and the recent development of the spectrometer MIMOS II prompted the authors to perform, for the first time, *in situ* Mössbauer measurements on rock paintings. The paintings were selected based on their colours and apparent thickness, and six spots were measured. We have used a prototype of the instrument to investigate the iron oxide composition of rock paintings in the field. The instrument was mounted on a tripod, and the power was supplied by battery (see Fig. 14). The data suggest that hematite and goethite are the pigments responsible for the red and yellow colours observed (see Fig. 15).

Raman spectroscopy has been used widely in recent years for laboratory and *in situ* study of materials related to the historical and cultural heritage (Veneranda *et al.*, 2014). Here, a recent example has been selected to illustrate the capabilities of the technique for *in situ* analysis of materials in prehistoric paintings in caves.

Figure 14. The MIMOS II instrument is mounted on a tripod and positioned at the picture of interest (left); part of the wall showing several distinct paintings with different colours (right).

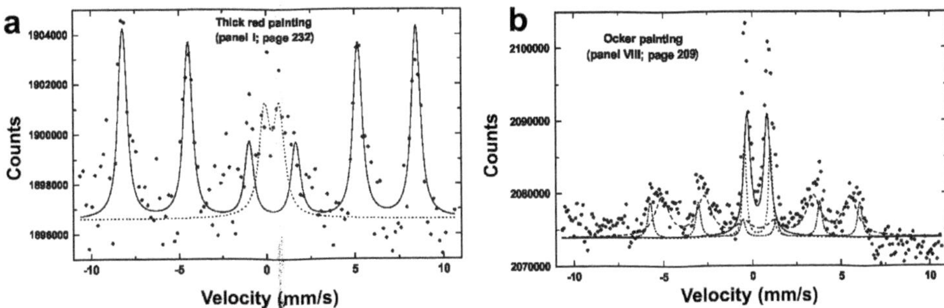

Figure 15. Mössbauer spectra of (a) the thick red painting located in panel I, and (b) the Ochre painting located in panel VIII (mentioned by Klingelhöfer *et al.*, 2001, and references therein).

The Altamira Cave is referred to as the 'Sistine Chapel' of the Palaeolithic and contains several of the most outstanding representations of prehistoric rock art (up to 22,000 y old). *In situ* Raman analysis was performed using a portable instrument with specific characteristics (related mainly to the optical Raman probe-head rather than the spectrometer. The optical Raman head was adapted specifically to a micrometric double stage. This was very important in order to avoid any physical contact with the pigments, to overcome the tiny humidity layer and to focus appropriately on the selected spot. The fine focus adjustment and spectral optimization were performed, in general, with the upper stage). A double micrometric XYZ positioning system was designed to focus precisely the laser beam on the mineral surface taking into account the fine humidity layer and avoiding direct contact with the painted surface. The whole system was mounted on a tripod allowing the samples to be focused at ~6 mm distance with a spatial resolution of ~300 μm (spot of 150 μm) and a depth of field of ±100 μm.

The Raman analysis revealed mainly the presence of calcite as the surface matrix and Fe-bearing minerals, including hematite and goethite forming part of the ochre-coloured bisons of the famous 'Polychromes Hall' (Fig. 16). These results concur with those obtained from mineralogical analysis of pigment remainders and painting tools found previously at the Altamira archaeological site (Rull *et al.*, 2014b). In addition, Raman spectroscopy can supply interesting information about the possible alteration processes that affect the painting. The Raman spectrum shown in Fig. 17 is of a droplet collected from the ceiling of Altamira Cave. Besides the clear hematite bands, two broad features of the D and G bands of disordered carbon at 1315 cm^{-1} and 1600 cm^{-1}, respectively, are observed. The presence of carbon could be associated with the black pigment (mainly charcoal) used but it might also indicate biological activity that has caused alteration of pigments. In this last case, the presence of calcium oxalate (Rull *et al.*, 2001) as a bio-product is an interesting feature that warrants further study.

6. Concluding remarks

Mössbauer and Raman spectroscopy have been demonstrated to be very powerful techniques for the characterization of iron-bearing minerals. Regarding Mössbauer

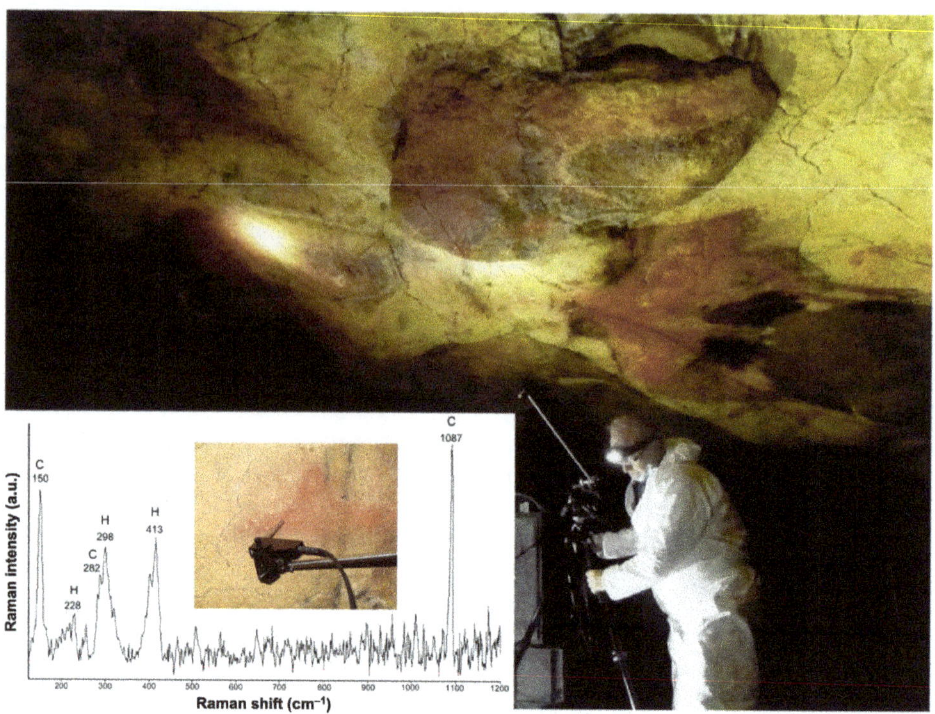

Figure 16. In situ Raman analysis of prehistoric rock art in Altamira Cave (northern Spain). The ochre colour of the paintings is due to hematite. Black lines are composed mainly of charcoal.

spectroscopy, the mineralogical results obtained on Mars by the MER Rovers and in the Martian analogue sites show its ability to distinguish between iron compounds, and especially to identify iron oxidation states in minerals.

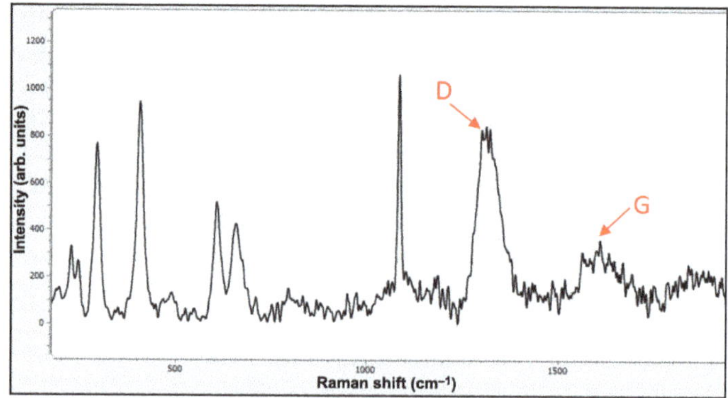

Figure 17. Raman spectrum of a droplet from the ceiling of Altamira Cave. In addition to the clear hematite Raman bands, the D and G bands of disordered carbon are also identified.

On the other hand, Raman spectroscopy which will be part of the next ESA mission to Mars, is a powerful technique for identifying precise mineral phases, including hydrated sulfate minerals like those found the Jaroso Ravine and Rio Tinto (most of them also described from Mars). Remarkably, the presence of such minerals clearly points to aqueous processes which occurred in the early geological history of Mars.

Moreover, Mössbauer spectroscopy produces a wider spot size (~1 cm) and penetrates several hundreds of microns into the sample compared to Raman spectroscopy which analyses only the surface of the sample (Raman penetrating depth is ~150 μm). As a result, Mössbauer spectroscopy provides average information about the mineral assemblage, whilst Raman spectroscopy, with a spot size of ~100 μm, typical of the grain size in these samples, can supply information about single mineral phases and their settings in the sample surfaces (context information). The combination of these techniques provides great advantages for precise *in situ* mineral identification in a non-destructive mode because the analysis requires only a minimal contact in the case of Mössbauer or no contact (the laser focus is at ~10 mm from the Raman optical head) in the case of Raman. In addition, the ability of Raman spectroscopy to detect organic compounds turns this technique into a potential discoverer of present or ancient life in Mars. Given the obvious advantages of these techniques, combining Mössbauer and Raman spectroscopy will surely be considered for future missions to Mars.

In situ Mössbauer and Raman studies of cultural heritage features have also facilitated detection of a variety of pigments, mainly Fe-bearing minerals, in a Lekythos funerary vase, rock-art paintings in Santana do Riacho and Altamira archaeological sites. The advantages of non-destructive and *in situ* techniques is obvious in these cases, in which removal of material for analysis in the laboratory could cause irreparable damages in such well preserved cultural heritage sites. Future work includes the combined use of these Raman and Mössbauer prototypes in the Spanish caves and other important historical sites including those in Castilla and Leon (Spain). The research team is particularly interested in using a Raman-LIBS prototype because it has been proven to be a powerful technique in providing structural and chemical information about mineralogical samples which are complementary to Mössbauer analysis (Rull *et al.*, 2011).

Acknowledgements

Financial support for this work was made available through Projects AYA2008-04529-4 and AYA2011-30291-C02-02 (Ministry of Science and Innovation, Spain and FEDER funds of the EU) and the ESP2013-48427-C3-2-R Project of the MINECO. The authors thank Carmen de las Heras and Jose Antonio Lasheras for providing access to Altamira Cave and allowing us to use their facilities. The Mössbauer investigations were supported by the German Space Agency DLR under contracts 50QM9902, 50QM1102, and 50QX0603.

References

Boulter, C.A. (1996) Extensional tectonics and magmatism as drivers of convection leading to Iberian Pyrite Belt massive sulphide deposits? *Journal of the Geological Society of London*, **153**, 181–184.

Bell, J. (editor) (2008) *The Martian Surface*. Cambridge University Press, Cambridge, UK.

Bigham, J.M., Schwertmann, U., Traina, S.J., Winland, R.L. and Wolf, M. (1996) Schwertmannite and the chemical modeling of iron in acid sulphate waters. *Geochimica et Cosmochimica Acta*, **60**, 2111–2121.

Burns, R.G. (1993) Mössbauer spectral characterization of iron in planetary surface materials. Pp. 539–556 in: *Remote Geochemical Analysis: Elemental and Mineralogical Composition* (C.M. Pieters and P.A.J. Englert, editors). Cambridge University Press, Cambridge, UK.

Burns, R.G. and Solberg, T.C. (1990) ^{57}Fe-bearing oxide, silicate, and aluminosilicate minerals, crystal structure trends in Mössbauer spectra. Pp. 262–283 in: *Spectroscopic Characterization of Minerals and their Surfaces* (L.M. Coyne, D.F. Blake and S.W.S. McKeever, editors). American Chemical Society, Washington, D.C.

Clark, B.C., Morris, R.V., McLennan, S.M., Gellert, R., Jolliff, B., Knoll, A.H., Squyres, S.W., Lowenstein, T.K., Ming, D.W., Tosca, N.J., Yen, A., Christensen, P.R., Gorevan, S., Bruckner, J., Calvin, W., Dreibus, G., Farrand, W., Klingelhoefer, G., Waenke, H., Zipfel, J., Bell III J.F., Grotzinger, J., McSween, H.Y. and Rieder, R. (2005) Chemistry and mineralogy of outcrops at Meridiani Planum. *Earth Planetary and Science Letters*, **240**, 73–94.

Cockell, C. (2015) *Astrobiology: Understanding Life in the Universe*. Wiley-Blackwell, Oxford, UK.

Edwards, H.G.M., Vandenabeele, P., Jorge-Villar, S.E., Carter, E.A., Rull-Pérez, F. and Hargreaves, M.D. (2007) The Rio Tinto Mars Analogue site: An extremophilic Raman spectroscopic study. *Spectrochimica Acta Part A*, **68**, 1133–1137.

Fernandez-Remolar, D., Gomez-Elvira, J., Gomez, F., Sebastian, E., Martin, J., Manfredi, J.A., Torres, J., Gonzalez-Kesler, C. and Amils, R. (2004) The Tinto River, an extreme acidic environment under control of iron, as an analog of the Terra Meridiani hematite site of Mars. *Planetary and Space Science*, **52**, 239–248.

Fernandez-Remolar, D., Morris, R., Gruener, J.E., Amils, R. and Knoll, A.H. (2005) The Rio Tinto Basin, Spain: mineralogy, sedimentary geobiology, and implications for interpretation of outcrop rocks at Meridiani Planum, Mars. *Earth and Planetary Science Letters*, **240**, 149–167.

Frost, R.L., Weier, M.L., Kloprogge, J.T., Rull, F. and Martinez-Frias, J. (2005) Raman spectroscopy of halotrichite from Jaroso, Spain. *Spectrochimica Acta Part A*, **62**,166–180.

García-Guinea, J., Martínez-Frías, J., López Ruiz, J., López García, J.A. and Benito García, R. (1989) Las mineralizaciones epitermales de Sierra Almagrera y de la cuenca de Herrerías (Cordilleras Béticas). *Boletín de la Sociedad Espaola de Mineralogía*, **12**, 261–271.

Gibson, E.K., Pillinger, C.T., Wright, I.P., Morgan, G.H., Yau, D., Stewart, J.L.C., Leese, M.R., Praine, I.J., Sheridan, S., Morse, A.D., Barber, S.J., Ebert, S., Goesmann, F., Roll, P., Rosenbauer, H. and Sims M.R. (2004) Beagle 2: Mission to Mars – current status. *35th Lunar and Planetary Science Conference*, **35**, 1845.

Gonser, U. (1986) Mössbauer spectroscopy. Pp. 409–448 in: *Microscopic Methods in Metals* (U. Gonser, editor). Springer Verlag, New York.

Gonser, U., Aubertin, F., Stenger, S., Fischer, H., Smirnov, G. and Klingelhöfer, G. (1991) Polarization and thickness effects in Mössbauer spectroscopy. *Hyperfine Interactions*, **67**, 701–710.

Greenwood, N.N. and Gibb, T.C. (1971) *Mössbauer Spectroscopy*. Chapman & Hall Ltd, London.

Guerrero, J. (2012) *Estudio espectroscópico y mineralógico de los sulfatos precipitados en Río Tinto a partir de aguas ácidas. Aplicaciones medioambientales*. PhD Thesis, Servicio de publicaciones, Universidad de Valladolid, Spain, 291 pp.

Haskin, L.A., Wang, A., Rockow, K.M., Jolliff, B.L., Korotev, R.L. and Viskupic, K.M. (1997) Raman spectroscopy for mineral identification and quantification for in situ planetary surface analysis: A point count method. *Journal of Geophysical Research*, **102**, 19293–19306.

Klingelhöfer, G., Held, P., Teucher, R., Schlichting, F., Foh, J. and Kankeleit, E. (1995) Mössbauer spectroscopy in space. *Hyperfine Interactions*, **95**, 305–339.

Klingelhöfer, G., Fegley Jr., B., Morris, R.V., Kankeleit, E., Held, P., Evlanov, E. and Priloutskii, O. (1996) Mineralogical analysis of Martian soils and rocks by a miniaturized backscattering MB spectrometer. *Planetary and Space Science*, **44**, 1277–1288.

Klingelhöfer, G., da Costa, G.M., Prous, A. and Bernhardt, B. (2001) Rock Paintings from Minas Gerais, Brasil, investigated by in situ Mössbauer Spectroscopy. Pp. 423–426 in: *Hyperfine Interactions (C), Vol.* **5** (M.F. Thomas, J.M. Williams and T.C. Gibb, editors). International Conference on the Applications of the Mössbauer Effect (ICAME 2001), Oxford, UK. Kluwer Academic Publishers, Dordrecht, The Netherlands.

Klingelhöfer, G., Morris, R.V., Bernhardt, B., Rodionov, D., de Souza Jr. P.A, Squyres, S.W., Foh, J., Kankeleit, E., Bonnes, U., Gellert, R., Schröder, C., Linkin, S., Evlanov, E., Zubkov, B. and Prilutski, O. (2003) Athena MIMOS II Mössbauer spectrometer investigation. *Journal of Geophysical Research*, **108**, (E12), 8067–8084.

Klingelhöfer, G., Morris R.V., Bernhardt, B., Schröder,C., Rodionov, D.S., de Souza Jr. P.A., Yen, A., Gellert, R., Evlanov, E.N., Zubkov, B., Foh, J., Bonnes, U., Kankeleit, E., Gütlich, P., Ming, D.W., Renz, F., Wdowiak, T., Squyres, S.W. and Arvidson, R.E. (2004) Jarosite and hematite at Meridiani Planum from Opportunity's Mössbauer Spectrometer. *Science*, **306**, 1740–1745.

Klingelhöfer, G., Rull, F., Venegas, G., Fleischer, I., Martínez Frías, J., Blumers, M., Medina, J., Sansano, A., Navarro, R. and Henrich, C. (2011) *In-situ* mineralogical analysis with the miniaturised Mössbauer spectrometer MIMOS II and Raman Spectrometer at the terrestrial Martian analogues Rio Tinto and Jaroso Ravine. *Workshop at Université Cady Ayyad, Ibn Battuta Centre – Morocco.*

Lane, M.D., Darby Dyar, M. and Bishop, J.L. (2004) Spectroscopic evidence for hydrous iron sulfate in the Martian soil. *Geophysical Research Letters*, **31**, L19702, 1–4.

Leistel, J.M., Marcoux, E., Thiéblemont, D., Quesada, C., Sánchez, A., Almodóvar, G.R., Pascual, E. and Sáez, R. (1997) The volcanic-hosted massive sulphide deposits of the Iberian Pyrite Belt Review and preface to the Thematic Issue. *Mineralium Deposita*, **33**, 2–30.

López-Archilla, A.I., Marín, I. and Amils, R. (1993) Bioleaching and interrelated acidophilic microorganisms from Río Tinto, Spain. *Geomicrobiology Journal*, **11**, 223–233.

López-Archilla, A.I., Marín, I. and Amils, R. (2001) Microbial community composition and ecology of an acidic aquatic environment: the Tinto River, Spain. *Microbial Ecology*, **41**, 20–35.

López-Reyes, G., Rull, F., Venegas, G., Westall, F., Foucher, F., Bost, N., Sanz-Arranz, A., Catalá-Espí, A., Vegas A., Hermosilla, I., Sansano, A. and Medina. J. (2013) Analysis of the scientific capabilities of the ExoMars Raman Laser Spectrometer instrument. *European Journal of Mineralogy*, **25**, 721–733.

Marion, G.M., Kargel, J.S. and Calling, D.C. (2006) Modeling ferrous/ferric iron chemistry with application to Martian surface geochemistry. 37th *Lunar and Planetary Science Conference*, 1898.

Martínez-Frías, J. (1991) Sulphide and sulphosalt mineralogy and paragenesis from the Sierra Almagrera veins (Betic Cordillera). *Estudios Geológicos*, **47**, 271–279.

Martinez-Frías, J., Lunar, R., Rodriguez-Losada, J.A., Delgado, A. and Rull, F. (2004) The volcanism related multistage hydrothermal system of El Jaroso (SE Spain): Implications for the exploration of Mars. *Earth Planets Space*, **56**, v–viii.

Ming, D.W., Mittlefehldt, D.W., Morris, R.V., Golden, D.C., Gellert, R., Yen, A., Clark, B.C., Squyres, S.W., Farrand, W.H., Ruff, S.W. and Arvidson, R.E. (2006) Geochemical and mineralogical indicators for aqueous processes in the Columbia Hills of Gusev crater, Mars. *Journal of Geophysical Research: Planets*, 111(E2).

Rull, F. and Martinez-Frías, J. (2006) Raman spectroscopy goes to Mars. *Spectroscopy Europe*, **1**, 18–21.

Rull, F., Medina, J., Sansano, A., Sánchez-Moral, S. and Soler, V. (2001) Estudio y caracterización de minerales de interés arqueológico en la Cueva de Altamira por DRX y Raman. *Boletín de la Sociedad Española Mineralogía*, **24**, 119–120.

Rull, F., Martinez-Frías, J. and Medina, J. (2005) Surface mineral analysis from two possible Martian analogs (Rio Tinto and Jaroso Ravine, Spain) using micro-, macro-, and remote laser Raman spectroscopy. *European Geosciences Union, Geophysical Research Abstracts*, **7**, 09114.

Rull, F., Fleischer, I., Martinez-Frías, J., Sanz, A., Upadhyay, C. and Klingelhöfer, G. (2008) Raman and Mössbauer spectroscopic characterisation of sulphate minerals from the Mars analogue sites at Rio Tinto and Jaroso Ravine, Spain. *39th Lunar and Planetary Science Conference*, **1391**, 1616.

Rull, F., Klingelhöfer, G., Sansano, A., Fleischer, I., Sobrón, P., Blumers, M., Lafuente, A., Schmanke, D. and Maul, J. (2009) In-situ micro-Raman and Mössbauer spectroscopic study of evaporite minerals in Rio Tinto (Spain): applications for planetary exploration. *Conference on Micro-Raman Spectroscopy and Luminescence Studies in the Earth and Planetary Sciences*, **1473**, 68–69.

Rull, F., Vegas, A. and Barreiro, F. (2011) In-situ Raman-LIBS spectroscopy for surface mineral analysis at stand-off distances. *Proceedings 42nd Lunar and Planetary Science Conference 42*, Abstract#2275.

Sobrón, P., Sanz, A., Acosta, T. and Rull, F. (2009) A Raman spectral study of stream waters and efflorescent

salts in Rio Tinto, Spain. *Spectrochimica Acta Part A*, **71**, 1678–1682.
Squyres, S.W., Arvidson, R.E., Baumgartner, E.T., Bell III, J.F., Christensen, P.R., Gorevan, S., Herkenhoff, K.E., Klingelhöfer, G., Madsen, M.B., Morris, R.V., Rieder, R. and Romero R.A. (2003) The Athena Mars Rover science investigation. *Journal of Geophysic Research*, **108**, E12, 8062.
Squyres, S.W., Arvidson, R.E., Bell III, J.F., Calvin, W.M., Christensen, P.R., Clark, B.C., Crisp, J.A., Farrand, W.H., Herkenhoff, K.E., Johnson, J.R., Klingelhoefer, G., Knoll, A.H., McLennan, S.M., McSween, H.Y., Morris, R.V., Rice, J.W., Rieder, R. and Soderblom, L.A. (2004) In situ evidence for an ancient aqueous environment at Meridiani Planum, Mars. *Science*, **306**, 1709–1714.
Walton, M.S., Svoboda, M., Mehta, A., Webb, S. and Trentelman, K. (2010) Material evidence for the use of Attic white-ground Lektyhoi ceramics in cremation burials. *Journal of Archaeological Science*, **37**, 936–940.
Venegas, G. (2014) *Raman study of sulphates formed by hydrothermal, evaporitic and weathering processes in the south east of Spain: implications for the exploration of Mars.* PhD thesis Servicio de publicaciones. Universidad de Valladolid, Spain, 209 pp.
Venegas, G., Guerrero, J., Sansano, A., Sanz, A. and Rull, F. (2012) Raman study of mineralogical precipitation sequence of Rio Tinto "Mars analog". *European Planetary Science Congress*, 754.
Veneranda, M., Irazola, M., Pitarch, A., Olivares, M., Iturregui, A., Castro, K. and Madariaga, J.M. (2014) In-situ and laboratory Raman analysis in the field of cultural heritage: the case of a mural painting. *Journal of Raman Spectroscopy*, **45**, 228–237.
Zhu, M., Xie, H., Guan, H. and Smith, R.K. (2006) Mineral and lithologic mapping of Martian low-albedo regions using OMEGA data. *37th Lunar and Planetary Science Conference*, 2173.

EMU Notes in Mineralogy, Vol. 17 (2017), Chapter 4, 55–93

The timescales of mineral redox reactions

BENJAMIN GILBERT* and GLENN A. WAYCHUNAS

Energy Geosciences Division, Lawrence Berkeley National Laboratory, MS 74R316C, 1 Cyclotron Road, Berkeley, California 94720, USA, e-mail: bgilbert@lbl.gov
** Corresponding author*

Redox-reactive minerals can serve as electron donors or acceptors for abiotic reactants or microbial metabolic processes, and hence can play important roles in terrestrial and aquatic environments, particularly if their reaction rates are comparable to those of other biogeochemical processes. Under such circumstances, their reactions can control metal and contaminant bioavailability and change the permeability of soils and sediments. While the thermodynamic driving force for a mineral redox reaction is frequently a good predictor of relative rates of reaction, there are many examples in which kinetic factors limit reaction rate. Understanding the thermodynamic and kinetic aspects of mineral reaction rates, and their sensitivity to environmental conditions such as temperature or pH, is important for anticipating the biogeochemical evolution of natural environments subjected to change. Achieving this goal requires knowledge of the reaction pathway, the timescales of intermediate steps, and the lifetimes of metastable reaction states. Mineral redox reactions proceed through a combination of steps that can include electron and proton transfer, the breaking or formation of bonds, and mineral dissolution or phase transformation. The combination of conventional kinetics approaches, newly-developed ultrafast time-resolved methods and molecular simulation can provide elucidation of such complex reaction pathways. This chapter summarizes key concepts in mineral and interfacial redox reactions with illustrations from the rich geochemistry of iron and iron-bearing minerals.

1. Introduction

Minerals that are partly or fully composed of redox-active elements play important roles in near-surface biogeochemistry. For example, the oxidation or reduction of redox-reactive minerals typically cause a major structural change, such as dissolution or phase transformation. Such reactions are thus a form of chemical weathering (Zinder *et al.*, 1986; Stumm and Sulzberger, 1992) that can rapidly alter the mineralogy and aqueous geochemistry of an environment subjected to changing redox conditions. Moreover, the redox chemistry of minerals in the environment is inseparable from the planet's biochemistry, as the reducing or oxidizing substances (such as O_2, H_2 and diverse organic molecules) are typically produced directly or indirectly by living organisms. Consequently, mineral redox chemistry is often linked tightly to the carbon cycle (Torn *et al.*, 1997; Dubinsky *et al.*, 2010). Understanding the redox reactivity of minerals is thus crucial for anticipating the consequences of environmental change and for interpreting records of past change such as those captured in banded iron formations or in palaeosols (Hu *et al.*, 2013; Posth *et al.*, 2013).

DOI: 10.1180/EMU-notes.17.4

In addition, redox reactions involving minerals and contaminant ions can effectively sequester pollutants from groundwater and, as reviewed in other contributions to this volume, redox-reactive minerals or their engineered analogues may have considerable utility for environmental remediation. Finally, many redox-reactive minerals often absorb photons from sunlight, leading to environmentally important photochemical reactions (Sulzberger and Laubscher, 1995). Indeed, there is considerable interest in the use of such minerals or mineral-inspired clusters as abiotic charge-transfer catalysts for reactions as water oxidation or carbon dioxide fixation (Hocking *et al.*, 2011).

Mineral redox reactions can occur over a very wide range of timescales. The principal goal of this contribution is to give examples of important molecular-scale steps that are involved in mineral redox reactions, thereby providing possible answers to the question: what determines reaction rate? Although detailed answers to that question are currently possible only for a few systems studied in the laboratory, the general concepts are of wide relevance for understanding the changes that occur when natural systems are subjected to fluctuating redox conditions.

We shall introduce key concepts regarding the rates of mineral redox reactions using examples from the rich geochemistry of iron and iron-bearing minerals (Jickells *et al.*, 2005; Boyd and Ellwood, 2010). Iron is the most common redox-active transition metal, found in nature in the +2 or +3 oxidation states and, less often, as a metallic iron ore. Common redox-reactive iron-bearing minerals include bulk iron(III) oxides and oxyhydroxides (Schwertmann and Cornell, 2000), iron(II) sulfides that are important ore minerals (Rosso and Vaughan, 2006) and iron(II) carbonate that is considered a key phase for geological carbon sequestration (Palandri and Kharaka, 2005). In addition, low-temperature precipitation processes frequently generate nanoparticles of iron(III) oxyhydroxides that exhibit a large range of crystallinity and that can incorporate or sorb anions, cations and natural organic matter (Lalonde *et al.*, 2012; Riedel *et al.*, 2013). Figure 1 provides examples of iron-bearing minerals found in ocean waters and aerosol particles (Schroth *et al.*, 2009; von der Heyden *et al.*, 2012). Iron is an essential element for all life with the Fe^{2+}/Fe^{3+} redox couple harnessed for a number of metabolic functions. In particular, outer-membrane electron-transfer proteins incorporating iron bound by heme groups can enable microorganisms to couple biological and mineral redox processes.

At the molecular scale, mineral redox reactions are complex, involving many elementary processes among which an actual electron-transfer step is just the most obvious example. Understanding reaction mechanism and the constraints on reaction rate requires a model of how all such elementary steps combine for the overall reaction to proceed. As an example, Fig. 2 presents a scheme for a representative reaction, the reductive dissolution of an iron(III) (oxyhydr)oxide. This important reaction that mobilizes iron into solution has been studied extensively and Fig. 2 summarizes current understanding of key processes, which are discussed in more detail below. To develop and test such a model, we require a theoretical framework for the reaction rate and experimental approaches to observe dynamic changes in the chemical or mineralogical parameters relevant to the reaction. A powerful framework for understanding the rates

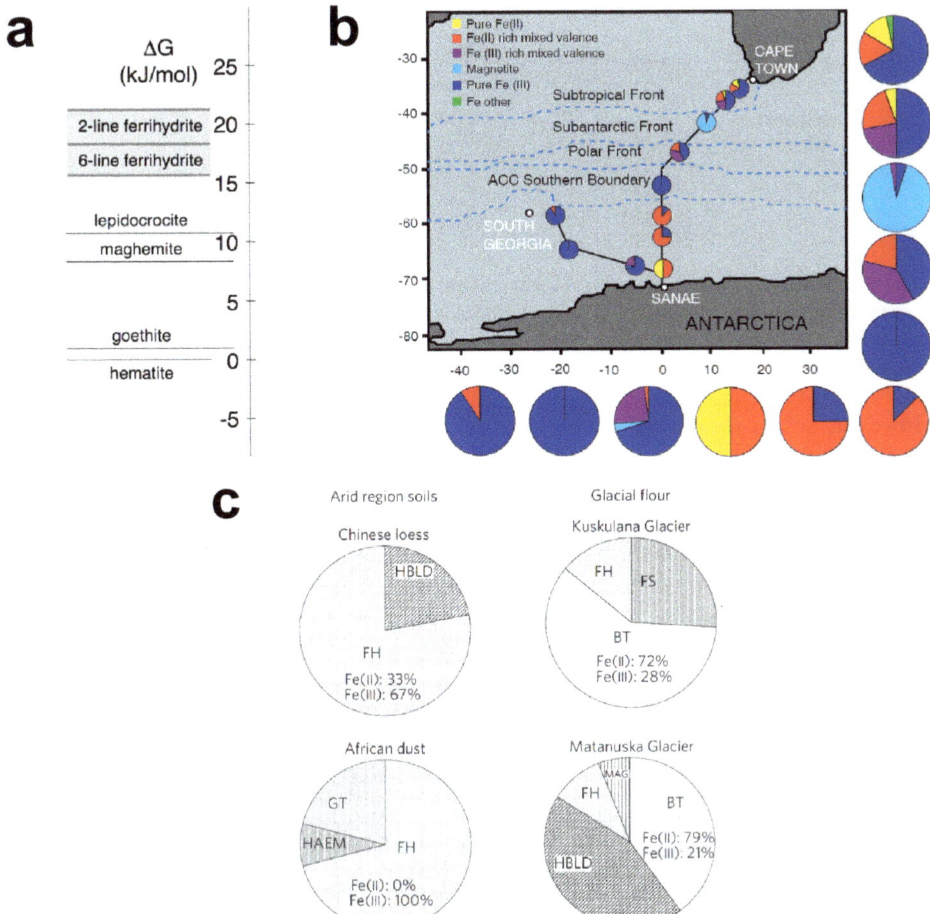

Figure 1. (a) Free energies of transformation of the listed iron oxide, hydroxide and oxyhydroxide phases relative to hematite plus water. For example, the ΔG value for goethite is calculated for the transformation α-FeOOH $\rightarrow$ $1/2\alpha$-Fe_2O_3 + $1/2H_2O$. The stoichiometry both of 2- and 6-line ferrihydrite is assumed to be $Fe(OH)_3$. Based on Majzlan *et al.* (2003, 2004). (b) Summaries of solid-phase iron speciation (pie charts) from surface water samples along the mapped cruise of the South Atlantic and Southern oceans. From von der Hayden *et al.* (2012) with permission of the American Association for the Advancement of Science. (c) Summaries of solid-phase iron speciation from sampled aerosols from the indicated sources. The labelled phases are biotite (BT), coquimbite (CQ), ferrihydrite (FH), ferrosmectite (FS), goethite (GT), hornblende (HBLD), magnetite (MAG) and hematite (HAEM). From Schroth *et al.* (2009) with permission of the Nature Publishing Group.

of chemical reactions, including electron transfer, is given by transition state theory (Laidler and King, 1983) and the concept of activation barriers. This concept helps conceptualize why a specific reaction step is fast or slow, and can provide a quantitative expression for the dependence of reaction rate on temperature (see section 2.2).

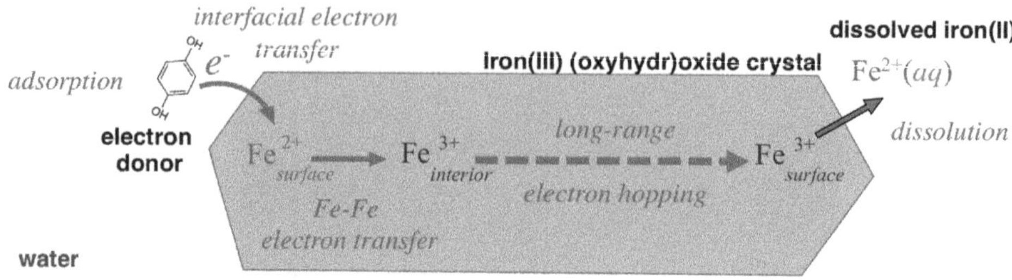

Figure 2. Schematic cartoon showing the key chemical steps involved in the chemical reduction and consequent dissolution of an iron(III) oxide, hydroxide or oxyhydroxide crystal. Each step indicated by arrows has an associated free energy change and a kinetic rate constant that contribute to the overall reaction.

However, such expressions have chemical meaning only if the reaction pathway is known confidently, a demanding task for both experimental and simulation approaches. We shall introduce some state-of-the art approaches for experimental methods, and refer to relevant literature for simulation methods.

Much of what we know about the fundamentals of redox reactions comes from studies of simple systems and thus comparisons between molecular and mineral redox reactions can be insightful[1]. Although we sought to make the material presented in the chapter self contained, familiarity with the concepts of chemical kinetics, as given in many excellent treatments, is recommended (Espenson, 1981; Upadhyay, 2006). Many of the conceptual and experimental difficulties in understanding the rates of mineral redox reactions closely match those encountered in the long-standing challenge of understanding mineral dissolution rates (Wehrli *et al.*, 1990; Ohlin *et al.*, 2010).

1.1. Timescales relevant to geochemical reaction kinetics

It is our experience that many solid-phase redox reactions, such as the corrosion of an iron bolt or the formation of an oxidation bloom on a mineral specimen, occur relatively slowly, often on the month-to-year timescale. Yet, our chemical intuition probably tells us that the molecular steps required for reaction, such as electron transfer, occur very quickly. An important goal of this chapter is to reconcile these ideas. Figure 3 illustrates the vast range of timescales that may be relevant. To help understand the molecular-scale steps that control mineral reaction rate, we propose a division into three classes.

(1) *Elementary* processes are the discrete molecular steps such as the exchange of a water neighbour or the transfer of an electron. Elementary processes are typically very fast (ps–ms) when they occur. However, even processes that are fast when they occur may have a low probability of occurring if there is a significant activation barrier.

---

[1] This chapter includes examples of reactions involving mononuclear atoms or ions (*e.g.* Fe^{2+}(aq)), molecules (*e.g.* O_2 or benzoquinone) and minerals (*e.g.* hematite). For brevity, all solid-phase reagents will be termed 'mineral' and all atomic or molecular reagents will be termed 'molecular'.

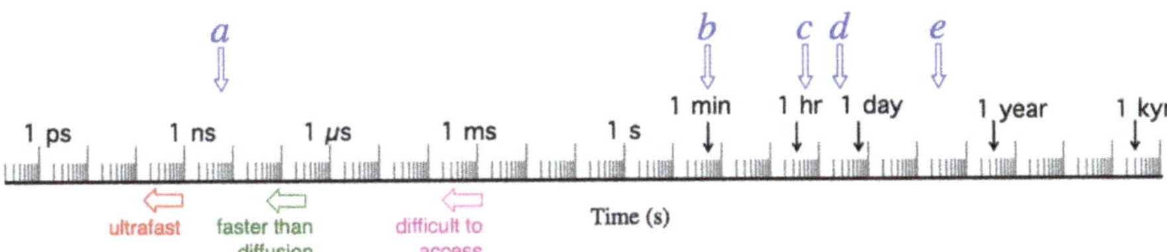

Figure 3. Logarithmic time axis illustrating the range of timescales relevant to mineral redox reactions. Characteristic timescales for a few environmental reactions, or intermediate steps, are indicated by labelled arrows. (a) 5 ns is the time constant for electron transfer between neighbouring iron sites in hematite (Katz *et al.*, 2004; Iordanova *et al.*, 2005). (b) 1 min is the half-life for the oxidation of nanomolar concentrations of Fe^{2+}(aq) in oxygen-saturated water at pH 7 (Morgan, 2007). (c) 2 h is the time required for Fe^{2+} to reduce a submonolayer of sorbed $(UO_2)^{2+}$ forming UO_2 nanoparticles at the (111) magnetite surface (Singer *et al.*, 2012). (d) 10 h is the average time after which a randomly chosen water molecule in pure pH 7 solution will dissociate spontaneously (Marx, 2006). (e) 3 months is the time to completely oxidize 8-nm diameter magnetite nanoparticles in oxygen-saturated water (Tang, 2003)

(2) *Ensemble* processes require an assortment of elementary processes to occur in a coordinated fashion. Ensemble processes represent an important, but intermediate step, in a more complex reaction. For example, dissolution – the release of an ion from a mineral surface into solution – requires multiple elementary processes that may include ion complexation by one or more water molecules (aquation); proton transfer or water hydrolysis reactions; and the breaking of bonds to the mineral surface. Other examples include the nucleation of a new mineral particle from solution and the nucleation of a new structure during a solid-state phase transformation. Certain redox reactions require significant valence changes such as $U^{6+} \rightarrow U^{4+}$ and $S^{2-} \rightarrow S^{6+}$. Such reactions involve a succession of single-electron transfers that are associated with changes in the binding of ligands to the reacting species, or in the binding of the species to a mineral surface. An example multi-electron redox reaction of the U^{6+} species is given by Skomurski *et al.* (2011).

(3) *Transport* processes involve the motion of reactive species to the site of reaction and the formation of an *encounter complex*. For example, the rate of reagent diffusion (or diffusion plus advective mixing) is an intrinsic limit on the rate of reaction in solution. Diffusion-limited reactions are those in which reaction always occurs upon an encounter between reagents. Radical species, including reactive oxygen species (ROS) or free radical forms of organic molecules, are important intermediates in many environmental redox reactions and reactions of these unstable species are often diffusion limited.

The timescales for diffusion-limited reactions can be estimated for a two-reagent solution system that reacts at the rate $v = k[C_1][C_2]$, where C_i is the concentration of the ith reagent. The rate constant, k, can be estimated using the diffusion coefficients for the species:

$$k_{diff} = 4\pi(D_1 + D_2)(r_1 + r_2)/1000 \approx 10^8 \ M^{-1}s^{-1} \quad (1)$$

The D values are the diffusion coefficients; the r values are the interaction radii, such that reaction occurs when the distance between the species is less than $(r_1 + r_2)$. This equation assumes that there is no electrostatic interaction between ions that may enhance or impede close interaction. If initially, $C_1 = C_2 = 1$ mM, we can estimate that 90% of the reagents are consumed within 10 μs.

Diffusion in solution is an important example of a transport process that can affect the rate of mineral redox reactions. While laboratory kinetic studies are performed typically in well mixed solutions, mineral reactions can be limited by the diffusion of reagents in soil and subsurface solutions. Geochemical reactive transport codes have been developed to calculate the combined effects of flow, diffusion and reaction in porous media (Steefel and Maher, 2009). As described below, solid-phase diffusion of atoms or electrons can also be important in mineral redox reactions. These transport processes connect a redox reaction inside a mineral with a surface reagent or adsorbate.

1.2. Thermodynamic basis for mineral redox reactions

The thermodynamic basis for redox reactions is well established, and an excellent succinct review relevant to iron minerals was given by Amonette (2002). Briefly, the standard free energy change for a given redox half reaction, ΔG^o, can be calculated from the free energy of formation of the reactants and products. Table 1 displays some ΔG_f^o values relevant to iron oxide redox chemistry (Majzlan *et al.*, 2003, 2004). For the reductive dissolution of goethite, α-FeOOH:

$$FeOOH + 3H^+ + e^- \rightarrow Fe^{2+}(aq) + 2H_2O \quad \Delta G^o = -63.3 \text{ kJ mol}^{-1} \tag{2}$$

Under standard conditions the temperature is taken to be 25°C and all solution species are assumed to be at unit activity (for protons, this corresponds to pH 0). A

Table 1. Thermodynamic data for calculations of half-reaction reduction potentials of iron oxide phases.

Mineral	Formula	ΔG_f^0 (kJ/mol, 298 K)	Refs
Hematite	α-Fe_2O_3	−742.7	Robie and Hemingway (1995)
Maghemite	γ-Fe_2O_3	−727.9	Majzlan *et al.* (2003)
Goethite	α-FeOOH	−489	Majzlan *et al.* (2003)
Lepidocrocite	γ-FeOOH	−480.1	Majzlan *et al.* (2003)
6-line Fhyd	$Fe(OH)_3$	−711.0 to −708.5	Majzlan *et al.* (2004)
2-line Fhyd	$Fe(OH)_3$	−708.5 to −705.2	Majzlan *et al.* (2004)
Magnetite	Fe_3O_4	−1017.4	Lilova *et al.*, (2012)
Fe^{2+}(1 M, aq)		−78.9	Stumm and Morgan (1981)
H_2O (l)		−237.1	Bard *et al.* (1985)

given half-reaction would generate a potential relative to a standard hydrogen electrode (SHE) that is given by $E^0 = -\Delta G^0/n\mathrm{F} = 0.656$ V *vs*. SHE. The Nernst equation describes the dependence of the electrochemical potential on the activities of the participating species away from standard conditions. For a one-electron reaction at 25°C, with the reagent activities given by the reaction quotient, Q:

$$E = E^0 + \frac{\mathrm{R}T}{n\mathrm{F}}\ln Q = E^0 + 0.2303\frac{\mathrm{R}T}{n\mathrm{F}}\ln Q = E^0 + 0.059\log Q \qquad (3)$$

For the reaction expressed in equation 3 using $\mathrm{pH} = -\log[\mathrm{H}^+]$:

$$E = E^0 - 0.059\log(a_{\mathrm{Fe}^{2+}}) - 0.177\mathrm{pH} \qquad (4)$$

A similar expression describes the electrochemical potential for a redox-active species such as ascorbate or uranium(VI) species.

Example: The reduction potential of surface adsorbed iron(II)

It has been frequently observed that iron(II) adsorbed to the surface of iron(III) oxide minerals exhibits faster rates of reaction with dissolved oxidizing species than free dissolved Fe^{2+}(aq) (White and Peterson, 1996; Elsner *et al.*, 2004). The coordination environment of iron probably plays a key role, and a proposed thermodynamic explanation is that the stability of the iron(III) product is enhanced at an oxide surface. As described below, general descriptions of thermally-activated reactions predict that stabilizing a product may increase the rate of reaction. However, there may be additional kinetic effects that enhance the rates of reactions at surfaces. To investigate this issue, Silvester *et al.* (2005) developed a thermodynamic surface complexation model to calculate the affinity of iron(II) to the surfaces of iron(III) (oxyhdr)oxides as a function of pH. Assuming that the Fe^{3+} species formed by the oxidation of surface-bound Fe^{2+} was added to the crystal lattice they predicted the E_{H} of particle suspensions equilibrated with Fe^{2+}(aq), and compared the prediction with experimental measurements of the rest potential at a platinum electrode. At pH below which sorption occurs, the suspension E_{H} is determined by the mineral reductive dissolution redox couple (*i.e.* equation 2, above). Above the sorption edge, E_{H} is dominated by the surface redox couple involving surface sites $\equiv Fe^{3+}OH$ and $\equiv Fe^{3+}OFe^{2+}$.

$$E = 0.814 - 0.059\log\frac{[\equiv \mathrm{Fe}^{3+}\mathrm{OH}]}{\equiv \mathrm{Fe}^{3+}\mathrm{OFe}^{2+}} - 0.118\mathrm{pH} - \psi_0 \quad (\mathrm{pH} >\sim 6) \qquad (5)$$

The sorption enthalpy is a function of the surface potential, ψ_0, that was derived from a model of the electrical double layer. As shown in Fig. 4, the model predicts that sorption enhances the reduction potential of a mixture of iron(III) oxides and Fe^{2+}(aq) and thus increases the thermodynamic driving force for the reduction of U(VI), nitrobenzene or nitrite.

1.3. Mechanistic descriptions of mineral redox reactions

While thermodynamic considerations provide a rigorous assessment of the possibility of reaction, the half-reactions do not reflect the complexity of either the mineral or

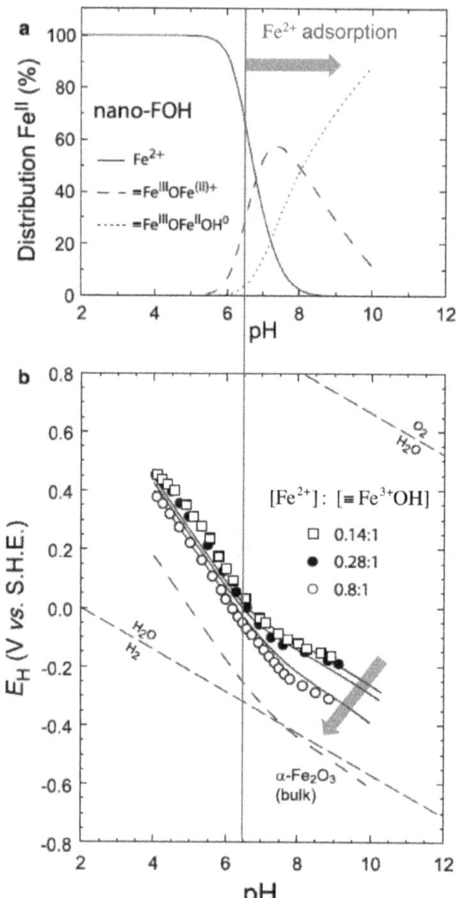

Figure 4. Silvester *et al.* (2005) modelled the sorption of Fe^{2+} to iron(III) oxyhydroxide nanoparticles (nano-FOH) and the corresponding effect on solution E_H. (a) Iron(II) distribution diagram for a concentration of oxyhydroxide surface sites, $\equiv Fe^{3+}OH$, of 0.2 mM and total Fe^{2+} concentration of 0.16 mM. (b) Calculated suspension E_H of three Fe^{2+} concentrations (lines) compared with experimental rest potentials measured at a platinum electrode. The red arrows illustrate how the onset of Fe^{2+} adsorption corresponds to a change in the measured reduction potential. Figure modified from Silvester *et al.* (2005) with permission from Elsevier.

molecular processes. The reductive dissolution of iron oxides (equation 2 and Fig. 2) clearly involves elementary, ensemble and transport processes. These steps must occur at successive points in time and may even be spatially separated.

As depicted in Fig. 3, this chapter describes complex chemical reactions (including mineral redox reactions) as a progression through a sequence of 'reaction intermediates' to arrive at a product. Reaction intermediates are often short-lived, and hence are challenging to identify and characterize directly. However, a core assumption is that the reaction intermediates are well defined chemical species. That is, they require a specific amount of free energy to be created, and they reproducibly adopt the same structural configuration. Such intermediate states are additionally characterized by their lifetime, which is determined by the number of additional reaction steps available to them (including reverse reaction steps) and the activation barrier for each step. Activation barriers are crossed *via* a configuration of all relevant atoms at the apex of the barrier which is called the 'transition state'.

1.3.1. Reductive dissolution of iron(III) oxides

It has been proposed by many authors that the reductive dissolution process can be divided into separate electron-transfer and iron(II) release (dissolution) steps.

$$Fe^{3+}(s) + e^{-} \rightarrow Fe^{2+}(s) \qquad \Delta G_{et}, k_{et} \qquad (6a)$$

$$Fe^{2+}(s) \rightarrow Fe^{2+}(aq) \qquad \Delta G_{diss}, k_{diss} \qquad (6b)$$

While this description is intuitive, key features were missing from early models. Most remarkable is the observation that electrons transferred to iron(III) oxides – semiconducting or insulating minerals not known in general for their electrical conductivity – are mobile inside the crystal. Although it is now well established that electron mobility in iron oxides proceeds *via* electron hopping between iron sites (Iordanova *et al.*, 2005; Katz *et al.*, 2012), the driving force for this process, and the consequences for mineral reactions, remain to be fully established. In addition, we still lack confident assessment of the relative contributions of successive steps to the observed reaction rates and overall free energy change.

(1) *Free energy changes during reaction steps.* The free energy changes for key steps, ΔG_{et} and ΔG_{diss}, respectively, are, in principle, well defined, but they are difficult to measure experimentally because they usually involve at least one reaction intermediate (as depicted schematically in Fig. 5). In general, reaction intermediates cannot be isolated experimentally and studied with conventional methods. Amonette (2002) discussed an approach to estimate ΔG_{et} based on the electronic structure of iron(II) sites. Below, we introduce approaches to identify such intermediates. If the identity of a reaction intermediate is known, such species can be created and examined using molecular simulations on the computer.

(2) *Rates of reaction steps.* An important goal in the study of any complex reaction is the determination of the rate-limiting step. With reference to Fig. 5, an equivalent goal is to determine which reaction intermediate has the longest lifetime. For reductive dissolution reactions, the dissolution step is widely assumed to be rate-limiting. Interfacial electron transfer reactions can be very fast, and it is reasonable that Fe^{2+} release, with many attendant bond-breaking and hydration steps, would generally be slower. However, there is little direct evidence for this. Pedersen *et al.* (2005) studied

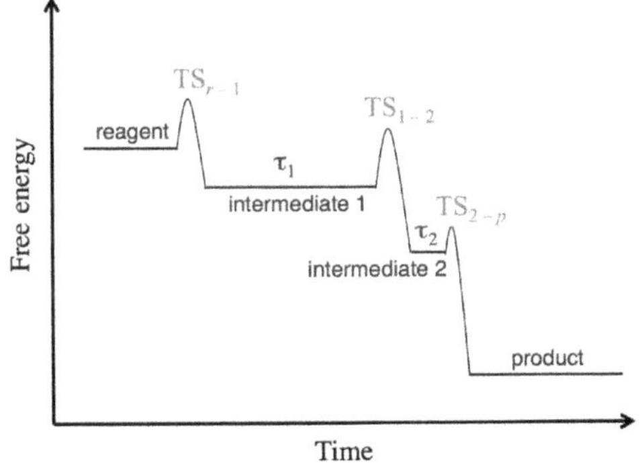

Figure 5. Conceptual model for multi-step chemical reactions in which a reagent is transformed into a product *via* a succession of reaction intermediates.

the addition of Fe^{2+}(aq) to suspensions of ^{55}Fe-enriched iron(III) (oxyhydr)oxides, finding that the rate of isotope exchange and release of $^{55}Fe^{2+}$ into solution was dependent on mineral structure. This important study suggests strongly that there is a relationship between mineral stability and the rate of Fe^{2+} release. However, such studies cannot distinguish directly between transport processes (*e.g.* electron hopping inside the particles) and chemical processes (*e.g.* the formation of a surface-leaving group involving a number of ligand exchanges).

1.3.2. Additional features of mineral redox reactions

Mineral redox reactions have some distinctions with respect to molecular reactions. For example, minerals that form in low-temperature environments are often not in the thermodynamically expected form. There are numerous pathways for the formation of metastable phases of iron(III) (oxyhydr)oxides such as lepidocrocite or maghemite at low temperatures. Transformation to the stable hydrated (goethite) or dehydrated (hematite) phase, depending on conditions, can take thousands of years. Moreover, aqueous precipitation reactions frequently generate nanoscale mineral particles that are metastable with respect to bulk phases because of the excess interfacial energy in such high-surface-area materials. Here too, dissolution-based growth rates (Ostwald ripening) can be slow if mineral solubility is low, and growth may be dominated by direct particle–particle joining (oriented aggregation). Metastable bulk or nanoscale iron (oxyhydr)oxides (Schroth *et al.*, 2009; von der Heyden *et al.*, 2012), including iron(II) bearing phases (Postma and Brockenhuusschack, 1987), can persist years or much longer in the environment. Several groups have considered surface energy contributions to ΔG_f^o for small mineral particles (Navrotsky *et al.*, 2008; Roden, 2006; Silvester *et al.*, 2005).

Mineral redox reactions are also typically not reversible. Adding or removing an electron from an atom in solution or in a crystal alters the ionic radius and local balance of charge. In water, these changes may be accommodated by changing the radius of the hydration shell or coordination by water and hydroxyl ions. In a bulk crystal, changes in atomic radius are less easily accommodated, and the energy penalty associated with loss of charge balance ultimately can be satisfied only by changing mineral stoichiometry, which requires the loss or gain of atoms from every unit cell in the structure. Very few crystal structures can undergo a transformation to a new, charge-balanced stoichiometry, which leads to changes that can include dissolution or disintegration[2]. Even isolated redox-reactive impurity iron sites in minerals such as clays typically do not react reversibly.

---

[2] This is a serious challenge for solid-state battery materials, and it can sometimes be solved by coupling cation transport (*e.g.* Li^+) to electron charging/discharging cycles. Hematite batteries can be made to function in this manner.

2. Chemical kinetics analysis of mineral redox reactions

By definition, the rate of any chemical reaction is measured experimentally by observing the appearance of the product as a function of time. Performing such 'chemical kinetics' experiments under varying conditions can be enormously informative about the nature of a reaction and can be used to distinguish between hypotheses for a reaction mechanism. For example, recording reaction kinetics as a function of solution ionic strength can indicate whether the reactions involve inner- or outer-sphere binding configurations of reagents. Studies as a function of pH can be illuminating for both solution-phase (homogeneous) or interfacial (heterogeneous) redox reactions. pH can effect the speciation of ions in aqueous solution, the affinity of a charged reagent to a mineral surface and other reaction steps. Many excellent reviews and monographs describe the elucidation of reaction mechanism using chemical kinetics analysis (Espenson, 1981; Upadhyay, 2006; Brantley *et al.*, 2008).

2.1. Experimental determination of reaction rates

Studies of the kinetics of mineral redox reactions have two requirements: initiating the reaction of interest and following the reaction progress. The techniques employed must have a temporal resolution significantly better than the rate of the full reaction or important reaction steps.

2.1.1. Initiating reactions by reagent mixing

Chemical kinetics studies are most accurate when the rate of reaction is slower than any transport limitations, such as mixing. Fast reactions can be studied by the rapid turbulent mixing of small volumes in continuous-flow or stopped-flow reactors. The precise start time of a reaction is limited by the rate of mixing.

2.1.2. Measuring reaction progress

The reaction extent may be determined through measurement of the structure and relative abundance of the reacting phase(s), the net oxidation state of mineral atoms or the concentration of a reagent or product. Studies are most conveniently performed by directly probing the reacting system *in situ* but reaction quenching and *ex situ* analysis may also be informative. Neutron or X-ray diffraction (XRD), vibrational spectroscopy (including infrared (IR) and Raman spectroscopy), and X-ray absorption near-edge spectroscopy (XANES) have all been used to observe mineral reactions (Magnien *et al.*, 2006; Ahmed *et al.*, 2010; Yang *et al.*, 2010). Using specially designed apparatus, millisecond timescale reactions can be studied (Haumann *et al.*, 2005). For reactions involving nanoparticles in fluids, the Penn group has pioneered the use of high-resolution imaging of statistically significant numbers of individual particles to obtain distributions of reaction rates (Burrows *et al.*, 2012). For certain semiconductor minerals, redox reactions can be controlled and rates measured using electrochemical methods (see below) (Ahmed *et al.*, 2010).

2.1.3. Electrochemical experiments

The mineral redox reactions addressed in this chapter involve electron transfer between a molecule and a hydrated mineral surface. Interfacial electron-transfer mechanisms and rates have been studied extensively in the field of electrochemistry which thus warrants a separate introduction. Theoretical and technical descriptions of the diverse range of possible electrochemical experiments are given elsewhere (Bard and Faulkner, 1980; Memming, 2001). Here we describe relevant concepts and key experimental approaches. Central to modern electrochemical methods is the use of a reference electrode that incorporates a redox couple such as Ag/AgCl. This electrode provides a stable reference potential used in three types of study.

(1) *Redox potential measurements*. The redox potential of solutions, suspensions or solid-phase samples containing redox-active species can be estimated from the potential difference that develops between a suitable working electrode and the reference without current flow between them. For example, the E_H of iron (oxyhydr)oxide suspensions in the presence of Fe^{2+}(aq) was determined by the 'rest potential' measured at a Pt electrode (Silvester *et al.*, 2005). The reduction potential of iron(0)- and iron(II)-bearing particles was estimated from a similar quantity (also called the 'self-induced' or 'open-circuit' potential) developed by an electrode composed of packed particles (White and Peterson, 1996; Gorski *et al.*, 2010).

(2) *Interfacial molecular redox reactions*. Oxidation or reduction reactions of dissolved reagents may be driven and monitored by setting the potential, E, of an inert, conductive working electrode relative to the reference electrode and recording the net current flow[3]. As described below, the reaction rate is a strong function of the difference between E and the solution E_H. Both metallic and (sufficiently conductive) semiconductor working electrodes may be used. For example, electrodes can be constructed from thin films of micro- or nanocrystalline metal oxides, including TiO_2 or αFe_2O_3 (Sivula *et al.*, 2010; Shimizu *et al.*, 2012). Thus, electrochemical studies with semiconductor electrodes can be used to investigate aspects of molecule–mineral electron transfer. However, there are important differences between reactions at metallic and semiconductor electrodes, and between electrochemically controlled electron-transfer processes and chemical redox reactions.

For both metallic and semiconductor electrodes, the net redox chemistry at the surface is the sum of reduction and oxidation processes that proceed until the average chemical potential of electrons in the solution (E_H) is equal to the average chemical potential of electrons in the electrode (called the Fermi energy, E_F). At a semiconductor surface, the current is usually a combination of reduction processes driven by electrons in the conduction band and oxidation processes driven by holes in the valence band. Certain

---

[3] A third electrode, called the auxiliary or counter electrode, is essential for stable electrochemical experiments. This electrode is responsible for providing or consuming electrons as required by the reaction at the working electrode, to avoid drifts in E, and may perform oxidative or reductive reactions of solutes, electrolytes or solvent molecules to do so.

highly reducing molecules may also transfer electrons to the conduction band. Changing E alters E_F in the working electrode but this change is manifested differently in metallic and semiconductor electrodes. Metal electrodes become polarized, increasing or decreasing the energy of electrons in the metal and of charged species at the surface. In contrast, ideal semiconductor electrode surfaces do not become polarized. As established by Gerischer (1990) and reviewed by Memming (2001), changes in E_F add or remove electrons or holes in the conduction and valence bands, respectively, but the electrochemical potentials of those bands remain fixed relative to the solution E_H.

Importantly, oxide surfaces in water can become polarized due to acid–base reactions involving oxygen atoms at the surface. Protonation changes the surface charge, directly changing the energy of electronic states at the surface (Lyon and Hupp, 1999). For many oxides, the bottom of the conduction band, E^{CB}, obeys a Nernstian relationship as a function of pH away from the pH of zero surface change, pH_{pzc}:

$$E^{CB}(pH) = E^{CB}_{pzc} + 0.059(pH_{pzc} - pH) \tag{7}$$

Of course, many redox reactions involve protons so that the solution E_H may also exhibit a pH dependence.

There are compilations of the conduction band (CB) and valence band (VB) energy positions for common semiconductors (Xu and Schoonen, 2000), and recent studies have used X-ray absorption and emission spectroscopy to probe these electronic states directly (Braun *et al.*, 2012; Gilbert *et al.*, 2013). Sherman (2005) used this approach to predict the solution pH values for which absorption of light by manganese and iron oxides created photoexcited electrons that were sufficiently reducing to drive mineral photoreduction

(3) *Electrochemically controlled mineral redox reactions*. The control and study of mineral redox reactions can be performed on conducting minerals. Studies of insulating minerals have been performed either by fabricating packed-powder electrodes (Nurmi *et al.*, 2004; Renock *et al.*, 2013) or by studying nanoscale particles at an inert electrode surface.

Mckenzie and Marken (2001) reported cyclic voltammetry of hematite nanoparticles that adsorbed to a conductive glass electrodes. The CV was consistent with a cathodic reduction of hematite followed by the anodic oxidation of Fe^{2+}(aq) ions released in the first sweep. Grygar (1998) measured the rates of the irreversible mineral redox transformations using 'abrasive stripping voltammetry'. This author proposed that differences in the trends in the rates of the oxidative dissolution of α-Cr_2O_3 and the reductive dissolution of γ-Fe_2O_3 point to differences in the rates or mechanisms of key interfacial reactions required for dissolution.

2.2. Activation barriers in mineral redox reactions

Chemical kinetics experiments often show that reactions are not diffusion limited and that the reaction rates increase with temperature. These observations are rationalized by the concept of an energy barrier that lies on the reaction pathway and hinders the reaction from proceeding. The barrier is overcome only if there is sufficient thermal energy for reagents and solvent molecules to access the reaction transition state – the unfavourable configuration of atoms at the barrier apex. Here we introduce three expressions for the rates of thermally activated reactions, which are summarized schematically in Fig. 6. They are introduced for simple, reversible reactions, but may also represent one rate-limiting step in a more complex reaction.

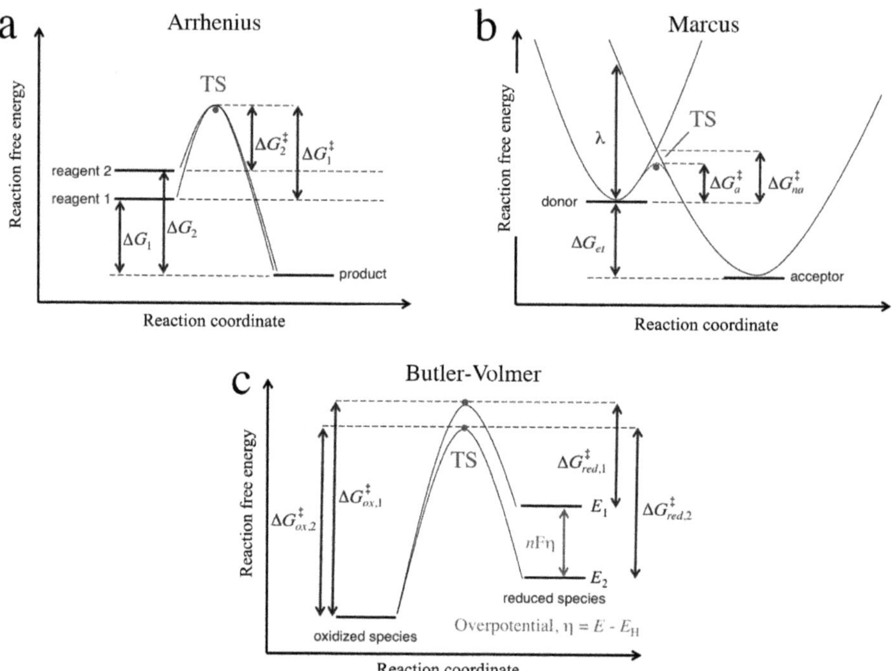

Figure 6. Summary of the free energy diagrams for thermally activated chemical reaction processes discussed in the text. The 'Reaction Coordinate' (the label of the horizontal axis) refers to any suitable parameter that can be used as the extent of the reaction progress, such as the distance between the reagents. The labelled 'Transition State' (TS) locations are the highest-free-energy points along reaction pathways. (a) The Arrhenius equation is a generic rate law for any reaction that requires thermal energy to proceed. If two similar molecular or mineral reagents (*e.g.* reagents 1 and 2) undergo a reaction to the same product(s), a free-energy relationship between reaction free energy and rate is expected. (b) The Marcus equation for electron transfer is derived by considering the free energy required to find the transition-state configuration by distorting the donor and acceptor species on two parabolic free-energy curves that intersect at the transition state. Electronic interactions between the donor and acceptor lower the activation free energy from $\Delta G_1^\ddagger$ to $\Delta G_2^\ddagger$, which can increase greatly the reaction rate. The activation free energy can be calculated from estimates of reaction free energy change ΔG_{et}, the reorganization energy, λ, and the electronic interaction energy. (c) The Butler-Volmer equation describes the rate of electron transfer between a redox-active species in solution and the surface of a conducting electrode. When the electrode potential is equal to the equilibrium potential of the solution, $E_1 = E_{\mathrm{H}}$, interfacial charge transfer rate reduces to the Marcus equation for a reaction with $\Delta G = 0$. Changing the electrode potential alters the rates of forward and reverse electron transfer.

2.2.1. *Arrhenius equation*

The Arrhenius equation is a simple, model-free kinetic rate equation in which the reaction rate, k, is expressed in terms of an enthalpy or free energy barrier

$$k = A\exp(-\Delta G^{\ddagger}/\mathrm{R}T) \quad (8)$$

where $\Delta G^{\ddagger}$ is the Gibbs free energy of activation, T is the temperature in Kelvin and R is the gas constant. If a plot of $\ln(k)$ *vs.* $1/T$ is linear, it is a good indication that the reaction rate is controlled by a single rate-limiting step with an activation barrier that can be estimated from the slope. The pre-exponential factor, A, also called the frequency factor and expressed in units of s^{-1}, accounts for the rate of transport and association of reagents that is required prior to reaction.

Example: The size dependence of the reduction of ferrihydrite by hydroquinone

Ferrihydrite is a common metastable iron(III) oxyhydroxide mineral that is found only as nanoparticles with a range of sizes and crystallinity. Ferrihydrite is more reactive with chemical and biological reducing agents than more crystalline bulk iron(III) phases, but the relationship between size, structure and reactivity has been difficult to establish. Erbs *et al.* (2008) studied the temperature dependence of the rate of reductive dissolution by hydroquinone of 6-line ferrihydrite nanoparticles with a range of average particle diameters from 3.4–5.9 nm and a relatively narrow size distribution (~10%). The initial rate of reduction at a single temperature exhibited a slight dependence on particle size once normalized to nanoparticle surface area. As shown in Fig. 7a, surface-area-normalized Arrhenius plots were highly linear, with no size dependence of the calculated activation energy. These results indicate a single common rate-limiting step at all temperatures and particle sizes. However, analysis of the Arrhenius plots revealed a striking size dependence of the pre-exponential factor, which varied sixfold, outpacing the change in surface area. This trend indicated that smaller nanoparticles had a greater frequency of opportunities to undergo reaction. Erbs *et al.* (2008) hypothesized that this could be caused by a change in the ability of hydroquinone molecules to diffuse through structured interfacial water and form a charge-transfer complex at the nanoparticle surface. Experiments and simulations have demonstrated clearly that water typically forms highly ordered layers at the surface of bulk oxides (Catalano *et al.*, 2009). As illustrated in Fig. 7b for hematite nanoparticles of two sizes, molecular simulations suggest that particle size and crystallinity may affect strongly the extent of water ordering (Spagnoli *et al.*, 2009). The effect of interfacial water structure on the transport of dissolved species to mineral surfaces is presently poorly understood, but it is plausible that it may contribute to the rates of interfacial reactions, including redox reactions, for certain systems.

2.2.2. *Marcus equation for electron transfer*

The development of a predictive, molecular description of electron-transfer reactions between reagents is one of the major accomplishments of chemical science. For the theoretical description of electron transfer in solution, Rudolph A. Marcus was awarded the Nobel prize in 1992. Marcus theory is a semi-classical theory for reaction

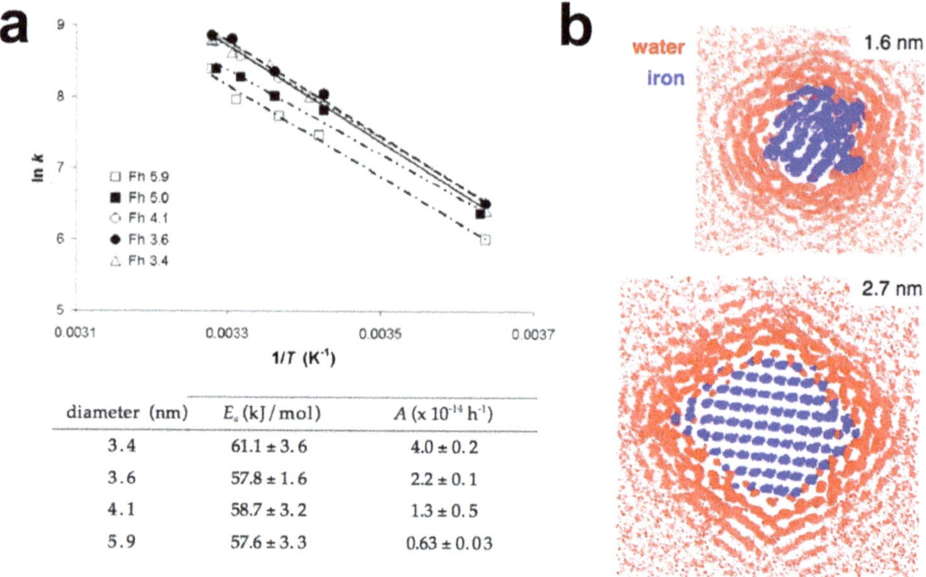

diameter (nm)	E_a (kJ/mol)	A (x 10^{-14} h^{-1})
3.4	61.1 ± 3.6	4.0 ± 0.2
3.6	57.8 ± 1.6	2.2 ± 0.1
4.1	58.7 ± 3.2	1.3 ± 0.5
5.9	57.6 ± 3.3	0.63 ± 0.03

Figure 7. (a) Results from a study of the size dependence of the rate of reductive dissolution of 6-line ferrihydrite nanoparticles by hydroquinone. Shown (top) are Arrhenius plots of the temperature dependence of the initial rate for the indicated particle diameters and (bottom) a table of fitted activation energies, E_A, and pre-exponential factors, A. Data from Erbs *et al.* (2008) with permission from the American Chemical Society. (b) Results of molecular dynamics simulations of the structure of water surrounding hematite nanoparticles of two particle sizes. The results are presented as cross-sectional views of the most frequent locations of water molecules (red) and iron atoms (blue). Surface water layers are predicted to be more highly structured around the larger, more crystalline nanoparticle. Data from Spagnoli *et al.* (2009) with permission from Elsevier.

in which the electron-transfer rate, k_{et}, is described by a thermally activated process (Marcus and Sutin, 1985).

$$k_{et} = \kappa A \exp(-\Delta G^{\ddagger}/RT) \qquad (9)$$

The pre-factor A has the same meaning as in equation 8 as it represents the frequency at which the electron donor and acceptor species associate and form a pre-charge-transfer complex (Hupp and Weaver, 1983b). The free energy of activation stems from the requirement for the system to adopt a configuration of reagent and solvent atoms that bring the energy levels for the electronic donor and acceptor states into alignment. A key aspect of the Marcus approach is the premise that the ground state of the reagents and products are at the bottom of a parabolic free energy well (Fig. 6b). This explicit functional form allows the activation barrier $\Delta G^{\ddagger}$ to be estimated from ΔG_{et}, in contrast to generic reaction rate expressions such as the Arrhenius equation. An additional term, κ, the electronic transfer coefficient, describes the probability that electron transfer occurs once the appropriate configuration has been attained.

There are many excellent reviews of the theory and its development to include quantum mechanical aspects. For the prediction of the rates of environmental redox reactions, Rosso and Dupuis (2006) developed simulation methods that calculate both inner- and outer-sphere bonding contributions to the Marcus equation using quantum chemical and molecular dynamics methods. This approach has enabled molecular-scale insights into many aspects of environmental electron-transfer reactions, including homogenous and heterogeneous reactions, and inorganic reagents as well as proteins. Here we summarize briefly the chemical factors that can affect electron-transfer rate, as they are pertinent both to molecular and mineral reactions.

(1) *Activation free energy*. For outer-sphere reagents in solution, electron transfer occurs through at least one hydration shell, and there are only weak direct interactions between donor and acceptor orbitals. In these systems, $\Delta G^{\ddagger}$ is a function of the overall reaction free energy, ΔG, and a reorganization free energy, λ, required to attain the transition state configuration. Because Marcus assumed a specific free-energy potential for electron transfer, an explicit expression could be derived for the activation energy:

$$\Delta G^{\ddagger} = (\Delta G + \lambda)^2/4\lambda \tag{10}$$

The term λ includes contributions from two physical factors. First, short-range distortion of bonds is required to attain the transition state configuration. Second, the electronic charge density of the transition state configuration interacts with a polarizable medium such as water. These quantities are typically intrinsic to a reaction pathway and may be approximately transferable between similar reactions.

(2) *Electronic interactions*. Electron transfer requires some overlap between donor and acceptor wavefunctions, but the extent of the interaction, and the impact on k_{et}, can vary substantially. Strong electronic interactions generally occur when reagents approach closely. In homogeneous aqueous solutions, inner-sphere configurations of reagents (in which each reagent co-penetrates the hydration shell of the other) generally lead to strong electronic interactions for electron transfer. Reagent adsorption to minerals can also involve inner- or outer-sphere binding, with equivalent impacts on the rates of interfacial electron transfer.

Strong interactions can increase k_{et} dramatically through two mechanisms. First, the height of the free energy activation barrier is lowered below that expected for weak coupling. This is due to a universal effect in quantum mechanics in which interaction between electronic states leads to the formation of a new lower-energy ground state.

Second, the probability for forward electron transfer (*i.e.* κ) also increases strongly with the extent of electronic wavefunction overlap between the reagents. For an outer-sphere configuration of hydrated species, such as the self-exchange reaction between the hexaaqua ions of M^{3+} and M^{2+} in solution (M = Ru, V, Cr), typically $\kappa < 1$. This self-exchange reaction proceeds at an anomalously faster rate for iron, possibly due to the formation of aqueous bridges between ions that enhance the electronic interactions (Hupp and Weaver, 1983a). For electron transfer between neighbouring and strongly interacting Fe^{3+} and Fe^{2+} sites in an oxide such as hematite, $\kappa \approx 1$, and in fact the electron can transfer back and forth between the sites many times when a favourable configuration is attained by spontaneous lattice vibrations (Iordanova *et al.*, 2005).

(3) *Wavefunction symmetry*. For atoms and molecules for which the occupied valence electron orbitals do not have spherical symmetry, electronic interaction is strongly orientation dependent. For some transition metals in solution, strong directional overlap between *3d* orbitals and reagent pi orbitals can lead to facile electron transfer (Rosso *et al.*, 2004; Luther, 2005).

As discussed recently by Renock and Becker (2010), there may be additional quantum mechanical constraints on certain mineral redox reactions if the initial and final electronic ground states of any reagents possess different symmetries that are not easily accessed by the electron-transfer reaction. For example, the electronic ground state of dioxygen is a triplet paramagnetic state, $^3O_2$, with two unpaired electrons. As reviewed by Miller *et al.* (1990), the ground state of most organic molecules is a singlet diamagnetic state, with fully paired valence electrons. Transfer of a single electron cannot be achieved while conserving the overall spin state. Thus direct oxidation (often called 'autoxidation') is forbidden. Oxidation may proceed *via* a spin transition to the $^1O_2$ singlet state, but this can add ~100 kJ/mol activation energy. Dissolved transition metals with more accessible spin states associated with partially occupied *3d* orbital can exchange electrons readily with organic molecules and dioxygen, and thus can catalyze the rates of organic oxidation. This phenomenon is also relevant to the oxidation of low-spin iron(II)-bearing minerals such as pyrite. Direct oxidation of surface iron(II) atoms by O_2 is kinetically limited, and pyrite oxidation is strongly catalyzed by dissolved iron that shuttles electrons from the surface to dissolve oxygen.

Example: Oxidation of iron(II) by dioxygen

It is estimated that primary productivity in 40% of the ocean is iron limited because insoluble iron(III) formed by the oxidation of soluble iron(II) is far less bioavailable for phytoplankton (Moore *et al.*, 2002). As Fe^{2+} is continually generated in aerobic surface waters by reactions with organic molecules and photochemical cycling, the rate of Fe^{3+}(aq) oxidation is a key constraint on ocean biogeochemistry. At pH ~3 or lower, the rate of oxidation of hexaaqua Fe^{2+}(aq) is relatively slow and is predicted quantitatively by first-principles simulation of outer-sphere electron transfer to O_2 to form superoxide, $O_2^{\bullet-}$. However, the oxidation rate increases rapidly with pH in a manner that follows the appearance of the iron(II) hydrolysis products (King *et al.*, 1995; Morgan and Lahav, 2007). Thus, the initial steps of iron(II) reaction involve at least three parallel processes:

$$
\begin{aligned}
&Fe^{2+} + O_2 \rightarrow Fe^{3+} + O_2^{\bullet-} && k_1 = 7.9\ \mathrm{m}^{-1}\ \mathrm{min}^{-1} \\
&Fe(OH)^{+} + O_2 \rightarrow Fe(OH)^{2+} + O_2^{\bullet-} && k_2 = 4.4 \times 10^{3}\ \mathrm{m}^{-1}\ \mathrm{min}^{-1} \\
&Fe(OH)_2^{0} + O_2 \rightarrow Fe(OH)^{+} + O_2^{\bullet-} && k_3 = 1.9 \times 10^{8}\ \mathrm{m}^{-1}\ \mathrm{min}^{-1} \qquad (11)
\end{aligned}
$$

However, the reasons for the considerable variation in reaction rates, taken from Millero (2009), remain incompletely understood. There is probably a thermodynamic component, as ligation by hydroxyl stabilizes the Fe^{3+} final state (Stumm and Sulzberger, 1996). Iron complexation by inorganic (*e.g.* CO_3^{2-}) and organic (*e.g.* citrate) anions is also observed to enhance oxidation rate (King, 1998; Pham and Waite, 2008). Accordingly, Wehrli (1990) demonstrated the existence of a linear free-energy relationship (see below) between the free energy of reaction for iron(II) species and the reaction rate, a possible indication that electron transfer occurs *via* the outer-sphere pathway for all hydrolysis species. However, while molecular simulation of dissolved iron oxidation rate, assuming outer-sphere pathway, gave excellent agreement for Fe^{2+}(aq), this approach greatly underestimated the rates for the hydrolysis product, indicating that an inner-sphere mechanism is more likely for those species (Rosso and Morgan, 2002).

2.2.3. Interfacial electron transfer

Similar to homogeneous redox reactions, electron transfer at a solid surface requires overcoming an intrinsic free energy of activation, $\Delta G^{\ddagger}$ (Hupp and Weaver, 1985). As above, rate equations for interfacial electron transfer have been derived that either make no specific assumption concerning reaction mechanism, or follow the Marcus approach (Henstridge *et al.*, 2012).

The Butler-Volmer equation is a widely used expression derived for metal electrodes (Bard and Faulkner, 1980). If a redox-active species at an electrode surface accepts n electrons, the energy of the reduced state is altered by the electrode potential, E. Using an *ad hoc* assumption that the transition state is affected partly by the electrode polarization (*cf.* Fig. 6c), the rates of both oxidative and reductive electron transfer are modified by E. Alternative derivations for the Butler-Volmer equation have been developed (Bazant, 2013) but it remains a phenomenological (although successful) treatment. The forward and reverse electron-transfer rates are

$$k_{ox} = k^0 \exp\left[\frac{1}{2}\frac{n\mathrm{F}}{\mathrm{R}T}\eta\right]; \qquad k_{red} = k^0 \exp\left[-\frac{1}{2}\frac{n\mathrm{F}}{\mathrm{R}T}\eta\right] \tag{12}$$

where the overpotential, $\eta = (E - E_H)$, and $k^0 = A\exp(-\Delta G^{\ddagger}/\mathrm{R}T)$ is the rate of interfacial electron transfer at equilibrium, $E = E_H$, when oxidative and reductive currents are equal and opposite.

Employing Marcus theory to describe redox reaction at a metal electrode introduces an alternative expression for the activation energy that now incorporates the reorganization energy, λ, as well as the reaction driving force, η.

$$k_{ox} = k^0_{et} \exp\left[-\frac{1}{4}\frac{\mathrm{F}}{\mathrm{R}T}(\eta + \lambda)^2\right]; \qquad k_{red} = k^0_{et} \exp\left[-\frac{1}{4}\frac{\mathrm{F}}{\mathrm{R}T}(\eta - \lambda)^2\right] \tag{13}$$

where k^0_{et} is also modified by the incorporation of $\Delta G^{\ddagger}$ from equation 10.

2.3. Free energy relationships

Above, we have sketched out how the concept of reaction activation barriers leads to expressions describing how the thermodynamic driving force and temperature can affect the rate of reactions. In fact, for many chemical reactions studied in the laboratory there is a clear correlation between ΔG and k. Such correlations are called 'free energy relationships' (FERs), and they may be used to test mechanistic hypotheses concerning a series of reactions (Hammet, 1937). Sverjensky and Molling (1992) appear to have been the first to extend the concept to minerals. All the activation relationships given above predict a specific FER under the correct conditions.

2.3.1. Linear free energy relationships

As illustrated in Fig. 6a, if a range of substances undergoes a type of reaction to the same final product, and the temperature dependence is given by the Arrhenius equation, a linear FER would be expected.

$$\ln k = \text{constant} + \Delta G/RT \quad (14)$$

This general expression can be used to test whether a series of reactions may proceed *via* a similar underlying mechanism.

Example: Biological and chemical reduction of iron oxides

Dissimilatory iron-reducing bacteria (DIRB) have the capability to respire organic compounds in the absence of O_2 by using solid-phase iron(III)-bearing minerals as the terminal electron acceptor (Lovley, 1991; Weber *et al.*, 2006). Work by several groups has identified the role of extracellular multi-heme *c*-type cytochromes in transferring electrons to iron (oxyhydr)oxides, causing iron reduction and Fe^{2+}(aq) release that may be accompanied by mineral transformation (Hansel *et al.*, 2005). As illustrated by Table 1, iron(III) minerals can vary significantly in their structure and stability, suggesting that there could be preferential utilization of particular phases with consequences for the biogeochemical evolution of anaerobic settings. However, it has proven surprisingly challenging to identify mineralogical controls on DIRB metabolism. As illustrated in Fig. 8a, Roden, (2003; 2006) compared the initial rates of chemical and microbial reduction of a range of iron (oxyhdr)oxides with the appropriate ΔG for reductive dissolution (*cf.* equation 2). A linear FER was obtained for chemical reduction but microbial reduction was dominated by mineral surface area not any thermodynamic parameter. Cutting *et al.* (2009) confirmed the importance of mineral surface area (and the effect of aggregation in reducing accessible surface area) but also noted differences in rate between poorly crystalline and more stable minerals. The most compelling evidence for a mineralogical control was presented by Bonneville *et al.* (2009), who found a linear FER connecting the initial reduction rate per cell with the free energy of formation of the iron(III) phase (Fig. 8b). This observation suggests that the Fe^{2+} release is rate-limiting and is controlled by the structure and stability of the reacting iron(III) phase. However, the authors note that the finding does not rule out alternative interpretations, such as a structure-dependent effect on protein–mineral electron transfer.

There is currently considerable experimental and theoretical work aimed at determining electron-transfer pathways and rates involving DIRB multi-heme cytochromes. One innovative approach is to embed purified outer membrane proteins in liposomes containing a chemical reductant and observe the rate of mineral particle reduction (White *et al.*, 2013).

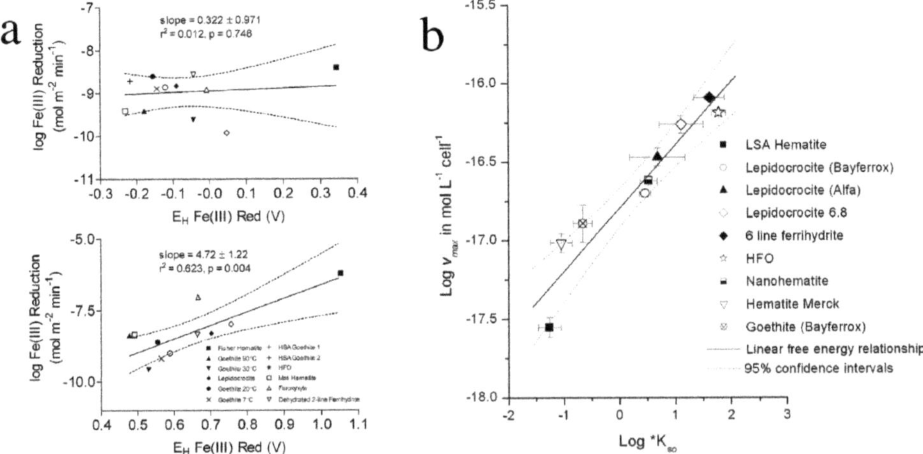

Figure 8. Examples of free energy relationships for the abiotic and microbial reductive dissolution of iron(III) oxides, hydroxides and oxyhydroxides. (a) Free energy relationships for Fe(III) (oxyhydr)oxide reduction by (top) *Shewanella putrefaciens* and (bottom) ascorbate. The logarithm of the initial rate is plotted against the electrochemical potential for the reductive dissolution of the oxide, *e.g.* $FeOOH + 3H^+ + e^- \rightarrow Fe^{2+} + 2H_2O$. Plot is figure 4 from Roden (2003) with permission from the American Chemical Society, with legend added from figure 1. (b) Linear free energy relationship for Fe(III) oxyhydroxide reduction by *S. putrefaciens*. The logarithm of the initial reduction rate per cell is plotted against the logarithm of the solubility product of the oxide, $^*KSO = a_{Fe^{3+}} a_{H^+}$, defined for reactions such as $1/2Fe_2O_3 + 3H^+ \leftrightarrow Fe^{3+}(aq) + 3/2H_2O$. From Bonneville *et al.* (2009) with permission from Elsevier.

2.3.2. Marcus theory free energy relationships

Marcus theory predicts a distinct FER functional form. For comparison with experimental data, the rate of formation of an encounter complex must be included. If the reagents combine at a diffusion rate, k_D, the rate of reaction, k, is a competition between disassociation, at rate $k_D{}^*$, and electron transfer, at rate k_{ET} given by equation 9. The steady-state approximation leads to the following expression, where $K_D = k_D/k_D{}^*$.

$$k = \frac{k_D}{1 + \frac{k_D}{K_D A k}\exp\left[\frac{\Delta G^\ddagger}{RT}\right]} \tag{15}$$

Figure 9 shows that this FER has been applied successfully to organic and inorganic molecular redox reactions.

2.4. Additional constraints on the rates of multistep reactions

Many dissolved species that drive mineral redox reactions exhibit complex redox chemistry that can make the prediction of mineral reaction rates very difficult. Very few environmental reagents undergo simple, reversible one-electron redox reactions and we briefly list common situations.

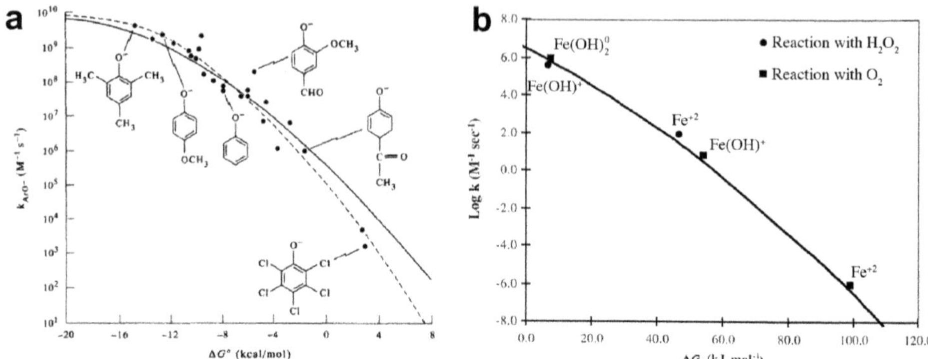

Figure 9. Examples of free energy relationships for two series of molecular electron transfer reactions. (a) Correlation between the second-order rate constants for oxidation of substituted phenoxide anions by chlorine dioxide and estimated values of reaction, ΔG. The solid line is a fit of Marcus theory FER in which only the reorganization energy, λ, is a free parameter. The dashed line fits a correction to the ΔG values to account for the irreversibility of the organic oxidation step. Plot taken from Tratnyek and Hoigné (1994) with permission from Elsevier. (b) Correlation between the second-order rate constant for oxidation of the Fe(II) species indicated and the calculated ΔG for oxidation by O_2 or H_2O_2. The solid line is a fit of Marcus theory FER in which only the reorganization energy, λ, is a free parameter. Plot is a modified version of figure 5 from King *et al.* (1995) with permission from Elsevier.

2.4.1. Multi-electron redox reactions

Many metals and metalloids, including arsenic, selenium, sulfur and chromium are found in only a few stable oxidation states separated by two- or three-electron redox reactions. Intermediate states are either never observed or exist only transiently in nature, such as uranium(V) or manganese(III) (unless complexed; see Madison *et al.*, 2013). Although there have been many mechanistic studies of two-electron redox reactions at metallic electrodes, very little knowledge exists about the pathways of environmental multi-electron processes occurring on mineral surfaces. Such reactions almost certainly occur in a sequential manner, frequently involving both electron and proton transfer to the reagent. There is probably an important role for the mineral surface in stabilizing the intermediate oxidation state with suggestions that this can confer a mineral face specificity to some reactions.

2.4.2. Formation of radical species

Many reagents do participate in one-electron redox reactions, but form unstable and reactive radical products, such as superoxide, $O_2^{\bullet-}$, or the sulfide radical anion, $S^{\bullet-}$, both potent reducing agents. Such species typically have short lifetimes in aqueous solution due to reaction with water, chloride or bicarbonate ions (King, 1998), or disproportionation. The electrochemical properties of many common radicals are well known. However, the rates at which such species undergo secondary reactions, and consequently their impact on the overall reaction progress, is far less well studied. Most organic molecules typically either undergo two-electron reactions or form radical species following one-electron transfer. Quinones are the best studied of the latter.

2.4.3. Redox reactions coupled to precipitation

Under circumneutral conditions, metal-atom oxidation-state change is frequently followed by precipitation of insoluble phases. Such precipitates can form a rind on a reactive mineral surface, hindering reagent transport and further reaction, a process often called passivation.

Example: Reaction of iron and magnetite nanoparticles with O_2

Magnetite is a mixed-valence iron(II/III) oxide that can be formed in the environment when iron(III) oxyhydroxides such as ferrihydrite are exposed to dissolved iron(II). Magnetite is an important environmental reductant and has application for remediation through contaminant-sorption (*e.g.* arsenate) and contaminant-reduction (*e.g.* uranium(VI) and organic pollutants) mechanisms. The oxidation of magnetite by O_2 and other species illustrates many examples of constraints on mineral reaction rate.

Below ~250ºC, the complete oxidation of magnetite forms maghemite, γFe_2O_3, through a solid-state reaction that preserves the spinel structure and the mineral morphology. However, crystal stoichiometry must change to preserve charge balance. As the following balanced half-reaction for magnetite oxidation suggests, electron loss must be coupled to considerable structural rearrangement. It has long been concluded that this is achieved through the diffusion of iron cations to the surface where oxidation occurs.

$$3Fe_3O_{4(\text{magnetite})} \rightarrow 4\gamma\text{-}Fe_2O_{3(\text{maghemite})} + Fe^{2+} + 2e^- \quad (16)$$

Tang *et al.* (2003) studied the oxidation of magnetite nanoparticles in aerobic aqueous solutions using near infrared spectroscopy to observe the proportion of the maghemite phase. The oxidation kinetics were well described by a kinetic rate law based on solid-state diffusion, and Arrhenius analysis of the temperature dependence furnished an activation energy attributed solid-state diffusion of Fe^{2+}. As a consequence, full oxidation of even nanoscale or microscale particles can be very slow at low temperatures (*cf.* Fig. 3).

Zerovalent (*i.e.* metallic) iron nanoparticles are stronger reducing agents than magnetite nanoparticles, and thus may be suited to a wider range of remediation applications. Exposure to air leads rapidly to the formation of a partially oxidized surface layer, but the particles retain their redox activity. Strikingly, for small particles full oxidation leads to the formation of a hollow oxide shell (Fig. 10A(c)) (Wang *et al.*, 2009). This phenomenon is known as the nanoscale Kirkendall effect (Yin *et al.*, 2004), and is explained by unequal rates of iron and oxygen diffusion through the reacting crystal lattice. Pratt *et al.* (2014) used scanning transmission electron microscopy to obtain high-resolution images of the oxidized shell of a cube-shaped iron nanoparticle. The oxide shell phase was either magnetite or maghemite, or a mixture of these phases. They observed that a strain field propagated from the iron core to the surface, and demonstrated that solid-state ion diffusion was probably enhanced by the strain.

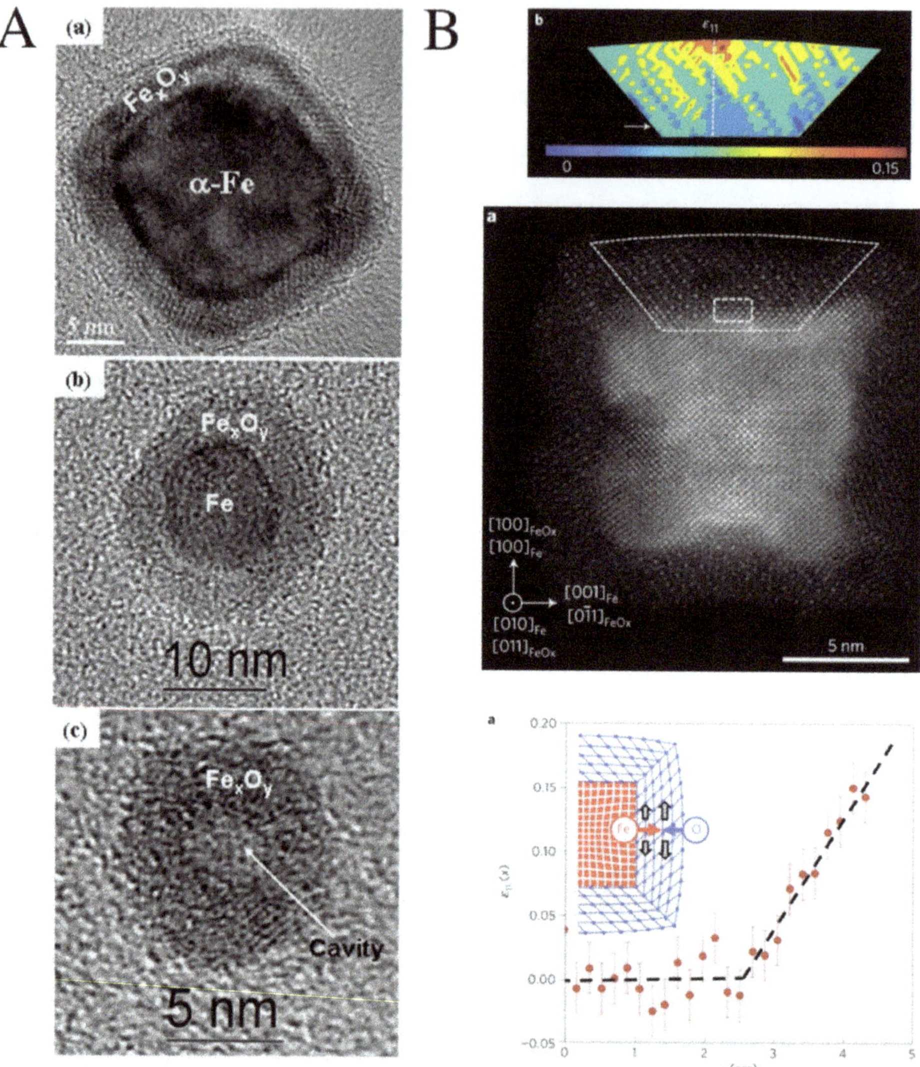

Figure 10. Transmission electron microscopy images showing oxidized metallic iron nanoparticles of different initial dimensions that developed a core-shell structure. (A) (a) Faceted core-shell particle, (b) spherical core-shell particles, and (c) a fully oxidized particle with a cavity at the centre. From Wang *et al.* (2009) with permission from the American Chemical Society. (B) High-resolution atomic-number contrast electron microscopy analysis of a partially oxidized metallic iron nanoparticle with a cubic morphology. Top image shows a strain field inferred from analysis of the indicated region of the shell region in the middle image. The bottom image shows the predicted enhancement of cation diffusion through the strained oxide shell. From Pratt *et al.* (2014) with permission from the Nature Publishing Group.

Example: Reaction of magnetite with substituted nitrobenzenes

Magnetite (and iron) nanomaterials have been proposed for reductive decontamination of organic molecules that often react less readily than O_2 either due to a lower thermodynamic driving force, a spin barrier or because the reaction requires successive electron transfer steps or molecular reconfiguration. Recently, Gorski *et al.* (2010) studied the reduction of nitrobenzene and two substituted analogues by ~11 nm magnetite nanoparticles prepared with a range in the stoichiometric ratio $x = [Fe^{2+}]/[Fe^{3+}]$. The maximum value, $x = 0.5$, represents pure magnetite. The reduction rate was a strong function of x (Fig. 11a) and of the nanoparticle open-circuit potential, E_{OCP}, measured using packed-powder electrodes. Gorski *et al.* (2010) found that trends in the reduction rates could be predicted from a Butler-Volmer-like expression based on the overpotentional between E_{OCP} and the molecular one-electron reduction potential (Fig. 11b). While best agreement was obtained for nitrobenzene, discrepancies for the substituted molecules suggested that other reactions steps, not only the first electron transfer step, contributed to the observed rate in those cases.

Example: Reaction of magnetite (111) surface with uranyl complexes

The transport of uranium in contaminated environments is strongly affected by the sorption of soluble uranyl, U(VI), species and by reduction reactions that form insoluble U(IV) oxides. Magnetite surfaces can participate in both reactions, but there is no firm consensus as to the conditions under which sorption or reduction predominate. Singer *et al.* (2012a,b) studied U(VI) interactions with the magnetite (111) surface. Although this surface is believed to contain both Fe^{2+} and Fe^{3+} sites

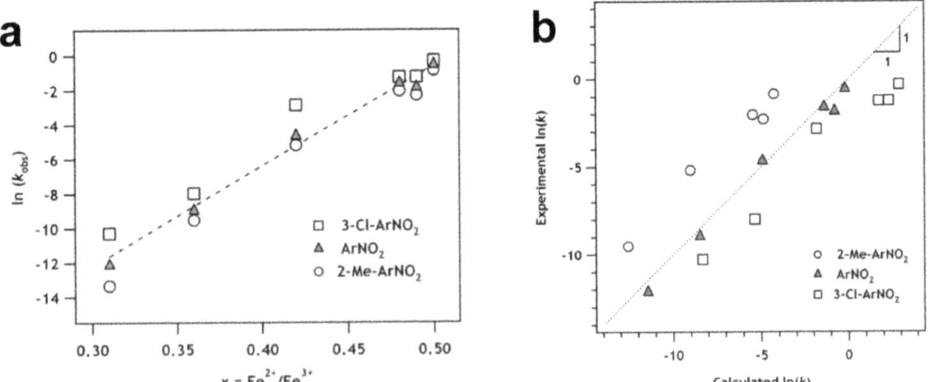

Figure 11. The oxidation state of ~11-nm magnetite nanoparticles, expressed as the stoichiometric ratio $x = Fe^{2+}/Fe^{3+}$, has a strong effect on the rate of reduction of nitrobenzene ($ArNO_2$), 2-methyl-nitrobenzene (2-Me) and 3-chlorine-nitrobenzen (3-Cl). (a) Natural log transformed observed kinetics for different stoichiometry magnetites. The fitted line shown is for all compounds *vs.* x (n = 15; R^2 = 0.955). (b) Experimental *vs.* predicted rates of reduction based on the magnetite open circuit potential and free energy changes for the molecular one-electron transfer. From Gorski *et al.* (2010) with permission from the American Chemical Society.

under anaerobic conditions, XPS and EXAFS studies showed that the uranyl ion sorbed to the surface is not readily reduced, and is stabilized as a carbonato-complex. However, slow reduction leads to the development of UO_2 nanoparticles only in the vicinity of magnetite surface defects associated with emergent screw dislocations and low-angle grain boundaries.

Singer *et al.* (2012) proposed that Fe^{2+} was released preferentially into solution from defect sites, revealing new terminal Fe^{3+} atoms. However, solid-state electron hopping between edge-sharing iron-oxygen octahedra along <110> directions could regenerate a surface Fe^{2+} site. The kinetics of this overall reaction have been measured but the underlying processes must include Fe^{2+} release, aqueous diffusion of the solvated Fe^{2+} to the vicinity of UO_2^{2-} ions, electron transfer to generate a U^{5+} valence complex, then a related sequence to produce a second electron transfer to U^{4+}. The nature of the hypothesized U^{5+} complex and the solvated Fe^{2+}–UO_2^{2-} ion association are both unknown.

If this model is correct, it is also unclear why surface Fe^{2+} does not drive sorbed uranyl ion reduction directly. It could be that this type of electron transfer is kinetically limited by the geometric arrangement of the ions involved, with the uranyl ion and Fe^{2+} octahedron too far apart. Hence a more favourable association and transfer may be between solvated Fe^{2+} and the uranyl ion in solution. As the UO_2^{2-} ion and other uranium complexes in solution exhibit strong optical absorption and fluorescence, this system may be amenable to pump-probe studies in the near future.

3. Chemical dynamics methods for studying mineral-reaction intermediates

A limitation of the chemical kinetics approach is that there is often more than one plausible reaction mechanism that could explain the data. Additional evidence may be obtained by acquiring direct characterization of intermediate species or molecular configurations. This is the goal of 'chemical dynamics' experiments, but it is challenging for two reasons. First, intermediate states often have a very short lifetime, much shorter than the timescales of reagent transport, and shorter than conventional methods for analysis. Second, in an ensemble of reagents, thermally-activated chemical reactions start essentially randomly. Thus, at any instant there will be a range of initial, final and intermediate species. Special techniques are required to circumvent relatively slow transport limits, to control reaction initiation and to characterize the intermediates while they exist.

3.1. The pump-probe method

The pump-probe approach has been most effective for studying reactions at the timescales of fast chemical steps. Energy is supplied to a system to initiate a reaction of interest (the 'pump') which is then interrogated by a suitable technique at a chosen time delay (the 'probe'). By probing the system at a range of delays after the pump, a picture of reaction pathway can be obtained. The development of high-powered lasers that deliver ultrafast (*i.e.* sub-nanosecond) pulses for both pump and probe steps has enabled the rapid growth of the field of 'femtochemistry', which has as its goal making three-dimensional molecular movies of chemical reactions. There are many excellent reviews of the technical and theoretical aspects of femtosecond timescale experiments (Chen, 2005; Nibbering *et al.*, 2005).

Almost every method of materials characterization can be performed with the pump-probe method and some examples are given below. However, we shall concentrate on so-called 'transient absorption' methods in which information as to the chemical evolution of a system is contained within a differential absorption spectrum, $\Delta A(\lambda,t)$, that is the mathematical difference between a ground-state absorption spectrum, $A_0(\lambda)$, and a transient absorption spectrum, $A(\lambda,t)$, where t is the time delay after the pump pulse.

$$\Delta A(\lambda,t) = A_0(\lambda) - A(\lambda,t) \tag{17a}$$

$$A(\lambda,t) = C_0 A_0(\lambda) + \Sigma_i C_i(t) A(\lambda) \tag{17b}$$

The concentration factors, C_i, sum to unity. Because the data are usually dominated by ground-state signal from non-excited species, and because C_i values are not known, even when there is a single reaction intermediate it is typically not possible to manipulate the ΔA data to obtain just the absorption spectrum of the reaction intermediate (Della-Longa *et al.*, 2009). Nevertheless, optical and X-ray transient absorption data can contain a great deal of interpretable information.

3.1.1. Optical pump-probe methods

Optical spectroscopies use electromagnetic radiation with a wavelength in the range from the infrared (IR) to the ultraviolet (UV) (the range for which lenses and other transmission optics can be fashioned). An important benefit of using sub-picosecond laser pulses for ultrafast optical spectroscopy is that extraordinarily high (if momentary) photon densities can be achieved. This is a condition for high-efficiency non-linear optical processes that allow tuning of the excitation or probe wavelength(s). Thus a single-wavelength output from an amplified femtosecond laser may be used to excite at a different energy and interrogate with a narrow or broadband probe from the IR to the UV.

Optical transient absorption (TA) spectroscopy reveals time-dependent changes to the UV-vis spectrum of a pumped system, and is a common method for studying transient kinetics. However, transient UV-vis spectra can be challenging to interpret because they reveal the electronic excitations of the material without directly determining the species that are excited or the nature of the excitation.

Example: The lifetime of electrons and holes in hematite solar cells

There is considerable interest in the development of materials that can capture energy from sunlight for the splitting of water into H_2 and O_2 gases. Earth-abundant mineral analogues may offer low cost and high stability, with hematite a potential material for a water-oxidation anode (Braun *et al.*, 2012). Ultrafast spectroscopy is proving vital for understanding chemical limitations on the mobility of electrons (and holes) in bulk and nanoscale crystals (Baxter *et al.*, 2014). We and others have studied the lifetime of band-gap excitation of hematite nanoparticles or thin films (Pendlebury *et al.*, 2011; Huang *et al.*, 2012; Gilbert *et al.*, 2013). A single lineshape in the transient difference spectrum is observed following 520 nm excitation. The signal decays to near zero within 200 ps, without change in lineshape, indicating that electron-hole recombination occurs on this timescale.

There has been considerable debate as to the origin of the transient signal, which represents additional absorption of light in the excited state. It has long been debated whether the light is absorbed by electrons newly promoted into the conduction band (CB), or by new electronic transitions in the valence band (VB) that are possible after one electron has been removed by photoexcitation. In TiO_2, electrons excited to the CB have an optical absorption feature in the 700–1000 nm range that can be used as a signature of interfacial electron transfer. In contrast, it appears that CB electrons in hematite do not absorb strongly and that the transient optical signals are dominated by VB holes. This limits the use of time-resolved optical spectroscopy for the study of iron oxide reduction reactions.

3.1.2. X-ray pump-probe methods

X-ray absorption spectroscopy (XAS) and X-ray emission spectroscopy (XES) are used widely for the determination of molecular-scale structure in geochemical systems. In particular, both methods are highly sensitive to the oxidation state of metals. Time-resolved XAS (Bressler and Chergui, 2004; Chen, 2005) and XES (Alonso-Mori *et al.*, 2012) are thus well suited to studying fast mineral redox reactions. In particular, XAS can provide a definitive determination of metal valence changes during the course of a chemical reaction even if no optical spectroscopic signatures for valence change have been established.

The essential components of a laser-pump/X-ray-probe study of a solution-phase reaction are depicted in Fig. 12. The sample is flowed in a liquid jet through the intersection point of the laser and X-ray beams. The laser excites the sample a short time prior to the X-ray probe. At typically synchrotron sources, the temporal resolution is determined by the X-ray puse width (60–120 ps) and the delay time can extend up to microsecond timescales. Time-resolved X-ray spectroscopy thus achieves far higher temporal resolution than quick-scanning methods (Ginder-Vogel *et al.*, 2009; Khalid *et al.*, 2010), but requires special sample preparation as described below. Technical descriptions of experimental approaches and facilities can be found elsewhere (Bressler and Chergui, 2004; Lima *et al.*, 2011; March *et al.*, 2011).

3.2. Other time-resolved methods

3.2.1. Relaxation methods

If a reversible process, such as a sorption–desorption reaction, is at equilibrium then interpretable kinetics data may be obtained by perturbing a parameter such as temperature or pressure (Grossl *et al.*, 1994) and measuring the rate at which the system reaches a new equilibrium. However, such methods are not well suited to irreversible processes, as is the case for most mineral reactions.

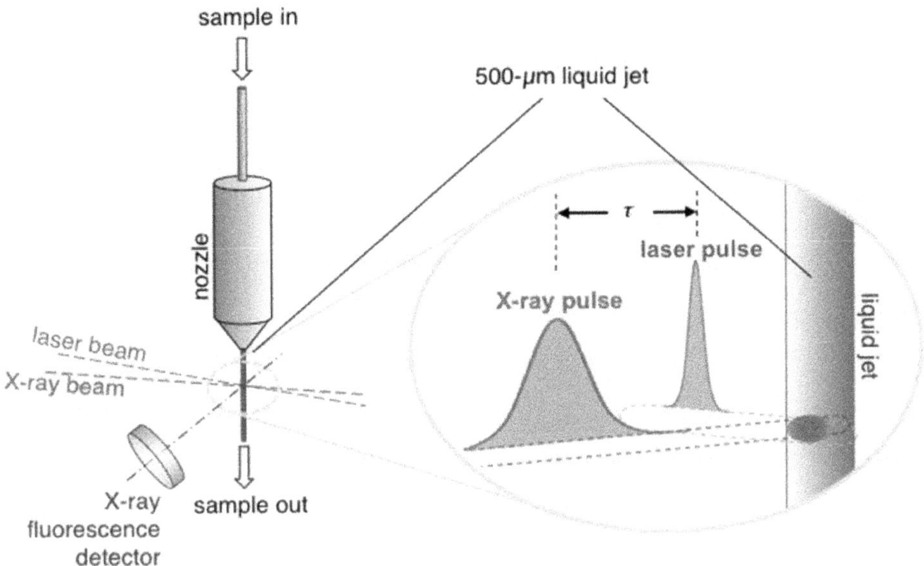

Figure 12. Scheme of experimental approach to perform laser-pump/X-ray-probe spectroscopy studies on suspensions of iron (oxyhydr)oxide nanoparticles. The suspension at 2–5 mM metal concentration is circulated through a thin nozzle to create a stable liquid jet flowing at a linear velocity of ~10 ms^{-1} in the X-ray beam. The laser focus is aligned to the X-ray spot. X-ray absorption spectra (XAS) are collected in fluorescence yield mode using a large-area, fast-response avalanche photodiode (APD) plus Z-1 filter and Soller slit (not shown). Time-resolved XAS signal at a single X-ray energy is acquired by exciting the sample with a laser pulse a short delay, τ, prior to the synchrotron X-ray pulse. After averaging many pulses, the X-ray energy is changed in order to acquire a full time-dependent spectrum. The APD detection is gated electronically to sort time-dependent data (after laser excitation) from the ground-state XAS data (no laser excitation).

3.2.2. Time-resolved electrochemical methods

Electrochemistry experiments can be modelled as electrical circuits with a characteristic time constant that is determined mainly by the solution resistance and the capacitance of the electrochemical cell. If voltage-scanning measurements are performed faster than the time constant, the redox (or Faradaic) contribution to the current is overwhelmed by the capacitive contribution. Reducing the capacitance by using a microscopic working electrode, such as the polished end of a 5 μm diameter gold wire, can partially but not fully alleviate this limitation. Recently, however, undistorted CV measurements acquired at voltage scanning speeds of up to 10^6 V/s have been demonstrated using an ultramicroelectrode and real-time electronic compensation of the drop in cell resistance (Amatore *et al.*, 2003; Guo and Lin, 2005). These approaches promise to enable CV studies of molecular redox reactions with nanosecond time resolution. It is unknown if they will be applicable to mineral redox reactions.

3.3. Pump-probe studies of photochemical reactions

Photochemical reactions are suited naturally to the pump-probe methodology. For example, the mechanism of the photolysis of ferrioxalate was studied by time-resolved X-ray and optical methods (Chen *et al.*, 2007). Photochemical reactions play a large role in atmospheric processes that can affect cloud formation and ozone depletion. Ultrafast methods can distinguish the mechanisms through which atmospheric molecules such as chlorine dioxide (OClO) break up into reactive products (Thogersen *et al.*, 1998). We are currently studying the surface photoreduction of iron oxide crystals (Gilbert *et al.*, 2012), photochemical cycling of manganese(VI) oxides (Femi Marafatto *et al.*, 2015) and the photocatalytic reduction of small organic acids at the surface of zinc sulfide minerals.

3.4. Initiating chemical reactions with light

3.4.1. Acid-base reactions

Acid–base reactions are fundamental processes in aqueous geochemistry. So-called photoacids or photobases are molecules that undergo rapid dissociation following light absorption and generate a short- or long-lived free proton or hydroxyl ion in solution. To date, such reagents have been used to study molecular aspects of aqueous solutions including the pathways and rates for H^+ or OH^- transfer (Cox and Bakker, 2008). There are few examples of use of such reagents to study geochemical reactions. Recently, Adamczyk *et al.* (2009) used a photoacid that was triggered optically to transfer a proton to bicarbonate (HCO_3^-) on an ultrafast timescale, and measured the deprotonation lifetime of carbonic acid, its conjugate acid (Adamczyk *et al.*, 2009).

3.4.2. Electron-transfer reactions

In order to perform ultrafast studies of interfacial redox reactions we implemented an approach that was inspired by the work of O'Regan and Gratzel (1991) and others in the field of solar energy capture and conversion. Many fluorescent dye molecules that bind to the surface of inert semiconductor oxides can act as efficient 'sensitizers', absorbing light at a lower energy than the semiconductor band gap and transferring ('injecting') the photoelectron into the CB. We showed that fluorescein-based dyes could sensitize redox-reactive iron(III) oxides, and used this approach to initiated iron reduction using a fast laser pulse (Katz *et al.*, 2010). Figure 13 illustrates some differences between chemical- and light-initiated reduction. Light-initiated interfacial electron transfer competed with other processes including band-gap absorption of the oxide and dye-to-oxide energy transfer that reduced the efficiency of iron reduction to ~25% of absorbed photons (Gilbert *et al.*, 2013). Nevertheless, iron reduction was fast enough (sub-ps) to enable time-resolved studies.

Example: Interfacial electron transfer processes in iron(III) (oxyhydr)oxides

We used pump-probe spectroscopy to investigate the molecular-scale processes that occur during the reductive dissolution of iron(III) oxides (hematite and maghemite) and oxyhydroxide (ferrihydrite). As described by Katz *et al.* (2012), this work provided the

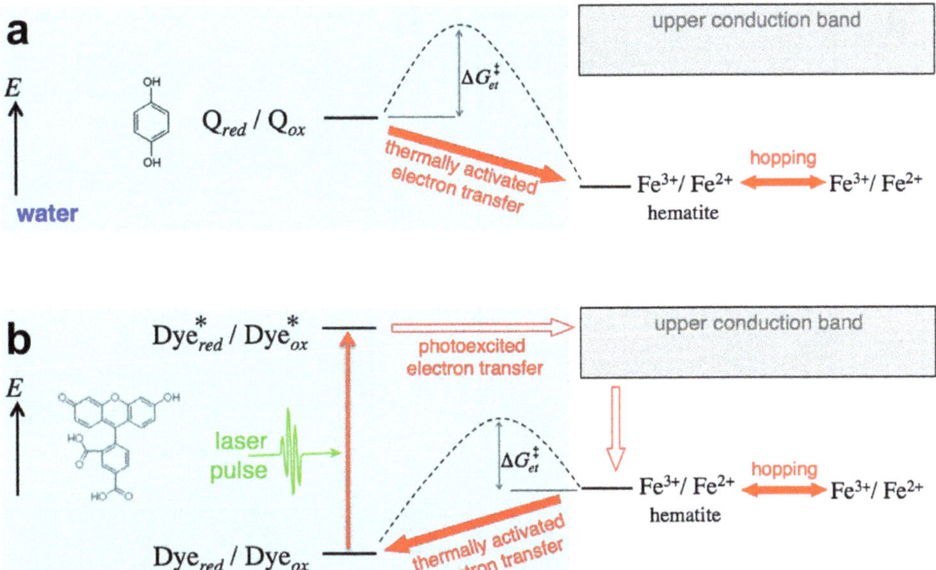

Figure 13. Energy diagrams illustrating the differences between (a) the chemical reduction of an iron(III) oxide and (b) the light-initiated reduction of an iron(III) oxide using a surface-bound dye molecule. In the first case, donor–mineral electron transfer is a thermally activated process that creates a reduced $Fe^{2+}(s)$ atom in its ground state. If ΔG_{et} is large, the electron will tend not to return to the donor. In the second case, donor–mineral electron transfer is light-activated, and the electron is transferred to a high-energy state at the mineral surface before rapidly decaying to the ground state of the surface $Fe^{2+}(s)$ atom. However, electron transfer back to the dye remains energetically favourable.

first direct experimental confirmation of the nature of the reduced Fe^{2+} sites created by interfacial electron transfer. The transient ΔXA spectrum acquired 150–200 ps after electron transfer was consistent with a change in both iron oxidation state and oxygen coordination geometry. The coupled change of valence state and coordination is an indication that the electron is localized at the iron site, a concept termed a 'small polaron'. Figures 14a and b illustrate the lifetime of the Fe^{2+} species in hematite and ferrihydrite. The initial fast loss was interpreted as a recombination with the dye. The long-lived state (up to >40 μs) was interpreted as electrons that hopped into the nanoparticle interior. Both recombination and hopping rates were mineral-phase dependent, and could be estimated using a simple kinetic model. The hopping rate in hematite was within the range predicted from prior simulation (Iordanova *et al.*, 2005).

Figure 14c presents a tentative scheme of the recombination process for hematite and ferrihydrite. For each case, the redox potential of the electron acceptor (the oxidized dye molecule) is identical. For each case, the electron donor is almost certainly a surface Fe^{2+} atom that is sixfold coordinated by oxygen in a distorted octahedral configuration. The pathway for electron transfer is then very similar in the two cases, and the difference in recombination rates has two possible explanations. The transition

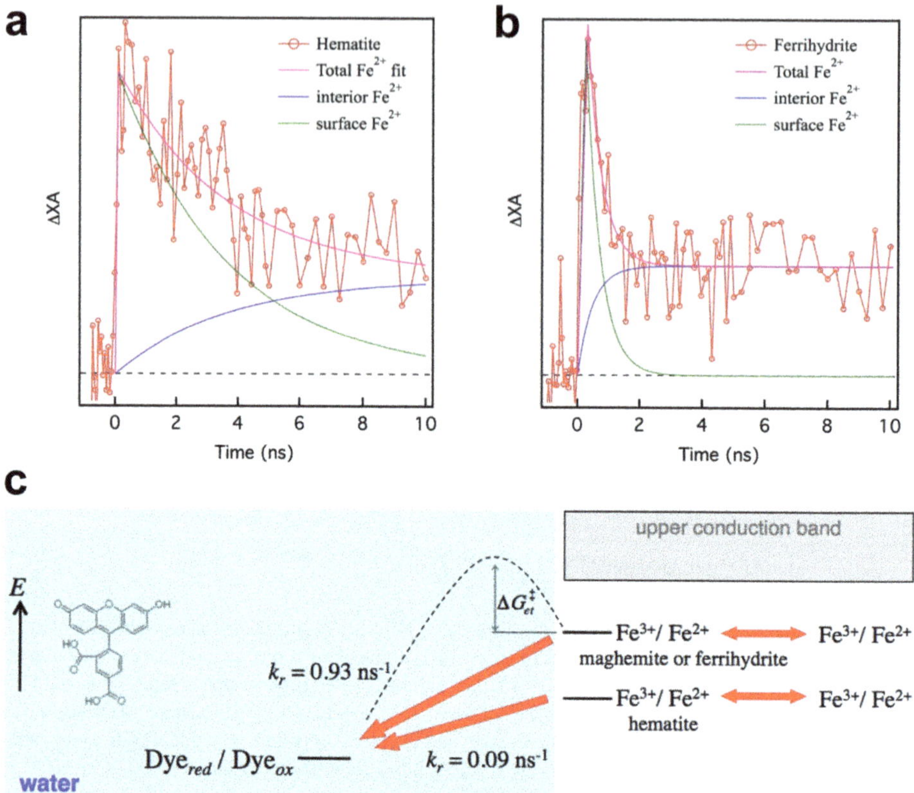

Figure 14. Top. Experimental pump-probe data showing the nanosecond timescale lifetime of iron(II) sites created in (a) 7 nm hematite and (b) 3 nm ferrihydrite nanoparticles coated with a fluorescein dye. The strength of the ΔXA signal at 7.125 keV is proportional to the iron(II) concentration. The kinetics are fitted by a three-site model for electron location: a surface iron atom, and interior iron atom, or back on the dye molecule. Both the surface-to-interior hopping rate and surface-to-dye recombination rate are faster for ferrihydrite *vs.* hematite. Data for maghemite (not shown) were identical to ferrihydrite. From Katz *et al.* (2012) with permission from the American Association for the Advancement of Science. (c) Tentative scheme to interpret the differences in the rate of electron recombination with the dye shown above.

state for the electron-transfer pathway may be less accessible for the hematite structure, leading to a higher activation barrier. This could occur if the vibrational states available to surface sites on hematite do not easily access the electron-transfer transition state. Alternatively, the Fe^{2+} site may be more stable in hematite relative to the other phases. This interpretation addresses the issue of the Fe^{2+}/Fe^{3+} couple at the surface (*cf.* equation 6a), and leads to the prediction that iron(II) sites at the surface of ferrihydrite (or maghemite) are stronger reducing agents than iron(II) sites on hematite. We are not aware of experimental studies that have evaluated this comparison. Thus, time-resolved methods have the potential to characterize the chemical properties of intermediate states that are unstable on the timescale of conventional experiments, and we are working to better establish the potential for these approaches.

4. Outlook

Mineral redox reactions are complex processes, involving one or more intermediate states and chemical steps that can occur over an exceedingly large range of timescales. It is difficult to establish a mechanistic description of such reactions that confidently identifies specific reaction steps as being rate limiting, and that provides insight into the factors that control reaction rate. Ultrafast time-resolved methods have considerable potential for obtaining insight into these reactions, and all geochemical processes, because experiments can be designed to probe the reaction at the timescale appropriate to a selected chemical process. There is presently rapid progress in developing versions of almost every method for chemical analysis that can be used in such studies, including electron and X-ray microscopy, and spectroscopies from the infrared to hard-X-ray regimes. Advances in two main areas will be needed. First, it is essential to develop ways to control reaction initiation so that reactions of interest may be performed using the pump-probe approach. Second, molecular simulation will continue to be crucial in order to test the identification of intermediate species inferred from experiment, and to calculate properties of the intermediates that remain inaccessible.

Acknowledgements

We thank the many colleagues and collaborators who have contributed to our understanding of mineral redox reactivity, including Kevin Rosso, Thilo Behrends and Jordan Katz. Ultrafast X-ray spectroscopy was performed at beamline 6.0.1 at the Advanced Light Source (ALS) and at beamline 11-ID-D at the Advanced Photon Source (APS). The authors were supported by the Director, Office of Science, Office of Basic Energy Sciences, of the U.S. Department of Energy, hereby abbreviated to DOE-BES, under Contract No. DE-AC02-05CH11231. Use of the ALS and the APS is supported by DOE-BES under Contract Numbers DE-AC02-05CH11231 and W-31-109-ENG-38, respectively.

References

Adamczyk, K., Premont-Schwarz, M., Pines, D., Pines, E. and Nibbering, E.T.J. (2009) Real-time observation of carbonic acid formation in aqueous solution. *Science*, **326**, 1690–1694.

Ahmed, I.A.M., Benning, L.G., Kakonyi, G., Sumoondur, A.D., Terrill, N.J. and Shaw, S. (2010) Formation of green rust sulfate: A combined *in situ* time-resolved X-ray scattering and electrochemical study. *Langmuir*, **26**, 6593–6603.

Alonso-Mori, R., Kern, J., Gildea, R.J., Sokaras, D., Weng, T.-C., Lassalle-Kaiser, B., Rosalie, T., Hattne, J., Laksmono, H., Hellmich, J., Gloeckner, C., Echols, N., Sierra, R.G., Schafer, D.W., Sellberg, J., Kenney, C., Herbst, R., Pines, J., Hart, P., Herrmann, S., Grosse-Kunstleve, R.W., Latimer, M.J., Fry, A.R., Messerschmidt, M.M., Miahnahri, A., Seibert, M.M., Zwart, P.H., White, W.E., Adams, P.D., Bogan, M.J., Boutet, S., Williams, G.J., Zouni, A., Messinger, J., Glatzel, P., Sauter, N.K., Yachandra, V.K., Yano, J. and Bergmann, U. (2012) Energy-dispersive X-ray emission spectroscopy using an X-ray free-electron laser in a shot-by-shot mode. *Proceedings of the National Academy of Sciences of the United States of America*, **109**, 19103–19107.

Amatore, C., Bouret, Y., Maisonhaute, E., Abruna, H. D. and Goldsmith, J.I. (2003) Electrochemistry within molecules using ultrafast cyclic voltammetry. *Comptes Rendus Chimie*, **6**, 99–115.

Amonette, J.E. (2002) Iron redox chemistry of clays and oxides: Environmental applications. Pp. 90–147 in: *Electrochemical Properties of Clays* (A. Fitch, editor). The Clay Minerals Society, Aurora, Colorado, USA.

Bard, A.J. and Faulkner, L.R. (1980) *Electrochemical methods*. John Wiley, New York.

Bard, A.J., Parsons, R. and Jordan, J. (1985) *Standard Potentials in Aqueous Solution*. Marcel Dekker, New York.

Baxter, J.B., Richter, C. and Schmutternmaer, C.A. (2014) Ultrafast carrier dynamics in nanostructures for solar fuels. *Annual Review of Physical Chemistry*, **65**, 423–447.

Bazant, M.Z. (2013) Theory of chemical kinetics and charge transfer based on nonequilibrium thermodynamics. *Accounts of Chemical Research*, **46**, 1144–1160.

Bonneville, S., Behrends, T. and Van Cappellen, P. (2009) Solubility and dissimilatory reduction kinetics of iron(III) oxyhydroxides: A linear free energy relationship. *Geochimica et Cosmochimica Acta*, **73**, 5273–5282.

Boyd, P.W. and Ellwood, M.J. (2010) The biogeochemical cycle of iron in the ocean. *Nature Geoscience*, **3**, 675–682.

Brantley, S., Kubicki, J. and White, A. (2008) *Kinetics of Water–Rock Interaction*. Springer Verlag, Berlin.

Braun, A., Sivula, K., Bora, D.K., Zhu, J., Zhang, L., Graetzel, M., Guo, J. and Constable, E.C. (2012) Direct observation of two electron holes in a hematite photoanode during photoelectrochemical water splitting. *Journal of Physical Chemistry C*, **116**, 16870–16875.

Bressler, C. and Chergui, M. (2004) Ultrafast X-ray absorption spectroscopy. *Chemical Reviews*, **104**, 1781–1812.

Burrows, N.D., Hale, C.R.H. and Penn, R.L. (2012) Effect of ionic strength on the kinetics of crystal growth by oriented aggregation. *Crystal Growth and Design*, **12**, 4787–4797.

Catalano, J.G., Fenter, P. and Park, C. (2009) Water ordering and surface relaxations at the hematite (110)–water interface. *Geochimica et Cosmochimica Acta*, **73**, 2242–2251.

Chen, J., Zhang, H.H., Tomax, I.V., Wolfsburg, M., Ding, Z. and Rentzepis, P.M. (2007) Transient structures and kinetics of the ferrioxalate redox reaction studies by time-resolved EXAFS, optical spectroscopy and DFT. *Journal of Physical Chemistry A*, **111**, 9326–9335.

Chen, L.X. (2005) Probing transient molecular structures in photochemical processes using laser-initiated time-resolved X-ray absorption spectroscopy. *Annual Review of Physical Chemistry*, **56**, 221–254.

Cox, M.J. and Bakker, H.J. (2008) Parallel proton transfer pathways in aqueous acid-base reactions. *Journal of Chemical Physics*, **128**.

Cutting, R.S., Coker, V.S., Fellowes, J.W., Lloyd, J.R. and Vaughan, D.J. (2009) Mineralogical and morphological constraints on the reduction of Fe(III) minerals by *Geobacter sulfurreducens*. *Geochimica et Cosmochimica Acta*, **73**, 4004–4022.

Della-Longa, S., Chen, L.X., Frank, P., Hayakawa, K., Hatada, K. and Benfatto, M. (2009) Direct deconvolution of two-state pump-probe X-ray absorption spectra and the structural changes in a 100 ps transient of Ni(II)-tetramesitylporphyrin. *Inorganic Chemistry*, **48**, 3934–3942.

Dubinsky, E.A., Silver, W.L. and Firestone, M.K. (2010) Tropical forest soil microbial communities couple iron and carbon biogeochemistry. *Ecology*, **91**, 2604–2612.

Elsner, M., Schwarzenbach, R.P. and Haderlein, S.B. (2004) Reactivity of Fe(II)-bearing minerals toward reductive transformation of organic contaminants. *Environmental Science & Technology*, **38**, 799–807.

Erbs, J.J., Gilbert, B. and Penn, R. L (2008) Influence of size on reductive dissolution of six-line ferrihydrite. *Journal of Physical Chemistry B*, **112**, 12127–12133.

Espenson, J.H. (1981) *Chemical Kinetics and Reaction Mechanisms*. McGraw-Hill Book Company, New York.

Femi Marafatto, F., Strader, M.L., Gonzalez-Holguera, J., Schwartzberg, A., Gilbert, B. and Pea, J. (2015) Rate and mechanism of birnessite (δ-MnO_2) nanosheet photoreduction. *Proceedings of the National Academy of Sciences*, **112**, 4600–4605.

Gerischer, H. (1990) The impact of semiconductors on the concepts of electrochemistry. *Electrochimica Acta*, **35**, 1677–1699.

Gilbert, B., Katz, J.E., Rude, B., Glover, T.E., Herlein, M., Kurtz, C. and Zhang, X. (2012) Thin water film formation on metal oxide crystal surfaces. *Langmuir*, **28**, 14308–14312.

Gilbert, B., Katz, J.E., Huse, H.E., Zhang, X., Frandsen, C., Falcone, R.W. and Waychunas, G. (2013) Ultrafast energy and electron transfer in dye-sensitized iron oxide nanoparticles. *PCCP*, **15**, 17303–17313.

Ginder-Vogel, M., Landrot, G.L., Fischel, J.S. and Sparks, D.L. (2009) Quantification of rapid environmental redox processes with quick-scanning X-ray absorption spectroscopy. *Proceedings of the National Academy of Sciences*, **106**, 16124–16128.

Gorski, C.A., Nurmi, J.T., Tratnyek, P.G., Hofstetter, T.B. and Scherer, M.M. (2010) Redox behavior of magnetite: Implications for contaminant reduction. *Environmental Science & Technology*, **44**, 55–60.

Grossl, P.R., Sparks, D.L. and Ainsworth, C.C. (1994) Rapid kinetics of Cu(II) adsorption-desorption in goethite. *Environmental Science & Technology*, **28**, 1422–1429.

Grygar, T. (1998) Phenomenological kinetics of irreversible electrochemical dissolution of metal-oxide microparticles. *Journal of Solid State Electrochemistry*, **2**, 127–136.

Guo, Z. and Lin, X. (2005) Ultrafast sinusoidal voltammetry. *Analytical Letters*, **38**, 1007–1017.

Hammet, L.P. (1937) The effect of structure upon the reactions of organic compounds. Benzene derivatives. *Journal of the American Chemical Society*, **59**, 96–103.

Hansel, C.M., Benner, S.G. and Fendorf, S. (2005) Competing Fe(II)-induced mineralization pathways of ferrihydrite. *Environmental Science & Technology*, **39**, 7147–7153.

Haumann, M., Muller, C., Liebisch, P., Neisius, T. and Dau, H. (2005) A novel bioXAS technique with sub-millisecond time resolution to track oxidation state and structural changes at biological metal centers. *Journal of Synchrotron Radiation*, **12**, 35–44.

Henstridge, M.C., Laborda, E., Rees, N.V. and Compton, R.G. (2012) Marcus-Hush-Chidsey theory of electron transfer applied to voltammetry: A review. *Electrochimica Acta*, **84**, 12–20.

Hocking, R.K., Brimblecombe, R., Chang, L.-Y., Singh, A., Cheah, M.H., Glover, C., Casey, W.H. and Spiccia, L. (2011) Water-oxidation catalysis by manganese in a geochemical-like cycle. *Nature Chemistry*, **3**, 461–466.

Hu, P., Liu, Q., Torrent, J., Barron, V. and Jin, C. (2013) Characterizing and quantifying iron oxides in Chinese loess/paleosols: Implications for pedogenesis. *Earth and Planetary Science Letters*, **369**, 271–283.

Huang, Z., Lin, Y., Xiang, X., Rodriguez-Cordoba, W., McDonald, K.J., Hagen, K.S., Choi, K.-S., Brunschwig, B.S., Musaev, D.G., Hill, C.L., Wang, D. and Lian, T.Q. (2012) *In situ* probe of photocarrier dynamics in water-splitting hematite (α-Fe_2O_3) electrodes. *Energy and Environmental Science*, **5**, 8923–8926.

Hupp, J.T. and Weaver, M.J. (1983a) Electrochemical and homogeneous exchange kinetics for transition-metal aquo couples: Anomalous behavior of hexaaquoiron(III/II). *Inorganic Chemistry*, **22**, 2557–2564.

Hupp, J.T. and Weaver, M.J. (1983b). The frequency factor for outer-sphere electrochemical reactions. *Journal of Electroanalytical Chemistry*, **152**, 1–14.

Hupp, J.T. and Weaver, M.J. (1985) Prediction of electron-transfer reactivities from contemporary theory – unified comparisons for electrochemical and homogeneous reactions. *Journal of Physical Chemistry*, **89**, 2795–2804.

Iordanova, N., Dupuis, M. and Rosso, K.M. (2005) Charge transport in metal oxides: A theoretical study of hematite α-Fe_2O_3. *Journal of Chemical Physics*, **122**, 144305.

Jickells, T.D., An, Z.S., Andersen, K.K., Baker, A.R., Bergametti, G., Brooks, N., Cao, J.J., Boyd, P.W., Duce, R.A., Hunter, K.A., Kawahata, H., Kubilay, N., laRoche, J., Liss, P.S., Mahowald, N., Prospero, J.M., Ridgwell, A.J., Tegen, I. and Torres, R. (2005) Global iron connections between desert dust, ocean biogeochemistry, and climate. *Science*, **307**, 67–71.

Katz, J.E., Gilbert, B., Zhang, X.Y., Attenkofer, K., Falcone, R.W. and Waychunas, G.A. (2010) Observation of transient iron(II) formation in dye-sensitized iron oxide nanoparticles by time-resolved X-ray spectroscopy. *Journal of Physical Chemistry Letters*, **1**, 1372–1376.

Katz, J.E., Zhang, X.Y., Attenkofer, K., Chapman, K., Frandsen, C., Zarzycki, P., Rosso, K.M., Falcone, R.W., Waychunas, G.A. and Gilbert, B. (2012) Electron small polarons and their mobility in iron (oxyhydr)oxide nanoparticles. *Science*, **337**, 1200–1203.

Khalid, S., Caliebe, W., Siddons, P., Clay, I.S., Lenhard, B.T., Hanson, J., Wang, Q., Frenkel, A.I., Marinkovic, N., Hould, N., Ginder-Vogel, M., Landrot, G.L., Sparks, D.L. and Ganjoo, A. (2010) Quick extended X-ray absorption fine structure instrument with millisecond time scale, optimized for *in situ* applications. *Review of Scientific Instruments*, **81**, 015105.

King, D.W. (1998) Role of carbonate speciation on the oxidation rate of Fe(II) in aquatic systems. *Environmental Science & Technology*, **32**, 2997–3003.

King, D.W., Lounsbury, H.A. and Millero, F.J. (1995) Rates and mechanism of Fe(II) oxidation at nanomolar total iron concentrations. *Environmental Science & Technology*, **29**, 818–824.

Laidler, K.J. and King, M.C. (1983) The development of the transition-state theory. *Journal of Physical Chemistry*, **87**, 2657–2664.

Lalonde, K., Mucci, A., Ouellet, A. and Gelinas, Y. (2012) Preservation of organic matter in sediments promoted by iron. *Nature*, **483**, 198–200.

Lilova, K.I., Xu, F., Rosso, K.M., Pearce, C.I., Kamali, S. and Navrotsky, A. (2012) Oxide melt solution calorimetry of Fe^{2+}-bearing oxides and application to the magnetite–maghemite (Fe_3O_4–$Fe_{8/3}O_4$) system. *American Mineralogist*. **97**, 164–175.

Lima, F.A., Milne, C.J., Amarasinghe, D.C.V., Rittman-Frank, M.H., van der Veen, R.M., Reinhard, M., Pham, V.-T., Karlsson, S., Johnson, S.L., Grolimund, D., Borca, C., Huthwelker, T., Janousch, M., van Mourik, F., Abela, R. and Chergui, M. (2011) A high-repetition rate scheme for synchrotron-based picosecond laser pump/X-ray probe experiments on chemical and biological systems in solution. *Review of Scientific Instruments*, **82**, 063111.

Lovley, D.R. (1991) Dissimilatory Fe(III) and Mn(IV) reduction. *Microbiological Reviews*, **55**, 259–287.

Luther, G.W. (2005) Manganese(II) oxidation and Mn(IV) reduction in the environment – two one-electron transfer steps versus a single two-electron step. *Geomicrobiology Journal*, **22**, 195–203.

Lyon, L.A. and Hupp, J.T. (1999) Energetics of the nanocrystalline titanium dioxide aqueous solution interface: Approximate conduction band edge variations between $H_0 = -10$ and $H_- = +26$. *Journal of Physical Chemistry B*, **103**, 4623–4628.

Madison, A.S., Tebo, B.M., Mucci, A., Sundby, B. and Luther, G.W., III (2013) Abundant porewater Mn(III) is a major component of the sedimentary redox system. *Science*, **341**, 875–878.

Magnien, V., Neuville, D.R., Cormier, L., Roux, J., Hazemann, J.L., Pinet, O., and Richet, P. (2006) Kinetics of iron redox reactions in silicate liquids: A high-temperature X-ray absorption and Raman spectroscopy study. *Journal of Nuclear Materials*, **352**, 190–195.

Majzlan, J., Grevel, K.D. and Navrotsky, A. (2003) Thermodynamics of Fe oxides: Part II. Enthalpies of formation and relative stability of goethite (alpha-FeOOH), lepidocrocite (gamma-FeOOH), and maghemite (gamma-Fe_2O_3). *American Mineralogist*, **88**, 855–859.

Majzlan, J., Navrotsky, A. and Schwertmann, U. (2004) Thermodynamics of iron oxides: Part III. Enthalpies of formation and stability of ferrihydrite (~$Fe(OH)_3$), schwertmannite (~$FeO(OH)_{3/4}(SO_4)_{1/8}$), and epsilon-$Fe_2O_3$. *Geochimica et Cosmochimica Acta*, **68**, 1049–1059.

March, A.M., Stickrath, A., Doumy, G., Kanter, E.P., Krässig, B., Southworth, S.H., Attenkofer, K., Kurtz, C.A., Chen, L.X. and Young, L. (2011) Development of high-repetition-rate laser pump/X-ray probe methodologies for synchrotron facilities. *Review of Scientific Instruments*, **82**, 073110.

Marcus, R.A. and Sutin, N. (1985) Electron transfers in chemistry and biology. *Biochimica et Biophysica Acta*, **811**, 265–322.

Marx, D. (2006) Proton transfer 200 years after von Grotthuss: Insights from *ab initio* simulations. *ChemPhysChem*, **7**, 1848–1870.

McKenzie, K.J. and Marken, F. (2001) Direct electrochemistry of nanoparticulate Fe_2O_3 in aqueous solution and adsorbed onto tin-doped indium oxide. *Pure and Applied Chemistry*, **73**, 1885–1894.

Memming, R. (2001) *Semiconductor Electrochemistry*. Wiley-VCH, Weinheim, Germany.

Miller, D.M., Buettner, G.R. and Aust, S.D. (1990) Transition metals as catalysts of "autoxidation" reactions. *Free Radical Biology and Medicine*, **8**, 95–108.

Millero, F.J. (2009) Thermodynamic and kinetic properties of natural brines. *Aquatic Geochemistry*, **15**, 7–41.

Moore, J.K., Doney, S.C., Glover, D.M. and Fung, I.Y. (2002) Iron cycling and nutrient-limitation patterns in surface waters of the world ocean. *Deep-Sea Research Part II – Topical Studies in Oceanography*, **49**, 463–507.

Morgan, B. and Lahav, O. (2007) The effect of pH on the kinetics of spontaneous Fe(II) oxidation by O_2 in aqueous solution – basic principles and a simple heuristic description. *Chemosphere*, **68**, 2080–2084.

Navrotsky, A., Mazeina, L. and Majzlan, J. (2008) Size-driven structural and thermodynamic complexity in iron

oxides. *Science*, **319**, 1635–1638.
Nibbering, E.T.J., Fidder, H. and Pines, E. (2005) Ultrafast chemistry: Using time-resolved vibrational spectroscopy for interrogation of structural dynamics. *Annual Review of Physical Chemistry*, **56**, 337–367.
Nurmi, J.T., Bandstra, J.Z. and Tratnyek, P.G. (2004) Packed powder electrodes for characterizing the reactivity of granular iron in borate solutions. *Journal of the Electrochemical Society*, **151**, B347–B353.
O'Regan, B. and Gratzel, M. (1991) A low-cost, high-efficiency solar-cell based on dye-sensitized colloidal TiO_2 films. *Nature*, **353**, 737–740.
Ohlin, C.A., Villa, E.M., Rustad, J.R. and Casey, W.H. (2010) Dissolution of insulating oxide materials at the molecular scale. *Nature Materials*, **9**, 11–19.
Palandri, J.L. and Kharaka, Y.K. (2005) Ferric iron-bearing sediments as a mineral trap for CO_2 sequestration: Iron reduction using sulfur-bearing waste gas. *Chemical Geology*, **217**, 351–364.
Pedersen, H.D., Postma, D., Jakobsen, R. and Larsen, O. (2005) Fast transformation of iron oxyhydroxides by the catalytic action of aqueous Fe(II). *Geochimica et Cosmochimica Acta*, **69**, 3967–3977.
Pendlebury, S.R., Barroso, M., Cowan, A.J., Sivula, K., Tang, J., Grätzel, M., Klug, D.R. and Durrant, J.R. (2011) Dynamics of photogenerated holes in nanocrystalline α-Fe_2O_3 electrodes for water oxidation probed by transient absorption spectroscopy. *Chemical Communications*, **47**, 716–718.
Pham, A.N. and Waite, T.D. (2008) Oxygenation of Fe(II) in the presence of citrate in aqueous solutions at ph 6.0–8.0 and 25 degrees C: Interpretation from an Fe(II)/citrate speciation perspective. *Journal of Physical Chemistry A*, **112**, 643–651.
Posth, N.R., Koehler, I., Swanner, E.D., Schroeder, C., Wellmann, E., Binder, B., Konhauser, K.O., Neumann, U., Berthold, C., Nowak, M. and Kappler, A. (2013) Simulating precambrian banded iron formation diagenesis. *Chemical Geology*, **362**, 66–73.
Postma, D. and Brockenhuusschack, B.S. (1987) Diagenesis of iron in progracial sand deposits in late- and post-Weichselian age. *Journal of Sedimentary Petrology*, **57**, 1040–1053.
Pratt, A., Lari, L., Hovorka, O., Shah, A., Woffinden, C., Teat, S.P., Binns, C. and Kröger, R. (2014) Enhanced oxidation of nanoparticles through strain-mediated ionic transport. *Nature Materials*, **13**, 26–30.
Renock, D. and Becker, U. (2010) A first principles study of the oxidation energetics and kinetics of realgar. *Geochimica et Cosmochimica Acta*, **74**, 4266–4284.
Renock, D., Mueller, M., Yuan, K., Ewing, R.C. and Becker, U. (2013) The energetics and kinetics of uranyl reduction on pyrite, hematite, and magnetite surfaces: A powder microelectrode study. *Geochimica et Cosmochimica Acta*, **118**, 56–71.
Riedel, T., Zak, D., Biester, H. and Dittmar, T. (2013) Iron traps terrestrially derived dissolved organic matter at redox interfaces. *Proceedings of the National Academy of Sciences of the United States of America*, **110**, 10101–10105.
Robie, R.A. and Hemingway, B.S. (1995) Thermodynamic properties of minerals and related substances at 298.15 K and 1 bar (10^5 Pascals) and at higher temperatures. *US Geological Survey Bulletin*, **2131**, 461 pp.
Roden, E.E. (2003) Fe(III) oxide reactivity toward biological versus chemical reduction. *Environmental Science & Technology*, **37**, 1319–1324.
Roden, E.E. (2006) Geochemical and microbiological controls on dissimilatory iron reduction. *Comptes Rendus Geoscience*, **338**, 456–467.
Rosso, K.M. and Dupuis, M. (2006) Electron transfer in environmental systems: A frontier for theoretical chemistry. *Theoretical Chemistry Accounts*, **116**, 124–136.
Rosso, K.M. and Morgan, J.J. (2002) Outer-sphere electron transfer kinetics of metal ion oxidation by molecular oxygen. *Geochimica et Cosmochimica Acta*, **66**, 4223–4233.
Rosso, K.M. and Vaughan, D.J. (2006) Reactivity of sulfide mineral surfaces. Pp. 557–608 in: *Sulfide Mineralogy and Geochemistry* (D.J. Vaughan, editor). Reviews in Mineralogy and Geochemistry, **61**. Mineralogical Society of America and the Geochemical Society, Chantilly, Virginia, USA.
Rosso, K.M., Smith, D.M.A. and Dupuis, M. (2004) Aspects of aqueous iron and manganese (II/III) self-exchange electron transfer reactions. *Journal of Physical Chemistry A*, **108**, 5242–5248.
Schroth, A.W., Crusius, J., Sholkovitz, E.R. and Bostick, B.C. (2009) Iron solubility driven by speciation in dust sources to the ocean. *Nature Geoscience*, **2**, 337–340.
Schwertmann, U. and Cornell, R.M. (2000) *Iron Oxides in the Laboratory: Preparation and Characterization.*

Wiley-VCH Verlag, Hoboken, New Jersey, USA.

Sherman, D.M. (2005) Electronic structures of iron(III) and manganese(IV) (hydr)oxide minerals: Thermodynamics of photochemical reductive dissolution in aquatic environments. *Geochimica et Cosmochimica Acta*, **69**, 3249–3255.

Shimizu, K., Lasia, A. and Boily, J.-F. (2012) Electrochemical impedance study of the hematite/water interface. *Langmuir*, **28**, 7914–7920.

Silvester, E., Charlet, L., Tournassat, C., Gehin, A., Greneche, J.M. and Liger, E. (2005) Redox potential measurements and Mössbauer spectrometry of Fe-II adsorbed onto Fe-III (oxyhydr)oxides. *Geochimica et Cosmochimica Acta*, **69**, 4801–4815.

Singer, D.M., Chatman, S.M., Ilton, E.S., Rosso, K.M., Banfield, J.F. and Waychunas, G.A. (2012a). Identification of simultaneous U(VI) sorption complexes and U(IV) nanoprecipitates on the magnetite (111) surface. *Environmental Science & Technology*, **46**, 3811–3820.

Singer, D.M., Chatman, S.M., Ilton, E.S., Rosso, K.M., Banfield, J.F. and Waychunas, G.A. (2012b) U(VI) sorption and reduction kinetics on the magnetite (111) surface. *Environmental Science & Technology*, **46**, 3821–3830.

Sivula, K., Zboril, R., Le Formal, F., Robert, R., Weidenkaff, A., Tucek, J., Frydrych, J. and Grätzel, M. (2010) Photoelectrochemical water splitting with mesoporous hematite prepared by a solution-based colloidal approach. *Journal of the American Chemical Society*, **132**, 7436–7344.

Skomurski, F.N., Ilton, E.S., Engelhard, M.H., Arey, B.W. and Rosso, K.M. (2011) Heterogeneous reduction of U^{6+} by structural Fe^{2+} from theory and experiment. *Geochimica et Cosmochimica Acta*, **75**, 7277–7290.

Spagnoli, D., Gilbert, B., Waychunas, G.A. and Banfield, J.F. (2009) Prediction of the effects of size and morphology on the structure of water around hematite nanoparticles. *Geochimica et Cosmochimica Acta*, **73**, 4023–4033.

Steefel, C.I. and Maher, K. (2009) Fluid–rock interaction: A reactive transport approach. Pp. 485–532 in: *Thermodynamics and Kinetics of Water–Rock Interaction* (E.H. Oelkers and J. Schott, editors). Reviews in Mineralogy and Geochemistry, **70**. Mineralogical Society of America and the Geochemical Society, Chantilly, Virginia, USA.

Stumm, W. and Morgan, J.J. (1981) *Aquatic Chemistry*. Wiley, New Jersey, USA.

Stumm, W. and Sulzberger, B. (1996) The cycling of iron in natural environments: Considerations based on laboratory studies of heterogeneous redox processes. *Geochimica et Cosmochimica Acta*, **56**, 3233–3257.

Sulzberger, B. and Laubscher, H. (1995) Reactivity of various types of iron(III) (hydr)oxides towards light-induced dissolution. *Marine Chemistry*, **50**, 103–115.

Sverjensky, D. and Molling, P.A. (1992) A linear free energy relationship for crystalline solids and aqueous ions. *Nature*, **356**, 231–234.

Tang, J., Myers, M., Bosnik, K.A. and Brus, L.E. (2003) Magnetite Fe_3O_4 nanocrystals: Spectroscopic observation of aqueous oxidation kinetics. *Journal of Physical Chemistry B*, **107**, 7501–7506.

Thogersen, J., Thomsen, C.L., Poulsen, J.A. and Keiding, S.R. (1998) Chemical reactions in liquids: Photolysis of OClO in water. *Journal of Physical Chemistry A*, **102**, 4186–4191.

Torn, M.S., Trumbore, S.E., Chadwick, O.A., Vitousek, P.M. and Hendricks, D.M. (1997) Mineral control of soil organic carbon storage and turnover. *Nature*, **389**, 170–173.

Tratnyek, P.G. and Hoigné, J. (1994) Kinetics of reactions of chlorine dioxide (OClO) in water – II. Quantitative structure-activity relationships for phenolic compounds. *Water Research* **28**, 57–66.

Upadhyay, S.K. (2006) *Chemical Kinetics and Reaction Dynamics*. Springer, Berlin.

von der Heyden, B.P., Roychoudhury, A.N., Mtshali, T.N., Tyliszczak, T. and Myneni, S.C.B. (2012) Chemically and geographically distinct solid-phase iron pools in the southern ocean. *Science*, **338**, 1199–1201.

Wang, C., Baer, D.R., Amonette, J.E., Engelhard, M.H., Antony, J. and Qiang, Y. (2009) Morphology and electronic structure of the oxide shell on the surface of iron nanoparticles. *Journal of the American Chemical Society*, **131**, 8824–8832.

Weber, K., Achenbach, L.A. and Coates, J.D. (2006) Microorganisms pumping iron: Anaerobic microbial iron oxidation and reduction. *Nature Reviews Microbiology*, **4**, 752–764.

Wehrli, B. (1990) Redox reactions of metal ions at mineral surfaces. Cha. 11 in: *Aquatic Chemical Kinetics:*

Reaction Rates of Processes in Natural Waters (W. Stumm, editor). Wiley-Interscience, New York.

Wehrli, B., Wieland, E. and Furrer, G. (1990) Chemical mechanisms in the dissolution kinetics of minerals – the aspect of active-sites. *Aquatic Sciences*, **52**, 3–31.

White, A.F. and Peterson, M.L. (1996) Reduction of aqueous transition metal species on the surfaces of Fe(II)-containing oxides. *Geochimica et Cosmochimica Acta*, **60**, 3799–3814.

White, G.F., Shi, Z., Shi, L., Wang, Z., Dohnalkova, A.C., Marshall, M.J., Fredrickson, J.K., Zachara, J.M., Butt, J.N., Richardson, D.J. and Clarke, T.A. (2013) Rapid electron exchange between surface-exposed bacterial cytochromes and Fe(III) minerals. *Proceedings of the National Academy of Sciences of the United States of America*, **110**, 6346–6351.

Xu, Y. and Schoonen, M.A.A. (2000) The absolute energy positions of conduction and valence bands of selected semiconducting minerals. *American Mineralogist* **85**, 543–556.

Yang, L., Steefel, C.I., Marcus, M.A. and Bargar, J.R. (2010) Kinetics of Fe(II)-catalyzed transformation of 6-line ferrihydrite under anaerobic flow conditions. *Environmental Science & Technology*, **44**, 5469–5475.

Yin, Y., Rioux, R.M., Erdonmez, C.K., Hughes, S., Somorjai, G.A. and Alivisatos, A.P. (2004) Formation of hollow nanocrystals through the nanoscale kirkendall effect. *Science*, **304**, 711–714.

Zinder, B., Furrer, G. and Stumm, W. (1986) The coordination chemistry of weathering: II. Dissolution of Fe(III) oxides. *Geochimica et Cosmochimica Acta*, **50**, 1861–1869.

EMU Notes in Mineralogy, Vol. 17 (2017), Chapter 5, 95–119

Biogeochemical redox processes of sulfide minerals

DAVID J. VAUGHAN and VICTORIA S. COKER

Williamson Research Centre for Molecular Environmental Science, and School of Earth, Atmospheric and Environmental Sciences, University of Manchester, Manchester M13 9PL, UK, e-mail: david.vaughan@manchester.ac.uk; vicky.coker@manchester.ac.uk

The crystal structures and chemical compositions of sulfide minerals are summarized briefly before going on to review their redox chemistries, particularly the roles played by bacteria. In the formation of sulfide minerals, two processes need to be considered; one applicable to all sulfide systems is the microbial reduction of sulfate, the other is microbial reduction of metals, especially iron. Sulfate-reducing prokaryotes (SRP) can supply reactive sulfide ions for the formation of sulfide minerals. The SRP, of which there are more than 120 species, are ubiquitous in many anaerobic environments, although marine sediments are the most important. The SRP are able to grow under extreme conditions of pH and temperature. Bacteria can also conserve energy by reducing metals, such as reduction of Fe(III) coupled to the oxidation of organic matter. Biological processes also mediate the dissolution of sulfides under acid mine drainage conditions, and there are a diversity of acidophilic (pH <3) metal sulfide-oxidizing microorganisms. The oxidation reactions of pyrite, galena, arsenopyrite and chalcopyrite are discussed in detail before considering how toxic metals may be bound to the surfaces of sulfides such as pyrite and mackinawite. The applications of sulfide bacterial redox processes in clean technologies, such as bioleaching and biomining, are discussed briefly.

1. Introduction

Sulfide minerals are the major source of world supplies of a wide range of metals and are the most important group of ore minerals. They are found concentrated in ore deposits and areas of mineralization, and a limited number of sulfides are found as accessory minerals in rocks. However, only pyrite (FeS_2), pyrrhotite ($Fe_{1-x}S$), galena (PbS), sphalerite ((Zn,Fe)S) and chalcopyrite ($CuFeS_2$) can be classed as rock-forming minerals (Bowles *et al.*, 2011). Pyrite is by far the most abundant sulfide mineral, and the very fine-particle iron sulfides found in reducing environments beneath the surfaces of recent sediments and soils, although transient species, are also volumetrically important. Formerly called amorphous iron sulfide, these phases are now known to be composed largely of fine-particle mackinawite (tetragonal FeS) and, in some cases, the sulfospinel mineral greigite (Fe_3S_4). Both phases are metastable in relation to pyrite and pyrrhotite. Sulfide minerals are also potential sources of pollution. In particular, the release of sulfur through the weathering of sulfides in

DOI: 10.1180/EMU-notes.17.5

natural rocks or in mine wastes generates sulfuric acid, resulting in acid rock drainage (ARD) or acid mine drainage (AMD). The minerals arsenopyrite (FeAsS) and, to a lesser extent, enargite (Cu_3AsS_4) are also the primary mineral sources of arsenic, a highly toxic element when mobilized.

The redox chemistry of sulfur is complex, given its wide range of possible oxidation states, and this has consequences for our understanding of sulfide oxidation reactions. Redox processes involving the cation components of sulfide minerals may also be important, especially in the case of iron, the most important cation in sulfide minerals. Bacteria have a very important role to play in the redox chemistry of sulfide minerals, both in their formation and their dissolution. The activities of sulfate-reducing organisms in the formation, and metal-oxidizing organisms in the dissolution of sulfides, are central to understanding these processes.

In this chapter, the redox chemistry of sulfide minerals is reviewed with particular reference to the role played by bacteria. Included here is a brief introduction to sulfide minerals, their compositions, crystal structures and relevant aspects of their chemistries. For much fuller accounts of sulfide mineralogy and geochemistry, readers are referred to Vaughan and Craig (1978), Vaughan and Lennie (1991), Vaughan (2006) and Bowles *et al.* (2011). The development of microbial leaching technologies for recovery of metals from sulfide ores is very long established and will be discussed in terms of clean technologies at the end of in this chapter; however, see the recent review of Brierley and Brierley (2013) for details.

2. Structures and compositions

Sulfide minerals can be separated into a series of groups based on major crystal structure types, or through having key structural features in common (Table 1). Many are well known structures of crystalline solids, such as the rock-salt structure of the galena group (see Fig. 1a), the sphalerite and wurtzite forms of zinc sulfide (Fig. 1b,c), and the nickel arsenide (niccolite) structure (Fig. 1e). The disulfides are a group characterized by the presence of dianion (S–S, S–As, As–As, *etc.*) units in the structure. As well as the pyrite structure, in which FeS_6 octahedral units share corners along the *c*-axis direction, there is the marcasite form of FeS_2, in which octahedra share edges to form chains along the *c* axis (see Fig. 1d). The structures of loellingite ($FeAs_2$) and arsenopyrite (FeAsS) are variants of the marcasite structure that have shorter and alternate long and short, respectively, metal–metal distances across the shared octahedral edge (see Fig. 1d). Sulfides such as molybdenite, covellite (Fig. 1f) and mackinawite have layer structures and a small number exhibit structures that are best characterized as containing rings or chains of linked atoms (*e.g.* realgar, AsS, which is essentially a puckered As_4S_4 ring). Finally, a diverse group of sulfides is characterized by compositions with excess in the ratio of metals to sulfur, *e.g.* pentlandite ($(Ni,Fe)_9S_8$; see Fig. 1g).

Many sulfides share the same structure types; however, there are also sulfide minerals that have structures which are related to one of the common sulfides and can be thought

Table 1. Sulfide mineral structural groups.

Disulfide group			
Pyrite structure	Marcasite structure	Arsenopyrite structure	Loellingite structure
FeS_2 pyrite	FeS_2 marcasite	FeAsS arsenopyrite	$CoAs_2$ safflorite
CoS_2 cattierite		FeSbS gudmundite	$FeAs_2$ loellingite
NiS_2 vaesite			$NiAs_2$ rammelsbergite
Galena group			
PbS galena			
α-MnS alabandite			
Sphalerite group			
Sphalerite structure	Derived by ordered substitution		Stuffed derivatives
β-ZnS sphalerite	$CuFeS_2$ chalcopyrite		$Cu_9Fe_8S_{16}$ talnakhite
CdS hawleyite	Cu_2FeSnS_4 stannite		$Cu_9Fe_9S_{16}$ mooihoekite
Hg(S,Se) metacinnabar	Cu_2ZnSnS_4kesterite		$Cu_4Fe_5S_8$ haycockite
Wurtzite group			
Wurtzite structure	Composite structure derivatives		Derived by ordered substitution
α-ZnS wurtzite	$CuFe_2S_3$ cubanite		Cu_3AsS_4 enargite
CdS greenockite	$AgFe_2S_3$ argentopyrite		
Nickel arsenide group			
NiAs structure	Distorted derivatives		Ordered omission derivatives
NiAs niccolite	FeS troilite		Fe_7S_8 monoclinic pyrrhotite
NiSb breithauptite	CoAs modderite		Fe_9S_{10}, $Fe_{11}S_{12}$ hexagonal pyrrhotites
Thiospinel group			
Thiospinel structure			
Co_3S_4 linnaeite			
$FeNi_2S_4$ violarite			
$CuCo_2S_4$ carrollite			
Fe_3S_4 greigite			
Layer sulfides group			
Molybdenite structure	Tetragonal PbO structure		Covellite structure
MoS_2 molybdenite	FeS mackinawite		CuS covellite
WS_2 tungstenite			~Cu_3FeS_4 idaite
Metal excess group			
Pentlandite structure	Argentite structure	Chalcocite structure	Digenite structure
$(Ni,Fe)_9S_8$ pentlandite	Ag_2S argentite	Cu_2S chalcocite	Cu_9S_5digenite
		Ag_2S acanthite	Cu_5FeS_4 bornite
Nickel sulfide structures			
Ni_3S_2 heazlewoodite			
NiS millerite			
Ring or chain structure group			
Stibnite structure	Realgar structure		Cinnabar structure
Sb_2S_3 stibnite	As_4S_4 realgar		HgS cinnabar

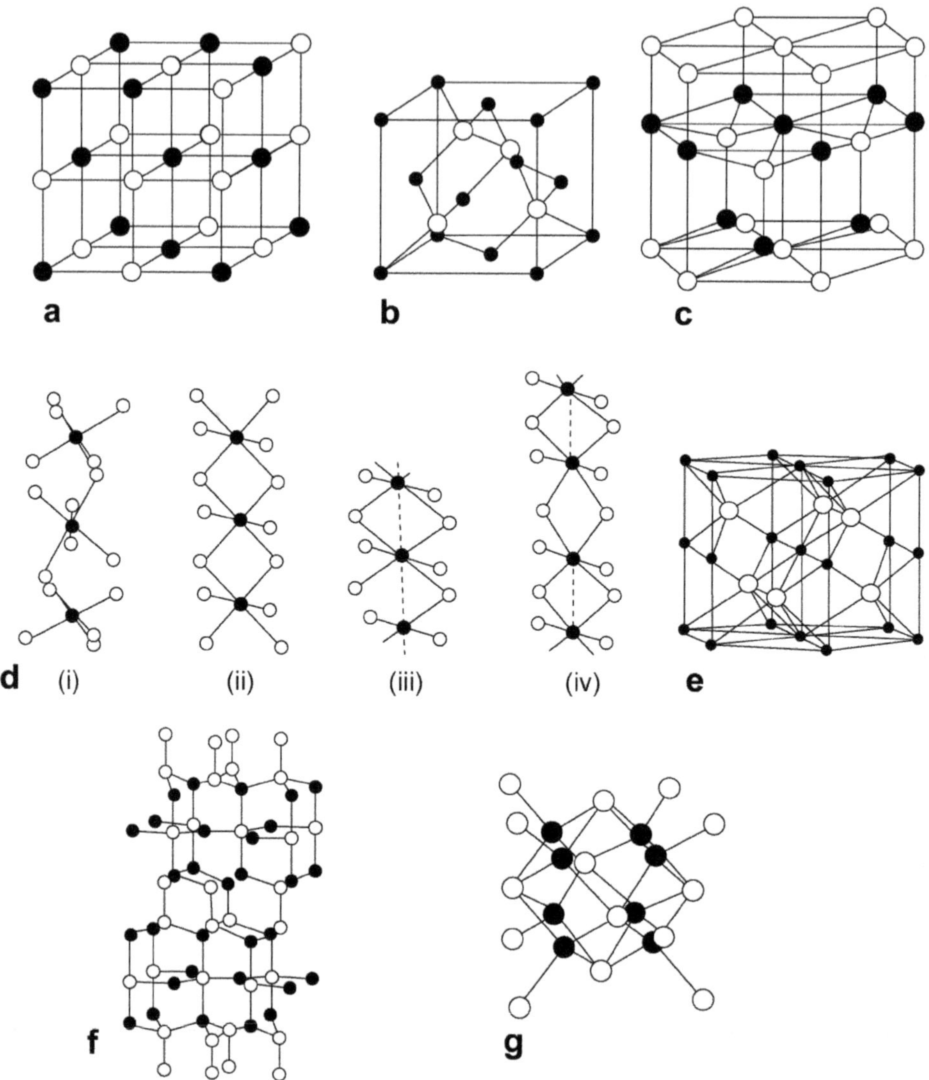

Figure 1. Crystal structures of some major sulfide minerals. Shown here are the structures of: (a) galena; (b) sphalerite; (c) wurtzite; (d) pyrite illustrated just to show the linkage of metal–sulfur octahedra along the *c*-axis direction (in (i)) compared with (ii) marcasite, (iii) loellingite and (iv) arsenopyrite; (e) niccolite; (f) covellite, and (g) pentlandite showing the cube cluster of octahedrally coordinated metals

of as derived from the 'parent' structure. The relationship between a derivative and parent structure may involve:

(1) ordered substitution, as where the structure of chalcopyrite ($CuFeS_2$) is derived from that of sphalerite (ZnS) by alternate replacement of zinc atoms with copper and iron, resulting in an enlarged (tetragonal) unit

cell; the structure of stannite (Cu_2FeSnS_4) results from further ordered substitution of half of the iron atoms in chalcopyrite by tin;

(2) a stuffed derivative, as where talnakhite ($Cu_9Fe_8S_{16}$) is derived from chalcopyrite by the occupation of additional normally empty cavities in the structure;

(3) ordered omission, as where monoclinic pyrrhotite (Fe_7S_8) is derived from the nickel arsenide structured troilite (FeS) by removal of iron atoms, leaving holes (vacancies) that are ordered;

(4) distortion, as where the troilite form of FeS is simply a distortion of the parent nickel arsenide structure form.

The compositions of the sulfide minerals given in Table 1 are for the pure endmembers whereas in natural samples, various substitutions are common. For example, most natural sphalerite samples contain substantial amounts of other metals substituting for Zn; iron can be as much as 27 wt.% and a wide range of other metals (Mn, Ni, Cu, Ga, Ge, Cd, In) may occur at concentrations ranging from ppm to percent levels. Certain pyrite samples may contain as much as 8 wt.% arsenic and there is extensive solid solution with NiS_2 and CoS_2 in some natural samples. Further details of these and similar substitutions are provided by Hudson-Edwards and Vaughan (2017). Many metal sulfides also show evidence of non-stoichiometry, *i.e.* that the elements comprising them are not combined in simple whole-number ratios. For example, the mineral pyrrhotite is commonly given the general formula $Fe_{1-x}S$ where $0 < x < 0.125$; here, the varying compositions correspond to varying concentrations of vacancies in sites that would otherwise be occupied by iron atoms. However, in systems such as these, ordering of the vacancies occurs at low temperatures, and the result may be a series of stoichiometric phases (a 'family' of pyrrhotite minerals all of slightly different compositions). Other examples of non-stoichiometry involve relatively small deviations from the simple ratio. For example, galena (PbS) is reported to exhibit a range of non-stoichiometry of 0.1 atomic percent. Such subtle variations in composition can have a significant effect on certain physical properties such as electrical conductivity, and have an influence on mineral reactivity that is still poorly understood.

3. Formation of sulfide minerals through biological processes

Many sulfide minerals can form at Earth surface temperatures by biologically induced mineralization processes, although the precipitation of iron sulfides is by far the most important. Rickard and Morse (2005) have reviewed iron sulfide formation; sedimentary pyrite formation has also been reviewed by Schoonen (2004) and other aspects of the biologically induced formation of sulfides were discussed by Rickard and Luther (2006). Two processes need to be considered here; one applicable to all sulfide systems is the microbial reduction of sulfate, the other is the microbial reduction of metals, especially iron. Microorganisms play a critical role in the global cycling of a variety of elements including carbon, iron, sulfur and nitrogen. The sulfur cycle is extremely complex due to the range of sulfur oxidation states (−2 to +6), and the ability of microorganisms to both oxidize and reduce the sulfur species, resulting in biogeochemical cycling of this important element and the formation of new minerals. The sulfur cycle is closely related to the carbon and nitrogen cycles as, for example, dissimilatory sulfate-reducing microorganisms are thought to account for >50% of the organic carbon mineralization in marine sediments (Jorgensen, 1982).

3.1. Microbial sulfate reduction

Dissimilatory sulfate-reducing microorganisms, those using sulfate as a terminal electron acceptor, are present in both the Bacteria and Archaea and as such can be referred to as sulfate-reducing prokaryotes (SRP). SRP can supply reactive sulfide ions for the formation of sulfide minerals (Frankel and Bazylinski, 2003) through the oxidation of common sulfate minerals such as gypsum ($CaSO_4.2H_2O$) (Muyzer and Stams, 2008). The ability of bacteria to cycle sulfur is also implicated as the oldest form of microbial metabolism (Canfield and Raiswell, 1999).

Microorganisms can oxidize organic carbon using a variety of terminal electron acceptors, ranging from O_2 under aerobic conditions to SO_4^{2-} in anoxic environments. The SRP are a morphologically and phylogenetically heterogeneous group of anaerobes that oxidize simple organic compounds or hydrogen using sulfate ions; *e.g.* through the reaction:

$$2CH_2O + SO_4^{2-} \rightarrow 2HCO_3^- + H_2S \quad (1)$$

in which the sulfur in the sulfate is reduced completely to sulfide, which is released into the environment, where it is available to combine with iron or other metals to form sulfide minerals. Dissolved sulfate is widely available in natural waters, particularly in seawater. The SRP use molecular hydrogen and relatively small organic molecules such as alcohols and fatty acids (*e.g.* acetate, butyrate and propionate) as electron donors. However, extensive studies have resulted in a host of additional organic compounds that support sulfate reduction by microorganism including methane and aromatic hydrocarbons. SRP commonly depend on coexisting fermentative microbes that can degrade polymeric substrates (Muyzer and Stams, 2008).

Overall there are >120 species of SRP (Thauer *et al.*, 2007) and, physiologically, these can be separated into two main groups: one is a group that tends to grow quickly and incompletely oxidizes organic substrates into acetate (*e.g. Desulfovibrio*, *Desulfotomaculum*, *Desulfomonas*, *Desulfobulbus*), and the other is a group that grows more slowly but is able to oxidize organic matter completely to CO_2 (*e.g. Desulfococcus*, *Desulfosarcina*, *Desulfonema*). The SRP are ubiquitous in many anaerobic environments including: lakes, swamps, soils, waste ponds, hydrothermal systems and even deep in the lithosphere (Lovley and Chapelle, 1995). In terms of global biogeochemical cycles, the SRP that occur in marine sediments are by far the most important (Schoonen, 2004).

The mechanism of sulfate reduction by SRP is complex. As the redox couple between sulfate and sulfite is unfavourable (−516 eV) for direct bacterial reduction, first the sulfate has to be transported into the cell and activated by adenosine triphosphate (ATP) to form adenosine phosphosulfate (APS), which can then be reduced directly to sulfite (SO_3^{2-}). The sulfite is then, in turn, reduced to sulfide; however, whether the process proceeds *via* a direct 6-electron transfer step, or in smaller 2-electron steps is, as yet, unclear (for reviews see Grein *et al.* (2013) and Muyzer and Stams (2008)). As well as sulfate, SRP are able to reduce other sulfur compounds to sulfide including thiosulfate, sulfite and sulfur, as well as other terminal electron acceptors such as nitrate and Fe(III), resulting in their ubiquitous occurrence within a host of natural environments, not all of them rich in sulfate (Lovley, 1993; Muyzer and Stams, 2008).

SRP are able to grow under extremes of both pH and temperature. The optimum growth conditions for the majority of SRP is a pH in the range of 5–9. Bacteria that are able to tolerate pH 10 have been reported; however, the focus of previous work in this area has predominantly been on the more acidic pH range (Tang *et al.*, 2009). This is due to the presence of SRP in acidic, sulfur-rich environments (as low as pH < 3) that can be associated with natural geothermal systems. However, a pure culture of anacidophile SRP remains elusive (Dopson and Johnson, 2012; Tang *et al.*, 2009). As well as mesophiles, there are thermophilic SRP strains able to reduce sulfate at optimum temperatures of 60°C (Tang *et al.*, 2009).

Another role played by microbes in the formation of sulfide minerals is in providing nucleation surfaces, as reviewed by Gilbert *et al.* (2005). At near-neutral pH, the outer surfaces of most bacterial cell walls are negatively charged; they attract positive ions from solution, which can initiate the nucleation of metal sulfides. The complexities of many cell surfaces, with their irregular arrays of proteins, their stalks and filaments, can also promote nucleation of metal sulfides.

3.2. Microbial metal reduction

A large number of Bacteria and Archaea are able to conserve energy by reducing a wide variety of metal and metalloid species through dissimilatory metabolism (Lloyd, 2003). One process of most relevance to the sulfur cycle is the reduction of Fe(III) coupled to the oxidation of organic matter (Balashova and Zavarzin, 1980). Many different Fe(III)-reducing bacteria have been discovered that are able to reduce both soluble, and solid-phase Fe(III) in the form of oxides and oxyhydroxides (Lovley *et al.*, 1997; Vargas *et al.*, 1998; Cutting *et al.*, 2009). One of the first to be isolated was *Geobacter metallireducens*, able to couple the reduction of amorphous ferric oxyhydroxide to the oxidation of organic matter (Lovley *et al.*, 1987). *G. metallireducens* resides in the δ-subdivision of the class Proteobacteria (Lovley *et al.*, 1993) where it is closely related to *Desulfuromonasacetoxidans*, a sulfur-reducing bacterium (Pfennig and Biebl, 1976; Roden and Lovley, 1993) also capable of reducing Fe(III) when grown on Fe(III)-oxyhydroxides (Nealson and Saffarini, 1994). Since then many other closely related Fe(III)-reducers have been isolated that are part of the family *Geobacteraceae*, and are capable of complete oxidation of organic compounds coupled to metal reduction (Lonergan *et al.*, 1996). As well as strict anaerobes in the δ-subdivision of Proteobacteria, there are also Fe(III)-reducing organisms in the γ-subdivision. These bacteria have a limited ability to use organic electron donors (Lonergan *et al.*, 1996) as they are only able to oxidize multi-carbon compounds as far as acetate and not all the way to CO_2, unlike the *Geobacter* species and close relatives (Coates *et al.*, 1999). The ability to respire using ferric iron is also found in other prokaryotic groups, *e.g.* *Geothrixfermentans*, in the *Holophaga-Acidobacterium* phylum (Coates *et al.*, 1999), and *Geovibrio ferrireducens* (Caccavo Jr. *et al.*, 1996).

Dissimilatory Fe(III)-reducing bacteria have putative ferric iron reductases located on the outer membrane of the cell (Myers and Myers, 1992, 1997; Lloyd, 2003) that may act as the terminal reductases of an electron transport chain that is linked to the cytoplasmic membrane. The terminal reductases have not yet been identified unequivocally in any organism, but are thought to involve *c*-type cytochromes in electron transport (Lloyd, 2005). Electrons are transferred through the chain to the surface of the cell where the exposed reductases transfer them to the Fe(III)-bearing mineral surface. This process conserves energy through the generation of ATP (Wilkins *et al.*, 2006).

For the reduction of structural Fe(III) to occur there needs to be a method for the bacteria to transfer electrons from the reductase to the mineral. This can be achieved either through: (1) direct contact between the bacterial cell surface and Fe(III)-bearing mineral; (2) release of chelating compounds to increase the solubility of the Fe(III)-bearing mineral to allow reduction; (3) release and/or utilization of exogenous extracellular electron shuttling compounds that act as electron transfer mediator compounds between bacterial cell surface and mineral; or (4) through 'nanowires' or electrically conducting pili (Kappler and Straub, 2005; Reguera *et al.*, 2005). *G. metallireducens* has also been shown to use flagella and pili to aid movement and attachment to insoluble Fe(III) oxyhydroxides. Their 'chemotaxis' towards Fe(II) has been demonstrated (Childers *et al.*, 2002), facilitating access to metal oxides in anoxic environments (Lovley *et al.*, 2004). Electron shuttles also provide a method for bacteria to access Fe(III)-oxides that are physically inaccessible, *e.g.* sub-micron scale fractures in rocks, or to reduce relatively recalcitrant crystalline forms of Fe(III)-bearing minerals, such as hematite (α-Fe_2O_3), goethite (α-FeOOH) and the Fe-bearing clay, smectite (Lovley *et al.*, 1998; Pentráková *et al.*, 2013).

3.3. Iron sulfide formation in marine sediments

Microbially generated iron sulfide minerals are ubiquitous in modern anoxic sediments, particularly marine sediments, and also in certain lakes, swamps and soils. This topic has been reviewed most recently by Rickard and Morse (2005). The main pathways of sedimentary iron sulfide formation are summarized in Fig. 2. The rate of iron sulfide formation depends particularly on the rate of microbial sulfate reduction (which also depends on the availability of organic carbon) and on the quantities of competing electron acceptors, including reactive Fe(III)-bearing minerals. The very fine-particle black precipitate that forms when dissolved sulfide produced by SRP reacts with Fe^{2+} is now known, from X-ray diffraction and spectroscopy experiments, to be poorly ordered mackinawite or mixtures of mackinawite and greigite. These are metastable phases; thermodynamically, only pyrite should be expected under equilibrium conditions in

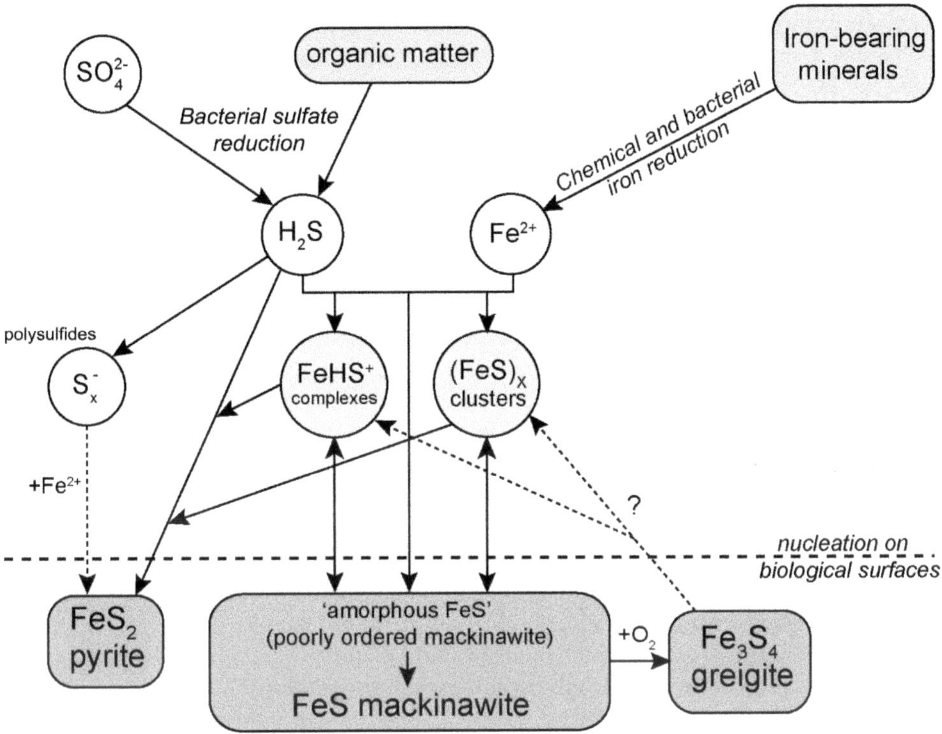

Figure 2. Pathways for sedimentary iron sulfide formation. (Reproduced from Posfai and Dunin-Borkowski, 2006, with the permission of the Mineralogical Society of America.)

low-temperature sedimentary environments. The formation of pyrite from these metastable precursors has been discussed in terms of three pathways, (see Schoonen (2004), for original references) which are: [1] FeS oxidation by a polysulfide species; [2] FeS oxidation by H_2S; and [3,4] conversion of FeS by iron loss through an intermediate greigite phase. The reactions involved are:

$$[1]\ FeS + S_n^{2-} \rightarrow FeS_2 + S_{n-1}^{2-} \quad (2)$$

$$[2]\ FeS + H_2S \rightarrow FeS_2 + H_2 \quad (3)$$

$$[3]\ 4\ FeS + \tfrac{1}{2}\ O_2 + 2\ H^+ \rightarrow Fe_3S_4 + Fe^{2+} + H_2O \quad (4)$$

$$[4]\ Fe_3S_4 + 2\ H^+ \rightarrow FeS_2 + Fe^{2+} + H_2 \quad (5)$$

The importance of reaction [2] is supported indirectly by the persistence and the large proportion of iron monosulfide in euxinic sediments. In such an environment, reactive iron is available in abundance and the lack of dissolved sulfide prevents the further reaction of FeS to form FeS_2. The conversion of mackinawite into greigite *via* iron loss (reaction [3]) was observed by Lennie *et al.* (1997). On the other hand, as summarized by Luther and Rickard (2005), highly reactive aqueous metal sulfide (FeS_{aq}) clusters may be key intermediaries in pyrite formation, as they react with either H_2S or polysulfide species to nucleate pyrite. Here, the conversion of mackinawite or greigite into pyrite would not be a solid-state transformation, but would involve their partial dissolution to form aqueous FeS clusters. Because FeS clusters can form by other routes, the presence of mackinawite and greigite would then not be a necessary condition for pyrite formation (Rickard and Morse, 2005). Bacterial activity may not only be involved in the precipitation of FeS, but also in its conversion to pyrite; *e.g.* Grimes *et al.* (2001) found that organic matter can provide nucleation sites for reaction of FeS to FeS_2.

4. Dissolution of sulfide minerals through biological processes

Sulfide minerals are unstable when exposed to the Earth's atmosphere and oxygenated surface waters. Hence, oxidation reactions are extremely important and dominate sulfide dissolution. Biological processes mediate such reactions, and biologically-mediated metal sulfide oxidation has been studied at low pH in terrestrial ecosystems, such as sites of acid mine drainage where the release of toxic metals (*e.g.* arsenic) can be associated with mineral sulfide breakdown, at terrestrial geothermal sites (*e.g.* Yellowstone National Park, USA), and in the mineral processing (bioleaching) of sulfide ores. In these cases, prevalent conditions are aerobic and the oxidation of the sulfide minerals generates acidic conditions that retard the oxidation of Fe(II) by abiotic processes creating an abundant availability of reduced Fe for use by acidophilic Fe(II)-oxidizing microorganisms (Johnson, 2011). However, there are also important mechanisms present in the biosphere for oxidizing metal sulfides under anaerobic, neutral pH regimes, such as those found in marine and subsurface environments. Phototrophs and denitrifying microorganisms are recognized as key Fe(II)-oxidizing microorganisms under these conditions, but they have been less well studied and are not discussed further here (see Weber *et al.*, 2001; Kappler and Newman, 2004).

There is a wide diversity of acidophilic ($pH < 3$) metal sulfide-oxidizing microorganisms spread across the Bacteria and Archaea that are able to oxidize Fe(II) and/or sulfur compounds (Hedrich *et al.*, 2011; Johnson, 2011; Vera *et al.*, 2013). These prokaryotes can be delineated by their temperature tolerance into: mesophiles (*e.g. Acidithiobacillus*, *Leptospirillum*), moderate thermophiles (*e.g. Alicyclobacillus*) and thermophiles (*e.g. Sulfobus* and *Acidianusbrierleyi*). The best studied of these prokaryotes is the bacterium *Acidithiobacillus ferrooxidans* (previously *Thiobacillus ferrooxidans* (Kelly and Wood, 2000)) which grows autotrophically and oxidizes Fe(II), sulfur or reduced sulfur compounds while using dissolved oxygen as the electron acceptor; curiously, it also has the ability to oxidize sulfur anaerobically coupled to the reduction of Fe(III) (Pronk *et al.*, 1992; Gleisner *et al.*, 2006).

The biologically mediated oxidation of sulfide minerals, particularly pyrite, has been described previously as having both a direct and indirect mechanism (Silvermann and Ehrlich, 1964); however, molecular-scale interactions are still under debate. The direct mechanism utilizes the enzymatic oxidation of the bound sulfide within the mineral phase, resulting in a breakdown of the structure, whereas the indirect mechanism is described as the result of oxidation of the mineral sulfide by Fe(III) ions together with enzymatic re-oxidation of the Fe(II) ions. It is generally considered that the indirect mechanism is the more likely process for mineral sulfide oxidation (Rawlings and Johnson, 2007). Bacterial production of extracellular polysaccharide (EPS) during biofilm attachment and growth on metal sulfides is also implicated in having a key role in the oxidation of metal sulfides. Under acidic conditions Schippers and Sand (1999) describe sulfide minerals as either being chemically attacked by Fe(III) as above and producing thiosulfate (FeS_2, WS_2 and MoS_2) or as being susceptible to attack by protons and Fe(III), resulting in elemental sulfur being formed *via* intermediate polysulfides. Further discussions on mechanisms can be found in the publications by Sand *et al.* (1995), Schippers and Sand (1999), Sand *et al.* (2001), Ehrlich and Newman (2009) and Vera *et al.* (2013).

In what follows, a brief account is given of acidophilic, biologically mediated metal sulfide oxidation processes and how these relate to a small number of key sulfides, namely: pyrite, as the overwhelmingly dominant sulfide in natural systems; galena and arsenopyrite, because of their potential for releasing toxic elements (As and Pb), and chalcopyrite, both as a mineral containing two redox sensitive metals, and as the most important copper ore mineral. Our focus will be on aqueous oxidation and processes involving microbes. We will not discuss extensive literature on the oxidation of these sulfide minerals in air.

4.1. Pyrite oxidation

Under acid mine drainage conditions, pyrite oxidation can be accelerated through aerobic respiration by chemolithoautotrophic bacteria able to oxidize Fe(II) (Singer and Stumm, 1970), such as *Leptospirillum sp. and Acidithiobacillus sp.* (Hallberg and Johnson, 2001; Johnson, 2010). The redox chemistry of pyrite in aqueous solution is complex. Rimstidt and Vaughan (2003) note that the key controls of mechanisms and rates of oxidation remain poorly understood. This is because the processes of aqueous oxidation involve a complex series of elementary reactions. The elementary steps almost always involve the transfer of only one electron at a time so that oxidation of a disulfide to release sulfate requires transfer of seven electrons and, therefore, up to seven elementary steps. The minerals are semiconductors so the reactions are electrochemical and various reactions can happen at different sites with electron transfer through the mineral. Following Rimstidt and Vaughan (2003), it is suggested that pyrite aqueous oxidation is an electrochemical process with three steps, as illustrated in Fig. 3. These are: (1) cathodic reaction, (2) electron transport and (3) anodic reaction. The cathodic reaction involves an aqueous species that accepts electrons from an Fe^{2+} site on the mineral surface, such as O_2 (equation 6) or Fe(III) (equation 7). The continued formation of Fe^{3+} is then mediated by oxygen or bacterial oxidation (equation 8).

$$FeS_2 + 3.5\ O_2 + H_2O \rightarrow Fe^{2+} + 2\ H^+ + 2\ SO_4^{2-} \quad (6)$$

$$FeS_2 + 14\ Fe^{3+} + 8\ H_2O \rightarrow 15\ Fe^{2+} + 16\ H^+ + 2\ SO_4^{2-} \quad (7)$$

$$2\ Fe^{2+} + 0.5\ O_2 + 2\ H^+ \rightarrow 2\ Fe^{3+} + H_2O \quad (8)$$

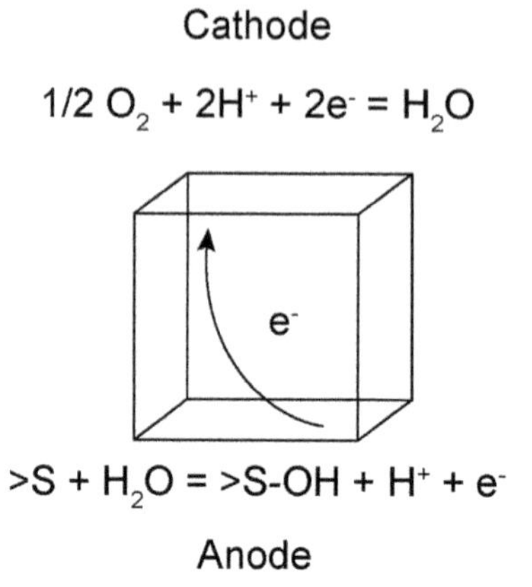

Cathode

$1/2\ O_2 + 2H^+ + 2e^- = H_2O$

e^-

$>S + H_2O = >S\text{-}OH + H^+ + e^-$

Anode

Figure 3. Diagram of aqueous pyrite oxidation showing the three stages of cathodic reaction, electron transport and anodic reaction. (Reproduced from Rimstidt and Vaughan (2003) with the permission of Elsevier.)

It is now clear that the cathodic reaction is the rate-determining step for sulfide oxidation. Williamson and Rimstidt (1994) showed that the pyrite oxidation rate depends on concentration of O_2 or another oxidant (such as Fe^{3+}), and this is the case for other sulfide minerals including galena, sphalerite, chalcopyrite and arsenopyrite. It is suggested that the anodic reaction involves a multistep sulfur atom oxidation (electron removal) process with a sequence of surface reactions of the type:

$$\text{py–S–S} = \text{py–S–S}^+ + e^- \quad (9)$$
$$\text{py–S–S}^+ + \text{H}_2\text{O} = \text{py–S–S–OH} + \text{H}^+ \quad (10)$$
$$\text{py–S–S–OH} = \text{py–S–SO} + e^- + \text{H}^+ \quad (11)$$
$$\text{py–S–SO} = \text{py–S–SO}_2 + 2e^- + 2\text{H}^+ \quad (12)$$
$$\text{py–S–SO}_2 = \text{py–S–SO}_3 + 2e^- + 2\text{H}^+ \quad (13)$$

At this point, there would be a tendency for this species to break away from the surface to form a thiosulfate complex:

$$\text{py–S–S–O}_3 = \text{py} + M\text{S}_2\text{O}_3 \quad (14)$$

(where M is a metal ion in solution). The final step is suggested to be pH dependent. If the pH is high, the terminal $S\text{–}SO_3$ ionizes completely, making the S–S bond stronger than the Fe–S bond, so much of the sulfur is released into solution as $S_2O_3^{2-}$. However, at low pH, the majority of terminal $S\text{–}SO_3$ groups would retain a proton (to be $S\text{–}SO_3H$), which encourages transfer of electrons into the S–S bond where they are more easily transferred to the cationic site, leaving the sulfur with a very positive charge. This would lead to further nucleophilic attack by a water molecule to produce SO_4^{2-}.

The ultimate products of pyrite oxidation are dependent on solution chemistry, with evidence of reaction products coming from a wide range of analytical techniques. For example, England *et al.* (1999) used glancing-angle X-ray absorption spectroscopy to probe the surface (~2–3 nm) region of pyrite oxidized at pH 9.2, finding a goethite-like surface species. Todd *et al.* (2003) used X-ray absorption spectroscopy (of the oxygen K- and sulfur and iron L-edges) to determine the products of oxidation in aqueous electrolytes at pH 2 to pH 10 in the top 2–5 nm region of the surface. Below pH 4, a ferric (hydroxyl) sulfate is the main product whereas at higher pH, an Fe^{3+} oxyhydroxide is also found. The presence of Fe^{3+} in solution autocatalyzes the oxidation process, thereby promoting ferric oxyhydroxide formation at low pH. Under the most alkaline conditions, the O *K*-edge spectrum resembles that of goethite (*cf.* England *et al.*, 1999). As discussed above, under acidic conditions, biologically mediated breakdown of pyrite is thought to proceed through attack by biogenic Fe(III), which is recycled for further oxidative breakdown of pyrite by bacteria. During bacterial oxidation of pyrite there can be a build-up of elemental sulfur if the bacterium, such as *Leptospirillum ferrooxidans*, is able to oxidize Fe(II) but not sulfur species.

4.2. Galena oxidation

Like pyrite, galena has been a much studied sulfide because of its widespread natural occurrence and potential as a source of pollution as well as its relevance to commercial bioleaching of sphalerite (ZnS) as the minerals are often co-associated (da Silva, 2004). The important early work on the oxidation of galena in air and in aqueous solution was summarized by Tossell and Vaughan (1987); later work on aqueous oxidation has involved electrochemical methods (Richardson and Odell, 1984; Fornasiero *et al.*, 1994), Raman spectroscopy (Turcotte *et al.*, 1993), thermodynamic methods (Acharya and Paul, 1991; Janczuk *et al.*, 1992a,b) and the use of bacteria (Bang *et al.*, 1995). X-ray photoelectron spectroscopy (XPS) has been important in providing insights into the oxidation processes of galena, particularly work using synchrotron radiation (SR-XPS). For example, the low temperature (143–298 K), *ex situ* examination of the S2*p* spectra of an electrochemically oxidized PbS (100) surface showed a layer structure of elemental sulfur and metal polysulfides (Laajalehto *et al.*, 1997).

In a different approach, Chernyshova (2003) used *in situ* FTIR to study the products of electrochemical oxidation of galena formed at the electrode/electrolyte interface in solution at pH 9.2. It was suggested that, under these conditions, the oxidation of galena starts with the reaction:

$$PbS + 2x\, h^+ \rightarrow Pb_{1-x}S + x\, Pb^{2+} \qquad (15)$$

followed by the reaction:

$$PbS + 2\, h^+ \rightarrow Pb^{2+} + S^0 \qquad (16)$$

where h^+ denotes a hole in the valence band of this semiconductor. Lead sulfate, thiosulfate and polythionate ions are formed from the elementary sulfur produced at the first oxidation stage, and $Pb(OH)_2$ is precipitated.

Under acidic conditions, *Acidithiobacillus sp.* are known to produce lead sulfate as an end product (Torma and Subramanian, 1974). This relatively insoluble mineral is undesirable in bioleaching processes making galena biooxidation far less well studied than, *e.g.* pyrite. Several studies have looked at the mechanism of biogenic galena oxidation (Attia and El Zeky, 1990; Bang *et al.*, 1995; da Silva, 2004). Da Silva (2004) concluded that both elemental sulfur and lead sulfate are the products of bacterial oxidation of galena and that the process is controlled by diffusion due to the layers of insoluble products which build up on the PbS surface. In the presence of Fe, the primary role of the bacteria is in the production of soluble Fe(III) which then facilitates PbS oxidation. In the absence of Fe, as galena is considered to be acid-soluble (unlike pyrite), it is therefore susceptible to proton attack. In theory the lead–sulfur bonds can be disrupted, leading to production of hydrogen sulfide and, therefore, mineral dissolution (Schippers and Sand, 1999; Johnson, 2010).

4.3. Arsenopyrite oxidation

Laboratory studies of arsenopyrite oxidation have been reviewed comprehensively by Corkhill and Vaughan (2009). Here we will focus on aqueous oxidation, particularly that involving microbial mediation. Jones *et al.* (2003) studied acidic (pH 2.3) oxidative leaching of arsenopyrite in the presence of *Thiobacillus ferrooxidans* and the essential salts required to sustain bacterial growth. Reactions between arsenopyrite and *T. ferrooxidans* produced a uniform, solid $FePO_4$ overlayer (~0.2 μm thick) within 1 week. The phosphate component was derived from the essential salts – no such layer was seen in an abiotic control experiment. The layer continues to thicken with time; reaction continues with oxidation, diffusion and dissolution of arsenopyrite beneath the overlayer. The bacterial cells are therefore separated from the arsenopyrite surface and thus it appears that they do not need to be in contact with the arsenopyrite in order to promote rapid reaction.

Corkhill *et al.* (2008) studied acidic (pH 1.8) oxidative dissolution of arsenopyrite both in the presence and the absence of the acidophilic microorganism *Leptospirillum ferrooxidans*. Biologically mediated oxidation was seen to proceed more extensively than abiotic oxidation. The formation of Fe(III) oxyhydroxides, ferric sulfate and arsenate was observed. Environmental Scanning Electron Microscope (ESEM) studies showed extensive coating of extracellular polymeric substances associated with the *L. ferrooxidans* on the surface. Bacterial leach pits suggest a direct biological oxidation mechanism, although a combination of direct and indirect bioleaching cannot be ruled out. It is interesting that the cells of *L. ferrooxidans* appear able to withstand concentrations of several hundred ppm of As(III) or As(V) in solution.

In their review, Corkhill and Vaughan (2009) observe that oxidation in acid solution appears to be more rapid than in water or alkaline solution, and that the oxidation products reported include: Fe(III) oxide, As(III), As(V), SO_3^{2-} and SO_4^{2-}. Electrochemical studies have highlighted the formation of elemental S at the surface, which contrasts with the sulfur oxyanions reported from XPS studies. Kinetic studies suggest that O_2 and Fe^{3+} are the dominant inorganic agents causing oxidative dissolution of arsenopyrite, and that bacterially mediated oxidation involving acidophillic Fe- and S-oxidizing bacteria is more extensive than abiotic oxidation. Corkhill and Vaughan (2009) suggested a multi-step molecular-scale mechanism for arsenopyrite oxidation, essentially the same as that proposed by Rimstidt and Vaughan (2003) for the oxidation of pyrite (as described above; see equations 9 to 14). It was also pointed out, however, that although the elemental constituents of arsenopyrite appear to oxidize at different rates; there is no consensus as to which is fastest or slowest and the literature is divided with regards to reaction stoichiometry and the composition and layering of surface overlayers.

4.4. Chalcopyrite oxidation

Studies of the aqueous oxidation of chalcopyrite have commonly used electrochemical techniques; a brief summary of the results obtained from many of these investigations was provided by Yin *et al.* (2000). Those authors combined cyclic voltammetry with surface analysis by X-ray photoelectron spectroscopy (XPS) to determine the oxidation reactions leading to the breakdown of chalcopyrite in an alkaline solution (pH 9.2), conditions typical of systems used in processing sulfide ores by froth flotation. In this case, combining electrochemical and spectroscopic data enabled the authors to propose a sequence of oxidation reactions as a function of increasing electrode potential (*i.e.* 'intensity of oxidation') as follows:

$$2\ CuFeS_2 + 6x\ OH^- \rightarrow 2\ CuFe_{1-x}S_2 + x\ Fe_2O_3 + 3x\ H_2 + 6x\ e^- \quad (17)$$

$$2\ CuFeS_2 + 6\ OH^- \rightarrow 2\ CuS_2^* + Fe_2O_3 + 3\ H_2O + 6\ e^- \quad (18)$$

$$CuFeS_2 + 3\ OH^- \rightarrow CuS_2^* + Fe(OH)_3 + 3\ e^- \quad (19)$$

These reactions suggest that, with initial oxidation, a monolayer of $Fe_2O_3/Fe(OH)_3$ is formed, leaving Cu and S unoxidized in the original chalcopyrite structure as a metastable phase of CuS_2 stoichiometry (and designated CuS_2^*). Together, these phases would cause passivation of the surface. A very simple model of the development of the surface during oxidation is shown in Fig. 4. With increasing potential, these reactions continue, removing Fe from deeper within the chalcopyrite. Here, solid-state diffusion is the likely rate-controlling mechanism; this is further evidenced by the marked differences in oxidation rates shown by the metal-enriched 'stuffed derivatives' of chalcopyrite such as talnakhite (Vaughan *et al.*, 1995). Above a critical potential, the metastable CuS_2^* phase decomposes (Yin *et al.*, 2000). The oxidation of chalcopyrite in solution at pH 4 seems to proceed in a similar fashion to that under alkaline conditions, with an iron oxide/hydroxide 'layer' forming at the surface (Farquhar *et al.*, 2003). However, AFM studies of the oxidized surface, both *ex situ* and in the presence of the fluid, show the presence of islands (< 0.15 μm wide) of reaction products. In electrochemical experiments, coverage of the surface with these islands increases with the amount of charge passed.

Acidiphilic bacteria have been implicated in chalcopyrite oxidation, again, particularly *Acidithiobacillus ferrooxidans*. Passivating mineral species can form on the chalcopyrite during biooxidation, as with the abiotic processes mentioned above, *e.g.* polysulfides and jarosite, and these can also encrust the bacterial biofilms that form on mineral surfaces (Harneit *et al.*, 2006; Lei *et al.*, 2009; Lara *et al.*, 2013; Vera *et al.*, 2013).

5. Toxic metals bound to sulfide mineral surfaces

Sulfide minerals, particularly those formed biogenically, commonly have surfaces which are highly reactive and can take up contaminant species from aqueous solutions

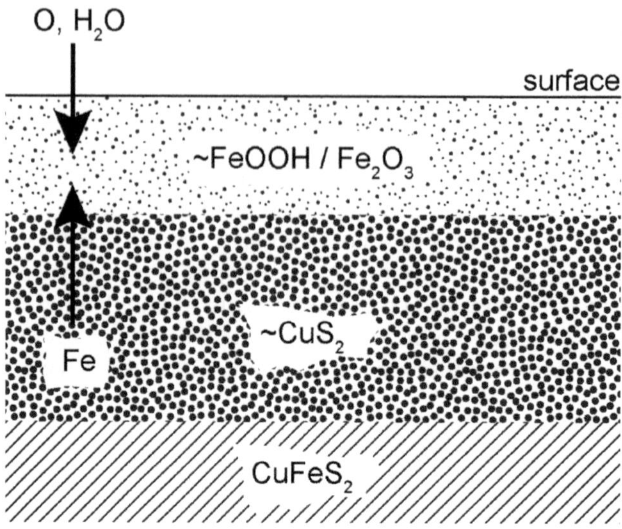

Figure 4. A model for chalcopyrite surface oxidation. (Reproduced from Rosso and Vaughan, 2006, with the permission of the Mineralogical Society of America.)

including natural waters. A comprehensive review of this topic has been provided by Rosso and Vaughan (2006). Here, we focus on a few illustrative examples with heavy metals and metalloids, and radioactive elements as contaminants, and pyrite and mackinawite as sulfide substrates. It is important to recall that uptake of species, such as metal ions in solution, by mineral surfaces can involve one or more of several processes. The ion may retain its hydration sphere and be weakly held to the surface as an 'outer sphere complex', or it may bond directly and more strongly to the surface as an 'inner sphere complex'. It may replace other ions at the surface (or even diffuse into the solid matrix) in an exchange reaction, or undergo a more extensive reaction (possibly a redox reaction) with the surface that may result in precipitation or co-precipitation of a new phase, or wholesale replacement of the substrate.

5.1. Uptake by pyrite

There have been numerous studies of surface interactions of pyrite with metal ions in solution. For example, Parkman *et al.* (1999) used bulk measurements of uptake as a function of initial concentration combined with X-ray absorption spectroscopy (EXAFS and XANES) to study uptake of Cu^{2+} and Cd^{2+} ions in solution at circum-neutral pH. In these experiments, the bulk uptake of Cd contrasted markedly with that of Cu. Whereas Cd shows decreasing efficiency of uptake with increasing concentration, Cu shows constant uptake, which is an order of magnitude greater than that of Cd. The behaviour of Cu is interpreted as due to formation (by precipitation/replacement) of a copper sulfide at the surface; this is in line with spectroscopic data showing bonding of Cu to an average of four sulfur atoms at 2.3 Å. The spectroscopic data for Cd indicate bonding to six oxygen atoms at 2.28 Å over all measured concentrations. The bulk uptake behaviour suggests limited availability of surface sites for 'sorption' of the Cd; hence, the simplest interpretation is formation of an outer sphere surface complex. Parkman *et al.* (1999), however, suggested that this is unlikely given the high capacity and relatively strong binding of Cd at low concentrations. A more complex interaction process involving oxidation is suggested. Bostick *et al.* (2000) tackled this problem using methods including Raman spectroscopy, electron microscopy and STM; they proposed complex reactions involving pyrite surface reconstruction and disproportionation, leading to a mixture of sulfide and oxide surface products (elemental sulfur, iron hydroxide and CdS).

Examples of other studies of metal uptake by pyrite include work on Ni^{2+} (Hacquard *et al.*, 1999), Hg^{2+} (Ehrhardt *et al.*, 2000) and As^{3+} and As^{5+} (Bostick and Fendorf, 2003; Farquhar *et al.*, 2002). The work on Ni^{2+} was at pH 10 under conditions where "the mineral surface is rapidly covered by a layer of Fe^{3+} oxides". The first step of the interaction with nickel is the formation of a hydroxylated surface complex (Hacquard *et al.*, 1999). The 'sorption' of Hg^{2+} was investigated as a function of pH and using XPS for analysis of exposed pyrite surfaces (Ehrhardt *et al.*, 2000). The surfaces showed partial oxidation, with islands of Fe^{3+} oxyhydroxides as well as pyrite present. The XPS measurements were interpreted in terms of (inner sphere) surface complexation involving both pyritic functional groups and hydroxyl groups on the oxyhydroxide islands. No evidence was found for the formation of Hg sulfide, sulfate or thiosulfate species.

Farquhar *et al.* (2002) used EXAFS and XANES to study the uptake of As^{3+} and As^{5+} from aqueous solutions at pH 5.5–6.5. For both arsenic species, the primary coordination of arsenic sorbed at the pyrite surface was to four oxygen atoms (with As–O distances of 1.69 Å for As^{5+} and 1.73 Å for As^{3+}). Spectroscopic evidence for Fe atoms at 3.35–3.40 Å lends further weight to the suggestion that both arsenite and arsenate species form outer sphere complexes under the conditions studied (Farquhar *et al.*, 2002). Bostick and Fendorf (2003) studied arsenite sorption onto pyrite using bulk sorption measurements, X-ray absorption spectroscopy and XPS; they showed patterns of uptake that generally follow the Langmuir isotherms characteristic of interaction with a limited number of surface sites, which eventually become saturated. Such behaviour is in accord with the surface complexation suggested above. However, the spectra reported by Bostick and Fendorf (2003) were interpreted in terms of formation of an "FeAsS-like surface precipitate". These discrepancies could arise from differences in starting materials and sample handling, although another explanation could be that Farquhar *et al.* (2002) worked at acidic pH, where Bostick and Fendorf (2003) found lower sorption levels and closer adherence to Langmuir isotherm uptake behaviour as a function of concentration.

5.2. Uptake by mackinawite

As with pyrite, and using similar techniques, interactions between the mackinawite surface and dissolved metal cations have been studied for Cu and Cd (Parkman *et al.*,

1999). Bulk uptake experiments show substantial removal of the metal up to high concentrations without any apparent limitation (*i.e.* a linear uptake rather than the decreasing sorption associated with a Langmuir isotherm). The results of spectroscopic studies point to formation of a CdS precipitate in the case of cadmium, whereas for copper, a precipitate is again formed, but with clear evidence of formation of the ternary sulfide, chalcopyrite ($CuFeS_2$). This is an unexpected result, but it is supported by work showing chalcopyrite formation in natural Cu-rich sulfidic sediments (Parkman *et al.*, 1996).

Farquhar *et al.* (2002) and Wolthers *et al.* (2003, 2005) have studied arsenic uptake by mackinawite. Farquhar *et al.* (2002) used XAS to investigate the mechanisms whereby As^{3+} and As^{5+} in aqueous solution (pH 5.5–6.5) interact with the mackinawite surface and found evidence for the formation of outer sphere complexes for both species. Wolthers *et al.* (2005) conducted similar experiments using a "poorly crystalline mackinawite" regarded as akin to the material found naturally in Recent sediments. Their results also point to outer sphere complex formation as the mechanism of interaction between arsenic solution species and the mackinawite surface.

Uranium and neptunium provide contrasting examples of the behaviour of dissolved radioactive elements on interaction with the mackinawite surface. Uranium, introduced as the uranyl ion, shows bulk behaviour similar to that exhibited by copper, being very effectively removed from solution by mackinawite (Moyes *et al.*, 2000). However, the environment of the metal at the FeS surface, as determined from EXAFS data, contrasts with that of copper. The uranium is clearly bonded to oxygen with U–O distances characteristic of those expected for the uranyl ion in the first shell, and a further four bonded oxygen atoms at 2.1–2.3 Å. Beyond these shells are further oxygen atoms at 2.9 Å, and possibly iron at ~4 Å (Moyes *et al.*, 2000). Great care was taken to avoid introduction of contaminant oxygen in these experiments; nevertheless, it was suggested that a redox reaction takes place at the FeS surface on interaction with the uranyl ion, leading to precipitation of a U_3O_8-type phase. Support for this suggestion comes from studying the spectra as a function of increasing total uranium concentrations; a systematic decrease in relative intensity of the peak at 1.79 Å, characteristic of the uranyl species, occurs and there is also a systematic decrease in the second shell U–O distance from a value of ~2.2 Å to 2.1 Å, characteristic of reduced uranium (Moyes *et al.*, 2000).

The interaction of dissolved pentavalent neptunium with the surface of mackinawite (Moyes *et al.*, 2002) provides an interesting contrast with uranium. The uptake of neptunium from solution is relatively low (~10%) and independent of solution concentration over the range 1000–20,000 ppm. The X-ray absorption spectra for the Np *L*-edge are also essentially the same over this range of concentrations. The best fits to these data involve neptunium being coordinated to four oxygen atoms at ~2.25 Å, two sulfur atoms at ~2.6 Å, and iron atoms at ~3.9 and 4.1 Å. The interpretations of these data are that, on interaction with the sulfide surface, a reduction of neptunium (V) to neptunium (IV) occurs, accompanied by a loss of axial oxygen atoms. The neptunium then bonds directly to sulfur atoms at the FeS surface. This formation of an inner sphere surface complex would explain the limited capacity of FeS to take up neptunium, in contrast to the behaviour of uranium, a result with important implications for the mobility of this element in environments where FeS occurs.

6. Applications in clean technologies

The areas of clean technologies associated with sulfides and bacterial redox processes are diverse and range from the large and well established areas of bioleaching and biomining of sulfide ores through to the novel technological applications of biologically produced nano-sulfides. Metal sulfides as nanomaterials have interesting and useful electronic and optical properties, making them useful in applications such as light-emitting diodes, gas sensors or solar cells (Lai *et al.*, 2012) and nanotubes of arsenic sulfide, which have previously been synthesized using *Shewanella* sp. HN-41, from the reduction of As(V) and $S_2O_3^{2-}$, and found to have novel electrical and photoconductive properties (Lee *et al.*, 2007).

Bioleaching of sulfides as a simple process of metal recovery from mined ores has been practiced for many centuries. In fact, long before the role of microbes in the process was properly understood, metals such as copper were being recovered in this way (Ehrlich, 2001). In contrast to this treatment of ores which have already been mined, the more recently established industry of biomining involves using the ability of bacteria to oxidize metals in the mining itself (Johnson, 2010; Brierley and Brierley, 2013). Biomining was originally based on the discovery in the 1940s that *Acidithiobacillus ferrooxidans* could oxidize Fe(III) to Fe(II) under acidic conditions. Hence, the notion that metal sulfide ores could be 'mined' through oxidative dissolution by bacteria, and which has now been employed by the mining industry for over half a century (Johnson, 2010; Brierley and Brierley, 2013). The sulfur cycle under biomining conditions is shown in Fig. 5 and shows that, initially, the role of bacteria is in the oxidation of Fe(III) to Fe(II), followed by oxidation of sulfur species to sulfate (Johnson, 2010). Biomining of ores is particularly economically effective on ores of low grade (low metal content), or refractory ores which are difficult to process, or those difficult to mine. It is increasingly important as high-grade ore bodies become exhausted. In addition, biomining has lower environmental impact than more traditional methods of mining. Biomining encompasses bioleaching, which is the dissolution of base metals such as zinc and copper, as well as biooxidation, which applies to the release of precious metals such as gold from sulfide minerals (Brierley and Brierley, 2013). Copper is the most extensively biomined base metal, and there are more than 18 active copper biomines in operation today. The majority are situated in South America where there is a prevalence of deposits well suited to bioleaching technologies; see Brierley and Brierley (2013) for further details.

Biogenic sulfate reduction, to produce hydrogen sulfide and capture toxic metals such as cadmium through the formation of insoluble sulfides, is also an established industrial process (Hulshoff Pol *et al.*, 1998). This process has been used to remediate mining site run-off, particularly in regard to AMD situations. Here, sulfate-reducing

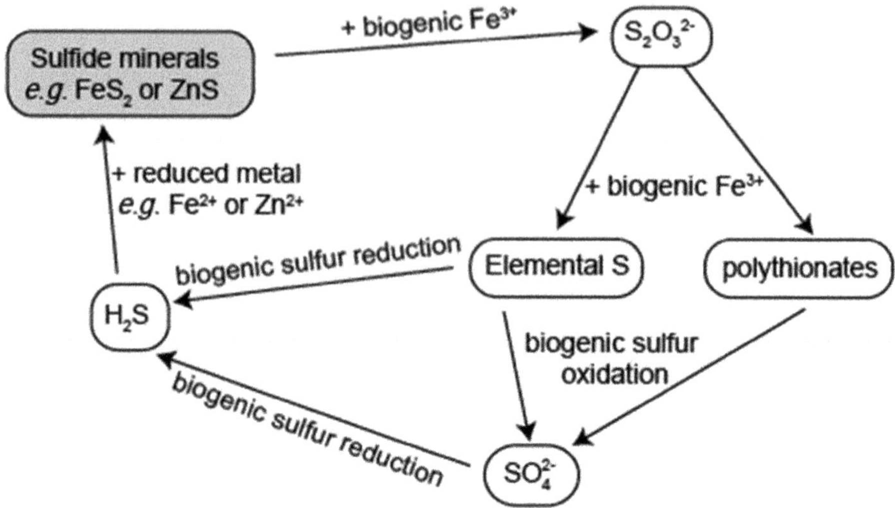

Figure 5. The sulfur cycle under biomining acidic conditions (after Johnson, 2010).

bacteria are stimulated with a suitable carbon source in order to generate sulfide. The sulfide is then re-combined with the AMD waters containing heavy metal contamination causing a rise in pH and the precipitation of metal sulfides. Finally, any excess sulfide is oxidized to sulfur using S-oxidizing bacteria (Muyzer and Stams, 2008; Tang *et al.*, 2009). This process can also be used to process leachate from biomining operations, and an example of this is the THIOTEQ™ system patented by the Dutch company Paques (Johnson, 2010).

Overall, the importance of the redox-cycling of sulfur using sulfate reducing and both metal oxidizing and reducing bacteria is of significant environmental consequence, and involves processes of considerable complexity. Sulfide minerals are potential sources of pollution, and also sources of economically and strategically important metals. Bacteria play an important role in the processes causing pollution but they also can play a major role in the remediation of such pollution and in clean technologies for metal extraction from complex orebodies.

References

Acharya, H.N. and Paul, A. (1991) Thermodynamic analysis of selective reduction of lead-oxide in galena. *Journal of Materials Science Letters*, **10**, 257–259.

Attia, Y. and El Zeky, M. (1990) Bioleaching of non-ferrous sulfides with adapted thiophilic bacteria. *Chemical Engineering Journal*, **44**, B31–B40.

Balashova, V.V. and Zavarzin, G.A. (1980) Anaerobic reduction of ferric iron by hydrogen bacteria. *Microbiology*, **48**, 635–639.

Bang, S.S., Deshpande, S.S. and Han, K.N. (1995) The oxidation of galena using *Thiobacillus ferrooxidans*. *Hydrometallurgy*, **37**, 181–192.

Bostick, B.C. and Fendorf, S. (2003) Arsenite sorption on troilite (FeS) and pyrite (FeS_2). *Geochimica et Cosmochimica Acta*, **67**, 909–921.

Bostick, B.C., Fendorf, S. and Fendorf, M. (2000) Disulfide disproportionation and CdS formation upon cadmium sorption on FeS_2. *Geochimica et Cosmochimica Acta*, **64**, 247–255.

Bowles, J.F.W., Howie, R.A., Vaughan, D.J. and Zussman, J. (2011) *Non-silicates*. Rock-forming Minerals. Vol. 5A. Geological Society, London

Brierley, C.L. and Brierley, J.A. (2013) Progress in bioleaching. Part B: applications of microbial processes by the minerals industries. *Applied Microbiology and Biotechnology*, **97**, 7543–7552.

Caccavo Jr, F., Coates, J.D., Rosello-Mora, R.A., Ludwig, W., Schleifer, K.H., Lovley, D.R. and McInerney, M.J. (1996) *Geovibrio ferrireducens*, a phylogenetically distinct dissimilatory Fe(III)-reducing bacterium. *Archives of Microbiology*, **165**, 370–376.

Canfield, D.E. and Raiswell, R. (1999) The evolution of the sulfur cycle. *American Journal of Science*, **299**, 697–723.

Chernyshova, I.V. (2003) An in situ FTIR study of galena and pyrite oxidation in aqueous solution. *Journal of Electroanalytical Chemistry*, **558**, 83–98.

Childers, S.E., Ciufo, S. and Lovley, D.R. (2002) *Geobacter metallireducens* accesses insoluble Fe(III) oxide by chemotaxis. *Nature*, **416**, 767–769.

Coates, J.D., Ellis, D.J., Gaw, C.V. and Lovley, D.R. (1999) *Geothrix fermentans* gen. nov., sp. nov., a novel Fe(III)-reducing bacterium from a hydrocarbon-contaminated aquifer. *International Journal of Systematic Bacteriology*, **49**, 1615–1622.

Corkhill, C.L. and Vaughan, D.J. (2009) Arsenopyrite oxidation – a review. *Applied Geochemistry*, **24**, 2342–2361.

Corkhill, C.L., Wincott, P.L., Lloyd, J.R. and Vaughan, D.J. (2008) The oxidative dissolution of arsenopyrite (FeAsS) and enargite (Cu_3AsS_4) by *Leptospirillum ferrooxidans*. *Geochimica et Cosmochimica Acta*, **72**, 5616–5633.

Cutting, R.S., Coker, V.S., Fellowes, J.W., Lloyd, J.R. and Vaughan, D.J. (2009) Mineralogical and morphological constraints on the reduction of Fe(III) minerals by *Geobacter sulfurreducens*. *Geochimica et Cosmochimica Acta*, **73**, 4004–4022.

da Silva, G. (2004) Kinetics and mechanism of the bacterial and ferric sulphate oxidation of galena. *Hydrometallurgy*, **75**, 99–110.

Dopson, M. and Johnson, D.B. (2012) Biodiversity, metabolism and applications of acidophilic, sulfur-metabolizing microorganisms. *Environmental Microbiology*, **14**, 2620–2631.

Ehrhardt, J.J., Behra, P., Bonnissel-Gissinger, P. and Alnot, M. (2000) XPS study of the sorption of Hg(II) onto pyrite FeS_2. *Surface and Interface Analysis*, **30**, 269–272.

Ehrlich, H. (2001) Past, present and future of biohydrometallurgy. *Hydrometallurgy*, **51**, 127–134.

Ehrlich, H. and Newman, D. (2009) *Geomicrobiology* 5th Edition. CRC Press Taylor Francis Group, Boca Raton, Florida, USA.

England, K.E.R., Charnock, J.M., Pattrick, R.A.D. and Vaughan, D.J. (1999) Surface oxidation studies of chalcopyrite and pyrite by glancing-angle X-ray absorption spectroscopy (REFLEXAFS). *Mineralogical Magazine*, **63**, 559–566.

Farquhar, M.L., Charnock, J.M., Livens, F.R. and Vaughan, D.J. (2002) Mechanisms of arsenic uptake from aqueous solution by interaction with goethite, lepidocrocite, mackinawite, and pyrite: An X-ray absorption spectroscopy study. *Environmental Science & Technology*, **36**, 1757–1762.

Farquhar, M.L., Wincott, P.L., Wogelius, R.A. and Vaughan, D.J. (2003) Electrochemical oxidation of the chalcopyrite surface: an XPS and AFM study in solution at pH 4. *Applied Surface Science*, **218**, 34–43.

Fornasiero, D., Li, F.S. and Ralston, J. (1994) Oxidation of galena. 2. Electrokinetic study. *Journal of Colloid and Interface Science*, **164**, 345–354.

Frankel, R.B. and Bazlinski, D.A. (2003) Biologically induced mineralisation by bacteria. Pp. 95–114 in: *Biomineralization* (P.M. Dove, J.J. De Yoreo and S. Weiner, editors). Reviews in Mineralogy and Geochemistry, **54**. Mineralogical Society of America and The Geochemical Society, Washington D.C.

Gilbert, P., Abrecht, M. and Frazer, B.H. (2005) The organic-mineral interface in biominerals. *Molecular Geomicrobiology*, **59**, 157–185.

Gleisner, M., Herbert Jr, R.B. and Frogner Kockum, P.C. (2006) Pyrite oxidation by *Acidithiobacillus ferrooxidans* at various concentrations of dissolved oxygen. *Chemical Geology*, **225**, 16–29.

Grein, F., Ramos, A.R., Venceslau, S.S. and Pereira, I.A.C. (2013) Unifying concepts in anaerobic respiration: Insights from dissimilatory sulfur metabolism. *Biochimica et Biophysica Acta (BBA) - Bioenergetics*, **1827**, 145–160.

Grimes, S.T., Brock, F., Rickard, D., Davies, K.L., Edwards, D., Briggs, D.E.G. and Parkes, R.J. (2001) Understanding fossilization: Experimental pyritization of plants. *Geology*, **29**, 123–126.

Hacquard, E., Bessiere, J., Alnot, M. and Ehrhardt, J.J. (1999) Surface spectroscopic study of the adsorption of Ni(II) on pyrite and arsenopyrite at pH 10. *Surface and Interface Analysis*, **27**, 849–860.

Hallberg, K. and Johnson, D. (2001) Biodiversity of acidophilic prokaryotes. *Advances in Applied Microbiology*, **49**, 37–84.

Harneit, K., Goksel, A., Kock, D., Klock, J., Gehrke, T. and Sand, W. (2006) Adhesion to metal sulfide surfaces by cells of *Acidithiobacillus ferrooxidans*, *Acidithiobacillus thiooxidans* and *Leptospirillum ferrooxidans*. *Hydrometallurgy*, **83**, 245–254.

Hedrich, S., Schlomann, M. and Johnson, D. (2011) The iron-oxidizing proteobacteria. *Microbiology*, **157**, 1551–1564.

Hudson-Edwards, K. and Vaughan, D.J. (2017) *Environmental Minerals*. Wiley-Blackwell, Oxford, UK, in press.

Hulshoff Pol, L., Lens, P.L., Stams, A.M. and Lettinga, G. (1998) Anaerobic treatment of sulphate-rich wastewaters. *Biodegradation*, **9**, 213–224.

Janczuk, B., Wojcik, W., Zdziennicka, A. and Gonzalezcaballero, F. (1992a) Components of surface free-energy of galena. *Journal of Materials Science*, **27**, 6447–6451.

Janczuk, B., Wojcik, W., Zdziennicka, A. and Gonzalezcaballero, F. (1992b) Determination of the galena free-energy components from contact-angle measurments. *Materials Chemistry and Physics*, **31**, 235–241.

Johnson, D.B. (2010) The biogeochemistry of biomining. Pp. 401–426 in: *Geomicrobiology: Molecular and Environmental Perspective* (L.L. Barton, A. Loy and M. Mandl, editors). Springer, Berlin.

Johnson, D.B. (2011) Geomicrobiology of extremely acidic subsurface environments. *FEMS Microbiology Ecology*, **81**, 2–12.

Jones, R.A., Koval, S.F. and Nesbitt, H.W. (2003) Surface alteration of arsenopyrite (FeAsS) by *Thiobacillus ferrooxidans*. *Geochimica et Cosmochimica Acta*, **67**, 955–965.

Jorgensen, B.B. (1982) Mineralization of organic-matter in the sea bed – the role of sulfate reduction. *Nature*, **296**, 643–645.

Kappler, A. and Newman, D.K. (2004) Formation of Fe(III)-minerals by Fe(II)-oxidizing photoautotrophic bacteria. *Geochimica et Cosmochimica Acta*, **68**, 1217–1226.

Kappler, A. and Straub, K.L. (2005) Geomicrobiological cycling of iron. Pp. 85–108 in: *Molecular Geomicrobiology* (J.F. Banfield, J. Cervini-Silva and K.N. Nealson, editors). Reviews in Mineralogy and Geochemistry, **59**. Mineralogical Society of America and The Geochemical Society, Washington D.C.

Kelly, D. and Wood, A. (2000) Reclassification of some species of *Thiobacillus* to the newly designated genera *Acidithiobacillus* gen. nov., *Halothiobacillus* gen. nov. and *Thermithiobacillus* gen. nov. *International Journal of Systematic and Evolutionary Microbiology*, **50**, 511–516.

Laajalehto, K., Kartio, I. and Suoninen, E. (1997) XPS and SR-XPS techniques applied to sulfide mineral surfaces. *International Journal of Mineral Processing*, **51**, 163–170.

Lai, C.-H., Lu, M.-Y. and Chen, L.-J. (2012) Metal sulfide nanostructures: synthesis, properties and applications in energy conversion and storage. *Journal of Materials Chemistry*, **22**, 19–30.

Lara, R.H., Garcia-Meza, J.V., Gonzalez, I. and Cruz, R. (2013) Influence of the surface speciation on biofilm attachment to chalcopyrite by *Acidithiobacillus thiooxidans*. *Applied Microbiology and Biotechnology*, **97**, 2711–2724.

Lee, J.H., Kim, M.G., Yoo, B.Y., Myung, N.V., Maeng, J.S., Lee, T., Dohnalkova, A.C., Fredrickson, J.K., Sadowsky, M.J. and Hur, H.G. (2007) Biogenic formation of photoactive arsenic-sulfide nanotubes by Shewanella sp strain HN-41. *Proceedings of the National Academy of Sciences of the United States of America*, **104**, 20410–20415.

Lei, J., Huaiyang, Z., Xiaotong, P. and Zhonghao, D. (2009) The use of microscopy techniques to analyze microbial biofilm of the bio-oxidized chalcopyrite surface. *Minerals Engineering*, **22**, 37–42.

Lennie, A.R., Redfern, S.A.T., Champness, P.E., Stoddart, C.P., Schofield, P.F. and Vaughan, D.J. (1997)

Transformation of mackinawite to greigite: An in situ X-ray powder diffraction and transmission electron microscope study. *American Mineralogist*, **82**, 302–309.

Lloyd, J.R. (2003) Microbial reduction of metals and radionuclides. *FEMS Microbiology Reviews*, **27**, 411–425.

Lloyd, J.R. (2005) Mechanisms and environmental impact of microbial metal reduction. Pp. 273–302 in: *Micro-organisms and Earth systems – Advances in Geomicrobiology* (G.M. Gadd, K.T. Semple and H.M. Lappin-Scott, editors). Symposia of the Society of General Microbiology, **65**, Cambridge University Press, New York.

Lonergan, D.J., Jenter, H., Coates, J.D., Phillips, E.J.P., Schmidt, T.M. and Lovley, D.R. (1996) Phylogenetic analysis of dissimilatory Fe(III)-reducing bacteria. *Journal of Bacteriology*, **178**, 2402–2408.

Lovley, D.R. (1993) Dissimilatory metal reduction. *Annual Reviews in Microbiology*, **47**, 263–290.

Lovley, D.R., Stoltz, J.F., Nord Jr, G.L. and Phillips, E.J.P. (1987) Anaerobic production of magnetite by a dissimilatory iron-reducing microorganism. *Nature*, **330**, 252–254.

Lovley, D.R., Giovannoni, S.J., White, D.C., Champine, J.E., Phillips, E.J.P., Gorby, Y.A. and Goodwin, S. (1993) *Geobacter metallireducens* gen. nov. sp. nov., a microorganism capable of coupling the complete oxidation of organic coumpounds to the reduction of iron and other metals. *Archives of Microbiology*, **159**, 336–344.

Lovley, D.R., Coates, J.D., Saffarini, D. and Lonergan, D.J. (1997) Diversity of dissimilatory Fe(III)-reducing bacteria. Pp. 187–215 in: *Iron and Related Transition Metals in Microbial Metabolism* (G. Winkelman and C.J. Carrano, editors). Harwood Academic Publishers, Chur, Switzerland.

Lovley, D.R., Fraga, J.L., Blunt-Harris, E.L., Hayes, L.A., Phillips, E.J.P. and Coates, J.D. (1998) Humic substances as a mediator for microbially catalyzed metal reduction. *Acta Hydrochimica et Hydrobiologica*, **26**, 152–157.

Lovley, D.R., Holmes, D.E. and Nevin, K.P. (2004) Dissimilatory Fe(III) and Mn(IV) reduction. *Advances in Microbial Physiology*, **49**, 219–286.

Luther, G.W. and Rickard, D.T. (2005) Metal sulfide cluster complexes and their biogeochemical importance in the environment. *Journal of Nanoparticle Research*, **7**, 389–407.

Moyes, L.N., Parkman, R.H., Charnock, J.M., Vaughan, D.J., Livens, F.R., Hughes, C.R. and Braithwaite, A. (2000) Uranium uptake from aqueous solution by interaction with goethite, lepidocrocite, muscovite and mackinawite: An X-ray absorption spectroscopy study. *Environmental Science & Technology*, **34**, 1062–1068.

Moyes, L.N., Jones, M.J., Reed, W.A., Livens, F.R., Charnock, J.M., Mosselmans, J.F.W., Hennig, C., Vaughan, D.J. and Pattrick, R.A.D. (2002) An X-ray absorption spectroscopy study of neptunium (V) reactions with mackinawite (FeS). *Environmental Science & Technology*, **36**, 179–183.

Muyzer, G. and Stams, A.J.M. (2008) The ecology and biotechnology of sulphate-reducing bacteria. *Nature Reviews Microbiology*, **6**, 441–454.

Myers, C.R. and Myers, J.M. (1992) Localisation of cytochromes to the outer membrane of anaerobically grown *Shewanella putrefaciens* MR-1. *Journal of Bacteriology*, **174**, 3429–3438.

Myers, C.R. and Myers, J.M. (1997) Outer membrane cytochromes of *Shewanella putrefaciens* MR-1: spectral analysis, and purification of the 83-kDa c-type cytochrome. *Biochimica et Biophysica Acta*, **1326**, 307–318.

Nealson, K.H. and Saffarini, D. (1994) Iron and manganese in anaerobic respiration: Environmental significance, physiology, and regulation. *Annual Reviews in Microbiology*, **48**, 311–343.

Parkman, R.H., Curtis, C.D., Vaughan, D.J. and Charnock, J.M. (1996) Metal fixation and mobilisation in the sediments of the Afon Goch estuary – Dulas Bay, Anglesey. *Applied Geochemistry*, **11**, 203–210.

Parkman, R.H., Charnock, J.M., Bryan, N.D., Livens, F.R. and Vaughan, D.J. (1999) Reactions of copper and cadmium ions in aqueous solution with goethite, lepidocrocite, mackinawite, and pyrite. *American Mineralogist*, **84**, 407–419.

Pentráková, L., Su, K., Pentrák, M. and Stucki, J. W. (2013) A review of microbial redox interactions with structural Fe in clay minerals. *Clay Minerals*, **48**, 543–560.

Pfennig, N. and Biebl, H. (1976) *Desulfuromonas acetoxidans* gen. nov. and sp. nov., a new anaerobic sulfur reducing, acetate oxidising bacterium. *Archives of Microbiology*, **110**, 3–12.

Pronk, J., de Bruyn, J., Bos, P. and Kuenen, J. (1992) Anaerobic growth of *Thiobacillus ferrooxidans*. *Applied and Environmental Microbiology*, **58**, 2227–2230.

Rawlings, D.E. and Johnson, P.B. (2007) *Biomining*. Springer-Verlag, Berlin, 314 pp.

Reguera, G., McCarthy, K.D., Mehta, T., Nicoll, J.S., Tuominen, M.T. and Lovley, D.R. (2005) Extracellular electron transfer via microbial nanowires. *Nature*, **435**, 1098–1101.

Richardson, P.E. and Odell, C.S. (1984) Semiconducting characteristics of galena electrodes. *Journal of the Electrochemical Society*, **131**, C99–C99.

Rickard, D. and Luther, G.W. (2006) Metal sulfide complexes and clusters. Pp. 421–504 in: *Sulfide Mineralogy and Geochemistry* (D.J. Vaughan, editor). Reviews in Mineralogy and Geochemistry, **61**. Mineralogical Society of America and The Geochemical Society, Washington D.C.

Rickard, D. and Morse, J.W. (2005) Acid volatile sulfide (AVS). *Marine Chemistry*, **97**, 141–197.

Rimstidt, J.D. and Vaughan, D.J. (2003) Pyrite oxidation: A state-of-the-art assessment of the reaction mechanism. *Geochimica et Cosmochimica Acta*, **67**, 873–880.

Roden, E.E. and Lovley, D.R. (1993) Dissimilatory Fe(III) reduction by the marine microorganism, *Desulfuromonas acetoxidans*. *Applied and Environmental Microbiology*, **59**, 734–742.

Rosso, K.M. and Vaughan, D.J. (2006) Reactivity of sulfide mineral surfaces. Pp. 557–607 in: *Sulfide Mineralogy and Geochemistry* (D.J. Vaughan, editor). Reviews in Mineralogy and Geochemistry, **61**. Mineralogical Society of America and The Geochemical Society, Washington D.C.

Sand, W., Gehrke, T., Hallman, R. and Schippers, A. (1995) Sulfur chemistry, biofilm, and the indirect attack mechanism – a critical evaluation of bacterial leaching. *Applied Microbiology and Biotechnology*, **43**, 961–966.

Sand, W., Gehrke, T., Jozsa, P. and Schippers, A. (2001) (Bio)chemistry of bacterial leaching – direct vs. indirect bioleaching. *Hydrometallurgy*, **59**, 159–175.

Schippers, A. and Sand, W. (1999) Bacterial leaching of metal sulfides proceeds by two indirect mechanisms via thiosulfate or via polysulfides and sulfur. *Applied and Environmental Microbiology*, **65**, 319–321.

Schoonen, M.A.A. (2004) Mechanisms of sedimentary pyrite formation. Pp. 117–134 in: *Sulfur Biogeochemistry – Past and Present* (J. Amend, K. Edwards. and T. Lyons, editors). **379**, Geological Society of America Special Paper, **379**.

Silvermann, M. and Ehrlich, H. (1964) Microbial formation and degradation of minerals. *Advanced and Applied Microbiology*, **6**, 153–206.

Singer, P.C. and Stumm, W. (1970) Acid mine drainage: the rate determining step. *Science*, **167**, 1121–1123.

Tang, K., Baskaran, V. and Nemati, M. (2009) Bacteria of the sulfur cycle: An overview of microbiology, biokinetics and their role in petroleum and mining industries. *Biochemical Engineering Journal*, **44**, 73–94.

Thauer, R.K., Stackebrandt, E. and Hamilton, W.A. (2007) Energy metabolism and phylogenetic diversity of sulphate-reducing bacteria. Pp. 1–37 in: *Sulphate-Reducing Bacteria: Environmental and Engineered Systems* (L.L. Barton and W.A. Hamilton, editors). Cambridge University Press, Cambridge, UK.

Todd, E.C., Sherman, D.M. and Purton, J.A. (2003) Surface oxidation of pyrite under ambient atmospheric and aqueous (pH=2 to 10) conditions: Electronic structure and mineralogy from X-ray absorption spectroscopy. *Geochimica et Cosmochimica Acta*, **67**, 881–893.

Torma, A. and Subramanian, K. (1974) Selective bacterial leaching of a lead sulphide concentrate. *International Journal of Mineral Processing*, **1**, 125–134.

Tossell, J.A. and Vaughan, D.J. (1987) Electronic-structure and the chemical-reactivity of the surface of galena. *The Canadian Mineralogist*, **25**, 381–392.

Turcotte, S.B., Benner, R.E., Riley, A.M., Li, J., Wadsworth, M.E. and Bodily, D.M. (1993) Surface-analysis of electrochemically oxidized metal sulfides using Raman spectroscopy. *Journal of Electroanalytical Chemistry*, **347**, 195–205.

Vargas, M., Kashefi, K., Blunt-Harris, E.L. and Lovley, D.R. (1998) Microbiological evidence for Fe(III) on early Earth. *Nature*, **395**, 65–67.

Vaughan, D.J. (editor) (2006) *Sulfide Mineralogy and Geochemistry*. Reviews in Mineralogy and Geochemistry, **61**. Mineralogical Society of America and The Geochemical Society, Washington D.C.

Vaughan, D.J. and Craig, J.R. (1978) *Mineral Chemistry of Metal Sulfides*. Cambridge University Press, Cambridge, UK, 493 pp.

Vaughan, D.J. and Lennie, A.R. (1991) The iron sulfide minerals – their chemistry and role in nature. *Science Progress*, **75**, 371–388.

Vaughan, D.J., England, K.E.R., Kelsall, G.H. and Yin, Q. (1995) Electrochemical oxidation of chalcopyrite ($CuFeS_2$) and the related metal-enriched derivatives $Cu_4Fe_5S_8$, $Cu_9Fe_9S_{16}$ and $Cu_9Fe_8S_{16}$. *American Mineralogist*, **80**, 725–731.

Vera, M., Schippers, A. and Sand, W. (2013) Progress in bioleaching: fundamentals and mechanisms of bacterial metal sulfide oxidation – part A. *Applied Microbiology and Biotechnology*, **97**, 7529–7541.

Weber, K.A., Picardal, F.W. and Roden, E.E. (2001) Microbially catalyzed nitrate-dependent oxidation of biogenic solid-phase Fe(II) compounds. *Environmental Science & Technology*, **35**, 1644–1650.

Wilkins, M.J., Livens, F.R., Vaughan, D.J. and Lloyd, J.R. (2006) The impact of Fe(III)-reducing bacteria on uranium mobility. *Biogeochemistry*, **78**, 125–150.

Williamson, M.A. and Rimstidt, J.D. (1994) The kinetics and electrochemical rate-determining step of aqueous pyrite oxidation. *Geochimica et Cosmochimica Acta*, **58**, 5443–5454.

Wolthers, M., Charlet, L. and van der Weijden, C.H. (2003) Arsenic sorption onto disordered mackinawite as a control on the mobility of arsenic in the ambient sulphidic environment. *Journal de Physique IV*, **107**, 1377–1380.

Wolthers, M., Charlet, L., Van der Weijden, C.H., Van der Linde, P.R. and Rickard, D. (2005) Arsenic mobility in the ambient sulfidic environment: Sorption of arsenic(V) and arsenic(III) onto disordered mackinawite. *Geochimica et Cosmochimica Acta*, **69**, 3483–3492.

Yin, Q., Vaughan, D.J., England, K.E.R., Kelsall, G.H. and Brandon, N.P. (2000) Surface oxidation of chalcopyrite ($CuFeS_2$) in alkaline solutions. *Journal of the Electrochemical Society*, **147**, 2945–2951.

Iron minerals as archives of Earth's redox and biogeochemical evolution

CAROLINE L. PEACOCK[1], STEFAN V. LALONDE[2] and KURT O. KONHAUSER[3]

[1] University of Leeds, School of Earth and Environment, Leeds LS2 9JT, UK, e-mail: C.L.Peacock@leeds.ac.uk
[2] European Institute for Marine Studies, CNRS-UMR6538 Laboratoire Domaines Océaniques UMR 6538, Brest, France, e-mail: Stefan.Lalonde@univ-brest.fr
[3] University of Alberta, Department of Earth and Atmospheric Sciences, Edmonton T6G 2E3 Canada, e-mail: KurtK@ualberta.ca

Iron minerals provide sedimentary repositories of chemical information pertaining to Earth's redox and biogeochemical evolution, from before the Great Oxidation Event some 2.5 billion years ago, to more recent events occurring up to and into the Cenozoic Era. The most powerful chemical information recorded in iron minerals comes in the form of trace-element signatures, most notably their concentrations and stable isotope compositions. Here we provide an introduction to iron mineralogy and the processes responsible for the accumulation and preservation of trace-element signatures in iron minerals, focusing on the deposition of iron minerals in three key ancient sedimentary archives: banded iron formations, ferromanganese crusts and black shales. We introduce the theory and practical use of non-traditional trace-element stable-isotope systems in redox and biogeochemical research, focusing on the recent use of iron, molybdenum and chromium stable isotopes to shed light on the redox and biogeochemical information stored in iron-rich sediments. By analysing both trace-element concentrations and stable-isotope compositions recorded in iron minerals, iron-rich sedimentary archives are providing a unique window into the past, where changes in trace-element signatures shed light on major transitions in Earth's redox and biogeochemical evolution.

1. Introduction

This chapter provides a review of the use of iron minerals as repositories of chemical information pertaining to Earth's redox and biogeochemical evolution, from before the Great Oxidation Event some 2.5 billion years ago, to more recent events occurring up to and into the Cenozoic Era. The most powerful chemical information recorded in iron minerals comes in the form of trace-element signatures, most notably their concentrations and stable isotope compositions. We begin our review by providing an introduction to iron mineralogy and the processes responsible for the accumulation and preservation of trace-element signatures in iron minerals, focusing on the deposition of these minerals in three key ancient sedimentary archives: banded iron formations (BIF), ferromanganese crusts and black shales. We then review the theory

DOI: 10.1180/EMU-notes.17.14

and practical use of non-traditional trace element stable isotope geochemistry, and in particular the use of emerging non-traditional stable isotope systems, in palaeo redox and biogeochemical research. Finally, we focus on how different trace element signatures, recorded in those ancient sedimentary archives, have helped to provide a window into the past, where changes in trace element signatures can be interpreted to reflect transitions in Earth's redox and biogeochemical evolution.

2. Iron minerals as archives of chemical information

Iron (Fe) minerals exist over a very wide range of redox conditions. They are extremely efficient scavengers of other elements, and effectively archive chemical information about the fluids from which they form. In the marine realm, Fe minerals form chemical sediments and cements under a variety of aqueous conditions and geological settings; these range from rapid water-column precipitation in Fe-rich hydrothermal plumes upon mixing with bulk seawater to the slow formation of concretions and cements at redox boundaries in sediments, nodules and hydrothermal chimneys. The ubiquity of Fe oxyhydroxides in the marine environment, both modern and ancient, and the fact that their composition is intimately linked to the chemistry of the waters from which they precipitate, make them especially promising recorders of the geochemical history of their formation waters. Specifically, the elemental and isotopic enrichments recorded in Fe minerals provide information on the redox state of the contemporaneous environment and the biogeochemical cycling of bio-essential metals, both of which are linked to primary productivity, carbon cycling, and ultimately the evolution of Earth's surface environments through time.

2.1. Iron mineralogy as a function of redox

Iron is the fourth most abundant element in the Earth's crust, where it exists predominantly as ferrous iron, Fe(II), in Fe-bearing silicate minerals (*e.g.* pyroxenes, amphiboles). As these minerals are weathered, Fe(II) is released into solution, largely oxidized to ferric iron, Fe(III), and then delivered to the oceans. Transport involves a continuum of different size fractions, defined as aqueous (<0.02 μm, mainly organically bound Fe), nanoparticulate (0.02–1 μm, mainly Fe(III) oxyhydroxide minerals that are also closely associated with organic species), and larger particles (>1 μm); for further explanation of Fe size fractions see Raiswell and Canfield (2012). There is a myriad of different processes delivering Fe to the oceans, and the relative magnitudes of the associated Fe fluxes vary both spatially and temporally (recently reviewed by Boyd and Ellwood, 2010; Raiswell and Canfield, 2012). In order of importance, processes supplying Fe to the global ocean include deposition of windborne dust (*e.g.* Jickells *et al.*, 2005), melting of glaciers and icebergs (*e.g.* Raiswell *et al.*, 2008), remobilization of Fe from submarine shallow-shelf systems (*e.g.* Poulton and Raiswell, 2002), Fe transport *via* rivers (*e.g.* MacKenzie *et al.*, 1979), and submarine hydrothermal emissions (*e.g.* Tagliabue *et al.*, 2010; Table 1). A significant proportion of the riverine Fe flux is probably removed *via* biogeochemical processes in

Table 1. Fe sources and estimated fluxes to the global ocean, compiled and discussed extensively by Raiswell and Canfield (2012).

Fe sources to the global ocean	Fe flux (Tg/y)
Aeolian	1.53–3.03
Icebergs	0.9–1.38
Sediment recycling	0.05–0.25
Riverine	0.14
Hydrothermal	0.05
Total	2.62–4.8

estuaries, while the remainder is deposited on the continental shelf (*e.g.* Boyle *et al.*, 1977; Boyd and Ellwood, 2010).

Once Fe reaches the oceans, and depending on prevailing geochemical conditions, it can be (re)precipitated *via* a number of different mechanisms as a variety of hydrogeneous and typically nanoparticulate Fe(II/III) minerals (for a review see Raiswell and Canfield, 2012). Because Fe is redox active, these minerals can form under almost all water-column redox conditions. In addition, many of these minerals have considerable surface area with an amphoteric surface charge (*i.e.* can be either positively or negatively charged depending on aqueous pH), meaning that they are highly reactive towards both dissolved cations and anions. As such, hydrogeneous Fe minerals make ideal sedimentary archives of trace-element concentrations and stable isotope compositions in contemporaneous seawater, recording elemental signatures over a wide range of redox regimes.

Under fully oxic water column conditions, when the concentration of dissolved O_2 is greater than ~10^{-6} molar, Fe(II) is oxidized to Fe(III), which precipitates primarily as the nanoparticulate and poorly crystalline Fe(III) oxyhydroxide, ferrihydrite $[Fe^{3+}_{4-5}(OH,O)_{12}]$. Ferrihydrite is metastable and ages to the more crystalline minerals goethite [α-FeOOH] or hematite $[Fe_2O_3]$ (Cornell and Schwertmann, 2003). Under anoxic conditions, when the concentration of dissolved O_2 is $<\sim 10^{-6}$ molar, a variety of different Fe(II/III) minerals are formed. In non-sulfidic anoxic conditions (total dissolved sulfide $<10^{-6}$ molar) a variety of non-sulfidized Fe minerals can precipitate, including ferrihydrite and/or ferrous carbonates (*e.g.* siderite $[FeCO_3]$) and ferrous phosphates (*e.g.* vivianite $[Fe_3(PO_4)_2 \cdot 8H_2O]$). In addition, a mixed Fe(II/III) layered double hydroxide can form, referred to as green rust (general formula $[Fe^{II}_{(1-x)}Fe^{III}_x(OH)_2]^{x+} \cdot [(x/n)A^{n-} \cdot (m/n)H_2O]^{x-}$ where $A^{n-} = CO_3^{2-}, SO_4^{2-}, Cl^-, OH^-$). While relatively rare today, recent research suggests that green rust might be an important reactive Fe phase in anoxic water columns that are devoid of free sulfide and rich in dissolved Fe(II) (termed ferruginous) (*e.g.* Zegeye *et al.*, 2012). Under sulfidic anoxic conditions where free sulfide is present (total dissolved sulfide $\geqslant 10^{-6}$ molar;

termed euxinic) reactive Fe is precipitated as Fe sulfides, first as metastable makinawite (FeS), which then ages to pyrite (FeS_2). The classification of oxic and anoxic environments, in terms of dissolved O_2 and sulfide, is taken from Berner (1981) who provided an extended discussion of Fe (and other) minerals that are thermodynamically stable in these geochemical regimes.

In ocean sediments Fe(II/III) minerals are subject to diagenesis, during which they can be reductively dissolved and reprecipitated. Reductive dissolution proceeds *via* two competing pathways (*e.g.* Canfield *et al.*, 1993a,b; Van Cappellen and Wang, 1996; Raiswell and Canfield, 1998; Haese, 2000), the first involving microbial reduction (dissimilatory Fe reduction, DIR) and the second involving abiotic reduction by dissolved sulfide, which diffuses upwards through the sediment column from a deeper zone of microbial sulfate reduction (*e.g.* Berner, 1964; Staubwasser *et al.*, 2006). If Fe(III) oxyhydroxides are reduced by DIR, reactive Fe can go through a repeated cycle of reductive dissolution, upward diffusion, oxidation and reprecipitation as Fe(III) oxyhydroxides or, with partial oxidation, reprecipitation as a myriad of other Fe(II/III) minerals as described above (*e.g.* Canfield, 1989; Staubwasser *et al.*, 2006). Ultimately, reduction of Fe(III) oxyhydroxides by dissolved sulfide removes Fe from the diagenetic cycle (*e.g.* Raiswell and Canfield, 1998). The diagenetic formation of Fe sulfide minerals constitutes a major sulfur and Fe exit channel that occurs at anoxic depths in the sedimentary pile where both reduction of Fe(III) and sulfate occur.

Use of Fe-rich sediments as archives of palaeo-seawater composition assumes that the Fe minerals provide a faithful record of seawater trace-element concentrations and/or stable isotope compositions, despite their diagenetic history. In some sedimentary systems, where Fe(III) concentrations exceed organic carbon, Fe(III) oxyhydroxides can be preserved in the sediments without reductive dissolution (*e.g.* Konhauser *et al.*, 2005). Under different scenarios, where reductive dissolution is likely to have occurred, it is important to demonstrate that the diagenetic processes have not significantly effected the original trace-element concentrations and/or stable isotope compositions at the bulk scale, *i.e.* that there has been minimal loss of Fe and elements of interest from the bulk archive over time, and minimal redistribution of the isotopes between the bulk archive and sedimentary pore-waters (*e.g.* Poulton and Canfield, 2006; Frierdich *et al.*, 2011; Robbins *et al.*, 2015). Indeed, in the case of some geochemical proxies, the environmental signal recorded by Fe minerals may be largely a diagenetic one (*e.g.* the case of Fe isotope compositions; Section 3.4.1).

2.2. Uptake of trace elements by iron minerals

2.2.1. Iron oxyhydroxides

Iron oxyhydroxides are highly effective sorbents, a property born out of the functional group and electrostatic charging characteristics of their surfaces. Some minerals (*e.g.* clays) possess permanent surface charge as the result of structural defects or isomorphous substitution. Others, including most hydrated oxyhydroxide minerals (whether they be Fe oxyhydroxides, silicon oxides, titanium oxides, *etc.*), develop surface charge as the result of pH-dependent surface functional group protonation/

deprotonation reactions. These minerals develop a positive surface charge at low pH as surface sites are neutralized or rendered positively charged with the surface sorption of protons and a negative charge at high pH as progressive deprotonation occurs (*i.e.* they are amphoteric). Adopting a 1pK formalism for the representation of surface sites, then surface sites of Fe oxyhydroxides (including hydrous ferric oxide, HFO, which is synonymous with ferrihydrite) develop charge through the following protonation/ deprotonation reaction:

$$>FeOH^{-0.5} + H^{+} \leftrightarrow >FeOH_2^{+0.5} \qquad pK_a = \sim 8 \tag{1}$$

where >FeOH denotes surface Fe hydroxyl functional groups (log acidity (pK_a) value for ferrihydrite given as an example and taken from *e.g.* Davis and Leckie (1978)). At pH values above and below ~8, surface sites, and thus the net charge on the mineral surface, will be predominantly negatively or positively charged as $>FeOH^{-0.5}$ or $>FeOH_2^{+0.5}$, respectively. At pH ~8, surface sites of opposite charge are present in equal amounts and this is deemed the point of zero charge (PZC). Crucially, ferrihydrite, and other pure Fe oxyhydroxide surfaces in general, are characterized by PZC values around 8, and thus possess abundant negatively and positively charged sites that are strongly reactive under many natural conditions (*e.g.* Sverjensky and Sahai, 1996). Considering their ubiquity and propensity to form nm-scale aggregates with large surface areas (in some cases upwards of 600 m^2 per dry gram, Roden *et al.*, 2003), Fe oxyhydroxides provide the most prevalent positively charged reactive mineral surface found in nature (Cornell and Schwertmann, 2003). In contrast, the surfaces of silicate minerals generally possess PZC values around 2 (*e.g.* Ihler, 1979), and are thus strongly negatively charged under most natural conditions.

The surface chemistry of Fe oxyhydroxides has long been studied in relation to the sorption of contaminants in natural waters. Equilibrium surface complexation models (SCM), employing experimentally- and/or theoretically-derived parameters (*e.g.* stability constants for (de)protonation of the surface sites, surface area, surface site density, surface capacitance), have been developed to account for Fe oxyhydroxide sorption behaviour under a wide variety of pH values, ionic strengths, and sorbent-to-sorbate ratios (for an introduction to SCM see Davies and Kent, 1990). Among the early works, Dzombak and Morel (1990) provide a comprehensive database of thermodynamic parameters that describe the experimentally observed sorption of a variety of anions and cations to HFO. These surface sorption reactions may proceed to complete surface coverage, and in some cases, sorption that exceeds surface-site saturation can lead to surface precipitation (*e.g.* Ler and Stanforth, 2003).

More recently, research on the sorption of trace elements to Fe oxyhydroxides has focused on determining the speciation and precise molecular coordination environment (*e.g.* mono- *vs.* bi-dentate, mono- *vs.* bi-nuclear) of sorbed trace elements using synchrotron-based spectroscopic techniques (for a review see Brown and Parks, 2001). Following the application of synchrotron spectroscopy to metal sorption systems, a new class of SCM has been developed by studies that first determine the molecular mechanisms of trace-element sorption, then invoke this directly observed surface complex to describe experimental trace-element sorption data (*e.g.* Peacock and

Sherman, 2004a,b; Moon and Peacock, 2011, 2012, 2013). These SCM are better able to predict trace element partitioning between solution and solid compared to those based solely on macroscopic sorption experiments. By properly defining the stoichiometry of the surface complex (*i.e.* the necessary input reaction that represents the formation of the surface complex in the model), and thus its complex-specific dependence on pH, aqueous trace element concentration and surface site density, it is possible to obtain an accurate description of its formation as a function of these same parameters. In order to apply model predictions to both modern and, in particular, ancient natural systems, where pH, trace-element concentrations and surface-site density (*i.e.* mineral crystallinity) vary over time and space, it is increasingly desirable to constrain SCM with molecular-level information whenever possible (*e.g.* for micronutrient trace element concentrations in the modern oceans, Peacock and Sherman, 2007; Sherman and Peacock, 2010).

Significant progress has been made in testing predictive models using natural Fe oxyhydroxides (*e.g.* Tessier *et al.*, 1996; Kennedy *et al.*, 2003; Lalonde *et al.*, 2007). In some cases, highly parameterized SCM may not be practical for the description of natural systems; critical information such as mineral surface area and surface site density may be difficult to determine. In these cases, lumped-process distribution coefficients, based on the Freundlich isotherm approach, are often applied. The distribution coefficient approach, which employs a linear or exponential relationship to describe the distribution of a trace element between solid and solution, benefits from simplicity and ease of execution, but, because it does not determine the speciation or coordination environment of the sorbed trace element, it cannot properly describe the formation of the surface complex and thus the sorption of the trace element as a function of the important environmental parameters (pH, trace-element concentration, *etc.*). Partition coefficients derived *via* this approach, either experimentally or by measuring trace element concentrations in contemporaneous solids and solutions are, therefore, limited to the particular aqueous conditions (fixed pH, trace-element concentration and ionic strength) present in the experiments or at the time of trace element uptake (Langmuir, 1997). Nonetheless, as we shall see below, distribution coefficient models have proven effective for the description of trace element sequestration by natural Fe oxyhydroxides, especially in marine systems (*e.g.* Feely *et al.*, 1998; Edmonds and German, 2004), and with consideration of these limitations, they have been used successfully to relate trace-element concentrations recorded in Fe mineral archives to contemporaneous seawater chemistry (*e.g.* Konhauser *et al.*, 2009).

2.2.2. Iron sulfides

Like their oxide counterparts, Fe sulfide minerals are generally insoluble, and in anoxic waters and sediment porewaters where Fe^{2+} and S^{2-} (predominantly as HS^- at marine pH) are present, they precipitate rapidly. Taking the form Fe_xS_y, seven stable minerals composed entirely of Fe and S have been described, of which four may be found in low-temperature aqueous and sedimentary environments: mackinawite (meta-stable, amorphous to tetragonal FeS), pyrite (cubic FeS_2), marcasite (orthorhombic FeS_2), and greigite (cubic Fe_3S_4). Precipitation of amorphous mackinawite proceeds faster than the other phases, such that it is often considered a precursor for its more crystalline counterparts. However, mackinawite is not a strict precursor in the sense that it does not appear to experience solid-state transformation during diagenesis, but rather undergoes dissolution and re-precipitation (*cf.* Rickard and Luther, 2007). Pyrite is the most abundant Fe sulfide mineral that is stable at Earth's surface, and like Fe oxides, its tendency to concentrate trace elements from its formation waters makes pyrite-rich sediments both an important marine trace-element sink and a rich palaeo-environmental chemical archive (*e.g.* Large *et al.*, 2014).

Similar to Fe oxides, Fe sulfide minerals present a reactive surface upon which significant trace-element sorption may occur. In the case of disordered mackinawite, the reactive surface appears to be composed of strongly acidic mono-coordinated sulfur sites ($>FeSH^0$) and weakly acidic tri-coordinated sulfur sites ($>Fe_3SH^0$), with a specific surface area of ~350 $m^2\ g^{-1}$ (Wolthers *et al.*, 2005). Few thermodynamic surface stability constants exist for trace elements at Fe-sulfide mineral surfaces; however, partition coefficients for the adsorption of metal-sulfide-forming elements to the mackinawite surface appear to scale with the solubility products of their respective metal sulfides (Morse and Arakaki, 1993). Despite mackinawite being an unstable phase, the linear free-energy relationship that characterizes its trace-element uptake appears reflected in the trace-element compositions of sedimentary pyrites.

In pyrite-bearing marine sediments, the degree of trace-element pyritization (DTMP), defined for any given trace element as the percent of that metal in a sediment that is found in pyrite (Raiswell and Canfield, 1998), also appears to scale with the respective metal sulfide solubility products (Morse and Luther, 1999; Fig. 1). In other words, the more pyrite a sediment contains, the greater the proportion of the total trace-element load will be hosted in pyrite. However, not all metals display this thermodynamically controlled behaviour; for those which may precipitate rapidly as

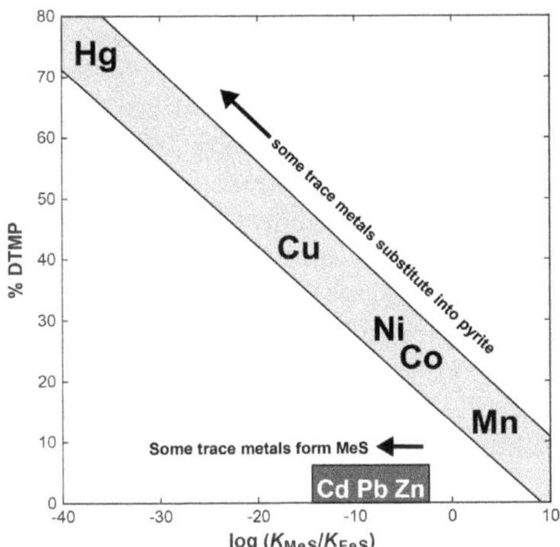

Figure 1. The degree to which different trace elements in anoxic sediments are concentrated in pyrite (% DTMP) *vs.* their own solubility as a metal sulfide (relative to FeS; log K_{MeS}/K_{FeS}). This relationship speaks to the affinity of different metals for sulfide as well as the kinetics of MeS *vs.* FeS precipitation. Some metals precipitate slowly as sulfides and thus are more likely to be found substituted into pyrite in anoxic sediments (metals on the pyritization trend), whereas others rapidly form their own metal sulfides and fall off this trend (Pb, Zn, Cd). Redrawn from Morse and Luther (1999).

their own amorphous metal sulfide, notably, lead (Pb), zinc (Zn) and cadmium (Cd), the kinetics of water exchange are faster than that of Fe^{2+}, and these elements tend to form their own metal sulfide phases in pyritized sediments, and DTMP for these elements remains low (Morse and Lurther, 1999). As a consequence, sedimentary pyrite is generally considered an important sink for arsenic (As), mercury (Hg), and molybdenum (Mo), moderately important for cobalt (Co), copper (Cu), manganese (Mn), and nickel (Ni), but generally unimportant for Pb, Zn and Cd (Huerta-Diaz and Morse, 1992).

While predictive techniques, such as partition coefficient and SCM model calculations, are relatively undeveloped compared to those advanced for Fe oxide minerals, trace element concentrations in sedimentary pyrite nonetheless appear to scale with aqueous concentrations in their local environment, such that pyrite-rich sediments (*e.g.* black shales) represent a rich archive of information on Earth's surface geochemical evolution (Section 4.3). Indeed, focusing on just pyrite grains from shales, Large *et al.* (2014) documented the temporal variability of several trace elements, including Mo, Co, Ni, As and Se, throughout geological time. Those authors suggested that in the same way that hydrothermal pyrite tracks the chemistry of ore-forming fluids (Large *et al.*, 2009), sedimentary pyrite can be used to track changes in seawater composition, and ultimately the redox state of Earth's surface environment. This view has recently found support in the work by Swanner *et al.* (2014) who provided pyrite, shale, and BIF trace element data that all suggested an expansion of the palaeo-marine Co reservoir between 2.48 and 1.84 Ga, coincident with anoxic marine conditions and expanded hydrothermal activity. Similarly, Gallagher *et al.* (2015) recently reported a suite of trace element data from Precambrian to Ordovician carbonate-hosted pyrites centred on several key geological events such as the Great Oxidation Event (GOE) ~2.45 Ga (billion years ago), that were in most cases argued to be largely consistent with previous pyrite and shale records.

2.2.3. Exemplar molybdenum

Molybdenum is the most abundant transition metal in seawater (~105 nmol/kg; Anbar and Knoll, 2002) and can exist in a variety of different oxidation states depending on ocean redox. In oxic conditions, Mo is present as molybdate ($Mo(VI)O_4^{2-}$) and its concentration in seawater is though to be controlled by an equilibrium reaction in which molybdate sorbs to Fe and manganese (Mn) oxyhydroxide minerals present in ocean sediments (Barling *et al.*, 2004; Goldberg *et al.*, 2009, 2012). Under anoxic non-sulfidic conditions, where dissolved Fe is commonly present in the water column (termed ferruginous), Mo is probably drawn down in association with other minerals, such as green rust (Zegeye *et al.*, 2012). In the presence of dissolved sulfide, molybdate is converted into a series of thiomolybdate complexes, starting with oxythiomolybdates [$Mo(VI)O_{4-x}S_x{}^{2-}$] (when H_2S ~0.1–11 μM), which progressively dominate Mo speciation and possibly culminate in tetrathiomolybdate [$Mo(VI)S_4^{2-}$] (when H_2S >11 μM) (Helz *et al.*, 1996; Erickson and Helz, 2000). These thiomolybdates are extremely reactive and Mo might, therefore, be drawn down *via* the scavenging of these complexes onto Fe minerals and sulfidized organic particles (likely at low concentrations of dissolved sulfide), or perhaps *via* the precipitation of Mo-sulfide phases (at higher concentrations of dissolved sulfide (~50–250 μM) and presumably Mo) (Helz *et al.*, 1996; Zheng *et al.*, 2000). On the other hand, these complexes might react with zero-valent sulfur present in natural sulfidic waters to produce Mo(IV)-polysulfides, which might similarly be scavenged by Fe minerals and/or organic matter (Dahl *et al.*, 2010). As such, the concentration and speciation of Mo recorded in Fe mineral archives can be used to infer prevailing ocean redox conditions at the time of sediment deposition.

Understanding how the amount of oxygen in seawater has changed through time is key to elucidating links between the evolution of ocean redox chemistry, climate and life. Oxygen provides an important control on seawater chemistry because many elements present in seawater are critically sensitive to redox conditions – existing in oxic waters either in solution as aqueous ions/organic carbon complexes or as solid forms that may be (depending on the precise chemical nature of the redox environment) sequestered into the sediments under reducing conditions. This drawdown of redox-sensitive elements, such as Mo, provides a means of reconstructing past ocean redox *via* careful analyses of the rock record. Indeed, the determination of Mo concentrations in marine shales is one of the most commonly used means to identify the most extreme water column redox condition (highly sulfidic, termed euxinic) throughout Earth history (Scott *et al.*, 2008).

The availability of Mo in seawater also has important implications for the biosphere because it is essential for various cellular metabolisms. For example, it is required for nitrate assimilation by eukaryotes and some prokaryotes, and its enhanced drawdown under widespread euxinic conditions in the mid-Proterozoic ocean (~1.8 Ga) has been postulated widely as an explanation for the maintenance of low levels of atmospheric oxygen and the protracted pace of biological evolution during Earth's 'middle age' (Anbar and Knoll, 2002). As such, intimate links exist between ocean redox, biological activity, atmospheric chemistry and climate, and these can be identified and evaluated by detailed understanding of the environmental behaviour of important redox-sensitive species such as Fe and Mo.

2.3. Iron mineral archives

The most commonly utilized Fe mineral archives are those found in BIF and marine ferromanganese crusts, in which the concentrations and/or stable isotope compositions of trace elements present in the constituent Fe (and Mn) minerals can be interpreted to reflect contemporaneous seawater trace-element signatures. In addition to these Fe-rich archives, scavenging of trace elements onto discrete Fe minerals, which are then deposited into marine sediments as relatively dispersed Fe phases, might also provide repositories for contemporaneous trace-element seawater signatures. Indeed, trace element enrichment in black shales, which provide an important archive for redox-sensitive trace elements (*e.g.* Arnold *et al.*, 2004; Scott *et al.*, 2008), probably results from the scavenging of particle reactive trace element species by Fe minerals (*e.g.* Helz *et al.*, 1996; Zheng *et al.*, 2000; Dahl *et al.*, 2013).

2.3.1. Banded iron formations

Banded iron formations represent vast quantities of Fe-rich sediments that were precipitated in the oceans between ~3.8 and 1.8 Ga. Trace element signatures recorded in BIF can therefore provide information on contemporaneous seawater chemistry during early Earth evolution. The name BIF stems from the appearance of these enigmatic deposits, composed of alternating bands of Fe-rich (~20–40% Fe) and Si-rich (~40–50% SiO_2) layers. Banding can often be observed on a wide range of scales, from coarse macrobands (metres in thickness) to mesobands (centimetre-thick units) to millimetre and submillimetre layers. Among the latter is the wide variety of varve-like repetitive laminae, known as microbands (Trendall and Blockley, 1970). The mineralogy of BIF from the best preserved successions is remarkably uniform, comprising mostly quartz (in the form of chert), magnetite, hematite, Fe-rich silicate minerals (stilpnomelane, minnesotaite, greenalite and riebeckite), carbonate minerals (siderite, ankerite, calcite and dolomite), and minor sulfides (pyrite and pyrrhotite); the presence of both ferric and ferrous minerals give BIF an average oxidation state of $Fe^{2.4+}$ (Klein and Beukes, 1992). However, it is generally agreed that none of the minerals in BIF are primary in origin. Instead, the minerals reflect significant post-depositional alteration under diagenetic and metamorphic conditions (including, in some cases, post-depositional fluid flow). The effect of increasing temperature and pressure is manifested by the progressive change in mineralogy through replacement and recrystallization, increase in crystal size and obliteration of primary textures (Klein, 2005). For instance, the alternating layers of magnetite and hematite are interpreted to have formed from an initial Fe oxyhydroxide phase, *e.g.* ferrihydrite, that precipitated in the photic zone when dissolved Fe^{2+} was oxidized and hydrolysed to insoluble ferric Fe. The Fe oxyhydroxide particles then sank through the water column and were deposited on the seafloor where they eventually formed: (1) magnetite or Fe carbonates when organic remineralization was coupled with Fe(III) reduction; (2) hematite, when organic material was lacking; or (3) Fe silicates, possibly in the form of a precursor mineral such as greenalite, when Si-sorbed ferric oxyhydroxides reacted with other cationic species in the sediment pore waters (Morris, 1993). The chert is widely considered to have precipitated from the water column as colloidal Si co-precipitated with Fe-rich particles (Fischer and Knoll, 2009) given that the Archaean ocean had significantly elevated concentrations of dissolved Si, at least as high as at saturation with cristobalite (0.67 mM at 40°C in seawater), and possibly even amorphous Si (2.20 mM) (Siever, 1992; Maliva *et al.*, 2005; Konhauser *et al.*, 2007).

There is still considerable effort towards understanding how BIF formed but almost all researchers now agree that BIF were precipitated from a predominantly anoxic water column (*e.g.* Holland, 1984). The duration of BIF deposition spans major evolutionary changes in the Earth's surface composition, from an early anoxic atmosphere dominated by CO_2 and CH_4 to an atmosphere that became partially oxygenated. Therefore, it is likely that BIF formed *via* different mechanisms throughout the Precambrian. The traditional model of BIF precipitation assumes the oxidation of dissolved Fe(II) *via* abiotic oxidation by cyanobacterially produced O_2 (Cloud, 1973;

Klein and Beukes, 1989) and/or biotic oxidation by chemolithotrophic bacteria (Holm, 1989; Planavsky *et al.*, 2009). Both models suggest the presence of free molecular oxygen in the Precambrian ocean and, hence, require the presence of oxygenic photosynthesis at that time of Earth history. Given the recent evidence for oxidative processes having already existed at 3.0 Ga (Crowe *et al.*, 2013; Planavsky *et al.*, 2014), and perhaps even at 3.2 Ga (Satkowski *et al.*, 2015), it might be plausible that O_2-driven reactions in the photic zone led to BIF deposition since the Mesoarchaean. An alternative biological mechanism, and one that might explain BIF deposition prior to the Mesoarchaean, involves Fe(II) oxidation *via* anoxygenic photosynthesis (Czaja *et al.*, 2013; Jones *et al.*, 2015; Pecoits *et al.*, 2015). Phototrophic Fe(II)-oxidizing bacteria were discovered two decades ago (*e.g.* Widdel *et al.*, 1993), and their metabolism involves light-energy fuelled CO_2 fixation coupled to the microbial oxidation of Fe^{2+}. The attractiveness of this concept is that it explains BIF deposition in the absence of molecular oxygen using the abundant availability of Fe^{2+}, light and CO_2 at that time (Garrels *et al.*, 1973; Hartman, 1984; Konhauser *et al.*, 2002). It has even been demonstrated by eco-physiological lab experiments in combination with modelling that these phototropic bacteria would have been capable of oxidizing enough Fe(II) to explain the large expansion of BIF deposits (Kappler *et al.*, 2005). There is, in addition, one abiological mechanism, *i.e.* ultraviolet photo-oxidation of Fe^{2+} in the surface waters (*e.g.* Braterman *et al.*, 1983). However, this view has been challenged on the basis that those early experiments were not conducted in complex solutions mimicking seawater (Konhauser *et al.*, 2007). Whatever the mechanism for their precipitation, BIF today have probably undergone repeated episodes of diagenesis and metamorphism, and their Fe mineralogy is now dominated by magnetite and hematite (for reviews on BIF deposition and mineralogy see Klein, 2005 and Bekker *et al.*, 2010).

2.3.2. Ferromanganese crusts

Marine ferromanganese crusts are found throughout the world's oceans and are formed of subequal amounts of Fe(III) oxyhydroxide and Mn(III,IV) minerals that have precipitated from seawater and deposited as pavements and coatings on submarine rock outcrops and topographic highs (*e.g.* Glasby, 1977). Crusts are arguably the slowest growing geological precipitates, with only a few mm deposited per million years (Ma) (*e.g.* Hein *et al.*, 2000). Accordingly, crust formations of ~10 cm thickness can provide trace element archives stretching over the last ~65 Ma of Earth history. Because ferromanganese crusts precipitate directly from ambient seawater (termed "hydrogenetic") these archives can be interpreted to reflect contemporaneous seawater chemistry throughout the Cenozoic. Crusts are restricted to submarine rock outcrops and topographic highs because these environments are often kept sediment free, *via*, for example, local currents generated by obstructional upwelling along seamount flanks. These currents also promote enhanced turbulent mixing and upwelling, leading to increased primary productivity and development of an oxygen-minimum zone (OMZ) that increases the concentrations of Fe^{2+} and Mn^{2+} which can be scavenged into crusts during oxic conditions resulting from turbulence (Muios *et al.*, 2013). As a result of these conditions, Fe and Mn oxyhydroxides precipitated from seawater can deposit onto the hard-rock surfaces and build up in layers over time, as opposed to elsewhere in the oceans where the same precipitates are deposited and dispersed into seafloor sediments. Coupled sorption and redox reactions occurring between dissolved trace elements present in contemporaneous seawater and the mineral precipitates then sequester metals as the minerals precipitate (*via* co-precipitation and structural incorporation) and during their transit through the marine water column (*via* adsorption onto mineral surfaces) (for a general model of crust formation see Koschinsky and Halbach, 1995). Rare and critical trace elements, in particular, are often concentrated to economically valuable concentrations in marine ferromanganese crusts, making the deep-sea mining of crusts and nodules a potentially attractive endeavour (*e.g.* Hein *et al.*, 2013).

The Fe mineralogy of crusts is dominated by ferrihydrite, with older crust layers sometimes featuring small ($< 10\%$) amounts of goethite, typically attributed to aging of the authigenic ferrihydrite (*e.g.* Hein *et al.*, 2000), or a switch in precipitation from authigenic ferrihydrite to goethite during a temporary change in palaeo-seawater conditions (*e.g.* Koschinsky *et al.*, 1997). Manganese mineralogy is dominated by poorly crystalline phyllomanganates, namely δ-MnO_2 (often termed vernadite) and birnessite (*e.g.* Burns and Burns, 1977). Minor quartz, feldspar and other detrital minerals, along with sometimes significant (~20%) carbonate fluorapatite in older crusts subject to diagenesis and/or precipitated during palaeo-seawater transition, are also present (*e.g.* Hein *et al.*, 1993).

Marine ferromanganese nodules are also prevalent in the deep oceans, and, despite their similarly slow growth rates, they do accumulate and comprise a significant component of some marine and freshwater sediment. For instance, in the Pacific Ocean, it has been estimated that 10^{12} tons of nodules exist, predominantly in pelagic sediments, with an annual rate of formation of 6×10^6 tons (Mero, 1962), while in Oneida Lake, New York, for example, within a 20 km^2 area of lake, 10^6 tons of nodules, that average 15 cm in diameter, exist at the bottom sediment (Dean and Greeson, 1979). Diagenetic nodule formation involves a series of microbially-catalysed reactions. The process begins with cyanobacterial and algal plankton concentrating Fe^{2+} and Mn^{2+} from solution. Upon their death the microbial biomass and those bound metals are transported to the bottom sediment where anaerobic respiratory processes in the suboxic layers liberates Fe^{2+} and Mn^{2+} back into the sediment pore waters. Concurrently, reduction of particulate ferric oxyhydroxides and Mn(III,IV) oxides occurs. Upward diffusion to the sediment–water interface facilitates microbial re-oxidation and incorporation of metals onto some form of nucleus, which can be any solid mineral or organic substrate. With continued accretion of Fe and Mn, a nodule, that may or may not display concentric laminations, forms. Plankton contributes to metal oxidation by producing high-pH, oxygenated surface waters that are conducive to the re-oxidation (and hydrolysis) reactions (*e.g.* Richardson *et al.*, 1988). Because nodules grow at the sediment–water interface (termed 'diagenetic') it is difficult to determine to what extent their trace-element signatures reflect contemporaneous seawater chemistry and their use as archives of seawater chemical information is somewhat limited.

2.3.3. Black shales

Shales are an example of marine sediments that, although not technically classed as Fe mineral archives, can never-the-less record contemporaneous seawater trace-element signatures, probably as a result of trace element drawdown in association with Fe (and sulfide and organic) phases (*e.g.* for Mo, Helz *et al.*, 1996; Zheng *et al.*, 2000; Dahl *et al.*, 2013). They comprise fine-grained clastic sediments predominantly composed of clay minerals (*e.g.* kaolinite, montmorillonite and illite) and silt-sized fragments of other minerals (*e.g.* quartz and calcite). Shales are typically grey in colour, but addition of variable amounts of minor constituents can alter their appearance significantly; most notably perhaps in the case of black shales which contain >1% unoxidized carbonaceous material, and display an Fe mineralogy dominated by amorphous Fe sulfides and pyrite (for a summary of the geochemistry of black shale deposits see Vine and Tourtelot, 1970). This relatively high unoxidized carbon content and the presence of Fe sulfides suggests that black shales were deposited under an anoxic water column, where organic carbon was not subject to aerobic oxidative degradation and ferrous Fe was thermodynamically stable. The presence of sedimentary fabrics also indicates a lack of physical disruption by burrowing animals (at least since the Neoproterozoic). This, combined with depositional occurrences that span most of the geological record (*e.g.* Tourtelot, 1979), means that trace element-signatures recorded in black shale archives can represent seawater chemistry during episodes of ocean anoxia occurring throughout Earth history.

3. Stable isotope fractionation of trace elements

Subtle physicochemical discrimination between the different stable isotopes of any given element may impart a robust geochemical fingerprint of the processes that the element has experienced, making stable isotope geochemistry a powerful tool for palaeo-environmental reconstruction. This has been recognized and exploited by geochemists for over 60 years, with early studies naturally focusing on the stable isotopes of lighter elements (H, C, N, O and S) whose greater relative mass differences imparted stronger, more easily measured fractionations. These 'traditional' stable isotope systems have since been supplemented by 'non-traditional' ones whose smaller natural fractionations have now become accessible due to advances in mass spectrometry technology. Natural stable isotope variability has now been described and quantified for an increasing number of non-traditional elements, including Li, B, Mg, Si, Cl, Ca, V, Cr, Fe, Ni, Cu, Zn, Ge, Se, Mo, Cd, Sn, Ce and Eu, and the list continues to grow (see Johnson *et al.*, 2004 dedicated to the subject).

3.1. Stable isotope nomenclature and conventions

Most stable isotope data are reported in delta notation where the isotope composition of a sample is defined as the relative deviation of the sample isotope ratio to the same isotope ratio in a known reference material (*e.g.* O'Neil, 1986):

$$\delta^{i/j}E = 1000\left(\frac{R_{i/j}\ \text{Sample}}{R_{i/j}\ \text{Reference}} - 1\right) \tag{2}$$

where i and j are the particular isotopes used in ratio R of element E. Units for $\delta^{i/j}E$ are in parts per thousand, *i.e.* per mil (‰). For the traditional stable isotope systems (H, C, N, O and S), the isotope ratio $R^{i/j}$ is expressed as the abundance of the minor isotope over the abundance of the major isotope, *e.g.* D/H and $^{13}C/^{12}C$. This convention has ensured a consistent point of reference for traditional stable isotope enrichments and depletions, where a positive or negative value of $\delta^{i/j}E$ indicates that a sample is relatively enriched or depleted in the heavy isotope, respectively, compared to the standard. For several non-traditional stable isotope systems, however, $R^{i/j}$ expressed as rare over major isotope does not satisfy the heavy over light convention, because the minor isotope is the isotopically light. This is the case for the Fe and Mo stable isotope systems, where, for example, ^{54}Fe and ^{95}Mo are more rare than ^{56}Fe and ^{98}Mo. In order for stable isotope enrichments and depletions to be expressed in the same direction (heavy enrichments as positive delta values, light enrichments as negative delta values) between traditional and non-traditional systems, in non-traditional systems the convention heavy over light is usually preferred. Considering ^{54}Fe and ^{56}Fe and ^{95}Mo and ^{98}Mo, for the Fe and Mo stable isotope systems, respectively, $R^{i/j}$ is typically expressed as $^{56}Fe/^{54}Fe$ and $^{98}Mo/^{95}Mo$, with corresponding delta notation of $\delta^{56}Fe$ and $\delta^{98}Mo$.

For traditional stable isotope systems the choice of reference material for each isotope system has been standardized internationally, where, for example, the reference used for reporting the isotope compositions of H and O is Standard Mean Ocean Water (SMOW), and Air for N. For some of the non-traditional systems the reference material is yet to be agreed upon and is therefore not standardized across the field. In more recent work, Fe stable isotope measurements are usually expressed with reference to the isotope composition of the international standard IRMM-14 (Belshaw *et al.*, 2000), however, for the Mo system an international standard has only recently been proposed (NIST SRM 3134; Wen *et al.*, 2010; Greber *et al.*, 2012), and the Mo isotope community has employed various in-house Mo isotope standards over the years to normalize results (for an inter-comparison see Goldberg *et al.*, 2013). In many non-traditional stable isotope systems there is also a choice of isotope ratios that can be measured, because several of the non-traditional elements have multiple naturally occurring stable isotopes. For example, Fe has four naturally occurring stable isotopes ^{54}Fe (5.84%), ^{56}Fe (91.7%), ^{57}Fe (2.12%) and ^{58}Fe (0.28%), where stable isotope data are commonly reported as $^{56}Fe/^{54}Fe$, while Mo has seven naturally occurring stable isotopes ^{92}Mo (14.84%), ^{94}Mo (9.25%), ^{95}Mo (15.92%), ^{96}Mo (16.68%), ^{97}Mo (9.55%), ^{98}Mo (24.13%) and ^{100}Mo (9.63%), and stable isotope data have been reported for both $^{97}Mo/^{95}Mo$ and $^{98}Mo/^{95}Mo$. As both the reference material, *REF*, and the stable isotopes measured in $R^{i/j}$ can be different for the same non-traditional stable isotope system, care is required when comparing $\delta^{i/j}E$ values between different studies. For an extended review of stable isotope nomenclature and conventions the reader is directed to Johnson *et al.* (2004).

3.2. Stable isotope fractionation theory

3.2.1. Mass-dependent stable isotope fractionation

Mass-dependent stable isotope fractionation refers to any physical or chemical process that acts to partially separate stable isotopes of the same element between two different reservoirs, where the amount of separation scales in proportion with the difference in the masses of the isotopes. For example, using the O stable isotope system, the mass difference between ^{18}O and ^{16}O is +2, while the mass difference between ^{18}O and ^{17}O is +1, so considering the partitioning of O isotopes between two different reservoirs, $\delta^{18}O$ will be approximately twice that of $\delta^{17}O$ as the result of mass-dependent fractionation. Separation of stable isotopes of the same element between two different reservoirs, and thus two separate substances, occurs because chemical bond strengths are mass-dependent. In the case of the two most abundant isotopes of O, ^{18}O and ^{16}O, in an H_2O molecule, there is a difference in bond strength between ^{18}O–H and ^{16}O–H. Specifically, ^{18}O–H has a lower vibrational frequency, and hence a lower ground state energy (or zero point energy, ZPE), than ^{16}O–H. This means that the bond between ^{18}O–H is stronger than that between ^{16}O–H. Differences in chemical bond strength lead to differences in reaction rate constants, where species with the stronger bonds will react more slowly than species with the weaker bonds. This effect on reaction rate constants is the origin of kinetic isotope fractionation, which acts to separate isotopes during unidirectional or incomplete processes, perhaps best illustrated during evaporation of water where light water molecules (with ^{16}O) will evaporate more rapidly than heavier water molecules (with ^{18}O). Differences in chemical bond strength also lead to differences in equilibrium constants for reactions that are otherwise identical. This is the origin of equilibrium isotope fractionation, involving the partial separation of isotopes between two substances that are in chemical equilibrium. The theory of stable isotope fractionation pre-dates the advent of multi-collector inductively coupled plasma mass spectrometry (MC-ICP-MS) and the subsequent discovery of stable isotope fractionation in the heavy trace elements, and includes a few very early studies (Lindemann and Aston, 1919; Lindemann, 1919; Urey and Greiff, 1935). The modern theoretical basis for calculating mass-dependent fractionation was established over 60 years ago in a series of seminal studies by Harold Urey, Jacob Bigeleisen and Maria Mayer (*e.g.* Urey, 1947; Bigeleisen and Mayer, 1947). A rigorous treatment of mass-dependent stable isotope theory can be found in several texts, including a more recent review of fractionation fundamentals by Criss (1999).

Equilibrium stable isotope fractionation describes the partial separation of isotopes between two or more substances that are in, or are approaching, chemical equilibrium. There are a number of general qualitative chemical rules governing equilibrium stable isotope fractionations which can be used to predict the magnitude and direction of isotope fractionation between two separate substances (for a complete review see Schauble, 2004). In the first instance, equilibrium isotope fractionations usually decrease as temperature increases, and, all else being equal, fractionations are largest for light elements and for isotopes with very different masses (*e.g.* Criss, 1999; Schauble, 2004). In addition to these rules, a further consideration of equilibrium thermodynamics provides a means to predict which substances will be enriched in heavy isotopes in a given geochemical system. In equilibrium thermodynamics, all systems strive to achieve their lowest energy state. Because chemical bonds involving heavier isotopes have a lower ground state energy than otherwise identical bonds involving lighter isotopes of the same element, then during equilibrium isotope fractionation the heavy isotope will tend to concentrate in the substance that confers the lowest ground state energy for that particular bond, *i.e.* in the substance that confers the strongest bonding environment. This process is well demonstrated in the isotope exchange of ^{18}O and ^{16}O between liquid water and water vapour:

$$H_2^{16}O\ (l) + H_2^{18}O\ (g) = H_2^{18}O\ (l) + H_2^{16}O\ (g) \quad (3)$$

where at, or approaching, chemical equilibrium, ^{18}O will concentrate in the stronger O–H bonds of water relative to the weaker O–H bonds of steam. The formation of strong bonding environments, *i.e.* strong chemical bonds, correlates with several physicochemical properties, including high oxidation state in the element of interest and low coordination number (for a full list see Schauble, 2004). Thus with a choice of bonding environment between substance A with element E in low oxidation state, and substance B with element E in high oxidation state, the heavier isotopes will tend to concentrate in substance B, all else being equal. Similarly, with a choice of bonding environment between substance A with element E in high coordination, and substance B with element E in low coordination, the heavier isotopes will again tend to concentrate in substance B, all else being equal. Overall, the general rules for equilibrium isotope fractionation predict that large equilibrium fractionations are most likely to occur at low temperature between substances with significantly different oxidation states, coordination numbers, bonding partners and/or electronic configurations (Schauble, 2004). Furthermore, these rules suggest that stable isotope signatures of redox-active trace elements, preserved in the sedimentary record, can provide a record of the palaeo redox conditions present in the environment during the time of sediment deposition. Specifically, the extent of the measured fractionation in the sedimentary record should reflect the extent of oxidation/reduction of that redox-active species in the depositional system, which is itself related to the overall redox state of the contemporaneous environment. In essence, this is the basis for utilizing redox-active trace element stable isotopes recorded in sedimentary archives as tracers of past redox and biogeochemical conditions.

3.2.2. Fractionation factors

The fractionation factor is a convenient quantification of the contrast in isotope compositions between two substances and is defined as (*e.g.* O'Neil, 1986):

$$\alpha_{A-B} = \frac{R_{i/j}A}{R_{i/j}B} \tag{4}$$

where A and B are two distinct substances, and $R_{i/j}A$ and $R_{i/j}B$ are the isotope ratios of the particular element of interest in substances A and B, respectively. The fractionation factor may also be cast in terms of $\delta^{i/j}E$ values:

$$\alpha_{A-B} = \frac{1000 + \delta^{i/j}E_A}{1000 + \delta^{i/j}E_B} \tag{5}$$

A fractionation factor of 1 indicates that the isotopes of element E are distributed evenly between substances A and B and there is no isotope fractionation. A fractionation factor > 1 indicates isotope ^iE (for non-traditional isotope systems ^iE is usually the heavy isotope) is concentrated in substance A, while a fractionation factor of < 1 indicates isotope ^iE is concentrated in substance B. The fractionation factor does not allow differentiation between the physical processes that lead to different isotope compositions in different substances, and thus may reflect equilibrium or non-equilibrium isotope partitioning.

3.2.3. Processes producing isotopically distinct reservoirs

There are generally two types of process that can lead to the partial separation of isotopes between two reservoirs, namely, closed-system equilibrium and Rayleigh fractionation. Closed-system equilibrium is best illustrated by considering the slow reaction of two substances A and B, where A and B remain open to complete isotope exchange throughout the reaction. In this process the isotope composition of A and B is a function of the fraction of A remaining and the fraction of B produced. As the reaction proceeds, the shifting mass balances of A and B requires that their $\delta^{i/j}E$ values shift in order to maintain a constant offset in fractionation between A and B. As such, $\delta^{i/j}E_A$ and $\delta^{i/j}E_B$ will progress along a straight line as a function of the fraction of B produced (Fig. 2).

During Rayleigh fractionation the products of a reaction do not continue to exchange with other phases in the system, because, for example, they are constantly isolated or removed. So as the reaction proceeds, product B is in thermodynamic and isotope equilibrium with reactant A, but is then instantaneously removed from the system. In this case the isotope compositions of A and B evolve along an exponential trajectory, and because the product is progressively isolated, large changes in δ^iE values in the remaining components can occur (Fig. 2). An example would be the Fe isotope fractionations associated with ferrihydrite precipitating from seawater during the Precambrian, such that the BIF which formed in the water column and then settled out to the seafloor became progressively heavy in ^{56}Fe, while the residual Fe(II) pool in solution became isotopically more depleted (*e.g.* Rouxel *et al.*, 2005).

In addition to changes in the redox conditions of the contemporaneous environment and changes in the biogeochemical processes that ultimately control the source and sink fluxes of trace elements to and from the oceans, there are a number of independent

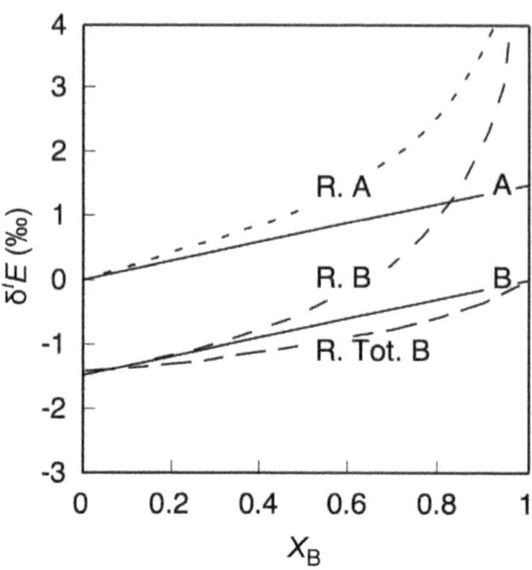

Figure 2. Comparison of the stable isotope fractionation produced by closed-system equilibrium *vs.* Rayleigh fractionation processes, during reaction of substance A to substance B with a fractionation factor α_{A-B} of 1.0015, as a function of the proportion of B produced (X_B). The solid lines labelled A and B reflect the isotopic evolution of the reactant A and product B during closed-system fractionation, where isotopic equilibrium is maintained. The dashed lines labelled R. A and R. B reflect the isotopic evolution of the reactant A and instantaneous product B during Rayleigh fractionation, where the product B is immediately isolated from isotopic exchange with reactant A after its formation. The dashed line labelled R. Tot. B shows the isotopic fractionation for the total product B during Rayleigh fractionation. Adapted from Johnson *et al.*, 2004.

physicochemical processes that occur in seawater and which, according to the general rules governing equilibrium stable isotope fractionation (Section 3.2.1), should be expected to cause stable isotope fractionation of trace elements and thus affect the isotope signal recorded in oceanic sedimentary archives. The first concerns the fact that almost all dissolved trace elements in seawater have varied speciation chemistry involving both inorganic and organic ligands. Physicochemical differences between the different dissolved aqueous species of element E in seawater, including differences in redox state, coordination number, bonding partners and electronic configurations, are expected to induce stable isotope fractionation of E between the relevant aqueous reservoirs (Section 3.2.1). Isotope fractionation between different aqueous species have been studied for several trace element stable isotope systems, including Fe and Mo (*e.g.* Anbar *et al.*, 2000; Roe *et al.*, 2003; Welch *et al.*, 2003; Anbar *et al.*, 2005; Tossell, 2005; Weeks *et al.*, 2007). Secondly, because there can be differences in coordination number and bonding partners (*i.e.* bonding environment) between the dissolved aqueous species of E in seawater and the inorganic and organic particulate phases that sequester E into the sediments, more often than not, the act of sequestration itself can induce an isotope fractionation. In particular the stable isotope fractionation of trace elements upon sorption to Fe and Mn oxyhydroxides has been measured for Mo (Barling and Anbar, 2004; Kashiwabara *et al.*, 2011; Wasylenki *et al.*, 2011), for Ni (Wasylenki *et al.*, 2015), for Cu and Zn (Pokrovsky *et al.*, 2005; Balistrieri *et al.*, 2008; Pokrovsky *et al.*, 2008) and for Cd (Wasylenki *et al.*, 2014). In addition, for Fe and Mn oxyhydroxides in particular, sorption of element E can be coupled with its oxidation and the concomitant reductive dissolution of the sorbent mineral phase (*e.g.* for chromium (Cr), Manceau and Charlet, 1992; for Tl, Peacock and Moon, 2012). According to the rules governing equilibrium stable isotope fractionation, this redox reaction can also be expected (*e.g.* for Cr, Schauble *et al.*, 2004), or in select systems has in fact been shown (*e.g.* for Tl, Peacock and Moon, 2012; Nielsen *et al.*, 2013), to induce stable isotope fractionation of *E* during sequestration to the sediments. Despite a number of studies now documenting trace element stable isotope fractionation occurring during sorption to sedimentary mineral phases, relatively few seek to determine the molecular mechanism(s) responsible for the fractionation, and thus provide a fundamental framework for understanding, interpreting and predicting fractionations between palaeo-seawater and contemporaneous trace element sedimentary archives (notable exceptions include, for Mo, Wasylenki *et al.*, 2011; for Ni, Wasylenki *et al.*, 2015; for Cu and Zn, Juliot *et al.*, 2008; Pokrovsky *et al.*, 2008; Little *et al.*, 2014; for Tl, Peacock and Moon, 2012). These studies are crucial for understanding the isotope signals measured in ancient sedimentary archives and subsequently how these archives relate to contemporaneous seawater. In particular, the isotope fractionation of trace elements during sequestration can be the dominant isotope effect in the modern oceans; *e.g.* ferromanganese-rich sediments are known to be the main sedimentary sink for Mo in modern seawater, and during sorption to ferromanganese minerals, light Mo is sequestered preferentially leaving seawater with a heavy isotope signature of 2.3‰; this sequestration process controls the modern global seawater Mo isotope composition (*e.g.* Barling *et al.*, 2001; Siebert *et al.*, 2003; Barling and Anbar, 2004). Similar sequestration processes are also likely to be important in controlling the global seawater isotope compositions of a range of other trace elements, which are now emerging as potentially important tracers of past biogeochemical processes (*e.g.* Cu and Zn, Little *et al.*, 2014).

Recent research now indicates that the magnitude and direction of isotope fractionation that occurs during sequestration of trace elements to Fe (and Mn) oxyhydroxide minerals is controlled by the precise molecular mechanism by which the metals are sorbed from solution (*e.g.* for Ni, Wasylenki *et al.*, 2015). This, in turn, depends on an often complex combination of geochemical, physicochemical and mineralogical factors (*e.g.* for Ni; Peacock and Sherman, 2007; Peacock, 2009; Pea *et al.*, 2010). Ultimately, isotope fractionation that occurs as a result of physicochemical processes in seawater must be considered when using sedimentary trace element isotope signals as tracers of contemporaneous seawater isotope composition and, ultimately, as proxies for contemporaneous environmental conditions and processes.

3.3. Predicted and measured fractionations of trace elements in geological systems

Stable isotope fractionation of the trace elements can be predicted from the theoretical basis for mass-dependent fractionation, established by Urey (1947) and Bigeleisen and Mayer (1947). These calculations are important for the new trace element systems because they provide limits to the fractionations that we might expect to measure in nature, and are particularly useful for systems that cannot be reproduced experimentally (*e.g.* Johnson *et al.*, 2004). Several of the trace element stable isotope systems have been explored theoretically, including Fe (*e.g.* Hill and Schauble,

2008), Mo (*e.g.* Weeks *et al.*, 2007, 2008; Wasylenki *et al.*, 2008, 2011), Cr (*e.g.* Schauble *et al.*, 2004), Tl (*e.g.* Schauble, 2007), and Ce (*e.g.* Nakada *et al.*, 2013). Ideally, calculated fractionations should be verified experimentally, and this is an active research area where several groups are working with both theory and experiment to shed light on measured natural isotope values and isotope variations between reservoirs (*e.g.* for Mo, Wasylenki *et al.*, 2011; for Cu and Zn, Little *et al.*, 2014).

The Fe and Mo stable isotope systems have, perhaps, received the most attention in terms of theoretical studies that predict fractionation between different Fe-Fe and Mo-Mo species and experimental studies quantitatively constraining that fractionation in different Fe and Mo systems. As detailed below, both elements experience significant change in their geochemical behaviour with changes in environmental redox conditions, both are important bio-essential metals linked to modern and ancient biogeochemical cycling, and both have revealed a rich sedimentary record of evolving conditions at Earth's surface.

3.4. Emerging non-traditional stable isotope systems

3.4.1. Iron

Compared to the lower-mass traditional stable isotopes (H, C, N, O and S), fractionations for higher-mass elements are not expected to exceed 10‰ at temperatures $\leqslant 100°C$ (Johnson *et al.*, 2004). Unlike other more volatile isotopic systems (*e.g.* C and O), Fe isotopes are geochemically robust, retaining their environmental signature even in the face of extreme metamorphic conditions (*e.g.* Frost *et al.*, 2007; Dauphas *et al.*, 2007), making them a useful tool in deciphering the biogeochemical history of highly deformed and metamorphosed rocks.

Iron has four naturally occurring stable isotopes ^{54}Fe (5.84%), ^{56}Fe (91.7%), ^{57}Fe (2.12%) and ^{58}Fe (0.28%), where stable isotope data are commonly reported as $^{56}Fe/^{54}Fe$. Many natural materials tend to have distinctively narrow $\delta^{56}Fe$ ranges, where most igneous rocks fall within the 0.00 ± 0.05 ‰ range, as do riverine and marine sediments, shales, and modern wind-blown sediments (Beard *et al.*, 2003; Fig. 3). In contrast, Fe-rich sediments have characteristically large variations in $\delta^{56}Fe$ (Fig. 3). This large variation is interpreted to have been caused by a combination of factors, including mineral-specific equilibrium isotope fractionation, variations in the original Fe isotope composition of the fluids from which these minerals precipitated, and bacterial metabolic processes which might have influenced their formation (Fig. 4; Johnson *et al.*, 2003, 2008b). Metabolically processed Fe may also retain a 'vital effect' due to the redox-related changes driving Fe isotope fractionation during microbially influenced Fe mineral precipitation (Fig. 4; Johnson *et al.*, 2002).

While Fe isotope systematics has gained widespread attention in applications ranging from chemical weathering, soil formation, the formation of ore deposits, and modern marine metal cycling, their principal application to date in terms of Earth's redox and biogeochemical evolution has focused on the Precambrian record of Fe isotopes archived in BIF (Section 4.1.2).

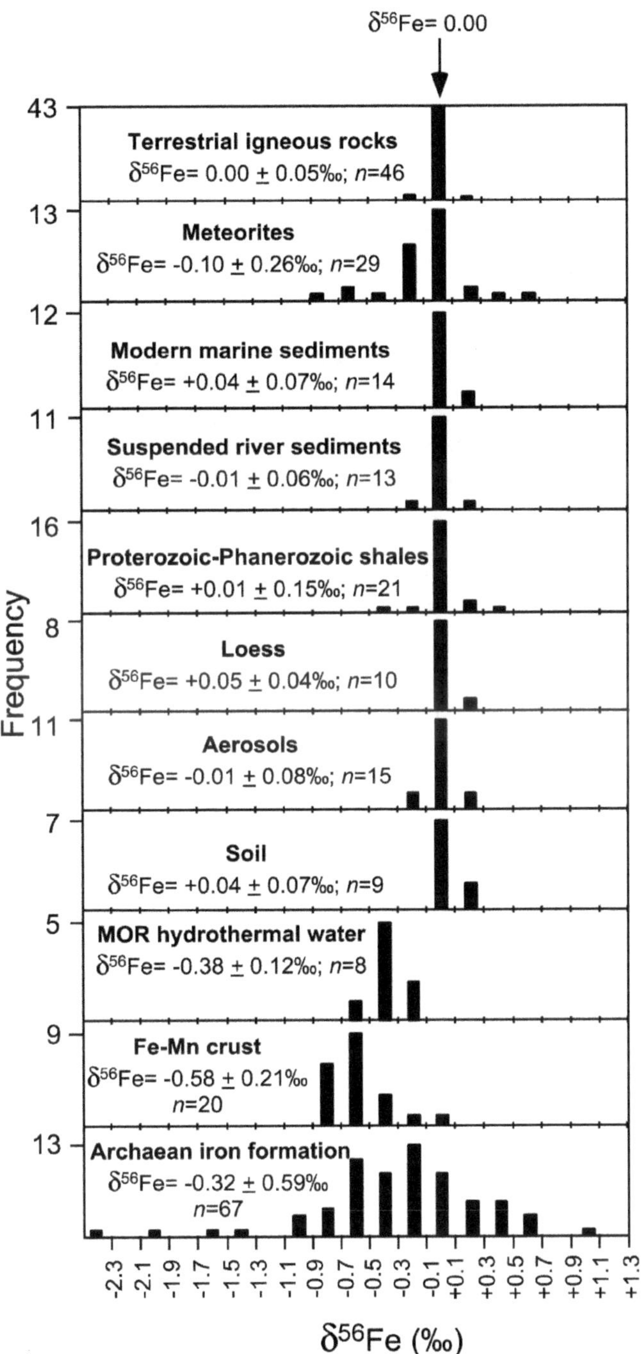

Figure 3. Iron isotope composition of natural materials. Adapted from Beard *et al.* (2003).

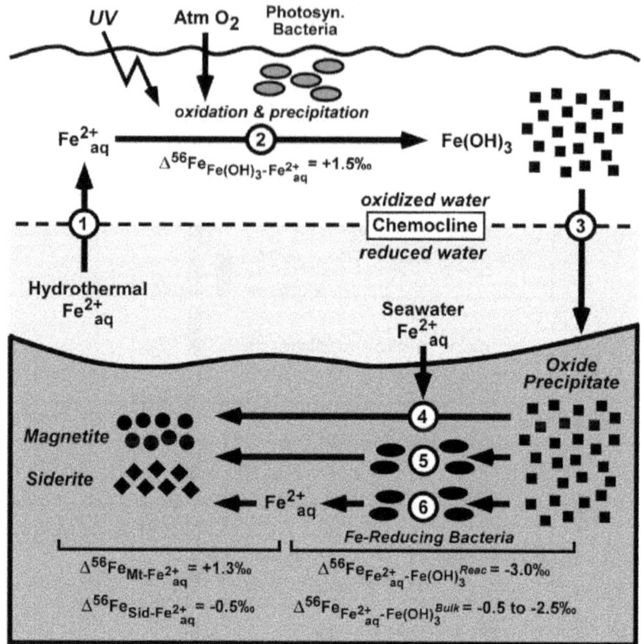

Figure 4. Pathways of iron minerals cycling and associated isotopic fractionations during the deposition and early diagenesis of iron formation. Adapted from Johnson *et al.* (2007).

3.4.2. Molybdenum

The Mo isotope composition of modern and ancient Fe mineral archives has received significant attention due to the strong potential for palaeo-redox reconstruction (Barling *et al.*, 2001; McManus *et al.*, 2002; Siebert *et al.*, 2003; Arnold *et al.*, 2004; Barling and Anbar, 2004; Archer and Vance, 2008; Voegelin *et al.*, 2009, 2010; Planavsky *et al.*, 2014). Molybdenum has seven naturally occurring stable isotopes ^{92}Mo (14.84%), ^{94}Mo (9.25%), ^{95}Mo (15.92%), ^{96}Mo (16.68%), ^{97}Mo (9.55%), ^{98}Mo (24.13%) and ^{100}Mo (9.63%) and Mo isotopic compositions represented in the conventional δ notation have been primarily reported in two different ways in the literature. The most common reporting is $\delta^{98/95}$Mo with respect to the isotope standard "Bern-Mo" (Johnson Matthey ICP standard, lot 602332B; McManus *et al.*, 2002; Siebert *et al.*, 2003; Voegelin *et al.*, 2009, 2010). The second most common reporting is $\delta^{97/95}$Mo with respect to Roch-Mo2 (Johnson Matthey Specpure Molybdenum Plasma Standard, lot 7024991; Barling *et al.*, 2001; Barling and Anbar, 2004). Despite this apparent complexity, it is relatively easy to convert from $\delta^{98/95}$Mo to $\delta^{97/95}$Mo with a simple multiplication factor of 2/3, corresponding to the relative mass difference between these notations. The NIST SRM 3134 Mo standard has since been proposed by Greber *et al.* (2012) in order to enable easier data comparisons between studies. Goldberg *et al.* (2013) calibrated the NIST SRM 3134 standard relative to Roch-Mo2 and Bern-Mo as follows: $\delta^{98/95}\mathrm{Mo}_{\mathrm{SRM\ 3134}} = \delta^{98/95}\mathrm{Mo}_{\mathrm{Roch\text{-}Mo2}} + 0.34 \pm 0.05$ ‰ and $\delta^{98/95}\mathrm{Mo}_{\mathrm{SRM\ 3134}} = \delta^{98/95}\mathrm{Mo}_{\mathrm{Bern\text{-}Mo}} + 0.27 \pm 0.06$‰.

Siebert *et al.* (2003) were the first to constrain mean ocean $\delta^{98/95}$Mo, reporting an isotopically heavy value of 2.63‰ ± 0.1 (2SD). An even heavier isotopic composition was reported by McManus *et al.* (2002) for deep marine sediment pore waters, where isotopically light Mo (relative to seawater) accumulates in reducing sediments (the most important marine Mo sink), leaving a low concentration of heavy residual Mo in associated pore waters.

All oceanic Mo sources and sinks have a lighter isotopic composition than seawater (Fig. 5). The isotopic offset between oxic and anoxic sediments is fixed by the isotopic composition of the most important Mo source, river water, which may vary in dissolved Mo isotope composition between 0 and 2.3‰ and averages around 0.7‰ (Archer and Vance, 2008). The heavy Mo isotopic composition of seawater is largely determined by the areal extent of sediments underlying an anoxic water column, which have a light isotopic composition and are the most important marine sink of Mo (McManus *et al.*, 2002; Siebert *et al.*, 2003). The second most important sink, Fe and Mn oxide-rich sediments, represents an even lighter Mo isotope exit channel, but is less important today than the anoxic sedimentary sink (Arnold *et al.*, 2004). Thus, isotopically heavier

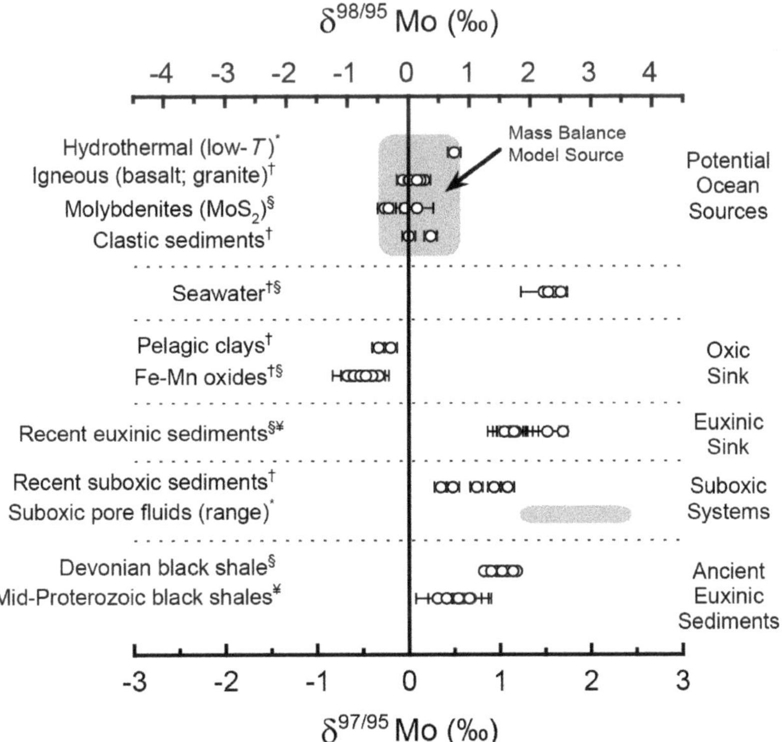

Figure 5. Mo isotope compositions of marine sediments deposited under diverse water column redox conditions. Adapted from Anbar (2004).

seawater values reflect a rigorous oxic sink removing light Mo isotopes, and isotopically lighter seawater values reflect a more important role for anoxic and euxinic sinks that sequester the heavier isotopes. Uptake of Mo into carbonates is assumed to have an insignificant effect on the Mo isotopic composition of seawater because they contain very little Mo and are thus relatively unimportant from a mass-balance perspective compared to the major sources and sinks identified.

3.4.3. Chromium

The mobile chromium(VI) anion ($HCrO_4^-$) is the most thermodynamically stable form of Cr in equilibrium with modern atmospheric oxygen. Chromium(III) is the most stable form under reduced conditions, and its mobilization is only prevalent under strongly acidic conditions. The oxidation to Cr(VI) depends upon the solubilization of primary Cr(III)-bearing mineral phases, such as chromite ($FeCr_2O_4$), and the subsequent reaction of dissolved Cr(III) with Mn(IV) oxides (reaction 6). The latter is the only naturally occurring oxidant of Cr(III) at $pH<9$ (Eary and Rai, 1987), but because it requires the presence of elevated O_2 levels to oxidize Mn(II) to Mn(IV), Cr mobility is ultimately tied to environmental oxygenation. Chromium(VI) can be reduced to Cr(III) by microbes (Sikora *et al.*, 2008) and by Fe^{2+} (reaction 7) or Fe(II)-bearing minerals (Ellis *et al.*, 2002). The oxidation of dissolved Fe^{2+} by Cr(VI) proceeds on the order of minutes, and is much faster than with oxygen, even under well aerated, high-pH conditions (Eary and Rai, 1989). This means that in the presence of Fe(II), Cr(IV) is efficiently reduced to Cr(III). The Cr(III) is subsequently and effectively scavenged into Fe(III)-Cr(III) oxyhydroxides owing to the very low solubility of $(Fe,Cr)(OH)_3$ (Sass and Rai, 1987). Moreover, the formation of a solid phase greatly reduces the potential for transformation back to Cr(VI) (Fendorf, 1995).

$$Cr^{3+} + 1.5MnO_2 + H_2O \rightarrow HCrO_4^- + 1.5Mn^{2+} + H^+ \quad (6)$$

$$3HCrO_4^- + 9Fe^{2+} + 15H_2O \rightarrow 3(Fe,Cr)(OH)_3 + 6Fe(OH)_3 + 15H^+ \quad (7)$$

Chromium has four stable isotopes of masses 50 (4.35%), 52 (83.8%), 53 (9.50%) and 54 (2.37%). After Cr(III) oxidation, the Cr(VI) anion is enriched by up to 7‰ in ^{53}Cr compared to coexisting compounds containing Cr(III), leaving the aqueous phase with positive $\delta^{53}Cr$ values and the residual soils and weathered rock with negative $\delta^{53}Cr$ values (Schauble *et al.*, 2004; Zink *et al.*, 2010). This positive Cr(VI) signal should be transferred to the oceans because subsequent sorption of Cr(VI) onto soil and sediment particles apparently produces no isotope effect (Ellis *et al.*, 2004). The microbial reduction of Cr(VI) further generates isotopic shifts of up to 4.1‰, comparable to those produced during abiotic reduction (Ellis *et al.*, 2002). The overall effects of Cr redox reactions are, therefore, to enrich the heavier isotope in the remaining dissolved Cr(VI). Chemical sediments may then capture the signal of oxidative continental weathering if the Cr(VI) pool is large enough to alter the Cr isotope composition of seawater.

4. Iron-based records of Earth's redox and biogeochemical evolution

4.1. Banded iron formations (BIF)

One of the unique sedimentological features of the Archaean oceans was the deposition of banded iron formation (BIF). As reviewed below, several aspects of the trace element signatures recorded in these enigmatic deposits are employed to reconstruct Earth's ancient redox and biogeochemical evolution.

4.1.1. Rare earth elements

Rare earth element (*REE*) studies in BIF focus on two central facets, namely understanding the sources of Fe to the ancient oceans, and the ancient water column redox conditions. At the heart of *REE* studies in BIF is the assumption that there is minimal fractionation of *REE* during precipitation of the primary ferric oxyhydroxides. Iron formations are, therefore, inferred to have trapped a *REE* signature of seawater at the site of Fe precipitation.

For over half of Earth's history, seawater was characterized by its high Fe concentrations, leading to the notion that the Archaean oceans (4.0–2.5 Ga) were "ferruginous" throughout much of the anoxic water column. Early studies invoked a continental source of Fe, since Fe(II) was much more mobile during weathering in the absence of atmospheric O_2 (*e.g.* James, 1954; Lepp and Goldich, 1964) and potentially since continents had a more mafic composition than today (Condie, 1993). Holland (1973) later suggested that Fe was instead sourced from deep marine waters and supplied to the depositional settings *via* upwelling. With the discovery of modern seafloor-hydrothermal systems (*e.g.* Corliss *et al.*, 1978) and the subsequent recognition that modern hydrothermal systems may contribute up to 75% of dissolved Fe to the Fe budget in the deep oceans (Carazzo *et al.*, 2013), submarine venting has become the generally agreed upon source for seawater Fe.

Perhaps the strongest support for a submarine volcanic source for BIF comes from its europium (Eu) enrichment, which indicates a strong influence of high-temperature hydrothermal fluids on the seawater dissolved *REE* load (*e.g.* Klinkhammer *et al.*, 1983; Derry and Jacobsen, 1988, 1990). The disparate behaviour of Eu from neighbouring *REE* in hydrothermal fluids is linked with Eu(III) reduction at high-temperature and low Eh conditions (Klinkhammer *et al.*, 1983; Sverjensky, 1984). It is generally assumed that Fe and *REE* will not be fractionated during transport from spreading ridges or other exhalative centres, and, therefore, a strong positive Eu anomaly indicates that the Fe in the precursor sediment was hydrothermally derived (*e.g.* Slack *et al.*, 2007). In this regard, secular trends in the magnitude of Eu anomalies in BIF has historically been assumed to indicate variations in hydrothermal flux, with a long-term decrease in hydrothermal activity from the Eoarchaean to Palaeoproterozoic (*e.g.* Huston and Logan, 2004; Sreenivas and Murakami, 2005). This *REE* trend is probably linked with an overall decline in the delivery of reductants from Earth's interior.

Neodymium (Nd) isotopes have also been shown to be highly valuable in tracing the source input of Fe to BIF. For example, a number of BIF record mixing of $^{143}Nd/^{144}Nd$

from primarily two sources: (1) anoxic deep water with generally positive initial εNd (εNd(i)) values reflecting submarine mid-ocean ridge basalt (MORB)-like depleted sources; and (2) shallower waters with generally lower εNd(i), reflecting riverine input from evolved continental sources (*e.g.* Alibert and McCulloch, 1993; Bau *et al.*, 1997; Frei and Polat, 2007). Newly compiled $^{143}Nd/^{144}Nd$ data by Alexander *et al.* (2009) of BIF older than 2.7 Ga suggest that bulk anoxic Archaean seawater was dominated by high-temperature hydrothermal alteration of mantle-derived oceanic crust, presumably also responsible for delivering the Fe to the BIF. However, deviation from this mantle Nd signal has been recorded in the 2.9 Ga-old Pongola BIF in South Africa, which was deposited in a near-shore, shallow-water environment where continental fluxes of Fe and other solutes contributed greatly to the unit's composition (Alexander *et al.*, 2009). Similarly, Haugaard *et al.* (2013) reported positive εNd values in the Si-rich bands of the 2.9 Ga Itilliarsuk BIF of West Greenland, suggesting that those *REE* were controlled by a local, depleted continental crust, whereas the negative ε_{Nd}(i) values found in the Fe-rich layers were hydrothermally sourced. Nevertheless, from Archaean to Proterozoic, there is a gradual decrease in hydrothermal activities as the Earth cooled, and a stronger continental contribution of Nd to the ocean (Miller and O'Nions, 1985; Jacobsen and Pimentel-Klose, 1988).

Anomalies in cerium (Ce) abundance in BIF have been used to constrain water column redox conditions. In general, oxygenated marine settings display a strong negative Ce anomaly when normalized to shale composite ($Ce_{(SN)}$), while suboxic and anoxic waters lack significant negative $Ce_{(SN)}$ anomalies and can even show positive anomalies (*e.g.* German and Elderfield, 1990; Byrne and Sholkovitz, 1996). Oxidation of Ce(III) to Ce(IV) greatly reduces Ce solubility, resulting in its preferential removal onto Mn(IV)-Fe(III)-oxyhydroxides, organic matter and clay particles. In contrast, suboxic and anoxic waters lack significant negative $Ce_{(SN)}$ anomalies due to reductive dissolution of settling Mn(IV)-Fe(III)-oxyhydroxide particles. Similarly, light *REE* depletion develops in oxygenated waters due to preferential removal of light *vs.* heavy *REE* onto Mn(IV)-Fe(III)-oxyhydroxides and other particle-reactive surfaces. As a result, the ratio of light-to-heavy *REE* markedly increases across redox boundaries due to reductive dissolution of Mn(IV)-Fe(III)-oxyhydroxides (German *et al.*, 1991; Byrne and Sholkovitz, 1996). In many Archaean and early Palaeoproterozoic BIF there are no Ce anomalies, and thus no deviation from trivalent Ce behaviour (*e.g.* Alexander *et al.*, 2008; Frei *et al.*, 2008), suggesting that the water column from which ferric oxyhydroxides precipitated was anoxic (Bau and Dulski, 1996). In support of this model, a recent survey of 18 different Palaeoproterozoic and Archaean BIF did not display significant Ce anomalies until after atmospheric oxygenation at ~2.5–2.4 Ga (Planavsky *et al.*, 2010a).

There also appear to be differences in trivalent *REE* behaviour in BIF before and after the permanent rise of atmospheric oxygen. Late Palaeoproterozoic BIF show significant ranges in light-to-heavy *REE* ($Pr/Yb_{(SN)}$) ratios, both below and above the shale composite value (Planavsky *et al.*, 2010a). This range of light-to-heavy *REE* and Y/Ho ratios in late Palaeoproterozoic BIF probably reflects variable fractionation of *REE* + Y by Mn(IV)-Fe(III)-oxyhydroxide precipitation and dissolution. This interpretation implies deposition of late Palaeoproterozoic BIF at ~1.9 Ga in basins with varying redox conditions and a strong redoxcline separating an upper oxic water column from deeper waters that were suboxic to anoxic (Planavsky *et al.*, 2009). These ranges are also similar to those seen in modern anoxic basins. In contrast, most Archaean and early Palaeoproterozoic BIF deposited before the rise of atmospheric oxygen are characterized by consistent light *REE* depletion (Planavsky *et al.*, 2010a). This consistent depletion in light *REE* suggests the lack of a discrete redoxcline and points towards O_2-independent Fe(II) oxidation mechanisms.

4.1.2. Trace element isotope compositions

In addition to *REE* signatures recorded in BIF, Fe isotope signatures archived in these deposits are also used to shed light on ancient redox conditions. The primary Fe minerals that ultimately sedimented to yield BIF were probably formed in the water column as the result of oxidation of $Fe(II)_{aq}$, where overall fractionation appears to be independent of the oxidative pathway, whether by molecular oxygen, anaerobic photoferrotrophy or UV photo-oxidation (Fig. 4; *e.g.* Bullen *et al.*, 2001; Croal *et al.*, 2004; Balci *et al.*, 2006; Staton *et al.*, 2006). While complete oxidation of an $Fe(II)_{aq}$ pool will result in ferric oxyhydroxides with a $\delta^{56}Fe$ composition equal to that of the initial pool, partial oxidation will result in ferric oxyhydroxides that are up to +4‰ higher (*e.g.* Wu *et al.*, 2012). Reactions between these ferric oxyhydroxide precipitates and various $Fe(II)_{aq}$ pools as they descend through the water column and their diagenesis upon burial result in their conversion to important BIF forming minerals (*e.g.* hematite, magnetite, siderite, ankerite).

In Precambrian BIF, positive $\delta^{56}Fe$ signatures are thought to occur *via* the interaction of ferric oxyhydroxides formed by partial oxidation of hydrothermally sourced $Fe(II)_{aq}$, with more pristine pools of hydrothermally sourced $Fe(II)_{aq}$ contributing near-zero $\delta^{56}Fe$ values. Such a scenario is easily envisioned for oxidative mechanisms restricted to the upper water column, where isotopically heavy precipitates formed in a zone of partial oxidation rain into a deeper, isotopically pristine hydrothermal $Fe(II)_{aq}$ pool. Alternatively, positive signatures may be generated *via* the partial or near-complete reduction of ferric oxyhydroxides of nearly any initial isotope composition. Dissimilatory Fe reduction (DIR) by Fe-reducing bacteria releases isotopically light $Fe(II)_{aq}$ and leaves the residual oxide isotopically heavy (Fig. 4). This appears reflected in the Fe isotope composition of mineral separates from Precambrian BIF, where hematite is systematically heavier than magnetite, the latter probably formed by reaction of primary Fe oxyhydroxides with isotopically light Fe released by DIR (Johnson *et al.*, 2008a). With few electron acceptors available for anaerobic respiration in marine sediments on the O_2-poor early Earth, DIR would have probably constituted a major remineralization pathway during sedimentary diagenesis (Konhauser *et al.*, 2005). The Fe isotope record in both sedimentary pyrites and Precambrian BIF (Johnson *et al.*, 2008b) demonstrates a profound excursion to extremely negative Fe isotope values ~2.7 Ga that would appear to reflect increasingly important DIR in the run-up to Earth surface oxygenation (Fig. 6). In this model, extremely light Fe isotope

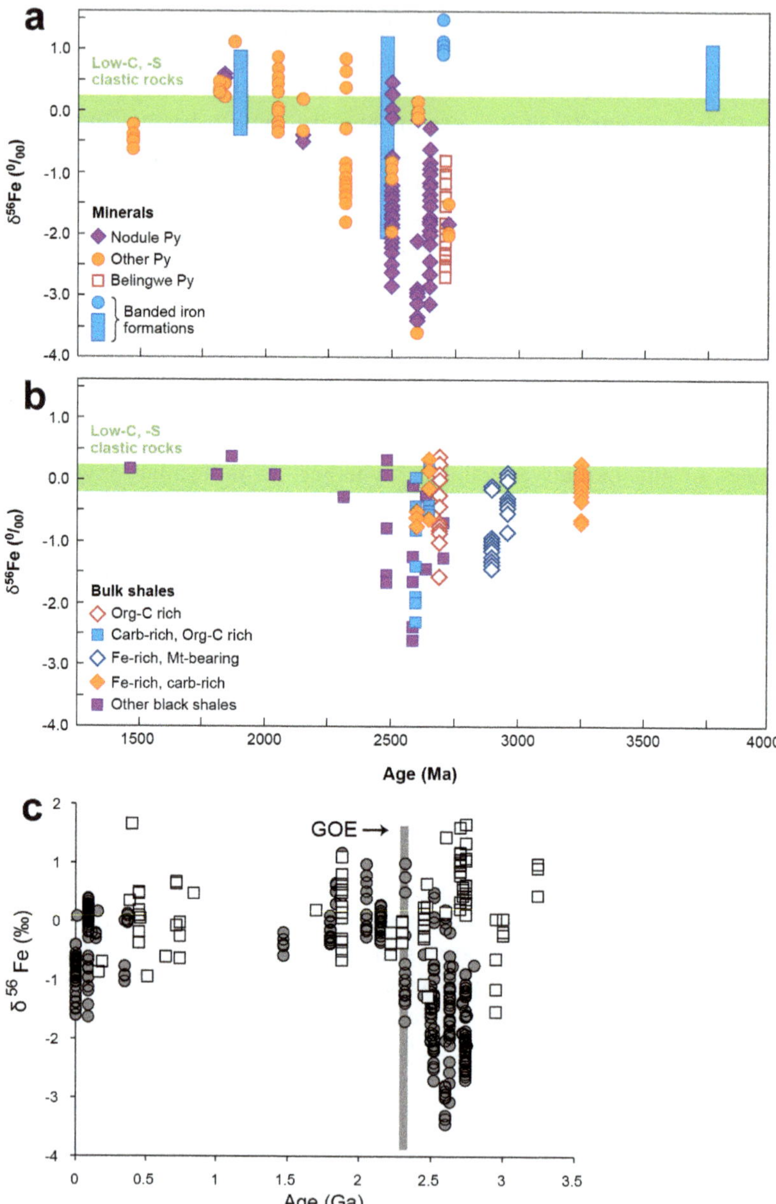

Figure 6. Iron isotope compositions of sedimentary sulfide minerals (a), iron formations (b) and whole-rock data from shales (c) through the Archaean to the Mesoproterozoic. Adapted from Johnson *et al.* (2008) and Planavsky *et al.* (2012).

compositions reflect a period in Earth's history where reduction of increasingly abundant sedimentary Fe oxyhydroxides was one of the most important organic carbon remineralization pathways in deep marine settings on the O_2- and SO_4-poor early Earth.

The apparently sensitive character of Mo isotope redox proxies has also been exploited in Precambrian Fe formations in the search for the oldest traces of free O_2 on Earth. Examining a suite of Precambrian Fe formations of various ages, Planavsky *et al.* (2014) revealed a trend between Mo isotope compositions and Fe/Mn ratios (Fig. 7), with isotopically lighter Mo isotope compositions occurring as Fe/Mn decreases (and thus Mn increases). This is interpreted as a redox array where under more reducing conditions, when conditions are prohibitive to Mn(II) oxidation and Mn(III,IV) oxide deposition, Mo isotope compositions of BIF tend towards values expected for equilibrium isotope fractionation between Fe oxyhydroxides and seawater (with Mo in Fe oxyhydroxides being ~1‰ lighter than in coeval seawater); when Mn(III,IV) oxides are deposited under more oxidizing conditions, the greater equilibrium Mo isotope fractionation between Mn oxides and seawater is expressed (with Mo in Mn oxides being ~2‰ lighter than coeval seawater). Molybdenum isotope compositions as light as −0.9‰ in the 2.95 Ga Sinqeni Fe formation (Pongola Supergroup, S. Africa) could be

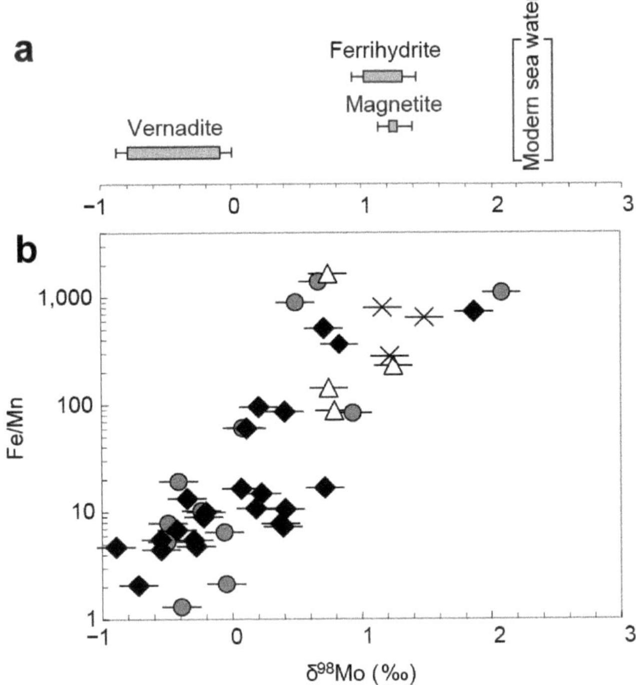

Figure 7. Mo isotope compositions of diverse Precambrian iron formations whose trend as a function of Fe/Mn ratios indicate adsorption of Mo onto Mn oxides, and by consequence the presence of appreciable free O_2, as far back as 2.96 Ga. Adapted from Planavsky *et al.* (2014).

generated only by Mo adsorption onto Mn(III,IV) oxides during deposition of this deposit; in the study of Planavsky *et al.* (2014), the Mo isotope data appear to be signalling the ancient presence of Mn(III,IV) oxides, and thus significant O_2, in the environment despite the fact that no Mn(III,IV) oxides are currently found in the deposit (probably having been the subject of dissimilatory Mn(III,IV) reduction, leaving the Fe formation enriched only in Mn(II)). These data are amongst Earth's earliest evidence for free O_2 in the environment, and place important constraints on the timing of the evolution of oxygenic photosynthesis. Mo isotope proxies have also been applied widely in the study of evolving marine palaeo-redox using the more extensive shale record (Section 4.3.2).

4.1.3. Trace-element concentrations

In natural systems where trace-element sequestration by authigenic ferric oxyhydroxides results from a continuum of adsorption and co-precipitation reactions, lumped-process distribution coefficient models can be used to relate the concentration of an element in the Fe oxide to the dissolved concentration present at the time of precipitation, *i.e.* they reflect contemporaneous seawater trace element signatures. This predictive aspect of metal sorption reactions has been exploited to better understand the BIF record with respect to ancient seawater composition and nutrient limitations on Precambrian primary productivity.

The phosphorus (P) content of BIF has received significant attention as ferric oxyhydroxides possess a strong affinity for dissolved phosphate, and phosphate is a bio-essential nutrient the availability of which can limit phytoplankton productivity in modern oceans. Today, particles of ferric oxyhydroxide that precipitate out of some marine hydrothermal vent fluids upon mixing with oxygenated seawater have been noted to strongly adsorb phosphate from seawater, to the point where the resulting particles attain molar P/Fe(III) ratios of >0.2 (Feely *et al.*, 1990, 1998). Seawater phosphate concentrations at hydrothermally active spreading centres may be strongly affected by adsorption onto ferric oxyhydroxides; at the Juan de Fuca ridge for example, significant negative phosphate anomalies have been noted at depth, with phosphate declining most appreciably (by ~60 nM) nearest to the ridge axis, coincident with increased sediment Fe and P concentrations (Feely *et al.*, 1990). Later work by Wheat *et al.* (1996) indicated that marine hydrothermal systems annually remove ~50% of the pre-industrial dissolved riverine phosphate flux, and that the co-precipitation of phosphate with ferric oxyhydroxides in seawater accounts for 36–66% of that removal rate. The extent to which phosphate is incorporated into the particles depends on the available phosphate concentration, in a manner that can be described using a distribution coefficient (K_D):

$$K_D = \frac{P/Fe(III)}{[P]} \tag{8}$$

where *P/Fe(III)* represents the particle's solid phase molar ratio, and [*P*] represents the dissolved phosphate concentration at the time of particle precipitation. Both experimental studies (Konhauser *et al.*, 2007) and field data (Feely *et al.*, 1998) demonstrate that this relationship is well described at pH ~8 by a K_D value of ~0.06–0.075 mM^{-1}.

Assuming that the ferric oxyhydroxides precipitates originally constituting Precambrian BIF behaved similarly, Bjerrum and Canfield (2002) applied a K_D value of 0.06 mM^{-1} to extrapolate Precambrian ocean phosphate concentrations from the concentrations of Fe(III) and P retained in BIF. Their results indicated that ocean phosphate concentrations might have been low enough (10–25% of present-day values) to significantly limit biological productivity prior to 1.8 Ga.

Konhauser *et al.* (2007) re-examined the BIF palaeo-proxy for P by evaluating the role of Si as a ferric oxyhydroxide co-precipitate and phosphate adsorption competitor. Throughout the Precambrian, the oceans had a significantly higher concentration of dissolved Si, at least as high as at saturation with cristobalite (0.67 mM at 40°C in seawater), and possibly even amorphous silica (2.20 mM) (Maliva *et al.*, 2005). Silica concentrations close to or at saturation with respect to amorphous silica probably persisted until the early Cretaceous, when the disappearance of VMS-associated hematitic jasper deposits indicate that diatoms or other Si-secreting organisms began actively scavenging Si (Grenne and Slack, 2003), eventually to reduce its concentration to a modern average that is <0.10 mM (Tréguer *et al.*, 1995). Crucially, Si strongly adsorbs to ferric oxyhydroxides and effectively competes with phosphate for available adsorption sites. This was demonstrated experimentally by Konhauser *et al.* (2007), who used a revised K_D approach accounting for Fe-Si co-precipitation indicates that with sufficiently high Si concentrations, the scavenging of phosphate by Fe-rich hydrothermal fluid precipitates is minimal (Fig. 8). The BIF P record, interpreted in this light, might indicate that phosphate was more abundant in Precambrian oceans than today. Additionally, high levels of carbonate saturation during the Precambrian (*e.g.* Grotzinger, 1990) would have inhibited carbonate fluorapatite (CFA) formation, which today is the largest burial flux for phosphorous in modern oceans (Ruttenberg and Berner, 1993). Combined, these two factors suggest higher, rather than lower dissolved Archaean phosphorous concentrations. A subsequent increase in P content of BIF in the Neoproterozoic, following Snowball Earth deglaciations, may then have led to enhanced cyanobacterial photosynthesis, which, in turn, produced enough oxygen to facilitate the evolution of animal life (Planavsky *et al.*, 2010b; Fig. 9a). Jones *et al.* (2015) recently re-evaluated the partitioning of P in solutions even more closely approximating seawater, and found that beyond Si, Ca and Mg can have equally important effects on P partitioning to the surfaces of Fe oxyhydroxide minerals, and the subject of calibration of adsorption-dependent Fe oxyhydroxide proxies remains an area of active investigation.

The Ni content of BIF has also received significant attention because it has been shown that Ni concentrations in BIF have changed dramatically over time, and that a drop in Ni availability in the oceans around 2.7 Ga would have had profound consequences for microorganisms that depended on it, that being methane-producing bacteria called methanogens (Konhauser *et al.*, 2009, 2015; Fig. 9b). These bacteria

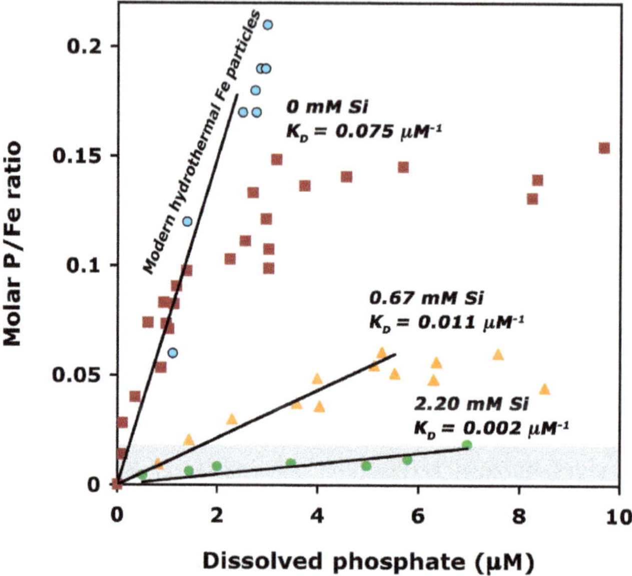

Figure 8. Experimentally determined partitioning coefficients (K_D; black lines and bold text) for partitioning of phosphate between simulated seawater and ferrihydrite under different ambient dissolved silica conditions (red squares, orange triangles, and green circles represent 0, 0.67, and 2.2 mM dissolved Si, respectively). Also shown are data for modern hydrothermal plumes worldwide (blue circles). This experiment demonstrates that suppressed partitioning of P into iron oxyhydroxides under Precambrian marine Si concentrations may be responsible for the low P/Fe ratio of BIF (area shaded grey). Adapted from Konhauser *et al.* (2007).

have a unique Ni requirement for their methane-producing enzymes, and crucially, these bacteria have been implicated in controlling oxygen levels on the ancient Earth as the methane they produced was reactive with oxygen and kept atmospheric oxygen levels low. It is possible that a Ni famine eventually led to a cascade of events that began with reduced methane production, the expansion of cyanobacteria into shallow-water settings previously occupied by methanogens, and ultimately increased oxygenic photosynthesis that tipped the atmospheric balance in favour of oxygen, the so-called GOE at 2.45 Ga. Perhaps it is no coincidence that at this time stromatolites expanded and diversified into shallow-water settings (McNamara and Awramik, 1992), and with increased consumption of CO_2, indiscriminate sedimentation of massive amounts of calcium carbonate took place *in situ* on the sea floor, leading to decimetre- to metre-thick beds that extended over thousands of square kilometres (Grotzinger and Knoll, 1999).

A recent compilation of Cr enrichment in BIF shows a profound enrichment coincident with the GOE around 2.45 Ga (Konhauser *et al.*, 2011; Fig. 9c). After that, Cr enrichments started to increase in shallow-water BIF and peaked synchronous with the permanent loss at ~2.32 Ga of mass-independent fractionation of sulfur isotopes that defines the GOE (Bekker *et al.*, 2004, Guo *et al.*, 2009). Given the insolubility of Cr

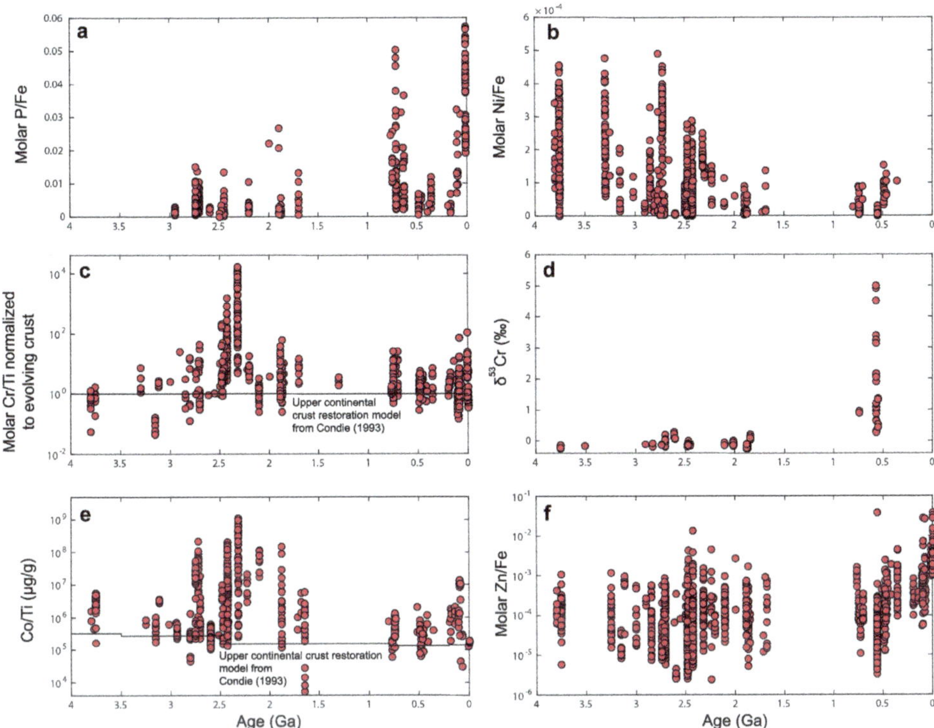

Figure 9. Several notable BIF trace element enrichment and isotope records: (a) phosphorus as molar P/Fe ratios from Planavsky *et al.* (2010b), (b) Ni as molar Ni/Fe ratios from Konhauser *et al.* (2015), (c) Cr as molar Cr/Ti ratios from Konhauser *et al.* (2011), (d) Cr isotope compositions from Frei *et al.* (2009), (e) Co as molar Co/Ti ratios from Swanner *et al.* (2015), and (f) Zn as molar Zn/Fe ratios from Robbins *et al.* (2013). See text for detailed explanations of each.

minerals, its mobilization and incorporation into BIF indicates enhanced chemical weathering at that time, probably associated with the evolution of aerobic continental pyrite oxidation. As the GOE commenced, aerobic chemolithoautotrophic weathering of pyrite on land produced significant acidity, which would have increased sulfate and nutrient fluxes to seawater (Reinhard *et al.*, 2009; Konhauser *et al.*, 2011; Bekker and Holland, 2012). This profound shift in weathering regimes constitutes Earth's first acid continental drainage system, and accounts for independent evidence of increased supply of sulfate (Bekker *et al.*, 2004) and sulfide-hosted trace elements to the oceans at that time (Scott *et al.*, 2008). With increased sulfate supply to the oceans, marine sulfate-reducing bacteria would have begun to dominate the anoxic layers of the water column, leading to the further marginalization of methanogens to the sulfate-poor bottom sediments, as is the case today. The end result of the progressive decline in methane production (a potent greenhouse gas) was a cooling climate, eventually culminating in the onset of major Palaeoproterozoic glaciations (Zahnle *et al.*, 2006).

During the past few years, Cr isotopes in BIF have also been used to track the progressive oxygenation of Earth's surface environment during the Archaean and Palaeoproterozoic (Fig. 9d). Frei *et al.* (2009) measured slight positive Cr fractionations (+0.04 to +0.29 ‰) in samples deposited between 2.8 and 2.6 Ga, which they suggested was a first-order proxy for the presence of Cr(VI) in seawater at that time. Because of the efficient reduction of Cr(VI) by Fe^{2+} (which was abundant in the highly ferruginous Archaean oceans), and the near instantaneous precipitation of $(Fe,Cr)(OH)_3$ at seawater pH, the stable Cr isotope signatures of BIF should thus mirror the initial reaction of Cr(III) from mineral dissolution on land with MnO_2. Interestingly, BIF deposited between 2.5 and 1.9 Ga show minimal positive fractionations in Cr isotopes, suggesting that oxygen levels then dropped in the aftermath of the GOE. This is an unusual finding in that it was previously believed that atmospheric O_2 levels progressively increased post GOE (Kump, 2008), but the Cr isotope record in BIF instead documents transient fluctuations in atmospheric oxygenation. Most recently, Crowe *et al.* (2013) reported ^{53}Cr-depletion in 2.96 Ga palaeosols (ancient soil horizons) and ^{53}Cr-enrichment in overlying 2.96 Ga BIF. These near-contemporaneous units thus seem to have captured the Cr(VI) pool originating from oxidative continental weathering some 500 Ma before the GOE. Based on the assumption that export of Cr(VI) from the weathering environment to the oceans necessitates minimal encounter with Fe(II) in rivers or groundwater, those authors further calculated that oxygen concentrations may have been as high as 3×10^{-4} times present atmospheric levels.

Frei *et al.* (2009, 2012) reported further strong positive fractionations (δ^{53}Cr up to +4.9‰) in the late Neoproterozoic, suggestive of increased surface oxygenation at that time. Importantly, this work provided supporting evidence that linked a significant rise in global oxygenation to the emergence of early animals (*e.g.* Planavsky *et al.*, 2010b). If positive δ^{53}Cr values are indicative of oxidative weathering, then the lack of such fractionations suggests minimal O_2 (to produce the MnO_2 catalyst). Indeed, in a recent study of granular Fe formations (reworked BIF) that were deposited between 1.7 and 0.9 Ga, Planavsky *et al.* (2014) observed δ^{53}Cr of −0.12‰. They hypothesized that the lack of Cr redox cycling must reflect atmospheric O_2 levels of <0.1% present atmospheric levels, values below the theoretical estimates for the minimum O_2 requirements of the earliest animals. In other words, after the GOE oxygen levels plummeted, and it was not until the Neoproterozoic that O_2 levels increased sufficiently to facilitate animal evolution.

Cobalt is a bio-essential trace element and co-limiting (*e.g.* with Fe) nutrient in some regions of the modern oceans. It has been proposed that Co was more abundant in poorly ventilated Precambrian oceans based on the greater utilization of Co by anaerobic microbes relative to plants and animals. Based on the sedimentary record of Co enrichments and a model of marine Co sources and sinks (Fig. 9e), Swanner *et al.* (2014) proposed that before ~1.8 Ga the large volume of hydrothermal fluids circulating through abundant submarine ultramafic rocks and the predominance of anoxic sediments with a negligible capacity for Co burial resulted in a large marine Co reservoir. A decrease in marine Co concentrations in the Middle Proterozoic resulted from waning hydrothermal Co sources and the expansion of oxic and euxinic sediments that buried Co with Mn(III,IV) oxides or pyrite, respectively. It is interesting to note that the initial diversification of eukaryotes, and thus eukaryotic Co-utilizing genes, occurred in the middle to late Proterozoic (*e.g.* Knoll *et al.*, 2007), at a time when marine Co concentrations were declining. Along similar lines, marine cyanobacteria probably utilize Co-binding ligands as a strategy for Co acquisition (Saito and Moffett, 2001; Saito *et al.*, 2005), which may help to explain their adaptation to low-Co environments despite high physiological demand. Perhaps this suggests that the number of Co-utilizing genes encoded by an organism's genome do not always mirror the metal's availability.

The lack of a clear link between metabolic metal demands and environmental concentrations is most clearly demonstrated by Zn. The relatively late proliferation of Zn metallomes by eukaryotes had previously been linked to marine Zn biolimitation (Williams and da Silva, 1996; Saito *et al.*, 2003; Dupont *et al.*, 2010). However, a recent examination of the BIF and shale records (Robbins *et al.*, 2013; Scott *et al.*, 2012, respectively; Fig. 9f) indicates that Zn may have been near modern abundances and was probably bioavailable to eukaryotes throughout the Precambrian, casting doubt on the coupled geochemical and eukaryotic evolutions with respect to Zn utilization. A novel possibility stemming from this work is that the late proliferation of Zn metalloenzymes in eukaryotes could have been a biologically intrinsic process related to the regulation of increasingly complex genomes, rather than solely dependent on the dramatic changes in the marine bioavailability of aqueous Zn species.

4.2. Ferromanganese crusts

Ferromanganese crusts have received significant attention due their ability to sorb certain trace elements from seawater, in particular Mo and Co. Molybdenum is supplied to the oceans primarily by the riverine flux of molybdate (MoO_4^{2-}) derived from the oxidative weathering of sulfide minerals (*e.g.* Archer and Vance, 2008). The molybdate anion is highly conservative, with a long residence time, in today's well mixed oceans. As a result, Mo is the most abundant transition element in the modern oxygenated oceans, at an average concentration of ~100 nM (Bertine and Turekian, 1973). The only significant sink for molybdate in well oxygenated environments is believed to be co-precipitation and burial with Mn oxyhydroxides, but this is a very slow process that removes only minor amounts of Mo from the ocean today (Bertine and Turekian, 1973). The adsorption of Co(II) and Co(III) to surface sites on ferric oxyhydroxides and Mn(III,IV) oxides, respectively, is an important pathway for scavenging of Co under oxic conditions (Koschinsky and Hein, 2003; Takahashi *et al.*, 2007; Stockdale *et al.*, 2010). By contrast, soluble Co maxima occur below the O_2-H_2S chemocline in modern euxinic basins in conjunction with both the soluble Fe and Mn peaks (Viollier *et al.*, 1995), which indicates that Co is released *via* the reductive dissolution of both Fe(III) oxyhydroxides and Mn(III,IV) oxides.

4.3. Black shales

4.3.1. Trace element concentrations

In modern sedimentary environments, black shales are used to examine palaeo-productivity based on trace element enrichments that correspond to increased organic carbon burial (*e.g.* Tribovillard *et al.*, 2006) or the concentrations of trace elements relative to total organic carbon (Me/TOC) (*e.g.* Alego and Rowe, 2012). This linkage is mainly due to trace elements with multiple redox states being precipitated and enriched during periods of reducing conditions, *e.g.* uranium (U) and Mo under sulfidic conditions or Cr under ferruginous conditions (Arnold *et al.*, 2004; Dahl *et al.*, 2013; Partin *et al.*, 2013; Reinhard *et al.*, 2013) or incorporated into microbial biomass, *e.g.* Ni, Cu, Zn and Cd (Twining and Baines, 2013). Indeed, the utility of shales as palaeo-marine proxies has grown in recent years as an indicator of palaeo-environmental conditions on the ocean floor (*i.e.* oxic, ferruginous or euxinic) with the advent of Fe speciation analyses that assess the degree of pyritization by comparing the ratio of pyrite Fe to reactive Fe (*e.g.* Raiswell and Canfield, 1998; Poulton and Canfield, 2005).

Black shales represent an important fraction of all ancient marine sediments preserved, and because of this, their trace element and isotope compositions have been used to track major biogeochemical events in Earth's history (*e.g.* Lyons *et al.*, 2009). This includes, amongst others: (1) the rise of atmospheric oxygen; (2) the onset of oxidative continental weathering; (3) the establishment of aerobic marine ecosystems; (4) the post-GOE drop in atmospheric O_2; and (5) the extent of Proterozoic marine euxinia. Arguably the smoking gun for the rise of atmospheric O_2 (*i.e.* the GOE) comes from the loss of sulfur isotope mass-independent fractionation (S-MIF) (Farquhar *et al.*, 2000). The processes that lead to S-MIF are restricted to the photochemical dissociation of SO_2 in the upper atmosphere into elemental and water-soluble S species. They then get rained out and incorporated into diagenetic pyrites, such as those associated with black shales. The only way to preserve the S-MIF signal is under low atmospheric O_2 levels because O_2 erases the unique signatures of these reactions and the presence of ozone prevents the penetration of ultraviolet radiation that generates the signals. The last appearance of S-MIF can be dated to pyrites in 2.32 Ga-old black shales of the Rooihoogte/Duitschland and Timeball Hills formations in South Africa, marking this point in Earth's history as a time of atmospheric oxygenation (Bekker *et al.*, 2004; Guo *et al.*, 2009).

In the past decade, several geochemical studies have been suggestive of oxidative continental weathering having occurred prior to the GOE. For example, increases in Mo and rhenium (Re) concentrations, as well as the first trend towards loss of S-MIF, comes from the 2.5 Ga Mount McRae Shale of Western Australia (Anbar *et al.*, 2007; Kaufman *et al.*, 2007). The trace element enrichments were interpreted as reflecting oxidative weathering of terrestrial sulfide minerals (but at much lower levels than at present), their transport to the oceans as dissolved oxyanion forms (*e.g.* MoO_4^{2-} and ReO_4^-), and then their being scavenged into black shales. The shales lack U enrichment, which, because it is predominantly hosted in silicates, would be less affected by rising atmospheric oxygen levels. Interestingly, Mo and Re enrichments were not observed in the upper sections of the shales, suggesting a subsequent drop in atmospheric O_2, or as coined by Anbar *et al.* (2007), the transient O_2 increase reflected a "whiff" of oxygen. Similar patterns of these metals had been reported in the 2.64–2.50 Ga Ghaap Group in South Africa (Wille *et al.*, 2007). Most recently, Stueken *et al.* (2012) suggested that an increase in the total sulfur and Mo supply to marine shales at 2.8 Ga was best explained by the biological oxidation of crustal sulfide minerals.

Evidence in support of transient rises in atmospheric O_2 prior to the GOE implies the evolution and establishment of cyanobacteria in the ocean-water column. The presence of dissolved O_2 in seawater as early as 2.7 Ga is inferred from (1) black shales based on shifts toward heavier $\delta^{15}N$ values in kerogen consistent with the onset of an oxic nitrogen cycle (Godfrey and Falkowski, 2009); (2) C isotope data in kerogens indicating methanotrophy (Eigenbrode and Freeman, 2006; Thomazo *et al.*, 2013); and (3) coupled Fe and Mo isotopes that indicate the formation of ferric oxyhydroxides in the upper water column which were probably linked to cyanobacteria rather than photoferrotrophs (Czaja *et al.*, 2012): by 2.56 Ga, O_2 levels may have already reached several hundred metres on the continental slope (Kendall *et al.*, 2010). Based on the presence of 10 wt.% organic content in the 3.2 Ga Gorge Creek Group in northwestern Australia, and the lack of Fe and sulfur, it is even possible that oxygenic photosynthesis may have evolved then. As Buick (2008) suggested, it is difficult to explain another mode of primary productivity that would have had enough available substrates over so wide an area for such lengths of time.

Until a few years ago, most researchers were of the opinion that once O_2 accumulated in the oceans and atmosphere it would have continued to rise until modern values were reached, sometime in the late Neoproterozoic (*e.g.* Kump, 2008). However, a study focusing on the temporal trends of U enrichment in black shale demonstrated that the initial rise of atmospheric O_2 at ~2.45 Ga was followed by a dramatic decline to less oxidizing conditions only several hundred million years after the GOE itself (Partin *et al.*, 2013). Somewhat similar trends were observed in the Mo and Mo/TOC records in black shales, but with evidence for a large oceanic Mo inventory occurring only at 2.2 Ga and then dropping to 10–20% of modern ocean levels until the Phanerozoic (Scott *et al.*, 2008). The smaller seawater Mo inventory, in the immediate aftermath of the GOE, was interpreted to reflect the Mo distribution in the continental crust, as has been argued for Re by Hannah *et al.* (2004). The present continental flux of Mo, like that of Re, is largely derived from organic matter-rich sulfidic shales; at the time of the GOE, the pre-GOE organic matter-rich sulfidic shales were not significantly enriched in Mo above crustal levels (Yamaguchi, 2002; Scott *et al.*, 2008). The predominant crustal reservoir of Mo that oxidized to form molybdate at this time would have been crystalline crustal rocks with generally low Mo content. This would result in a small supply of Mo to the oceans in the immediate aftermath of the GOE. The time for ingrowth of a large reservoir of Mo-enriched shales may have been comparable to the crustal residence time (200 My), and the process would ultimately lead to greater oceanic Mo contents as its concentration increased in the source rocks (*e.g.* black shales) that were being weathered.

The solubility and burial of U and Mo also depend on the sulfide concentration because both are chalcophilic. Therefore, in modern euxinic environments, such as the Black Sea, Mo sinks outpace Mo supply. Low Mo concentrations develop in the water column, and the magnitude of Mo enrichment and Mo/TOC ratios in sediments drop precipitously (Algeo and Lyons 2006). This feature of Mo geochemistry forms the basis for the 'bio-inorganic bridge' hypothesis, whereby changes in the chemistry of the early oceans directly affected the bioavailability of trace elements, and in turn, the evolution of life (Anbar and Knoll, 2002). Those authors suggested that the drawdown of Mo under euxinic conditions limited primary photosynthetic productivity in the photic zone of the oceans for over a billion years between the late Palaeoproterozoic and Neoproterozoic (the so-called 'boring billion'; Buick *et al.*, 1995). The cause for the global euxinia was attributed to the progressive rise in atmospheric O_2 after the GOE that increased pyrite oxidation on land, which in turn led to a greater supply of dissolved sulfate to the oceans, enhanced sulfate reduction in the oceans, and the generation of sufficient dissolved sulfide that the total dissolved Fe flux in the oceans was titrated out by 1.8 Ga (Canfield, 1998). It was not until the Neoproterozoic, in association with a second large oxidation of the Earth's surface, that the deep oceans became oxygenated. However, the observed Mo drawdown and complementary Mo isotope data during that time interval are inconsistent with ocean-wide euxinia (*e.g.* Arnold *et al.*, 2004), and a recent investigation of sedimentary Mo and Cr enrichments in black shales, using mass-balance box models, demonstrated that only ~1–10% of the modern seafloor area needed to be anoxic at that time (Reinhard *et al.*, 2013).

4.3.2. Molybdenum isotopes in shales

When Mo removal from seawater is quantitative, the Mo isotope composition of reducing sediments may be linked directly to the isotopic composition of contemporaneous seawater, and in turn, to the relative importance of reducing and oxic areas of the seafloor. This property of the Mo isotope system has been exploited extensively to reconstruct palaeo-redox from black shales, the organic-rich nature of which testifies to depletion of available electron acceptors in the sedimentary pile and the establishment of reducing conditions (either anoxic or euxinic) amenable to recording the Mo isotope composition of contemporaneous seawater (Section 3.4.2). Using a conservative assumption of 70% Mo removal by euxinic black shales in the mid-Proterozoic, Arnold *et al.* (2004) used simple isotope mass balance to calculate the Mo isotopic composition of mid-Proterozoic seawater to be ~0.8‰. This isotopically light value for seawater was interpreted to reflect expanded euxinia and a suppressed oxic sink. Assuming a riverine source near 0‰, Arnold *et al.* (2004) initially estimated a ratio between the fractional importance of oxic (f_{ox}) and euxinic (f_{eux}) sinks as being 10 times lower than today (f_{ox}/f_{eux} <0.4 in the mid-Proterozoic *vs.* ~3 today). Using a newly determined modern riverine source value of ~0.7‰, Archer and Vance (2008) re-interpreted the low isotopic composition of mid-Proterozoic black shales (relative to modern seawater) as reflecting near total drawdown of the marine Mo reservoir as the result of expanded euxinia, with sedimentary Mo isotope compositions tending towards the source composition (mostly river water). In other words, Mo isotope data from shales reveal that mid-Proterozoic oceans were characterized by important euxinic Mo drawdown and little evidence for any important role for oxic sedimentary Mo exit channels. Similarly light (relative to modern seawater) Mo sedimentary isotope compositions indicating near total absence of oxic Mo exit channels have been reported for black shales from the 2.05 Ga Palaeoproterozoic Shunga carbon burial event in the aftermath of the GOE (Asael *et al.*, 2013), during the mid-Neoproterozoic in the run-up to the appearance of the Ediacaran fauna (Dahl *et al.*, 2011), and during Phanerozoic ocean anoxic events (OAE) such as the Toarcian (early Jurassic) OAE (Pearce *et al.*, 2008). Using a mass-balance model with spatially variable metal burial rates and by also considering Cr (whose marine exit channels appear insensitive to the distinction between anoxia and euxinia), Reinhard *et al.* (2013) refined sedimentary Mo exit channels under low-O_2 (anoxic and euxinic) conditions, finding that a relatively small spatial extent of euxinia (1–10% of the seafloor, *vs.* 0.1% today) would be sufficient to crash the marine dissolved Mo reservoir, and by inference, draw seawater Mo isotope compositions towards riverine sources. Early models indicating widespread global euxinia throughout mid-Proterozoic oceans (*e.g.* Canfield *et al.*, 1998) are thus better characterized by expanded but spatially restricted euxinia combined with widespread anoxia.

Molybdenum isotope-based redox proxies appear to be very sensitive to even the smallest amounts of free oxygen, and are now being used to examine trace amounts of dissolved Mo that may have been released prior to the ~2.4 Ga GOE and sustained oxidation of crustal Mo sources. Wille *et al.* (2007) demonstrated small but non-negligible Mo isotope deviations from crustal values in 2.6–2.5 Ga black shales from the Transvaal Supergroup, South Africa, indicating the production of some free O_2 at that time. In only slightly younger (2.50 Ga) black shales in the Hamersley basin, Western Australia, work by Anbar and colleagues (Anbar *et al.*, 2007; Duan *et al.*, 2010) has revealed coupled Mo abundance and isotope excursions occurring while the atmosphere remained anoxic, reinforcing the idea that small, local oxygen-production events (*i.e.* "whiffs") probably preceded global oxygenation of the atmosphere. Kurzweil *et al.* (2015) extended the Hamersley Mo isotope dataset back by nearly 100 Ma to reveal a long-term trend of increasingly heavy Mo isotope compositions upwards in the basin. This would indicate a gradual increase in the prevalence of oxic Mo exit channels for seawater Mo prior to the GOE, consistent with the early occurrence of non-negligible Mo isotope fractionation as reported by Wille *et al.* (2007) for the contemporaneous Griqualand West basin, S. Africa.

5. Conclusions and future directions

By nature of their physicochemical surface properties, redox-reactive Fe minerals are extremely efficient scavengers of trace elements from natural waters and so can be used to provide a rich archive of trace-element concentrations and stable isotope compositions in terrestrial and marine environments over the entire history of their formation. These trace element signals can then be interpreted to shed light on environmental conditions in these environments at the time of mineral precipitation. Three Fe mineral archives are currently employed to help understand the evolution of Earth's redox and biogeochemical conditions, namely banded iron formations (BIF), ferromanganese crusts and black shales, where although shales are not technically classed as Fe-rich sediments, it is likely that many of their trace elements are sequestered in association with Fe minerals. In these sediments Fe oxyhydroxides, including ferrihydrite, goethite, hematite and magnetite, and Fe sulfides, including pyrite, provide the primary Fe-rich phases for trace-element sorption.

To date Fe mineral archives have provided a wealth of trace=element concentration and stable-isotope data which have been interpreted to reflect environmental conditions during Earth's earliest oceans and atmosphere, during the Archaean in the run up to and events after the Great Oxidation Event (GOE), through subsequent Ocean Anoxic Events (OAE), and into the relatively recent geological history of the Cenozoic. In light of these studies, it is certain that these enigmatic deposits have revolutionized our understanding of the Earth system. To continue forward with their investigation and bring new insights into Earth's redox and biogeochemical evolution, it is apparent that we must continue to improve our understanding of both the Fe mineral archives and the behaviour of their trace element impurities, both at the small-scale within the archives and at the larger-scale within the environment as a whole. By continuing to investigate the precipitation, deposition and diagenesis of Fe minerals, and both the macro- and microscopic uptake of trace elements to these minerals, we can be increasingly confident in the fidelity of trace-element signals recorded in Fe mineral archives, and better able to relate these signals to the contemporaneous environment. Similarly, by continuing to investigate the role and function of trace elements in the environment, we can continue to make links between trace element signals recorded in Fe mineral archives and biogeochemical processes that reflect contemporaneous environmental conditions.

Acknowledgements

Financial support to CP from the Science and Technology Facilities Council (STFC) UK, and to KK from the Natural Sciences and Engineering Research Council of Canada (NSERC) is gratefully acknowledged.

References

Algeo, T.J. and Lyons, T.W. (2006) Mo-total organic carbon covariation in modern anoxic marine environments: Implications for analysis of paleoredox and paleo-hydrographic conditions. *Paleoceanography*, **21**, 1016, DOI: 10.1029/2004PA001112.

Algeo, T.J. and Rowe, H. (2012) Paleoceanographic applications of trace-metal concentration data. *Chemical Geology*, **324**, 6–18.

Alexander, B.W., Bau, M., Andersson, P. and Dulski, P. (2008) Continentally-derived solutes in shallow Archean seawater: Rare earth element and Nd isotope evidence in iron formation from the 2.9 Ga Pongola Supergroup, South Africa. *Geochimica et Cosmochimica Acta*, **72**, 378–394.

Alexander, B.W., Bau, M. and Andersson, P. (2009) Neodymium isotopes in Archean seawater and implications for the marine Nd cycle in Earth's early oceans. *Earth and Planetary Science Letters*, **283**, 144–155.

Alibert, C. and McCulloch, M.T. (1993) Rare-earth element and neodymium isotopic compositions of the banded iron-formations and associated shales from Hamersley, Western-Australia. *Geochimica et Cosmochimica Acta*, **57**, 187–204.

Anbar, A.D. and Knoll, A.H. (2002) Proterozoic ocean chemistry and evolution: A bioinorganic bridge? *Science*, **297**, 1137.

Anbar, A.D., Roe, J.E., Barling, J. and Nealson, K.H. (2000) Non-biological fractionation of iron isotopes. *Science*, **288**, 126–128.

Anbar, A.D., Jarzecki, A.A. and Spiro, T.G. (2005) Theoretical investigation of iron isotope fractionation between $Fe(H_2O)(^{3+})_{(6)}$ and $Fe(H_2O)(^{2+})_{(6)}$: Implications for iron stable isotope geochemistry. *Geochimica et Cosmochimica Acta*, **69**, 825–837.

Anbar, A.D., Duan, Y., Lyons, T.W., Arnold, G.L., Kendall, B., Creaser, R.A., Kaufman, A.J., Gordon, G.W., Scott, C., Garvin, J. and Buick, R. (2007) A whiff of oxygen before the great oxidation event? *Science*, **317**, 1903–1906.

Archer, C. and Vance, D. (2008) The isotopic signature of the global riverine molybdenum flux and anoxia in the ancient oceans. *Nature Geoscience*, **1**, 597–600.

Arnold, G.L., Anbar, A.D., Barling, J. and Lyons, T.W. (2004) Molybdenum isotope evidence for widespread anoxia in mid-Proterozoic oceans. *Science*, **304**, 87–90.

Asael, D., Tissot, F.L.H., Reinhard, C.T., Rouxel, O., Dauphas, N., Lyons, T.W., Ponzevera, E., Liorzou, C. and Chéron, S. (2013) Coupled molybdenum, Fe and uranium stable isotopes as oceanic palaeo-redox proxies during the Palaeoproterozoic Shunga Event. *Chemical Geology*, **362**, 193–210.

Balci, N., Bullen, T., Witte-Lien, K., Shanks, W., Motelica, M. and Mandernack, K. (2006) Iron isotope fractionation during microbially stimulated Fe(II) oxidation and Fe(III) precipitation. *Geochimica et Cosmochimica Acta*, **70**, 622–639.

Barling, J. and Anbar, A.D. (2004) Molybdenum isotope fractionation during adsorption by manganese oxides. *Earth and Planetary Science Letters*, **217**, 315–329.

Barling, J., Arnold, G.L. and Anbar, A.D. (2001) Natural mass-dependent variations in the isotopic composition of molybdenum. *Earth and Planetary Science Letters*, **193**, 447–457.

Balistrieri, L.S., Borrok, D.M., Wanty, R.B. and Ridley, W.I. (2008) Fractionation of Cu and Zn isotopes during adsorption onto amorphous Fe(III) oxyhydroxide: Experimental mixing of acid rock drainage and ambient river water. *Geochimica et Cosmochimica Acta*, **72**, 311–328.

Bau, M. and and Dulski, P. (1996) Distribution of yttrium and rare-earth elements in the Penge and Kuruman iron-formations, Transvaal Supergroup, South Africa. *Precambrian Research*, **79**, 37–55.

Bau, M., Hohndorf, A., Dulski, P. and Beukes, N.J. (1997) Sources of rare-earth elements and iron in paleoproterozoic iron-formations from the Transvaal Supergroup, South Africa: Evidence from neodymium isotopes. *Journal of Geology*, **105**, 121–129.

Beard, B., Johnson, C., Damm, von K. and Poulson, R. (2003) Iron isotope constraints on Fe cycling and mass balance in oxygenated Earth oceans. *Geology*, **31**, 629–632.

Bekker, A. and Holland, H.D. (2012) Oxygen overshoot and recovery during the early Paleoproterozoic. *Earth and Planetary Science Letters*, **317**, 295–304.

Bekker, A., Slack, J.F., Planavsky, N., Krapez, B., Hofmann, A., Konhauser, K.O. and Rouxel, O.J. (2010) Iron formation: The sedimentary product of a complex interplay among mantle, tectonic, oceanic, and biospheric processes. *Economic Geology*, **105**, 467–508.

Bekker, A., Holland, H.D., Wang, P.L., Rumble, D., Stein, H.J., Hannah, J.L., Coetzee L. L. and Beukes, N.J. (2004) Dating the rise of atmospheric oxygen. *Nature*, **427**, 117–120.

Belshaw, N.S., Zhu, X.K., Guo, Y. and O'Nions, R.K. (2000) High precision measurement of iron isotopes by plasma source mass spectrometry. *International Journal of Mass Spectrometry*, **197**, 191–195.

Berner, R.A. (1964) An idealized model of dissolved sulphate distribution in recent sediments. *Geochimica et Cosmochimica Acta*, **28**,1497–1503.

Berner, R.A. (1981) A new geochemical classification of sedimentary environments. *Journal of Sedimentary Petrology*, **51**, 359–365.

Bertine, K.K. and Turekian, K.K. (1973) Molybdenum in marine deposits. *Geochimica et Cosmochimica Acta*, **37**, 1415–1434.

Bigeleisen, J. and Mayer, M.G. (1947) Calculation of equilibrium constants for isotopic exchange reactions. *Journal of Chemical Physics*, **15**, 261–267.

Bjerrum, C.J. and Canfield, D.E. (2002) Ocean productivity before about 1.9 Gyr ago limited by phosphorus adsorption onto iron oxides. *Nature*, **417**, 159–162.

Boyd, P.W. and Ellwood, M.J. (2010) The biogeochemical cycle of iron in the ocean. *Nature Geoscience*, **3**, 675–682.

Boyle, E.A., Edmond, J.M. and Sholkovitz, E.R. (1977) The mechanism of iron removal in estuaries. *Geochimica et Cosmochimica Acta*, **41**, 1313–1324.

Braterman, P.S., Cairnssmith, A.G. and Sloper, R.W. (1983) Photooxidation of hydrated Fe^{2+} – Significance for banded iron formations. *Nature*, **303**, 163–164.

Brown, G.E. and Parks, G.A. (2001) Sorption of trace elements on mineral surfaces: Modern perspectives from spectroscopic studies, and comments on sorption in the marine environment. *International Geological Review*, **43**, 963–1073.

Buick, R. (2008) When did oxygenic photosynthesis evolve? *Philosophical Transactions of the Royal Society B*, **363**, 2731–2743.

Buick, R., Marais D.J.D. and Knoll, A.H. (1995) Stable isotope compositions of carbonates from the Mesoproterozoic Bangemall Group, northwestern Australia. *Chemical Geology*, **123**, 153–171.

Bullen, T.D., White, A.F., Childs, C.W. and Vivit, D.V. (2001) Demonstration of significant abiotic Fe isotope fractionation in nature. *Geology*, **29**, 699–702.

Burns, R.G. and Burns, V.M. (1977) Mineralogy of ferromanganese nodules. Pp. 185–248 in *Marine Manganese Deposits* (G.P. Glasby, editor). Elsevier, Amsterdam.

Byrne, R.H. and Sholkovitz, E.R. (1996) Marine chemistry and geochemistry of the lanthanides. Pp. 497–592 in: *Handbook on the Physics and Chemistry of Rare Earths*. Vol. **23** (K.A. Gschneider Jr. and L. Eyring, editors). Elsevier.

Canfield, D.E. (1998) A new model for Proterozoic ocean chemistry. *Nature*, **396**, 450–453.

Canfield, D.E., Thamdrup, B. and Hansen, J.W. (1993a) The anaerobic degradation of organic-matter in Danish coastal sediments – Iron reduction, manganese reduction and sulfate reduction. *Geochimica et Cosmochimica Acta*, **57**, 3867–3883.

Canfield, D.E., Jorgensen, B.B., Fossing, H., Glud, R., Gundersen, J., Ramsing, N.B., Thankdrup, B., Hansen, J.W., Nielsen, L.P. and Hall, P.O.J. (1993b) Pathways of organic-carbon oxidation in 3 continental-margin sediments. *Marine Geology*, **113**, 27–40.

Carazzo, G., Jellinek, A.M. and Turchyn, A.V. (2013) The remarkable longevity of submarine plumes: Implications for the hydrothermal input of iron to the deep-ocean. *Earth and Planetary Science Letters*, **382**, 66–76.

Cloud, P. (1973) Paleoecological significance of banded iron-formation. *Economic Geology*, **68**, 1135–1143.

Condie, K.C. (1993) Chemical composition and evolution of the upper continental crust – Contrasting results from surface samples and shales. *Chemical Geology*, **104**, 1–37.

Corliss, J.B., Lyle, M., Dymond, J. and Crane, K. (1978) Chemistry of hydrothermal mounds near Galapagos Rift. *Earth and Planetary Science Letters*, **40**, 12–24.

Cornell, R.M. and Schwertmann, U. (2003) *The Iron Oxides: Structure, Properties, Reactions, Occurrences and Uses, Second Edition*, Wiley-VCH Verlag GmbH and Co. KGaA, Weinheim, Germany.

Criss, R.E. (1999) *Principles of Stable Isotope Distribution*. Oxford University Press, Oxford, UK.

Croal, L., Johnson, C., Beard, B. and Newman, D. (2004) Iron isotope fractionation by Fe (II)-oxidizing photoautotrophic bacteria 1. *Geochimica et Cosmochimica Acta*, **68**, 1227–1242.

Crowe, S.A., Dossing, L.N., Beukes, N.J., Bau, M., Kruger, S.J., Frei, R. and Canfield, D.E. (2013) Atmospheric oxygenation three billion years ago. *Nature*, **501**, 535.

Czaja, A.D., Johnson, C.M., Beard, B.L., Roden, E.E., Li, W.Q. and Moorbath, S. (2013) Biological Fe oxidation controlled deposition of banded iron formation in the *ca.* 3770 Ma Isua Supracrustal Belt (West Greenland). *Earth and Planetary Science Letters*, **363**, 192–203.

Dahl, T.W., Canfield, D.E., Rosing, M.T., Frei, R.E., Gordon, G.W., Knoll, A.H. and Anbar, A.D. (2011)

Molybdenum evidence for expansive sulfidic water masses in ~750 Ma oceans. *Earth and Planetary Science Letters*, **311**, 264–274.

Dahl, T.W., Chappaz, A., Fitts, J.P. and Lyons, T.W. (2013) Molybdenum reduction in a sulfidic lake: Evidence from X-ray absorption fine-structure spectroscopy and implications for the Mo paleoproxy. *Geochimica et Cosmochimica Acta*, **103**, 213–231.

Dauphas, N., Cates, N.L., Mojzsis, S.J. and Busigny, V. (2007) Identification of chemical sedimentary protoliths using Fe isotopes in the >3750 Ma Nuvvuagittuq supracrustal belt, Canada. *Earth and Planetary Science Letters*, **254**, 358–376.

Davis, J.A. and Leckie, J.O. (1978) Surface ionization and complexation at the oxide/water interface. II. Surface properties of amorphous Fe oxyhydroxide and adsorption of metal ions. *Journal of Colloid and Interface Science*, **67**, 90–105.

Dean, W.E. and Greeson, P.E. (1979) Influences of algae on the formation of freshwater ferromanganese nodules, Onedia Lake, New York. *Archiv für Hydrobiologie*, **86**, 181–192.

Derry, L.A. and Jacobsen, S.B. (1988) The Nd and Sr isotopic evolution of Proterozoic seawater. *Geophysical Research Letters*, **15**, 397–400.

Derry, L.A. and Jacobsen, S.B. (1990) The chemical evolution of Precambrian seawater – Evidence from REEs in banded iron formations. *Geochimica et Cosmochimica Acta*, **54**, 2965–2977.

Duan, Y., Anbar, A.D., Arnold, G.L., Lyons, T.W., Gordon, G.W. and Kendall, B. (2010) Molybdenum isotope evidence for mild environmental oxygenation before the Great Oxidation Event. *Geochimica et Cosmochimica Acta*, **74**, 6655–6668.

Dupont, C.L., Butcher, A., Valas, R.E., Bourne, P.E. and Caetano-Anolles, G. (2010) History of biological metal utilization inferred through phylogenomic analysis of protein structures. *Proceedings of the National Academy of Sciences USA*, **107**, 10567–10572.

Dzombak, D.A. and Morel, F.M. (1990) *Surface Complexation Modelling: Hydrous Ferric Oxide.* Wiley-Interscience, New York.

Eary, L.E. and Rai, D. (1987) Kinetics of chromium(III) oxidation to chromium(VI) by reaction with manganese-dioxide. *Environmental Science & Technology*, **21**, 1187–1193.

Eary, L.E. and Rai, D. (1989) Kinetics of chromate reduction by ferrous-ions derived from hematite and biotite at 25-degrees-C. *American Journal of Science*, **289**, 180–213.

Edmonds, H.N. and German, C.R. (2004) Particle geochemistry in the Rainbow hydrothermal plume, Mid-Atlantic Ridge. *Geochimica et Cosmochimica Acta*, **68**, 759–772.

Eigenbrode, J.L. and Freeman, K.H. (2006) Late Archean rise of aerobic microbial ecosystems. *Proceedings of the National Academy of Sciences USA*, **103**, 15759–15764.

Ellis, A.S., Johnson, T.M. and Bullen, T.D. (2002) Chromium isotopes and the fate of hexavalent chromium in the environment. *Science*, **295**, 2060–2062.

Ellis, A.S., Johnson, T.M. and Bullen, T.D. (2004) Using chromium stable isotope ratios to quantify Cr(VI) reduction: Lack of sorption effects. *Environmental Science & Technology*, **38**, 3604–3607.

Erickson, B.E. and Helz G.R. (2000) Molybdenum(VI) speciation in sulfidic waters: Stability and lability of thiomolybdates. *Geochimica et Cosmochimica Acta*, **64**, 1149–1158.

Farquhar, J., Bao, H.M. and Thiemens, M. (2000) Atmospheric influence of Earth's earliest sulfur cycle. *Science*, **289**, 756–758.

Feely, R.A., Massoth, G.J., Baker, E.T., Cowen, J.P., Lamb, M.F. and Krogslund, K.A. (1990) The effect of hydrothermal processes on midwater phosphorus distributions in the northeast Pacific. *Earth and Planetary Science Letters*, **96**, 305–318.

Feely, R.A., Trefry, J.H., Lebon, G.T. and German, C.R. (1998) The relationship between P/Fe and V/Fe ratios in hydrothermal precipitates and dissolved phosphate in seawater. *Geophysical Research Letters*, **25**, 2253–2256.

Fendorf, S. (1995) Surface-reactions of chromium in soils and waters. *Geoderma* 67, 55–71.

Fischer, W.W. and Knoll, A.H. (2009) An iron shuttle for deepwater silica in Late Archean and early Proterozoic iron formation. *Geological Society of America Bulletin*, **121**, 222–235.

Frei, R. and Polat, A. (2007) Source heterogeneity for the major components of similar to 3.7 Ga Banded Iron Formations (Isua Greenstone Belt, Western Greenland): Tracing the nature of interacting water masses in

BIF formation. *Earth and Planetary Science Letters*, **253**, 266–281.
Frei, R., Dahl, P.S., Duke, E.F., Frie, K.M., Hansen, T.R., Frandsson, M.M. and Jensen, L.A. (2008) Trace element and isotopic characterization of Neoarchean and Paleoproterozoic iron formations in the Black Hills (South Dakota, USA): Assessment of chemical change during 2.9–1.9 Ga deposition bracketing the 2.4–2.2 Ga first rise of atmospheric oxygen. *Precambrian Research*, **162**, 441–474.
Frei, R., Gaucher, C., Poulton, S.W. and Canfield, D.E. (2009) Fluctuations in Precambrian atmospheric oxygenation recorded by chromium isotopes. *Nature*, **461**, 250–253.
Frierdich, A.J., Luo, Y. and Catalano, J.G. (2011) Trace element cycling through iron oxide minerals during redox-driven dynamic recrystallization. *Geology*, **39**, 1083–1086.
Frost, C., Blanckenburg, F. von, Schoenberg, R., Frost, B. and Swapp, S. (2007) Preservation of Fe isotope heterogeneities during diagenesis and metamorphism of banded Fe formation. *Contributions to Mineralogy and Petrology*, **153**, 211–235.
Gallagher, M., Turner, E.C. and Kamber, B.S. (2015) *In situ* trace element analysis of Neoarchean – Ordovician shallow-marine microbial-carbonate-hosted pyrites. *Geobiology*, doi: 10.1111/gbi.12139.
Garrels, R.M., Perry, E.A. and Mackenzi, F.T. (1973) Genesis of Precambrian iron-formations and development of atmospheric oxygen. *Economic Geology*, **68**, 1173–1179.
German, C.R. and Elderfield, H. (1990) Rare-earth elements in the NW Indian-Ocean. *Geochimica et Cosmochimica Acta*, **54**, 1929–1940.
German, C.R., Holliday, B.P. and Elderfield, H. (1991) Redox cycling of rare-earth elements in the suboxic zone of the Black Sea. *Geochimica et Cosmochimica Acta*, **55**, 3553–3558.
Glasby, G.P. (1977) *Marine Manganese Deposits*. Elsevier, Amsterdam.
Godfrey, L.V. and Falkowski, P.G. (2009) The cycling and redox state of nitrogen in the Archean ocean. *Nature Geoscience*, **2**, 725–729.
Goldberg, T., Archer, C., Vance, D. and Poulton, S.W. (2009) Mo isotope fractionation during adsorption to Fe (oxyhydr)oxides. *Geochimica et Cosmochimica Acta*, **73**, 6502–6516.
Goldberg, T., Archer, C., Vance, D., Thamdrup, B., McAnena, A. and Poulton, S.W. (2012) Controls on Mo isotope fractionations in a Mn-rich anoxic marine sediment, Gullmar Fjord, Sweden. *Chemical Geology*, **296–297**, 73–82.
Goldberg, T., Gordon, G., Izon, G., Archer, C., Pearce, C.R., McManus, J., Anbar, A.D. and Rehkamper, M. (2013) Resolution of inter-laboratory discrepancies in Mo isotope data: an intercalibration. *Journal of Analytical and Atomic Spectroscopy*, **28**, 724–735.
Greber, N.D., Siebert, C., Nägler, T.F. and Pettke, T. (2012) $\delta^{98/95}$Mo values and Molybdenum Concentration Data for NIST SRM 610, 612 and 3134: Towards a Common Protocol for Reporting Mo Data. *Geostandards and Geoanalytical Research*, **36**, 291–300.
Grenne, T. and Slack, J.F. (2003) Paleozoic and Mesozoic silica-rich seawater: Evidence from hematitic chert (jasper) deposits. *Geology*, **31**, 319–322.
Grotzinger, J.P. (1990) Geochemical model for Proterozoic stromatolite decline. *American Journal of Science*, **290A**, 80–103.
Grotzinger, J.P. and Knoll, A.H. (1999) Stromatolites in Precambrian carbonates: Evolutionary mileposts or environmental dipsticks? *Annual Reviews in Earth and Planetary Science*, **27**, 313–358.
Guo, Q.J., Strauss, H., Kaufmann, A.J., Schroder, S., Gutzmer, J., Wing, B., Baker, M.A., Bekker, A., Jin, Q.S., Kim, S.T. and Farquhar, J. (2009) Reconstructing Earth's surface oxidation across the Archean-Proterozoic transition. *Geology*, **37**, 399–402.
Haese, R.R. (2000) The reactivity of iron. Pp. 233–261 in: *Marine Geochemistry*. Springer Berlin, Heidelberg, Berlin.
Hannah, J.L., Bekker, A., Stein, H.J., Markey, R.J. and Holland, H.D. (2004) Primitive Os and 2316 Ma age for marine shale: implications for Paleoproterozoic glacial events and the rise of atmospheric oxygen. *Earth and Planetary Science Letters*, **225**, 43–52.
Hartman, H. (1984) The evolution of photosynthesis and microbial mats: A speculation on the banded iron formations. Pp. 449–454 in: *Microbial Mats: Stromatolites* (Y. Cohen, R.W. Castenholz and H.O. Halvorson, editors). Alan R. Liss, Inc., New York.
Haugaard, R., Frei, R., Stendal, H. and Konhauser, K.O. (2013) Petrology and geochemistry of the similar to

2.9 Ga Itilliarsuk banded iron formation and associated supracrustal rocks, West Greenland: Source characteristics and depositional environment. *Precambrian Research*, **229**, 150–176.

Hein, J.R., Yeh, H.W., Gunn, S.H., Sliter, W.V., Benninger, L.m. and Wang, C.H. (1993) Two major episodes of phosphogenesis recorded in equatorial Pacific seamount deposits. *Paleoceanography*, **8**, 293–311.

Hein, J.R., Koschinsky, A., Bau, M., Manheim, F.T., Kang, J.-K. and Roberts, L. (2000) Cobalt-rich ferromanganese crusts in the Pacific. Pp. 239–279 in: *Handbook of Marine Mineral Deposits* (D.S. Cronan, editor). CRC Press, Boca Raton, Florida, USA.

Hein, J.R., Mizell, K., Koschinsky, A. and Conrad, T.A. (2013) Deep-ocean mineral deposits as a source of critical metals for high- and green-technology applications: Comparison with land-based resources. *Ore Geology Review*, **51**, 1–14.

Helz, G.R., Miller, C.V., Charnock, J.M., Mosselmans, J.F.W., Pattrick, R.A.D., Garner, C.D. and Vaughan, D.J. (1996) Mechanism of molybdenum removal from the sea and its concentration in black shales: EXAFS evidence. *Geochimica et Cosmochimica Acta*, **60**, 3631–3642.

Hill, P.S. and Schauble, E.A. (2008) Modeling the effects of bond environment on Fe isotope fractionation in ferric aquo-chloro complexes. *Geochimica et Cosmochimica Acta*, **72**, 1939–1958.

Holland, H.D. (1973) Oceans – Possible source of iron in iron-formations. *Economic Geology*, **68**, 1169–1172.

Holland, H.D. (1984) *The Chemical Evolution of the Atmosphere and Oceans*. Princeton University Press, Princeton, New Jersey, USA.

Holm, N.G. (1989) The C-13 C-12 ratios of siderite and organic-matter of a modern metalliferous hydrothermal sediment and their implications for banded iron formations. *Chemical Geology*, **77**, 41–45.

Huerta-Diaz, M.A. and Morse, J.W. (1992) Pyritization of trace elements in anoxic marine sediments. *Geochimica et Cosmochimica Acta*, **56**, 2681–2702.

Huston, D.L. and Logan, G.A. (2004) Barite, BIFs and bugs: Evidence for the evolution of the Earth's early hydrosphere. *Earth and Planetary Science Letters*, **220**, 41–55.

Ihler, R.K. (1979) *Chemistry of Silica*. Wiley-Interscience, New York

Jacobsen, S.B. and Pimentel-Klose, M.R. (1988) Nd isotopic variations in Precambrian banded iron formations. *Geophysical Research Letters*, **15**, 393–396.

James, H.L. (1954) Sedimentary facies of iron formations. *Economic Geology*, **49**, 235–293.

Jickells, T.D., An, Z.S., Andersen, K.K., Baker, A.R., Bergametti, G., Brooks, N., Cao, J.J., Boyd, P.W., Duce, R.A., Hunter, K.A., Kawahata, H., Kubilay, N., laRoche, J., Liss, P.S., Mahowald, N., Prospero, J.M., Ridgwell, A.J., Tegen, I. and Torres, R. (2005) Global iron connections between desert dust, ocean biogeochemistry, and climate. *Science*, **308**, 67–71.

Johnson, C.M., Skulan, J.L., Beard, B.L., Sun, H., Nealson, K.H. and Braterman, P.S. (2002) Isotopic fractionation between Fe(III) and Fe(II) in aqueous solutions. *Earth and Planetary Science Letters*, **195**, 141–153.

Johnson, C.M., Beard, B.L., Beukes, N.J., Klein, C. and O'Leary, J.M. (2003) Ancient geochemical cycling in the Earth as inferred from Fe isotope studies of banded iron formations from the Transvaal Craton. *Contributions to Mineralogy and Petrology*, **144**, 523–547.

Johnson, C.M., Beard, B.L. and Albarede, F. (2004) Overview and general concepts. Pp. 1–24 in: *Geochemistry of Non-Traditional Stable Isotopes* (C.M. Johnson, B.L. Beard and F. Albarede, editors). Reviews in Mineralogy and Geochemistry, **55**. Mineralogical Society of America and the Geochemical Society, Washington, D.C.

Johnson, C.M., Beard, B.L., Klein, C., Beukes, N.J. and Roden, E.E. (2008a) Iron isotopes constrain biologic and abiologic processes in banded Fe formation genesis. *Geochimica et Cosmochimica Acta*, **72**, 151–169.

Johnson, C.M., Beard, B.L. and Roden, E.E. (2008b) The iron isotope fingerprints of redox and biogeochemical cycling in modern and ancient Earth. *Annual Reviews in Earth and Planetary Science*, **36**, 457–493.

Jones, C., Nomosatryo, S., Crowe, S.A., Bjerrum, C.J. and Canfield, D.E. (2015) Iron oxides, divalent cations, silica, and the early Earth phosphorus crisis. *Geology*, **43**, 135–138.

Kappler, A., Pasquero, C., Konhauser, K.O. and Newman, D.K. (2005) Deposition of banded iron formations by anoxygenic phototrophic Fe(II)-oxidising bacteria. *Geology*, **33**, 865–868.

Kashiwabara, T., Takahashi Y. and Tanimizu, M. (2011) Molecular-scale mechanisms of distribution and isotopic fractionation of molybdenum between seawater and ferromanganese oxides. *Geochimica et*

Cosmochimica Acta, **75**, 5762–5784.

Kaufman, A.J., Johnston, D.T., Farquhar, J., Masterson, A.L., Lyons, T.W., Bates, S., Anbar, A.D., Arnold, G.L., Garvin, J. and Buick, R. (2007) Later Archean biospheric oxygenation and atmospheric evolution. *Science*, **317**, 1900–1903.

Kendall, B., Reinhart, C.T., Lyons, T., Kaufman, A.J., Poulton, S.W. and Anbar, A.D. (2010) Pervasive oxygenation along late Archean ocean margins. *Nature Geoscience*, **3**, 647–652.

Kennedy, C.B., Martinez, R.E., Scott, S.D. and Ferris, F.G. (2003) Surface chemistry and reactivity of bacteriogenic iron oxides from Axial Volcano, Juan de Fuca Ridge, north-east Pacific Ocean. *Geobiology*, **1**, 59–69.

Klein, C. (2005) Some Precambrian banded iron-formations (BIFs) from around the world: Their age, geologic setting, mineralogy, metamorphism, geochemistry, and origin. *American Mineralogist*, **90**, 1473–1499.

Klein, C. and Beukes, N.J. (1989) Geochemistry and sedimentology of a facies transition from limestone to iron-formation deposition in the early Proterozoic Transvaal Supergroup, South-Africa. *Economic Geology*, **84**, 1733–1774.

Klein, C., and Beukes, N.J., 1992. Time distribution, stratigraphy, and sedimentologic setting, and geochemistry of Precambrian iron-formations. Pp. 139–146 in: *The Proterozoic Biosphere: A Multidisciplinary Study* (J.W. Schopf and C. Klein, editors). Cambridge University Press. Cambridge, UK.

Klinkhammer, G., Elderfield, H. and Hudson, A. (1983) Rare-earth elements in seawater near hydrothermal vents. *Nature*, **305**, 185–188.

Knoll, A.H., Summons, R.E., Waldbauer, J.R. and Zumberge, J.E. (2007) The geological succession of primary producers in the oceans. Pp. 133–163 in: *Evolution of Primary Producers in the Sea* (P.G. Falkowski and A.H. Knoll, editors). Elsevier, Amsterdam.

Konhauser, K., Hamade, T., Raiswell, R., Morris, R.C., Ferris, F.G., Southam, G. and Canfield, D.E. (2002) Could bacteria have formed the Precambrian banded iron formations? *Geology*, **30**, 1079–1082.

Konhauser, K.O., Newman, D.K. and Kappler, A. (2005) The potential significance of microbial Fe(III) reduction during deposition of Precambrian banded iron formations. *Geobiology*, **3**, 167–177.

Konhauser, K.O., Lalonde, S.V., Amskold L. and Holland, H.D. (2007) Was there really an Archean phosphate crisis? *Science*, **315**, 1234.

Konhauser, K.O., Pecoits, E., Lalonde, S.V., Papineau, D., Nisbet, E.G., Barley, M.E., Arndt, N.T., Zahnle, K. and Kamber, B.S. (2009) Oceanic Ni depletion and a methanogen famine before the great oxidation event. *Nature*, **458**, 750–753.

Konhauser, K.O., Lalonde, S.V., Planavsky, N.J., Pecoits, E., Lyons, T.W., Mojzsis, S.J., Rouxel, O.J., Barley, M.E., Rosiere, C., Fralick, P.W., Kump, L.R. and Bekker, A. (2011) Aerobic bacterial pyrite oxidation and acid rock drainage during the Great Oxidation Event. *Nature*, **478**, 369.

Konhauser, K.O., Robbins L.J., Pecoits, E., Peacock, C., Kappler, A. and Lalonde, S.V. (2015) The Archean nickel famine revisited. *Astrobiology*, **15**, 804–815.

Koschinsky, A. and Halbach, P. (1995) Sequential leaching of marine ferromanganese precipitates: Genetic implications. *Geochimica et Cosmochimica Acta*, **59**, 5113–5132.

Koschinsky, A. and Hein, J.R. (2003) Uptake of elements from seawater by ferromanganese crusts: Solid-phase associations and seawater speciation. *Marine Geology*, **198**, 331–351.

Koschinsky, A., Stascheit, A., Bau, M. and Halbach, P. (1997) Effects of phosphatization on the geochemical and mineralogical composition of marine ferromanganese crusts. *Geochimica et Cosmochimica Acta*, **61**, 4079–4094.

Kump, L.R. (2008) The rise of atmospheric oxygen. *Nature*, **451**, 277–278.

Kurzweil, F., Wille, M., Schoenberg, R., Taubald, H. and Van Kranendonk, M.J. (2015) Continuously increasing δ^{98}Mo values in Neoarchean black shales and iron formations from the Hamersley Basin. *Geochimica et Cosmochimica Acta*, **164**, 523–542.

Lalonde, S., Konhauser, K.O., Amskold, L., McDermott, T. and Inskeep, B.P. (2007) Chemical reactivity of microbe and mineral surfaces in hydrous ferric oxide depositing hydrothermal springs. *Geobiology*, **5**, 219–234.

Langmuir, D. (1997) *Aqueous Environmental Geochemistry*. Prentice Hall, New Jersey.

Large, R.R., Danyushevsky, L., Hollit, C., Maslennikov, V., Meffre, S., Gilbert, S., Bull, S., Scott, R., Emsbo, P.,

Thomas, H., Singh, B. and Foster, J. (2009) Gold and trace element zonation in pyrite using a laser imaging technique: Implications for the timing of gold in orogenic and carlin-style sediment-hosted deposits. *Economic Geology*, **104**, 635–668.

Large, R.R., Halpin, J.A., Danyushevsky, L.V., Maslennikov, V.V., Bull, S.W., Long, J.A., Gregory, D.D., Lounejeva, E., Lyons, T.W., Sack, P.J., McGoldrick, P.J . and Claver, C.R. (2014) Trace element content of sedimentary pyrite as a new proxy for deep-time ocean-atmosphere evolution. *Earth and Planetary Science Letters*, **389**, 209–220.

Ler, A. and Stanforth, R. (2003) Evidence for surface precipitation of phosphate on goethite. *Environmental Science & Technology*, **37**, 2694–2700.

Lepp, H. and Goldich, S.S. (1964) Origin of Precambrian iron formations. *Economic Geology*, **59**, 1025–1061.

Lindemann, F.A. (1919) Note on the vapour pressure and affinity of isotopes. *Philosophical Magazine*, **38**, 173–181.

Lindemann, F.A. and Aston, F.W. (1919) The possibility of separating isotopes. *Philosophical Magazine*, **37**, 530.

Little, S.H., Sherman, D.M., Vance, D. and Hein, J.R. (2014) Molecular controls on Cu and Zn isotope fractionation in Fe-Mn crusts. *Earth and Planetary Science Letters*, **396**, 213–222.

Lyons, T.W., Anbar, A.D., Severmann, S., Scott, C. and Gill, B.C. (2009) Tracking euxinia in the ancient ocean: A multiproxy perspective and Proterozoic case study. *Annual Reviews in Earth and Planetary Science*, **37**, 507–534.

Mackenzie, F.T., Lantzy, R.J. and Paterson, V. (1979) Global trace element cycles and predictions. *Mathematical Geology*, **11**, 99–142.

Maliva, R.G., Knoll, A.H. and Simonson, B.M. (2005) Secular change in the Precambrian silica cycle: insights from chert petrology. *Geological Society of America Bulletin*, **117**, 835–845.

Manceau, A. and Charlet, L. (1992) X-ray absorption spectroscopic study of the sorption of Cr(III) at the oxide water interface. 1. Molecular mechanism of Cr(III) oxidation on Mn oxides. *Journal of Colloid and Interface Science*, **148**, 425–442.

McNamara, K.J. and Awramik, S.M. (1992) Stromatolites – A key to understanding the early evolution of life. *Science Progress*, **76**, 345–364.

McManus, J., Nagler, T.F., Siebert, C., Wheat, C.G. and Hammond, D.E. (2002) Oceanic molybdenum isotope fractionation: Diagenesis and hydrothermal ridge-flank alteration. *Geochemistry Geophysics Geosystems*, **3**, 1078.

Mero, J.L. (1962) Ocean floor manganese nodules. *Economic Geology*, **57**, 747–767.

Miller, R.G. and O'Nions, R.K. (1985) Source of Precambrian chemical and clastic sediments. *Nature*, **314**, 325–330.

Moon, E.M. and Peacock, C.L. (2011) Adsorption of Cu(II) to Bacillus subtilis: A pH-dependent EXAFS and thermodynamic modelling study. *Geochimica et Cosmochimica Acta*, **75**, 6705–6719.

Moon, E.M. and Peacock, C.L. (2012) Adsorption of Cu(II) to ferrihydrite and ferrihydrite-bacteria composites: Importance of the carboxyl group for Cu mobility in natural environments. *Geochimica et Cosmochimica Acta*, **92**, 203–219.

Moon, E.M. and Peacock, C.L. (2013) Modelling Cu(II) adsorption to ferrihydrite and ferrihydrite-bacteria composites: Deviation form additive adsorption in the composite sorption system. *Geochimica et Cosmochimica Acta*, **104**, 148–164.

Morris, R.C. (1993) Genetic modelling for banded iron-formation of the Hamersley Group, Pilbara Craton, Western Australia. *Precambrian Research*, **60**, 243–286.

Morse, J.W. and Arakaki, T. (1993) Adsorption and coprecipitation of divalent metals with mackinawite (FeS). *Geochimica et Cosmochimica Acta*, **57**, 3635–3640.

Morse, J.W. and Luther, G.W. (1999) Chemical influences on trace element-sulfide interactions in anoxic sediments. *Geochimica et Cosmochimica Acta*, **63**, 3373–3378.

Muios, S.B., Hein, J.R., Frank, M., Monteiro, J.H., Gaspar, L., Conrad, T., Pereira, H.G. and Abrantes, F. (2013) Deep-sea Fe-Mn crusts from the Northeast Atlantic Ocean: Composition and resource considerations. *Marine Georesources and Geotechnology*, **31**, 40–70.

Nakada, R., Takahashi, Y. and Tanimizu, M. (2013) Isotopic and speciation study on cerium during its solid-water distribution with implication for Ce stable isotope as a palaeo-redox proxy. *Geochimica et Cosmochimica Acta*, **103**, 49–62.

Nielsen, S.G., Wasylenki, L.E., Rehkamper, M., Peacock, C.L., Xue, Z.C. and Moon, E.M. (2013) Towards and understanding of thallium isotope fractionation during adsorption to manganese oxides. *Geochimica et Cosmochimica Acta*, **117**, 252–265.

O'Neil, J.R. (1986) Theoretical and experimental aspects of isotopic fractionation. Pp. 1–40 in: *Stable Isotopes in High Temperature Geological Processes* (J.W. Valley, H.P. Taylor and J.R. O'Neil, editors). Reviews in Mineralogy, **16**. Mineralogical Society of America, Washington, D.C.

Partin, C.A., Bekker, A., Planavsky, N.J., Scott, C.T., Gill, B.C., Li, C., Podkovyrov, V., Maslov, A., Konhauser, K.O., Lalonde, S.V., Love, G.D., Poulton, S.W. and Lyons, T.W. (2013) Large-scale fluctuations in Precambrian atmospheric and oceanic oxygen levels from the record of U in shales. *Earth and Planetary Science Letters*, **369**, 284–293.

Peacock, C.L. (2009) Physiochemical controls on the crystal-chemistry of Ni in birnessite: Genetic implications for ferromanganese precipitates. *Geochimica et Cosmochimica Acta*, **73**, 3568–3578.

Peacock, C.L. and Moon, E.M. (2012) Oxidative scavenging of thallium by birnessite: Explanation for thallium enrichment and stable isotope fractionation in marine ferromanganese precipitates. *Geochimica et Cosmochimica Acta*, **84**, 297–313.

Peacock, C.L. and Sherman, D.M. (2004a) Vanadium adsorption onto goethite (a-FeOOH) at pH 1.5 to 12: A surface complexation model based on ab inito molecular geometries and EXAFS spectroscopy. *Geochimica et Cosmochimica Acta*, **68**, 1723–1733.

Peacock, C.L. and Sherman, D.M. (2004b) Copper (II) sorption onto goethite, hematite and lepidocrocite: A surface complexation model based on ab inito molecular geometries and EXAFS spectroscopy. *Geochimica et Cosmochimica Acta*, **68**, 2623–2637.

Peacock, C.L. and Sherman, D.M. (2007) Sorption of Ni by birnessite: equilibrium controls on Ni in seawater. *Chemical Geology*, **238**, 94-106.

Pearce, C.R., Cohen, A.S., Coe, A.L. and Burton, K.W. (2008) Molybdenum isotope evidence for global ocean anoxia coupled with perturbations to the carbon cycle during the Early Jurassic. *Geology*, **36**, 231–234.

Pecoits, E., Smith, M.L., Catling, D.C., Philippot, P., Kappler A. and Konhauser, K.O. (2015) Atmospheric hydrogen peroxide and Eoarchean iron formations. *Geobiology*, **13**, 1–14.

Pea, J., Kwon, K.D., Refson, K., Bargar, J.R. and Sposito, G. (2010) Mechanisms of nickel sorption by a bacteriogenic birnessite. *Geochimica et Cosmochimica Acta*, **74**, 3076–3089.

Planavsky, N., Rouxel, O., Bekker, A., Shapiro, R., Fralick, P. and Knudsen, A. (2009) Iron-oxidizing microbial ecosystems thrived in late Paleoproterozoic redox-stratified oceans. *Earth and Planetary Science Letters*, **286**, 230–242.

Planavsky, N., Bekker, A., Rouxel, O.J., Kamber, B., Hofmann, A., Knudsen, A. and Lyons, T.W. (2010a) Rare Earth Element and yttrium compositions of Archean and Paleoproterozoic Fe formations revisited: New perspectives on the significance and mechanisms of deposition. *Geochimica et Cosmochimica Acta*, **74**, 6387–6405.

Planavsky, N.J., Rouxel, O.J., Bekker, A., Lalonde, S.V., Konhauser, K.O., Reinhard, C.T. and Lyons, T.W. (2010b) The evolution of the marine phosphate reservoir. *Nature*, **467**, 1088–1090.

Planavsky, N.J., Asael, D., Hofmann, A., Reinhard, C.T., Lalonde, S.V., Knudsen, A., Wang, X., Ossa, F.O., Pecoits, E., Smith, A.J.B., Beukes, N.J., Bekker, A., Johnson, T.M., Konhauser, K.O., Lyons, T.W. and Rouxel, O.J. (2014) Evidence for oxygenic photosynthesis half a billion years before the Great Oxidation Event. *Nature Geoscience*, **7**, 283–286.

Pokrovsky, O.S., Viers J. and Freydier, R. (2005) Zinc stable isotope fractionation during its adsorption on oxides and hydroxides. *Journal of Colloid and Interface Science*, **291**, 192–200.

Pokrovsky, O.S., Viers, J., Emnova, E.E., Kompantseva, E.I. and Freydier, R. (2008) Copper isotope fractionation during its interaction with soil and aquatic microorganisms and metal oxy(hydr)oxides: Possible structural control. *Geochimica et Cosmochimica Acta*, **72**, 1742–1757.

Poulton, S. and Canfield, D. (2005) Development of a sequential extraction procedure for iron: implications for iron partitioning in continentally derived particulates. *Chemical Geology*, **214**, 209–221.

Poulton, S.W. and Canfield, D.E. (2006) Co-diagenesis of iron and phosphorus in hydrothermal sediments from the southern East Pacific Rise: Implications for the evaluation of paleoseawater phosphate concentrations. *Geochimica et Cosmochimica Acta*, **70**, 5883–5898.

Poulton, S.W. and Raiswell, R. (2002) The low-temperature geochemical cycle of iron: From continental fluxes to marine sediment deposition. *American Journal of Science*, **302**, 774–805.

Raiswell, R. and Canfield, D.E. (1998) Sources of iron for pyrite formation in marine sediments. *American Journal of Science*, **298**, 219–245.

Raiswell, R. and Canfield, D.E. (2012) The Fe biogeochemical cycle past and present. *Geochemical Perspectives*, **1**, 1–220.

Raiswell, R., Benning, L.G., Tranter, M. and Tulaczyk, S. (2008) Bioavailable iron in the Southern Ocean: the significance of the iceberg conveyor belt. *Geochemical Transactions*, **9**, 7.

Reinhard, C.T., Raiswell, R., Scott, C., Anbar, A.D. and Lyons, T.W. (2009) A late Archean sulfidic sea stimulated by early oxidative weathering of the continents. *Science*, **326**, 713–716.

Reinhard, C.T., Planavsky, N.J., Robbins, L.J., Partin, C.A., Gill, B.C., Lalonde, S.V., Bekker, A., Konhauser, K.O. and Lyons, T.W. (2013) Proterozoic ocean redox and biogeochemical stasis. *Proceedings of the National Academy of Science*, **110**, 5357–5362.

Richardson, L.L., Aguilar, C. and Nealson, K.H. (1988) Manganese oxidation in pH and O_2 microenvironments produced by phytoplankton. *Limnology and Oceanography*, **33**, 352–363.

Rickard, D. and Luther, G.W. III. (2007) Chemistry of iron sulfides. *Chemical Reviews*, **107**, 514–562.

Robbins, L.J., Lalonde, S.V., Saito, M.A., Planavsky, N.J., Mloszewska, A.M., Pecoits, E., Scott, C., Dupont, C.L., Kappler, A. and Konhauser, K.O. (2013) Authigenic iron oxide proxies for marine zinc over geological time and implications for eukaryotic metallome evolution. *Geobiology*, **11**, 295–306.

Robbins, L.J., Swanner, E.D., Lalonde, S.V., Eickhoff, M., Paranich, M.L., Reinhard, C.T., Peacock, C.L., Kappler, A. and Konhauser, K.O. (2015) Limited Zn and Ni mobility during simulated iron formation diagenesis. *Chemical Geology*, **402**, 30–39.

Roden, E.E. (2003) Fe(III) Oxide reactivity toward biological versus chemical reduction. *Environmental Science & Technology*, **37**, 1319–1324.

Roe, J.E., Anbar, A.D. and Barling, J. (2003) Nonbiological fractionation of Fe isotopes: evidence of an equilibrium isotope effect. *Chemical Geology*, **195**, 69–85.

Rouxel, O., Bekker, A. and Edwards, K. (2005) Iron Isotope Constraints on the Archean and Paleoproterozoic Ocean Redox State. *Science*, **307**, 1088–1091.

Ruttenberg, K. and Berner, R. (1993) Authigenic apatite formation and burial in sediments from non-upwelling, continental margin environments. *Geochimica et Cosmochimica Acta*, **57**, 991–1007.

Saito, M.A. and Moffett, J.W. (2001) Complexation of cobalt by natural organic ligands in the Sargasso Sea as determined by a new high-sensitivity electrochemical cobalt speciation method suitable for open ocean work. *Marine Chemistry*, **75**, 49–68.

Saito, M.A., Sigman, D.M. and Morel, F. M.M. (2003) The bioinorganic chemistry of the ancient ocean: the co-evolution of cyanobacterial metal requirements and biogeochemical cycles at the Archean-Proterozoic boundary? *Inorganica Chimica Acta*, **356**, 308–318.

Saito, M.A., Rocap, G. and Moffett, J.W. (2005) Production of cobalt binding ligands in a Synechococcus feature at the Costa Rica upwelling dome. *Limnology and Oceanography*, **50**, 279–290.

Sass, B.M. and Rai, D. (1987) Solubility of amorphous chromium(III)-iron(III) hydroxide solid solutions. *Inorganic Chemistry*, **26**, 2228–2232.

Satkowski, A.M., Beukes, N.J., Li, W., Beard, B.L. and Johnson, C.M. (2015) A redox-stratified ocean 3.2 billion years ago. *Earth and Planetary Science Letters*, **430**, 43–53.

Schauble, E.A. (2004) Applying stable isotope fractionation theory to new systems. Pp 65–111 in: *Geochemistry of Non-Traditional Stable Isotopes* (C.M. Johnson, B.L. Beard and F. Albarede, editors). Reviews in Mineralogy and Geochemistry, **55**. Mineralogical Society of America and the Geochemical Society, Washington, D.C.

Schauble, E.A. (2007) Role of nuclear volume in driving equilibrium stable isotope fractionation of mercury, thallium, and other very heavy elements. *Geochimica et Cosmochimica Acta*, **71**, 2170–2189.

Schauble, E., Rossman, G.R. and Taylor, H.P., Jr. (2004) Theoretical estimates of equilibrium chromium-isotope fractionations. *Chemical Geology*, **205**, 99–114.

Scott, C., Lyons, T.W., Bekker, A., Shen, Y., Poulton, S.W., Chu, X. and Anbar, A.D. (2008) Tracing the stepwise oxygenation of the Proterozoic ocean. *Nature*, **452**, 456–459.

Scott, C., Planavsky, N.J., Dupont, C.L., Kendall, B., Gill, B.C., Robbins, L.J., Husband, K.F., Arnold, G.L., Wing, B.A., Poulton, S.W., Bekker, A., Anbar, A.D., Konhauser, K.O. and Lyons, T.W. (2012) Bioavailability of zinc in marine systems through time. *Nature Geoscience*, **6**, 125–128.

Sherman, D.M. and Peacock, C.L. (2010) Surface complexation of Cu on birnessite (δ-MnO_2): controls on Cu in the deep ocean. *Geochimica et Cosmochimica Acta*, **74**, 6721–6730.

Siebert, C., Nägler, T.F., von Blanckenburg, F. and Kramers, J.D. (2003) Molybdenum isotope records as a potential new proxy for paleoceanography. *Earth and Planetary Science Letters*, **211**, 159–171.

Siever, R. (1992) The silica cycle in the Precambrian. *Geochimica et Cosmochimica Acta*, **56**, 3265–3272.

Sikora, E.R., Johnson, T.M. and Bullen, T.D. (2008) Microbial mass-dependent fractionation of chromium isotopes. *Geochimica et Cosmochimica Acta*, **72**, 3631–3641.

Slack, J.F., Grenne, T., Bekker, A., Rouxel, O.J. and Lindberg, P.A. (2007) Suboxic deep seawater in the late Paleoproterozoic: Evidence from hematitic chert and iron formation related to seafloor-hydrothermal sulfide deposits, central Arizona, USA. *Earth and Planetary Science Letters*, **255**, 243–256.

Sreenivas, B. and Murakami, T. (2005) Emerging views on the evolution of atmospheric oxygen during the Precambrian. *Journal of Mineralogical and Petrological Science*, **100**, 184–201.

Staton, S.J.R., Amskold, L., Gordon, G., Anbar, A.D. and Konhauser, K.O. (2006) Iron isotope fractionation during photo-oxidation of aqueous ferrous iron. *Astrobiology*, **6**, 215–216.

Staubwasser, M., von Blanckenburg, F. and Schoenberg, R. (2006) Iron isotopes in the early marine diagenetic iron cycle. *Geology*, **34**, 629–632.

Stockdale, A., Davison, W., Zhang, H. and Hamilton-Taylor, J. (2010) The association of cobalt with iron and manganese (oxyhydr)oxides in marine sediment. *Aquatic Chemistry*, **16**, 575–585.

Stüeken, E.E., Catling, D.C. and Buick, R. (2012) Contributions to late Archaean sulphur cycling by life on land. *Nature Geoscience*, **5**, 722–725.

Sverjensky, D.A. (1984) Europium redox equilibria in aqueous solution. *Earth and Planetary Science Letters*, **67**, 70–78.

Sverjensky, D.A. and Sahai, N. (1996) Theoretical prediction of single-site surface-protonation equilibrium constants for oxides and silicates in water. *Geochimica et Cosmochimica Acta*, **60**, 3773–3797.

Swanner, E.D., Planavsky, N.J., Lalonde, S.V., Robbins, L.J., Bekker, A., Rouxel, O.J., Saito, M.A., Kappler, A., Mojzsis, S.J. and Konhauser, K.O. (2014) Cobalt and marine redox evolution. *Earth and Planetary Science Letters*, **390**, 253–263.

Tagliabue, A., Bopp, L., Dutay, J.-C., Bowie, A.R., Chever, F., Jean-Baptiste, P., Bucciarelli, E., Lannuzel, D., Remenyi, T., Sarthou, G., Aumont, O., Gehlen, M. and Jeandel, C. (2010) Hydrothermal contribution to the oceanic dissolved iron inventory. *Nature Geoscience*, **3**, 252–256.

Takahashi, Y., Manceau, A., Geoffroy, N., Marcus, M.A. and Usui, A. (2007) Chemical and structural control of the partitioning of Co, Ce, and Pb in marine ferromanganese oxides. *Geochimica et Cosmochimica Acta*, **71**, 984–1008.

Tessier, A., Fortin, D., Belzile, N., DeVitre, R.R. and Leppard, G.G. (1996) Metal sorption to diagenetic iron and manganese oxyhydroxides and associated organic matter: Narrowing the gap between field and laboratory measurements. *Geochimica et Cosmochimica Acta*, **60**, 387–404.

Thomazo, C., Nisbet, E.G., Grassineau, N.V., Peters, M. and Strauss, H. (2013) Multiple sulfur and carbon isotope composition of sediments from the Belingwe Greenstone Belt (Zimbabwe): A biogenic methane regulation on mass independent fractionation of sulfur during the Neoarchean? *Geochimica et Cosmochimica Acta*, **121**, 120–138.

Tossell, J.A. (2005) Calculating the partitioning of the isotopes of Mo between oxidic and sulfidic species in aqueous solution. *Geochimica et Cosmochimica Acta*, **69**, 2981–2993.

Tourtelot, H.A. (1979) Black shale – its deposition and diagenesis. *Clays and Clay Minerals*, **27**, 313–321.

Tréguer, P., Nelson, D.M., Van Bennekom, A.J., DeMaster, D.J., Leynaert, A. and Queguiner, B. (1995) The silica balance in the world ocean: a reestimate. *Science*, **268**, 375–379.

Trendall, A.F. and Blockley, J.G. (1970) The iron formations of the Precambrian Hamersley Group Western Australia with special reference to the associated crocidolite. *Geological Survey of Western Australia*

Bulletin, **119**, 1–366.

Tribovillard, N., Algeo, T.J., Lyons, T. and Riboulleau, A. (2006) Trace elements as paleoredox and paleoproductivity proxies: An update. *Chemical Geology*, **232**, 12–32.

Twining, B.S. and Baines, S.B. (2013) The trace element composition of marine phytoplankton. *Annual Reviews in Marine Science*, **5**, 191–215.

Urey, H.C. (1947) The thermodynamic properties of isotopic substances. *Journal of the Chemical Society*, 562–580.

Urey, H.C. and Greiff, L.J. (1935) Isotopic exchange equilibria. *Journal of the American Chemical Society*, **57**, 321–327.

van Cappellen, P. and Wang, Y. (1996) Cycling of iron and manganese in surface sediments; a general theory for the coupled transport and reaction of carbon, oxygen, nitrogen, sulfur, iron, and manganese. *American Journal of Science*, **296**, 197–243.

Vine, J.D. and Tourtelot, E.B. (1970) Geochemistry of black shale deposits; a summary report. *Economic Geology*, **65**, 253–272.

Viollier, E., Jezequel, D., Michard, G., Pepe, M., Sarazin, G. and Alberic, P. (1995) Geochemical study of a crater lake (Pavin Lake, France) – Trace-element behaviour in the monimolimnion. *Chemical Geology*, **125**, 61–72.

Voegelin, A.R., Nägler, T.F., Samankassou, E. and Villa, I.M. (2009) Molybdenum isotopic composition of modern and Carboniferous carbonates. *Chemical Geology*, **265**, 488–498.

Voegelin, A.R., Nägler, T.F., Beukes, N.J. and Lacassie, J.P. (2010) Molybdenum isotopes in late Archean carbonate rocks: Implications for early Earth oxygenation. *Precambrian Research*, **182**, 70–82.

Wasylenki, L.E., Rolfe, B.A., Weeks, C.L., Spiro, T.G. and Anbar, A.D. (2008) Experimental investigation of the effects of temperature and ionic strength on Mo isotope fractionation during adsorption to manganese oxides. *Geochimica et Cosmochimica Acta*, **72**, 5997–6005.

Wasylenki, L.E., Weeks, C.L., Bargar, J.R., Spiro, T.G., Hein, J.R. and Anbar, A.D. (2011) The molecular mechanism of Mo isotope fractionation during adsorption to birnessite. *Geochimica et Cosmochimica Acta*, **75**, 5019–5031.

Wasylenki, L.E., Swihart, J.W. and Romaniello, S.J. (2014) Cadmium isotope fractionation during adsorption to Mn oxyhydroxide at low and high ionic strength. *Geochimica et Cosmochimica Acta*, **140**, 212–226.

Wasylenki, L.E., Howe, H.D., Spivak-Birndorf, L.J. and Bish, D.L. (2015) Ni isotope fractionation during sorption to ferrihydrite: Implications for Ni in banded iron formations. *Chemical Geology*, **400**, 56–64.

Weeks, C.L., Anbar, A.D., Wasylenki, L.E. and Spiro, T.G. (2007) Density functional theory analysis of molybdenum isotope fractionation. *Journal of Physical Chemistry, A*, **111**, 12434–12438.

Weeks, C.L., Anbar, A.D., Wasylenki, L.E. and Spiro, T.G. (2008) Density functional theory analysis of molybdenum isotope fractionation. *Journal of Physical Chemistry, A*, **112**, 10703–10704.

Welch, S.A., Beard, B.L., Johnson, C.M. and Braterman, P.S. (2003) Kinetic and equilibrium Fe isotope fractionation between aqueous Fe(II) and Fe(III). *Geochimica et Cosmochimica Acta*, **67**, 4231–4250.

Wen, H., Carignan, J., Cloquet, C., Zhu, X. and Zhang, Y. (2010) Isotopic delta values of molybdenum standard reference and prepared solutions measured by MC-ICP-MS: Proposition for delta zero and secondary references. *Journal of Analytical and Atomic Spectroscopy*, **25**, 716–721.

Wheat, C.G., Feely, R.A. and Mottl, M.J. (1996) Phosphate removal by oceanic hydrothermal processes: An update of the phosphorus budget in the oceans. *Geochimica et Cosmochimica Acta*, **60**, 3593–3608.

Widdel, F., Schnell, S., Heising, S., Ehrenreich, A., Assmus, B. and Schink, B. (1993) Ferrous iron oxidation by anoxygenic phototrophic bacteria. *Nature*, **362**, 834–836.

Wille, M., Kramers, J., Nägler, T., Beukes, N., Schröder, S., Meisel, T., Lacassie, J. and Voegelin, A. (2007) Evidence for a gradual rise of oxygen between 2.6 and 2.5 Ga from Mo isotopes and Re-PGE signatures in shales. *Geochimica et Cosmochimica Acta*, **71**, 2417–2435.

Williams, R. and Da Silva, J. (1996) The natural selection of the chemical elements. *Chemistry in Britain*. Clarendon Press, Oxford, UK.

Wolthers, M., Charlet, L., van Der Linde, P.R., Rickard, D. and van Der Weijden, C.H. (2005) Surface chemistry of disordered mackinawite (FeS). *Geochimica et Cosmochimica Acta*, **69**, 3469–3481.

Wu, L., Percak-Dennett, E.M., Beard, B.L., Roden, E.E. and Johnson, C.M. (2012) Stable Fe isotope fractionation between aqueous Fe(II) and model Archean ocean Fe-Si coprecipitates and implications for Fe isotope variations in the ancient rock record. *Geochimica et Cosmochimica Acta*, **84**, 14–28.

Yamaguchi, K.E. (2002) *Geochemistry of Archean-Paleoproterozoic black shales: the early evolution of the atmosphere, oceans, and biosphere*. Ph.D. dissertation, Pennsylvania State University.

Zahnle, K., Claire, M. and Catling, D. (2006) The loss of mass-independent fractionation in sulfur due to a Palaeoproterozoic collapse of atmospheric methane. *Geobiology*, **4**, 271–283.

Zegeye, A., Bonneville, S., Benning, L.G., Sturm, A., Fowle, D.A., Jones, C., Canfield, D.E., Ruby, C., MacLean, L.C., Nomosatryo, S., Crowe, S.A. and Poulton, S.W. (2012) Green rust formation controls nutrient availability in a ferruginous water column. *Geology*, **40**, 599–602.

Zheng, Y., Anderson, R.F., van Geen, A. and Kuwabara, J. (2000) Authigenic molybdenum formation in marine sediments: a link to pore water sulfide in the Santa Barbara Basin. *Geochimica et Cosmochimica Acta*, **64**, 4165–4178.

Zink, S., Schoenberg, R. and Staubwasser, M. (2010) Isotopic fractionation and reaction kinetics between Cr(III) and Cr(VI) in aqueous media. *Geochimica et Cosmochimica Acta*, **74**, 5729–5745.

EMU Notes in Mineralogy, Vol. 17 (2017), Chapter 7, 173–196

Formation of manganese oxide minerals by bacteria

SUNG-WOO LEE, MATTHEW JONES, CHRISTINE ROMANO and BRADLEY M. TEBO

Division of Environmental & Biomolecular Systems, Institute of Environmental Health, Oregon Health & Science University, Portland, Oregon 97239, USA, e-mail: sungwlz@gmail.com

Manganese is an element that is relatively abundant and can be found in three oxidation states (II, III and IV) in the environment. Mn(II) is generally found in the soluble phase while Mn(IV), normally found as Mn(IV) oxide is relatively insoluble. Mn(III) is unstable in the environment and will disproportionate into Mn(II) and Mn(IV) unless it is complexed by a ligand, which allows Mn(III) to be present in the soluble phase. Mn(IV) oxide is a strong oxidant and sorbent that plays an important role in the cycling and mobility of various elements and organic compounds. In the environment Mn(IV) oxides are predominantly formed by bacteria (and fungi) through the oxidation of Mn(II) *via* a Mn(III) intermediate. Although the reason(s) why bacteria oxidize Mn(II) is still unknown, recent studies have revealed the nature of the enzymes that carry out oxidation of Mn(II) to Mn(IV). Studies have shown that depending on the bacterium, a multicopper oxidase and/or a heme peroxidase is involved in bacterial Mn(II) oxidation, similar to fungi. Studies also have reported the Mn oxides formed by Mn(II)-oxidizing bacteria to be layer-type birnessites. This chapter aims to introduce what is currently known about how bacteria form Mn oxides and also the structures of bacteriogenic Mn oxides.

1. Introduction

Manganese (Mn) is the second most abundant redox active transition metal in the Earth's crust. It is ubiquitous in the environment, where it is found in three oxidation states: II, III and IV. Manganese geochemistry has often been defined operationally using filtration: soluble Mn that passes through certain sized filters, usually <0.45 or <0.2 μm, is considered to be Mn(II) (as the divalent cation, Mn^{2+}) while the particulate fraction is composed of Mn(III), Mn(IV) and mixed Mn(III, IV) oxyhydroxide/oxide phases (collectively referred to in this paper as Mn oxides) which occur in a variety of Mn minerals. Unless complexed with organic or inorganic ligands, aqueous Mn(III) is unstable and will disproportionate to Mn(II) and Mn(IV). Mn oxides are effective sorbents; their presence can result in a decreased mobility of various other elements (Nelson *et al.*, 1999; Foster *et al.*, 2003; Tebo *et al.*, 2004; Zhu *et al.*, 2009; Pena *et al.*, 2010; Wang *et al.*, 2012b). Additionally, Mn oxides are strong oxidants, oxidizing other elements such as U(IV), Cr(III) and As(III) and also complex organics (Stone and

DOI: 10.1180/EMU-notes.17.13

Morgan, 1984a,b; Sunda and Kieber, 1994; Fendorf and Zasoski, 2002; Chinni *et al.*, 2008; Lafferty *et al.*, 2010):

$$UO_2 + MnO_2 + 4H^+ \leftrightarrow UO_2^{2+} + Mn^{2+} + 2H_2O \quad (1)$$

$$2Cr(OH)_3 + 3MnO_2 \leftrightarrow 3Mn^{2+} + 2CrO_4^{2-} + 2H_2O + 2OH^- \quad (2)$$

$$H_3AsO_3 + MnO_2 \leftrightarrow HAsO_4^{2-} + Mn^{2+} + H_2O \quad (3)$$

During these reactions Mn oxides are reduced to form Mn(II). Mn(III) can form complexes with a variety of ligands both inorganic and organic, *e.g.* OH^- and pyrophosphate and citrate and desferrioxamine B (Duckworth and Sposito, 2005; Faulkner *et al.*, 1994; Harrington *et al.*, 2012; Klewicki and Morgan, 1998; Parker *et al.*, 2004, 2007) and recent studies have observed soluble Mn(III) species in various environments: the Black Sea, Chesapeake Bay, and sediment porewaters of the lower St Lawrence River Estuary (Trouwborst *et al.*, 2006; Madison *et al.*, 2011, 2013; Oldham *et al.*, 2015). In some of these redox stratified environments, soluble Mn(III), Mn(III) stabilized by forming complexes with varying ligands, is the predominant species of the soluble Mn phase. Considering that soluble Mn(III) can function as a dissolved oxidant, the occurrence of soluble Mn(III) becomes a critical component in how manganese cycling affects the fate and transport of other elements and compounds (Kostka *et al.*, 1995; Yakushev *et al.*, 2009). In addition to their role of being chemical oxidants, Mn(III) and Mn(IV) can serve as terminal electron acceptors for bacterial respiration when other energetically favourable electron acceptors such as O_2 and NO_3^- are limited (Kostka *et al.*, 1995; Myers and Nealson, 1988).

In many environments manganese oxide formation has been attributed to microbial activities as the biotic oxidation of manganese has been found to be faster than abiotic processes (Hastings and Emerson, 1986; Nealson *et al.*, 1988). Rates of manganese oxidation in the absence of a catalytic surface (1.1×10^{-13} nM h^{-1}) are far slower than in the presence of a surface, such as MnO_x (5.4×10^{10} nM^{-1} h^{-1} (pH 8, 25°C and 1 atm)); a real time perspective can be provided with FeOOH, at which surface the half-life of Mn(II) is ~1 month (Morgan, 2000). Rates of biotic oxidation in marine/estuarine systems range from 3.5 to 12.1 nM h^{-1} in the seasonally anoxic fjord, Saanich Inlet, Vancouver Island, British Columbia, Canada (Tebo and Emerson, 1986) to 0.62–3.5 nM h^{-1} in a mid-east coast USA estuary in North Carolina (Sunda and Huntsman, 1987) to the slowest measured rates in the open ocean 4×10^{-4}– 3.7×10^{-3} nM h^{-1} in the Sargasso Sea, North Atlantic Ocean (Sunda and Huntsman, 1988). To understand the fate and transport of the various elements affected by manganese in the environment, research is needed to elucidate the role of microbial activity on its cycling. This goal requires: (1) identification of organisms that form manganese oxides; (2) understanding processes of microbially mediated oxidation of manganese; (3) determining environmental factors that influence how these microbes initiate Mn oxidation; and (4) exploring how the Mn oxides are formed. In this chapter, we review the mechanisms by which bacteria oxidize Mn and form Mn oxide minerals. We also summarize briefly the properties of the bacteriogenic Mn oxides and propose why bacterial Mn oxidation systems may have some useful environmental applications.

2. Mn(II) oxidation

2.1. Mn(II)-oxidizing bacteria

The ability to form manganese oxides has been found in both bacteria and fungi. Mn(II)-oxidizing bacteria are a phylogenetically diverse group (Fig. 1). Strains from various lineages have been identified across the bacterial domain and in some cases the genes involved in Mn oxidation are known (Table 1). Within the well studied model organisms and/or organisms that have undergone genome sequencing, there are a

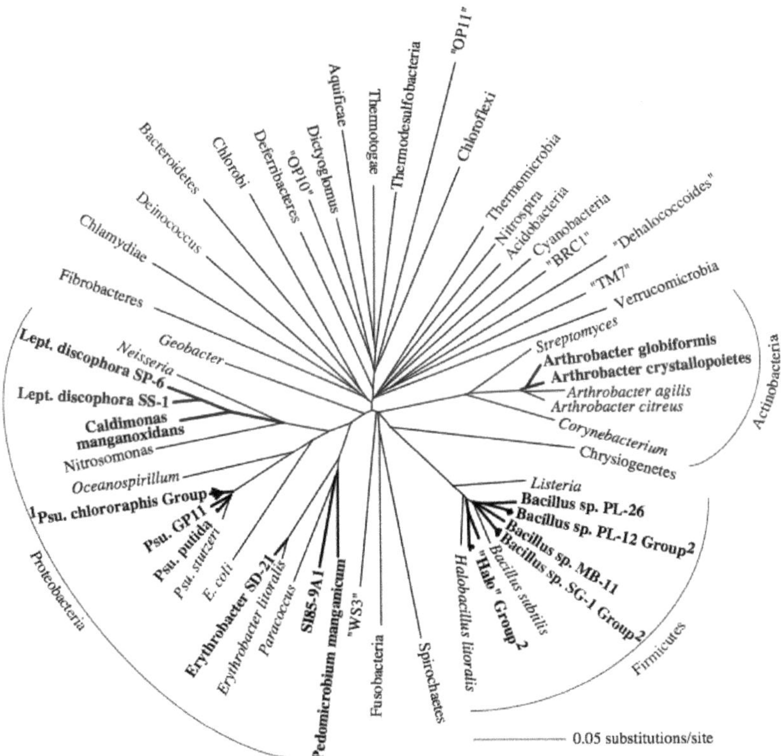

Figure 1. Neighbour-joining, unrooted phylonetic tree of the domain bacteria depicting well studied Mn(II)-oxidizing bacteria. 16S rRNA sequences aligned over 1536 positions were obtained from the Ribosomal Database Project (RDP; http://rdp.cme.msu.edu) and analysed using PAUP (phylogenetic analysis using *parsimony;* ver. 4.0b10). Mn(II)-oxidizing bacterial strains for which sequences appear in the RDP are shown in bold. Sequence similarity is only an estimate of evolutionary relationships and is expressed here as the distance between any two branch points; sequences that appear next to each other are not necessarily close relatives. Phyla in quotes are considered to be taxa of uncertain affiliation and have few (if any) cultivated members. Abbreviations: Lept., *Leptothrix*; Psu., *Pseudomonas.* (1) The *Psu. chlororaphis* group includes the Mn(II)-oxidizing strains ISO1, ISO6, GB-13, MG1 and PCP but not all similarly close relatives of *Psu. chlororaphis* are known to oxidize Mn(II). (2) Bacillus species groups are labelled according to Francis and Tebo (2001). Reproduced from Tebo *et al.* (2004).

Table 1. Examples of Mn(II)-oxidizing bacteria and the genes implicated to be involved in Mn oxidation.

Lineage	Organism	Gene	References
Firmicutes	*Bacillus* sp. SG-1	*mnxG*	(Rosson and Nealson, 1982; van Waasbergen *et al*., 1993; van Waasbergen *et al*., 1996)
Actinobacteria	*Arthrobacter globiformis* BKM 661	–	(Dubinina and Zhdanov, 1975)
Alphaproteobacteria	*Aurantimonas manganoxydans* SI85-9A1	*mopA*	(Anderson *et al*., 2009a; Anderson *et al*., 2009b; Caspi *et al*., 1996)
	Erythrobacter sp. SD-21	*mopA*	(Anderson *et al*., 2009b; Francis *et al*., 2001; Johnson and Tebo, 2008)
	Pedomicrobium sp. ACM 3067	*moxA*	(Larsen *et al*., 1999; Ridge *et al*., 2007)
	Roseobacter sp. AzwK-3b	*ahp*	(Andeer *et al*., 2015; Hansel and Francis, 2006)
	Fulvimarina pelagi HTCC2506	–	(Kang *et al*., 2010)
Betaproteobacteria	*Leptothrix discophora* SS-1	*mofA*	(Corstjens *et al*., 1997; de Vrind-de Jong *et al*., 1990)
	Caldimonas manganoxidans HS^T	–	(Takeda *et al*., 2002)
Gammaproteobacteria	*Pseudomonas putida* strains GB-1 and MnB1	*mnxG*, *mcoA*, *mopA*	(Caspi *et al*., 1998; Geszvain *et al*., 2013; Geszvain *et al*., 2016; de Vrind *et al*., 1998)
	Marinobacter manganoxydans MnI7-9	–	(Wang *et al*., 2012a)

variety of lineages including the Firmicutes, Actinobacteria and alpha-, beta- and gamma-proteobacteria, that have all shown an ability to form manganese oxides. Interestingly, even though the ability to oxidize Mn(II) appears to be widespread, it is not yet known why these organisms possess this ability. Although oxidation of Mn(II) by bacteria may be fortuitous or a by-product or side reaction of another process, in the model organisms that have been studied in greatest detail, there appear to be specific enzyme(s) that are involved in Mn(II) oxidation. Speculation on what biological function Mn(II) oxidation serves includes, but is not limited to, energy generation (Ehrlich and Salerno, 1990), protection from reactive oxygen species and UV (Ghiorse, 1984; Banh *et al.*, 2013) and the breakdown of organic molecules to be utilized as metabolic substrates (Sunda and Kieber, 1994). The potential biological functions of Mn(II) have been discussed elsewhere (Tebo *et al.*, 1997)

2.2. Mn(II) oxidases

Based on the orbital properties of the electron configuration of Mn(II), its oxidation is considered to be carried out through two single electron transfers with Mn(III) forming as an intermediate (Luther, 2005). Microorganisms, both fungi and bacteria, facilitate this oxidation enzymatically. While more is known about the Mn oxidizing capabilities of fungi, research into the bacterially produced enzymes responsible for oxidation of Mn(II) to Mn(IV) oxides is in its initial stages. In fungi, the enzymes that carry out oxidation of Mn(II) have been found to be in two groups, the multicopper oxidases (MCO) and the peroxidases. Laccase, which is a member of the MCO (blue coloured enzymes) family of proteins, is an enzyme that has a broad range of substrate specificities, from phenolic compounds to Mn(II) (Baldrian, 2006) and carries out oxidations coupled with the four-electron reduction of O_2 to H_2O (Solomon *et al.*, 1996). Laccases generally have one type-1 (T1) copper ion in the active site along with three other copper ions (one type-2 (T2) and two type-3 (T3)) forming a trinuclear cluster (TNC) (Baldrian, 2006). In laccases, the T1 copper receives the electron generated from the oxidation of substrate while the TNC is responsible for the reduction of O_2 (Schlosser and Hofer, 2002).

The manganese peroxidases that have been characterized thus far are part of the heme peroxidase group and examples can be found within the lignin degrading system in white rot and litter decaying fungi (Schlosser and Hofer, 2002). In the manganese-containing heme peroxidase from *Phanerochaete chrysosporium*, the manganese is bound to heme propionate and three other amino acids along with two water ligands (Sundaramoorthy *et al.*, 1994). The active centres of manganese peroxidases are oxidized by H_2O_2, and reduced to the resting state by transferring electrons from Mn^{2+} to form Mn^{3+} (Schlosser and Hofer, 2002; Sundaramoorthy *et al.*, 1994). The resulting Mn(III) when chelated with a ligand becomes a diffusive oxidant that can oxidize various organic compounds.

Similar to the fungal manganese oxidases the bacterially produced variants have been found to be either an MCO or a heme peroxidase. Studies on *Bacillus* sp. SG-1, *Pseudomonas putida* GB-1, *Leptothrix discophora* SS-1, and *Pedomicrobium* sp. ACM

3067 have provided direct or indirect information suggesting that these organisms use an MCO to form manganese oxides. *Aurantimonas manganoxydans* SI85-9A1 and *Erythrobacter* sp. SD-21 utilize heme peroxidases. *P. putida* GB-1 also has a heme peroxidase that is expressed under certain conditions that oxidizes Mn(II) (Geszvain *et al.*, 2013). Interestingly, a study on *Roseobacter* sp. AzwK-3b, another Mn(II) oxidizing bacterium showed that Mn(II) oxidation was enhanced in the presence of light (Hansel and Francis, 2006). More recently it has been shown that *Roseobacter* sp. AzwK-3b oxidizes Mn(II) by enzymatically forming superoxide which subsequently oxidizes Mn(II) (Learman *et al.*, 2011a,b, 2013). Collectively, it appears that bacteria have various enzymatic pathways to oxidize Mn(II) to form manganese oxides. The following parts of this chapter will focus on what is currently known about the manganese oxidases in bacteria and their product, manganese oxides, which have properties suitable for a variety of biotechnological applications.

2.2.1. MCOs of Bacillus species

Most information on Mn(II) oxidation *via* putative MCOs are based on studies of *Bacillus* sp. strains SG-1 and PL-12, though other bacteria are also thought to utilize MCOs for Mn(II) oxidation. Spores of *Bacillus* sp. SG-1 were found to be responsible for Mn oxide formation in that species (Nealson and Ford, 1980; Rosson and Nealson, 1982; de Vrind *et al.*, 1986). Later studies showed that spores of various *Bacillus* species form manganese oxides (Dick *et al.*, 2006; Francis and Tebo, 2002). Bacillus spores are now known to play an important role in manganese cycling and ultimately to alter the speciation of various elements. Genetic studies revealed that the putative Mn(II) oxidase in *Bacillus* sp. SG-1 is encoded by *mnxG* located within the *mnx* operon with other genes located upstream, *mnxA-F* (van Waasbergen *et al.*, 1996). It is currently unknown why spores would form Mn oxides although it has been suggested that the Mn oxides could serve as an electron acceptor for anaerobic respiration or for further protection against reactive oxygen species, UV light or other environmental factors (Tebo *et al., 1997*). The deduced amino acid sequence of *mnxG* contained multiple copper binding regions and showed high similarity to MCOs (van Waasbergen *et al.*, 1996). Further studies have shown that the Mn(II) oxidase is localized in the outer layers of the spores, specifically the exosporium (Francis *et al.*, 2002). Activity of the Mn(II) oxidase, MnxG, is inhibited by the addition of azide and o-phenanthroline, both known to inhibit MCOs (Francis *et al.*, 2002). Azide inhibits MCOs by bridging T2 and T3 copper sites (Solomon *et al.*, 1996) while o-phenanthroline is a copper chelator. The protein was identified by *in gel* Native PAGE (polyacrylamide gel electrophoresis) assays. When a native gel is incubated in a solution of Mn(II) the formation of Mn oxides can be detected visually by the brown to black coloration that occurs where the Mn(II) oxidase is present. For *Bacillus* sp. SG-1, the band on the gel that produced Mn oxides was excised and tandem mass spectrometry was used to identify the proteins as MnxG and MnxF (Dick *et al., 20*08b). MnxF is a small hydrophobic protein of unknown function that contains a copper binding site (Dick *et al.*, 2008b). Recently, the MnxG complex of *Bacillus* sp. PL-12, has been successfully overexpressed in *E. coli* using a *mnxDEFG* expression construct enabling a faster collection of relatively large amounts of MnxG (Butterfield *et al.*, 2013). The expressed product was analyzed by *in gel* assay and tandem mass spectrometry. The band consisted of MnxG, MnxE and MnxF. The presence of three proteins suggested that this Mn(II) oxidase is a multi-protein complex, a phenomenon which has not been observed in other MCOs (Butterfield *et al.*, 2013). Recently, a study on MnxE and MnxF has revealed that these proteins bind copper and heme but will not oxidize Mn(II) in the absence of MnxG (Butterfield *et al.*, 2015).

Although bacterial Mn(II) oxidation was proposed to be carried out *via* two single electron-transfer processes (Luther, 2005), supporting experimental evidence came only recently. Initially, a study reported that *Bacillus* sp. SG-1 formed hausmannite (Mn_3O_4) which transformed to MnO_x (x = 1.9) with aging and concluded that this conversion required two single electron transfers (Hastings and Emerson, 1986). Other studies also reported the formation of manganese oxides with lower valence which transform to higher-valence manganese oxides with aging (Mann *et al.*, 1988; Mandernack *et al.*, 1995). In a later study, Mn(III) formation during Mn(II) oxidation to Mn(IV) oxides by the exosporium of *Bacillus* sp. SG-1 was shown using a Mn(III) chelator to trap the otherwise unstable and transient Mn(III) (Webb *et al.*, 2005a). A *Bacillus* sp. SG-1 mutant with an exosporium lacking MnxG exhibited no Mn(III) formation from Mn(II) and no Mn(IV) oxide from Mn(III), suggesting that MnxG plays a role in both the oxidation of Mn(II) to Mn(III) and Mn(III) to Mn(IV) oxides (Webb *et al.*, 2005a). A more recent study verified these results and demonstrated that both steps, *i.e.* Mn(II) to Mn(III) and Mn(III) to Mn(IV), are enzyme mediated, inhibited by sodium azide, and dependent on O_2 (Soldatova *et al.*, 2012). The study proposed a mechanism for how a single multicopper oxidase could oxidize its substrate *via* two single electron transfer processes. This mechanism was later modified based on observations using electron paramagnetic resonance (EPR) (Fig. 2), which found evidence of a dimeric Mn(II, II) species present in Mnx (Tao *et al.*, 2015). The first step in Mn oxidation, Mn(II) to Mn(III), was proposed to be similar to that observed in Fe(II) oxidation by human ceruloplasmin (Lindley *et al.*, 1997; Machonkin and Solomon, 2000; Quintanar *et al.*, 2004) or yeast Fet3p (Quintanar *et al.*, 2004; Stoj *et al.*, 2006; Taylor *et al.*, 2005) in that a 'holding site' may bind a partially oxidized intermediate. A mononuclear Mn(II) would be near the T1 copper site, where it could be oxidized to Mn(III). The mononuclear Mn(III) then receives an electron from a binuclear Mn(II, II) site (Tao *et al.*, 2015). Multiple electron transfers could occur between the mononuclear and binuclear substrate binding sites, resulting gradually in the formation of Mn(III) and Mn(IV) intermediates. The electrons from reduced manganese would gradually be transferred to the TNC, which would transfer them to dioxygen to form water (Tao *et al.*, 2015). Formation of hydroxide or oxide bridges using water would provide a driving force for the successive oxidation of each dinuclear intermediate to form $[Mn(IV)_2O_2]^{4+}$. In the final step, the $[Mn(IV)_2O_2]^{4+}$ could be moved outside of the protein. If enough such MnO_2 particles accumulate, they could form a 'manganese wire' through which electrons could be transferred from an external Mn(II) atom to the

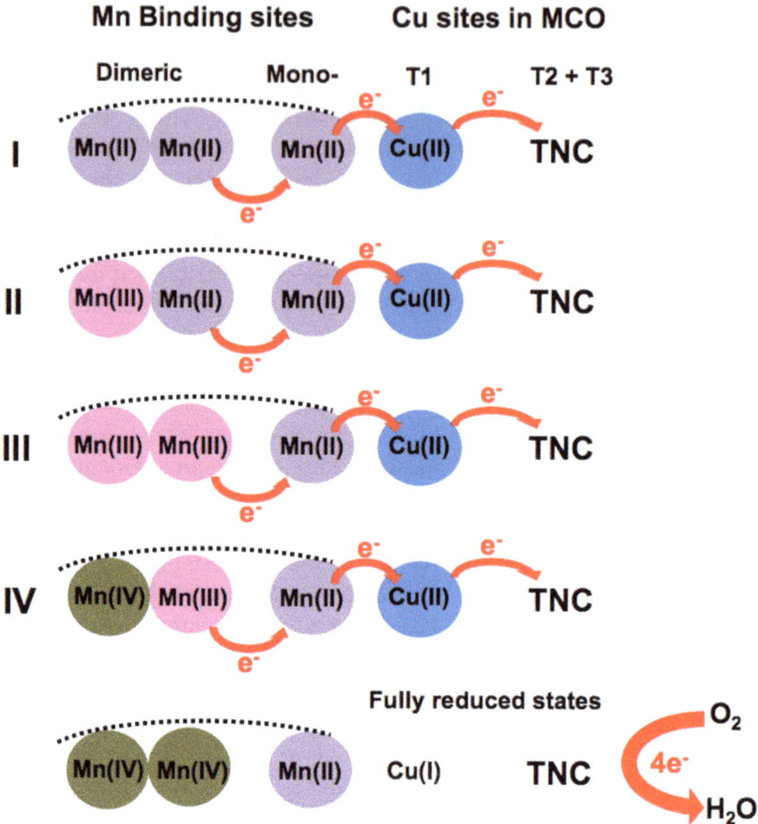

Figure 2. Proposed mechanism for Mn(II) binding and oxidation by the MCO in the Mnx protein complex of *Bacillus* sp. strain PL-12. Reproduced from Tao *et al.* (2016) after Soldatova *et al.* (2012) with the permission of the American Chemical Society.

redox active copper sites of the MCO. These particles could also nucleate to form the MnO_2 mineral (Soldatova *et al.*, 2012; Tao *et al.*, 2015), similar to how Fe oxide nucleation occurs in the Fe storage protein, ferritin (Liu and Theil, 2005). Because relatively large amounts of MnxG can now be purified, further biochemical studies should be able to test this proposed mechanism of mineral formation.

2.2.2. MCOs in other Mn(II)-oxidizing bacteria

In addition to *Bacillus* species, MCOs have been implicated in Mn(II) oxidation in other microorganisms. Evidence for MCO-like enzymes involved in Mn(II) oxidation also exists for *Pedomicrobium* sp. ACM3067, *Leptothrix discophora* SS-1, and *P. putida* GB-1. *P. putida* GB-1 and *P. putida* MnB1 are closely related Mn(II) oxidizing bacteria that form manganese oxides late in their exponential growth phase (Okazaki *et al.*, 1997). The Mn(II) oxidation activity from *P. putida* GB-1 was found within whole-cell extracts and its activity was inhibited by o-phenanthroline and azide, strengthening the hypothesis that an MCO carries out Mn(II) oxidation for *P. putida* strains (Okazaki *et al.*, 1997). Through transposon mutagenesis the *cumA* gene, which encodes a putative MCO, was initially proposed as the Mn oxidase (Brouwers *et al.*, 1999). However, a later study showed that the *cumA* mutant strain created in the early work had a second mutation in a putative sensor histidine kinase (Geszvain and Tebo, 2010). When only *cumA* was deleted from *P. putida* GB-1, the strain was still capable of forming manganese oxides, indicating *cumA* was not required for Mn(II) oxidation in *P. putida* (Geszvain and Tebo, 2010). The genome of *P. putida* GB-1 contains multiple genes encoding homologues of putative MCOs such as *mnxG*, *mofA* and *moxA* (Geszvain *et al.*, 2013). *moxA* and *mofA* encode the putative Mn(II) oxidases in *Pedomicrobium* ACM 3067 (Ridge *et al.*, 2007) and *L. discophora* SS-1 (Corstjens *et al.*, 1997), respectively. A homologue of *mofA*, is found in the genome of *P. putida* GB-1; however, due to its low homology to and large size difference from *L. discophora* SS-1 *mofA*, the *P. putida* GB-1 gene was given the annotation *mcoA* (Geszvain *et al.*, 2013). The deletion of either *mnxG* or *mcoA* from *P. putida* GB-1 decreases manganese oxide formation to a certain degree but these mutants still retained the ability to form manganese oxides. Only a mutant with both genes deleted was no longer able to form manganese oxides under the conditions examined. This result suggested that both genes encode a Mn(II) oxidase, an intriguing phenomenon which has yet to be reported in other Mn(II) oxidizing organisms (Geszvain *et al.*, 2013). The genes *mcoA* and *mnxG* have been found in known Mn(II) oxidizing pseudomonads but not in known non-Mn(II) oxidizing pseudomonads. Therefore, it is highly likely that these two MCOs have critical roles in Mn(II) oxidation. Further study such as heterologous expression of the genes encoding the two MCOs would shed light on the mechanism of Mn(II) oxidation by these two enzymes. Interestingly, it was determined that a Δ*mnxG*Δ*mcoA* double knock-out of *P. putida* GB-1 can produce Mn oxides when a third gene, *fleQ*, is also deleted. It was hypothesized that the *fleQ* deletion allows activity of a third Mn oxidase, MopA (Geszvain *et al.*, 2016). The MopA protein is a putative animal heme peroxidase (see below; Geszvain *et al.*, 2016).

Leptothrix species belong to the beta-proteobacteria and are capable of forming sheaths wherein deposits of manganese or iron oxides form (van Veen *et al.*, 1978). Early studies indicated that *Leptothrix discophora* SS-1 secreted a factor that oxidized Mn(II) (Boogerd and de Vrind, 1987). Later work using an antibody raised against this secretion factor indicated *mofA*, a gene with multiple copper binding regions, encodes the Mn(II) oxidizing factor (Corstjens *et al.*, 1997). The *mofA* gene is not unique to *L. discophora* SS-1 but is found in other Mn(II) oxidizing *Leptothrix* species (Siering and Ghiorse, 1997). Evidence that an MCO is involved in Mn(II) oxidation was provided through the addition of copper to stimulate the oxidizing ability of either the active cell culture or spent media from *L. discophora* SS-1 grown in the presence of copper (Brouwers *et al.*, 2000; Zhang *et al.*, 2002). Mn(II) oxidation activity of *L. discophora* SS-1 is inhibited in cells grown in iron-limited conditions, and activity is not restored with added Cu(II) (El Gheriany *et al.*, 2009). However, Mn(II) oxidizing activity under

Fe-replete conditions is stimulated by the addition of Cu(II), although copper additions do not affect transcript levels of *mofA*. It is worth noting that iron also did not affect the transcript levels of *mofA* (El Gheriany *et al.*, 2009). Based on these studies, it was suggested that regulation of Mn(II) oxidase activity is not dependent on the metal present and perhaps the activity is post-transcriptionally regulated by Cu(II) (El Gheriany *et al.*, 2011). The possibility of having multiple Mn(II) oxidases in *L. discophora* SS-1 has been suggested (Boogerd and de Vrind, 1987; Corstjens *et al.*, 1992) and could be said to have set the precedent for the idea of multiple MCOs in *P. putida* GB-1 (Geszvain *et al.*, 2013). In the genome of *L. cholodnii* SP-6, an organism closely related to *L. discophora* SS-1, in addition to homologues of *mofA* and *mcoA*, five other genes were found that encode MCOs (El Gheriany *et al.*, 2011; Geszvain *et al.*, 2013). It is therefore possible that *L. discophora* SS-1 could, like *P. putida* GB-1, have multiple Mn(II) oxidases. Progress in understanding the genes encoding the Mn(II) oxidase in *L. discophora* SS-1 is hindered by the difficulty of doing genetic manipulations in this organism.

Pedomicrobium sp. ACM 3067 belongs to the alpha-proteobacteria and is unlike other Mn(II) oxidizing bacteria in that Mn(II) oxidation occurs during early stages of growth (Larsen *et al.*, 1999). Additions of copper did not affect Mn(II) oxidation by washed cells of *Pedomicrobium* sp. ACM 3067. This was attributed to an incomplete removal of copper from the cells (Larsen *et al.*, 1999). When diethyldithiocarbamic acid, a copper chelator, was added to the washed cells, Mn(II) oxidation was inhibited; the activity was restored through the re-addition of copper (Larsen *et al.*, 1999). The authors concluded that the results were evidence that *Pedomicrobium* sp. ACM 3067 utilizes an MCO to catalyse Mn(II) oxidation. In a subsequent study, an operon containing three genes, *moxCBA*, was identified within the genome of *Pedomicrobium* sp. ACM 3067 (Ridge *et al.*, 2007). One of the genes in this operon, *moxA*, has four copper binding motifs indicative of MCOs. Mutation of *moxA*, *via* integration of a plasmid, abolished the ability of *Pedomicrobium* sp. ACM 3067 to oxidize Mn(II). However, the study did not confirm whether complementation of *moxA* could restore Mn(II) oxidation. Coincidently, the copper binding motifs of MoxA in *Pedomicrobium* sp. ACM 3067 showed high homology to *CumA*, the multicopper oxidase in *P. putida* GB-1 previously considered but refuted to be the Mn(II) oxidase (Geszvain and Tebo, 2010). Therefore, although the role of MoxA may be crucial in Mn(II) oxidation by *Pedomicrobium* sp. ACM 3067, further studies are necessary to conclude that MoxA is a Mn(II) oxidase in this organism.

2.2.3. Mn(II) oxidation mediated by Ca^{2+}-binding animal heme peroxidases

Several organisms are believed to use a Ca^{2+}-binding animal heme peroxidase enzyme to oxidize manganese. In the marine alpha-proteobacteria *Aurantimonas manganoxydans* SI85-9A1 and *Erythrobacter* sp. SD-21, the enzyme has been called MopA for Mn(II)-oxidizing peroxidase (Anderson *et al.*, 2009b). Based on inhibition by azide (an inhibitor of metalloenzymes) and *o*-phenanthroline (a copper chelator) these organisms were thought to utilize MCOs to oxidize Mn(II), similar to *Bacillus* sp. SG-1, *P. putida* GB-1 and *L. discophora* SS-1 (Francis *et al.*, 2001). The genome of *A. manganoxydans*

SI85-9A1 also has multiple genes encoding MCOs: homologues of MoxA (similar to *Pedomicrobium* sp. ACM 3067) and an MCO exhibiting high similarity to a gene found in *P. putida* GB-1 (Dick *et al.*, 2008a). However, the role of an MCO in Mn(II) oxidation by *Erythrobacter* sp. SD-21 was questioned, as a partially purified Mn(II) oxidase was stimulated by the addition of quinones and not copper. The absorbance spectrum also suggested it contained quinones and not copper (Johnson and Tebo, 2008). The majority of the Mn(II) oxidation activity of *A. manganoxydans* SI85-9A1 was found in the supernatant of the growth media; peptides found in that fraction were identified as part of hemolysin-type Ca^{2+} binding peroxidase (Anderson *et al.*, 2009b). At the time, the study speculated that either (1) Mn(II) may act as a substrate that reduces this peroxidase and itself becomes oxidized to Mn(III) (Wariishi *et al.*, 1992) or (2) the heme peroxidase would be functioning in concert with an MCO similar to fungal systems where laccase oxidizes Mn(II) to Mn(III) and forms H_2O_2 which serve as a substrate for the Mn peroxidase along with Mn(II) to form Mn(III) (Schlosser and Hofer, 2002). However, Mn(II) oxidation by both *Erythrobacter* sp. SD-21 and *A. manganoxydans* SI85-9A1 is currently attributed directly to Ca^{2+} binding heme peroxidases based on their detection by mass spectrometry in the active fraction of the proteins that can carry out metal oxidation in the absence of manganese oxidizing MCOs (Anderson *et al.*, 2009b). To further strengthen the case for heme peroxidases, the Mn(II)-oxidizing activity of partially purified proteins from both *Erythrobacter* sp. SD-21 and *A. manganoxydans* SI85-9A1 were shown to be stimulated by the addition of Ca^{2+} and were stained by both in-gel Mn(II) oxidation and heme-staining assays (Johnson and Tebo, 2008; Anderson *et al.*, 2009b). Interestingly, it should be noted that the heme peroxidase in *A. manganoxydans* SI85-9A1 has ten predicted Ca^{2+} binding sites and two peroxidase domains while that in *Erythrobacter* sp. SD-21 has seven predicted Ca^{2+} binding sites and only one peroxidase domain (Anderson *et al.*, 2009b). However, it appears that the role of these multiple Ca^{2+} binding sites is not as critical in Mn(II) oxidation. A recent study showed that the heme peroxidase of *Erythrobacter* sp. SD-21 which lacks the Ca^{2+} binding sites can carry out oxidation of Mn(II) (Nakama *et al.*, 2014). On the other hand, unlike the heme peroxidase from the native host, addition of Ca^{2+} was required for the heterologously expressed heme peroxidase (Nakama *et al.*, 2014). At this time, it is unknown why Ca^{2+} is necessary for heme peroxidase to oxidize Mn(II).

During Mn(II) oxidation, both *A. manganoxydans* SI85-9A1 and *Erythrobacter* sp. SD-21 form Mn(III) (Johnson and Tebo, 2008; Anderson *et al.*, 2009b) but differences in the removal mechanism for Mn(III) were observed. In *A. manganoxydans* SI85-9A1, Mn(III) was formed transiently similar to that observed with *Bacillus* sp. SG-1, suggesting that Mn(III) is an intermediate of Mn(II) oxidation (Anderson *et al.*, 2009b). In contrast, *Erythrobacter* sp. SD-21 produced Mn(III) but it was not further oxidized to Mn(IV) oxides, suggesting that Mn(III) is the final product of Mn(II) oxidation (Johnson and Tebo, 2008). Therefore, although both *A. manganoxydans* SI85-9A1 and *Erythrobacter* sp. SD-21 employ heme peroxidase to catalyse Mn(II) oxidation, it appears that the heme peroxidase in *A. manganoxydans* SI85-9A1 either carries out two single electron transfers or has a second mechanism for oxidizing Mn(III) while that in *Erythrobacter* sp. SD-21 only facilitates one single electron transfer. Perhaps this difference is due to the number of Ca^{2+} binding sites and/or peroxidase domains in the two heme peroxidases. However, further genetic and/or biochemical studies are needed to elucidate how these heme peroxidases catalyse Mn(II) oxidation.

2.2.4. Combined enzymatic and non-enzymatic Mn(II) oxidation

In addition to utilizing multicopper oxidases or heme peroxidases, there appears to be yet another bacterially mediated pathway in the formation of manganese oxides through the enzymatic production of the reactive oxygen species, superoxide (O_2^-). In the marine α-proteobacterium, *Roseobacter* sp. AzwK-3b, Mn(II) oxidation by live cultures and cell-free filtrates was found to be enhanced in the presence of light, a process attributed to photo-catalysed superoxide production (Hansel and Francis, 2006). A subsequent study in the dark showed that superoxide (O_2^-) was also enzymatically produced and could react with Mn(II) to form Mn(III) with hydrogen peroxide (H_2O_2) as a byproduct (Learman *et al.*, 2011a). A similar process has been observed in abiotic systems following photochemical excitation of humic substances; the subsequent reactive intermediate species formed were found to oxidize Mn(II) resulting in the formation of Mn oxides (Nico *et al.*, 2002). In the *Roseobacter* experiments, additions of Cu(II), a scavenger of superoxide, resulted in inhibition of Mn(II) oxidation (Learman *et al.*, 2011a), in contrast to the Cu-stimulated oxidation observed in MCO-catalysed Mn(II) oxidation. Mn oxide formation mediated by abiotically produced superoxide was shown to be inhibited in the presence of H_2O_2, a product of Mn(II) oxidation by superoxide (Learman *et al.*, 2013). The inhibition is attributed to the effect of (1) reduction of Mn(III) or Mn(IV) by H_2O_2 and/or (2) thermodynamic inhibition due to increased levels of H_2O_2. *Roseobacter* sp. AzwK-3b contains homologs of *mopA* in its genome (Anderson *et al.*, 2011) and recently has been shown to produce Mn(II)-oxidizing animal heme peroxidases on its outer membrane and extracellularly (Andeer *et al.*, 2015). The removal of H_2O_2 by these peroxidases results in the production of superoxide which may oxidize both Mn(II) and Mn(III) ultimately allowing Mn oxides to form.

2.2.5. Bacteriogenic Mn oxides

The products of Mn(II) oxidation by bacteria, manganese oxides, are typically found as two different structures: (1) chain or tunnel structures, or (2) layer structures (Fig. 3; Post, 1999). Chain or tunnel structures are chains of edge-sharing MnO_6 octahedra with the chains sharing the corners to form tunnels of varying size. Tunnel size affects surface area and the accessibility of water and cations (Post, 1999; Feng *et al.*, 1999; Tebo *et al.*, 2004). Layer structures are layers of MnO_6 octahedron sheets with cations and water occupying space between the interlayer regions (Post, 1999). Manganese oxides with layer structures can be further divided into manganese oxides with either hexagonal layer symmetry or orthogonal layer symmetry (Gaillot *et al.*, 2003; Lanson *et al.*, 2000).

Earlier studies on bacteriogenic manganese oxides have suggested that MnO_x (x = 1.9) is the final product and hausmannite (Mn_3O_4) is an intermediate (Hastings and

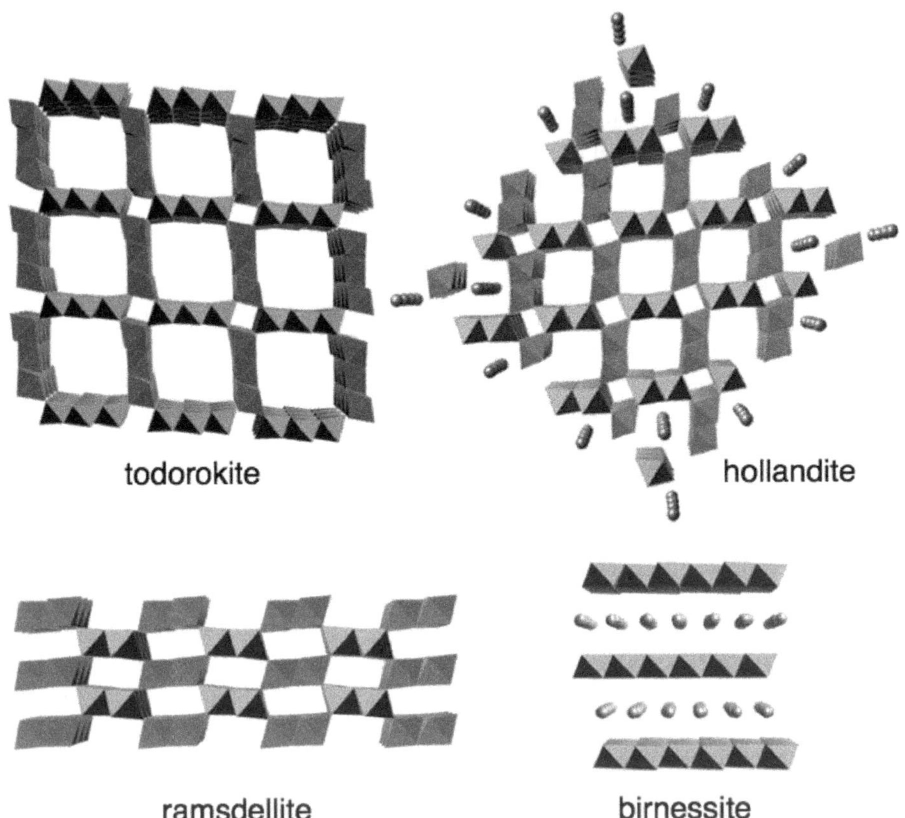

Figure 3. Examples of Mn oxides with tunnel structures (todorokite, hollandite and ramsdellite) or layer-type structures (birnessite). The Ba cations are not shown for hollandite and the charge-balancing cations are not shown for todorokite. Reproduced from Tebo *et al.* (2004).

Emerson, 1986; Mann *et al.*, 1988). Recent studies using synchrotron-based X-ray absorption spectroscopy (XAS), specifically X-ray absorption near edge spectroscopy (XANES) and extended X-ray absorption fine structure spectroscopy (EXAFS), have revealed detailed information on bacteriogenic manganese oxides and how certain geochemical parameters affect their structures (Bargar *et al.*, 2005; Jürgensen *et al.*, 2004; Learman *et al.*, 2011b; Villalobos *et al.*, 2003, 2006; Webb *et al.*, 2005c, 2006; Zhu *et al.*, 2010a). XANES and EXAFS provide information on the oxidation state and local structure of the solids. The technique enables bulk investigation of amorphous and poorly crystalline phases as well as wet samples. This becomes useful in studying biogenic manganese oxides as they may have varying oxidation states, tend to be amorphous and poorly crystalline and may change structure upon dehydration (Tebo *et al.*, 2004). Other techniques employed in studying the structures of bacteriogenic manganese oxides are Fourier transform infrared (FTIR) spectroscopy (Parikh and Chorover, 2005) and scanning transmission soft X-ray microscopy (STXM) (Pecher *et al.*, 2003). Among the model Mn(II) oxidizing bacteria, several strains have been used for structural studies: specifically, *Bacillus* sp. SG-1, *P. putida* GB-1, *Roseobacter* sp. AzwK-3b, and *Leptothrix discophora* SS-1 (Bargar *et al.*, 2005; Jürgensen *et al.*, 2004; Learman *et al.*, 2011b; Saratovsky *et al.*, 2006; Villalobos *et al.*, 2003, 2006; Webb *et al.*, 2006). They were studied in a defined medium to minimize environmental effects and to examine the effect of specific geochemical parameters, *e.g.* the presence of varying cations on the bacteriogenic manganese oxide structure. Most of these studies used spores of *Bacillus* sp. SG-1 and washed cells of *P. putida* GB-1 which have shown that certain geochemical parameters do significantly affect the structure of the manganese oxides (Webb *et al.*, 2005b,c, 2006; Zhu *et al.*, 2010a). Other studies of the structural properties of bacteriogenic Mn oxides using other organisms were undertaken in rich bacterial growth media (where organic carbon and many other nutrients are present), and unfortunately are difficult to evaluate meaningfully.

Studies on the structure of manganese oxides formed by both *Bacillus* sp. SG-1 and *P. putida* MnB1 have shown that they are similar to that of δ-MnO_2, a synthetic Mn oxide analogue of vernadite, or a poorly ordered layered hexagonal birnessite with an average oxidation state between 3.7 and 4.0 (Bargar *et al.*, 2005; Villalobos *et al.*, 2003, 2006). These bacteriogenic oxide products are different from crystalline triclinic birnessite, which has orthogonal layer symmetry, and have a large number of vacancies in the mineral lattice (Villalobos *et al.*, 2006). The initial Mn oxide product of *Roseobacter* sp. AzwK-3b is also a Mn oxide with hexagonal symmetry (Learman *et al.*, 2011b). Studies with *L. discophora* SS-1 and *L. cholodnii* SP-6 were performed in culture growth media containing various other components which may or may not have effects on the structure of the biogenic manganese oxides. Nonetheless, it was reported that both species also form manganese oxides with hexagonal symmetry (Jürgensen *et al.*, 2004; Saratovsky *et al.*, 2006). Therefore, in all these organisms the initial biogenic Mn oxide product appears primarily to be a nanoparticulate layered hexagonal birnessite.

These initial bacteriogenic Mn oxide products are extremely reactive and transform rapidly to more stable oxides depending on the geochemical conditions (Fig. 4; Bargar *et al.*, 2005). In the presence of low (1 μM) concentrations of Mn(II) these poorly ordered birnessites appear to be more stable. But in the presence of higher Mn(II) concentrations the bacteriogenic oxides catalyse abiotic Mn(II) oxidation and form other secondary Mn oxide products, such as feitknechtite (β-MnOOH, an Mn(III) mineral) or triclinic Na-birnessite (a mixed Mn(III,IV) phase with orthogonal symmetry). Because of the highly reactive nature of these bacteriogenic Mn oxides they may be so transient in nature that they may be practically undetectable (Templeton and Knowles, 2009).

Several cations have been tested in order to examine their impacts on the structure of bacteriogenic manganese oxides from the model organisms *Bacillus* sp. SG-1 and *P. putida* GB-1. When the medium was supplemented with Ca^{2+}, both *Bacillus* sp. SG-1 and *P. putida* GB-1 produced manganese oxides with apparent orthogonal layer symmetry similar to triclinic birnessite (Fig. 4) (Webb *et al.*, 2005c; Zhu *et al.*, 2010a).

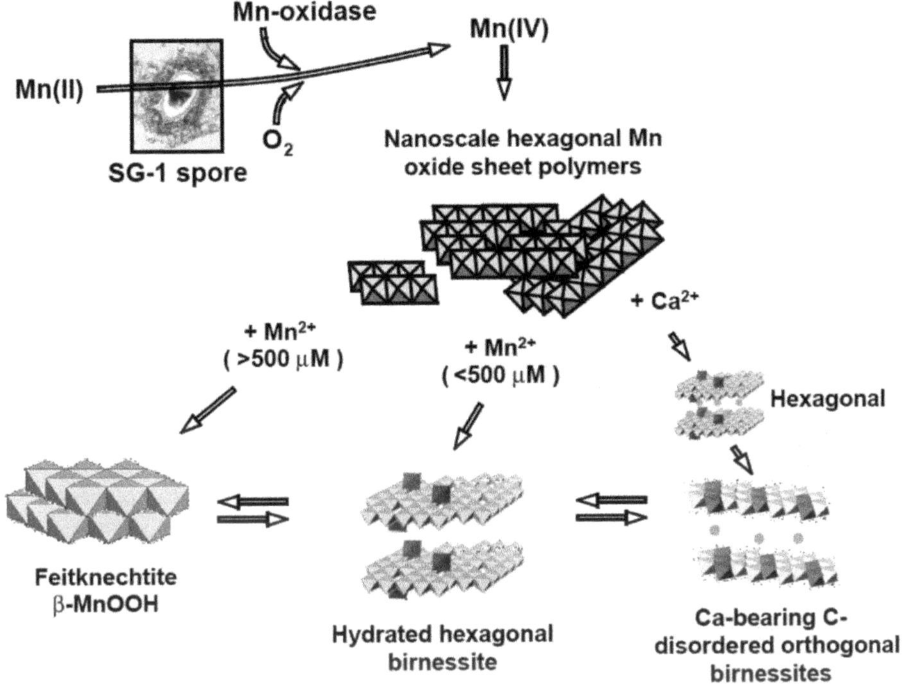

Figure 4. A conceptual image of the conditions allowing the biogenic Mn oxides to transform to other mineral structures depending on the environmental concentrations of cations such as Mn^{2+} and Ca^{2+} (after Templeton and Knowles, 2009), reproduced with the permission of '*Annual Reviews of Earth and Planetary Sciences*'.

Addition of Ca^{2+} is known to stimulate Mn oxide formation by *Bacillus* sp. SG-1, a phenomenon that can, at least in part, be attributed to an increase in the kinetics of the enzymatic reaction of MnxG (Toyoda and Tebo, 2013). Manganese oxides with orthogonal layer symmetry are shown to be less reactive (Learman *et al.*, 2011b). Due to this lower reactivity and a lack of vacancies, one can expect the formation of low-denticity (*i.e.* ligands with a small number of donor groups) inner sphere surface complexes or outer sphere surface complexes (Villalobos *et al.*, 2003). Although the abundance of manganese oxides would increase if their formation is driven by Ca^{2+}, the resulting products would possess a diminished reactive nature relative to Mn oxides with hexagonal layer symmetry. When *Bacillus* sp. SG-1 is introduced to a more complex system, *i.e.* seawater, the initial product of Mn(II) oxidation has hexagonal layer symmetry. Over time, however, manganese oxides with orthogonal layer symmetry are observed (Webb *et al.*, 2005b). Similarly, *Roseobacter* sp. AzwK-3b initially produces colloidal manganese oxides with hexagonal layer symmetry but these transform to particulate manganese oxides with orthogonal layer symmetry (Learman *et al.*, 2011b). In the case of *Bacillus* sp. SG-1, when manganese oxides are allowed to age, and depending on the amount of replenished Mn(II) in the system, secondary phases such as feiknechtite (β-MnOOH) and/or triclinic birnessite develop (Bargar *et al.*, 2005).

Biogenic manganese oxides which are generally found in hexagonal layer symmetry can also undergo structural changes to form tunnel structures. It is believed that manganese oxides with tunnel structures, *e.g.* todorokite, are formed *via* unknown geochemical transformation processes of layered manganese oxides (Feng *et al.*, 2010; Cui *et al.*, 2008, 2010). A study investigating the interaction between UO_2^{2+} and Mn(II) oxidation using *Bacillus* sp. SG-1 has shown that todorokite, a manganese oxide with a tunnel structure, can be formed (Webb *et al.*, 2006). In this study, under relatively low concentrations of UO_2^{2+} (< 4 μM) and Mn(II) (10 μM), manganese oxides did not differ from those formed in the absence of UO_2^{2+} (hexagonal symmetry layer structures). However, at the highest concentration of UO_2^{2+} tested (20 μM), manganese oxides with tunnel structures similar to todorokite but poorly ordered were formed. Another study examined the transformation of biogenic manganese oxide with hexagonal layer symmetry to todorokite using mild reflux under atmospheric pressure (Feng *et al.*, 2010). It was previously suggested that mild reflux under atmospheric pressure can simulate and accelerate the transformation processes observed in the environment (Cui *et al.*, 2008). This study showed that manganese oxide formed by *P. putida* GB-1 can be transformed into a todorokite-like phase and suggested that the hexagonal layer symmetry manganese oxide initially transforms into triclinic birnessite, a manganese oxide with orthogonal layer symmetry before it transforms into a todorokite-like phase (Feng *et al.*, 2010). Therefore, although bacterial processes may not result in the formation of tunnel-structured manganese oxides, once formed they can undergo changes to tunnel structures depending on environmental conditions.

To summarize, biogenic manganese oxides initially form as a highly reactive, poorly ordered, layer-type hexagonal birnessite. With time, conditioning occurs in the presence of ions such as Ca^{2+} or Mn^{2+}, and the highly reactive phases can transform into less reactive phases or into tunnel-type structures. In the environment, the reactive properties of the Mn oxides undoubtedly influence the activities of Mn(II) oxidizing bacteria because many reactions involving Mn oxides as the oxidant will lead to the production of Mn(II) which can then be re-oxidized by the bacteria. In this "cryptic" Mn cycle, reactive intermediates are short-lived and in low concentration but their importance to elemental cycling is greatly magnified (Hansel *et al.*, 2015).

3. Environmental applications of Mn-oxidizing enzymes and bacteriogenic Mn oxides

Mn oxides have long been recognized for their ability to sorb a variety of different elements and compounds. In fact, they have been referred to as the "scavengers of the sea" because of their ability to sorb trace elements and thereby control their distributions in the oceans (Goldberg, 1954). From a materials science perspective they have been recognized as a class of porous materials that can be used as ion and molecular sieves and as catalysts (Feng *et al.*, 1999). In the environment they are one of the strongest natural oxidants capable of oxidizing a wide variety of inorganic (*e.g.*

reduced forms of Fe, S, Cr and U) and organic (*e.g.* natural humic material) compounds (Remucal and Ginder-Vogel, 2014; Stone and Morgan, 1984a,b; Tebo *et al.*, 2004).

Mn-oxidizing bacteria, Mn oxidase enzymes, and biogenic Mn oxides have properties that make them potentially useful for a variety of environmental applications. Mn oxides are highly reactive due to their large surface area, abundance of surface charges and the vacancies in the mineral lattice. These properties account for the rapid binding, high affinity and sorption capacity of the biogenic oxides for metals like lead, zinc and nickel (Toner *et al.*, 2006; Villalobos *et al.*, 2005; Zhu *et al.*, 2010b). Additionally, Mn oxides are potent oxidants and oxidize a great variety of organic compounds including endocrine disruptors and antimicrobial compounds (Remucal and Ginder-Vogel, 2014); see Johnson *et al.* (2017, this volume). These sorptive and oxidizing properties of Mn oxides are currently exploited in water-treatment works where Mn oxides contribute to removal of other metals and organic compounds (Johnson *et al.*, 2015). Biogenic Mn oxides would be expected to be even better than the synthetic oxides typically studied and are probably the important phases in natural ecosystems, such as waters and soils. Thus, biogenic Mn oxides may find important applications in remediation as well as drinking and waste water treatment systems.

Mn-oxidizing organisms and their enzymes are capable of extensive substrate turnover, formation of fresh reactive Mn oxide surfaces, and, in cases when the Mn oxides are acting as oxidants, producing Mn(II) that will be recycled (re-oxidized) back to Mn oxide. The Mn oxidases typically have a very high affinity for Mn(II) so Mn has the potential to cycle rapidly between its reduced and oxidized forms and couple to other desirable reactions. The advantage of employing such an 'active' system over just the biogenic Mn oxides is that the oxidant (Mn oxides) is continuously being regenerated. Consequently, the rates are likely to be faster and more extensive because of the catalytic properties of the freshly formed Mn oxide minerals.

4. Conclusions

In the last decade, much has been learned about the mechanisms of bacterial Mn(II) oxidation and the structure and properties of the bacteriogenic Mn oxides. Clearly, the oxidative portion of the Mn cycle is more complicated than was once believed. In bacteria, there are now three distinct enzyme-mediated mechanisms of Mn oxide biomineralization: (1) direct oxidation by a novel type of multicopper oxidase; (2) oxidation by a Ca^{2+}-binding animal heme peroxidase; and (3) indirect Mn(II) oxidation through enzymatic production of superoxide, also carried out by an animal heme peroxidase. The bacteriogenic Mn oxides have now been fairly well characterized in terms of their structure, sorptive properties and the mechanisms of binding other metals. They are generally considered to be poorly crystalline nanoparticulate layered Mn oxides with hexagonal symmetry similar to δ-MnO_2 (vernadite). The high sorptive and oxidizing capacity of the bacteriogenic oxides relative to their synthetic analogues are probably due to the greater number of vacancies in the layers creating favourable surface charges for cations to bind. Mn(II)-oxidizing bacteria and their biominerals hold great promise for use in a variety of biotechnological applications.

Acknowledgements

The authors gratefully acknowledge the agencies which have funded their research on manganese oxidation over the years including the National Science Foundation (OCE-1154307 and -1558692 and CHE-1410688) and the Department of Energy (Subsurface Biogeochemical Research Program, DESC0005324). BMT would also like to acknowledge fellowship support from Durham University's (UK) interdisciplinary Institute of Advanced Studies. CAR was supported by an NSF Postdoctoral Research Fellowship in Biology Award ID: DBI-1202859.

References

Andeer, P.F., Learman, D.R., McIlvin, M., Dunn, J.A. and Hansel, C.M. (2015) Extracellular haem peroxidases mediate Mn(II) oxidation in a marine *Roseobacter* bacterium via superoxide production. *Environmental Microbiology*, **17**, 3925–3936. doi:10.1111/1462-2920.12893.

Anderson, C.R., Dick, G.J., Chu, M.L., Cho, J.C., Davis, R.E., Brauer, S.L. and Tebo, B.M. (2009a) *Aurantimonas manganoxydans*, sp. nov. and *Aurantimonas litoralis*, sp. nov.: Mn(II) oxidizing representatives of a globally distributed clade of alpha-Proteobacteria from the order *Rhizobiales*. *Geomicrobiology Journal*, **26**, 189–98.

Anderson, C.R., Johnson, H.A., Caputo, N., Davis, R.E., Torpey, J.W. and Tebo, B.M. (2009b) Mn(II) oxidation Is catalyzed by heme peroxidases in "*Aurantimonas manganoxydans*" strain SI85-9A1 and *Erythrobacter* sp. strain SD-21." *Applied and Environmental Microbiology*, **75**, 4130–4138. doi:10.1128/AEM.02890-08.

Anderson, C.R., Davis, R.E., Bandolin, N., Baptista, A. and Tebo, B.M. (2011) Analysis of *in situ* manganese(II) oxidation in the Columbia River and Offshore Plume: linking *Aurantimonas* and the associated microbial community to an active biogeochemical cycle. *Environmental Microbiology*, **13**, 1561–1576. doi:10.1111/j.1462-2920.2011.02462.x.

Baldrian, P. (2006) Fungal laccases – Occurrence and properties. *FEMS Microbiology Reviews*, **30**, 215–242.

Banh, A., Chavez, V., Doi, J., Nguyen, A., Hernandez, S., Ha, V., Jimenez, P., Espinoza, F. and Johnson, H.A. (2013) Manganese (Mn) oxidation increases intracellular Mn in *Pseudomonas putida* GB-1. *PLoS One*, **8**, e77835. doi:10.1371/journal.pone.0077835.

Bargar, J.R., Tebo, B.M., Bergmann, U., Webb, S.M., Glatzel, P., Chiu, V.Q. and Villalobos, M. (2005) Biotic and abiotic products of Mn(II) oxidation by spores of the marine *Bacillus* sp. strain SG-1. *American Mineralogist*, **90**, 143–54.

Boogerd, F.C. and de Vrind, J.P.M. (1987) Manganese oxidation by *Leptothrix discophora*. *Journal of Bacteriology*, **169**, 489–94.

Brouwers, G. -J., de Vrind, J.P. M., Corstjens, P.L., Cornelis, P., Baysse, C. and de Vrind-de Jong, E.W. (1999) *CumA*, a gene encoding a multicopper oxidase, is involved in Mn^{2+}-oxidation in *Pseudomonas putida* GB-1. *Applied and Environmental Microbiology*, **65**, 1762–1768.

Brouwers, G.-J., Corstjens, P.L.A.M., de Vrind, J.P. M., Verkamman, A., de Kuyper, M. and de Vrind-de Jong, E.W. (2000) Stimulation of Mn^{2+} oxidation in *Leptothrix discophora* SS-1 by Cu^{2+} and sequence analysis of the region flanking the gene encoding putative multicopper oxidase MofA. *Geomicrobiology Journal*, **17**, 25–33. doi:10.1080/014904500270468.

Butterfield, C.N., Soldatova A. V., Lee, S. -W., Spiro, T.G. and Tebo, B.M. (2013) Mn(II,III) Oxidation and MnO_2 mineralization by an expressed bacterial multicopper oxidase. *Proceedings of the National Academy of Sciences, USA*, **110**, 11731–11735. doi:10.1073/pnas.1303677110.

Butterfield, C.N., Tao, L., Chacon, K.N., Spiro, T.G., Blackburn, N.J., Casey, W.H., Britt, R.D. and Tebo, B.M. (2015) Multicopper manganese oxidase accessory proteins bind Cu and heme. *Biochimica et Biophysica Acta*, **1854**, 1853–1859.

Butterfield, C.N., Lee, S.-W. and Tebo, B.M. (2016) The role of bacterial spores in metal cycling and their potential application in metal contaminant bioremediation. *Microbiology Spectrum*, **4**, TBS-0018-2013

Caspi, R., Haygood, M.G. and Tebo, B.M. (1996) Unusual ribulose-1,5-bisphosphate carboxylase/oxygenase genes from a marine manganese-oxidizing bacterium. *Microbiology*, **142**, 2549–2559.

Caspi, R., Tebo, B.M. and Haygood, M.G. (1998) C-type cytochromes and manganese oxidation in *Pseudomonas putida* strain MnB1. *Applied and Environmental Microbiology*, **64**, 3549–3555.

Chinni, S., Anderson, C.R., Ulrich, K.-W., Giammar, D.E. and Tebo, B.M. (2008) Indirect UO_2 oxidation by Mn(II)-oxidizing spores of *Bacillus* sp. strain SG-1 and the effect of U and Mn concentrations. *Environmental Science & Technology*, **42**, 8709–8714.

Corstjens, P.L., de Vrind, J.P., Westbroek, P. and de Vrind-de Jong, E.W. (1992) Enzymatic iron oxidation by *Leptothrix discophora*: identification of an iron-oxidizing protein. *Applied and Environmental Microbiology*, **58**, 450–454.

Corstjens, P.L., de Vrind, J.P. M., Goosen, T. and de Vrind-de Jong, E.W. (1997) Identification and molecular analysis of the *Leptothrix discophora* SS-1 *mofA* gene, a gene putatively encoding a manganese-oxidizing protein with copper domains. *Geomicrobiology Journal*, **14**, 91–108.

Cui, H., Liu, X., Tan, W., Feng, X., Liu, F. and Ruan, H.D. (2008) Influence of Mn(III) availability on the phase transformation from layered buserite to tunnel-structured todorokite. *Clays and Clay Minerals*, **56**, 397–403. doi:10.1346/ccmn.2008.0560401.

Cui, H., Liu, F., Feng, X., Tan, W. and Wang, M. (2010) Aging promotes todorokite formation from layered manganese oxide at near-surface conditions. *Journal of Soils and Sediments*, **10**, 1540–1547.

de Vrind, J.P., de Vrind-de Jong, E.W., de Voogt, J.W., Westbroek, P., Boogerd, F.C. and Rosson, R.A. (1986) Manganese oxidation by spores and spore coats of a marine *Bacillus* species. *Applied and Environmental Microbiology*, **52**, 1096–1100.

de Vrind, J.P., Brouwers, G.J., Corstjens, P.L.A.M., Dulk, J.D. and de Vrind-de Jong, E.W. (1998) The cytochrome c maturation operon is involved in manganese oxidation in *Pseudomonas putida* GB-1. *Applied and Environmental Microbiology*, **64**, 3556–3562.

de Vrind-de Jong, E.W., Corstjens, P.L., Kempers, E.S., Westbroek, P. and de Vrind, J.P. (1990) Oxidation of manganese and iron by *Leptothrix discophora*: use of N,N,N',N'-tetramethyl-P-phenylenediamine as an indicator of metal oxidation. *Applied and Environmental Microbiology*, **56**, 3458–3462.

Dick, G.J., Lee, Y.E. and Tebo, B.M. (2006) Manganese(II)-oxidizing *Bacillus* spores in Guaymas Basin hydrothermal sediments and plumes. *Applied and Environmental Microbiology*, **72**, 3184–3190.

Dick, G.J., Podell, S. Johnson, H.A., Rivera-Espinoza, Y., Bernier-Latmani, R., McCarthy, J.K., Torpey, J.W., Clement, B.G., Gaasterland, T. and Tebo, B.M. (2008a) Genomic insights into Mn(II) oxidation by the marine alphaproteobacterium *Aurantimonas* sp. strain SI85-9A1. *Applied and Environmental Microbiology*, **74**, 2646–2658. doi:10.1128/AEM.01656-07.

Dick, G.J., Torpey, J.W., Beveridge, T.J. and Tebo, B.M. (2008b) Direct identification of a bacterial manganese(II) oxidase, the multicopper oxidase MnxG, from spores of several different marine *Bacillus* species. *Applied and Environmental Microbiology*, **74**, 1527–1534. doi:AEM.01240-07 [pii] 10.1128/AEM.01240-07.

Dubinina, G. and Zhdanov, A.V. (1975) Recognition of the iron bacteria "*Siderocapsa*" as Arthrobacters and description of *Arthrobacter siderocapsulatus* sp.nov. *International Journal of Systematic Bacteriology*, **25**, 340–350. doi:10.1099/00207713-25-4-340.

Duckworth, O.W. and Sposito, G. (2005) Siderophore-manganese(III) interactions. I. Air-oxidation of manganese(II) promoted by desferrioxamine B. *Environmental Science & Technology*, 39, 6037–6044.

Ehrlich, H.L. and Salerno, J.C. (1990) Energy coupling in Mn^{2+} oxidation by a marine bacterium. *Archives of Microbiology*, **154**, 12–17.

El Gheriany, I.A., Bocioaga, D., Hay, A.G., Ghiorse, W.G., Shuler, M.L. and Lion, W.L. (2009) Iron requirement for Mn(II) oxidation by *Leptothrix discophora* SS-1. *Applied and Environmental Microbiology*, **75**, 1229–1235. doi:AEM.02291-08 [pii] 10.1128/AEM.02291-08.

El Gheriany, I.A., Bocioaga, D., Hay, A.G., Ghiorse, W.G., Shuler, M.L. and Lion, L.W. (2011) An uncertain role for Cu(II) in stimulating Mn(II) oxidation by *Leptothrix discophora* SS-1. *Archives of Microbiology*, **193**, 89–93. doi:10.1007/s00203-010-0645-x.

Faulkner, K.M., Stevens, R.D. and Fridovich, I. (1994) Characterization of Mn(III) complexes of linear and cyclic desferrioxamines as mimics of superoxide dismutase activity. *Archives of Biochemistry and Biophysics*, **310**, 341–346.

Fendorf, S.E. and Zasoski, R.J. (2002) Chromium(III) oxidation by δ-manganese oxide (MnO_2). 1.

Characterization. *Environmental Science & Technology*, **26**, 79–85.

Feng, Q., Kanoh, H. and Ooi, K. (1999) Manganese oxide porous crystals. *Journal of Materials Chemistry*, **9**, 319–333. doi:10.1039/a805369c.

Feng, X.H., Zhu, M., Ginder-Vogel, M., Ni, C., Parikh, S.J. and Sparks, D.L. (2010) Formation of nano-crystalline todorokite from biogenic Mn oxides. *Geochimica et Cosmochimica Acta*, **74**, 3232–3245. doi:10.1016/j.gca.2010.03.005.

Foster, A.L., Brown, G.E. and Parks, G.A. (2003) X-Ray Absorption Fine Structure Study of As(V) and Se(IV) sorption complexes on hydrous Mn oxides. *Geochimica et Cosmochimica Acta*, **67**, 1937–1953. doi:http://dx.doi.org/10.1016/S0016-7037(02)01301-7.

Francis, C.A. and Tebo, B.M. (2001) *cumA* Multicopper oxidase genes from diverse Mn(II)-oxidizing and non-Mn(II)-oxidizing *Pseudomonas* strains. *Applied and Environmental Microbiology*, **67**, 4272–4278.

Francis, C.A. and Tebo, B.M. (2002) Enzymatic manganese(II) oxidation by metabolically dormant spores of diverse *Bacillus* species. *Applied and Environmental Microbiology*, **68**, 874–880.

Francis, C.A., Co, E.-M. and Tebo, B.M. (2001) Enzymatic Manganese(II) oxidation by a marine α-Proteobacterium. *Applied and Environmental Microbiology*, **67**, 4024–4029. doi:10.1128/aem.67.9.4024-4029.2001.

Francis, C.A., Casciotti, K.L. and Tebo, B.M. (2002) Localization of Mn(II)-oxidizing activity and the putative multicopper oxidase, MnxG, to the exosporium of the marine *Bacillus* sp. strain SG-1. *Archives of Microbiology*, **178**, 450–456. doi:10.1007/s00203-002-0472-9.

Gaillot, A. -C., Flot, D., Drits, V.A., Manceau, A., Burghammer, M. and Lanson, B. (2003) Structure of synthetic K-rich birnessite obtained by high-temperature decomposition of $KMnO_4$. I. Two-layer polytype from 800°C experiment. *Chemistry of Materials*, **15**, 4666–4678.

Geszvain, K. and Tebo, B.M. (2010) Identification of a two-component regulatory pathway essential for Mn(II) oxidation in *Pseudomonas putida* GB-1. *Applied and Environmental Microbiology*, **76**, 1224–1231. doi:10.1128/AEM.02473-09.

Geszvain, K., McCarthy, J.K. and Tebo, B.M. (2013) Elimination of manganese(II, III) oxidation in *Pseudomonas putida* GB-1 by a double knockout of two putative multicopper oxidase genes. *Applied and Environmental Microbiology*, **79**, 357–366. doi:10.1128/aem.01850-12.

Geszvain, K., Smesrud, L. and Tebo, B.M. (2016) Identification of a third Mn(II) oxidase enzyme in *Pseudomonas putida* GB-1. *Applied and Environmental Microbiology*, doi:10.1128/AEM00046-16.

Ghiorse, W C. (1984) Biology of iron-and manganese-depositing bacteria. *Annual Review of Microbiology*, **38**, 515–550. doi:10.1146/annurev.mi.38.100184.002503.

Goldberg, E.D. (1954) Marine geochemistry I. Chemical scavengers of the sea. *Journal of Geology*, **62**, 249–265.

Hansel, C.M. and Francis, C.A. (2006) Coupled photochemical and enzymatic Mn(II) oxidation pathways of a planktonic *Roseobacter*-like bacterium. *Applied and Environmental Microbiology*, **72**, 3543–3549. doi:10.1128/AEM.72.5.3543.

Hansel, C.M., Ferdelman, T.G. and Tebo, B.M. (2015) Cryptic cross-linkages among biogeochemical cycles: novel insights from reactive intermediates. *Elements*, **11**, 409–414. doi:10.2113/gselements.11.6.409.

Harrington, J.M., Parker, D.L., Bargar, J.R., Jarzecki, A.A., Tebo, B.M., Sposito, G. and Duckworth, O.W. (2012) Structural dependence of Mn complexation by siderophores: donor group dependence on complex stability and reactivity. *Geochimica et Cosmochimica Acta*, **88**, 106–119.

Hastings, D. and Emerson, S. (1986) Oxidation of manganese by spores of a marine *Bacillus* – Kinetic and thermodynamic considerations. *Geochimica et Cosmochimica Acta*, **50**, 1819–1824.

Johnson, H.A. and Tebo, B.M. (2008) I*n Vitro* studies indicate a quinone is involved in bacterial Mn(II) oxidation. *Archives of Microbiology*, **189**, 59–69.

Johnson, K., Purvis, G., Lopez-Capel, E., Peacock, C., Gray, N., Wagner, T., März, C. and six others (2015) Towards a mechanistic understanding of carbon stabilization in manganese oxides. *Nature Communications*, **6**, 7628. doi:10.1038/ncomms8628.

Johnson, K.L., McCann, C.M. and Clarke, C.E. (2017) Breakdown of organic contaminants in soil by manganese oxides: A short review. Pp. 313–356 in: *Redox-Reactive Minerals: Properties, Reactions and Applications in Clean Technologies* (I.A.M. Ahmed and K.A. Hudson-Edwards, editors). EMU Notes in

Mineralogy, **17**. European Mineralogical Union and Mineralogical Society of Great Britain & Ireland.

Jürgensen, A., Widmeyer, J.R., Gordon, R.A., Bendell-Young, L.I., Moore, M.M. and Crozier, E.D. (2004) The structure of the manganese oxide on the sheath of the bacterium *Leptothrix discophora*: An XAFS study. *American Mineralogist*, **89**, 1110–1118.

Kang, I., Oh, H.-M., Lim, S.-I., Ferriera, S., Giovannoni, S.J. and Cho, J.-C. (2010) Genome sequence of *Fulvimarina pelagi* HTCC2506T, a Mn(II)-oxidizing alphaproteobacterium possessing an aerobic anoxygenic photosynthetic gene cluster and xanthorhodopsin. *Journal of Bacteriology*, **192**, 4798–4799. doi:10.1128/jb.00761-10.

Klewicki, J.K. and Morgan, J.J. (1998) Kinetic behavior of Mn(III) complexes of pyrophosphate, EDTA, and citrate. *Environmental Science & Technology*, **32**, 2916–2922. doi:10.1021/es980308e.

Kostka, J.E., Luther, G.W. and Nealson, K.H. (1995) Chemical and biological reduction of Mn (III)-pyrophosphate complexes: potential importance of dissolved Mn (III) as an environmental oxidant. *Geochimica et Cosmochimica Acta*, **59**, 885–894. doi:10.1016/0016-7037(95)00007-0.

Lafferty, B.J., Ginder-Vogel, M., Zhu, M., Livi, K.J.T. and Sparks, D.L. (2010) Arsenite oxidation by a poorly crystalline manganese-oxide. 2. Results from X-Ray Absorption Spectroscopy and X-Ray Diffraction. *Environmental Science & Technology*, **44**, 8467–8472. doi:10.1021/es102016c.

Lanson, B., Drits, V.A., Silvester, E. and Manceau, A. (2000) Structure of H-exchanged hexagonal birnessite and its mechanism of formation from Na-rich monoclinic buserite at low pH. *American Mineralogist*, **85**, 826–838.

Larsen, E.I., Sly, L.I. and McEwan, A.G. (1999) Manganese(II) adsorption and oxidation by whole cells and a membrane fraction of *Pedomicrobium* sp. ACM 3067. *Archives of Microbiology*, **171**, 257–264.

Learman, D.R., Voelker, B.M., Vazquez-Rodriguez, A.I. and Hansel, C.M. (2011a) Formation of manganese oxides by bacterially generated superoxide. *Nature Geoscience*, **4**, 95–98. doi:10.1038/ngeo1055.

Learman, D.R., Wankel, S.D., Webb, S.M., Martinez, N., Madden, A.S. and Hansel, C.M. (2011b) Coupled biotic-abiotic Mn(II) oxidation pathway mediates the formation and structural evolution of biogenic Mn oxides. *Geochimica et Cosmochimica Acta*, **75**, 6048–6063. doi:10.1016/j.gca.2011.07.026.

Learman, D.R., Voelker, B.M., Madden, A.S. and Hansel, C.M. (2013) Constraints on superoxide mediated formation of manganese oxides. *Frontiers in Microbiology*, **4**, 1–12. doi:10.3389/fmicb.2013.00262.

Lindley, P.F., Card, G., Zaitseva, I., Zaitsev, V., Reinhammar, B., Selin-Lindgren, E. and Yoshida, K. (1997) An X-ray structural study of human ceruloplasmin in relation to ferroxidase activity. *Journal of Biological Inorganic Chemistry*, **2**, 454–463. doi:10.1007/s007750050156.

Liu, X. and Theil, E.C. (2005) Ferritins: dynamic management of biological iron and oxygen chemistry. *Accounts of Chemical Research*, **38**, 167–175. doi:10.1021/ar0302336.

Luther, G.W. (2005) Manganese(II) oxidation and Mn(IV) reduction in the environment – two one-electron transfer steps versus a single two-electron step. *Geomicrobiology Journal*, **22**, 195–203. doi:10.1080/01490450590946022.

Machonkin, T.E. and Solomon, E.I. (2000) The thermodynamics, kinetics, and molecular mechanism of intramolecular electron transfer in human ceruloplasmin. *Journal of the American Chemical Society*, **122**, 12547–12560. doi:10.1021/ja002339r.

Madison, A.S., Tebo, B.M. and Luther, G.W. (2011) Simultaneous determination of soluble manganese(III), manganese(II) and total manganese in natural (pore)waters. *Talanta*, **84**, 374–381.

Madison, A.S., Tebo, B.M., Mucci, A., Sundby, B. and Luther, G.W. (2013) Abundant porewater Mn(III) is a major component of the sedimentary redox system. *Science*, **341**, 875–878. doi:10.1126/science.1241396.

Mandernack, K.W., Post, J. and Tebo, B.M. (1995) Manganese mineral formation by bacterial spores of the marine *Bacillus*, strain SG-1: evidence for the direct oxidation of Mn(II) to Mn(IV)." *Geochimica et Cosmochimica Acta*, **59**, 4393–4408. doi:10.1016/0016-7037(95)00298-E.

Mann, S., Sparks, N.H. C., Scott, G.H.E. and de Vrind-de Jong, E.W. (1988) Oxidation of manganese and formation of Mn_3O_4 (hausmannite) by spore coats of a marine *Bacillus* sp. *Applied and Environmental Microbiology*, **54**, 2140–2143.

Morgan, J.J. (2000) Manganese in natural waters and Earth's Crust: Its availability to organisms. Pp. 1–34 in: *Manganese and Its Role in Biological Processes* (A. Sigel and H. Sigel, editors). *Metal Ions in Biological Systems*, **37**. Marcel Dekker, New York.

Myers, C.R. and Nealson, K.H. (1988) Bacterial manganese reduction and growth with manganese oxide as the sole electron acceptor. *Science*, **240**, 1319–1321. doi:10.2307/1701057.

Nakama, K., Medina, M., Lien, A., Ruggieri, J., Collins, K. and Johnson, H.A. (2014) Heterologous expression and characterization of the manganese-oxidizing protein from *Erythrobacter* sp. Strain SD21. *Applied and Environmental Microbiology*, **80**, 6837–6842.

Nealson, K.H. and Ford, J. (1980) Surface enhancement of bacterial manganese oxidation: implications for aquatic environments. *Geomicrobiology Journal*, **2**, 21–37. doi:10.1080/01490458009377748.

Nealson, K.H., Tebo, B.M. and Rosson, R.A. (1988) Occurrence and mechanisms of microbial oxidation of manganese. Pp. 279–318 in *Advances in Applied Microbiology*, **33** (I. Laskin Allen, editor). Academic Press. doi:10.1016/S0065-2164(08)70209-0.

Nelson, Y.M., Lion, L.W., Ghiorse, W.C. and Shuler, M.L. (1999) Production of biogenic Mn oxides by *Leptothrix discophora* SS-1 in a chemically defined growth medium and evaluation of their Pb adsorption characteristics. *Applied and Environmental Microbiology*, **65**, 175–180.

Nico, P.S., Anastasio, C. and Zasoski, R.J. (2002) Rapid photo-oxidation of Mn(II) mediated by humic substances. *Geochimica et Cosmochimica Acta*, **66**, 4047–4056. doi:10.1016/S0016-7037(02)01001-3.

Okazaki, M., Sugita, T., Shimizu, M., Ohode, Y., Iwamoto, K., de Vrind-de Jong, E.W., de Vrind, J.P. and Corstjens, P.L. (1997) Partial purification and characterization of manganese-oxidizing factors of *Pseudomonas fluorescens* GB-1. *Applied and Environmental Microbiology*, **63**, 4793–4799.

Oldham, V.E., Owings, S.M., Jones, M.R., Tebo, B.M. and Luther III, G.W. (2015) Evidence for the presence of strong Mn(III)-binding ligands in the water column of the Chesapeake Bay. *Marine Chemistry*, **171**, 58–66. doi:10.1016/j.marchem.2015.02.008.

Parikh, S.J. and Chorover, J. (2005) FTIR spectroscopic study of biogenic Mn-oxide formation by *Pseudomonas putida* GB-1. *Geomicrobiology Journal*, **22**, 207–218. doi:10.1080/01490450590947724.

Parker, D.L., Sposito, G. and Tebo, B.M. (2004) Manganese(III) binding to a pyoverdine siderophore produced by a manganese(II)-oxidizing bacterium. *Geochimica et Cosmochimica Acta*, **68**, 4809–4820. doi:10.1016/j.gca.2004.05.038.

Parker, D.L., Morita, T., Mozafarzadeh, M.L., Verity, R., McCarthy, J.K. and Tebo, B.M. (2007) Inter-relationships of MnO_2 precipitation, siderophore-$Mn^{(III)}$ complex formation, siderophore degradation, and iron limitation in $Mn^{(II)}$-oxidizing bacterial cultures. *Geochimica et Cosmochimica Acta*, **71**, 5672–5683.

Pecher, K., McCubbery, D., Kneedler, E., Rothe, J., Bargar, J., Meigs, G., Cox, L., Nealson, K. and Tonner, B. (2003) Quantitative charge state analysis of manganese biominerals in aqueous suspension using Scanning Transmission X-Ray Microscopy (STXM). *Geochimica et Cosmochimica Acta*, **67**, 1089–1098.

Pena, J., Kwon, K., Refson, K., Bargar, J.R. and Sposito G. (2010) Mechanisms of nickel sorption by a bacteriogenic birnessite. *Geochimica et Cosmochimica Acta*, **74**, 3076–3089. doi:10.1016/j.gca.2010.02.035.

Post, J.E. (1999) Manganese oxide minerals: Crystal structures and economic and environmental significance. *Proceedings of the National Academy of Sciences, USA*, **96**, 3447–3454.

Quintanar, L., Gebhard, M., Wang, T.P., Kosman, D.J. and Solomon, E.I. (2004) Ferrous binding to the multicopper oxidases *Saccharomyces cerevisiae* Fet3p and human ceruloplasmin: contributions to ferroxidase activity. *Journal of the American Chemical Society*, **126**, 6579–6589. doi:10.1021/ja049220t.

Remucal, C.K. and Ginder-Vogel, M. (2014) A critical review of the reactivity of manganese oxides with organic contaminants. *Environmental Science: Processes and Impacts*, **16**, 1247–1266. doi:10.1039/c3em00703k.

Ridge, J.P., Lin, M., Larsen, E.I., Fegan, M., McEwan, A.G. and Sly, L.I. (2007) A multicopper oxidase is essential for manganese oxidation and laccase-like activity in *Pedomicrobium* sp. ACM 3067. *Environmental Microbiology*, **9**, 944–953. doi:10.1111/j.1462-2920.2006.01216.x.

Rosson, R.A. and Nealson, K.H. (1982) Manganese binding and oxidation by spores of a marine *Bacillus*. *Journal of Bacteriology*, **151**, 1027–1034.

Saratovsky, I., Wightman, P.G., Pastén, P.A., Gaillard, J.F. and Poeppelmeier, K.R. (2006) Manganese oxides: parallels between abiotic and biotic Structures. *Journal of the American Chemical Society*, **128**, 11188–11198. doi:doi:10.1021/ja062097g.

Schlosser, D. and Hofer, C. (2002) Laccase-catalyzed oxidation of Mn^{2+} in the presence of natural Mn^{3+}

chelators as a novel source of extracellular H_2O_2 production and Its Impact on manganese peroxidase. *Applied and Environmental Microbiology*, **68**, 3514–3521.

Siering, P.L. and Ghiorse, W.C. (1997) PCR detection of a putative manganese oxidation gene (*mofA*) in environmental samples and assessment of *mofA* gene homology among diverse manganeseoxidizing bacteria. *Geomicrobiology Journal*, **14**, 109–125. doi:10.1080/01490459709378038.

Soldatova, A.V., Butterfield, C., Oyerinde, O.F., Tebo, B.M. and Spiro, T.G. (2012) Multicopper oxidase involvement in both Mn(II) and Mn(III) oxidation during bacterial formation of MnO_2. *Journal of Biological Inorganic Chemistry*, **17**, 1151–1158. doi:10.1007/s00775-012-0928-6.

Solomon, E.I., Sundaram, U.M. and Machonkin, T.E. (1996) Multicopper oxidases and oxygenases. *Chemical Reviews*, **96**, 2563–2606. doi:10.1021/cr950046o.

Stoj, C.S., Augustine, A.J., Zeigler, L., Solomon, E.I. and Kosman, D.J. (2006) Structural basis of the ferrous iron specificity of the yeast ferroxidase, Fet3p. *Biochemistry*, **45**, 12741–12749. doi:10.1021/bi061543+.

Stone, A.T. and Morgan, J.J. (1984a) Reduction and dissolution of manganese(III) and manganese(IV) oxides by organics. 1. Reaction with hydroquinone. *Environmental Science & Technology*, **18**, 450–456.

Stone, A.T. and Morgan, J.J. (1984b) Reduction and dissolution of manganese(III) and manganese(IV) oxides by organics. 2. Survey of the reactivity of organics. *Environmental Science & Technology* **18**, 617–624.

Sunda, W.G. and Huntsman, S.A. (1987) Microbial oxidation of manganese in a North Carolina estuary. *Limnology and Oceanography* **32**, 552–564. doi:10.4319/lo.1987.32.3.0552.

Sunda, W.G. and Huntsman, S.A. (1988) Effect of sunlight on redox cycles of manganese in the southwestern Sargasso Sea. *Deep-Sea Research Part A – Oceanographic Research Papers*, **35**, 1297–1317.

Sunda, W.G. and Kieber, D.J. (1994) Oxidation of humic substances by manganese oxides yields low-molecular-weight organic substrates. *Nature*, **367**, 62–64.

Sundaramoorthy, M., Kishi, K., Gold, M.H. and Poulos, T.L. (1994) The crystal structure of manganese peroxidase from *Phanerochaete chrysosporium* at 2.06-A resolution. *Journal of Biological Chemistry*, **269**, 32759–32767.

Takeda, M., Kamagata, Y., Ghiorse, W.C., Hanada, S. and Koizumi, J. (2002) *Caldimonas manganoxidans* gen. nov., sp. nov., a poly(3-hydroxybutyrate)-degrading, manganese-oxidizing thermophile." *International Journal of Systematic and Evolutionary Microbiology*, **52**, 895–900. doi:10.1099/ijs.0.02027-0.

Tao, L., Stich, T.A., Butterfield, C.N., Romano, C.A., Spiro, T.G., Tebo, B.M., Casey, W.H. and Britt, R.D. (2015) Mn(II) binding and subsequent oxidation by the multicopper oxidase MnxG investigated by Electron Paramagnetic Resonance Spectroscopy. *Journal of American Chemical Society*. **137**, 10563–10575.

Taylor, A.B., Stoj, C.S., Ziegler, L., Kosman, D.J. and Hart, P.J. (2005) The copper-iron connection in biology: structure of the metallo-oxidase Fet3p. *Proceedings of the National Academy of Sciences of the USA*, **102**, 15459–15464. doi:10.1073/pnas.0506227102.

Tebo, B.M. and Emerson, S. (1986) Microbial manganese(II) oxidation in the marine environment: a quantitative study. *Biogeochemistry*, **2**, 149–161. doi:10.1007/BF02180192.

Tebo, B.M., Ghiorse, W.C., van Waasbergen, L.G., Siering, P.L. and Caspi, R. (1997) Bacterially-mediated mineral formation: Insights into manganese(II) oxidation from molecular genetic and biochemical studies. Pp. 225–266 in: *Geomicrobiology: Interactions between Microbes and Minerals* (J.F. Banfield and K.H. Nealson, editors). Reviews in Mineralogy, **35**. Mineralogical Society of America. Washington, D.C.

Tebo, B.M., Bargar, J.R., Clement, B.G., Dick, G.J., Murray, K.J., Parker, D., Verity, R. and Webb, S.M. (2004) Biogenic manganese oxides: properties and mechanisms of formation. *Annual Review of Earth and Planetary Sciences*, **32**, 287–328.

Templeton, A.S. and Knowles, E. (2009) Microbial transformations of minerals and metals: recent advances in geomicrobiology derived from synchrotron-based X-Ray Spectroscopy and X-ray Microscopy. *Annual Review of Earth and Planetary Sciences*, **37**, 367–91. doi:10.1146/annurev.earth.36.031207.124346.

Toner, B., Manceau, A., Webb, S.M. and Sposito, G. (2006) Zinc sorption to biogenic hexagonal-birnessite particles within a hydrated bacterial biofilm. *Geochimica et Cosmochimica Acta*, **70**, 27–43. doi:10.1016/j.gca.2005.08.029.

Toyoda, K. and Tebo, B.M. (2013) The effect of Ca^{2+} ions and ionic strength on Mn(II) oxidation by spores of the marine *Bacillus* sp. SG-1. *Geochimica et Cosmochimica Acta*, **101**, 1–11.

Trouwborst, R.E., Clement, B.G., Tebo, B.M., Glazer, B.T. and Luther, G.W. (2006) Soluble Mn(III) in suboxic zones. *Science*, **313**, 1955–1957.

van Veen, W.L., Mulder, E.G. and Deinema, M.H. (1978) The *Sphaerotilus-Leptothrix* group of Bacteria." *Microbiological Reviews*, **42**, 329–356.

van Waasbergen, L.G., Hoch, J.A. and Tebo, B.M. (1993) Genetic analysis of the marine manganese-oxidizing *Bacillus* sp. strain SG- 1: protoplast transformation, Tn917 mutagenesis, and identification of chromosomal loci involved in manganese oxidation. *Journal of Bacteriology*, **175**, 7594–7603.

van Waasbergen, L.G., Hildebrand, M. and Tebo, B.M. (1996) Identification and characterization of a gene cluster involved in manganese oxidation by spores of the marine *Bacillus* sp. strain SG-1. *Journal of Bacteriology*, **178**, 3517–3530.

Villalobos, M., Toner, B., Bargar, J. and Sposito, G. (2003) Characterization of the manganese oxide produced by *Pseudomonas putida* strain MnB1. *Geochimica et Cosmochimica Acta*, **67**, 2649–2662. doi:10.1016/S0016-7037(03)00217-5.

Villalobos, M., Bargar, J. and Sposito, G. (2005) Mechanisms of Pb(II) sorption on a biogenic manganese oxide. *Environmental Science & Technology*, **39**, 569–576. doi:10.1021/es049434a.

Villalobos, M., Lanson, B., Manceau, A., Toner, B. and Sposito, G. (2006) Structural model for the biogenic Mn oxide produced by *Pseudomonas putida*. *American Mineralogist*, **91**, 489–502. doi:10.2138/am.2006.1925.

Wang, H., Li, H., Shao, Z., Liao, S., Johnstone, L., Rensing, C. and Wang, G. (2012a) Genome sequence of deep-sea manganese-oxidizing bacterium *Marinobacter manganoxydans* MnI7-9. *Journal of Bacteriology*, **194**, 899–900. doi:10.1128/jb.06551-11.

Wang, Z., Lee, S.-W., Catalano, J.G., Lezama-Pacheco, J.S., Bargar, J.R., Tebo, B.M. and Giammar, D.E. (2012b) Adsorption of uranium(VI) to manganese oxides: X-Ray Absorption Spectroscopy and Surface Complexation Modeling. *Environmental Science & Technology*, **47**, 850–858. doi:10.1021/es304454g.

Wariishi, H., Valli, K. and Gold, M.H. (1992) Manganese(II) oxidation by manganese peroxidase from the Basidiomycete *Phanerochaete chrysosporium*. Kinetic mechanism and role of chelators. *Journal of Biological Chemistry*, **267**, 23688–23695.

Webb, S.M., Dick, G.J., Bargar, J.R. and Tebo, B.M. (2005a) Evidence for the presence of Mn(III) intermediates in the bacterial oxidation of Mn(II). *Proceedings of the National Academy of Sciences, USA*, **102**, 5558–5563.

Webb, S.M., Tebo, B.M. and Bargar, J.R. (2005b) Structural characterization of biogenic Mn oxides produced in seawater by the marine *Bacillus* sp strain SG-1. *American Mineralogist*, **90**, 1342–1357.

Webb, S.M., Tebo, B.M. and Bargar, J.R. (2005c) Structural influences of sodium and calcium ions on the biogenic manganese oxides produced by the marine *Bacillus* sp., strain SG-1. *Geomicrobiology Journal*, **22**, 181–193. doi:10.1080/01490450590946013.

Webb, S.M., Fuller, C.C., Tebo, B.M. and Bargar, J.R. (2006) Determination of uranyl incorporation into biogenic manganese oxides using X-ray absorption spectroscopy and scattering. *Environmental Science & Technology*, **40**, 771–777.

Yakushev, E., Pakhomova, S., Sørenson, K. and Skei, J. (2009) Importance of the different manganese species in the formation of water column redox zones: observations and modeling. *Marine Chemistry*, **117**, 59–70. doi:10.1016/j.marchem.2009.09.007.

Zhang, J., Lion, L.W., Nelson, Y.M., Shuler, M.L. and Ghiorse, W.C. (2002) Kinetics of Mn(II) oxidation by *Leptothrix discophora* SS-1. *Geochimica et Cosmochimica Acta*, **66**, 773–781. doi:10.1016/S0016-7037(01)00808-0.

Zhu, M., Paul, K.W., Kubicki, J.D. and Sparks, D.L. (2009) Quantum chemical study of arsenic (III, V) adsorption on Mn-oxides: implications for arsenic(III) oxidation. *Environmental Science & Technology*, **43**, 6655–6661.

Zhu, M., Ginder-Vogel, M., Parikh, S.J., Feng, X.H. and Sparks, D.L. (2010a) Cation effects on the layer structure of biogenic Mn-oxides. *Environmental Science & Technology*, **44**, 4465–4471. doi:10.1021/es1009955.

Zhu, M., Ginder-Vogel, M. and Sparks, D.L. (2010b) Ni(II) sorption on biogenic Mn-oxides with varying Mn octahedral layer structure. *Environmental Science & Technology*, **44**, 4472–4478. doi:10.1021/es9037066.

EMU Notes in Mineralogy, Vol. 17 (2017), Chapter 8, 197–228

Bioconversion of Fe(III) oxides into magnetic nanoparticles: Processes and applications

VICTORIA S. COKER, MATHEW P. WATTS and JONATHAN R. LLOYD

School of Earth, Atmospheric and Environmental Sciences, University of Manchester, Manchester M13 9PL, UK, e-mail: vicky.coker@manchester.ac.uk

The formation of magnetic nanomaterials by biogenic processes is an untapped green chemistry approach to nanomaterial production. This chapter focuses on magnetite-based nanomaterials made through extracellular processes by dissimilatory Fe(III)-reducing and Fe(II)-oxidizing bacteria as these microbes have particular importance in global Fe cycling. Biogenic nano-magnetite is typically between 5 and 50 nm in size and as such exhibits different mechanical, electrical, magnetic and chemical behaviours compared to larger-scale equivalent material. Therefore, an environmentally benign process for the production of these materials, as offered by bacteria, is highly desirable. The formation of biogenic nano-magnetite is discussed, primarily focusing on Fe(III)-reducing bacteria as these are the most widely researched. The methods that can be used to tailor the magnetic nanoparticles to give enhanced reactivity involve using different mineral precursor phases, different ratios of biomass: precursor mineral and also the addition of different metals either within the magnetite structure or by arraying metals on the surface of the magnetite nanoparticles. The applications of these tailored magnetite nanoparticles are wide ranging and covered within this review are a variety of redox-sensitive organic and inorganic contaminants that can be stabilized through a reductive transformation or degraded by biogenic magnetite. There are also biomedical applications for biogenic nano-magnetite. Nano-sized magnetite particles are produced by a wide range of Bacteria and Archaea and have potential uses in environmental remediation strategies and the challenges of process scale-up and the utilization of waste materials in the production of these nanomaterials are the next steps to harnessing this green technology.

1. Introduction

Functional nanomaterials are employed in a variety of different technological and industrial applications and the search for new and unique particles and production methods is ongoing. The ability of microbes to precipitate nanomaterials represents a wide and untapped resource. Of particular appeal is the potential to offer a green-chemistry approach to nanomaterial production, compared to traditional processing methods, including those focused on palladium and gold nanocatalysts, antimicrobial silver nanoparticles, selenium, tellurium and sulfur-based semi-conductors and magnetic iron (Fe)-based nanomaterials. For a comprehensive review of the area see Lloyd *et al.* (2011); this chapter will focus on magnetite-based nanomaterials made through extracellular microbial processes.

DOI: 10.1180/EMU-notes.17.6

Biogenic magnetite is produced through the activity of both dissimilatory Fe(III)-reducing bacteria as well as Fe(II)-oxidizing bacteria. Understanding the formation of biogenic Fe minerals such as magnetite is particularly important as Fe redox-cycling bacteria are key participants in the global Fe cycle (Kappler and Straub, 2005). Biogenic magnetite is nanoparticulate in character (typically between 5 and 50 nm in size), and therefore has physical properties that differ from its larger-scale counterpart, related to a change in the surface-to-mass ratio. These differences range from changes in mechanical (Guo *et al.*, 2014), electrical and magnetic behaviours, to chemical properties such as solubility, reactivity and catalytic activity. There are chemical processes established for the synthesis of magnetite; however, there is considerable interest in using biotechnological approaches to achieve scalable, cost-effective bioproduction. The advantage of these alternative processes is that more environmentally benign conditions can be used, which generally lack toxic organic and inorganic reagents and the processing extremes of pressure or temperature associated with conventional synthetic routes.

This chapter describes the formation of biogenic magnetite, concentrating on Fe(III)-reduction as it is the most widely studied; but Fe(II)-oxidation mechanisms are considered also. The methods that can be used to tailor nanoparticles with exceptional physical properties will be discussed, and also how these particles can be utilized in remediation settings as well as in more high-tech industries, including medical applications.

2. Redox cycling

2.1. Fe(III) reduction

Dissimilatory Fe(III) reduction encompasses a wide area of research, and since its inception many different taxonomically diverse Fe(III)-reducing bacteria have been discovered (Lovley *et al.*, 1997, 2004; Vargas *et al.*, 1998). Many of the best known Fe(III)-reducers that have been isolated in pure culture belong to the family *Geobacteraceae* in the δ-*Proteobacteria*, such as the genera *Geobacter*, *Desulfuromonas*, *Pelobacter* and *Desulfomusa* (Lovley, 2013). The first isolated Fe(III)-reducing bacterium in this group was *Geobacter metallireducens*, which formed magnetite from the reduction of ferrihydrite (Lovley *et al.*, 1987). Many other closely related Fe(III)-reducers, again mostly in the family *Geobacteraceae*, have been isolated including *Geobacter sulfurreducens* (Caccavo Jr. *et al.*, 1994), *Desulfuromonas palmitatis* (Coates *et al.*, 1995) and *Pelobacter carbinolicus* (Stackebrandt *et al.*, 1989), although the latter can accomplish only Fe(III)-reduction coupled to the oxidation of H_2 (Lonergan *et al.*, 1996). Most of these species, however, are capable of coupling the complete oxidation of organic compounds to CO_2 to the reduction of metals (Lonergan *et al.*, 1996) and are commonly found in subsurface sedimentary environments where Fe(III) is a dominant electron-accepting process (Lovley, 2013). Within the γ-*Proteobacteria*, there are also facultative micro-organisms capable of the incomplete oxidation of multi-carbon compounds, *e.g.*

Shewanella oneidensis MR-1 (Lovley *et al.*, 1989) and *Shewanella alga* (Caccavo Jr. *et al.*, 1992), both of which are also able to produce extracellular magnetite from dissimilatory Fe(III)-reduction. The ability to respire using ferric Fe and form extracellular magnetite is also found in other prokaryotic groups and includes *Geothrix fermentans*, part of the *Acidobacterium* phylum (Coates *et al.*, 1999), *Geovibrioferrireducens* (Caccavo Jr *et al.*, 1996), *Deferribacterthermophilus* (Greene *et al.*, 1997), *Anaeromyxobacter dehalogenans* (Petrie *et al.*, 2003) and *Acidiphilium cryptum* JF-5, the latter isolated from acidic sediments (Küsel *et al.*, 1999). *Geothermobacteriumferrireducens* from a hydrothermal environment (Kashefi *et al.*, 2002; Wilkins *et al.*, 2006) and the Gram-positive *Thermoanaerobacter ethanolicus* (previously TOR-39) are also able to precipitate magnetite (Zhang *et al.*, 1998). The latter produces magnetite nanoparticles, where the majority are single-domain phases which are closer in structure to intracellular magnetite produced by magnetotactic bacteria, than the more poorly ordered extracellular magnetite formed by Fe(III)-reducing strains such as *Geobacter* and *Shewanella* (Zhang *et al.*, 1998; Frankel and Bazylinski, 2003). Although the thermophiles mentioned are part of the bacterial kingdom, some Archaeal strains are also of interest. *Pyrobaculum islandicum* is a thermophile that is able to produce magnetite nanoparticles at 100°C (Kashefi *et al.*, 2008). In addition, Strain-121 has raised the upper temperature limit for life by 8°C to 121°C and was cultured from sediment collected from an active black smoker hydrothermal vent in the northeastern Pacific Ocean (Kashefi and Lovley, 2003). Phylogenetic analyses placed the microbe in the Archaeal Kingdom and it is related to members of the genera *Pyrodictium* and *Pyrobaculum*. Strain-121 was cultured on solid amorphous Fe(III) oxide with formate as the electron donor, and biogenic extracellular magnetite was produced as the result of incubation at 100°C (Kashefi and Lovley, 2003).

A diverse group of microorganisms, including chemolithic bacteria, are also capable of synthesizing magnetite, or less commonly greigite (Fe_3S_4) nanoparticles, within intracellular vesicle structures, termed magnetosomes. Greigite is a magnetic mineral with a similar structure to magnetite; see Section 3.2 for details. Magnetosomes are generally aligned into chain structures within the bacterium and individually are each <150 nm in size (Lin *et al.*, 2014). Magnetosomes have been found in numerous species of aquatic prokaryotes within the *a*, *d* and *Nitrospira* lineages of *Proteobacteria* (Schüler, 2004) since the first discovery of *Magnetospirillum* by Blakemore (1975); however, greigite-producing magnetotactic bacteria appear to be confined to the δ-*Proteobacteria* (Lin *et al.*, 2014). This group of bacteria primarily uses magnetic intracellular structures for orientation according to the Earth's magnetic field for navigational purposes, and is therefore referred to as magnetotactic bacteria (Blakemore, 1982). Although poorly understood, the mechanism of intracellular magnetic particle production is thought to involve the secretion of siderophores (Schüler, 1999), uptake of chelated Fe(III) and intracellular reduction to Fe(II) (Bazylinski *et al.*, 2007), followed by particle oxidation forming magnetite (Frankel *et al.*, 1984) or greigite. This results in magnetic particles with a tightly constrained size range, chemical purity and crystal habit (Bazylinski *et al.*, 2000). However, a combination of the low yield of magnetite associated with intracellular reduction and the challenging microaerobic culturing conditions often required may limit its potential for scaling up production for biotechnology purposes.

2.1.1. Mechanisms of bacterial Fe(III) reduction

Microbial reduction of Fe(III) phases cause reductive dissolution of the initial Fe(III) phase, liberating Fe(II) and allowing the formation of secondary mineral phases, such as magnetite (Benner *et al.*, 2002). The precise mechanisms that dissimilatory Fe(III)-reducing bacteria use to transfer electrons to solid-phase Fe(III)-bearing mineral phases has been studied intensively for the last two decades, and the complex electron transfer chains that are involved are becoming better understood. These bacteria have a battery of putative ferric Fe reductases located on the outer membrane of the cell (Myers and Myers, 1992; 1997; Lloyd, 2003) that may act as the terminal reductases of an electron transport chain that is linked to the cytoplasmic membrane. The terminal reductases have not yet been unequivocally identified in any organism, but are thought to involve *c*-type cytochromes in an electron transport chain (Lloyd, 2005). Electrons are transferred through the chain to the surface of the cell where the exposed reductase transfers them to the Fe(III)-bearing mineral surface. This process conserves energy through the generation of ATP (Lovely *et al.*, 2004; Wilkins *et al.*, 2006). This is shown in the bacterial genomes for *Shewanella* spp. and *Geobacter* spp., where genes encoding for *c*-type cytochromes are found in abundance (Richter *et al.*, 2012). Recent studies have highlighted the involvement of OmcB (Leang *et al.*, 2003) and OmcS (Leang *et al.*, 2010) cytochromes as playing key roles in electron transfer to Fe(III) at the periphery of the *Geobacter* cell, with MtrC playing a similar role in *Shewanella* cells (Hartshorne *et al.*, 2007). The potential orientation of these proteins at the cell surface was discussed in detail by Liu *et al.* (2014).

Bacteria have developed several strategies for dealing with solid-phase minerals beyond the need for direct contact between the cell and the mineral in order to effect electron transfer; see Fig. 1 and, for a more detailed illustration of the bacterial membrane proteins involved, see Weber *et al.* (2006a). Some bacteria, such as *Shewanella* spp., are able to create soluble 'electron shuttling' compounds alleviating the need for direct contact with the mineral surface (Newman and Kolter, 2000). These can accelerate electron transfer processes and have been identified in *Shewanella* spp. as flavin compounds (Marsili *et al.*, 2008; von Canstein *et al.*, 2008). Fe(III)-reducing bacteria are not all capable of electron shuttle synthesis; however, naturally occurring organic materials, such as humics can be utilized as electron shuttling compounds by bacteria (Lovley *et al.*, 1996) and these reduced hydraquinone moieties can therefore transfer electrons abiotically to Fe(III)-containing minerals (Lloyd, 2005).

Additional mechanisms of Fe(III)-reduction have been identified (Fig 1). A study has shown that *G. sulfurreducens* produced pili when grown on insoluble ferric Fe but not when grown on soluble ferric citrate (Reguera *et al.*, 2005); the pili were useful not just for attachment to the mineral substrate, but were also highly conductive and could be

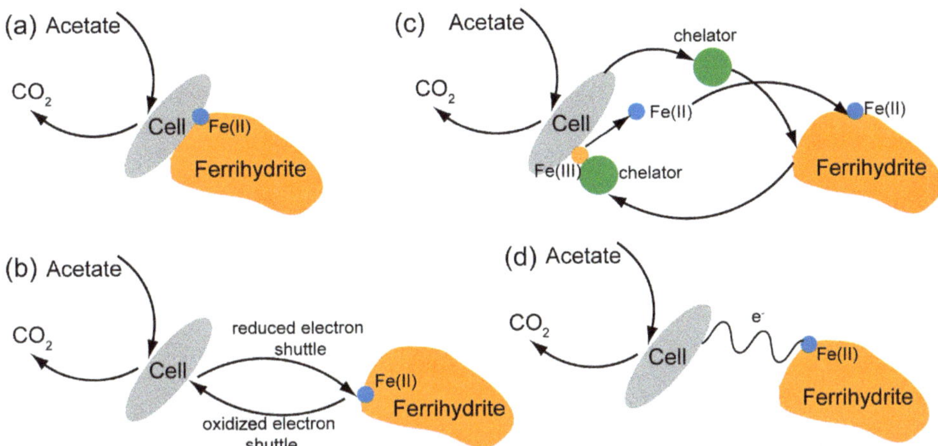

Figure 1. Possible mechanisms of reduction of solid-phase Fe(III) minerals utilized by Fe(III)-reducing microorganisms (a) direct contact; (b) electron shuttling compounds (both cell-derived and naturally occurring); (c) chelating agents; (d) nanowire electron transfer. Note acetate is given as an example electron donor while not all Fe(III)-reducing bacteria are capable of metabolizing this compound.

used for the transfer of electrons by acting as 'nanowires' (Kappler and Straub, 2005; Reguera *et al.*, 2005). Conductive extracellular appendages have also been described in *Shewanella* spp. (Gorby *et al.*, 2006). *G. metallireducens* has also been shown to use flagella and pili to aid movement and attachment to insoluble Fe(III) oxyhydroxides and is chemotactic towards ferric oxides (Childers *et al.*, 2002), facilitating access to metal oxides in anoxic environments (Lovley *et al.*, 2004). Recent studies have progressed the 'nanowire' concept in both model organisms dramatically. The structure in *Shewanella oneidensis* comprises extensions of the outer membrane and periplasm impregnated with outer membrane multiheme cytochromes MtrC and OmcA to facilitate electron transfer (Pirbadian *et al.*, 2014). The structure of nanowires in *Geobacter sulfurreducens* is very different, comprising proteinaceous pili, which are proposed to have intrinsic metallic-like conductivity due to overlapping π-orbitals of aromatic amino acids within the protein structure (Malvankar *et al.*, 2015).

Fe(III) oxides are also thought to be solubilized by chelating agents (Fig. 1), making Fe(III) more accessible to the cell (Lovley and Woodward, 1996). The addition of the chelating agent 2,2′,2″-Nitrilotriacetic acid was found to increase the rate of reduction of amorphous and crystalline Fe(III) minerals at circum-neutral pH (Lovley and Woodward, 1996). Chelators may also promote bioreduction by the increased solubilization of Fe(II), where adsorbed Fe(II) has been identified as an inhibitor to bioreduction of Fe(III) oxides *via* a surface passivation mechanism (Roden and Urrutia, 1999; Urrutia *et al.*, 1999).

2.2. Fe(II) oxidation

Bacterial oxidation of Fe(II) occurs *via* a variety of mechanisms; microaerophilic, nitrate-dependent and phototrophic, but magnetite formation has been identified primarily during anaerobic Fe(II) oxidation coupled to nitrate reduction (Benz *et al.*, 1998; Chaudhuri *et al.*, 2001). The first observation of this form of metabolism was made with a lithotrophic enrichment culture (Straub *et al.*, 1996). Most Fe(II)-oxidizing nitrate-reducers isolated so far require an organic substrate for growth, which means they grow only mixotrophically with Fe(II) (Straub *et al.*, 1996; Benz *et al.*, 1998; Kappler and Straub, 2005). However, *Azospira oryzae* (formally *Dechlorosoma suillum*) is one of the few species where it has been proven that the oxidation of ferrous Fe (in this case $FeCl_2$), coupled to dissimilatory nitrate reduction, supports growth. The formation pathway of the resulting biogenic magnetite proceeded *via* Fe(II)Fe(III)-hydroxides, identified as carbonate-containing green rust ($[Fe_6(OH)_{12}][4H_2O{\cdot}CO_3]$) (Chaudhuri *et al.*, 2001). Green rust is metastable and known to react with nitrate to form magnetite, so it is possible that the biogenic green rust was subsequently converted abiotically to magnetite (Chaudhuri *et al.*, 2001). *A. oryzae* can also oxidize the Fe(II) in siderite ($FeCO_3$) and almandine ($Fe_3Al_2(SiO_4)_3$), which might also form magnetite as the end-product (Chaudhuri *et al.*, 2001). Later work has also implicated the role of intermediates from nitrate reduction, NO_2^- and NO, in the abiotic oxidation of Fe(II), in the presence of the nitrate-reducing, Fe(II) oxidizer *Acidovorax* sp. BoFeN1 (Klueglein and Kappler, 2013). This bacterium has also been found to form magnetite *via* the oxidation of green rust, extending the number of nitrate-reducing Fe(II)-oxidizing bacteria capable of magnetite biomineralization (Miot *et al.*, 2014). Outside of the nitrate-reducing bacteria, few Fe(II)-oxidizing bacteria have been found to catalyse the formation of magnetite. One such bacterium, *Rhodopseudomonas palustris* strain TIE-1, was found to form magnetite by phototrophic Fe(II) oxidation above pH 7.2, forming goethite below pH 7 (Jiao *et al.*, 2005). Those authors proposed that the formation of magnetite at higher pH might be related to the increased adsorption of Fe(II) on to Fe(III) (hydr)oxides above the point of zero charge of these phases, promoting the formation of the mixed-valence Fe oxide.

Weber and colleagues (2006b) found that an enrichment culture from a wetlands sample, that previously had reduced Fe(III) in goethite, was then able, within the same experiment under anoxic conditions, to subsequently oxidize Fe(II), where Fe(II) oxidation was promoted by the addition of nitrate. Surprisingly, the bacteria that were the most likely candidates for the Fe(II) oxidation were the same as those that had caused the original Fe(III) reduction, and were members of the family *Geobacteraceae* (Weber *et al.*, 2006b).

The mechanistic details of anaerobic Fe(II) oxidation by bacteria are currently poorly understood, with a number of important questions remaining unanswered, *e.g.* the cellular location of Fe(II) oxidation, or how the bacteria cope with the poor solubility of the resultant ferric minerals (Kappler and Straub, 2005). However, recent studies have identified some of the enzymes utilized in the electron transfer from Fe(II) to oxidizing bacteria (Liu *et al.*, 2012a) This study identified a decaheme *c*-type cytochrome from the bacterium *Siberoxydans lithotrophicus* ES-1 capable of growing on siderite. Another study demonstrated that when *R. palustris* strain TIE-1 accepted electrons from a poised electrode Pio proteins were implicated in electron transfer with solid substrates (Bose *et al.*, 2014). This study also noted the increased expression of *ruBis*Co form I, coding for the expression of the RuBisCo enzyme of the Calvin cycle, during electron uptake in the presence of light.

3. Primary and secondary mineral phases

3.1. Primary Fe(III) phases

The formation of magnetite by dissimilatory Fe(III)-reducing bacteria (DIRB) is typically achieved by the reductive transformation of a variety of precursor phases, such as akaganeite (FeO(OH,Cl)), lepidocrocite (γ-FeO(OH)) and schwertmannite ($Fe_8O_8(OH)_6SO_4$). These Fe(III) phases are abundant in the natural environment and typically occur at concentrations in the hundreds of g kg^{-1} of aerobic soil and sediment (Hansel *et al.*, 2003; Cutting *et al.*, 2009). The formation of magnetite, by biologically-mediated Fe(III) reduction, is dependent on a complex range of variables, such as initial crystallinity and the presence of electron shuttle compounds (Cutting *et al.*, 2009). The rate of reductive transformation of the Fe phase is thought to be governed by the surface area and the crystallinity of the Fe(III)-bearing phase, and this, in turn, determines the mineralogical product of the reductive process (Roden and Zachara, 1996; Hansel *et al.*, 2003, 2005; Cutting *et al.*, 2009). The more crystalline Fe(III) phases, for example, do not result typically in the formation of magnetite upon incubation with DIRB (Cutting *et al.*, 2009).

As stated, ferrihydrite is the most widely studied Fe(III) precursor mineral to biogenic magnetite and is therefore discussed in detail here. This Fe(III) oxyhydroxide occurs principally as a nanoparticulate phase and when analysed using X-ray diffraction (XRD) it gives only two or six broad peaks. It is therefore often known as 2-line or 6-line ferrihydrite (Waychunas, 1991) or, alternatively, hydrous ferric oxide (HFO) (Schwertmann *et al.*, 1999). Ferrihydrite is the most bioavailable of the Fe(III)-containing minerals present in the subsurface (Lovley and Phillips, 1986; Glasauer *et al.*, 2003; Wilkins *et al.*, 2006) and can make up $>20\%$ of the total Fe in sediments (Thamdrup, 2000). Ferrihydrite is formed abundantly as the initial product of the hydrolysis and precipitation of dissolved Fe in natural geological systems, and is a strong scavenger of metal ions from aqueous solutions due to its small particle size (2–3 nm (Schwertmann *et al.*, 1999)) and large surface area (Waychunas, 1991). The structure or atomic arrangement of ferrihydrite remains controversial primarily due to the difficulty in characterizing the material using conventional techniques, due to its small particle size and structural disorder (Michel *et al.*, 2007). Michel *et al.* (2007) utilized pair-distribution function (PDF) from Fourier transformation of total X-ray scattering, to reveal hexagonal unit cells containing Fe arranged roughly in 20% tetrahedral and 80% octahedral sites, with an ideal formula of $Fe_{10}O_{14}(OH)_2$. However, this is not universally accepted because this structure does not agree necessarily with all the available experimental data (Manceau, 2009) and research is ongoing. Further discussion and references to this topic can be found in Manceau *et al.* (2014) and Cismasu *et al.* (2014). Ferrihydrite is thermodynamically unstable, and will transform to the crystalline goethite and hematite ferric phases (Schwertmann *et al.*, 1999). However, ferrihydrite has been found under specific environmental conditions at the Loihi Seamount in microbial mats (Toner *et al.*, 2012) and can also be found in a variety of different soils and sediments (Jambor and Dutrizac, 1998). However, crystalline Fe(III)-bearing minerals are, therefore, generally more abundant in sedimentary environments than poorly crystalline equivalents (Schwertmann and Taylor, 1989). *S. oneidensis* is reported to be capable of reducing Fe(III) in the crystalline phases goethite and hematite as well as ferrihydrite in pure culture experiments (Fredrickson *et al.*, 1998; Roden and Zachara, 1996; Zachara *et al.*, 1998), although reduction in 'static' systems does not go to completion due to passivation of Fe(III) reduction by Fe(II) surface-sorption to either the cell or the Fe(III)-mineral (Roden and Zachara, 1996; Cutting *et al.*, 2009).

The formation of extracellular magnetite from these Fe(III) starting phases requires suitable regulation of redox, pH and Fe chemistry (Mann *et al.*, 1990; Schwertmann and Fitzpatrick, 1992). Schwertmann and Fitzpatrick (1992) suggested that a requirement for microbial magnetite formation is a favourable rate of Fe(II) and Fe(III) supply to the growing crystal, combined with effective buffering capacity of the (natural or synthetic) system that allows bacteria to thrive. For any mineral to form, the solution must be supersaturated with respect to that mineral (Schwertmann and Fitzpatrick, 1992; Fredrickson *et al.*, 1998; Roh *et al.*, 2003), and rapidly changing conditions in natural systems may lead to sporadic magnetite formation. Abiotic factors that will determine whether magnetite will be precipitated or not include atmospheric composition, buffer type, solution composition, incubation temperature and time. For example, siderite instead of magnetite was produced using cultures of *S. oneidensis* under a headspace of CO_2-H_2 instead of N_2 (Roh *et al.*, 2003). Also, if phosphate instead of carbonate is used in the buffering system, then vivianite ($Fe_3(PO_4)_2$) can form (Islam *et al.*, 2005). Hansel *et al.* (2003) found that in a flow-through system, the first stage of mineralization was affected by bacterial metabolism, solution chemistry and kinetically regulated solid-phase precipitation, whereas the conversion of minerals formed during the first stage to magnetite was affected primarily by flow-regulated Fe(II) concentration, which was the main reason for the difference in mineral end-products (Hansel *et al.*, 2003).

3.2. Magnetite

Magnetite has a cubic spinel structure, with a quarter of the tetrahedral (T_d) and a half of the octahedral (O_h) sites filled by Fe (Fig. 2). The formula for magnetite is $(Fe^{3+})[Fe^{2+}Fe^{3+}]O_4^{2-}$, where Fe^{3+} is split equally between O_h [ ] and T_d ( ) sites and Fe^{2+} occupies only O_h sites. Hence, in stoichiometric magnetite the occupancy of the Fe d^6 $O_h : d^5 T_d : d^5$ O_h sites is 1 : 1 : 1 and the splitting of the Fe(III) between the two sites means magnetite is classified as an 'inverse' spinel. The Fe^{3+} in the O_h and T_d sub-lattices occurs as cations having antiparallel magnetic moments that cancel each other out, but there is a resulting net magnetization (ferrimagnetism) on the Fe(II) in the O_h site.

Oxidation of Fe^{2+} to Fe^{3+} in the O_h site in magnetite results in a charge imbalance and the release of Fe from the structure. The result is neo-vacancies in the structure giving

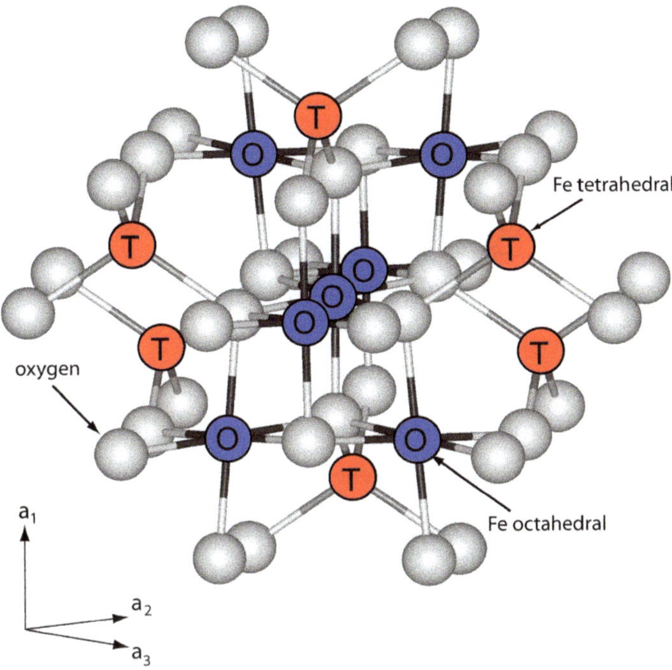

Figure 2. Structure of magnetite indicating the octahedral Fe sites (O, blue), tetrahedral Fe sites (T, red) and oxygen (grey).

rise to a chemical formula of $(Fe^{3+})[Fe^{2+}_{1-3\delta}Fe^{3+}_{1+2\delta}\Delta_\delta]O_4^{2-} = Fe_{3-\delta}O_4$, where δ is the number of vacancies Δ (Pearce *et al.*, 2006). The end-member of this oxidation process is the mineral maghemite, which has the same structure as magnetite but with no Fe(II) and a formula of γ-Fe_2O_3. Under some conditions, principally Fe(II) supply rate and magnitude, Fe(III)-reducing bacteria may favour the formation of this non-stoichiometric and more oxidized form of magnetite (Zachara *et al.*, 2002).

Because magnetite contains both Fe(III) as well as Fe(II) in a ratio of 2:1 it is possible to reduce magnetite chemically using dissolved sulfide. Although it was found that it is possible for *S. oneidensis* to couple growth to magnetite reduction in pure culture, at temperatures (22°C) and pH (6.2) values found in marine and aquatic systems, this study highlighted a thermodynamic barrier at higher Fe(II) concentrations (<100s μM) and pH >6.5 (Kostka and Nealson, 1995). Dong *et al.* (2000) conducted experiments in which *S. oneidensis* MR-1 and strain CN32 reduced Fe(III) in magnetite, precipitating siderite or vivianite, depending on whether the reduction was carried out in a carbonate or phosphate buffer. Byrne *et al.* (2015) also found that co-cultures of an Fe(II)-oxidizing (*Rhodopseudomonas palustris*) and Fe(III)-reducing bacteria (*G. sulfurreducens*) were able to cycle the oxidation state of the Fe within magnetite; showing that magnetite is bioavailable as both a sink and source of electrons under certain environmental conditions.

3.3. Greigite

Greigite (Fe_3S_4) is an iron sulfide mineral organized in an inverse spinel structure, isostructural with magnetite, with sulfur atoms in place of oxygen and therefore known as a thiospinel. This structure, which is similar to that of magnetite, also makes greigite ferrimagnetic, with an imbalance of Fe(II) in the octahedral site (Fassbinder and Stanjek, 1994); however, there are differences in the magnetic properties due to increased covalency caused by the presence of the sulfur atoms; see Chang *et al.* (2008). Like magnetite, greigite can be formed biogenically within magnetotactic bacteria (Kasama *et al.*, 2006), although it has received far less attention. However, it is more commonly associated with sulfate-reducing sediments, forming as a precursor to pyrite (FeS_2) (Benning *et al.*, 2000).

4. Tailoring magnetic minerals

The physical properties of biogenic magnetite can be altered and therefore tailored for optimal particle size, magnetic property and reactivity. Several variables can provide controls on these physical properties, the most significant being: chemistry of the culture medium; the quantity of bacterial biomass employed; and the chemistry and physical state of the precursor material. The details of the applications for these different physical attributes are discussed in more detail in Section 5.

As discussed previously, biogenic magnetite, produced by DIRB, can be formed by the transformation of a variety of Fe(III) starting phases. These starting phases have been found to provide a control on the physicochemical characteristics of the resulting magnetite. The biogenic magnetite formed from schwertmannite contains a greater quantity of Fe(II) at the surface of the nanoparticle, compared to ferrihydrite-derived magnetite, and it has been shown that this, in turn, leads to a greater reactivity towards contaminants such as hexavalent chromium (Cr(VI)) (Cutting *et al.*, 2010).

Several model bacterial systems have been used to produce biogenic magnetite with metal dopants substituting for some of the Fe cations within the spinel lattice of the magnetite. The most commonly used dopants for changing the physical attributes of biogenic magnetite are the top level transition metals including Co, Ni, Mn, Cr and Zn, and the lanthanides. All of these have been substituted successfully into the biogenic magnetite structure using Fe(III)-reducing bacteria, including *G. sulfurreducens, S. oneidensis* and *T. ethanolis* TOR-39 (Fredrickson *et al.*, 2001; Roh *et al.*, 2001, 2003, 2006; Moon *et al.*, 2007a,b,c; Coker *et al.*, 2008, 2009; Byrne *et al.*, 2013). The precursor mineral phases used to achieve a doped magnetite typically include akaganeite and ferrihydrite, where each is produced already containing a quantity of additional metal dopants, such as Co. Up to one third of the Fe in magnetite can be substituted for a different transitional metal using this method. For example, the addition of 23 atom% Co increased the magnetic coercivity (the intensity of applied magnetic field required to demagnetize a material) of the resulting spinel from 360 Oe (Oersted) for pure biogenic magnetite, to 7900 Oe, values equivalent to synthetic counterparts (Coker *et al.*, 2009). Controlling the quantity of Co substituted into biogenic magnetite also changes the size of the nanoparticles, with an increase in Co resulting in a subsequent decrease in particle size to 4 nm (from 8.5 nm), and an increase in the effective magnetic anisotropy, the directionality of a materials magnetic field (Byrne *et al.*, 2013). In Section 5.2 the application of these particles for hyperthermia cancer treatment is discussed.

The particle size of the biogenic magnetite can also be tailored, as is normally done during abiotic synthesis using *e.g.* organic solvents. The biogenic method described by Byrne *et al.* (2011) leads to a precise control on the nanoparticle size by simply adjusting the quantity of resting cells of *G. sulfurreducens* added to the ferrihydrite at the beginning of the experiment. Adding larger quantities of bacteria resulted in smaller particles and *vice versa*. The mechanism of the change in nanoparticle size is thought to be related to the rate of nucleation of the particles, with rapid nucleation rates due to increased Fe(II) flux from the greater biomass concentration resulting in the production of smaller particles (Byrne *et al.*, 2011).

5. Applications of magnetic biominerals

It is well established that Fe(II)-bearing biominerals are reactive towards a variety of redox-sensitive inorganic and reducible organic contaminants (Lovley, 1997). These interactions are often useful for environmental remediation; inducing reductive stabilization of inorganic pollutants or degradation of organic ones. Due to the prevalence of Fe(III)-reducing conditions in the polluted subsurface, of which magnetite is a common endmember, the reactivity of Fe(II) biominerals has important implications for the cycling of various contaminant species, see Fig. 3 (Lloyd, 2003). The biosynthesis of magnetic minerals, principally magnetite, offers potential for a biotechnological approach to contaminant remediation (Lloyd *et al.*, 2011).

A key aspect in the effectiveness of remedial applications of biogenic magnetite is related to the typically nano-scale particle size (Lloyd *et al.*, 2008), which is often associated with increased reactivity when compared to larger-scale equivalents (Hochella Jr., 2002; Hochella Jr. *et al.*, 2008). The harnessing of this increased reactivity has led to the application of various synthetic nano-scale particles for the remediation of contaminated soils and waters (Zhang, 2003; Tratnyek and Johnson, 2006). These have focused typically on highly reactive Fe(0) nano-particles alongside its reactive corrosion products, such as magnetite, which often form a shell-like structure at the particle surface (Nurmi *et al.*, 2004). Due to the prevalence of magnetite in subsurface environments, and its reactivity with several key contaminants, it has also been the focus of several studies aimed at contaminant-remediation processes.

This section will therefore address some of the aspects of magnetite interactions with a variety of key inorganic and organic contaminants. Also discussed are some additional aspects related to controls over its biosynthesis where they affect magnetite–contaminant interactions.

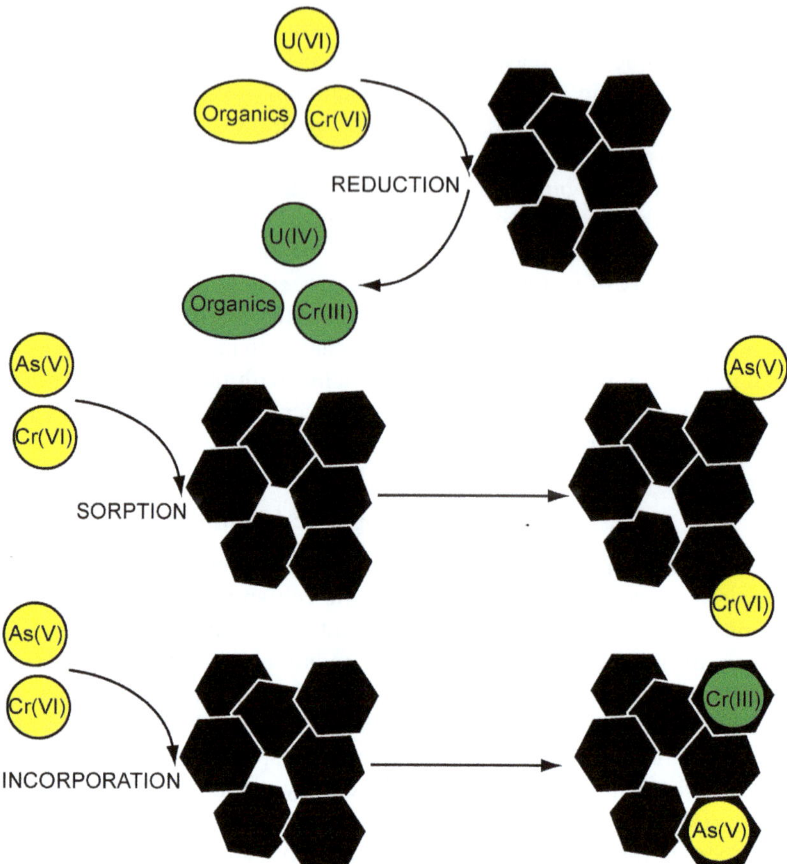

Figure 3. Remediation mechanisms facilitated by, for example, nanoscale biogenic magnetite contaminants. Oxidized contaminants (yellow, lighter grey) and reduced elements (green, darker grey).

5.1. Magnetite–contaminant interactions

5.1.1. Chromium(VI)

The common ground- and surface-water pollutant Cr(VI) is known to be susceptible to reductive transformation by Fe(II)-bearing phases to the benign Cr(III) phase (Fendorf, 1995). Under environmental conditions Cr occurs most commonly in the Cr(III) and Cr(VI) valence states. The oxidized Cr(VI) state occurs as hydrated species which are readily soluble and therefore mobile within most subsurface conditions (Rai *et al.*, 1989). Its mobile and oxidizing nature cause it to act as a toxicant to animals and plants (Costa, 1997; Shanker *et al.*, 2005). By contrast, the Cr(III) valence state occurs widely as insoluble oxides or hydroxides (Fendorf, 1995) and is regarded as non-toxic (Gad, 1989).

An early study demonstrated the ability of synthetic and natural magnetites to remove aqueous Cr(VI) from solution coupled to surface Fe(II) oxidation (White and

Peterson, 1996). The degree of oxidation of the magnetite surface, and therefore its surface Fe(II) density, was found to control the amount of Cr(VI) removal; with more removal using magnetite from reducing rather than oxic environments. Passivation of the reactive surface, by coating with an insulating layer of reduced Cr(III) (Fig. 3) and the oxidized Fe(III), was also found to be a limiting factor in Cr(VI) reduction (Peterson *et al.*, 1997). The thickness of this passivation layer, on synthetic magnetite, has been reported to be 10–20 Å from high-resolution transmission electron microscopy (HRTEM) images (Peterson *et al.*, 1997), and calculated from X-ray attenuation during X-ray photoemission spectroscopy (XPS) (Kendelewicz *et al.*, 1999). Near complete reduction to the Cr(III) state on the magnetite surface has been confirmed by X-ray absorption spectroscopy (XAS) analysis of the Cr *L*-edge (Kendelewicz *et al.*, 1999, 2000) and the Cr *K*-edge (Peterson *et al.*, 1997). The local structure of the passivation layer is inferred from X-ray absorption fine structure (EXAFS) analysis to be most similar to that of a Cr(III) and/or Fe(III)OOH (Peterson *et al.*, 1997). The pH of the system was also found to control the removal of Cr(VI); with a decrease in removal at higher pH conditions (He and Traina, 2005; Kendelewicz *et al.*, 2000). The latter study observed significant inhibition of Cr(VI) removal, ~20%, at pH 13 compared to those recorded at acidic to circumneutral pH.

Biogenic magnetite, synthesized by bioreduction of an Fe(III) starting phase by *G. sulfurreducens*, has also been demonstrated to be active towards Cr(VI) (Telling *et al.*, 2009; Cutting *et al.*, 2010; Byrne *et al.*, 2011; Crean *et al.*, 2012). The earlier study suggested, using X-ray magnetic circular dichroisim (XMCD) analysis of the Fe $L_{2,3}$-edge and Cr $L_{2,3}$-edge, Cr(III) incorporation into the magnetite octahedral sites in a spinel surface structure. A subsequent study was able to compare the reactivity of biogenic magnetite derived from bioreduction of schwertmannite and ferrihydrite with respect to a synthetic magnetite (Cutting *et al.*, 2010). The schwertmannite-derived magnetite had a larger surface Fe(II) content and a greater reactivity towards Cr(VI) than the synthetic and ferrihydrite-derived magnetite. Analyses of the post Cr(VI) reacted magnetite, again using XMCD, suggested incorporation of 14–20% of Cr(III) into the octahedral site of the magnetite. A later study sought to refine further the biogenic magnetite particles with regard to controlling their size, reactivity and magnetic properties by adjusting the total biomass during biosynthesis (Byrne *et al.*, 2011). The subsequent biogenic magnetites synthesized with increasing bacterial density were tested for their reactivity towards Cr(VI). The smaller particles yielded by greater bacterial density exhibited the highest reactivity, related probably to larger surface area. In addition, the reactivity of biogenic magnetite coated in palladium (Pd) nanoparticles has been examined using a flow-through column study. When formate as an electron donor was added to the column system with Pd and magnetite, the activity of the column was far greater against Cr(VI) than when biogenic magnetite was used alone (Crean *et al.*, 2012). More recent studies have focused on the successful use of biomagnetite (Watts *et al.*, 2015a) and biomagnetite coated with reactive Pd nanoparticles (Watts *et al.*, 2015b) to treat chromite-ore processing residue and associated alkaline Cr(VI) leachates.

5.1.2. Mercury(II)

Mercury (Hg) is also a common pollutant primarily through its release into the environment during the chloralkali process for the production of Cl and NaOH using a Hg cathode and mining activities (von Canstein *et al.*, 2002). It is highly toxic in its ionic Hg(II) state, where it is can bind and deactivate key metabolic enzymes (Barkey *et al.*, 2003). It can also form toxic organic Hg compounds, methyl Hg, which are readily bioavailable (O'Driscoll *et al.*, 2005). The toxicity of methyl Hg is, however, due to intracellular de-methylation yielding toxic Hg(II) within the cell (Morel *et al.*, 1998). Hg(0) in its reduced state is regarded as less toxic and, due to its low solubility and high vapour pressure, is easily expelled from contaminated soils and waters (Barkey and Wagner-Dobler, 2005).

Recent studies have also highlighted the ability of magnetite to reduce toxic Hg(II) to Hg(0) (Wiatrowski *et al.*, 2009; Pasakarnis *et al.*, 2013). The initial study found that, after reaction of a Hg(II) solution with 0.2 g/L of synthetic magnetite for 1 h, ~70% Hg was lost from solution, compared to minimal loss in the no-magnetite control experiments (Wiatrowski *et al.*, 2009). The Hg was recovered in a volatile trap of potassium permanganate, inferred as evidence for the reduction of Hg(II) to Hg(0) and its subsequent volatilization from solution. Analyses by Mössbauer and XPS noted a concurrent loss of Fe(II) and the presence of Hg(II) present on the magnetite surface, implying adsorption followed by Fe(II)-mediated reduction. This study also highlighted the kinetic impact of pH and Cl^- concentration. Cl^- is a common ion in chloralkali wastes, and the combined increase in its concentration and decreasing pH has been shown to slow the rate of Hg(II) removal from solution (Wiatrowski *et al.*, 2009). A later study (Pasakarnis *et al.*, 2013) found that in the absence of Cl^-, regardless of magnetite Fe stoichiometry between 0.29 and 0.50 Fe(II)/Fe(III), complete reduction of Hg(II) to Hg(0) was observed using Hg L_3-edge X-ray absorption near edge structure (XANES) analysis. However, in the presence of Cl^- the more oxidized magnetites, <0.42 Fe(II)/Fe(III), appeared to accumulate metastable Hg(I) species. This Hg(I) phase was absent in more reduced, >0.42 Fe(II)/Fe(III), reacted magnetites.

5.1.3. Arsenic(V)

Arsenic in groundwaters, particularly in areas such as West Bengal and Bangladesh, is known to have had extensive impacts on human health (Charlet and Polya, 2006). The exact mechanism for the release of As into aquifer systems remains under debate though it is understood that one of the main causes of As contamination is the activity of anaerobic metal-reducing bacteria, resulting in solubilization of As(V)-bearing mineral phases, releasing any adsorbed chemical species (Nickson *et al.*, 2000; Islam *et al.*, 2004). However, magnetite formed by metal-reducing bacteria can also be used as a remediation agent against As(V). Initial experiments focus on the sorption of As(V) and As(III) to Fe-minerals, including magnetite (Fig. 3), which was effective in removing both As species from solution (Dixit and Hering, 2003). Recently, data from EXAFS analysis suggest the formation of bidentate binuclear complexes for As(V) and tridentate hexanuclear complexes for As(III) adsorption to magnetite nanoparticles (Liu *et al.*, 2015).

When As(V) was sorbed to the Fe(III) precursor mineral, ferrihydrite, and prior to reductive transformation to magnetite, using *G. sulfurreducens*, it was found that As(V) was sequestered by the new mineral phase into the magnetite structure. As(III) was not sequestered, however, and remained as a sorbed species (Islam *et al.*, 2005; Coker *et al.*, 2006). This was also observed when As(V) was co-precipitated with schwertmannite and bioreduced (Cutting *et al.*, 2012). The incorporation of As(V) into nano-sized magnetite has been confirmed by Wang *et al.* (2011) and represents an As immobilization mechanism, which is potentially useful for water treatment.

The origin of the Fe(III) oxide is also an important factor in the subsequent Fe(III) bioreduction-related mobilization of As. Recently, a study sought to assess the mobility of As(III) and As(V) adsorbed on to abiotically synthesized ferrihydrite, in comparison to biogenic ferrihydrite with associated cell-derived organic matter, upon reduction by *S. oneidensis* MR-1 (Muehe *et al.*, 2013). Different behaviour was observed for abiotic and biotic ferrihydrite, where, upon reduction, As was more effectively immobilized upon reduction of the As(V)-loaded biotic than abiotic ferrihydrite and formation of magnetite.

Magnetite can also be used to catalyse the oxidation of As(III) to As(V), desirable due to the reduced toxicity of As(V), under aerated conditions through Fe(II)-mediated Fenton-like reactions (Ona-Nguema *et al.*, 2010).

5.1.4. Uranium(VI)

Uranium is one of the primary concerns regarding radionuclide contamination due to its presence at multiple stages of the nuclear fuel cycle (Kimber *et al.*, 2011). In the environment it occurs in two stable valence states, U(IV) and U(VI), typically forming relatively insoluble U(IV) oxides and soluble U(VI) carbonate complexes (Clark *et al.*, 1995). The reductive stabilization of mobile U(VI), preventing migration within the subsurface, has therefore received much attention particularly in regards to geodisposal scenarios, where magnetite is a common corrosion product of steel containers.

Early studies sought to assess the adsorption and reduction mechanism of the U(VI) reaction with a synthetic magnetite surface (el Aamrani *et al.*, 1999; Missana *et al.*, 2003; Scott *et al.*, 2005). The initial adsorption of U(VI) was observed to be relatively fast, and it was followed by a slower reduction step coupled to Fe(II) oxidation (el Aamrani *et al.*, 1999) and controlled strongly by pH (Missana *et al.*, 2003). The reduction process was probed further in a later study which analysed a series of U(VI) solutions (10 and 100 ppm) reacted, at pH 4–5, with magnetite surfaces (Scott *et al.*, 2005). XPS analysis observed, after 50 h, ~5 atomic % U on the magnetite surface determined to be of a mixed U(VI) and U(IV) phase. Interestingly, adsorbed U(VI) decreased over an exposure time, concomitant with a decrease in surface Fe(II), consistent with a fast sorption of U(VI) followed by a slower reduction step.

The fate of the reacted U(VI) on the surface of synthetic magnetite has also been the focus of several studies (Aamrani *et al.*, 2007; Ilton *et al.*, 2009). The former study reported the presence of a mixed U(IV) and U(VI) phase, from EXAFS analysis, with both the axial oxygen of schoepite and the high coordination number of UO_2 (Aamrani *et al.*, 2007). Using high-resolution XPS and EXAFS, reduction to the lesser-studied U(V) phase on the surface of magnetite was observed (Ilton *et al.*, 2009). The authors concluded that under the reaction conditions U(IV) should have been thermodynamically stable and the system was therefore not in redox equilibrium. Despite this, the occurrence of U(V) at the surface of magnetite has been confirmed using a combination of electrochemical and spectroscopic techniques (Yuan *et al.*, 2015a). Further to this it was found that although on a powdered magnetite electrode, U(V) and U(IV) formed, on a bulk magnetite, electrode only U(V) formed on the surface (Yuan *et al.*, 2015b)

The discrepancies observed in the degree of reduction of U(VI) to U(IV) by varying the stoichiometry of magnetite by the addition of aqueous Fe^{2+} have also been discussed (Latta *et al.*, 2012). Using Mössbauer and XAS analysis, a greater capacity of U(VI) to U(IV) reduction (>80%) in synthetic magnetite of $\geqslant 0.38$ (Fe^{2+}/Fe^{3+}) compared to more oxidized (<0.38) magnetite and maghemite (<20%) was observed. The ability to recharge the reductive capacity of the magnetite surface, by aqueous Fe(II) addition, thus increasing the removal of U(VI), was also noted. The chemical composition of the reaction solution, particularly the presence of common groundwater components Ca and HCO_3^-, have also been found to have a major control upon U(VI) remediation by magnetite (Singer *et al.*, 2012). The results for a continuous batch flow experiment with *in situ* U L_3-edge grazing incidence XANES (GI-XANES) indicated minimal U(VI) reduction when CO_3 and Ca are both present, with a proposed U(VI)-CO_3-Ca sorption complex inhibiting reduction. When absent or solely in the presence of CO_3, reduction was observed.

The reduction of U(VI) has also been assessed using the magnetic biomineral magnetite alongside biogenic vivianite (Law *et al.*, 2011; Veeramani *et al.*, 2011). The former study used the Fe(III)-reducing bacterium *G. sulfurreducens* to reduce ferrihydrite or ferric citrate for the synthesis of magnetite and vivianite, respectively. Interestingly, U(VI) reduction was dependent upon the stage of U(VI) addition, whereby minimal reduction ($\leqslant 10\%$) was observed when the U(VI) was added after biomineral formation. However, when the U(VI) was added to the precursor Fe(III) phase prior to inoculation with *G. sulfurreducens*, significant reduction to a U(IV) phase, recalcitrant re-oxidation to nitrate, was observed during co-precipitation. This study therefore questioned the role of Fe(II)-mediated reduction and instead appeared to indicate that, under the conditions tested, enzymatic reduction by *G. sulfurreducens* was the primary U(VI) reduction mechanism. A study also assessing the ability of the biominerals magnetite and vivianite to reduce U(VI) obtained contrasting results (Veeramani *et al.*, 2011). The minerals were synthesized using resting cells of *Shewanella putrefaciens* CN32 to reduce hydrous ferric oxide or ferric citrate to produce magnetite and vivianite, respectively. Rapid aqueous U(VI) removal (with rates increasing at higher mineral loadings), and near complete reduction of U(VI) to U(IV), were observed by XAS, at mineral:U(VI) ratios of 50:1, 20:1 and 10:1. EXAFS data suggested that magnetite favoured UO_2 formation, but the presence of phosphate was found to promote formation of a monomeric U(IV) phase.

5.1.5. Technetium(VII)

Technetium (Tc) isotopes are by-products of nuclear fission and due to the long half-life ($>2 \times 10^5$ years) of the β-emitting ^{99}Tc, it forms a major component of medium- to high-level wastes (Schulte and Scoppa, 1987; Icenhower *et al.*, 2010). When released into the environment under oxic conditions it is stable as the Tc(VII) valence state, primarily forming the highly soluble pertechnetate anion (TcO_4^-) (Rard, 1983). The Tc(VII) is however susceptible to reduction to the relatively insoluble Tc(IV), typically forming TcO_2 (Icenhower *et al.*, 2010).

The reduction of Tc(VII) in environmental samples such as soils and sediments is thought to be mediated primarily by biogenic Fe(II), with a competing enzymatic reduction mechanism shown to be less efficient (Lloyd *et al.*, 2000). An early study identified magnetite as an effective reductant of Tc(VII) when studied in comparison to Fe(II)-bearing geological material, in the form of an heterogenous mixture of chlorite, hornblende and magnetite, and dependent upon the ionic strength and pH of the media (Cui and Eriksen, 1996). A subsequent study assessed the Tc(VII) reduction kinetics mediated by synthetic nanoparticulate titanomagnetites with varying stoichiometry (Liu *et al.*, 2012b). The reduction kinetics were found to be fast (within minutes) and dependent upon the Fe(II)/Fe(III) ratio, as controlled by Ti content, with faster reduction rates associated with higher Fe(II) loadings. A model of $Tc(IV)O_2$ chains attached to surface Fe octahedra was proposed by fitting of EXAFS data. Analysis of XMCD data showed a decrease in surface Fe(II) coupled to Tc(VII) reduction. The authors of this study proposed a reaction mechanism of sustaining surface reactivity from a resupply of surface Fe(II) by outward migration of Fe(II) from the particle core, balanced by inward migration of charge-balancing cationic vacancies.

Biogenic magnetites have also been studied with respect to Tc(VII) removal and reduction (Cutting *et al.*, 2010; McBeth *et al.*, 2011). The former study assessed the efficiency of bio-magnetites for the removal of the radiotracer $^{99m}Tc(VII)$ (a metastable form of technetium used in medical imaging) in column experiments using a γ-camera. The nano-scale biogenic magnetite was synthesized by reduction of a ferrihydrite or schwertmannite starting phase by *G. sulfurreducens*. The schwertmannite-derived bio-magnetite, with greater surface Fe(II) abundances, exhibited greater retention of influent Tc(VII) than the ferrihydrite-derived biomagnetite, retaining 97.8% and 77.9%, respectively. A following study compared Tc(VII) removal by biomagnetites to that of microbially-synthesized vivianite and siderite, compared to an amorphous synthetic nanoparticulate Fe(II)-bearing phase (McBeth *et al.*, 2011). The addition of biomagnetite (7 mM Fe(II)) and the synthetic amorphous Fe(II) phase (20 mM Fe(II)) were most efficient, removing completely 1–5 μM TcO_4^-, while biosiderite (16 mM Fe(II)) and biovivianite (2 mM Fe(II)) removed 84% and 68%, respectively. The resulting phase, determined by extended EXAFS analysis, was air-stable $Tc(IV)O_2$.

5.1.6. Neptunium(V)

Due to its long half-life, 2.14×10^6 y, and its production from the decay of the short-lived ^{241}Am isotope, ^{237}Np is a long-term concern for intermediate- and high-level radioactive waste repositories (Kaszuba and Runde, 1999). It is also of great concern due to its chemical and radiological toxicity as it is a major α-emitting isotope (Taylor, 1989; Ruggiero *et al.*, 2005). Neptunium is stable typically in the Np(IV) and Np(V) valence states, where the highly soluble Np(V) dominates in oxic environments while the less soluble Np(IV) dominates under reducing conditions (Kaszuba and Runde, 1999). The reductive stabilization by biogenic Fe(II) in complex environmental samples, such as sediments, has recently been implicated as a major sink of mobile Np(V) (Law *et al.*, 2010).

A series of studies sought to assess the ability of synthetic magnetite to remove and reduce Np(V) under oxic and anoxic conditions (Nakata *et al.*, 2002, 2004). The earlier paper observed, after 7 days, greater sorption of Np(V) on to the magnetite under anoxic conditions, at a pH of 4–8, with minimal desorption in a KCl solution. For the reacted magnetite (pH 4.9 and 6.3) reduced Np(IV) was found in quantities >85% under anoxic conditions and ~13% under oxic conditions; determined using extraction with 0.5 M thenoyltrifluoroacetone (TTA) in xylene. A later study further probed the kinetics of Np(V) sorption and reduction using magnetite (Nakata *et al.*, 2004). As with other inorganic interactions, the study observed an initial fast adsorption step, followed by a second, slower, reduction step.

5.1.7. Plutonium(V) and plutonium(VI)

Environmental release of plutonium, through its use in nuclear weapons or from the nuclear fuel cycle, is of prime concern due to the long half-life of the predominant isotope ^{239}Pu; 2.4×10^4 y (Kimber *et al.*, 2012). It can be environmentally stable as Pu(III), Pu(IV), Pu(V) and Pu(VI), and due to similar values for their redox couples it can often occur in multiple valences within the same environment (Choppin, 1991). The solubilities of the valence states of Pu are strongly dependent upon extrinsic properties such as the presence of complexing ligands and humics (Schmeide *et al.*, 2006; Icopini *et al.*, 2009). Generally, the oxidized species Pu(V) and Pu(VI) are considered more soluble than the more reduced Pu(III) and Pu(IV) (Choppin and Morgenstern, 2001).

A recent study sought to assess the impact of synthetic magnetite upon the speciation of ^{239}Pu(V) in a 0.01 M NaCl aerobic solution with varying pH (3, 5 and 8) and magnetite additions (10 and 100 m^2/L) (Powell *et al.*, 2004). Adsorption increased with increasing pH, inferred as a result of increasing magnetite surface negativity. This was followed by rapid reduction to Pu(IV) in the solid phase. A further study sought to observe the effect of magnetite upon anoxic aqueous ^{242}Pu(III) and ^{242}Pu(V) (Kirsch *et al.*, 2011). Aqueous Pu (13 μM) was removed rapidly from solution and XANES analysis confirmed reduction to primarily Pu(III) (>90%). The reduced Pu(III) phase was shown, by EXAFS analysis, to be consistent with a stable tridentate sorption complex to the Fe-octahedral position.

5.1.8. Organic compounds

A variety of toxic organic compounds are also amenable to reduction by Fe(II), with halogenated solvents (Lee and Batchelor, 2002) and nitroaromatic compounds (Hofstetter *et al.*, 1999) receiving much attention. The chlorinated solvents are of particular concern due to their common occurrence at hazardous levels in groundwater resources (Moran *et al.*, 2006). This is due to their widespread use in commercial and industrial processes (Anderson *et al.*, 1992). Once in the subsurface environment, these dense, non-aqueous phase liquids can accumulate on top of low-permeability layers and, due to their relatively high solubility, pollute large areas of groundwater (Johnson and Pankow, 1992). Evidence of their toxic nature has been observed for a long time, primarily associated with damage to the central nervous system, liver and kidney toxicity and carcinogenicity (Ruder, 2006). The reductive remediation of chlorinated solvents is also complicated by the toxicity of intermediate daughter compounds (He *et al.*, 2003).

Nitroaromatic compounds are also commonly found as contaminants through their release associated with use as pesticides, explosives and as an industrial feedstock (Spain *et al.*, 2000). The initial nitroaromatic parent compounds and a proportion of their subsequent metabolite daughter compounds are often toxic, behaving as mutagens or carcinogens (Rodgers and Bunce, 2001). The toxic daughter compounds, aromatic amines, are typically the product of the initial reduction of the nitro group and for successful remediation further reduction is ideal (Agrawal and Tratnyek, 1995).

Remediation of organic azo dyes is also of concern as up to 50% of a dye can be lost to wastewater compounded by the high water solubility and characteristic brightness of the dyes. In the work of Pattrick *et al.* (2013), the reduction of the azo dye Remazol Black B was achieved using biogenic magnetite. A colourless solution was discernible after 31 days and spectroscopic analyses determined that the biogenic magnetite had reduced the azo chromophore to colourless amines. Fe(II) in the magnetite was partly consumed and an abiotic magnetite with a lower Fe(II) surface concentration was unable to decolourize the dye compound completely.

Earlier studies using synthetic magnetite did not observe direct magnetite-mediated reduction of nitroaromatics (Klausen *et al.*, 1995; Gregory *et al.*, 2004) and chlorinated solvents (Elsner *et al.*, 2004). These studies instead focused on the ability of the magnetite surface to catalyse the reduction of these organic contaminants by adsorbed Fe(II). Incomplete reduction of the chlorinated compounds 4-chloronitrobenzene (Klausen *et al.*, 1995), 4-chlorobenzene and hexachloroethane (Elsner *et al.*, 2004) resulted in the accumulation of partially chlorinated intermediates. While the nitroaromatic (Hexahydro-1,3,5-trinitro-1,3,5-triazine) tested by (Gregory *et al.*, 2004) was reduced to sequential nitroso-reduction metabolites and stable products. Commonly these studies noted an increase in reaction rate with increasing pH, inferred to correspond to further Fe(II) adsorption to the magnetite surface (Klausen *et al.*, 1995; Gregory *et al.*, 2004).

As with inorganic contaminants, the reactivity of magnetite towards organic contaminants has been studied with respect to Fe(II) stoichiometry (Gorski and Scherer, 2009; Gorski *et al.*, 2010). These studies initially observed that aqueous Fe(II), in the presence of magnetite, was oxidized and formed a partially reduced magnetite phase, reducing the octahedral Fe(III) to Fe(II) in the bulk magnetite (as opposed to forming an adsorbed Fe(II) species) (Gorski and Scherer, 2009). This method of controlling the stoichiometry was used to adjust the Fe(II) content of magnetite reacted with nitrobenzene, forming aniline. It was found that nearly-stoichiometric magnetite, with an Fe(II)/Fe(III) ratio of 0.48, could rapidly reduce nitrobenzene to aniline while non-stoichiometric magnetite (with a ratio of 0.31), did not exhibit appreciable reduction on the same timescale (60 min). This work was followed by a detailed study of magnetite reduction of three monosubstituted nitrobenzene compounds (Gorski *et al.*, 2010). The reduction rates of the three nitrobenzene compounds were found to be related linearly to magnetite stoichiometry, with reactivity dependent upon Fe(II) content.

Several subsequent studies have found magnetite to be active towards the dechlorination of carbon tetrachloride (CCl_4) (Danielsen and Hayes, 2004; Vikesland *et al.*, 2007). pH was found to control reactivity, increasing with increasing pH, and also the distribution of the products CO and $CHCl_3$, with greater accumulation of the chlorinated $CHCl_3$ at higher pH (Danielsen and Hayes, 2004). Both these pH dependencies are probably due to the increase in deprotonated surfaces at higher pH values, increasing electron density and affecting the ability to stabilize the surface $-CCl_3$ transitional phase. The particle size of the magnetite also exerted a control, with higher reactivity noted when smaller (9 nm) particles were employed compared to larger (80 nm) particles (Vikesland *et al.*, 2007). Interestingly, this study also found the degree of particle aggregation, increasing with higher ionic strength of the solution, controlled reaction rates, with lower rates observed with higher aggregation, due to an effective loss in reactive surface area.

Biogenic magnetite has also been shown to be reactive towards CCl_4 (McCormick *et al.*, 2002; McCormick and Adriaens, 2004). The latter study assessed the relative contributions of enzymatic- and biogenic magnetite-mediated reduction of CCl_4 during its incubation with a culture of *G. metallireducens* and the reducible phase hydrous ferric oxide. The rate of reduction by the magnetite was found to be far greater, 60–260 times faster, than enzymatic CCl_4 reduction and there was accumulation of the sole breakdown product, chloroform. A later study sought to investigate further the mechanism of the reaction between a biogenic magnetite with CCl_4 (McCormick and Adriaens, 2004). CCl_4 was found to degrade along three pathways; hydrogenolysis, carbine hydrolysis and carbine reduction. The first two pathways involve a variety of chlorinated intermediates while the latter results in the benign products CO and CH_4.

A further study using a biogenic magnetite synthesized by *G. metallireducens* was shown to be active towards a toxic groundwater explosives contaminant, hexahydro-1,3,5-trinitro-1,3,5-triazine (RDX) (Williams *et al.*, 2005). Despite a loss in total carbon mass balance, significant removal of the RDX was observed, along with the identification of three nitroso products. Interestingly, a decrease in reactivity was observed for magnetites that had undergone longer incubation (75 and 400 days), thought to be the result of cell lysis products affecting reactivity.

5.2. Biomedical applications

Magnetic nanoparticles have also been investigated with regard to their utilization in a variety of medical applications. These range from uses in targeted drug- and gene-delivery systems, magnetic hyperthermia cancer therapy, magnetic resonance imaging for cell labelling and cell separation (Pankhurst *et al.*, 2009). For these biomedical applications the magnetic properties, the size distribution and low cytotoxicity of the nanoparticles are of great importance (Gupta and Gupta, 2005; Shubayev *et al.*, 2009). As well as a variety of synthetic magnetite nanoparticles, several studies have addressed the use of biogenic nanoparticles for biomedical applications (Safarik and Safarikova, 2002).

The study by Albrecht *et al.* (2005) presented detailed characterization using atomic force microscopy and magnetic force microscopy of magnetosome magnetites produced by *Magnetospirillum gryphiswaldense*. These magnetites were found to be consistent in size and dimensions and were stable for extended periods, important features for biomedical applications. Another study also utilized biogenic magnetosome-derived magnetite from *M. gryphiswaldense* compared to a lipid-coated synthetic alternative for tracking of stem and dendritic cells (Schwarz *et al.*, 2009). The magnetic particles were used to label hematopoietic $Flt3^{+}$ stem cells and dendritic cells (DC) from mouse bone marrow, providing contrast for cell-migration assessment using magnetic resonance imaging. The magnetosome magnetite exhibited better cell retention, compared to synthetic particles, in the stem cells and similar retention was observed in DC cells. This study therefore proposed magnetosome magnetite as an alternative to synthetic magnetite for biomedical applications.

The majority of studies have focused on magnetite produced by magnetotactic bacteria; however, due to low yields, Fe(III)-reducing bacteria have been suggested as a higher-yielding alternative (Coker *et al.*, 2009). The bacterium *G. sulfurreducens* was used to synthesize magnetite from a ferrihydrite starting phase. Upon analysis, the particles were found to be of a narrow size-distribution and with magnetic properties equal to synthetic alternatives. A further development was the inclusion of Co into the spinel structure, confirmed by XMCD analysis, significantly enhancing the magnetic properties of the particles. Byrne *et al.* (2013) created a suite of Co-substituted biogenic magnetite for hyperthermia cancer treatment. In hyperthermia treatment a heating effect is induced in the magnetic nanoparticles (which are localized to the cancerous growth) by using an oscillating magnetic field, and the high temperatures created (~45°C) destroy the cancer. The heating effect is dependent on the size and the magnetic anisotropy of the particles. The heating effect, or specific loss power, of the nanoparticles was determined from the rate of change of temperature of a suspension containing a known quantity of particles and 20% Co substitution with a particle size of 10 nm was found to be the most effective (Byrne *et al.*, 2013). The ability to control the magnetic properties of the biogenic particles within a simple biosynthesis method represents a significant advance for biomedical application.

6. Conclusions

Magnetic minerals produced extracellularly by prokaryotes are produced under a wide range of environmental and laboratory conditions; the identity of the biomineral formed is dependent on the prevailing geochemical conditions and the bacterial metabolism producing the mineral. Nano-sized magnetite particles are produced by a wide range of Bacteria and Archaea and have potential uses in environmental remediation strategies. For the potential of these production methods to be attained a method for scaling up the processes to kilogram quantities will need to be realized, using large-scale reactors such as in Moon *et al.* (2010). In addition, there is scope for the use of waste Fe(III)-minerals as precursor phases to the high-value nanoparticles. These can be found at acid mine-drainage sites or from use in the water polishing steps within the water industry, and would add further green credentials to an already environmentally benign process. Magnetic biominerals are an important area of study as they improve knowledge of the biogeochemistry of early Earth and also on the environmental cycling of metals in the subsurface in contemporary Earth systems.

Acknowledgements

The authors acknowledge financial support from the European Union Seventh Framework Programme (FP7/2007– 2013) under Grant Agreement No. 309517 (NANOREM).

References

Aamrani, S.E., Gimenez, J., Rovira, M., Seco, F., Grive, M., Bruno, J., Duro, L. and de Pablo, J. (2007) A spectroscopic study of uranium(VI) interaction with magnetite. *Applied Surface Science*, **253**, 8794-8797.

Agrawal, A. and Tratnyek, P.G. (1995) Reduction of nitro aromatic compounds by zero-valent iron metal. *Environmental Science & Technology*, **30**, 153–160.

Albrecht, M., Janke, V., Sievers, S., Siegner, U., Schuler, D. and Heyen, U. (2005) Scanning force microscopy study of biogenic nanoparticles for medical applications. *Journal of Magnetism and Magnetic Materials*, **290**, 269–271.

Anderson, M.R., Johnson, R.L. and Pankow, J.F. (1992) Dissolution of dense chlorinated solvents into ground water: 1. dissolution from a well-defined residual source. *Ground Water*, **30**, 250–256.

Barkey, T. and Wagner-Dobler, I. (2005) Microbial transformations of mercury: potentials, challenges, and achievements in controlling mercury toxicity in the environment. *Advances in Applied Microbiology*, **57**, 1–52.

Barkey, T., Miller, S.M. and Summers, A.O. (2003) Bacterial mercury resistance from atoms to ecosystems. *FEMS Microbiology Reviews*, **27**, 355–384.

Bazylinski, D.A. and Frankel, R.B. (2000) Biologically-controlled mineralization of magnetic iron minerals by magnetotactic bacteria. *Environmental Microbe–Metal Interactions*. 109–144.

Bazylinski, D.A., Frankel, R.B. and Konhauser, K. (2007) Modes of biomineralization of magnetite by microbes, *Geomicrobiology Journal* **24**, 465–475.

Benner, S., Hansel, C.M., Wielinga, B.W., Barber, T.M. and Fendorf, S. (2002) Reductive dissolution and biomineralization of iron hydroxide under dynamic flow conditions. *Environmental Science & Technology*. **36**, 1705–1711.

Benning, L.G., Wilkin, R.T. and Barnes, H. (2000) Reaction pathways in the Fe–S system below 100ºC. *Chemical Geology*, **167**, 25–51.

Benz, M., Brune, A. and Schink, B. (1998) Anaerobic and aerobic oxidation of ferrous iron at neutral pH by

chemoheterotrophic nitrate-reducing bacteria. *Archives of Microbiology*, **169**, 159–165.
Blakemore, R.P. (1975) Magnetotactic bacteria. *Science*, **190**, 377–379.
Blakemore, R.P. (1982) Magnetotactic bacteria. *Annual Reviews in Microbiology*, **36**, 217–238.
Bose, A., Gardel, E.J., Vidoudez, C., Parra, E.A. and Girguis, P.R. (2014) Electron uptake by iron-oxidizing phototrophic bacteria. *Nature Communications*, **5**, 3391.
Byrne, J.M., Telling, N.D., Coker, V.S., Pattrick, R.A.D., van der Laan, G., Arenholz, E., Tuna, F. and Lloyd, J.R. (2011) Control of nanoparticle size, reactivity and magnetic properties during the bioproduction of magnetite by *Geobacter sulfurreducens*. *Nanotechnology*, **22**, 455709.
Byrne, J.M., Coker, V.S., Moise, S., Wincott, P.L., Vaughan, D.J., Tuna, F., Arenholz, E., van der Laan, G., Pattrick, R.A.D., Lloyd, J.R. and Telling, N.D. (2013) Controlled cobalt doping in biogenic magnetite nanoparticles. *Journal of the Royal Society Interface*, **10**, 20130134.
Byrne, J.M., Klueglein, N., Pearce, C., Rosso, K.M., Appel, E. and Kappler, A. (2015) Redox cycling of Fe(II) and Fe(III) in magnetite by Fe-metabolizing bacteria. *Science*, **347**, 1473–1476.
Caccavo Jr., F., Blakemore, R.P. and Lovley, D.R. (1992) A hydrogen-oxidizing, Fe(III)-reducing microorganism from the Great Bay Estuary, New Hampshire. *Applied and Environmental Microbiology*, **58**, 3211–3216.
Caccavo Jr, F., Lonergan, D.J., Lovley, D.R., Davis, M., Stolz, J.F. and McInerney, M.J. (1994) *Geobacter sulfurreducens* sp. nov., a hydrogen and acetate-oxidizing dissimilatory metal reducing bacterium. *Applied and Environmental Microbiology*, **60**, 3752–3759.
Caccavo Jr., F., Coates, J.D., Rosello-Mora, R.A., Ludwig, W., Schleifer, K.H., Lovley, D.R. and McInerney, M.J. (1996) *Geovibrio ferrireducens*, a phylogenetically distinct dissimilatory Fe(III)-reducing bacterium. *Archives of Microbiology*, **165**, 370–376.
Charlet, L. and Polya, D.A. (2006) Arsenic in shallow, reducing groundwaters in Southern Asia: An environmental health disaster. *Elements*, **2**, 91–96.
Chang, L., Roberts, A.P., Tang, Y., Rainford, B.D., Muxworthy, A.R. and Chen, Q. (2008) Fundamental magnetic parameters from pure synthetic greigite (Fe_3S_4). *Journal of Geophysical Research*, **113**, B06104.
Chaudhuri, S.K., Lack, J.G. and Coates, J.D. (2001) Biogenic magnetite formation through anaerobic biooxidation of Fe(II). *Applied and Environmental Microbiology*, **67**, 2844–2848.
Childers, S.E., Ciufo, S. and Lovley, D.R. (2002) *Geobacter metallireducens* accesses insoluble Fe(III) oxide by chemotaxis. *Nature*, **416**, 767–769.
Choppin, G. (1991) Redox speciation of plutonium in natural waters. *Journal of Radioanalytical and Nuclear Chemistry*, **147**, 109–116.
Choppin, G.R. and Morgenstern, A. (2001) Distribution and movement of environmental plutonium. *Radioactivity in the Environment*, **1**, 91–105.
Cismasu, A.C., Michel, M.F., Tcaciuc, A.P. and Brown Jr., G.E. (2014) Properties of impurity-bearing ferrihydrite III. Effects of Si on the structure of 2-line ferrihydrite. *Geochimica et Cosmochimica Acta*, **133**, 168–185.
Clark, D.L., Hobart, D.E. and Neu, M.P. (1995) Actinide carbonate complexes and their importance in actinide environmental chemistry. *Chemical Reviews*, **95**, 25–48.
Coates, J.D., Lonergan, D.J., Phillips, E.J.P., Jenter, H. and Lovley, D.R. (1995) *Desulfuromonas palmitatis* sp. nov., a marine dissimilatory Fe(III) reducer that can oxidise long-chain fatty acids. *Archives of Microbiology*, **164**, 406–413.
Coates, J.D., Ellis, D.J., Gaw, C.V. and Lovley, D.R. (1999) *Geothrix fermentans* gen. nov., sp. nov., a novel Fe(III)-reducing bacterium from a hydrocarbon-contaminated aquifer. *International Journal of Systematic Bacteriology*, **49**, 1615–1622.
Coker, V.S., Gault, A.G., Pearce, C.I., van der Laan, G., Telling, N.D., Charnock, J.M., Polya, D.A. and Lloyd, J.R. (2006) XAS and XMCD evidence for species-dependent partitioning of arsenic during microbial reduction of ferrihydrite to magnetite. *Environmental Science & Technology*, **40**, 7745–7750.
Coker, V.S., Pearce, C.I., Pattrick, R.A.D., van der Laan, G., Telling, N.D., Charnock, J.M., Arenholz, E. and Lloyd, J.R. (2008) Probing the site occupancies of Co, Ni and Mn substituted biogenic magnetite using XAS and XMCD. *American Mineralogist*, **93**, 1119–1132.
Coker, V.S., Telling, N.D., van der Laan, G., Pattrick, R.A.D., Pearce, C.I., Arenholz, E., Tuna, F., Winpenny,

R.E.P. and Lloyd, J.R. (2009) Harnessing the extracellular bacterial production of nanoscale cobalt ferrite with exploitable magnetic properties. *ACS Nano*, **3**, 1922–1928.

Costa, M. (1997) Toxicity and carcinogenicity of Cr(VI) in animal models and humans. *Critical Reviews in Toxicology*, **27**, 431–442.

Crean, D.E., Coker, V.S., van der Laan, G. and Lloyd, J.R. (2012) Engineering biogenic magnetite for sustained Cr(VI) remediation in flow-through systems. *Environmental Science & Technology*, **46**, 3352–3359.

Cui, D. and Eriksen, T.E. (1996) Reduction of pertechnetate in solution by heterogenous electron trnasfer from Fe(II)-containing geological material. *Environmental Science & Technology*, **30**, 2263–2269.

Cutting, R.S., Coker, V.S., Fellowes, J.W., Lloyd, J.R. and Vaughan, D.J. (2009) Mineralogical and morphological constraints on the reduction of Fe(III) minerals by *Geobacter sulfurreducens*. *Geochimica et Cosmochimica Acta*, **73**, 4004–4022.

Cutting, R., Coker, V.S., Telling, N.D., Kimber, R.L., Pearce, C.I., Ellis, B.L., Lawson, R.S., van der Laan, G., Pattrick, R.A.D., Vaughan, D.J., Arenholz, E. and Lloyd, J.R. (2010) Optimizing Cr(VI) and Tc(VII) remediation through nanoscale biomineral engineering. *Environmental Science & Technology*, **44**, 2577–2584.

Cutting, R.S., Coker, V.S., Telling, N.D., Kimber, R.L., van der Laan, G., Pattrick, R.A.D., Vaughan, D.J., Arenholz, E. and Lloyd, J.R. (2012) Microbial reduction of arsenic-doped schwertmannite by *Geobacter sulfurreducens*. *Environmental Science & Technology*, **46**, 12591–12599.

Danielsen, K.M. and Hayes, K.F. (2004) pH dependence of carbon tetrachloride reductive dechlorination by magnetite. *Environmental Science & Technology*, **38**, 4745–4752.

Dixit, S. and Hering, J.G. (2003) Comparison of arsenic(V) and arsenic(III) sorption onto iron oxide minerals: implications for arsenic mobility. *Environmental Science & Technology*, **37**, 4182–4189.

Dong, H., Fredrickson, J.K., Kennedy, D.W., Zachara, J.M., Kukkadapu, R.K. and Onstott, T.C. (2000) Mineral transformations associated with the microbial reduction of magnetite. *Chemical Geology*, **169**, 299–318.

el Aamrani, F., Casas, I., de Pablo, J., Duro, L., Grive, M. and Bruno, J. (1999) *Experimental and modelling study of the interaction between uranium(VI) and magnetite*. Svensk kärnbränslehantering AB/Swedish Nuclear Fuel and Waste Management.

Elsner, M., Schwarzenbach, R.P. and Haderlein, S.B. (2004) Reactivity of Fe(II)-bearing minerals toward reductive transformation of organic contaminants. *Environmental Science & Technology*, **38**, 799–807.

Fassbinder, J.W. and Stanjek, H. (1994) Magnetic properties of biogenic soil greigite (Fe_3S_4). *Geophysical Research Letters*, **21**, 2349–2352.

Fendorf, S.E. (1995) Surface reactions of chromium in soils and waters. *Geoderma*, **67**, 55–71.

Frankel, R.B. and Bazylinski, D.A. (2003) Biologically induced mineralization by bacteria. Pp. 95–114 in: *Biomineralization* (P.M. Dove, J.J. de Yoreo and S. Weiner, editors). Reviews in Mineralogy and Geochemistry, **54**, The Mineralogical Society of America, Washington DC.

Frankel, R.B., Blakemore, R.P. and Mackay, A.L. (1984) Precipitation of Fe_3O_4 in magnetotactic bacteria [and Discussion]. *Philosophical Transactions of the Royal Society of London. B, Biological Sciences*, **304**, 567–574.

Fredrickson, J.K., Zachara, J.M., Kennedy, D.W., Dong, H., Onstott, T.C., Hinman, N.W. and Li, S. (1998) Biogenic iron mineralization accompanying the dissimilatory reduction of hydrous ferric oxide by groundwater bacterium. *Geochimica et Cosmochimica Acta*, **62**, 3239–3257.

Fredrickson, J.K., Zachara, J.M., Kukkadapu, R.K., Gorby, Y.A., Smith, S.C. and Brown, C.F. (2001) Biotransformation of Ni-substituted hydrous ferric oxide by an Fe(III)-reducing bacterium. *Environmental Science & Technology*, **35**, 703–712.

Gad, S.C. (1989) Acute and chronic systemic chromium toxicity. *Science of the Total Environment*, **86**, 149–157.

Glasauer, S., Weidler, P.G., Langley, S. and Beveridge, T.J. (2003) Controls on Fe reduction and mineral formation by a subsurface bacterium. *Geochimica et Cosmochimica Acta*, **67**, 1277–1288.

Gorby, Y.A., Yanina, S., McLean, J.S., Rosso, K.M., Moyles, D., Dohnalkova, A., Beveridge, T.J., Chang, I.S., Kim, B.H., Kim, K.S., Culley, D.E., Reed, S.B., Romine, M.F., Saffarini, D.A., Hill, E.A., Shi, L., Elias, D.A., Kennedy, D.W., Pinchuk, G., Watanabe, K., Ishii, S., Logan, B., Nealson, K.H. and Fredrickson, J.K. (2006) Electrically conductive bacterial nanowires produced by Shewanella oneidensis strain MR-1 and other microorganisms. *Proceedings of the National Academy of Sciences*, **103**, 11358–11363.

Gorski, C.A. and Scherer, M.M. (2009) Influence of magnetite stoichiometry on Fe^{II} uptake and nitrobenzene reduction. *Environmental Science & Technology*, **43**, 3675–3680.

Gorski, C.A., Nurmi, J.T., Tratnyek, P.G., Hofstetter, T.B. and Scherer, M.M. (2010) Redox behavior of magnetite: Implications for contaminant reduction. *Environmental Science & Technology*, **44**, 55–60.

Greene, A.C., Patel, B.K.C. and Sheehy, A.J. (1997) *Deferribacter thermophilus* gen. nov. sp. nov., a novel thermophilic manganese- and iron-reducing bacterium isolated from a petroleum reservoir. *International Journal of Systematic Bacteriology*, **47**, 505–509.

Gregory, K.B., Larese-Casanova, P., Parkin, G.F. and Scherer, M.M. (2004) Abiotic transformation of hexahydro-1,3,5-trinitro-1,3,5-triazine by feII bound to magnetite. *Environmental Science & Technology*, **38**, 1408–1414.

Guo, D., Xie, G. and Luo, J. (2014) Mechanical properties of nanoparticles: basics and applications. *Journal of Physics D: Applied Physics* **47(1)**: 013001.

Gupta, A.K. and Gupta, M. (2005) Synthesis and surface engineering of iron oxide nanoparticles for biomedical applications. *Biomaterials*, **26**, 3995–4021.

Hansel, C.M., Benner, S.G., Neiss, J., Dohnalkova, A., Kukkadapu, R.K. and Fendorf, S. (2003) Secondary mineralization pathways induced by dissimilatory iron reduction of ferrihydrite under advective flow. *Geochimica et Cosmochimica Acta*, **67**, 2977–2992.

Hansel, C.M., Benner, S.G. and Fendorf, S. (2005) Competing Fe(II)-induced mineralization pathways of ferrihydrite. *Environmental Science & Technology*, **39**, 7147–7153.

Hartshorne, R.S., Jepson, B.N., Clarke, T.A., Field, S.J., Fredrickson, J., Zachara, J., Shi, L., Butt, J.N. and Richardson, D.J. (2007) Characterization of *Shewanella oneidensis* MtrC: a cell-surface decaheme cytochrome involved in respiratory electron transport to extracellular electron acceptors. *Journal of Biological and Inorganic Chemistry*, **12**, 1083–1094.

He, J.T., Ritalahti, K.M., Yang, K.-L., Koenigsberg, S.S. and Loffler, F.E. (2003) Detoxification of vinyl chloride to ethene coupled to growth of an anaerobic bacterium. *Nature*, **424**, 62–65.

He, Y.T. and Traina, S.J. (2005) Cr(VI) reduction and immobilization by magnetite under alkaline pH conditions: The role of passivation. *Environmental Science & Technology*, **39**, 4499–4504.

Hochella Jr., M.F. (2002) There's plenty of room at the bottom: nanoscience in geochemistry. *Geochimica et Cosmochimica Acta*, **66**, 735–743.

Hochella Jr., M.F., Lower, S.K., Maurice, P.A., Penn, R.L., Sahai, N., Sparks, D.L. and Twining, B.S. (2008) Nanominerals, mineral nanoparticles, and Earth systems. *Science*, **319**, 1631–1635.

Hofstetter, T.B., Heijman, C.G., Haderlein, S.B., Holliger, C. and Schwarzenbach, R.P. (1999) Complete reduction of TNT and other (poly)nitroaromatic compounds under iron-reducing conditions. *Environmental Science & Technology*, **33**, 1479–1487.

Icenhower, J.P., Qafoku, N.P., Zachara, J.M. and Martin, W.J. (2010) The biogeochemistry of technetium: A review of the behavior of an artificial element in the natural environment. *American Journal of Science*, **310**, 721–752.

Icopini, G., Lack, J.G., Hersman, L.E., Neu, M.P. and Boukhalfa, H. (2009) Plutonium(V/VI) reduction by the metal-reducing bacteria *Geobacter metallireducens* GS-15 and *Shewanella oneidensis* MR-1. *Applied and Environmental Microbiology*, **75**, 3641–3647.

Ilton, E.S., Boily, J.-F., Buck, E.C., Skomurski, F.N., Rosso, K.M., Cahill, C.L., Bargar, J.R. and Felmy, A.R. (2009) Influence of dynamical conditions on the reduction of UVI at the magnetite–solution interface. *Environmental Science & Technology*, **44**, 170–176.

Islam, F.S., Gault, A.G., Boothman, C., Polya, D.A., Chatterjee, D. and Lloyd, J.R. (2004) Role of metal-reducing bacteria in arsenic release from Bengal delta sediments. *Nature*, **430**, 68–71.

Islam, F.S., Pederick, R.L., Gault, A.G., Adams, L.K., Polya, D.A., Charnock, J.M. and Lloyd, J.R. (2005) Interactions between the Fe(III)-reducing bacterium *Geobacter sulfurreducens* and arsenate, and capture of the metalloid by biogenic Fe(II). *Applied and Environmental Microbiology*, **71**, 8642–8648.

Jambor, J.L. and Dutrizac, J.E. (1998) Occurrence and consititution of natural and synthetic ferrihydrite, a widespread iron oxyhydroxide. *Chemical Reviews*, **98**, 2549–2586.

Jiao, Y., Kappler, A., Croal, L.R. and Newman, D.K. (2005) Isolation and characterization of a genetically tractable photoautotrophic Fe (II)-oxidizing bacterium, *Rhodopseudomonas palustris* strain TIE-1. *Applied and Environmental Microbiology*, **71**, 4487–4496.

Johnson, R.L. and Pankow, J.F. (1992) Dissolution of dense chlorinated solvents into groundwater. 2. Source functions for the pools of solvent. *Environmental Science & Technology*, **26**, 896–901.

Kappler, A. and Straub, K.L. (2005) Geomicrobiological cycling of iron. Pp. 85–108 in: *Molecular Geomicrobiology* (J.F. Banfield, J. Cervini-Silva and K.N. Nealson, editors). Reviews in Mineralogy, **59**, Mineralogical Society of America and Geochemical Society, Washington DC.

Kasama, T., Pósfai, M., Chong, R.K., Finlayson, A.P., Buseck, P.R., Frankel, R.B. and Dunin-Borkowski, R.E. (2006) Magnetic properties, microstructure, composition, and morphology of greigite nanocrystals in magnetotactic bacteria from electron holography and tomography. *American Mineralogist*, **91**, 1216–1229.

Kashefi, K. and Lovley, D.R. (2003) Extending the upper temperature limit for life. *Science*, **301**, 934.

Kashefi, K., Holmes, D.E., Reysenbach, A.-L. and Lovley, D.R. (2002) Use of Fe(III) as an electron acceptor to recover previously uncultured hyperthermophiles: isolation and characterisation of *Geothermobacterium ferrireducens* gen. nov. sp. nov. *Applied and Environmental Microbiology*, **68**, 1735–1742.

Kashefi, K., Moskowitz, B.M. and Lovley, D.R. (2008) Characterization of extracellular minerals produced during dissimilatory Fe(III) and U(VI) reduction at 100°C by *Pyrobaculum islandicum*. *Geobiology*, **6**, 147–154.

Kaszuba, J.P. and Runde, W.H. (1999) The aqueous geochemistry of neptunium: dynamic control of soluble concentrations with applications to nuclear waste disposal. *Environmental Science & Technology*, **33**, 4427–4433.

Kendelewicz, T., Liu, P., Doyle, C.S., Brown, G.E., Nelson, E.J. and Chambers, S.A. (1999) X-ray absorption and photoemission study of the adsorption of aqueous Cr(VI) on single crystal hematite and magnetite surfaces. *Surface Science*, **424**, 219–231.

Kendelewicz, T., Liu, P., Doyle, C.S. and Brown Jr, G.E. (2000) Spectroscopic study of the reaction of aqueous Cr(VI) with Fe_3O_4(111) surfaces. *Surface Science*, **469**, 144–163.

Kimber, R., Livens, F.R. and Lloyd, J.R. (2011) Management of land contaminated by the nuclear legacy. Pp. 82–115 in: *Nuclear Power and the Environment* (R.M. Harrison and R.E. Hester, editors). The Royal Society of Chemistry, London.

Kimber, R.L., Boothman, C., Purdie, P., Livens, F.R. and Lloyd, J.R. (2012) Biogeochemical behaviour of plutonium during anoxic biostimulation of contaminated sediments. *Mineralogical Magazine*, **76**, 567–578.

Kirsch, R., Fellhauer, D., Altmaier, M., Neck, V., Rossberg, A., Fanghanel, T., Charlet, L. and Scheinost, A.C. (2011) Oxidation state and local structure of plutonium reacted with magnetite, mackinawite, and chukanovite. *Environmental Science & Technology*, **45**, 7267–7274.

Klausen, J., Troeber, S.P., Haderlein, S.B. and Schwarzenbach, R.P. (1995) Reduction of substituted nitrobenzenes by Fe(II) in aqueous mineral suspensions. *Environmental Science & Technology*, **29**, 2396–2404.

Klueglein, N. and Kappler, A. (2013) Abiotic oxidation of Fe (II) by reactive nitrogen species in cultures of the nitrate-reducing Fe (II) oxidizer *Acidovorax* sp. BoFeN1dation of Fe (II) bistence of enzymatic Fe (II) oxidation. *Geobiology*, **11**, 180–190.

Kostka, J.E. and Nealson, K.H. (1995) Dissolution and reduction of magnetite by bacteria. *Environmental Science & Technology*, **29**, 2535–2540.

Küsel, K., Dorsch, T., Acker, G. and Stackebrandt, E. (1999) Microbial reduction of Fe(III) in acidic sediments: isolation of *Acidiphilium cryptum* JF-5 capable of coupling the reduction of Fe(III) to the oxidation of glucose. *Applied and Environmental Microbiology*, **65**, 3633–3640.

Latta, D.E., Gorski, C.A., Boyanov, M.I., O'Loughlin, E.J., Kemner, K.M. and Scherer, M.M. (2012) Influence of magnetite stoichiometry on UVI reduction. *Environmental Science & Technology*, **46**, 778–786.

Law, G.T.W., Geissler, A., Lloyd, J.R., Livens, F.R., Boothman, C., Begg, J.D.C., Denecke, M.A., Rothe, J.R., Dardenne, K., Burke, I.T., Charnock, J.M. and Morris, K. (2010) Geomicrobiological redox cycling of the transuranic element neptunium. *Environmental Science & Technology*, **44**, 8924–8929.

Law, G.T.W., Geissler, A., Burke, I.T., Livens, F.R., Lloyd, J.R., McBeth, J.M. and Morris, K. (2011) Uranium redox cycling in sediments and biomineral systems. *Geomicrobiology Journal*, **28**, 497–506.

Lee, W. and Batchelor, B. (2002) Abiotic reductive dechlorination of chlorinated ethylenes by iron-bearing soil minerals. 1. Pyrite and magnetite. *Environmental Science & Technology*, **36**, 5147–5154.

Leang, C., Coppi, M.V. and Lovley, D.R. (2003) OmcB, a *c*-type polyheme cytochrome, involved in Fe(III) reduction in *Geobacter sulfurreducens*. *Journal of Bacteriology*, **185**, 2096–2103.

Leang, C., Qian, X., Mester, T. and Lovley, D.R. (2010) Alignment of the *c*-type cytochrome OmcS along pili of *Geobacter sulfurreducens*. *Applied and Environmental Microbiology*, **76**, 4080–4084.

Lin, W., Bazylinski, D.A., Xiao, T., W, L.-F. and Pan, Y. (2014) Life with compass: diversity and biogeography of magnetotactic bacteria. *Environmental Microbiology*, **16**, 2646–2658.

Liu, J., Wang, Z., Belchik, S.M., Edwards, M.J., Liu, C., Kennedy, D.W., Merkley, E.D., Lipton, M.S., Butt, J.N. and Richardson, D.J. (2012a) Identification and characterization of MtoA: a decaheme c-type cytochrome of the neutrophilic Fe (II)-oxidizing bacterium *Sideroxydans lithotrophicus* ES-1. *Frontiers in Geomicrobiology*, **3**, 37.

Liu, J.W., Pearce, C.I., Qafoku, O., Arenholz, E., Heald, S.M. and Rosso, K.M. (2012b) Tc(VII) reduction kinetics by titanomagnetite ($Fe_{3-x}Ti_xO_4$) nanoparticles. *Geochimica et Cosmochimica Acta*, **92**, 67–81.

Liu, Y., Wang, Z., Liu, J., Levar, C., Edwards, M.J., Babauta, J.T., Kennedy, D.W., Shi, Z., Beyenal, H., Bond, D.R., Clarke, T.A., Butt, J.N., Richardson, D.J., Rosso, K.M., Zachara, J.M., Fredrickson. J.K. and Shi, L. (2014) A trans-outer membrane porin-cytochrome protein complex for extracellular electron transfer by *Geobacter sulfurreducens*PCA. *Environmental Microbiology Reports*, **6**, 776–85.

Liu, C.-H., Chuang, Y.-H., Chen, T.-Y., Tian, Y., Li, H., Wang, M.-K. and Zhang, W. (2015) Mechanism of arsenic adsorption on magnetite nanoparticles from water: thermodynamic and spectroscopic studies. *Environmental Science & Technology*, **49**, 7726–7734.

Lloyd, J.R. (2003) Microbial reduction of metals and radionuclides. *FEMS Microbiology Reviews*, **27**, 411–425.

Lloyd, J.R. (2005) Mechanisms and environmental impact of microbial metal reduction. Pp. 273–302 in: *Micro-organisms and Earth Systems – Advances in Geomicrobiology* (G.M. Gadd, K.T. Semple and H.M. Lappin-Scott, editors). Symposia of the Society of General Microbiology, **65**, Cambridge University Press, New York.

Lloyd, J.R., Sole, V.A., van Praagh, C.V.G. and Lovley, D.R. (2000) Direct and Fe(II)-mediated reduction of technetium by Fe(III)-reducing bacteria. *Applied and Environmental Microbiology*, **66**, 3743–3749.

Lloyd, J.R., Pearce, C.I., Coker, V.S., Pattrick, R.A.D., van der Laan, G., Cutting, R., Vaughan, D.J., Paterson-Beedle, M., Mikheenko, I.P., Yong, P. and Macaskie, L.E. (2008) Biomineralization: linking the fossil record to the production of high value functional materials. *Geobiology*, **6**, 285–297.

Lloyd, J.R., Byrne, J.M. and Coker, V.S. (2011) Biotechnological synthesis of functional nanomaterials. *Current Opinion in Biotechnology*, **22**, 509–515.

Lonergan, D.J., Jenter, H., Coates, J.D., Phillips, E.J.P., Schmidt, T.M. and Lovley, D.R. (1996) Phylogenetic analysis of dissimilatory Fe(III)-reducing bacteria. *Journal of Bacteriology*, **178**, 2402–2408.

Lovley, D.R. (1997) Microbial Fe(III) reduction in subsurface environments. *FEMS Microbiology Reviews*, **20**, 305–313.

Lovley, D.R. (2013) Dissimilatory Fe(III)- and Mn(IV)-reducing prokaryotes. Pp. 287–308 in: *The Prokaryotes* (E. Rosenberg, E. DeLong, S. Lory, E. Stackebrandt and F. Thompson, editors). Springer, Berlin Heidelberg.

Lovley, D.R. and Phillips, E.J.P. (1986) Availability of ferric iron for microbial reduction in bottom sediments of the freshwater tidal Potomac River. *Applied and Environmental Microbiology*, **52**, 751–757.

Lovley, D.R. and Woodward, J.C. (1996) Mechanisms for chelator stimulation of microbial Fe(III)-oxide reduction. *Chemical Geology*, **132**, 19–24.

Lovley, D.R., Stoltz, J.F., Nord Jr, G.L. and Phillips, E.J.P. (1987) Anaerobic production of magnetite by a dissimilatory iron-reducing microorganism. *Nature*, **330**, 252–254.

Lovley, D.R., Phillips, E.J.P. and Lonergan, D.J. (1989) Hydrogen and formate oxidisation coupled to dissimilatory reduction of iron or manganese by *Alteromonas putrefaciens*. *Applied and Environmental Microbiology*, **55**, 700–706.

Lovley, D.R., Coates, J.D., Blunt-Harris, E.L., Phillips, E.J.P. and Woodward, J.C. (1996) Humic substances as electron acceptors for microbial respiration. *Nature*, **382**, 445–448.

Lovley, D.R., Coates, J.D., Saffarini, D. and Lonergan, D.J. (1997) Diversity of dissimilatory Fe(III)-reducing bacteria. Pp. 187–215 in: *Iron and Related Transition Metals in Microbial Metabolism* (G. Winkelman and C.J. Carrano, editors). Harwood Academic Publishers, Chur, Switzerland.

Lovley, D.R., Holmes, D.E. and Nevin, K.P. (2004) Dissimilatory Fe(III) and Mn(IV) reduction. *Advances in Microbial Physiology*, **49**, 219–286.

Malvankar, N.S., Vargas, M., Nevin, K., Tremblay, P.-L., Evans-Lutterodt, K., Nykypanchuk, D., Martz, E., Tuominen, M.T. and Lovley, D.R. (2015) Structural basis for metallic-like conductivity in microbial nanowires. *mBio*, **6**, e00084–15.

Manceau, A. (2009) Evaluation of the structural model for ferrihydrite derived from real-space modelling of high-energy X-ray diffraction data. *Clay Minerals*, **44**, 19–34.

Manceau, A., Skanthakumar, S. and Soderholm, L. (2014) PDF analysis of ferrihydrite: Critical assessment of the under-constrained akdalaite model. *American Mineralogist*, **99**, 102–108.

Mann, S., Sparks, N.H.C. and Wade, V.J. (1990) Crystallochemical control of iron oxide biomineralisation. Pp. 21–49 in: *Iron Biominerals* (R.B. Frankel and R.P. Blakemore, editors). Plenum Press, New York.

Marsili, E., Baron, D.B., Shikhare, I.D., Coursolle, D., Gralnick, J.A. and Bond, D.R. (2008) Shewanella secretes flavins that mediate extracellular electron transfer. *Proceedings of the National Academy of Sciences*, **105**, 3968–3973.

McBeth, J.M., Lloyd, J.R., Law, G.T.W., Livens, F.R., Burke, I.T. and Morris, K. (2011) Redox interactions of technetium with iron-bearing minerals. *Mineralogical Magazine*, **75**, 2419–2430.

McCormick, M.L. and Adriaens, P. (2004) Carbon tetrachloride transformation on the surface of nanoscale biogenic magnetite particles. *Environmental Science & Technology*, **38**, 1045–1053.

McCormick, M.L., Bouwer, E.J. and Adriaens, P. (2002) Carbon tetrachloride transformation in a model iron-reducing culture: Relative kinetics of biotic and abiotic reactions. *Environmental Science & Technology*, **36**, 403–410.

Michel, F.M., Ehm, L., Antao, S.M., Lee, P.L., Chupas, P.J., Liu, G., Strongin, D.R., Schoonen, M.A., Phillips, B.L. and Parise, J.B. (2007) The structure of ferrihydrite, a nanocrystalline material. *Science*, **316**, 1726–1729.

Miot, J., Li, J., Benzerara, K., Sougrati, M.T., Ona-Nguema, G., Bernard, S., Jumas, J.-C. and Guyot, F. (2014) Formation of single domain magnetite by green rust oxidation promoted by microbial anaerobic nitrate-dependent iron oxidation. *Geochimica et Cosmochimica Acta*, **139**, 327–343.

Missana, T., Garcia-Gutierrez, M. and Fernández, V. (2003) Uranium (VI) sorption on colloidal magnetite under anoxic environment: experimental study and surface complexation modelling. *Geochimica et Cosmochimica Acta*, **67**, 2543–2550.

Moon, J.-W., Roh, Y., Lauf, R.J., Vali, H., Yeary, L.W. and Phelps, T.J. (2007a) Microbial preparation of metal-substituted magnetite nanoparticles. *Journal of Microbial Methods*, **70**, 150–158.

Moon, J.-W., Roh, Y., Yeary, L.W., Lauf, R.J., Rawn, C.J., Love, L.J. and Phelps, T.J. (2007b) Microbial formation of lanthanide-substituted magnetites by *Thermoanaerobacter* sp. TOR-39. *Extremophiles*, **11**, 859–867.

Moon, J.-W., Yeary, L.W., Rondinone, A.J., Rawn, C.J., Kirkham, M.J., Roh, Y., Love, L.J. and Phelps, T.J. (2007c) Magnetic response of microbially synthesized transition metal- and lanthanide-substituted nanosized magnetites. *Journal of Magnetism and Magnetic Minerals*, **313**, 283–292.

Moon, J.-W., Rawn, C.J., Rondinone, A.J., Love, L.J., Roh, Y., Everett, S.M., Lauf, R.J. and Phelps, T.J. (2010) Large-scale production of magnetic nanoparticles using bacterial fermentation. *Journal of Industrial Microbiology and Biotechnology*, **37**, 1023–1031.

Moran, M.J., Zogorski, J.S. and Squillace, P.J. (2006) Chlorinated solvents in groundwater of the United States. *Environmental Science & Technology*, **41**, 74–81.

Morel, F.O.M.M., Kraepiel, A.M.L. and Amyot, M. (1998) The chemical cycle and bioaccumulation of mercury. *Annual Review of Ecology and Systematics*, **29**, 543–566.

Muehe, E.M., Scheer, L., Daus, B. and Kappler, A. (2013) Fate of arsenic during microbial reduction of biogenic versus abiogenic As-Fe(III)-mineral coprecipitates. *Environmental Science & Technology*, **47**, 8297–8307.

Myers, C.R. and Myers, J.M. (1992) Localisation of cytochromes to the outer membrane of anaerobically grown *Shewanella putrefaciens* MR-1. *Journal of Bacteriology*, 174, 3429–3438.

Myers, C.R. and Myers, J.M. (1997) Outer membrane cytochromes of *Shewanella putrefaciens* MR-1: spectral analysis, and purification of the 83-kDa c-type cytochrome. *Biochimica et Biophysica Acta*, **1326**, 307–318.

Nakata, K., Nagasaki, S., Tanaka, S., Sakamoto, Y., Tanaka, T. and Ogawa, H. (2002) Sorption and reduction of neptunium(V) on the surface of iron oxides. *Radiochimica Acta*, **90**, 665–669.

Nakata, K., Nagasaki, S., Tanaka, S., Sakamoto, Y., Tanaka, T. and Ogawa, H. (2004) Reduction rate of neptunium(V) in heterogenous solution with magnetite. *Radiochimica Acta*, **92**, 145–150.

Newman, D.K. and Kolter, R. (2000) A role for excreted quinones in extracellular electron transfer. *Nature*, **405**, 94–97.

Nickson, R.T., McArthur, J.M., Ravenscroft, P., Burgess, W.G. and Ahmed, K.M. (2000) Mechanism of arsenic release to groundwater, Bangladesh and West Bengal. *Applied Geochemistry*, **15**, 403–413.

Nurmi, J.T., Tratnyek, P.G., Sarathy, V., Baer, D.R., Amonette, J.E., Pecher, K., Wang, C., Linehan, J.C., Matson, D.W., Penn, R.L. and Driessen, M.D. (2004) Characterization and properties of metallic iron nanoparticles: spectroscopy, electrochemistry and kinetics. *Environmental Science & Technology*, **39**, 1221–1230.

O'Driscoll, N.J., Rencz, A. and Lean, D.R.S. (2005) The biogeochemistry and fate of mercury in the environment. *Biogeochemical Cycles of Elements*, **43**, 221–238.

Ona-Nguema, G., Morin, G., Wang, Y., Foster, A.L., Juillot, F., Calas, G. and Brown, G.E. (2010) XANES evidence for rapid arsenic(III) oxidation at magnetite and ferrihydrite surfaces by dissolved O_2 via Fe^{2+}-mediated reactions. *Environmental Science & Technology*, **44**, 5416–5422.

Pankhurst, Q.A., Thanh, N.T.K., Jones, S.K. and Dobson, J. (2009) Progress in applications of magnetic nanoparticles in biomedicine. *Journal of Physics D-Applied Physics*, **42**, 224001.

Pasakarnis, T.S., Boyanov, M.I., Kemner, K.M., Mishra, B., O'Loughlin, E.J., Parkin, G. and Scherer, M.M. (2013) Influence of chloride and Fe(II) content on the reduction of Hg(II) by magnetite. *Environmental Science & Technology*, **47**, 6987–6994.

Pattrick, R.A.D., Coker, V.S., Pearce, C.I., Telling, N.D., van der Laan, G. and Lloyd, J.R. (2013) Extracellular bacterial production of doped magnetite nanoparticles. Pp. 102–115 in: *Nanoscience: Volume 1: Nanostructures through Chemistry* (P. O'Brien, editor). The Royal Society of Chemistry.

Pearce, C.I., Henderson, C.M.B., Pattrick, R.A.D., van der Laan, G. and Vaughan, D.J. (2006) Direct determination of cation site occupancies in natural ferrite spinels by $L_{2,3}$ X-ray absorption spectroscopy and X-ray magnetic circular dichroism. *American Mineralogist*, **91**, 880–893.

Peterson, M.L., White, A.F., Brown Jr, G.E. and Parks, G.A. (1997) Surface passivation of magnetite by reaction with aqueous Cr(VI): XAFS and TEM results. *Environmental Science & Technology*, **31**, 1573–1576.

Petrie, L., North, N.N., Dollhopf, S.L., Balkwill, D.L. and Kostka, J.E. (2003) Enumeration and characterisation of iron(III)-reducing microbial communities from acidic subsurface sediments contaminated with uranium(VI). *Applied and Environmental Microbiology*, **69**, 7467–7479.

Pirbadian, S., Barchinger, S.E., Leung, K.M., Byun, H.S., Jangir, Y., Bouhenni, R.A., Reed, S.B., Romine, MF, Saffarini, D.A., Shi, L., Gorby, Y.A., Golbeck, J.H. and El-Naggar, M.Y. (2014) *Shewanella oneidensis* MR-1 nanowires are outer membrane and periplasmic extensions of the extracellular electron transport components. *Proceedings of the National Academy of Sciences*, **111**, 12883–12888.

Powell, B.A., Fjeld, R.A., Kaplan, D.I., Coates, J.T. and Serkiz, S.M. (2004) Pu(V)O_2^+adsorption and reduction by synthetic magnetite (Fe_3O_4). *Environmental Science & Technology*, **38**, 6016–6024.

Rai, D., Eary, L.E. and Zachara, J.M. (1989) Environmental chemistry of chromium. *Science of the Total Environment*, **86**, 15–23.

Rard, J.A. (1983) Critical review of the chemistry and thermodynamics of technetium and some of its inorganic compounds and aqueous species. Department of Energy, USA report UCRL-53440.

Reguera, G., McCarthy, K.D., Mehta, T., Nicoll, J.S., Tuominen, M.T. and Lovley, D.R. (2005) Extracellular electron transfer via microbial nanowires. *Nature*, **435**, 1098–1101.

Richter, K., Schicklberger, M. and Gescher, J. (2012) Dissimilatory reduction of extracellular electron acceptors in anaerobic respiration. *Applied and Environmental Microbiology*, **78**, 913–921.

Roden, E.E. and Urrutia, M.M. (1999) Ferrous iron removal promotes microbial reduction of crystalline iron(III) oxides. *Environmental Science & Technology*, **33**, 1847–1853.

Roden, E.E. and Zachara, J.M. (1996) Microbial reduction of crystalline iron (III) oxides: Influence of oxide surface area and potential for cell growth. *Environmental Science & Technology*, **30**, 1618–1628.

Rodgers, J.D. and Bunce, N.J. (2001) Treatment methods for the remediation of nitroaromatic explosives. *Water Research*, **35**, 2101–2111.

Roh, Y., Lauf, R.J., McMillan, A.D., Zhang, C., Rawn, C.J., Bai, J. and Phelps, T.J. (2001) Microbial synthesis and the characterisation of metal-substituted magnetites. *Solid State Communications*, **118**, 529–534.

Roh, Y., Zhang, C.-L., Vali, H., Lauf, R.J., Zhou, J. and Phelps, T.J. (2003) Biogeochemical and environmental factors in Fe biomineralisation: magnetite and siderite formation. *Clays and Clay Minerals*, **51**, 83–91.

Roh, Y., Vali, H., Phelps, T.J. and Moon, J.-W. (2006) Extracellular synthesis of magnetite and metal-substituted magnetite nanoparticles. *Journal of Nanoscience and Nanotechnology*, **6**, 3517–3520.

Ruder, A.M. (2006) Potential health effects of occupational chlorinated solvent exposure. *Annals of the New York Academy of Sciences*, **1076**, 207–227.

Ruggiero, C.E., Boukhalfa, H., Forsythe, J.H., Lack, J.G., Hersman, L.E. and Neu, M.P. (2005) Actinide and metal toxicity to prospective bioremediation bacteria. *Environmental Microbiology*, **7**, 88–97.

Safarik, I. and Safarikova, M. (2002) Magnetic nanoparticles and biosciences. *Monatshefte für Chemie*, **133**, 737–759.

Schmeide, K., Reich, T., Sachs, S. and Bernhard, G. (2006) Plutonium(III) complexation by humic substances studied by X-ray absorption fine structure spectroscopy. *Inorganica Chimica Acta*, **359**, 237–242.

Schüler, D. (1999) Formation of magnetosomes in magnetotactic bacteria. *Journal of Molecular Microbiology and Biotechnology*, **1**, 79–86.

Schüler, D. (2004) Molecular analysis of a subcellular compartment: the magentosome membrane in *Magnetospirillum gryphiswaldense*. *Archives of Microbiology*, **181**, 1–7.

Schulte, E.H. and Scoppa, P. (1987) Sources and behavior of technetium in the environment. *Science of the Total Environment*, **64**, 163–179.

Schwarz, S., Fernandes, F., Sanroman, L., Hodenius, M., Lang, C., Himmelreich, U., Schmitz-Rode, T., Schueler, D., Hoehn, M., Zenke, M. and Hieronymus, T. (2009) Synthetic and biogenic magnetite nanoparticles for tracking stem cells and dendritic cells. *Journal of Magnetism and Magnetic Materials*, **321**, 1533–1538.

Schwertmann, U. and Fitzpatrick, R.W. (1992) Iron minerals in surface environments. Pp. 7–30 in: *Biomineralisation, Processes of Iron and Manganese* (H.C.W. Skinner and R.W. Fitzpatrick, editors). Catena Verlag, Destedt, Germany.

Schwertmann, U. and Taylor, R.M. (1989) *Iron Oxides: Minerals in Soil Environments*. Soil Science Society of America, Madison, Wisconsin, USA.

Schwertmann, U., Freidl, J. and Stanjek, H. (1999) From Fe(III) ions to ferrihydrite and then to hematite. *Journal of Colloidal and Interface Science*, **209**, 215–223.

Scott, T.B., Allen, G.C., Heard, P.J. and Randell, M.G. (2005) Reduction of U(VI) to U(IV) on the surface of magnetite. *Geochimica et Cosmochimica Acta*, **69**, 5639–5646.

Shanker, A.K., Cervantes, C., Loza-Tavera, H. and Avudainayagam, S. (2005) Chromium toxicity in plants. *Environmental International*, **31**, 739–753.

Shubayev, V.I., Pisanic II, T.R. and Jin, S. (2009) Magnetic nanoparticles for theragnostics. *Advanced Drug Delivery Reviews*, **61**, 467–477.

Singer, D.M., Chatman, S.M., Ilton, E.S., Rosso, K.M., Banfield, J.F. and Waychunas, G.A. (2012) U(VI) sorption and reduction kinetics on the magnetite (111) surface. *Environmental Science & Technology*, **46**, 3821–3830.

Spain, J.C., Hughes, J.B. and Knackmuss, H.-J. (2000) *Biodegradation of Nitroaromatic Compounds and Explosives*. CRC Press, Boca Raton, Florida, USA

Stackebrandt, E., Wehmeyer, U. and Schink, B. (1989) The phylogenetic status of *Pelobacter acidigallici, Pelobacter venetianus*, and *Pelobacter carbinolicus*. *Systematic and Applied Microbiology*, **11**, 257–260.

Straub, K.L., Benz, M., Schink, B. and Widdel, F. (1996) Anaerobic, nitrate-dependant microbial oxidation of ferrous iron. *Applied and Environmental Microbiology*, **62**, 1458–1460.

Taylor, D.M. (1989) The biodistribution and toxicity of plutonium, americium and neptunium. *Science of the Total Environment*, **83**, 217–225.

Telling, N.D., Coker, V.S., Cutting, R.S., van der Laan, G., Pearce, C.I., Pattrick, R.A.D., Arenholz, E. and Lloyd, J.R. (2009) Remediation of Cr(VI) by biogenic magnetic nanoparticles: An X-ray magnetic circular

dichroism study. *Applied Physics Letters*, **95**, 163701.
Thamdrup, B. (2000) Microbial manganese and iron reduction in aquatic sediments. *Advances in Microbial Ecology*, **15**, 41–83.
Toner, B.M., Berquo, T.S., Michel, F.M., Sorensen, J.V., Templeton, A.S. and Edwards, K.J. (2012). Mineralogy of iron microbial mats from Loihi Seamount. *Frontiers in Microbiology*, **3**, 118.
Tratnyek, P.G. and Johnson, R.L. (2006) Nanotechnologies for environmental cleanup. *Nano Today*, 1, 44–48.
Urrutia, M.M., Roden, E.E. and Zachara, J.M. (1999) Influence of aqueous and solid-phase Fe(II) complexants on microbial reduction of crystalline iron(III) oxides. *Environmental Science & Technology*, **33**, 4022–4028.
Vargas, M., Kashefi, K., Blunt-Harris, E.L. and Lovley, D.R. (1998) Microbiological evidence for Fe(III) on early Earth. *Nature*, **395**, 65–67.
Veeramani, H., Alessi, D.S., Suvorova, E.I., Lezama-Pacheco, J.S., Stubbs, J.E., Sharp, J.O., Dippon, U., Kappler, A., Bargar, J.R. and Bernier-Latmani, R. (2011) Products of abiotic U(VI) reduction by biogenic magnetite and vivianite. *Geochimica et Cosmochimica Acta*, **75**, 2512–2528.
Vikesland, P.J., Heathcock, A.M., Rebodos, R.L. and Makus, K.E. (2007) Particle size and aggregation effects on magnetite reactivity toward carbon tetrachloride. *Environmental Science & Technology*, **41**, 5277–5283.
von Canstein, H., Li, Y.L., Leonhauser, J., Haase, E., Felske, A., Deckwer, W.-D. and Wagner-Dobler, I. (2002) Spatially oscillating activity and microbial succession of mercury-reducing biofilms in a technical-scale bioremediation system. *Applied and Environmental Microbiology*, **68**, 1938–1946.
von Canstein, H., Ogawa, J., Shimizu, S. and Lloyd, J.R. (2008) Secretion of flavins by Shewanella species and their role in extracellular electron transfer. *Applied and Environmental Microbiology*, **74**, 615–623.
Wang, Y., Morin, G., Ona-Nguema, G., Juillot, F., Calas, G. and Brown, G.E. (2011) Distinctive Arsenic(V) trapping modes by magnetite nanoparticles induced by different sorption processes. *Environmental Science & Technology*, **45**, 7258–7266.
Waychunas, G.A. (1991) Crystal chemistry of oxides and oxyhydroxides. Pp. 11–68 in: *Oxide Minerals: Petrologic and Magnetic Significance* (D.H. Lindsley, editor). Reviews in Mineralogy, **25**. Mineralogical Society of America, Chelsea, Michigan, USA.
Watts, M.P., Coker, V.S., Parry, S., Pattrick, R.A.D., Thomas, R., Kalin, R. and Lloyd, J.R. (2015a) Bionanoparticle treatment of alkaline Cr(VI) leachate and chromite ore processing residue. *Applied Geochemistry*, **54**, 27–42.
Watts, M.P., Coker, V.S., Parry, S., Thomas, R., Kalin, R. and Lloyd, J.R. (2015b) Effective treatment of alkaline Cr(VI) contaminated leachate using a novel Pd-bionanocatalyst; impact of electron donor and aqueous geochemistry. *Applied Catalysis B: Environmental*, **170–171**, 162–172.
Weber, K.A., Achenbach, L.A. and Coates, J.D. (2006a) Microorganisms pumping iron: Anaerobic microbial iron oxidation and reduction. *Nature Reviews Microbiology*, **4**, 752–764.
Weber, K.A., Urrutia, M.M., Churchill, P.F., Kukkadapu, R.K. and Roden, E.E. (2006b) Anaerobic redox cycling of iron by freshwater sediment microorganisms. *Environmental Microbiology*, **8**, 100–113.
White, A.F. and Peterson, M.L. (1996) Reduction of aqueous transition metal species on the surfaces of Fe(II)-containing oxides. *Geochimica et Cosmochimica Acta*, **60**, 3799–3814.
Wiatrowski, H.A., Das, S., Kukkadapu, R., Ilton, E.S., Barkay, T. and Yee, N. (2009) Reduction of Hg(II) to Hg(0) by magnetite. *Environmental Science & Technology*, **43**, 5307–5313.
Wilkins, M.J., Livens, F.R., Vaughan, D.J. and Lloyd, J.R. (2006) The impact of Fe(III)-reducing bacteria on uranium mobility. *Biogeochemistry*, **78**, 125–150.
Williams, A.G.B., Gregory, K.B., Parkin, G.F. and Scherer, M.M. (2005) Hexahydro-1,3,5-trinitro-1,3,5-triazine transformation by biologically reduced ferrihydrite: evolution of Fe mineralogy, surface area, and reaction rates. *Environmental Science & Technology*, **39**, 5183–5189.
Yuan, K., Ilton, E.S., Antonio, M.R., Li, Z., Cook, P.J. and Becker, U. (2015a) Electrochemical and spectroscopic evidence on the one-electron reduction of U(VI) to U (V) on magnetite. *Environmental Science & Technology*. **49**, 6206–6213.
Yuan, K., Renock, D., Ewing, R.C. and Becker, U. (2015b) Uranium reduction on magnetite: Probing for pentavalent uranium using electrochemical methods. *Geochimica et Cosmochimica Acta*, **156**, 194–206.
Zachara, J.M., Fredrickson, J.K., Li, S., Kennedy, D.W., Smith, S.C. and Gassman, P.L. (1998) Bacterial

reduction of crystalline Fe^{3+} oxides in single phase suspensions and subsurface materials. *American Mineralogist*, **83**, 1426–1443.

Zachara, J.M., Kukkadapu, R.K., Fredrickson, J.K., Gorby, Y.A. and Smith, S.C. (2002) Biomineralization of poorly crystalline Fe(III) oxides by dissimilatory metal reducing bacteria (DMRB). *Geomicrobiology Journal*, **19**, 179–207.

Zhang, C., Vali, H., Romanek, C.S., Phelps, T.J. and Liu, S.V. (1998) Formation of single-domain magnetite by a thermophilic bacterium. *American Mineralogist*, **83**, 1409–1418.

Zhang, W.-X. (2003) Nanoscale iron particles for environmental remediation: an overview. *Journal of Nanoparticle Research*, **5**, 323–332.

EMU Notes in Mineralogy, Vol. 17 (2017), Chapter 9, 229–272

Impact of iron redox chemistry on nuclear waste disposal

CAROLYN I. PEARCE[1,2], KEVIN M. ROSSO[1,3], RICHARD A. D. PATTRICK[3] and ANDREW R. FELMY[1]

[1] *Pacific Northwest National Laboratory, Richland, Washington 99352, USA, e-mail: carolyn.pearce@pnnl.gov*
[2] *School of Chemistry and Dalton Nuclear Institute, University of Manchester, Manchester M13 9PL, UK*
[3] *School of Earth, Atmospheric and Environmental Sciences, University of Manchester, Manchester M13 9PL, UK*

For the safe disposal of nuclear waste, the ability to predict the changes in oxidation states of redox active actinide elements and fission products, such as U, Pu, Tc and Np is a key factor in determining their long term mobility. Both in the Geological Disposal Facility (GDF) near-field and in the far-field subsurface environment, the oxidation states of radionuclides are closely tied to changes in the redox condition of other elements such as iron. Iron pervades all aspects of the waste-package environment, from the steel in the waste containers, through corrosion products, to the iron minerals present in the host rock. Over the long period required for nuclear waste disposal, the chemical conditions of the subsurface waste package will vary along the entire continuum from oxidizing to reducing conditions. This variability leads to the expectation that redox-active components such as Fe oxides can undergo phase transformations or dissolution; to understand and quantify such a system with respect to potential impacts on waste package integrity and radionuclide fate is clearly a serious challenge. Traditional GDF performance assessment models currently rely upon surface adsorption or single-phase solubility experiments and do not deal with the incorporation of radionuclides into specific crystallographic sites within the evolving Fe phases. In this chapter, we focus on the iron-bearing phases that are likely to be present in both the near and far-field of a GDF, examining their potential for redox activity and interaction with radionuclides. To support this, thermodynamic and molecular modelling is particularly important in predicting radionuclide behaviour in the presence of Fe-phases. Examination of radionuclide contamination of the natural environment provides further evidence of the importance of Fe-phases in far-field processes; these can be augmented by experimental and analogue studies.

1. Introduction and importance

One of the most important scientific challenges for the disposal of nuclear waste in a geological disposal facility (GDF) is the ability to predict the mobility or transport of actinide elements (*e.g.* U-238, Pu-238, Pu-239, Pu-240, Pu-241 and Np-237) and fission products (*e.g.* Cs-137, Sr-90 and Tc-99) in the subsurface environment over the long time spans required to ensure the safe disposal of high-level waste. The mobility of radionuclides in the environment depends on their physical state, oxidation state and

DOI: 10.1180/EMU-notes.17.7

complexation. Thus, understanding possible changes in oxidation state of actinides and fission products (AFP) is a key factor in predicting their long-term mobility in a GDF.

There are currently several countries with either proposed or operating GDFs including France, Germany, Sweden, Finland, USA and UK. The variable waste type, geology and depth associated with these facilities results in different temperatures, pressures, chemical potentials, water availability, radiation effects and microbiology. Some characteristics of different low-level waste (LLW) and intermediate level waste (ILW) disposal facilities are given in Table 1.

Of the total radioactive waste generated, the majority (>99%) is in the LLW and ILW category. The volume of high-level radioactive waste (HLW) is relatively small, but it contains >90% of the total radioactivity. To date there has been no practical need for final high-level waste (HLW) disposal facilities, as surface storage for the first 40–50 years is required to allow heat and radioactivity levels to decay to a sufficiently low level to allow handling, transport and final disposal of the waste. The international consensus is that geological disposal will also be the preferred method for the final disposal of HLW and the process of establishing a suitable site for long-term storage of containment of HLW, spent nuclear fuel (SNF), plutonium and uranium stocks in a GDF is underway in several countries, with Finland and Sweden at the most advanced stage (Miller, 2015). A suitable geological environment for the GDF must have restricted groundwater flow and host rock that can limit movement of radionuclides; inorganic sedimentary rocks or clays are considered the most suitable formations as they form an impermeable geological barrier (NDA, 2010a). The GDF must be located deep enough to protect against surface erosion, climate change, earthquakes and human intrusion for thousands of years; thus the waste package will be subjected to the elevated temperatures and pressures incumbent at depths of 200–1000 m (DECC, 2014). All proposed GDFs incorporate a multi-barrier system to isolate the HLW from the environment and to contain the radioactivity. The multi-barrier concept consists of radioactive waste encapsulated in an appropriate material (*e.g.* glass) poured into the waste container (*e.g.* stainless steel) and sealed to form a durable waste package contained in an engineered barrier or buffer (*e.g.* bentonite) which is complementary to the surrounding geological barrier (Fig. 1). The man-made barriers, consisting of the waste package and the engineered barrier, must be designed to immobilize completely highly active short-lived radionuclides, *e.g.* Cs-137 with a half-life of 30.2 years (Table 2), during the several hundred years that are required for them to decay. Over a longer timescale of thousands of years, it will become the safety function of the geological barrier to prevent release of long-lived radionuclides, *e.g.* Pu-242 with a half-life of 380,000 y (Table 2), into the surrounding subsurface.

Nature has already shown that, in the absence of human intervention, geological isolation is possible. At Oklo (Gabon, West Africa) ~2 billion years ago, spontaneous natural nuclear reactors operated for ~500,000 y as a result of the high concentration of uranium (>3% U-235), the absence of neutron absorbers and the presence of water as a moderator (Gauthier-Lafaye *et al.*, 1996). The naturally occurring reactor produced all the radionuclides found in HLW, including Pu-239 and its fission products. Isotopic

Table 1. Characteristics of some operating or closed GDFs.

Location	Waste type	Geology	Depth (m)
Finland			
Olkiluoto (VLJ)	LLW – packed into drums and compressed (fire-protected fabrics, plastic wrappings, protective clothing, machinery parts, pipes) ILW – dried, mixed with bitumen, and cast into drums (ion-exchange resin, washings, tank sludge)	Micaceous gneiss intercalated with sparsely fractured tonalite	60–100
Loviisa (VLJ)	LLW and ILW – cement solidification of wet waste. Solid waste packed into steel drums (as above)	Coarse-grained and porphyritic Rapakivi granite	120
Germany			
Lower Saxony (Asse II)	LLW and ILW	Salt Dome	750
Saxony-Anhalt (Morsleben)	LLW and ILW	Salt dome	630
Sweden			
Forsmark (SFR)	LLW and ILW	Granite	50
USA			
New Mexico (Waste Isolation Pilot Plant)	Transuranic waste	Salt Bed	655

studies of the Oklo fossil reactors show that that most of the fission products were retained within the reactor zones for ~2000 million years and eventually decayed into non-radioactive elements. There is also evidence of mobile fission products being retained close to reactor zones in Fe-rich clays, even in a highly porous environment, demonstrating many of the features of the multi-barrier concept and highlighting the importance of Fe in a GDF.

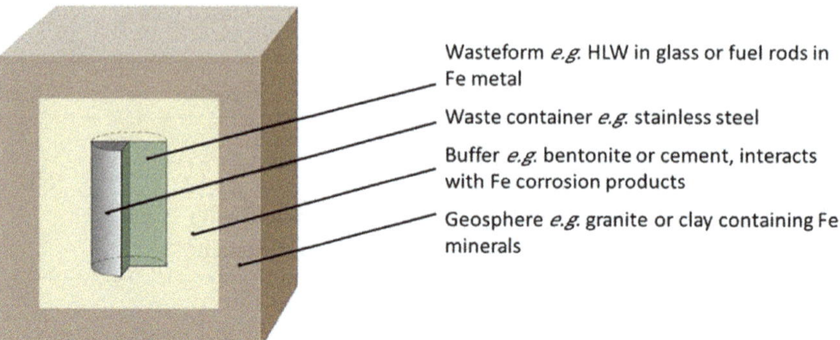

Figure 1. Waste-package concept and the roles of iron in a multi-barrier system for radionuclide containment. Adapted from NDA Report no. NDA/RWMD/033.

In both the GDF near-field and far-field, the oxidation state of the radionuclide is tied closely to the changes in Fe redox state. Fe is present in high concentrations in the waste package-environment in the form of steel or other components, and reacts to form ubiquitous coatings on most mineral phases in the subsurface. Over hundreds to millions of years, it is likely that the chemical environment both in the waste package near-field and far-field environment will vary along the entire continuum from oxidizing to reducing conditions. The presence of oxygen in a GDF in the saturated zone is expected during construction and emplacement of the waste, subsequent storage and eventual backfilling, decommissioning and closure, with reducing conditions dominating thereafter. Radiolysis in the vicinity of highly radioactive waste forms could create localized oxidizing conditions; however, recent studies have shown that that hydrogen produced by anoxic iron corrosion has the potential to prohibit oxidative corrosion processes (Wersin *et al.*, 1994). Variability in redox conditions can result in the conversion of Fe oxides from one phase to another or even the complete dissolution of the Fe phases. Traditional GDF performance assessment models rely upon surface adsorption or single-phase solubility experiments and do not address the long-term aging effects of radionuclide sorption to container corrosion phases, ion migration or incorporation of radionuclides into different crystallographic sites within the changing Fe phases (Vines and Beard, 2012). Unravelling the impact of such phase transformations on the oxidation state of the radionuclide is one of the central issues required to develop a long-term predictive capability for radionuclides in the subsurface. These processes can have an enormous impact on the migration potential of radionuclides, especially the redox-sensitive elements (U, Np, Pu and Tc) owing to the orders of magnitude differences in the solubility of the different oxidation states.

Radionuclides can interact with Fe phases by a range of mechanisms including adsorption, co-precipitation, incorporation into specific crystallographic sites, intra-crystalline diffusion and reductive precipitation with redox-active chemical species at the mineral surface (Fig. 2). The size and charge of the radionuclide ions, along with the pH and redox conditions in solution, are critical in determining which interaction

Table 2. Half-life of radionuclides relevant to geological disposal of high activity wastes. Data from NDA Report no. NDA/RWMD/034 (2010c).

Radionuclide		Half-life (y)
Actinides		
Uranium-238		4.5×10^9
Uranium-235		7.0×10^8
Thorium-230	U-238 daughter products	7.5×10^4
Radium-226	U-238 daughter products	1.6×10^3
Plutonium-239	Produced from U-238 by neutron capture in reactor cores	2.4×10^4
Plutonium-240	Produced from U-238 by neutron capture in reactor cores	6.6×10^3
Plutonium-242	Produced from U-238 by neutron capture in reactor cores	3.8×10^5
Neptunium-237	Produced from U-235 by neutron capture in reactor cores	2.1×10^6
Fission products		
Selenium-78		1.2×10^6
Zirconium-93		1.5×10^6
Technetium-99		2.1×10^5
Tin-126		1.0×10^5
Iodine-129		1.6×10^7
Cesium-137		30.2
Strontium-90		28.8
Activation products		
Chlorine-36	Generated within graphite	3.0×10^5
Carbon-14	Generated within graphite and steel	5.7×10^3
Nickel-63	Produced from irradiation of stable Ni	9.9×10^1
Niobium-94	Produced within steel	2.0×10^4

mechanism will dominate and multiple interaction mechanisms can occur simultaneously. For example, radionuclides can first adsorb to the Fe oxide surface and then be incorporated deeper into the structure as the solid phase ages or reconstructs in response to changing solution conditions. Radionuclides can also be incorporated directly during the precipitation of poorly crystalline Fe oxides that occur as a result of increasing pH or oxidation. Radionuclide incorporation can be either as a discrete phase or in specific crystallographic sites. The oxidation state of the radionuclides may also change even after substitution as a result of electron conduction occurring through the bulk or surface phases. In order to develop a long-term predictive capability it is necessary to determine how the speciation of the incorporated radionuclides change

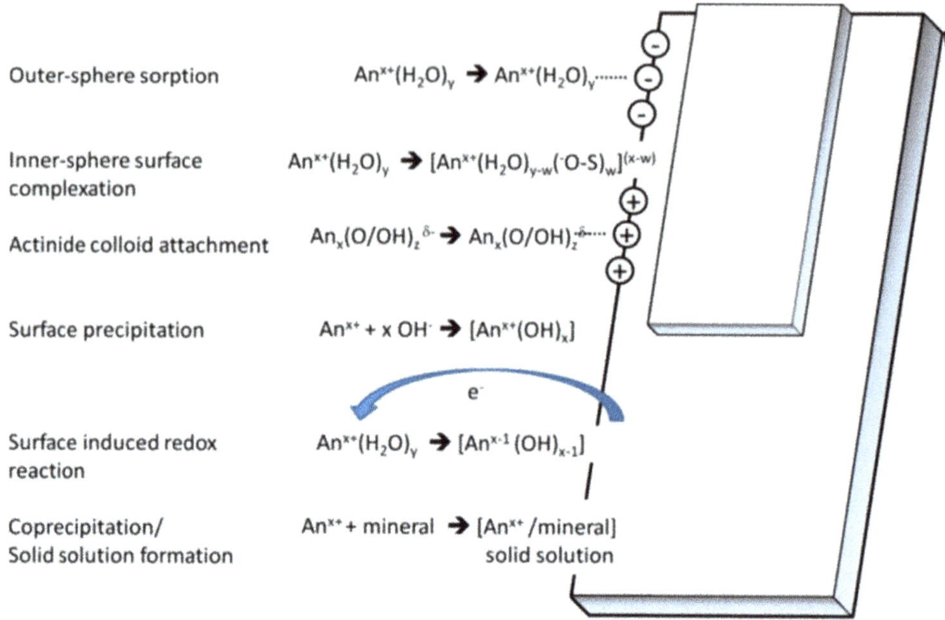

Figure 2. Radionuclide interaction with Fe phases (from Geckeis *et al.*, 2013).

over time in response to changing abiotic and/or biotic driven redox conditions. Ultimately, it will be necessary to develop molecular and macroscopic, thermodynamic and kinetic models to predict these changes.

2. Post-closure redox conditions in a GDF

Processes that begin after the closure of a GDF include heat generation, irradiation, degradation of engineered barriers, resaturation of the engineered zone with groundwater, waste-container corrosion, gas generation and microbial degradation. The thermodynamic equilibria and reaction kinetics of key chemical reactions involved in these processes will be affected by the extreme conditions of temperature, pressure and radiation associated with HLW. HLW can generate enough self-heat from decay of radionuclides to produce initial storage temperatures up to 600°C with temperatures remaining as high as 300°C after 100 y of storage; however, GDF concepts set temperature constraints for the canister/barrier interface of 200°C for evaporates and 100°C for higher-strength rocks and lower-strength sedimentary rocks. Even in the GDF far-field, temperatures in the host rock will be in the order of 10–25°C higher than at the Earth's surface as a result of the geothermal gradient (15–30°C/km). Pressure build up in the waste package can occur as a result of a number of process including: (1) metal corrosion resulting in the formation of corrosion products that occupy a greater volume; and (2) hydrogen generation and accumulation through corrosion of steel canisters, radiolysis of water and degradation of organics. The host rock surrounding the GDF will also experience pressures in the order of ~138 bar as a result of lithostatic pressure. Radiation damage processes, in addition to heat, associated with HLW include: (1) α-particle (He ion, 4.5–5.5 MeV) and α-recoil nucleus (70–100 keV) irradiation produced by α-decay of actinides (U, Np, Pu, Am and Cm); (2) β-particles (high energy electrons) produced by β-decay of fission products (Cs-137, Sr-90); and (3) γ-rays (ionizing radiation). All of these processes impact on the GDF redox state, which is the most important factor influencing radionuclide dissolution and release from the disposal facility, as well as subsequent transport through the surrounding host rock. As most radionuclides are less mobile in lower oxidation states, a reducing environment is a key feature for the safety function of the disposal facility and the corroding canister represents a strong redox buffer. However, oxidizing conditions can potentially occur at depth as a result of: (1) remnant oxygen from construction and operation for a short period after closure; and (2) oxidizing and diluted glacial meltwater intrusion through open fractures in high-strength rocks *e.g.* granite (Duro *et al.*, 2012). The host rock, in particular low-temperature Fe-containing minerals such as pyrite and iron oxyhydroxides, will also substantially influence redox conditions through reaction with the groundwater and, although these reactions are slow, over long time periods they will take on increased importance for the groundwater chemistry.

The redox state is a well defined thermodynamic property, but different techniques for redox measurement respond to different redox components and, given the timescales of environmentally significant reactions, it is unlikely that the overall system is in equilibrium. Consequently, the modelling of the system redox state, and its impact on the relevant radionuclides, for safety function assessment is challenging. In order to model the system, it is necessary to: (1) identify the components that govern the system redox chemistry; (2) characterize the kinetics of radionuclide redox reactions with near-field/far-field materials, including relevant intermediate phases; and (3) determine the deviation from equilibrium of different redox-sensitive components. In the waste package, near-field and far-field environments of the GDF, Fe(II) in magnetite as a corrosion product from steel canisters, Fe(II) in the clay buffer, natural Fe(II)-bearing minerals in the host rock and adsorbed Fe(II) in deep groundwater systems represent huge electron donor reservoirs (Geckeis *et al.*, 2013). Although the GDF has previously been considered to be an extreme environment where stresses including hyperalkalinity, high temperature, radiation and radionuclide toxicity may play a role in limiting microbial colonization, it is clear that microbes may tolerate such extreme conditions (Brown *et al.*, 2014; Bassil *et al.*, 2015). Thus, microbiological processes may play an as yet poorly understood role in controlling Fe redox in the GDF environment. The 'evolved' GDF, resaturated with groundwater, might be a potential niche for a variety of specialist microorganisms including Fe-reducing organisms that could lead to the formation of a 'bio-barrier' restricting the transport of radionuclides from the GDF (Lloyd, 2014).

Recently, an increasing array of spectroscopic techniques and quantum chemical calculations have provided molecular-level insight into redox reactivity between Fe phases and radionuclides in environments relevant to geological disposal (Felmy *et al.*, 2011a). These studies have shown that redox reactivity at Fe mineral surfaces can significantly impact the mobility of radionuclides including U, Np, Pu and Tc, with electron transfer (ET) from Fe(II) solid phases to surface-sorbed radionuclides kinetically preferred over reduction of dissolved species by Fe(II) in solution. In this chapter, we will discuss recent developments and future research needs relating to sorption processes, redox reactions and incorporation mechanisms that occur in association with Fe phases under conditions relevant to geological disposal, including their use in predictive models for radionuclide fate and transport.

3. Redox-active Fe components in the near field

The near field is defined as the wasteform itself (*e.g.* HLW glass), the canister (*e.g.* carbon steel) and the surrounding buffer zone (*e.g.* bentonite) extending into the Excavation Damaged Zone (EDZ) produced by excavation, disposal, backfilling and closure of a GDF. The key near field sources of Fe are: (1) corrosion products from waste canisters; (2) structural Fe and accessory Fe-bearing mineral phases in the buffer material; and (3) Fe-oxyhydroxide precipitation as a result of groundwater infiltration in the EDZ. Timescales for the sealing of discontinuities in the EDZ through mineral precipitation have been estimated as between several hundreds to more than 1 million y by hydrogeochemical modelling of EDZ evolution. The main precipitated mineral phases predicted to be responsible for the hydraulic sealing of the EDZ are calcite and ferric oxyhydroxides (Acero *et al.*, 2010).

3.1. Corrosion behaviour of Fe-bearing waste canisters

The properties of aged spent nuclear fuel will differ in terms of the composition, structure, surface area, radioactivity level and type of radioactive decay. Corrosion behaviour of the SNF will depend on fuel properties, container material and conditions in the near-field environment surrounding the spent fuel at the time of water contact. Experiments investigating alteration of irradiated UO_2 fuel revealed that spent fuel corrosion was inhibited due to the hydrogen overpressure generated in the near field by anaerobic steel corrosion (Fanghanel *et al.*, 2013). This is because low oxygen partial pressures (of the order of 10^{-65} atm at room temperature) ensure the thermodynamic stability of UO_2. Redox conditions inside the canister are controlled by both Fe(II) ions in solution and H_2 gas generated by corrosion of the Fe, producing a low Eh (<−100 mV) which limits the dissolution of spent nuclear fuel, and specifically UO_2. Anaerobic corrosion of steel produces magnetite (Fe_3O_4) as a final stable mineral phase and thereby consumes H_2O, according to equations 1 and 2 under neutral pH conditions:

$$\mathrm{Fe(II)} + 2\mathrm{H_2O_{(l)}} \rightarrow \mathrm{Fe(OH)_2} + \mathrm{H_{2\,(g)}} \quad (1)$$

$$3\mathrm{Fe(OH)_2} \rightarrow \mathrm{Fe_3O_4} + \mathrm{H_{2\,(g)}} + 2\mathrm{H_2O_{(l)}} \quad (2)$$

Due to different densities of steel (7.75−8.05 g/cm^3) and magnetite (5.21 g/cm^3), the volume of the container material increases by a factor of 54% as a result of corrosion and further reduces the available void volume in the near field and thus the maximum volume of water which may have access. Fe corrosion products including hematite ($Fe_2^{III}O_3$), 'green rust' ($Fe^{II}_{4.5}Mg_{1.5}Fe_2^{III}(OH)_{18} \cdot 4(H_2O)$), ferrous hydroxide ($Fe^{II}(OH)_2$) and siderite ($Fe^{II}CO_3$), as well as magnetite ($Fe^{II}Fe_2^{III}O_4$) offer considerable reactive surface area and demonstrate retention of radionuclides *via* surface sorption, coprecipitation or by reduction of the redox-sensitive radionuclides U, Np and Tc to the poorly soluble tetravalent states (Metz *et al.*, 2012). Both Pu(III) and Pu(IV) species can form in the presence of corrosion products, depending on pH and Pu concentration. Soluble Pu(III) dominates at low pH but can exist in equilibrium with solid Pu(IV) hydrous oxide up to pH10 in the presence of corroding iron powder (Altmaier *et al.*, 2009). Low soluble Pu concentrations can be maintained by strong sorption of Pu(III) to Fe_3O_4 surfaces, as shown by Kirsch *et al.* (2011), who found very low Pu concentrations in solution in contact with magnetite even though they proved the existence of trivalent species by XANES analysis. Because Fe corrosion and concomitant H_2 production are the dominant processes controlling redox in the GDF near-field environment, radiolytically produced oxidants have a limited effect and oxidative dissolution of irradiated fuel is inhibited (Metz *et al.*, 2012). For example, in the Swedish KBS-3 nuclear waste repository, the inhibitory effect of H_2 generated from Fe canister corrosion on SNF dissolution was demonstrated, along with minor effects produced by micromolar levels of oxidative species generated in aqueous solution by gamma irradiation during the early stages of disposal, highlighting the importance of using Fe-based canister material (Cui, 2011). Cast iron, copper, titanium and nickel-based alloys have all been proposed as waste-container materials, along with mild steel, stainless steel and carbon steel, in many geological disposal programmes worldwide (NDA, 2010b). The nature and structural properties of the corrosion products formed on these container materials will vary significantly with the storage environment, and particularly with the redox conditions. For example, structural characterization of corrosion products on carbon steel by X-ray diffraction (XRD) and confocal micro-Raman spectroscopy showed the formation of iron oxides (magnetite and maghemite) under aerobic conditions, whereas nanocrystalline mackinawite was identified as the first corrosion product on the steel surface under anaerobic conditions, followed by a fast transformation process into a monoclinic pyrrhotite phase (El Mendili *et al.*, 2013). Along with redox potential, other geochemical conditions including ionic strength and the concentrations of dissolved species such oxygen, sulfate and nitrate, will affect the rate of steel corrosion in a GDF. A series of electrochemical experiments showed that the corrosion rate of reinforcing steel was proportional to the concentration of Cl^- and dissolved oxygen, acting as a corrosive agent and an electron acceptor respectively. Measured corrosion rates were higher than those estimated from an empirical model based on dissolved oxygen diffusion (Jung *et al.*, 2011). In the United States, wastes for disposal in the Waste Isolation Pilot Plant (WIPP) and other salt dome repositories are contained in mild steel drums. The corrosion layer formed on steel surfaces after exposure to high ionic strength (up to 5 M) WIPP brines under anoxic conditions are porous and composed of iron oxides, principally green rust with some ionic substitutions from the brines (Wang *et al.*, 2001).

As well as characterizing the canister corrosion products in isolation, it is also necessary to understand matrix effects, including the interface between silicate glass used for SNF containment and Fe-bearing corrosion products. An investigation of this interface using synchrotron-based scanning transmission X-ray microscopy (STXM) revealed two types of Fe-Si interactions (Fig. 3); (1) precipitation of an outer layer of nanocrystalline Fe-silicates at the glass surface; and (2) migration of Fe within the inner altered porous gel layer and precipitation within the pores as Fe oxyhydroxide or Fe-silicates (Michelin *et al.*, 2013). More detailed multiscale characterization of the glass–Fe interface using electron microscopy and Raman micro-spectroscopy revealed an increase in glass alteration rates with decreasing distance between the glass and the Fe, and with increasing Fe content. Glass alteration kinetics and transport properties of the alteration layer were affected, respectively, by: (1) consumption of hydrolysed silica due to precipitation of Fe-silicates resulting in increased glass dissolution; and (2) a higher alteration rate, and penetration of Fe within the gel porosity in the form of precipitates, potentially resulting in pore clogging and a decrease in the diffusion coefficient of reactive species (Burger *et al.*, 2013). Several leaching tests of nuclear glass in the presence of Fe, under both aerobic and anaerobic conditions, have described the precipitation of Fe-silicates, either crystallized or in the form of amorphous nanocolloids (McVay and Buckwalter, 1983). Dissolution and precipitation of other crystalline phases including oxyhydroxides and nontronite clay have also been reported (Ildefonse *et al.*, 1994; Pelegrin *et al.*, 2010).

3.2. Corrosion of Fe-bearing phases at the interface between the waste package and the buffer material

Moving outwards from the waste package and into the near-field environment, a third component – the clay buffer material – must be added into the system, along with nuclear glass and Fe-bearing corrosion products, to understand the driving forces controlling the retention of HLW in a GDF. De Combarieu *et al.* (2011) investigated the

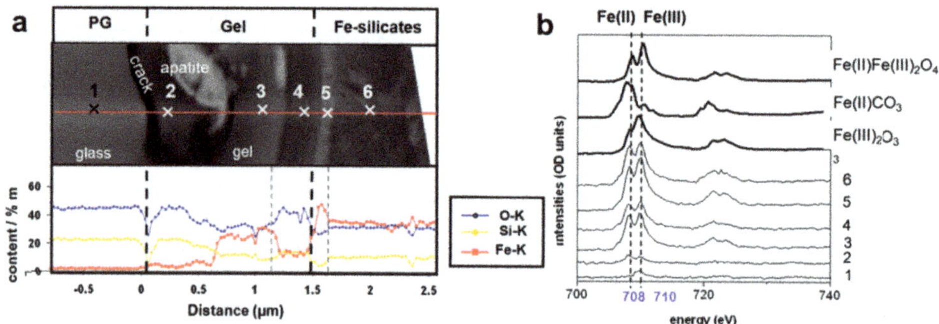

Figure 3. Silicate glass alteration enhanced by iron. (a) Micrograph of the glass/iron interface with the corresponding EDX profiles and STXM spot analyses (PG: pristine glass) and (b) $L_{2,3}$-edge XANES spectra extracted from different areas of the stack compared with the spectra of maghemite $Fe(III)_2O_3$, siderite $Fe(II)CO_3$ and magnetite $Fe(II)Fe(III)_2O_4$. Adapted from Burger *et al.* (2013).

coupled interactions between glass alteration, Fe corrosion and clay transformation at 90°C for up to 18 months, using a range of micro-scale techniques including Raman, XRD, XRF and XANES. An Fe foil (10 μm) was fully corroded in 10 months, leading to the formation of a layer containing magnetite, siderite ($FeCO_3$) and Fe-rich phyllosilicates. The glass, subsequently exposed to pore water, underwent alteration resulting in the formation of a gel layer, the thickness of which increased with reaction time (de Combarieu *et al.*, 2011). The chemical nature of this altered glass layer (AGL) suggested that it retained Al and Fe but was depleted in Si compared to the pristine glass, and that heavy elements released by glass leaching, *e.g.* Mo and Ln, were concentrated in a thin layer of secondary phases formed against the hard Fe foil at the expense of the AGL (de Combarieu *et al.*, 2011). Further investigation of Fe corrosion in water-saturated clay showed formation of an internal discontinuous thin magnetite layer, an external layer of Fe-rich phyllosilicate, and a clay transformation layer containing Ca-bearing siderite, with evidence for some Fe migration into the clay, where it could potentially interact with radionuclides released by alteration of nuclear glass (Schlegel *et al.*, 2010).

The interaction between metallic iron and kaolinite has also been studied under anoxic atmosphere at 90°C and in chloride solutions to simulate GDF conditions (Rivard *et al.*, 2013). Initially, the partial oxidation of metallic Fe resulted in an increase in solution pH and negative Eh values. Dissolved Fe was not found in solution but occurred as two new Fe-rich phases: magnetite in very small amounts and an Fe-rich clay phase, identified as the Fe-serpentine berthierine, $[(Fe^{2+}(Al,Fe^{3+})\square)]_3(Si,Al)_2O_5(OH)_4]$, by XRD and high-resolution TEM. The Si and Al contents of the berthierine are derived from the partial alkaline dissolution of kaolinite along crystal edge faces, with particles consisting of mixed berthierine and kaolinite layers. Oxygen gas was then introduced into the system at 90°C and the newly formed berthierine layers dissolved, mobilizing Fe to form secondary Fe oxides and oxyhydroxides while kaolinite layers recrystallized from released Al and Si (Rivard *et al.*, 2013).

Along with changing redox, temperature variations as a result of the evolving HLW within the canisters, will affect smectite transformations in contact with metallic Fe in aqueous solution. This is a particularly important consideration for ANDRA, the National Radioactive Waste Management Agency in France, where the design of the HLW GDF consists of steel canisters in a clay host rock. Jodin-Caumon *et al.* (2010) demonstrated that the initial bentonite, a dioctahedral, low-charge smectite, was destabilized towards trioctahedral products enriched in Fe and Mg as a function of temperature, including chlorite at 300°C and Fe-serpentine, Fe-saponite, mixed-layer chlorite-smectite and mixed-layer serpentine-smectite at lower temperatures (<150°C), as confirmed by XRD and TEM. Fe corrosion made the aqueous chloride medium basic and reductive, with redox equilibria affected by the thermal gradient resulting in more reduced conditions at the hottest point. Mass transport processes also occurred as a consequence of the thermal gradient, particularly affecting Mg migration and producing Mg-enriched clay minerals at the hottest point (300°C) (Jodin-Caumon *et al.*, 2012). These temperature-dependent reaction products reflect the existence of local thermodynamic equilibria and reaction pathways. The evolution of illite-smectite starting clay material towards serpentine or chlorite products will affect clay properties including swelling and adsorption, influencing the migration behaviour of radionuclides in the vicinity of the GDF (Jodin-Caumon *et al.*, 2012).

The French GDF involves disposal in an argillaceous environment, therefore steel/argillite interactions have been studied to determine how the release of Fe from the canisters will modify the properties of clay minerals in natural underground conditions of the Toarcian argillite layer (ANDRA, 2005a). Petrography and mineralogy characterization of modified argillite in contact with different steel samples for 6 y indicated no significant changes with the stainless steel but significant corrosion with carbon steel (Gaudin *et al.*, 2009). Iron diffusion from the corroded steel is accompanied by crystallization of significant amounts of goethite and lepidocrocite (FeOOH) and some traces of magnetite near the steel surface. The formation of Fe-rich phases within the argillite is associated with dissolution of calcite, phyllosilicate and pyrite and crystallization of some traces of gypsum and melanterite ($FeSO_4{\cdot}7H_2O$) (Gaudin *et al.*, 2009).

Lanson *et al.* (2012) also characterized Fe-rich 1:1 phyllosilicates formed as a result of interaction between metallic iron and smectites at 80°C and in the absence of O_2 using various spectroscopic, microscopic and diffraction-based techniques. Two Fe-rich 1:1 phyllosilicates products, cronstedtite ($(Fe_2^{2+}, Fe^{3+})(Si,Fe^{3+})_3O_5(OH)_4$) and odinite ($(Fe^{3+}Fe^{2+}Al,Mg)_3(Si,Al)_{2.5}O_5(OH)_4$), were formed with different relative compositions and different structural Fe valence states. Cronstedtite has a significantly higher Fe^{2+} content possibly enhancing its reactivity with respect to redox-active radionuclides (Lanson *et al.*, 2012). Fe(III)-montmorillonite has also been synthesized as a model clay buffer to study some of the basic properties of Fe-altered clay including osmotic swelling, tracer diffusion and thermal stability, in comparison with Na-montmorillonite (Manjanna *et al.*, 2009). The ferric form, rather than the ferrous form which would be the expected product under the reducing conditions of the GDF, was chosen due to ease of synthesis and handling. The osmotic swelling in water in compacted, water-saturated Fe-montmorillonite was eight times lower than in Na-montmorillonite (5 mL of sediment volume/g dry mass *vs.* >40 mL/g) due to the suppression of the electrical double layer around the highly charged interlayer Fe(III) cations. However, surprisingly, no significant difference in the migration behaviour of radioactive tracers ($^{22}Na^+$, HTO – tritiated H_2O, $^{36}Cl^-$) occurred when Na(I) ions were replaced with Fe(III) ions in the montmorillonite, possibly due to the use of different diffusion pathways depending on the charge, pore space, hydration and osmotic swelling properties of the clay (Manjanna *et al.*, 2009). The Fe(III)-montmorillonite continued to show reversible dehydration up to 190°C, indicating the presence of Fe^{3+} ions in the interlayer and supporting the thermal stability of Fe-altered clay in a GDF environment where temperatures of ~100°C are expected (Manjanna *et al.*, 2009).

3.3. Effect of microbial activity on Fe-bearing phases in the near field

In Sweden, where the Swedish Nuclear Fuel and Waste Management Company (SKB) is proposing to use copper-iron canisters for the disposal of SNF in a granitic GDF, surrounded by a compacted bentonite engineered barrier, model canisters in contact with bentonite have been investigated in the Äspö underground laboratory. Water analysis and electrochemical measurements indicated an increase in the corrosion rate of both Fe and Cu in contact with low-density bentonite, and of Fe only in experiments with no bentonite present. The corrosion may be attributed to microbial activity in the low-density bentonite (as opposed to fully compacted bentonite where microbial activity would be inhibited by swelling pressure and nm-sized pores with insufficient space for microbes to survive), resulting in sulfide production (Smart *et al.*, 2011). Over time, as the temperature decreases and groundwater from the host rock surrounding the GDF infiltrates the compacted clay, sulfate- and iron-reducing microbes will play a more significant role in the corrosion of metallic components. Urios *et al.* (2013) characterized the microbial diversity at the interface between re-compacted argillite and steel coupons after 10 years under *in situ* conditions in the Toarcian argillite layer. Sulfate-reducing bacteria, iron-reducing bacteria and thermophilic isolates were present, growing at the interface between the rock and the steel over very short time periods compared with the timescale of the GDF. Thus, the consequences of microbial activity on the evolution of metallic components in the GDF must be considered (Urios *et al.*, 2013). El Mendili *et al.* (2013) identified the corrosion products formed at 30°C on carbon steel under aerobic/anaerobic conditions, and their evolution due to enhanced microbial activity under environmental and geological conditions. Corrosion of the carbon steel under aerobic conditions was not influenced by the presence of the bacteria, with a surface Fe oxide formed as the corrosion product. Corrosion of carbon steel was more aggressive under anaerobic conditions and in the presence of sulfate-reducing bacteria nanocrystalline mackinawite was identified as the corrosion product, evolving into polycrystalline pyrrhotite over time. Thus the microbially enhanced corrosion of carbon steel under these conditions may not cause a significant problem due to the formation of this passive pyrrhotite layer on the surface (El Mendili *et al.*, 2013).

3.4. Effect of radiation damage on Fe-bearing phases in the near field

Once the shielding canister material of the primary containment has corroded away, Fe-bearing phases in the near field will no longer be protected from the waste package and will be exposed to radiation from the longer-lived radionuclides present in the waste. The engineered clay barrier will be exposed to ionizing gamma radiation and high-energy electrons produced by interaction of γ-rays with the waste container material. In the case of a leakage of transuranic elements from the radioactive waste form, in addition to gamma rays, radiation will come from beta emitters such as fission products (*e.g.* ^{137}Cs and ^{90}Sr) and alpha particle/recoil nuclei emitters such as actinides (*e.g.* U, Np, Pu, Am and Cm) (Ewing *et al.*, 1995). The mean penetration depth in silicates is nanometers for α recoil nuclei, ~20 μm for 5 MeV α-particles, and several centimetres in the case of β-particles. Allard *et al.* (2012) analysed the effect of radiation on several clay minerals (kaolinite, dickite, montmorillonite, illite and sudoite) using Electron Paramagnetic Resonance (EPR) Spectroscopy. Radiation damage produced defects consisting of electron holes associated with the O 2*p*-character molecular orbital on Si–O bonds oriented perpendicular to the silicate layer. Amorphization as a result of γ-radiation is unlikely but with its longer penetration range, γ-irradiation has the potential to affect surface properties of colloidal clay particles further out in the engineered barrier, resulting in increased colloid stability and changes in sedimentation behaviour due to an increase in surface potential through reaction with radical radiolysis products (Holmboe *et al.*, 2009). Whereas the radiation-induced defects produced by these electronic excitations tend to be point defects, ballistic interactions with heavy ions and recoil nuclei can cause atomic displacements and collision cascades resulting in extended defects and ultimately mineral amorphization. Thus, the effect of amorphization on the dissolution kinetics of smectite must be taken into account in the prediction of the long-term behaviour of engineered barriers.

Ionizing irradiation may also change the redox state of structural Fe in clay minerals. γ-irradiation of montmorillonite resulted in the reduction of 3% of the structural Fe^{3+} to Fe^{2+} as observed by Mössbauer spectroscopy and EPR (Allard *et al.*, 2012). Pattrick *et al.* (2013) also investigated α-particle effects on phyllosilicates revealing changes in both short- and long-range order in radiation damage haloes around thorium-containing monazite in Fe-rich biotite. X-ray absorption near edge structure (XANES) spectroscopy of the Fe K-edge showed a decrease in energy of the main absorption by up to 1 eV, revealing reduction of the Fe(III) components of the biotite by interaction with the $^{4}He^{2+}$. These radiation-induced Fe(III) reduction processes can be explained by electronic collisions that produce electronic holes and a cascade of secondary electrons, which gradually lose energy as they travel before becoming localized and trapped. Low- and thermal-energy electrons are highly reducing and can easily reduce Fe(III) centres to Fe(II). Alternatively, Fe(III) reduction may not be by direct electron transfer, but instead by radiolysis of water molecules in the phyllosilicate interlayer, or of hydroxyl groups present in the structure, to produce strongly reducing hydrogen radicals for localized Fe(III) reduction; however, such radiolysis would also result in the formation of OH radical oxidants. Radiation-induced Fe(III) reduction can result in an increase in the interlayer charge and may affect the interaction between phyllosilicate minerals and radionuclides released from the waste package (Allard *et al.*, 2012).

4. Fe-bearing minerals in candidate far-field lithologies

The far field will form the final barrier of a GDF where the complex geosphere will be required to retard and retain the long half-life AFP migrating out from the engineered barriers, for periods of 10^5–10^6 y. Because of its crustal abundance (5 wt.% Fe), complex mineralogy (Table 3, Fig. 4), redox geochemistry and involvement in

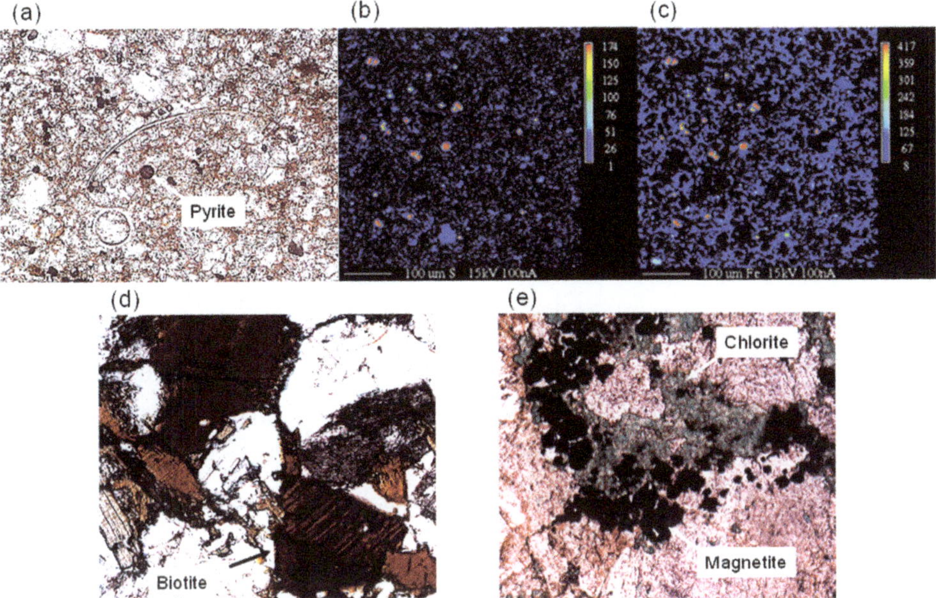

Figure 4. Lithologies demonstrating complex Fe mineralogy. Kimmeridge Clay (a) with elemental maps for (b) S and (c) Fe. Eskdale Granite (d) and altered Eskdale Granite (e).

bio-geological processes, it is essential to understand the behaviour of iron in the far field (Taylor and Konhuaser, 2011). In this section we focus on the mineralogical Fe-budget in key rock-types that might host a GDF and therefore their potential involvement in biotic and abiotic redox processes that would influence radionuclide migration. Candidate rock masses/lithologies being considered as GDF hosts are higher-geomechanical-strength crystalline rocks (igneous and metamorphic – already chosen by Sweden and Finland (SKB, 2005; Lahti *et al.*, 2009), lower strength sedimentary rocks (mudstones and clays – France, Belgium and Switzerland) and evaporites (some storage in the USA and Germany) (ANDRA, 2005b; NDA, 2010b).

Where lithological choices are available, mudstones and claystones are increasingly being considered as the likely GDF host (Enachescu *et al.*, 2002; ANDRA, 2005b). The Fe inventory of mudstones, especially the range of clay phyllosilicates, is complex and the fine scale lamination, small grain size (< 62.5 μm) and limited connectivity across layers leads to mm-scale geochemical heterogeneity. Mudstones are the most common sedimentary rock, forming successions, commonly in excess of 1000 m thick, during the filling of depositional sedimentary basins. They contain a diverse mixture of minerals dominated by clays (illite, smectites, kaolinite and chlorite), carbonates (siderite, calcite, dolomite) and silicates (quartz and feldspars) as well as pyrite, Fe-hydroxyoxides (goethite, limonite) and amorphous phases (ferrihydrite, Mn oxides), with varying amounts of organic matter (Aplin and Macquaker, 2011;

Table 3. Key iron-bearing phases in potential GDF host rocks. The table provides representative information on selected Fe phases from, often, complex mineral groups giving typical Fe contents, mineral formulae and Fe coordination.

Mineral	General formula	Fe coordination
Serpentines. Trioctahedral 1:1 silicates		
Serpentine ↔ *Cronstedtite*	$Mg_2Si_2O_5(OH)_4$ ↔ $(Fe_2^{2+}, Fe^{3+})(Si,Fe^{3+})_3O_5(OH)_4$	Fe^{3+} for Si, Fe^{3+}/Fe^{2+} for Mg. $Fe^{3+}O_h$, $Fe^{2+}T_d$, $Fe3^{+}T_d$. (Deer *et al.*, 2009; Hybler, 2014)
Berthierine 27 wt.% Fe	$[(Fe^{2+}(Al,Fe^{3+})\square)]_3$ $(Si,Al)_2O_5(OH)_4$	$Fe^{2+} >> Fe^{3+}$ $Fe^{3+}O_h$, $Fe^{2+}O_h$. (Brindley, 1982)
Odinite 20 wt.% Fe	$(Fe^{3},Fe^{2+},Al,Mg)_3$ $(Si,Al)_{2.5}O_5(OH)_4$	$Fe^{3+} > Fe^{2+}$, $Fe^{3+}O_h$, $Fe^{2+}O_h$. (Bailey, 1988)
Chlorites $(Mg,Fe)_5Al[(AlSi_3O_{10})(OH)_8]$ Trioctahedral phyllosilicates		
Chamosite Trioctahedral 2:1:1 30 wt.% Fe also *Fe clinochlore*	$(Fe^{2+},Fe^{3+}Mg)_3(Fe_2^{2+},Al)$ $[Si_3\ AlO_{10}](OH)_8$ 2:1 phyllosilicate layer with a brucite layer and Mg ↔ Fe substitution. $(Mg,Fe^{2+})_3[(MgFe^{2+})_2Al)]$ $[(Si_3Al)(O_{10})](OH)_8$	Fe^{2+} and Fe^{3+} in O_h, Fe^{2+} rarely in T_d Large range of Fe^{2+}/Fe^{3+} but $Fe^{2+} >> Fe^{3+}$ (Deer *et al.*, 2009; Brigatti *et al.*, 2011) Fe^{2+} usually >> Fe^{3+} but variable
Sudoite Di/trioctahedral 4 wt.% Fe	$Al_3[(Mg,Fe^{3+})_2]$ $[(Si_3Al)(O_{10})](OH)_8$	
Micas 2:1 phyllosilicates		
Glauconite 17 wt.% Fe	$(K,Na,Ca)_{0.6-1.0}$ $(Fe^{3+},Al,Fe^{2+},Mg)_{2.0}\square[Si_{3.5-3.8}$ $Al_{0.5-.02}]O_{10}(OH)_2.n(H_2O)$	Fe^{2+}/Fe^{3+} usually 0.12–0.14. Rarely >0.6 in ferro-glauconites. Mainly Fe^{3+} and Fe^{2+} O_h, some $Fe^{2+}T_d$ (Johnston and Cardile, 1987; Wilson, 2013).
Biotites Annite (Fe)–phlogopite (Mg) solid solution Typical 10–20 wt.% Fe	$K(Mg,Fe^{2+}, Fe^{3+})_2$ $[AlSi_3O_{10}](OH)_2$	Fe^{2+}/Fe^{3+} ratios vary with typical biotite 5–15% of ΣFe as Fe^{3+}. Fe mostly Fe^{2+} O_h, limited T_d. (Raeburn *et al.*, 1997; Fleet, 2003)

Table 3 (*contd.*)

Mineral	General formula	Fe coordination
Illite (mica-clay) Dioctahedral 3 wt.% Fe typical Up to 8.4 wt.% Fe	$K_{0.65}(Al,MgFe^{2+})_2$ $[Si_{3.35}Al_{0.65}]O_{10}(OH)_2$	Fe^{3+} O_h, Minor Fe^{2+}Oh and some $Fe^{2+}T_d$ in > 5 wt.% Fe compositions. Higher Fe^{3+} in higher Fe contents (Murad, 2010; Wilson, 2013)
Illite-smectites 3 wt.% Fe		Illite-smectites typically with Fe^{2+}/Fe^{3+} 0.13 (Środoń *et al.*, 1986; Wilson, 2013)
Smectites		
Nontronite Dioctahedral 15–30 wt.% Fe	$Na_{0.3}(Fe^{3+}Mg)_2$ $(SiAlFe^{3+})_4O_{10}(OH)_2.nH_2O$	$Fe^{3+}O_h$ with very minor T_d. Fe^{2+} can be present if >5 wt.% Fe (Johnston and Cardile, 1987; Gorski *et al.*, 2013; Wilson, 2013)
Saponite Trioctahedral 9 wt.% Fe	$(Na,K,Ca)_{0.3}(Mg,Fe^{2+}Fe^{3+})_3$ $(SiAl)_4O_{10}(OH)_2.nH_2O$	$Fe^{3+}O_h$ and $Fe^{2+}O_h$ in variable ratios (Andrews *et al.*, 1983)
Garnets		
Almandine	$Fe_3^{2+}Al_2(SiO_4)_3$	Fe^{2+} 8 fold
Andradite	$Ca_3Fe_2^{3+}(SiO_4)_3$	$Fe^{3+}O_h$ (Deer *et al.*, 1997a)
Amphiboles		
Hornblende Orthorhombic 20 wt.% Fe	$Ca_2[Fe_4^{2+}(Al,Fe^{3+})]$ $(Si_7Al)O_{22}(OH)_2$	Wide range of Fe^{3+}/Fe^{2+} Fe^{3+}/Fe^{2+} typically >0.2; Ferri-hornblendes can have $Fe^{3+}>Fe^{2+}$; $Fe^{2+}/Fe^{3+}O_h$
Grunerite Monoclinic Fe end-member has 39 wt.% Fe	$(Fe^{2+},Mg)_7Si_8O_{22}(OH)_2$	Mainly Fe^{2+} (Hawthorne *et al.*, 2007)
Epidote Group		
Epidote Monoclinic 24 wt.% Fe	$Ca_2(Al,Fe^{3+})_3Si_3O_{12}(OH)$	$Fe^{3+}O_h$ (Deer *et al.*, 1997b)

Table 3 (*contd.*)

Mineral	General formula	Fe coordination
Sulfides		
Pyrite 46 wt.% Fe	$Fe^{2+}S_2$	Fe^{2+} $O_{h.}$
Pyrrhotite 62 wt.% Fe	$Fe_{1-x}S$	Fe^{2+} O_h (Vaughan and Craig, 1978)
Carbonates		
Siderite-ankerite 48 wt.% Fe in siderite	$Fe^{2+}CO_3$ – $MgCO_3$	Fe^{2+} O_h in carbonates – oxidizes to Fe oxides during weathering (Pablo Gil *et al.*, 1992)
Ferroan calcite 2 wt.% Fe in calcite	$(Ca,Fe^{2+})CO_3$	
Ferroan dolomite 10 wt.% Fe	$(Mg,Ca,Fe^{2+})CO_3$	
Fe-(hydroxy)oxides		
Goethite 63 wt.% Fe	$Fe^{3+}O.OH$	Fe^{3+} O_h (Bowles *et al.*, 2011)
Ferrihydrite 62 wt.% Fe	$Fe^{3+}_{4\text{-}5}(OH,O)_{12}$	Fe^{3+} O_h 0.8, Fe^{3+} T_d 0.2. (Manceau and Drits, 1993; Maillot *et al.*, 2011)
Hematite 70 wt.% Fe	$Fe^{3+}_2O_3$	Fe^{3+} O_h (Lindsley, 1976; Bowles *et al.*, 2011)
Magnetite 72 wt.% Fe	$Fe^{2+}Fe^{3+}_2O_4$	$Fe^{3+}O_h$, $Fe^{2+}O_h$, $Fe^{2+}T_d$ (Pattrick *et al.*, 2002; Pearce *et al.*, 2006)

Taylor and Macquaker, 2011). The importance of clays in radionuclide retention is emphasized by many authors, *e.g.* Bradbury and Baeyens (2011), Baeyens *et al.* (2014) and the review of Brookshaw *et al.* (2012).

The relative amounts of the minerals present in mudstones, their Fe contents and Fe^{2+}/Fe^{3+} ratios depend on their sedimentary provenance and petrogenesis, especially depositional conditions and diagenetic history. Mudstone sequences typically have Fe contents between 2 and 6 wt.% Fe although rock sequences with an erosional hinterland of basic/intermediate igneous sources comprising easily weathered ferromagnesian minerals will have greater Fe contents of >7 wt.% Fe (Perri *et al.*, 2012). For many mudstones, clay minerals will be >50% of the mineral volume and, with clay grain sizes <4 μm, these will dominate the Fe-redox budget and reactivity of the bulk rock mass. The complex chemistry and multiple solid solutions in the clay minerals and other phyllosilicates mean that Fe contents (ΣFe) of individual mineral species can vary enormously. Although minerals may have characteristic Fe(II)/Fe(III) signatures, these too can be highly variable in different lithologies with different geo-histories.

Diagenetic history is a major factor in defining mudstone mineralogy. For instance, in immature mudstones/claystones (such as the Liassic Clays of northern Europe) which have suffered little burial and no tectonic alteration, the clay minerals will include smectite and interlayered smectite-illites (Peltonen *et al.*, 2008). Early, near-surface diagenesis is dominated by redox processes involving microbial oxidation of organic matter using sulfate and Fe-oxides as terminal electron acceptors, leading to the production of H_2S and Fe^{2+}, and the formation of pyrite/marcasite, ferroan carbonates, siderite and berthierine (Lovley and Phillips, 1988; Aplin and Macquaker, 2011; Pentrakova *et al.*, 2013). With increasing burial diagenesis and as temperatures reach >80°C, illite is produced and the creation of chlorite, *via* reactions involving K, such as:

$$\text{smectite} + \text{K-feldspar} = \text{illite} + \text{quartz} + \text{chlorite} + H_2O \quad (3)$$

(Hurst, 1985; Bjorlykke, 1998; Aplin and Macquaker, 2011). The Opalinus Clay, studied extensively in a GDF context at Mount Terri, Switzerland, is a typical mudstone (Table 4) with 66% clay minerals and significant illite, smectite and chlorite components, with overall iron contents of 2.85 wt.% FeO and 1.74 wt.%

Table 4. Mineral analysis of Opalinus Clay (Pearson *et al.*, 2004).

Mineral	Volume (%)
Total clay minerals	66.0
Clay minerals <2 μm	53.0
Illite	23.0
Chlorite	10.0
Smectite	11.0
Quartz	13.7
K-feldspar	1.0
Na-feldspar	1.0
Calcite	13.0
Dolomite/Ankerite	0.5
Siderite	3.0
Pyrite	1.1
Gypsum	0.20

Fe_2O_3; lower ΣFe and higher FeO is found in the in the more calcareous facies (Pearson *et al.*, 2004). Authigenic chlorite may also develop during diagenesis as a result of the dissolution of Fe- and Mg-rich detrital grains and volcanic rock fragments (Dowey *et al.*, 2012) and from mud intraclasts (Worden and Morad, 2003). These processes can lead to a wide range of chlorite compositions. Chlorite, generally, becomes an increasingly important Fe-bearing phyllosilicate as diagenesis progresses in mudstones, and remains the main stable Fe-carrier in mudstones during low-temperature regional metamorphism with the formation of slates and phyllites.

An example of small-scale heterogeneity in mudstones is shown by the Jurassic Kimmeridge Clay of the UK which is organic-rich and formed in marine basins of moderate and variable depth (10–100 m) under dysaerobic/anoxic conditions. In the relatively immature sections of the Kimmeridge Clay (containing 2–7 wt.% total Fe), the organic-rich mudstone layers contain kaolinite, illite, pyrite and/or siderite; higher pyrite concentrations represent depositional conditions that involved dysoxia/anoxia above the sediment–water interface (Fig. 4a–c). However, in organic-poor, carbonate-rich horizons of the same formation, ferroan calcite and dolomite are the dominant Fe phases (Matthews *et al.*, 2004). Ironstones are also found in close association with the Kimmeridge and Liassic Clays. These originally comprised layers rich in ferric iron oxyhydroxides that were wholly or partly replaced by chamosite/berthierine under reducing conditions during early diagenesis, and have been partially re-oxidized to goethite, under the influence of relatively recent ground waters (Palmer and Wilson, 1990; Taylor and Curtis, 1995; Matthews *et al.*, 2004; Mücke and Farshad, 2005). Similar deposits are forming in marine shelf systems such as the Amazonian Basin discharge where iron (rather than sulfate) reduction dominates and siderite and berthierine are the main diagenetic products. In these low-sulfur mudstone sequences, discrete ironstone layers (mm to m and <30 wt.% Fe) develop, with early ferric iron hydroxides replaced by chamosite and bethierine during reductive, early diagenesis (Taylor and Macquaker, 2011).

The Fe-bearing phyllosilicate minerals in mudstones have characteristic Fe(II)/Fe(III) ratios although these are, in part, controlled by petrogenesis. Smectites, such as montmorillonite, found in immature mudstones often have typical Fe contents in the range 3–7 wt.% with >95% of the iron as Fe(III). Nontronite, the Fe-smectite, is found in mudstones with significant ferruginous content and can typically contain 30 wt.% Fe, with the majority or all of the iron also Fe(III) (Gorski *et al.*, 2013). Fe(II) is, however, recorded in the smectite, saponite, where the protolith was basic igneous in origin (Andrews *et al.*, 1983). Both *in situ* bioreduction of smectites and their production as weathering products of Fe(II)-bearing chlorite and biotite yields smectites with Fe(II)/Fe(III) near to 1 (Bristow *et al.*, 2009; Jaisi *et al.*, 2011; Brookshaw *et al.*, 2012). Illites typically contain 1–5 wt.% Fe, with some as high as 12 wt.% Fe – the iron is present usually as Fe(III) with <10% of the ΣFe as Fe(II), although in some low-Fe illites (<2 wt.% Fe) the Fe(II) can be greater than Fe(III) (Wilson, 2013). Chlorite is the end product of many diagenetic reactions in mudstones and the chlorite group has a range of Mg ↔ Fe and Al ↔ Si substitutions which define a range of mineral species (Deer *et al.*, 2009). Iron contents of chlorites provide a full spectrum, from 0 wt.% Fe in pure clinochlore to >35 wt.% Fe in typical chamosites. The most important Fe-bearing chlorite-group minerals are chamosite and ferroan clinochlore in which Fe^{2+} dominates but 10–35% of the ΣFe is commonly Fe(III).

Although phyllosilicates will often contain the bulk of the Fe in mudstone, many other Fe phases might play important roles in GDF far-field performance. Pyrite and marcasite (FeS_2 polymorphs), are Fe(II) minerals, and are omnipresent components of mudstones that have suffered diagenesis – formed as a product of both abiotic and biotic iron reduction of seawater sulfate. In mudstones, pyrite is present as framboidal aggregates of micron-sized crystals that provide a large surface available to cause local reduction of permeating fluids, and will break down to Fe-oxyhydroxides if those fluids are oxidizing. Magnetite, $Fe(II)Fe_2(III)O_4$, can be present in mudstones as a refractory detrital phase or produced by bioreduction of Fe-oxide species during early diagenesis. Iron in carbonates is present as Fe(II), forming a major component in siderite and ankerite, and also up to 1000s ppm in calcite. The preservation of primary oxidized iron species such as ferrihydrite, lepidocrocite, limonite, goethite, maghemite and hematite in mudstones depends on subsequent diagenesis, when most or all of these Fe(III) phases are consumed, forming the sources of Fe in phyllosilicates and other more reduced species. Given their ability to adsorb ions from solutions and their importance as electron acceptors in bioreduction, these Fe^{3+} phases will play a key role in far-field performance. As 'later' migrating oxidizing fluids will lead to oxidation of the sulfides, carbonates and phyllosilicates in a rock mass and the production of ferrihydroxyoxide and ferrioxide species, these could be the key Fe-species in far-field sequences. This is especially true if this oxidation is concentrated in the higher permeability zones (especially fractures) surrounding the engineered GDF.

Crystalline, higher-strength rocks are produced where igneous or metamorphic petrogenesis has resulted in impervious rocks with limited porosity and fracture-dominated permeability. Coarsely crystalline granitic plutons such as the I-type plutons (*e.g.* Cordilleran batholiths and Caledonian granitoids) are GDF candidates. These have typical bulk iron contents of ~2.1 wt.% Fe, an Fe(III)/ΣFe of 0.44 with the main Fe-bearing phase being biotite. S-type granites (such as the Cornubian batholith, UK) are developed from melts derived from crustal sequences and have a lower oxygen fugacity (f_{O_2}) and are reduced with Fe(III)/ΣFe = 0.24 (Chappell and Hine, 2006). In more basic lithologies (granodiorite/diorite/andesite) containing greater Fe contents of 5.5 wt.% Fe and Fe(III)/ΣFe = 0.39, amphiboles (hornblende) become an increasingly important Fe-bearing phase. Basaltic rocks with 8.0 wt.% Fe (Fe(III)/ΣFe = 0.35) have pyroxene/olivine as the ferro-magnesium phases but these lithologies are not generally being considered as GDF candidates. In crystalline rocks, most of the iron in biotite is present as Fe(II) although almost all biotites contain between 5 and 15% Fe(III) of the ΣFe (Fleet, 2003; Brigatti *et al.*, 2008). The Fe-bearing amphiboles (*e.g.* grunerite and hornblende) contain iron mainly as Fe(II) with typical ratios Fe(III)/ΣFe = 0.2 in dioritic rocks although higher Fe(III) has been found in oxyamphiboles where Fe(II) ↔ OH^- exchange has taken place (Hawthorne *et al.*, 2007).

An example of a crystalline GDF host is the Swedish GDF site at Forsmark which is situated in the highly deformed and metamorphosed crystalline gneisses of the 1.87 billion year old Fennoscandanavian Shield. Most ancient shield terrains or greenstone belts are a complex series of gneisses with coarse grain size and mineralogy dependent on protolith. The Swedish GDF will be hosted in a less deformed granitic gneiss containing 5 vol.% biotite (with minor chlorite, pyrite and epidote) and a whole-rock Fe(III)/ΣFe = 0.44. Granodioritic rocks in the area contain equal amounts of biotite + amphibole and an Fe(III)/ΣFe = 0.3.In common with other metamorphosed terrains, the dominant Fe-bearing phase in meta-basic rocks is amphibole (hornblende) (up to 36 vol.%), along with several percent biotite in specific layers (Stephens, 2010). The area is extensively fractured (300 fractures at a depth of 500 m) where later fluids have precipitated a contrasting mineralogy, dominated by chlorite and calcite, with lesser amounts of clay phases, hematite, pyrite and minor goethite and epidote. The Fe(III)/ΣFe of the chlorite in these fractures is in the range 0.12–0.33.

Many crystalline terrains have suffered retrograde metamorphism, often involving hydrous fluids that alter the biotite and hornblende to phyllosilicates such as chlorite or illite, with magnetite as a by-product. This can affect the whole-rock mass or be confined to areas adjacent to fractures. The Fe^{3+} of the resulting illite, chlorites and sudoites ($Mg_{1.9}Fe^{2+}_{0.1}Al_{2.9}Fe^{3+}_{0.2}Si_3AlO_{10}(OH)_{7.9}$) varies considerably, dependent on the oxidation state of their 'parent' fluids and can show local variations. For instance, in the Canadian Shield, rocks affected by fluids involved in uranium ore-forming systems, sudoite and clinochlore with Fe^{3+}/ΣFe ratios of 0.3–0.4 formed from relatively reducing Fe^{2+}-bearing basement fluids. This contrasted with the rock mass interaction with relatively oxidized basinal fluids where the Fe^{3+}/ΣFe of the phases was raised to 0.50 (Kister *et al.*, 2005).

Lower grades of hydrous metamorphism can have a major control on Fe-mineralogy of candidate lithologies. The Ordovician Borrowdale Volcanic Series (BVS) of the Lake District has been a candidate for a geodisposal rock laboratory near Sellafield, UK, and mainly comprises a sequence of calc-alkaline, andesitic pyroclastics. However, little remains of the primary ferromagnesian mineralogy of amphibole-pyroxene-biotite and low-grade regional metamorphism has resulted in chlorite being the dominant Fe-phase in a chlorite + white mica + quartz ± calcite ± epidote ± actinolite ± prehnite ± pumpellyite assemblage, with significant pyrite and Fe-Ti oxides sporadically developed (Oliver *et al.*, 1984). Furthermore nine stages of fracture-controlled mineralization, divisable into early silicate-sulfide mineralizing fluids and later oxidized sulfate-hematite fluids have pervaded the BVS, contributing to its alteration (Meller, 1998; Milodowski *et al.*, 1998; Engelberg *et al.*, 2012).

Any modelling of GDF performance will necessarily account for the relative accessibility of near-field fluids to the far-field iron mineralogy – the fracture-based permeability of crystalline rocks being an example of the bulk lithology being dominated by biotite while fluids will preferentially access fractures lined with chlorite. In these fracture-controlled systems, 'rock matrix diffusion' of fluids through the limited pore space in the bulk rock is a factor, but the reactions in the high-permeability (fracture) zones will dominate. Even in mudstones, variations in porosity, grain size and burial history mean permeabilities can vary by 10 orders of magnitude (Dewhurst *et al.*, 1999). The permeability of the Jurassic Callovian-Oxfordian argillites being examined for the French GDF at Bure are low, with permeabilities of 10^{-12} to 10^{-14} m/s. However, within one rock sequence the different layers can have a very wide range of permeabilities, and transmissivity of fluids through any rock mass along layers will be orders of magnitude larger than across layers. Indeed in a mudstone sequence the ironstone layers with coarser grain sizes, ooidal textures could form the most permeable zones and be important in overall far-field performance, and Fe redox activity, especially where they intersect the engineered barrier.

In addition, modelling of far-field GDF performance must include microbial effects on Fe mineral–radionuclide interactions. Microbial activity may have direct or indirect effects on the host-rock minerals, as well as the radionuclides themselves. With appropriate amendments, microbially mediated processes could facilitate reducing subsurface conditions conducive to radionuclide immobilization. Brookshaw *et al.* (2012) provide a detailed review focussing on the reactions between microbes and mineral-radionuclide systems, including the role that microbes play in mineral dissolution and precipitation.

Generic studies of the reactions involving the Fe-bearing components of the geosphere provide an important blueprint for understanding of the diverse geological and microbiological processes involved in the far field of a GDF, but the natural inhomogeneity of geological terrains make it very difficult to produce robust, holistic geochemical models, until potential rocks masses that will host a specific GDF are identified.

5. Radionuclide interaction with redox-active Fe components: natural samples and synthetic analogues

Once the redox-active Fe components in both near and far field environments have been identified and characterized (Table 3), their interaction with radionuclides can be studied using single-phase natural samples and synthetic analogues. These studies provide a molecular-level understanding of adsorption, diffusion, growth, dissolution, precipitation and electron transfer processes at the mineral–groundwater interface, which is required to model radionuclide migration as reaction complexity at these small scales has a significant effect on geochemical cycling of elements, groundwater composition, and contaminant transport.

In countries with a long history of nuclear fission, such as the UK and the USA, large volumes of contaminated land exist as a result of controlled or accidental release of effluents from spent fuel reprocessing and storage facilities. The molecular and microscopic characterization of field samples from these sites has been a critical factor in predicting radionuclide migration behaviour in the safe disposal of nuclear waste. The Hanford nuclear reservation in Washington State, USA occupies 1518 km^2 and produced 67 tons of Pu for USDOE weapons programmes (1943–1989). 200 million L of high-level radioactive and chemical waste (60% of all US HLW) is stored at Hanford in 177 huge underground tanks. A third of the tanks are suspected to have leaked 1.9–3.8 million L of tank waste containing actinides (Pu, U, Am) and fission products (Cs, Sr, Tc) into the subsurface, making it the most contaminated US nuclear site with the nation's largest environmental clean-up activity. Technetium-99 is a fission product of uranium-235 and plutonium-239 and has been introduced into the environment through the processing of irradiated nuclear fuel. Because of its long half-life ($t_{1/2}$ = 2.13×10^5 y), Tc is considered to be one of the most hazardous radioactive contaminants. About 2000 kg of Tc was produced at the Hanford Site and, to date, >400 Ci (14800 GBq) of Tc has been released to the vadose zone.

Hanford sediments generally have basaltic provenance and are rich in Fe-bearing minerals of mixed valence including magnetite, titanomagnetite, ilmenite, Fe(II)/Fe(III) phyllosilicates, and Fe(III) oxides. Peretyazhko *et al.* (2012) characterized the Fe(II) mineral suite in a fine-grained unit of Ringold formation sediments collected from a 55 m borehole that sampled a semi-confined aquifer at the Hanford Site, and contained a dramatic redox transition zone. Fe(II)-bearing phyllosilicates, pyrite, magnetite and siderite were present in the sediments with the Fe(II) mineral phase distribution differing between the oxic and anoxic facies (Fig. 5). The redox reactivity of these mineralogically heterogeneous, Fe(II)-containing sediments was assessed with respect to abiotic Tc(VII) reduction. Technetium valence state is a crucial factor determining Tc mobility in the environments. In oxic conditions, anionic pertechnetate $[Tc(VII)O_4]^-$ is stable and mobile, as it is sorbed only weakly at circumneutral and basic pH. The reduced form of technetium, Tc(IV), is stable in anoxic environments (Peretyazhko *et al.*, 2012). The mechanisms of Tc(VII) to Tc(IV) reduction, particularly by reactive Fe(II) in naturally occurring mineral phases, are of interest because of their implications for subsurface migration in the GDF far-field environments. Technetium(VII) reduction occurred in all anoxic Ringold sediments at depths >18.3 m and reaction time differed significantly between the sediments (8–219 days). Analysis of the Tc-reacted sediment revealed changes in the concentrations of solid-phase Fe(II) and Fe(III), with a decrease in siderite and Fe(II)-containing phyllosilicates contributions as a result of oxidation during reaction with Tc(VII). Spectroscopic analysis of the Tc associated with sediments showed that it was present in the Tc(IV) valence state and immobilized as clusters of a $TcO_2.nH_2O$-like phase (Peretyazhko *et al.*, 2012).

From the overlying vadose zone, coarse unconsolidated sediments of the Hanford formation were sampled from several locations on Hanford's central plateau (Pearce *et al.*, 2014). Natural Fe(II)/Fe(III)-bearing microparticles were isolated by magnetic separation, and this mineral fraction was found to represent up to 1 wt.% of the total sediment and be composed of 90% magnetite with minor ilmenite and hematite, as determined by X-ray diffraction. The magnetite contained variable amounts of transition metals consistent with alio- and isovalent metal substitutions for Fe. X-ray microprobe analysis showed that Ti was the most significant substituent, and that these grains could be described with the titanomagnetite formula $Fe_{3-x}Ti_xO_4$, which falls

Figure 5. Redox boundary in 300 A Ringold sediments ~2.5 m below the Hanford-Ringold contact (from Williams *et al.*, 2007, used with the permission of PNNL, Washington, USA).

between endmember magnetite ($x = 0$) and ulvöspinel ($x = 1$); the dominant composition was determined to be $x = 0.15$. Reaction of the magnetically separated natural phases from Hanford sediments with Tc(VII) in solution showed that they were able to reductively immobilize Tc(VII) with concurrent oxidation of Fe(II) to Fe(III) at the mineral surface (Fig. 6). A series of mixed-valent titanium-doped magnetite nanoparticles, structurally and chemically analogous to the natural titanomagnetites, were subsequently synthesized to investigate further heterogeneous reduction of highly soluble $[Tc(VII)O_4]^-$ to sparingly soluble Tc(IV)-bearing solids. Reaction with Tc(VII) in solution yielded fast exponentially decaying reduction kinetics with rates that increased with increasing solid-state Fe(II)/Fe(III) ratio in the nanoparticles, a characteristic systematically controlled by the Ti content (Liu *et al.*, 2012). Thus various mineralogical constituents at the Hanford site bear reactive Fe(II), both in oxidizing vadose zone sediments and in reducing environments at depth near

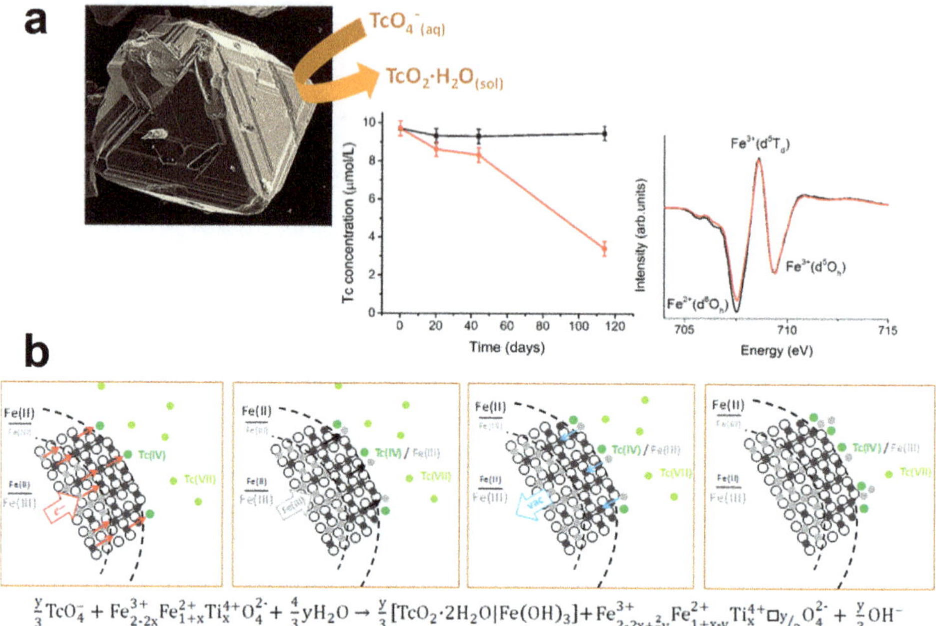

Figure 6. (a) Sample without titanomagnetite (black line) and sample with natural titanomagnetite (grey line, red in online version) showing reduction of Tc(VII) to Tc(IV) with corresponding Fe L-edge XMCD difference spectra before (black line) and after (red line) Tc(VII) reduction. (b) Conceptual model for TC(VII)(aq) reactivity with $Fe_{3-x}Ti_xO_4$ nanoparticles.

groundwater, and have the capacity to reductively immobilize contaminants such as technetium (Cui and Eriksen, 1996a,b).

The actinides, uranium and plutonium, are also abundant radioactive contaminants in the subsurface at the Hanford site. As with technetium, oxidation state plays a major role in determining U and Pu mobility in the environment. Under reducing conditions where U(IV) is thermodynamically stable, U forms sparingly soluble minerals such as uraninite (UO_2). Under oxidizing conditions, U(VI) is stable as the uranyl ion $[UO_2]^{2+}$ and is generally more mobile than U(IV). Latta *et al.* (2013) found that titanomagnetites ($Fe_{3-x}Ti_xO_4$) could also reduce U(VI) to U(IV). Redox reactivity of U(VI) with Fe(II) was controlled directly by the Fe(II)/Fe(III) ratio but was otherwise independent of Ti content. However, in contrast to previous studies with synthetic end-member magnetite, where U(VI) was reduced to nanocrystalline uraninite (UO_2) (Latta *et al.*, 2012), the presence of structural Ti resulted in the formation of U(IV) species that lacked the bidentate $U{-}O_2{-}U$ bridges of uraninite and instead formed a binuclear corner sharing adsorbed or incorporated monomeric U(IV) complex with the solid phase. This speciation distinction based on mineralogical composition yielding higher-affinity surface sites could profoundly influence the long-term stability of U(IV) phases formed in sediments (Latta *et al.*, 2013, 2014).

Singer and co-workers (Singer *et al.*, 2012a,b) measured sorption of U(VI) onto the (111) surface of end-member magnetite (Fe_3O_4) as a function of time and solution composition under continuous batch-flow conditions. A combination of atomic force microscopy (AFM) and U L-III-edge grazing-incidence X-ray absorption spectroscopy near-edge structure (GI-XANES) spectroscopy indicated that uranium uptake was divided into three-stages; (1) initial adsorption of U(VI); (2) reduction of U(VI) to UO_2 nanoprecipitates at surface-specific sites after 2–3 h of exposure; and (3) completion of U(VI) reduction after 6–8 h. Reaction stoichiometry was consistent with U(VI) reduction coupled with enhanced release of Fe(II) from magnetite (Singer *et al.*, 2012b). Uranium uptake pathways on magnetite as a function of U(VI) aqueous speciation was further investigated to determine the role of U(VI)-CO_3-Ca complexes in inhibiting U(VI) reduction. In the presence of CO_3, adsorption of U(VI) surface species occurred concurrently with reduction to U(IV) but in the presence of both Ca and CO_3, only U(VI) adsorption occurred. Nanoparticulate UO_2 – the product of U reduction – formed only within and adjacent to crystal boundaries and cracks, suggesting that U reduction is limited to defect-rich surface regions (Singer *et al.*, 2012a). The reactivity of both sorbed and structural Fe(II) with respect to heterogeneous reduction of U(VI) to U(IV) on magnetite has also been studied using XAS and X-ray photoelectron spectroscopy (XPS). The results suggested an important role for the formation of intermediate oxidation states in determining reaction rates, with the reaction pathway involving incorporation and stabilization of U(V) and U(VI) into secondary phases (Ilton *et al.*, 2010). The stability of pentavalent U was further investigated by locking U(VI) in the hematite structure at circum-neutral pH, exposing it to organic reductants, and comparing it to the reduction of non-incorporated U(VI). A combination of electron microscopy and X-ray spectroscopy showed that U(VI) was incorporated in hematite and reduced to U(V), whereas non-incorporated U(VI) was reduced to U(IV), suggesting that incorporation of U in Fe(III) (hydr)oxides is a distinct possibility and revealing U(V) as a possibly important species of uranium in subsurface environments (Ilton *et al.*, 2012).

Fe-containing phyllosilicates such as micas are also important mineralogical components of the Hanford subsurface. Like magnetite, these naturally occurring phases can host mixed-valence Fe, usually within the dioctahedral or trioctahedral sheets, and therein possess moderate charge-carrier mobilities at room temperature that enable these phases to serve as a source or sink of electron density for redox reactions with radionuclides (Rosso and Ilton, 2003, 2005; Alexandrov *et al.*, 2013; Alexandrov and Rosso, 2013). For example, Ilton *et al.* (2006) investigated the role of the interlayer region of micas containing 2.8, ~0.02 and 0.01 Fe(II) atomic % in the reduction of U(VI). Appreciable reduction of uranyl in all three micas occurred under anaerobic conditions and U was still partially reduced under aerobic conditions, although a limited amount of re-oxidation occurred over time. The results showed that reduction of U(VI) in the interlayer of Fe(II)-containing micas is viable and that under aerobic conditions reduction is initially faster than re-oxidation. This extends the range of possible environments where heterogeneous reduction of U(VI) by micas might occur (Ilton *et al.*, 2006)

As well as the heterogeneous reduction of actinides in higher, more soluble oxidation states to lower, more insoluble oxidation states by Fe(II), it is also possible for Fe(II) to increase the mobility of radionuclides, such as by inducing the reduction of sparingly soluble Pu(IV) to the more soluble, lower oxidation state Pu(III). A recent study of this topic laid out the importance of understanding factors that control the form of the oxidized Fe(III) minerals produced because of its thermodynamic consequences for the overall driving force for AFP reduction by Fe(II) (Felmy *et al.*, 2011b). The differences in the free energy of solid-phase Fe(III) oxides produced by this reaction was shown to greatly influence the extent of amorphous PuO_2 reduction by Fe(II) and thus the concentration of aqueous Pu(III). This was demonstrated by the enhanced reduction of PuO_2(am) to Pu(III) by Fe(II) when the Fe(III) mineral goethite (FeO(OH)) was spiked into the reaction. Reduction of PuO_2(am) was also studied in the presence of aqueous Fe(II) and magnetite (Fe_3O_4) nanoparticles. The amount of dissolved Pu mobilized from PuO_2(am) was much less than that predicted by the measured pH and redox potentials coupled to equilibrium thermodynamic modelling, which suggested the presence of sorbed Pu(III). Sorbed Pu(III) was displaced by competitive cations, *e.g.* Eu(III), but similar aqueous Pu concentrations were measured with and without magnetite, indicating that the Pu was solubilized from PuO_2(am), not from the nanoparticulate magnetite (Felmy *et al.*, 2012). The observed Fe(III) reaction products were formed from the direct oxidation of Fe(II) to Fe(III) as a result of solution phase radiolysis by the $^{239}PuO_2$(am) suspensions, without a direct link to the underlying Pu redox reaction. However, reaction of aqueous Pu-242(III) and Pu-242(V) with magnetite, under anoxic conditions, did reveal a highly specific Pu(III)-sorption complex on the magnetite surface (Kirsch *et al.*, 2011). These results highlight some of the complexities associated with developing quantitative models to predict redox-active radionuclide chemistry and mobility in the environment, especially in association with common mineral phases. Issues related to geochemical predictions of the behaviour of redox-active elements and mineralogical species, including the difficulties associated with measurement of redox potential in the environment are discussed by Ahmed *et al.* (2017, this volume).

Besides redox-active coupling with radionuclide oxidation state, Fe-bearing minerals are also important adsorbents. Sorption reactions at the mineral–aqueous solution interface have been researched extensively to obtain solid/liquid distribution constants (K_D-values). K_D values quantify the link between the concentration of radionuclide sorbed to particles (C_S) and the concentration of radionuclide in solution (C_w) (Boyle and Birks, 1999):

$$K_D = \frac{C_s}{C_w}(\mathrm{m^3kg^{-3}}) \tag{4}$$

Single distribution coefficients such as these are often used as input data for performance assessment modelling relating to GDF development and contaminated land remediation (Geckeis *et al.*, 2013); however, they are applicable only at trace concentration ranges and for reversible sorption reactions, thus are of limited use for complex, redox-reactive, Fe-bearing mineral surfaces. Sorption of radionuclides is strongly mineral-specific due to surface complexation considerations such as surface structure and charge density, *e.g.* Boily and Rosso (2011), as well as being sensitive to contact time (Ilton *et al.*, 2012). The adsorption and speciation of U(VI) on Fe-bearing minerals present in Hanford sediments (chlorite – $(Mg,Fe)_3(Si,Al)_4O_{10}$ and ferrihydrite – $(Fe^{3+})_2O_3{\cdot}0.5H_2O$)) have been investigated (Wang *et al.*, 2011). U(VI) adsorption K_d values varied with mineral type and U concentration with 100% adsorption on 6-line ferrihydrite and three different chlorites displaying high and variable K_d values. For three chlorites with different compositions, the K_d decreased with increasing Fe, suggesting that U(VI) was not reduced by structural Fe(II) (Wang *et al.*, 2011).

Whilst these studies involving contaminated land provide extremely useful information on the redox reactivity of Fe-bearing mineral phases in the environment, it is also important to look more specifically at radionuclide interaction with redox-reactive Fe components that are present in the candidate rock masses/lithologies being considered as GDF hosts. In countries such as the UK, where the siting process for a HLW GDF is ongoing, it is necessary to study radionuclide interaction with generic natural mineral samples that are present in a variety of potential host rocks. Hallam *et al.* (2011) investigated the sorption behaviour of technetium with respect to a range of materials likely to be found in the vicinity of a UK GDF, including hematite (Fe_2O_3) and goethite (α-FeO(OH)). The aerobic aqueous chemistry of Tc is dominated by the highly mobile pertechnetate anion (TcO_4^-), which is virtually non-sorbing in the geosphere. However, given the low Eh in the near field, most of the Tc is likely to be in the lower oxidation state Tc(IV). Batch ^{95m}Tc(IV) sorption experiments with goethite and hematite at pH 3–13 followed a rising trend from mildly acidic to a peak at circumneutral pH (pH 6.6–6.8), with a subsequent decrease in the degree of sorption as a function of pH to ~pH 11–12 before rising again in the hyperalkaline region (> pH 12). The peak partition coefficient (Rd) for goethite (2.15×10^4 cm^3 g^{-1} at pH 6.8) was ~three times larger than that for hematite. Modelling of the sorption data suggested that the binding mechanism involved monodentate or bidentate surface complexation (Hallam *et al.*, 2011).

In Belgium, the Boom Clay Formation is studied as a reference host formation for the installation of GDF. The major redox-controlling minerals in Boom Clay are considered to be pyrite, FeS_2 (1–5 wt.%), and, to a lesser extent, siderite, $FeCO_3$ (0–1 wt.%) (Bruggeman and Maes, 2010), with the measured solution redox potential (Eh) controlled by the equilibrium of these mineral phases:

$$FeS_2 + 9H_2O + CO_2 \leftrightarrow 18H^+ + 2SO_4^{2-} + FeCO_3 + 14e^- \quad (5)$$

Pyrite is a constituent of host rocks and bentonite backfills and is considered the most important iron sulfide with respect to radioactive waste disposal. X-ray absorption spectroscopy has been used to elucidate the speciation of U in these systems and to gain insight into the major reaction mechanisms between U and FeS_2. Results show that U(VI) was partly reduced to a UO_2-like precipitate, although the predominant valence state of U in solution remained U(VI), with the solid-liquid distribution governed by both reduction and adsorption processes. With dissolved organic matter present in the system, the fraction of U(IV) in the solid phase decreased and the concentration of U in solution increased but colloidal U complexed with the natural organic matter was not observed (Bruggeman and Maes, 2010).

The long-lived radioactive isotope Se-79 (half-life 4.8×10^5 y) is produced by fission of U-235 in nuclear power plants and thus its presence is of concern for the safe enclosure of nuclear waste repositories. Se forms anions and has a low sorption coefficient therefore exhibits limited retention in the repository system and high mobility in the geosphere; thus this fission product may dominate the radiation dose from geological disposal of high-level nuclear wastes for long periods of time (Finck *et al.*, 2012). Se mobility is controlled by the oxidation state: the oxidized species (Se(IV) and Se(VI)) are highly mobile, whereas the reduced species (Se(0) and Se(-II)) form low-solubility solids. To evaluate the long-term behaviour of Se in a GDF, it is important to investigate the solubility of Se-79 and its solubility-limiting solid phase(s) under repository conditions. The formation of Fe-Se solids is expected due to the release of Fe(II) by Fe or steel canister corrosion under reducing conditions. Se(-II) and Se(IV)-doped pyrite and mackinawite have been synthesized *via* direct precipitation under anoxic and acidic conditions. Rapid precipitation resulted in substitution of S(-I) by Se(-I) in Se-doped pyrite and of S(-II) by Se(-II) in Se-doped mackinawite. However, slower precipitation of Se-doped pyrite resulted in Se(-II) and Se(IV) retention by incorporation coupled with an oxidation-state change and Se is incorporated as Se0 into pyrite without structural bonding. ^{79}Se would thus be effectively immobilized in an anoxic pyritiferous environment (Diener *et al.*, 2012).

In the reducing Boom Clay environment Se(-II), with HSe^- as the main aqueous species, is the dominant thermodynamically stable form of Se, thus Se(-II) retention by coprecipitation with and by adsorption on mackinawite (FeS) has been further investigated. The Se did not significantly influence the FeS precipitate morphology and X-ray absorption spectroscopy revealed that coprecipitation resulted in the Se located in a mackinawite-like sulfide environment (Finck *et al.*, 2012). Dynamic dissolution-recrystallization of the highly reactive FeS in the presence of adsorbed Se(-II) also resulted in incorporation of the Se. Thus, both these interaction mechanisms (sorption on pre-existing substrate and coprecipitation) with FeS represent an effective retention potential for radioactive Se by forming a (meta)stable solid solution (Finck *et al.*, 2012). The products of the coprecipitation of $[Tc(VII)O_4]^-$ and Tc(IV) with FeS have also been characterized (Wharton *et al.*, 2000). $[Tc(VII)O_4]^-$ was reduced upon coprecipitation with FeS and, although it was not possible to confirm the incorporation of Tc in the FeS phase due to the poorly ordered nature of the precipitates, XAS suggested that the Tc was present as a TcS_2-like cluster.

6. Radionuclide interaction with redox-active Fe components: Thermodynamic and molecular modelling

There are many theory and simulation tools which are useful for understanding the chemistry of actinide and fission-product species in subsurface environments. First and foremost is the necessity to understand the equilibrium thermodynamics of speciation, phase stabilities, and reaction free energies for specific conditions. Developing reliable thermodynamic models for predicting the oxidation state, solubility and speciation of radionuclides is of critical importance to predicting their long-term fate and transport in the environment. Application of advanced spectroscopic techniques and computational molecular simulation in combination then often provides a powerful level of detailed insight into AFP structure and reaction processes that can be used to refine thermodynamic models. Modern computational chemistry methods are robust and at the cost of increasing computational expense can, in many cases, yield reliable estimates for thermodynamic quantities for direct inclusion in macroscopic models. Below, we overview select recent examples of both thermodynamic and molecular simulation approaches.

There has been a great deal of progress in developing accurate thermodynamic data for radionuclides over the past two decades, much of the progress relates to efforts by the OECD Nuclear Energy Agency (NEA) documented in a series of critical reviews of the thermodynamic data for many redox active actinide elements and fission products (Grenthe *et al.*, 1992; Rard *et al.*, 1999; Lemire *et al.*, 2001; Guillaumont *et al.*, 2003; Hummel *et al.*, 2005; Olin *et al.*, 2005; Bruno *et al.*, 2007). For example, the room-temperature solubility of amorphous, hydrous technetium(IV) oxide ($TcO_2.xH_2O$) was studied across a broad range of pH values extending from 1.5 to 12 and in oxalate concentrations from dilute (10^{-6} mol. kg^{-1}) to complete saturation with respect to sodium bioxalate at lower pH values, and to saturation with respect to sodium oxalate at higher pH values. The solubility was measured to very long equilibration times (*i.e.* as long as 1000 days or longer). The thermodynamic modelling results show that the dominant species in solution must have at least one more hydroxyl moiety present in the complex than proposed by previous investigators (*e.g.* TcO(OH)-oxalate$^-$ rather than TcO(oxalate)(aq)). Inclusion of the single previously unidentified species TcO(OH)-oxalate$^-$ in the aqueous thermodynamic model explains a wider range of observed solubility data for $TcO_2.xH_2O$(am) in the presence of oxalate and over a broad range of pH values. Inclusion of this species is also supported by the recently proposed thermodynamic data for the $TcO(OH)^+$ hydrolysis species which indicates that this species is stable at pH values as low as one (Hess *et al.*, 2008).

However, with the exception of a limited number of solid solutions (Bruno *et al.*, 2007), the vast majority of such thermodynamic data for radionuclides is for solids of uniform or ideal composition or species formed in the aqueous phase. Very little information is available for the redox-active radionuclides incorporated into bulk or natural materials, such as the iron oxides, the importance of which is described above. In addition, and also of critical importance, is the chemical form of the Fe(III) reaction product formed during the radionuclide reduction reaction. It is the oxidation of Fe(II) to Fe(III) that provides the energy for redox-active radionuclide reduction. Hence both the chemical form of the radionuclides before and after reaction as well as the chemical form of the Fe(II) before reaction and the chemical form of the Fe(III) after reaction must be known to accurately predict the overall extent of radionuclide reduction (Felmy *et al.*, 2011a,b, 2012).

As an example, the reduction of U(VI) to UO_2(am) by Fe(II) *via* the overall reaction,

$$\mathrm{U(VI)\ (aq) + 2Fe(II)(aq) \leftrightarrow UO_2(am) + Fe_2O_3(hematite),} \qquad (6)$$

becomes thermodynamically possible at pH values just above 6 whereas if the reaction product is an aqueous Fe(III) hydrolysis species, the reaction is only possible above pH 9 (Fig. 7). These calculations illustrate that the free energy of the Fe(III) reaction products is a key factor in determining the range of conditions under which it is possible to reduce the redox active radionuclides by Fe(II).

Unfortunately, several factors make identifying the structural and thermodynamic stability of both the radionuclide and the Fe(III) reaction product in environmental materials rather difficult. As regards the Fe(III) reaction product, Navrotsky (2008) has pointed out that variations in the surface area alone of different iron oxide phases has a large impact on the thermodynamic stability of the pure phase components. An even more complex situation is evident when one examines the reduction of iron-containing phyllosilicates. Iron-containing clay minerals are nearly ubiquitous in the subsurface and the Fe(III)/Fe(II) transformations in such phases often control the effective redox potential in the subsurface by serving as electron acceptors/donors for abiotic or biotic reduction reactions. Developing predictive models for the thermodynamics of Fe(II) to Fe(III) transitions in such materials is difficult owing to the wide range of Fe-containing clay minerals, the coupling of the Fe(III)/Fe(II) reduction processes to other charge compensation mechanisms in the mineral structure, the lack of equilibrium in the system, and potential issues with reaction reversibility (see Neumann *et al.* (2011) for a discussion). As a result, the treatment of the thermodynamic relations for the clay

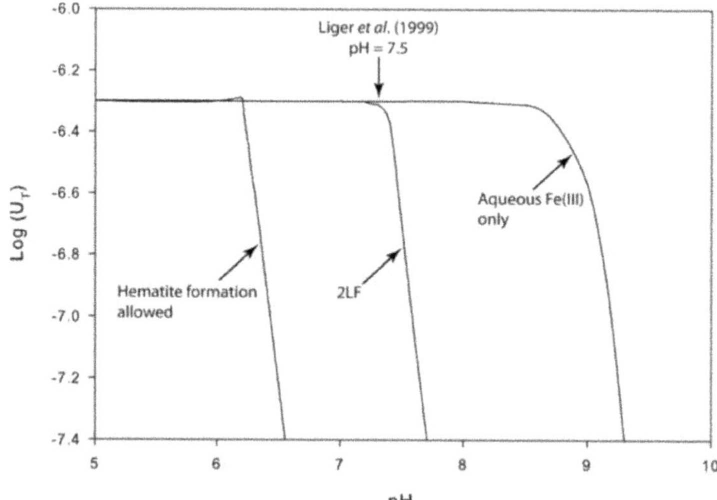

Figure 7. Calculated U(VI) concentrations in 0.1 M $NaNO_3$ allowing the formation of UO_2(am) and initial total concentrations of 1.66×10^{-4} M for Fe(II) and 5×10^{-7} M for U(VI). The symbol 2LF represents use of a solubility product for 2-line ferrihydrite given by Stefansson *et al.* (2007). Reproduced from Felmy *et al.*, 2011a with the permission of the Mineralogical Society of Great Britain & Ireland.

minerals in general has been limited to the either models of the end-member solid phases or limited to the treatment of hydration or dehydration reactions (Ransom and Helgeson, 1994; Vidal and Dubacq, 2009).

Furthermore, the radionuclide can be incorporated into a variety of crystallographic sites within the iron oxides themselves, and these different sites can have different thermodynamic stability and also inhibit electron transfer reactions to the substituted radionuclide. An example of the latter effect is the recent work by Ilton *et al.* (2012) on the different crystallographic sites of U(V) in substituted hematite. The final form of the radionuclide and the Fe(III) reaction product are often difficult to identify because they are present in miniscule concentrations, relative to other bulk phases. Therefore, understanding microscopic or molecular-scale factors that control nucleation and growth of specific types of Fe(III) reaction products and the substitution of radionuclides into bulk phases present in the environment constitutes a challenge for the future.

Molecular theory and simulation has been an essential tool for interpreting spectroscopic data on local coordination, bond distances, valence and electron transfer processes for actinides and fission products adsorbed or incorporated within mineral structures. These atomic-level modelling approaches range generally from crude but efficient empirical interatomic potential methods for thousands of atoms to rigorous quantum mechanical methods for smaller numbers of atoms. The latter entails solving the system electronic wavefunction self-consistently according to first principles. Although these approaches are now capable of dealing with complex mineral surfaces or defective bulk lattice structures, this kind of work is just beginning. Some representative studies are outlined below. For example, quantum mechanical methods were used to evaluate mechanisms for possible incorporation of Tc species into the model iron oxide, hematite (α-Fe_2O_3) (Skomurski *et al.*, 2010b). Using periodic supercell models, energies for charge-neutral incorporation of Tc(IV) or $[Tc(VII)O_4]^-$ ions, either as substituents for Fe(III) or as vacancy-based interstitials, were calculated using a Tc(IV)/Fe(II) substitution scheme on the Fe sub-lattice, or by insertion of $[Tc(VII)O_4]^-$ as an interstitial species within a hypothetical $Fe_3O_4^+$ cluster vacancy charge-balanced by co-incorporation of two protons, respectively. Although charge neutral, incorporation of the pertechnetate anion is found to be invariably unfavourable, consistent with its stable electronic configuration and low bonding affinity. However, incorporation of Tc(IV) along with charge-neutralizing electrons, *i.e.* Fe(II), as replacements for Fe(III) cations was predicted to be energetically feasible up to ~2.6 wt.%. The calculations predict that incorporated Tc(IV) and Fe(II) will favour clustering in the hematite lattice, attributed to less local net Coulombic repulsion relative to that of Fe(III)-Fe(III). These charge-neutralizing electrons are sufficiently mobile within the hematite lattice to diffuse by small polaron hopping to Fe sites nearest Tc(IV) sites to reduce the total energy (Rosso *et al.*, 2003; Iordanova *et al.*, 2005; Kerisit and Rosso, 2007)). Predicted Tc(IV) incorporation will require either Eh-pH conditions where both Tc(IV) and hematite are stable, such as in reducing basic environments, or a mechanism to produce Tc(IV) within the hematite stability field under oxidizing conditions. One such mechanism is reduction of $[Tc(VII)O_4]^-$ to Tc(IV) *via* interaction with radiolytic species, a process possible in, for example, the oxidizing high pH conditions of the residual tank waste environments at the Hanford site, where selective association of Tc with iron oxides has been observed experimentally.

Molecular modelling has also been helpful for understanding relationships between uranium valence, speciation, and incorporation in minerals. Very little is known about the elementary processes involved in uranium reduction from U(VI) to U(V) to U(IV) in general, although the importance of the local bonding environment around the uranium cation is increasingly apparent. For example, while U(VI) in the form of the hydrated uranyl cation is readily reduced by a range of natural reductants, uranyl complexation by ligands such as carbonate can greatly reduce its reduction potential, impose increased electron transfer (ET) distances, and thereby stabilize its more positive oxidation states, having significant impact on its fate and transport. A combination of empirical potential molecular dynamics (MD) simulations and density functional theory (DFT) electronic structure calculations were used to examine the theoretical kinetics of ET from aqueous ferrous iron to triscarbonato uranyl (Wander *et al.*, 2006). MD simulations predict that two hydrated ferrous iron cations will bind in an inner-sphere fashion to the three-membered carbonate ring of triscarbonato uranyl, forming a stable charge-neutral ternary $Fe_2UO_2(CO_3)_3(H_2O)_8$ complex. This complex thus possesses two possible electron donors in the form of two ferrous cations, bridged to the uranyl group in the center of the complex by carbonate anions. The complex is predicted to have a favourable first electron transfer step from one of the bound Fe(II) cations to the U(VI), yielding U(V). Through a sequential proton-coupled electron-transfer mechanism (PCET), this first ET step converting U(VI) to U(V) is predicted by DFT to occur with an electronic barrier that corresponds to a rate on the order of 1 s^{-1}. The second ET step converting U(V) to U(IV) is predicted to be significantly endergonic. Therefore, the intermediate oxidation state of U(V) is predicted to be a stabilized end product in the presence of soluble Fe(II) in this system.

This situation can be compared with the case of heterogeneous reduction of U(VI) by reductive adsorption on an Fe(II)-bearing mineral phase such as magnetite. Like the case above, reduction of U(VI) nominally to stable product U(IV) is a process that, in principle, entails transfer of up to two Fe(II) electrons, in this case from within the magnetite structure. Magnetite is a low band gap semiconductor at room temperature with a relatively high charge-carrier density provided by the Fe(II) electrons in the lattice. These electrons are very mobile in the bulk, migrating by small polaron hopping along the octahedral Fe sub-lattice at ~10^{12} s^{-1} as predicted by first-principles calculations (Skomurski *et al.*, 2010a). This provides for the interesting prospect of multi-electron transfer from the lattice to adsorbed U species and efficient reductive immobilization. Slower electron migration rates are predicted at magnetite (100) surfaces, by several orders of magnitude, suggesting that the rate of surface redox reactions could be controlled by both Fe(II) availability at the surface and its replenishment rate by electron migration from deeper within the solid.

The adsorption of U(VI) to various iron oxide surfaces in a biatomic bidentate oxo-bridged fashion is often suggested based on surface X-ray spectroscopy studies. For magnetite, molecular cluster Fe-Fe-U models representing a hydrated uranyl species adsorbed in this fashion to a neutral (100) surface iron group were constructed for first-principles calculations of the local bonding, coordination, and energetics of electron transfer (Skomurski *et al.*, 2011). Total energy minimizations of the cluster atom positions showed that this adsorption configuration is structurally robust. Various possible initial valence distributions for the Fe-Fe-U cations within this model surface complex were evaluated in terms of their relative stabilities, including a reduced initial state involving two Fe(II) cations with U(VI), and more oxidized initial state of an Fe(II) and Fe(III) with U(VI). The rate calculated for the first electron transfer step from Fe(II) to U(VI) (10^4 s^{-1}) is orders of magnitude faster than predicted for that within the aqueous triscarbonato complex. Furthermore, comparison with the Fe(II) to Fe(III) exchange rate ($>10^5$ s^{-1}) indicated that sequential multi-electron uranium reduction to U(IV) should not be limited kinetically by conductive electron resupply to the adsorption site. These calculations also suggest plausible stability of U(V), depending on both the availability of Fe(II) and conditions that favour U(V) coordination. These findings are generally consistent with associated experiments, in which freshly cleaved, single crystals of magnetite with different initial Fe(II)/Fe(III) ratios were exposed to uranyl-nitrate solution (pH similar to 4) for 90 h (Skomurski *et al.*, 2011). X-ray photoelectron spectroscopy and electron microscopy indicated the presence of a mixed U(VI)/U(V) precipitate heterogeneously nucleated and grown on stoichiometric magnetite surfaces, but only the presence of sorbed U^{6+} and no precipitate on sub-stoichiometric (partially oxidized) magnetite surfaces. Both theory and experiment point to structural Fe^{2+} density, taken as a measure of thermodynamic reducing potential, and sterically accessible uranium coordination as key controls on uranium reduction extent and rate. This impacts the prospect of adsorbed reduced uranium ultimately being incorporated into the magnetite structure through local recrystallization, where, if the uranium coordination is allowed to obtain the uranate type upon reduction, it should favour uranium incorporation. If uranium cannot acquire 8-fold coordination then reduction may proceed to U^{5+} but not necessarily U^{4+}. In both cases adsorption and incorporation into magnetite should widen the stability field of U^{5+}.

Classical atomistic simulations were carried out recently by Kerisit *et al.* (2011) to characterize and compare stable coordination environments of U incorporated in three Fe-(hydr)oxide minerals: goethite, magnetite and hematite. The simulations provided information on U–O and U–Fe distances, coordination numbers, and lattice distortion for U incorporated in different sites (*e.g.* unoccupied *vs.* occupied sites, octahedral *vs.* tetrahedral) as a function of the oxidation state of U and charge-compensation mechanisms (*i.e.* deprotonation, vacancy formation, or reduction of Fe(III) to Fe(II)). For goethite, deprotonation of first-shell hydroxyls enables substitution of U for Fe(III) with a minimal amount of lattice distortion, whereas substitution in unoccupied octahedral sites induced appreciable distortion to 7-fold coordination regardless of U oxidation states and charge-compensation mechanisms. Importantly, U–Fe distances similar to 3.6 Å were associated with structural incorporation of U and cannot be considered diagnostic of simple adsorption to goethite surfaces. For magnetite, the octahedral site accommodates U(V) or U(VI) with little lattice distortion. U substituted for Fe(III) in hematite maintained octahedral coordination in most cases. In general, comparison of the simulations with available experimental data provides further evidence for and constraints influencing the structural incorporation of U in iron (hydr)oxide minerals (Kerisit *et al.*, 2011).

7. Future research needs

An understanding of the plethora of iron redox reactions is an important component of a GDF safety case and this will require an iterative approach involving the integration of information from natural systems, experimentation and modelling. Whereas the redox state of many of the Fe-bearing engineered and natural components of a potential GDF is known and our understanding of key reactions is advanced, our knowledge of the spatial and temporal evolution of GDF redox is poorly constrained. In particular, the influence of chemically evolving aqueous fluids and gases, and microbial activity on redox reactions and radionuclide fate needs definition. A major challenge is that for most existing studies, establishment of thermodynamic equilibrium is assumed although redox and mineralization reactions are characterized partly by slow kinetics and time frames in the order of years. Data on actinide reaction kinetics are scarce and can be difficult to acquire, although they are essential for the reliable prediction of actinide fate and transport in natural systems over GDF time scales. Most existing data for actinide interaction with, for instance, mineral surfaces are available for uranium because of its low specific radioactivity and the ease of use. For transuranic elements, such as plutonium, however, safety requirements usually are so strict that they only can be handled in dedicated laboratories. This is notably true for *in situ* synchrotron-based X-ray spectroscopic studies, as only a few beamlines dedicated to actinide research are available worldwide.

Radionuclide interactions at interfaces and surfaces is the controlling factor of their mobility and the observation of "selective competition" of metal ions for sorption sites in minerals is certainly remarkable, and also worrying. It is obvious that geochemical calculations applying surface complexation constants for individual metal ions without knowledge of sorption competition bear large uncertainties. Defining the exact nature of mineral surface sites and their charges still represents a central challenge for quantifying actinide sorption and other safety case breaker radioactive species by geochemical sorption modelling. Multisite complexation models derive surface sites and their acid/base properties from the known structure of ideal crystal planes and are extremely successful in describing well defined systems. They are, however, difficult to apply to real-world, natural systems although some progress is being made. The latter are heterogeneous on various scales, all of which will be found in an evolving GDF. Therefore, pragmatic approaches are used for natural sediments, rocks and soils where generic surface sites need to be defined. Further simplifications have been proposed for clay rock systems, where non-electrostatic models have been applied successfully to describe metal ion sorption. As a consequence, various types of sorption models exist, which limits the use of the model-specific (thermodynamic) data that are available (Geckeis *et al.*, 2013).

This review describes the importance of Fe mineralogy, including total Fe content and Fe(II)/Fe(III) ratio. Future research should emphasize understanding the Fe distribution and reactivity, and its evolution over time, across the near and far fields of a GDF. Advanced imaging and spectroscopic techniques combined with computational molecular simulation will ultimately help to make the necessary inroads in this area. In particular, these tools should be focused on understanding the kinetics of redox-catalysed phase transformations and radionuclide incorporation, and the effect of these processes on radionuclide transport over time.

Acknowledgements

Professor C.M.B. Henderson is acknowledged for his insightful discussions and suggestions. The authors also thank Miss J. Quirke and Dr J. Charnock for providing the images in Fig. 4. KMR gratefully acknowledges support from the Geosciences Research Program at Pacific Northwest National Laboratory from the U.S. Department of Energy, Office of Science, Office of Basic Energy Sciences, Division of Chemical Sciences, Geosciences, and Biosciences.

References

Acero, P., Auque, L.F., Gimeno, M.J. and Gomez, J.B. (2010) Evaluation of mineral precipitation potential in a spent nuclear fuel repository. *Environmental Earth Sciences*, **59**, 1613–1628.

Ahmed, I.A.M., Acero, P. and Auqué, L.F. (2017) Biogeochemical modelling of redox processes in low-temperature natural systems. Pp. 273–311 in: *Redox-Reactive Minerals: Properties, Reactions and Applications in Clean Technologies* (I.A.M. Ahmed and K.A. Hudson-Edwards, editors). EMU Notes in Mineralogy, **17**. European Mineralogical Union and Mineralogical Society of Great Britain & Ireland.

Alexandrov, V., Neumann, A., Scherer, M.M. and Rosso, K.M. (2013) Electron exchange and conduction in nontronite from first-principles. *Journal of Physical Chemistry C*, **117**, 2032–2040.

Alexandrov, V. and Rosso, K.M. (2013) Insights into the mechanism of Fe(II) adsorption and oxidation at Fe-clay mineral surfaces from first-principles calculations. *Journal of Chemical Physics*, **117**, 22880–22886.

Allard, T., Balán, E., Calas, G., Fourdrin, C., Morichon, E. and Sorieul, S. (2012) Radiation-induced defects in clay minerals: A review. *Nuclear Instruments and Methods in Physics Research Section B – Beam Interactions with Materials and Atoms*, **277**, 112–120.

Altmaier, M., Neck, V., Lützenkirchen, J. and Fanghänel, T. (2009) Solubility of plutonium in $MgCl_2$ and $CaCl_2$ solutions in contact with metallic iron. *Radiochimica Acta*, **97**, 187.

ANDRA (2005a) Andra research on the geological disposal of high-level long-lived radioactive waste. ANDRA, Châtenay-Malabry, France, pp. 38

ANDRA (2005b) Synthesis: Evaluation of the feasibility of a geological repository in an argillaceous formation Meuse/Haute-Marne site. ANDRA, Châtenay-Malabry, France. 236 pp.

Andrews, A.J., Dollase, W.A. and Fleet, M.E. (1983) A Mössbauer Study of Saponite in Layer-2 Basalt, Deep-Sea Drilling Project Leg. *Initial Reports of the Deep Sea Drilling Project*, **69**, 585–588.

Aplin, A.C. and Macquaker, J.H.S. (2011) Mudstone diversity: Origin and implications for source, seal, and reservoir properties in petroleum systems. *AAPG Bulletin*, **95**, 2031–2059.

Baeyens, B., Marques Fernandes, M. and Bradbury, M.H. (2014) Comparison of sorption measurements on argillaceous rocks and bentonite with predictions using the SGT-E2 approach to derive sorption data bases.

Nagra Technical Report, NTB 12-05, Nagra, Wettingen, Switzerland, 74 pp.

Bailey, S.W. (1988) Odinite, a new dioctahedral-trioctahedral Fe^{3+}-rich 1:1 clay mineral. *Clay Minerals*, **23**, 237–247.

Bassil, N.M., Bryan, N.D. and Lloyd, J.R. (2015) Microbial degradation of isosaccharinic acid at high pH. *The ISME Journal*, **9**, 310–320.

Bjorlykke, K. (1998) Clay mineral diagenesis in sedimentary basins: A key to the prediction of rock properties – Examples from the North Sea Basin. *Clay Minerals*, **33**, 15–34.

Boily, J.F. and Rosso, K.M. (2011) Crystallographic controls on uranyl binding at the quartz/water interface. *Physical Chemistry Chemical Physics*, **13**, 7845–7851.

Bowles, J.F.W., Howie, R.A., Vaughan, D.J. and Zussman, J. (2011) *Non-Silicates: Oxides, Hydroxides and Sulphides*. Rock Forming Minerals, 5A, The Geological Society, London.

Boyle, J.F. and Birks, H.J.B. (1999) Predicting heavy metal concentrations in the surface sediments of Norwegian headwater lakes from atmospheric depostion: An application of a simple sediment-water partitioning model. *Water, Air, and Soil Pollution*, **114**, 27–51.

Bradbury, M.H. and Baeyens, B. (2011) Predictive sorption modelling of Ni(II), Co(II), Eu(IIII), Th(IV) and U(VI) on MX-80 bentonite and Opalinus Clay: A "bottom-up" approach. *Applied Clay Science*, **52**, 27–33.

Brigatti, M.F., Guidotti, C.V., Malferrari, D. and Sassi, F.P. (2008) Single-crystal X-ray studies of trioctahedral micas coexisting with dioctahedral micas in metamorphic sequences from western Maine. *American Mineralogist*, **93**, 396–408.

Brigatti, M.F., Malferrari, D., Laurora, A. and Elmi, C. (2011) Structure and mineralogy of layer silicates: recent perspectives and new trends. Pp. 1–71 in: *Layered Mineral Structures and their Application in Advanced Technologies* (M.F. Brigatti and A. Mottana, editors). EMU Notes in Mineralogy, **11**. The European Mineralogical Union and the Mineralogical Society of Great Britain & Ireland, London.

Brindley, G.W. (1982) Chemical compositions of berthierines – A review. *Clays and Clay Minerals*, **30**, 153–155.

Bristow, T.F., Kennedy, M.J., Derkowski, A., Droser, M.L., Jiang, G. and Creaser, R.A. (2009) Mineralogical constraints on the paleoenvironments of the Ediacaran Doushantuo Formation. *PNAS*, **106**, 13190–13195.

Brookshaw, D.R., Pattrick, R.A.D., Lloyd, J.R. and Vaughan, D.J. (2012) Microbial effects on mineral–radionuclide interactions and radionuclide solid-phase capture processes. *Mineralogical Magazine*, **76**, 777–806.

Brown, A.R., Wincott, P.L., LaVerne, J.A., Small, J.S., Vaughan, D.J., Pimblott, S.M. and Lloyd, J.R. (2014) The impact of γ-radiation on the bioavailability of Fe(III) minerals for microbial respiration. *Environmental Science & Technology*, **48**, 10672–10680.

Bruggeman, C. and Maes, N. (2010) Uptake of Uranium(VI) by pyrite under Boom Clay conditions: Influence of dissolved organic carbon. *Environmental Science & Technology*, **44**, 4210–4216.

Bruno, J., Bosbach, D., Kulik, D. and Navrotsky, A. (2007) *Chemical Thermodynamics of Solid Solutions of Interest in Nuclear Waste Management*. A State-of-the-Art Report. Nuclear Energy Agency Data Bank, Organisation for Economic Co-operation and Development, Paris, France.

Burger, E., Rebiscoul, D., Bruguier, F., Jublot, M., Lartigue, J.E. and Gin, S. (2013) Impact of iron on nuclear glass alteration in geological repository conditions: A multiscale approach. *Applied Geochemistry*, **31**, 159–170.

Chappell, B.W. and Hine, R. (2006) The Cornubian Batholith: an Example of magmatic fractionation on a crustal scale. *Resource Geology*, **56**, 203–244.

Cui, D.Q. (2011) A review of beneficial effects of reducing environment at the near-field of KBS-3 Repository. *Progress in Chemistry*, **23**, 1411–1428.

Cui, D.Q. and Eriksen, T.E. (1996a) Reduction of pertechnetate by ferrous iron in solution: Influence of sorbed and precipitated Fe(II). *Environmental Science & Technology*, **30**, 2259–2262.

Cui, D.Q. and Eriksen, T.E. (1996b) Reduction of pertechnetate in solution by heterogeneous electron transfer from Fe(II)-containing geological material. *Environmental Science & Technology*, **30**, 2263–2269.

de Combarieu, G., Schlegel, M.L., Neff, D., Foy, E., Vantelon, D., Barboux, P. and Gin, S. (2011) Glass-iron-clay interactions in a radioactive waste geological disposal: An integrated laboratory-scale experiment. *Applied Geochemistry*, **26**, 65–79.

DECC (2014) *Implementing Geological Disposal: A Framework for the long-term management of higher activity radioactive waste*. Department of Environment and Climate Change, London, pp. 55.

Deer, W.A., Howie, R.A. and Zussman, J. (1997a) *Orthosilicates*. Rock-forming Minerals, vol. **1A**, The Geological Society, London.

Deer, W.A., Howie, R.A. (1997b) *Disilicates and Ring Silicates*. Rock-forming Minerals, vol. **1B**, The Geological Society, London.

Deer, W.A., Howie, R.A. and Zussman, J. (2009) *Layered Silicates Excluding Micas and Clay Minerals*. The Rock-forming Minerals, vol. **3B**, The Geological Society, London.

Dewhurst, D.N., Yang, Y.L. and Aplin, A.C. (1999) Permeability and fluid flow in natural mudstones. Pp. 23–43 in: *Muds and Mudstones: Physical and Fluid-Flow Properties* (A.C. Aplin, A.J. Fleet and J.H.S. Macquaker, editors). Special Publication, **158**, The Geological Society, London, .

Diener, A., Neumann, T., Kramar, U. and Schild, D. (2012) Structure of selenium incorporated in pyrite and mackinawite as determined by XAFS analyses. *Journal of Contaminant Hydrology*, **133**, 30–39.

Dowey, P.J., Hodgson, D.M. and Worden, R.H. (2012) Pre-requisites, processes, and prediction of chlorite grain coatings in petroleum reservoirs: A review of subsurface examples *Marine and Petroleum Geology*, **32**, 63–75.

Duro, L., C. Domènech, C., Grivé, M., Roman-Ross, G., Bruno, J. and Källström, K. (2012) Assessment of the evolution of the redox conditions in the SKB ILW-LLW SFR-1 repository (Sweden). *MRS Proceedings*, **1518**, 70.

El Mendili, Y., Abdelouas, A. and Bardeau, J.F. (2013) Insight into the mechanism of carbon steel corrosion under aerobic and anaerobic conditions. *Physical Chemistry Chemical Physics*, **15**, 9197–9204.

Enachescu, C., Blümling, P., Castelao, A. and Steffen, P. (2002) Mont Terri GP-A and GS experiments synthesis report. *Nagra Internal Report NIB 02-51*,Nagra, Wettingen, Switzerland, pp. 32.

Engelberg, D.L., Pattrick, R.A.D., Wilson, C., McCrea, R. and Withers, P.J. (2012) 3D imaging of inhomogeneous lithologies using X-ray computed tomography – characterisation of drill core from the Borrowdale Volcanic Group. *Mineralogical Magazine*, **76**, 2931–2938.

Ewing, R.C., Weber, W.J. and Clinard, F.W.J. (1995) Radiation effects in nuclear waste forms for high-level radioactive waste. *Progress in Nuclear Energy*, **29**, 63–127.

Fanghanel, T., Rondinella, V.V., Glatz, J.P., Wiss, T., Wegen, D.H., Gouder, T., Carbol, P., Serrano-Purroy, D. and Papaioannou, D. (2013) Reducing uncertainties affecting the assessment of the long-term corrosion behavior of spent nuclear fuel. *Inorganic Chemistry*, **52**, 3491–3509.

Felmy, A.R., Ilton, E.S., Rosso, K.M. and Zachara, J.M. (2011a) Interfacial reactivity of radionuclides: emerging paradigms from molecular-level observations. *Mineralogical Magazine*, **75** 2379–2391.

Felmy, A.R., Moore, D.A., Rosso, K.M., Qafoku, O., Rai, D., Buck, E.C. and Ilton, E.S. (2011b) Heterogeneous reduction of PuO_2 with Fe(II): Importance of the Fe(III) reaction product. *Environmental Science & Technology*, **45**, 3952–3958.

Felmy, A.R., Moore, D.A., Pearce, C.I., Conradson, S.D., Qafoku, O., Buck, E.C., Rosso, K.M. and Ilton, E.S. (2012) Controls on soluble Pu concentrations in PuO_2/magnetite suspensions. *Environmental Science & Technology*, **46**, 11610–11617.

Finck, N., Dardenne, K., Bosbach, D. and Geckeis, H. (2012) Selenide retention by mackinawite. *Environmental Science & Technology*, **46**, 10004–10011.

Fleet, M.E. (2003) *Micas* (W.A. Deer, R.A. Howie and J. Zussman, editors). Rock-forming Minerals, vol. **3B**, The Geological Society, London.

Gaudin, A., Gaboreau, S., Tinseau, E., Bartier, D., Petit, S., Grauby, O., Foct, F. and Beaufort, D. (2009) Mineralogical reactions in the Tournemire argillite after in-situ interaction with steels. *Applied Clay Science*, **43**, 196–207.

Gauthier-Lafaye, F., Holliger, P. and Blanc, P.-L. (1996) Natural fission reactors in the Franceville basin, Gabon: A review of the condition and results of a "critical event" in a geologic system. *Geochimica et Cosmochimica Acta*, **60**, 4831–4852.

Geckeis, H., Lützenkirchen, J., Polly, R., Rabung, T. and Schmidt, M. (2013) Mineral–water interface reactions of actinides. *Chemical Reviews*, **113**, 1016–1062.

Gorski, C.A., Kluepfel, L.E., Voegelin, A., Sander, M. and Hofstetter, T.B. (2013) Redox properties of structural

Fe in clay minerals: 3. Relationships between smectite redox and structural properties. *Environmental Science & Technology*, **47**, 13477-13485

Grenthe, I., Fuger, J., Konings, R., Lemire, R., Muller, A., Nguyen-Trung, C. and Wanner, H. (1992) *Chemical Thermodynamics of Uranium* (H. Wanner and I. Forest, editors). Organisation for Economic Co-operation and Development/Nuclear Energy Agency, Elsevier, Amsterdam.

Guillaumont, R., Fanghänel, T., Fuger, J., Grenthe, I., Neck, V., Palmer, D. and Rand, M. (2003) *Update on The Chemical Thermodynamics of Uranium, Neptunium, Plutonium, Americium and Technetium.* Organisation for Economic Co-operation and Development/Nuclear Energy Agency, Elsevier, Amsterdam.

Hallam, R.J., Evans, N.D.M. and Jain, S.L. (2011) Sorption of Tc(IV) to some geological materials with reference to radioactive waste disposal. *Mineralogical Magazine*, **75**, 2439–2448.

Hawthorne, F.C., Oberti, R., Ventura, G.D. and Mottana, A., editors (2007) *Amphiboles: Crystal Chemistry, Occurrence and Health Issues.* Reviews in Mineralogy and Geochemistry, **67**. Mineralogical Society of America and the Geochemical Society, Chantilly, Virginia, USA.

Hess, N.J., Qafoku, O., Xia, Y., Moore, D.A. and Felmy, A.R. (2008) Thermodynamic model for the solubility of $TcO_2.xH_2O$ in aqueous oxalate systems. *Journal of Solution Chemistry*, **37**, 1471–1487.

Holmboe, M., Wold, S., Jonsson, M. and García-García, S. (2009) Effects of γ-irradiation on the stability of colloidal Na^+-Montmorillonite dispersions. *Applied Clay Science*, **43**, 86–90.

Hummel, W., Anderegg, G., Puigdomenech, I., Rao, L. and Tochiyama, O. (2005) *Chemical thermodynamics of complexes and compounds of U, Np, Pu, Am, Tc, Zr, Ni and Se with selected organic ligands.* OECD Nuclear Energy Agency Data Bank, Amsterdam.

Hurst, A. (1985) Diagenetic chlorite formation in some Mesozoic shales from the Sleipner area of the North Sea. *Clay Minerals*, **20**, 69–79.

Hybler, J. (2014) Refinement of cronstedtite-1M. *Acta Crystallographica*, **B70**, 963–972.

Ildefonse, P., Kirkpatrick, R.J., Montez, B., Calas, G., Flank, A.M. and Lagarde, P. (1994) ^{27}Al MAS NMR and aluminium X-ray absorption near edge structures study of natural imogolite and allophanes. *Clays and Clay Minerals*, **42**, 276–287.

Ilton, E.S., Heald, S.M., Smith, S.C., Elbert, D. and Liu, C. (2006) Reduction of uranyl in the interlayer region of low iron micas under anoxic and aerobic conditions. *Environmental Science & Technology*, **40**, 5003–5009.

Ilton, E.S., Boily, J.-F., Buck, E.C., Skomurski, F.N., Rosso, K.M., Cahill, C.L., Bargar, J.R. and Felmy, A.R. (2010) Influence of dynamical conditions on the reduction of U-VI at the magnetite–solution interface. *Environmental Science & Technology*, **44**, 170–176.

Ilton, E.S., Pacheco, J.S.L., Bargar, J.R., Shi, Z., Liu, J., Kovarik, L., Engelhard, M.H. and Felmy, A.R. (2012) Reduction of U(VI) incorporated in the structure of hematite. *Environmental Science & Technology*, **46**, 9428–9436.

Iordanova, N., Dupuis, M. and Rosso, K.M. (2005) Charge transport in metal oxides: A theoretical study of hematite α-Fe_2O_3. *Journal of Chemical Physics*, **122**, 144305.

Jaisi, D.P., Dong, H.L., Plymale, A.E., Fredrickson, J.K., Zachara, J.M., Heald, S. and Liu, C.X. (2011) Reduction and long-term immobilization of technetium by Fe(II) associated with clay mineral nontronite. *Chemical Geology*, **264**, 127–138.

Jodin-Caumon, M.C., Mosser-Ruck, R., Rousset, D., Randi, A., Cathelineau, M. and Michau, N. (2010) Effect of a thermal gradient on iron-clay interactions. *Clays and Clay Minerals*, **58**, 667–681.

Jodin-Caumon, M.C., Mosser-Ruck, R., Randi, A., Pierron, O., Cathelineau, M. and Michau, N. (2012) Mineralogical evolution of a claystone after reaction with iron under thermal gradient. *Clays and Clay Minerals*, **60**, 443–455.

Johnston, J.H. and Cardile, C.M. (1987) Iron substitution in montmorillonite, illite, and glauconite by ^{57}Fe Mössbauer Spectroscopy. *Clays and Clay Minerals*, **35**, 170–176.

Jung, H., Kwon, K.J., Lee, E., Kim, D.G. and Kim, G.Y. (2011) Effect of dissolved oxygen on corrosion properties of reinforcing steel. *Corrosion Engineering Science and Technology*, **46**, 195–198.

Kerisit, S., Felmy, A.R. and Ilton, E.S. (2011) Atomistic simulations of uranium incorporation into iron (hydr)oxides. *Environmental Science & Technology*, **45**, 2770–2776.

Kerisit, S. and Rosso, K.M. (2007) Kinetic Monte Carlo model of charge transport in hematite (α-Fe_2O_3). *Journal of Chemical Physics*, **127**, 124706.

Kirsch, R., Fellhauer, D., Altmaier, M., Neck, V., Rossberg, A., Fanghanel, T., Charlet, L. and Scheinost, A.C. (2011) Oxidation state and local structure of plutonium reacted with magnetite, mackinawite, and chukanovite. *Environmental Science & Technology*, **45**, 7267–7274.

Kister, P., Vieillard, P., Cuney, M., Quirt, D. and Laverret, E. (2005) Thermodynamic constraints on the mineralogical and fluid composition evolution in a clastic sedimentary basin: the Athabasca Basin (Saskatchewan, Canada). *European Journal of Mineralogy*, **17**, 325–342.

Lahti, M., Ahokas, T., Nordback, N., Paananen, M., Paulamaki, S. and Vaittinen T., (2009) The ONKALO Area Model. *Model Working Report -2009-113*, Posiva Oy, Eurajok, Olkiluoto, Finland, pp. 130

Lanson, B., Lantenois, S., van Aken, P.A., Bauer, A. and Plancon, A. (2012) Experimental investigation of smectite interaction with metal iron at 80 degrees C: Structural characterization of newly formed Fe-rich phyllosilicates. *American Mineralogist*, **97**, 864–871.

Latta, D.E., Gorski, C.A., Boyanov, M.I., O'Loughlin, E.J., Kemner, K.M. and Scherer, M.M. (2012) Influence of magnetite stoichiometry on U^{VI} reduction. *Environmental Science & Technology*, **46**, 778–786.

Latta, D.E., Mishra, B., Cook, R.E., Kemner, K.M. and Boyanov, M.I. (2014) Stable U(IV) Complexes Form at High-Affinity Mineral Surface Sites. *Environmental Science & Technology*, **48**, 1683–1691.

Latta, D.E., Pearce, C.I., Rosso, K.M., Kemner, K.M. and Boyanov, M.I. (2013) Reaction of U^{VI} with titanium-substituted magnetite: Influence of Ti on U^{IV} speciation. *Environmental Science & Technology*, **47**, 4121–4130.

Lemire, R.J., Fuger, J., Nitsche, H., Potter, P., Rand, M., Rydberg, J., Spahiu, K., Sullivan, J., Ullman, W., Vitorge, P. and Wanner, H. (2001) *Chemical Thermodynamics of Neptunium and Plutonium*. Organisation for Economic Co-operation and Development/Nuclear Energy Agency, Elsevier, Amsterdam.

Liger, E., Charlet, L. and Van Capellan, P. (1999) Surface catalysis of uranium(VI) reduction by iron(II). *Geochimica et Cosmochimica Acta*, **63**, 2939–2955.

Lindsley D.H., (1976) The crystal chemistry and structure of oxide minerals as exemplified by the Fe-Ti oxides. Pp. 1–60 in: *Oxide Minerals, Vol.* **3** (D. Rumble III, editor). Mineralogical Society of America Short Course Notes, Chantilly, Virginia, USA.

Liu, J., Pearce, C.I., Qafoku, O., Arenholz, A., Heald, S.M. and Rosso, K.M. (2012) Tc(VII) reduction kinetics by titanomagnetite ($Fe_{3-x}Ti_xO_4$) nanoparticles. *Geochimica et Cosmochimica Acta*, **92**, 67–81.

Lloyd, J.R. (2014) *The impact of microbial metabolism on the geodisposal of radwaste in multibarrier systems*. IGD-TP Geodisposal, Manchester, UK.

Lovley, D.R.P. and Phillips, E.J.P. (1988) Novel mode of microbial energy-metabolism-organic-carbon oxidation coupled to dissimilatory reduction of iron or manganese. *Applied and Environmental Microbiology*, **54**, 1472–1480.

Maillot, F., Morin, G., Wang, Y., Bonnin, D., Ildefonse, P., Chanaec, C. and Calas, G. (2011) New insight into the structure of nanocrystalline ferrihydrite: EXAFS evidence for tetrahedrally coordinated iron(III). *Geochimica et Cosmochimica Acta*, **75**, 2708–2720.

Manceau, A. and Drits, V.S. (1993) Local structure of ferrihydrite and ferroxhite by EXAFS spectroscopy. *Clay Minerals*, **28**, 165–184.

Manjanna, J., Kozaki, T. and Sato, S. (2009) Fe(III)-montmorillonite: Basic properties and diffusion of tracers relevant to alteration of bentonite in deep geological disposal. *Applied Clay Science*, **43**, 208–217.

Matthews, A., Morgans-Bell, H.S., Emmanuel, S., Jenkyns, H.C., Erel, Y. and Halicz, L. (2004) Controls on iron-isotope fractionation in organic-rich sediments (Kimmeridge Clay, Upper Jurassic, southern England). *Geochimica et Cosmochimica Acta*, **68**, 3107–3123.

McVay, G.L. and Buckwalter, C.Q. (1983) Effect of iron on waste-glass leaching. *Journal of the American Ceramic Society*, **66**, 170–174.

Meller, N. (1998) The metamorphic history of the Borrowdale Volcanic Group; a petrographic study. *Proceedings of the Yorkshire Geological Society*, **52**, 243–254.

Metz, V., Geckeis, H., Gonzalez-Robles, E., Loida, A., Bube, C. and Kienzler, B. (2012) Radionuclide behaviour in the near-field of a geological repository for spent nuclear fuel. *Radiochimica Acta*, **100**, 699–713.

Michelin, A., Burger, E., Rebiscoul, D., Neff, D., Bruguier, F., Drouet, E., Dillmann, P. and Gin, S. (2013) Silicate glass alteration enhanced by iron: Origin and long-term implications. *Environmental Science &*

Technology, **47**, 750–756.

Miller, B. (2015) *Overveiw of Knowledge Coverage for Geological Disposal*. AMEC Report Reference: 200094. Prepared for Office for Nuclear Regulation, AMEC, Warrington, UK.

Milodowski, A.E., Gillespie, M.R., Naden, J., Fortey, N.J., Shepherd, T.J., Pearce, J.M. and Metcalfe, R. (1998) The petrology and paragenesis of fracture mineralization in the Sellafield area, West Cumbria. *Proceedings of the Yorkshire Geological Society*, **52**, 215–241.

Mücke, A. and Farshad, F. (2005) Whole-rock and mineralogical composition of Phanerozoic ooidal ironstones: Comparison and differentiation of types and subtypes. *Ore Geology Reviews*, **26**, 227–262.

Murad, E. (2010) Mössbauer spectroscopy of clays, soils and their mineral constituents. *Clay Minerals*, **45**, 413–430.

Navrotsky, A. (2008) Size-driven structural and thermodynamic complexity in iron oxides. *Science*, **319**, 1635.

NDA (2010a) *Geological Disposal: Near-field evolution status report*. NDA Report No. NDA/RWMD/033, NDA, Didcot, UK.

NDA (2010b) *Geological Disposal: Package evolution status report*. NDA Report No. NDA/RWMD/031, NDA, Didcot, UK.

NDA (2010c) *Geological Disposal: Radionuclide behaviour status report*. NDA Report No. NDA/RWMD/034, NDA, Didcot, UK.

Neumann, A., Sander, M. and Hofstetter, T.B. (2011) Redox properties of Fe in smectite clay minerals. Pp. 361–379 in: *Aquatic Redox Chemistry* (P. Tratnyek, editor). ACS Symposium Series, American Chemical Society, Washington, D.C.

Olin, Å., Nolång, B., Öhman, L., Osadchii, E. and Rosén, E. (2005) *Chemical Thermodynamics of Selenium*. Organisation for Economic Co-operation and Development/Nuclear Energy Agency, Elsevier, Amsterdam.

Oliver, G.J.H., Smellie, J.L., Thomas, L.J., Casey, D.M., Kemp, A.E.S., Evans, L.J., Baldwin, J.R. and Hepworth, B.C. (1984) Early Palaeozoic metamorphic history of the Midland Valley, the Southern Uplands-Longford-Down massif and the Lake District, British Isles. *Transactions of the Royal Society of Edinburgh: Earth Science*, **75**, 259–273.

Pablo Gil, P., Pesquera, A. and Velasco, F. (1992) X-ray diffraction, infrared and Mössbauer studies of Fe-rich carbonates. *European Journal of Mineralogy*, **4**, 521–526.

Palmer, T.J. and Wilson, M.A. (1990) Growth of ferruginous oncoliths in the Bajocian (Middle Jurassic) of Europe. *Terra Nova*, **2**, 142–147.

Pattrick, R.A.D., van der Laan, G., Henderson, C.M.B., Kuiper, P., Dudzik, E. and Vaughan, D.J. (2002) Cation site occupancy in spinel ferrites studied by X-ray magnetic circular dichroism: developing a method for mineralogists. *European Journal of Mineralogy*, **14**, 1095–1102.

Pattrick, R.A.D., Charnock, J.M., Geraki, T., Mosselmans, J.F.W., Pearce, C.I., Pimblott, S. and Droop, G.T.R. (2013) Alpha particle damage in biotite characterized by microfocus X-ray diffraction and Fe K-edge X-ray absorption spectroscopy. *Mineralogical Magazine*, **77**, 2867–2882.

Pearce, C.I., Henderson, C.M.B., Pattrick, R.A.D., Van der Laan, G. and Vaughan, D.J. (2006) Direct determination of cation site occupancies in natural ferrite spinels by $L_{2,3}$ X-ray absorption spectroscopy and X-ray magnetic circular dichroism. *American Mineralogist*, **91**, 880–893.

Pearce, C.I., Liu, J., Baer, D., Qafoku, O., Heald, S.M., Arenholz, A., Grosz, A.E., McKinley, J.P., Resch, C.T., Bowden, M.E., Engelhard, M.H. and Rosso, K.M. (2014) Characterization of natural titanomagnetites ($Fe_{3-x}Ti_xO_4$) for studying heterogeneous electron transfer to Tc(VII) in the Hanford subsurface. *Geochimica et Cosmochimica Acta*, **128**, 114–127

Pearson, F.J., Arcos, D., Bath, A., Boisson, J.-Y., Fernández, A.M., Gäbler, H.-E., Gaucher, E., Gautschi, A., Griffault, L., Hernán, P. and Waber, H.N. (2004) Mont Terri Project – Geochemistry of water in the Opalinus Clay Formation at the Mont Terri Rock Laboratory. *Reports of the FOWG, Geology Series 5*, Bern, Switzerland, pp. 319.

Pelegrin, E., Calas, G., Ildefonse, P., Jollivet, P. and Galoisy, L. (2010) Structural evolution of glass surface during alteration: Application to nuclear waste glasses. *Journal of Non-Crystalline Solids*, **356**, 2497–2508.

Peltonen, C., Marcussen, O., Bjørlykke, K. and Jahren, J. (2008) Mineralogical control on mudstone compaction: a study of Late Cretaceous to Early Tertiary mudstones of the Vøring and Møre basins, Norwegian Sea. *Petroleum Geology*, **14**, 127–138.

Pentrakova, L., Sui, K., Pentrák, M. and Stucki, J.W. (2013) A review of microbial redox interactions with structural Fe in clay minerals. *Clay Minerals*, **48**, 543–560.

Peretyazhko, T.S., Zachara, J.M., Kukkadapu, R.K., Heald, S.M., Kutnyakov, I.V., Resch, C.T., Arey, B.W., Wang, C.M., Kovarik, L., Phillips, J.L. and Moore, D.A. (2012) Pertechnetate (TcO_4^-) reduction by reactive ferrous iron forms in naturally anoxic, redox transition zone sediments from the Hanford Site, USA. *Geochimica et Cosmochimica Acta*, **92**, 48–66.

Perri, F., Critelli, S., Cavalcante, F., Mongelli, G., Dominici, R., Sonnino, M. and de Rosa, R. (2012) Provenance signatures for the Miocene volcaniclastic succession of the Tufiti di Tusa Formation, southern Apennines, Italy. *Geological Magazine*, **149**, 423–442.

Raeburn, S.P., Ilton, E.S. and Veblen, D.R. (1997) Quantitative determination of the oxidation state of iron in biotite using X-ray photoelectron spectroscopy: II. In situ analyses *Geochimica et Cosmochimica Acta*, **61**, 4531–4537.

Ransom, B. and Helgeson, H. (1994) A chemical and thermodynamic model for aluminous dioctahedral 2:1 layer clay minerals in diagenetic processes: Regular solution representation of interlayer iehydration of smectite. *American Journal of Science*, **294**, 449.

Rard, J., Rand, H., Anderegg, G. and Wanner, H. (1999) *Chemical Thermodynamics of Technetium.* Organisation for Economic Co-operation and Development/Nuclear Energy Agency, Elsevier, Amsterdam.

Rivard, C., Pelletier, M., Michau, N., Razafitianamaharavo, A., Bihannic, I., Abdelmoula, M., Ghanbaja, J. and Villieras, F. (2013) Berthierine-like mineral formation and stability during the interaction of kaolinite with metallic iron at 90 degrees C under anoxic and oxic conditions. *American Mineralogist*, **98**, 163–180.

Rosso, K.M. and Ilton, E.S. (2003) Charge transport in micas: The kinetics of $Fe^{II/III}$ electron transfer in the octahedral sheet *Journal of Chemical Physics*, **119**, 9207.

Rosso, K.M. and Ilton, E.S. (2005) Effects of compositional defects on small polaron hopping in micas *Journal of Chemical Physics*, **122**, 244709.

Rosso, K.M., Smith, D.M.A. and Dupuis, M. (2003) An ab initio model of electron transport in hematite (α-Fe_2O_3) basal planes *Journal of Chemical Physics*, **118**, 6455.

Schlegel, M.L., Bataillon, C., Blanc, C., Pret, D. and Foy, E. (2010) Anodic activation of iron corrosion in clay media under water-saturated conditions at 90 degrees C: Characterization of the corrosion interface. *Environmental Science & Technology*, **44**, 1503–1508.

Singer, D.M., Chatman, S.M., Ilton, E.S., Rosso, K.M., Banfield, J.F. and Waychunas, G.A. (2012a) Identification of simultaneous U(VI) sorption complexes and U(IV) nanoprecipitates on the magnetite (111) surface. *Environmental Science & Technology*, **46**, 3811–3820.

Singer, D.M., Chatman, S.M., Ilton, E.S., Rosso, K.M., Banfield, J.F. and Waychunas, G.A. (2012b) U(VI) sorption and reduction kinetics on the magnetite (111) surface. *Environmental Science & Technology*, **46**, 3821–3830.

SKB (2005) Extended consultations according to the environmental code. *EnaInfo/Edita Norstedts Tryckeri.* Swedish Nuclear Fuel and Waste Management Co, Stockholm, Sweden, pp. 45.

Skomurski, F.N., Ilton, E.S., Engelhard, M.H., Arey, B.W. and Rosso, K.M. (2011) Heterogeneous reduction of U^{6+} by structural Fe^{2+} from theory and experiment. *Geochimica et Cosmochimica Acta*, **75**, 7277–7290.

Skomurski, F.N., Kerisit, S.N. and Rosso, K.M. (2010a) Structure, charge distribution, and electron hopping dynamics in magnetite (Fe_3O_4) (100) surfaces from first principles. *Geochimica et Cosmochimica Acta*, **74**, 4234–4248.

Skomurski, F.N., Rosso, K.M., Krupka, K.M. and McGrail, B.P. (2010b) Technetium incorporation into hematite (alpha-Fe_2O_3). *Environmental Science & Technology*, **44**, 5855–5861.

Smart, N., Rance, A., Reddy, B., Lydmark, S., Pedersen, K. and Lilja, C. (2011) Further studies of in situ corrosion testing of miniature copper-cast iron nuclear waste canisters. *Corrosion Engineering Science and Technology*, **46**, 142–147.

Środoń, J., Morgan, D.J. and Eslinger, E.V. (1986) Chemistry of illite/smectite and end-member illite. *Clays and Clay Minerals*, **34**, 368–378.

Stephens, M.B. (2010) Forsmark site investigation: Bedrock geology – overview and excursion guide. SKB R-10-04, Swedish Nuclear Fuel and Waste Management Co, Stockholm, Sweden, 52 pp.

Taylor, K.G. and Curtis, C.D. (1995) Stability and facies association of early diagenetic mineral assemblages; an example from a Jurassic ironstone-mudstone succession, U.K. *Journal of Sedimentary Research*, **65**, 358–368.

Taylor, K.G. and Konhauser, K. (2011) Iron in Earth surface systems: A major player in chemical and biological processes. *Elements*, **7**, 83–88.

Taylor, K.G. and Macquaker, J.H.S. (2011) Iron minerals in marine sediments record chemical environments. *Elements*, **7**, 113–118.

Urios, L., Marsal, F., Pellegrini, D. and Magot, M. (2013) Microbial diversity at iron-clay interfaces after 10 years of interaction inside a deep argillite geological formation (Tournemire, France). *Geomicrobiology Journal*, **30**, 442–453.

Vaughan, D.J. and Craig, J.R. (1978) *Mineral Chemistry of Metal Sulfides*. Cambridge University Press, Cambridge, UK.

Vidal, O. and Dubacq, B. (2009) Thermodynamic modeling of clay dehydration, stability and composition evolution with temperature, pressure and water activity. *Geochimica et Cosmochimica Acta*, **73**, 6544–6564.

Vines, S. and Beard, R. (2012) An overview of radionuclide behaviour research for the UK geological disposal programme. *Mineralogical Magazine*, **76**, 3373–3380.

Wander, M.C.F., Kerisit, S., Rosso, K.M. and Schoonen, M.A.A. (2006) Kinetics of triscarbonato uranyl reduction by aqueous ferrous iron: A theoretical study. *Journal of Physical Chemistry A*, **110**, 9691–9701.

Wang, Z., Moore, R.C., Felmy, A.R., Mason, M.J. and Kukkadapu, R.K. (2001) A study of the corrosion products of mild steel in high ionic strength brines. *Waste Management*, **21**, 335–341.

Wang, Z.M., Zachara, J.M., Boily, J.F., Xia, Y.X., Resch, T.C., Moore, D.A. and Liu, C. (2011) Determining individual mineral contributions to U(VI) adsorption in a contaminated aquifer sediment: A fluorescence spectroscopy study. *Geochimica et Cosmochimica Acta*, **75**, 2965–2979.

Wersin, P., Spahiu, K. and Bruno, J. (1994) Time evolution of dissolved oxygen and redox conditions in a HLW repository. SKB TR 94-02, Swedish Nuclear Fuel and Waste Management Co, Stockholm, Sweden.

Wharton, M.J., Atkins, B., Charnock, J.M., Livens, F.R., Pattrick, R.A.D. and Collison, D. (2000) An X-ray absorption spectroscopy study of the coprecipitation of Tc and Re with mackinawite (FeS). *Applied Geochemistry*, **15**, 347–354.

Williams, B.A., Brown, C.F., Um, W., Nimmons, M.J., Peterson, R.E., Bjornstad, B.N., Lanigan, D.C., Serne, R.J., Spane, F.A. and Rockhold, M.L. (2007) *Limited Field Investigation Report for Uranium Contamination in the 300-FF-5 Operable Unit at the 300 Area, Hanford Site, Washington*. PNNL-16435, Pacific Northwest National Laboratory, Richland, Washington, USA.

Wilson, M.J. (2013) *Sheet Silicates* (W.A. Deer, R.A. Howie, and J. Zussman, editors). Rock-forming Minerals, vol. **3C**, The Geological Society, London.

Worden, R.H. and Morad, S. (2003) Clay minerals in sandstones: controls on formation, distribution and evolution. Pp. 1–14 in: *Clay Mineral Cements in Sandstones* (R.H. Worden and S. Morad, editors). Blackwell Publishing Ltd., Oxford, USA.

EMU Notes in Mineralogy, Vol. 17 (2017), Chapter 10, 273–311

Biogeochemical modelling of redox processes in low-temperature natural systems

IMAD A. M. AHMED[1], PATRICIA ACERO[2] and LUIS F. AUQUÉ[3]

[1] *Department of Earth Sciences, University of Oxford, South Parks Road, Oxford OX1 3AN, UK, e-mail: imad.ahmed@earth.ox.ac.uk*
[2] *Department of Earth and Planetary Sciences, Birkbeck, University of London, Malet St., London WC1E 7HX, UK, e-mail: patri.acero@gmail.com*
[3] *Department of Earth Sciences, University of Zaragoza, c/Pedro Cerbuna 12, 50009 Zaragoza, Spain, e-mail: lauque@unizar.es*

Oxidation-reduction (redox) reactions, acid-base reactions and complex formation along with hydrolysis, precipitation and adsorption, account for the vast majority of chemical reactions and processes that occur in environmental and engineering systems. This chapter provides an overview of the basic principles of the thermodynamic and kinetic treatments of redox reactions in natural systems. Modelling redox transformations and predicting the concentrations of chemical species involved in redox reactions are the focus of this chapter. The practical utility of geochemical modelling applied to redox processes and some of its limitations are discussed and some practical modelling examples of geochemical problems are provided.

1. Introduction

Chemists use the term elemental 'species' to mean a specific form of an element defined by its isotopic composition, electronic or oxidation state, complexation or molecular structure (Brummer, 1986; Förstner, 1993; Templeton *et al.*, 2000). This is the IUPAC definition of speciation. The chemical forms ^{63}Cu, ^{65}Cu, ^{67}Cu, Cu^{I}, Cu^{II}, $CuCl^{+}$, $CuCl_2^0$, $CuSO_4^0$, $CuOH^{+}$, $Cu(OH)_2$ and $CuEdta^{2-}$ are examples of copper chemical species covered by the above definition. The concept 'speciation' was first introduced by Goldberg (1954) to provide a better understanding of the biogeochemical cycling of trace elements in seawater. However, some of the dramatic effects of speciation on metal toxicity have long been recognized. For example, organic mercury compounds have been known since the 1800s with the first reported fatal cases of methylmercury (CH_3Hg^{+}) poisoning in 1865. The high toxicity of methylmercury is linked to its high solubility in lipids compared to inorganic forms of mercury. This is a good example to demonstrate that the solubility of different forms of an element plays a crucial role in differential bioaccessibility and toxicity.

Note that the above IUPAC definition of speciation is perfectly appropriate for speciation in solution where species are defined specifically (*i.e.* specific chemical

DOI: 10.1180/EMU-notes.17.8

formula or oxidation state). This criterion is difficult to apply fully in soils and sediments because in these systems the majority of element species are found associated with mineralogical and biological components. In such systems, groups of molecules (not specific or singly identified species) are commonly defined functionally or operationally. The functional definition specifies the function or impact of species (groups) in living organisms such as uptake and essentiality of selenium to plants or its toxicity at certain concentrations. Operationally defined species refer to extractable clusters of molecules characterized or isolated by a particular analytical procedure (operation) such as the commonly used sequential extraction procedures in soil science (see Tessier and Campbell, 1987; Tessier and Turner, 1995; Young *et al.*, 2005).

The formation of different forms and compounds of an element in the environment is highly dependent on medium conditions such as redox potential, pH, composition, temperature and ionic strength. Redox potential, pH and the availability of complexing ligands (both organic and inorganic) are the most important variables controlling element speciation in aqueous systems. For example, arsenate ($H_2AsO_4^-$) is dominant at pH < 7 whereas arsenite ($HAsO_4^{2-}$) predominates at alkaline conditions (Cullen and Reimer, 1989). Uranium is another example that occurs naturally in four oxidation states: U^{III}, U^{IV}, $U^{V}$and U^{VI}. While the uranyl ion (UO_2^{2+}) is dominant under oxidizing conditions, the uranous ion (U^{IV}) predominates under reducing conditions. The availability of carbonate and natural organic matter as complexing agents are known to determine the fate and mobility of uranium in natural water (Choppin, 2007; Markich, 2002).

Speciation of an element affects not only its chemical (*e.g.* solubility, complexation and polymerization) or physical properties (*e.g.* volatilization and diffusivity), but also its ecotoxicity and bioavailability. For example, tellurite (Te^{IV}) is ~10 times more toxic than tellurate (Te^{VI}) and similarly the water-soluble Cr^{VI} is more toxic than the sparingly soluble Cr^{III} (Karlson and Frankenberger, 1993). Therefore, research of element speciation has been an active area in environmental ecotoxicology for decades and knowledge of chemical speciation is becoming of primary importance in environmental risk assessment and the development of environmental guidelines and regulations.

The calculations and numerical methods used to predict the geochemical speciation of elements from the analytical data in environmental aqueous systems are referred to as 'geochemical speciation modelling' or simply geochemical speciation. These calculations are usually carried out using sophisticated codes to solve complex thermodynamic and kinetic equations in order to predict the concentrations or activities of specific chemical species. Speciation calculations provide baseline data for other more complex geochemical problems; *e.g.* prediction of reaction pathways, mineral weathering rates or multi-element reactive transport calculations (*e.g.* Zhu and Anderson, 2002). Geochemical speciation can be used to simplify complex processes and to facilitate our understanding of the driving forces and pathways of reactions that may take place over time scales that are not easily ascertained by laboratory or field experimentation. Geochemical speciation has also become a standard research tool for interpreting and predicting a wide range of abiotically and biotically driven processes occurring at the water–mineral interface (Ahmed *et al.*, 2014; Nordstrom and Campbell, 2014; Magalhães *et al.*, 2015). It has also been used widely as a tool in the optimization of the remediation of contaminated soils and groundwaters, environmental risk assessment of toxic elements and performance assessment of high-level nuclear waste or of CO_2 storage in deep saline aquifers (Davis *et al.*, 1986; Hunter *et al.*, 1998; Prommer *et al.*, 2000; Amme, 2002; Marini, 2007; Auqué *et al.*, 2008). In all of these processes geochemical speciation enables key remediation parameters to be identified. If geochemical speciation is to provide useful results, however, it must be based on, and coupled with, suitable thermochemical, physical and chemical empirical data. Detailed information on model development and related concepts in geochemical context are not the scope of this chapter but the reader is directed to the following references (among many others) for further reading: Zhu and Anderson, 2002; Chien *et al.*, 2003; Bethke, 2008; Nordstrom and Campbell, 2014).

Geochemical processes range from sorption, ion exchange, dissolution and precipitation to more complex reactions in multicomponent systems (*e.g.* biogeochemical reactions, isotopic exchange, multiphase processes, *etc.*). Reduction-oxidation is an important class of chemical reactions which involves the transfer of electrons between dissolved and/or solid species. At the Earth's surface, oxidation (electron removal) is commonly driven by oxygen in the atmosphere while reduction (electron addition) is frequently driven by organic matter or is mediated by microorganisms. There is strong experimental evidence that the biogeochemical cycles of redox-sensitive transition metals (*e.g.* Fe, Mn, Mo, Ti, Cu, Cr, V), metalloids (*e.g.* As, Te, and Sb) and radionuclides (*e.g.* U, Tc and Np) are controlled mainly by redox processes (Morford and Emerson, 1999; Zhang and Brady, 2002; Ginder-Vogel and Fendorf, 2007; Borch *et al.*, 2010). For both terrestrial and aquatic environments reduction-oxidation is also a key constraint on the cycles of carbon, nitrogen, sulfur and phosphorus (Cappellen and Ingall, 1996; Dean, 1999; Hayes and Waldbauer, 2006; Godfrey and Falkowski, 2009; Borch *et al.*, 2010; Li *et al.*, 2012). Most of the redox-driven reactions are complex, often occur over a range of temporal scales ranging from milliseconds to years and often involve many chemical and mineralogical components, which makes it difficult to conduct sufficiently realistic laboratory experiments. For such cases, speciation modelling becomes a valuable tool in geochemical research.

This chapter provides an overview of redox speciation and also some modelling examples of redox chemistry in natural waters, sediments and soils using the *PHREEQC* geochemical code (Parkhurst and Appelo, 2013). The chapter focuses primarily on modelling the formation and transformation of redox-reactive minerals and their interactions with redox-sensitive elements. Particular attention is also given to the limitations, uncertainty and challenges of modelling redox geochemistry. The theoretical chemical basis for redox thermodynamics and some basic principles of related thermodynamic relationships and concepts are discussed briefly here but several other texts are also recommended for further reading (Mattigod and Zachara, 1996; Stumm and Morgan, 1996; Langmuir, 1997; Ganguly, 2009; Oelkers and Schott, 2009; Fegley, 2013).

2. Theoretical background

2.1. Equilibrium redox potentials

Many elements can exist in multiple oxidation states with each state displaying distinct properties and behaviour. For example, sulfur forms chemical species with oxidation states which range from −2 to +6. These include H_2S (−2), polysulfides S^{2-} (−2, 0), elemental S^0 (0), sulfite (SO_3^{2-}) (+4), thiosulfate ($S_2O_3^{2-}$) (+2), and sulfate (SO_4^{2-}) (+6). Each of these individual sulfur species displays a distinctive chemical behaviour.

The oxidation state of a system is an intensive thermodynamic variable, similar to temperature, pressure and concentration, meaning that it is independent of the size of the system or amount of materials in that system. The redox potential (Eh) is the most commonly used measure of oxidation state in low-temperature geosciences (*i.e.* T = 0−100°C). At higher temperatures, oxygen fugacity (f_{O_2}) is used more frequently as a measure of oxidation state. Experimentally, Eh is the measured potential (in volts) of a platinum electrode corrected to the Standard Hydrogen Electrode (SHE) which is given the arbitrary value of zero. Values of Eh in natural systems commonly range from −0.6 to +0.9 V. Throughout this chapter, we follow the Gibbs-Stockholm convention in which the positive direction of electrode potential implies an increasingly oxidizing condition whereas the negative direction implies an increasingly reducing condition at the electrode. Hence, a positive Eh value indicates an oxidizing condition while a negative Eh value indicates a reducing condition. A value of zero Eh signifies a system with no drive or, put another way, no capability to either oxidize or reduce.

The basic treatment of redox processes may be examined by taking a common redox reaction at the Earth's surface; *e.g.* the oxidation of magnetite (Fe_3O_4 or $Fe^{II}O.Fe_2^{III}O_3$) to hematite (α-Fe_2O_3 or $Fe_2^{III}O_3$) in the presence of water (equation 1):

$$2Fe^{II}O.Fe_2^{III}O_3 + H_2O \leftrightarrow 3Fe_2^{III}O_3 + 2H^+ + 2e^- \quad (1)$$

The essence of the above reaction is the oxidation of ferrous to ferric ion ($Fe^{II} \rightarrow Fe^{III} + e^-$). Unlike protons, hydrated electrons cannot exist as free species in nature (Hostettler, 1984; Thorstenson, 1984) and therefore they must be consumed in a complementary reaction (*e.g.* $2H^+ + 2e^- \rightarrow H_2$). At redox equilibrium, the net current is zero and the electrode is known to adopt a potential (E_{eq}, commonly referred to as Eh) that can be described by the Nernst equation (Bard and Faulkner, 2001) as follows:

$$E_{eq} = \mathrm{Eh(V)} = E^o + 2.3026 \times \frac{\mathrm{R}T}{n\mathrm{F}} \log_{10}\left(\frac{\prod_i^{O_i} \alpha_i(\text{Oxidized species})}{\prod_j^{R_j} \alpha_j(\text{Reduced species})}\right) = \frac{-\Delta G^o}{n\mathrm{F}} \quad (2)$$

where Eh is the equilibrium redox potential (in mV), E^o is the standard potential, R is the gas constant (8.314 J mol^{-1} K^{-1}), T is temperature (in Kelvin), F is the Faraday constant (9.6485×10^4 J V^{-1} mol^{-1}) and n is the number of electrons being transferred. The term $\prod_i^{O_i} \alpha_i$ is the activity product of the i chemical species in the oxidized assemblage and $\prod_j^{R_j} \alpha_j$ is the activity product of the j species in the reduced assemblage, where O_i and R_j refer to the stoichiometries of these species, respectively. ΔG^o is the free Gibbs energy under standard conditions.

The Nernst equation allows for calculation of the comparative activities of the individual components of a coupled redox pair at any specific value of the redox potential (Eh), given knowledge of the standard redox potential (E^o). It therefore enables the determination of which redox-specific species is dominant at any selected non-standard redox potential. Alternatively, if the activities and stoichiometries of the reduced and oxidized species are known (again, given knowledge of the standard redox potential) then quantification of the prevailing redox potential at the point in time is possible through application of the Nernst equation.

The use of the dimensionless variable pe as a measure of redox intensity of an environment is a more convenient approach both computationally and conceptually than the directly measured Eh. By analogy to pH ($-\log_{10}a_{H^+}$), pe is defined as the negative logarithm of a relative electron activity:

$$pe = -\log_{10}a_{e^-} \tag{3}$$

At equilibrium, the pe can be described in terms of the equilibrium constant K_{eq}, where pe^o is the relative electron activity when all species other than electrons are at unit activity:

$$pe = pe^o + 1/n\log_{10}K_{eq} \tag{4}$$

Conveniently, pe can be calculated directly from Eh as follows:

$$pe = \frac{F \times Eh}{2.3026RT} = \frac{5040 \times Eh}{T} \tag{5}$$

Incidentally, both hydrogen and electron activities are measured indirectly by use of specific pH and redox electrodes in solution. However, unlike pH, pe is a completely theoretical expression and does not have any physicochemical significance. This is because there is no physical meaning of the activity of aqueous electrons (a_{e-}) whereas the activity of protons (a_{H+}) is a meaningful chemical quantity. However, pe is still a useful concept and, by analogy with the acid-base chemistry, logarithmic plots of species activities as a function of pe can be constructed. Taking the nitrogen–water system as an example, the nitrogen speciation can be obtained by combining reactions 6–10 (Sillén *et al.*, 1964):

$$NO_3^- + 2H^+ + 2e^- \leftrightarrow NO_2^- + H_2O \qquad pe^o_{25°C} = +14.2 \tag{6}$$

$$NO_3^- + 10H^+ + 8e^- \leftrightarrow NH_4^+ + 3H_2O \qquad pe^o_{25°C} = +14.9 \tag{7}$$

$$2NO_3^- + 12H^+ + 10e^- \leftrightarrow N_{2(g)} + 6H_2O \qquad pe^o_{25°C} = +21.0 \tag{8}$$

$$NO_2^- + 8H^+ + 6e^- \leftrightarrow NH_4^+ + 2H_2O \qquad pe^o_{25°C} = +15.0 \tag{9}$$

$$N_{2(g)} + 8H^+ + 6e^- \leftrightarrow 2NH_4^+ \qquad pe^o_{25°C} = +4.68 \tag{10}$$

Figure 1 shows the distribution of nitrogen species as a function of pe at pH 7 and using a total nitrogen concentration of 1 mmol L^{-1}. At these conditions, note that $N_{2(g)}$ is the most stable N species which occupies most of the pe aqueous range. Ammonia (NH_3) and ammonium (NH_4^+) are the predominant reduced species at negative pe whereas nitrate (NO_3^-) and nitrite (NO_2^-) dominate the systems at pe > +8. Between NH_4^+

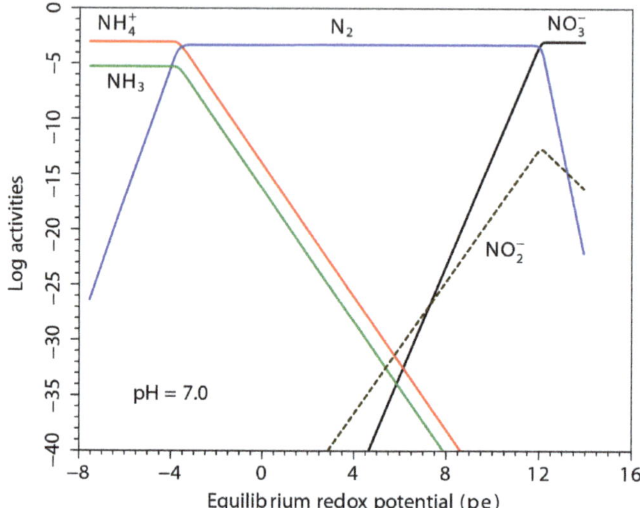

Figure 1. Log-transformed activities of nitrogen species as a function of pe at pH 7 and 25°C. This system was modelled in *PHREEQC* using the *WATEQ4F* thermodynamic database (Ball and Nordstrom, 1991) using a total N concentration of 1 mmol L^{-1}.

and NO_3^- there is an eight-electron shift in oxidation state but it is remarkable that the shifts in the relative predominance of the N species occur within a rather narrow pe range. The incomplete oxidation of $N_{2(g)}$ into NO_3^-/NO_2^- even under prevailing aerobic conditions may be explained by the lack of efficient biological catalysts in most surface waters.

Reaction pairs involved in redox reactions (*e.g.* magnetite and hematite in reaction 1) are termed 'redox buffers'. The resistance to change in Eh is the basis of redox buffering which is analogous to pH buffering but they refer to species other than H^+ such as O_2, SO_4^{2-} and NO_3^-. Redox-active elements such as Fe, Mn, S and Ti present in rocks may provide redox buffering under certain conditions.

2.2. Eh-pH stability diagrams and the range of redox potentials in natural systems

At Earth's surface environments, Eh (or pe), pH and the activities of dissolved species are the most important variables in redox reactions. Therefore, it is common practice in low-temperature geochemistry to study phase equilibria using Eh-pH (or pe-pH) stability diagrams at specified activities of dissolved species. In these diagrams, the relative stabilities of different oxidized or reduced components are represented as a function of pH and Eh (or pe). For their construction, thermodynamic equilibrium is implicitly assumed and, therefore, their use and interpretation should be made with caution for most natural systems.

Even though water does not, at first glance, seem to participate in redox reactions, it may become unstable under extreme oxidizing or reducing conditions. Under very oxidizing conditions, the partial pressure of oxygen in equilibrium with water is greater than 1 atm; similarly, the partial pressure of hydrogen is also greater than 1 atm under highly reducing conditions. Since water in natural systems is stable only at a total pressure of less than 1 atm, all the relevant redox processes in aqueous systems take place at or below that total pressure and, therefore, hyper oxidative and reductive conditions are excluded.

In an Eh-pH diagram, the stability boundaries of water may be obtained by using the Nernst equation. The upper boundary ($H_2O_{(l)}-O_{2(g)}$) is defined by reaction 11:

$$0.5O_{2(g)} + 2H^+_{(aq)} + 2e^- \leftrightarrow H_2O_{(l)} \tag{11}$$

At $T = 298.15$ K, the Eh of the above reaction can be represented according to equation 12:

$$\mathrm{Eh}_{H_2O(l)} = \mathrm{Eh}^{o}_{H_2O(l)} + \frac{0.0592}{n}\log_{10}\left(\frac{(\alpha_{H^+})^2(P_{O_2})^{0.5}}{\alpha_{H_2O_{(l)}}}\right) \tag{12}$$

for which $n = 2$. The potential $\mathrm{Eh}^{o}_{H_2O(l)}$ can be calculated easily to be 1.23 V from the Gibbs free energy for water at standard conditions ($\Delta G^o_f = -nfE^o$), which is -237.129 kJ mol^{-1} (Robie and Hemingway, 1995). Setting $P_{O_2} = 1$ atm and rearranging the terms in equation 12, an equivalent equation describing the upper boundary of water stability in terms of Eh and pH is obtained:

$$\mathrm{Eh} = 1.23 - 0.0592\ \mathrm{pH} \tag{13}$$

Analogously, the lower boundary of water stability ($H_{2(g)}-H_2O_{(l)}$) can be derived from reaction 14:

$$H^+_{(l)} + e^- \leftrightarrow 0.5H_{2(g)} \tag{14}$$

In this case, the following simple linear equation is obtained:

$$\mathrm{Eh} = -0.0592\mathrm{pH} \tag{15}$$

These boundaries enclose the field of possible combinations of pe and pH in natural hydrogeochemical systems (Fig. 2a). Within this field, the largest pe values correspond to waters in contact with atmospheric conditions, which are frequently grouped as 'oxic' environments. Regarding groundwaters, their pe or Eh values are usually more reducing (*i.e.* lower pe) because they are not in direct contact with atmospheric conditions and they are frequently termed as 'suboxic' or 'anoxic' environments. The range of redox potentials for this type of environment is comparatively wide and may even extend to values close to the lower boundary (Fig. 2a).

As displayed in Fig. 2b, every range of pH-pe values corresponds to the dominance of different redox reactions and redox couples, which are clearly segregated in the pH-pe field. For any given pH value, a unique and characteristic sequence of redox couples corresponding to different redox potential values can be obtained, as will be explained in more detail in section 3.2.

According to Drever (1997), four different ranges of pH-pe values and associated redox couples can be differentiated for natural waters (Fig. 2b). Range 1 corresponds to

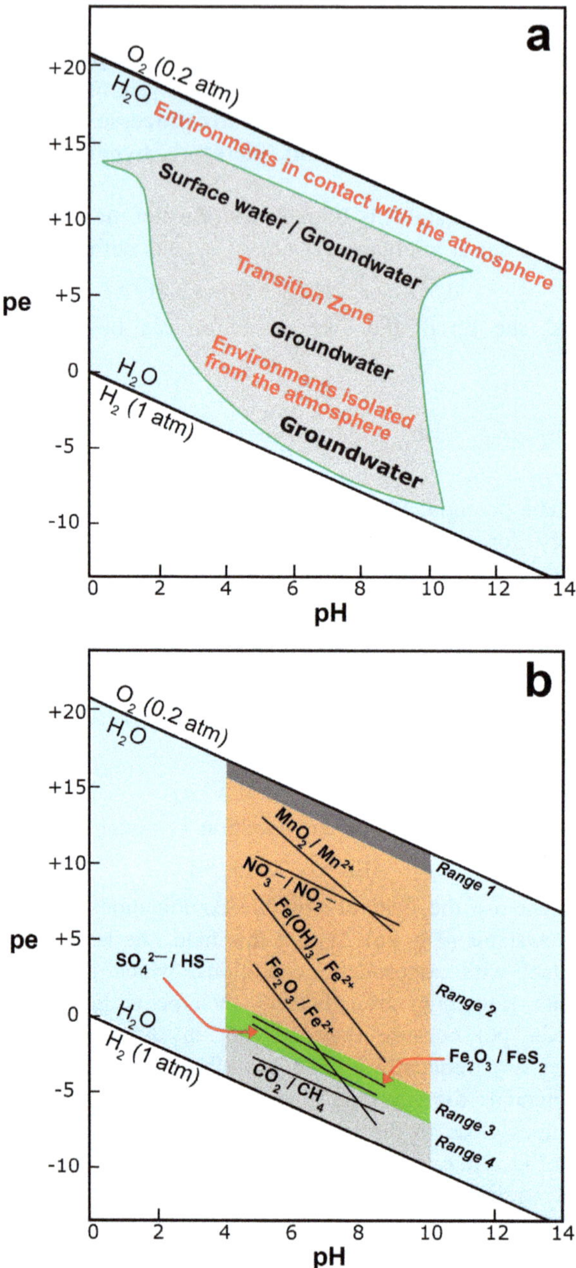

Figure 2. (a) pH-pe diagram showing approximate positions of some natural environments adapted from Garrels and Christ (1965) and Baas Becking *et al.* (1960). (b) pH-pe stability field diagram of principal redox buffering pairs in hydrochemical environments (soils, sediments, surface waters and groundwaters), adapted from Drever (1997).

oxygen-bearing waters. Range 2 contains waters without any free oxygen (anoxic environments) or dissolved sulfide. This range is the widest in the pH-pe space and the main corresponding redox processes are those associated with iron and manganese oxides, which favour (and buffer) the presence of dissolved Mn(II) and Fe(II) in this range. The range of redox potentials in Range 3 is much narrower and, in most cases, clearly reducing. This range corresponds to waters buffered mainly by sulfate reduction processes, usually accompanied by FeS or MnS precipitation. Additionally, crystalline iron oxides (*e.g.* hematite) could also participate in the control of dissolved Fe(II) at neutral or slightly basic pH conditions in this range. Overall, sulfur and iron redox couples (*e.g.* SO_4^{2-}/HS^-, SO_4^{2-}/FeS, SO_4^{2-}/FeS_2) are dominant under Range 3 conditions. Finally, Range 4 includes the lowest and most reducing redox potential values, frequently associated with methanic environments.

A wide variability in the Eh (pe) values can be observed in sediments, soils or groundwaters. In the case of groundwater aquifers, such variability is mainly determined by the following factors (Drever, 1997):

1. Dissolved oxygen concentration in the recharge waters which, in turn, may be controlled by factors such as precipitation and rate of the degradation of soil organic matter.
2. The amount, type, distribution and reactivity of organic matter and other reducing components (*e.g.* ascorbate). These factors influence mostly the microbial metabolic processes and, therefore, the development of more or less complex redox evolution sequences.
3. Presence, type, distribution and amount of redox buffers (predominantly mineral phases such as MnO_2, Fe_2O_3 and $Fe(OH)_2$) in the aquifer.
4. Hydrological processes, such as the flow rate. Such processes are the main controls on the residence time of the groundwaters in the aquifer and, therefore, the evolution time of the redox processes under the aquifer conditions. The hydraulic properties of the materials involved play a major role in the rate of supply and depletion of oxygen and hence introduce variability in the groundwater redox chemistry.

2.3. Modelling redox reactions

Various modelling frameworks have been developed to perform geochemical calculations in water–rock interaction systems (soils, sediments, aquifers, *etc.*). These include, among many others, the codes *PHREEQC* (Parkhurst and Appelo, 2013), *EQ3/6* (Wolery, 1992), *The Geochemist's Workbench* (Bethke and Yaekel, 2014), *ORCHESTRA* (Meeussen, 2003) which incorporates the NICA-DONNAN model (Kinniburgh *et al.*, 1996; Kinniburgh *et al.*, 1999), *ECOSAT* (Keizer and van Riemsdijk, 1994), *MINEQL+* (Schecher and McAvoy, 2001), GEM-Selektor code (Kulik *et al.*, 2013), *GEOCHEM* (Parker *et al.*, 1995), *SOILCHEM* (Sposito and Coves, 1988), Visual *MINTEQ* (Gustafsson, 2009), *CHESS* (van der Lee, 1993), *GEOSURF* (Sahai and Sverjensky, 1998) and *WHAM* VI and VII (Tipping, 1998; Tipping *et al.*, 2011). Most of these geochemical codes are equilibrium-based and use a wide range of input variables, including multi-element concentrations (or activities), alkalinity, pH, temperature and dissolved organic matter (DOM). *PHREEQC*, *ORCHESTRA*, *ECOSAT*, *EQ3/6*, *Geochemist's Workbench* and *MINEQL+* are among the most commonly used computer codes in simulating redox reactions under the Earth's surface conditions.

In the discussion and modelling exercises which follow we will be using the *PHREEQC* code (version 3.1.7) as the main modelling platform, together with the phreeqc.dat thermodynamic database distributed with the package, unless otherwise stated. *PHREEQC* is a computer program written in the C language based on the Fortran code *PHREEQE* (Parkhurst *et al.*, 1980) for modelling chemical reactions and transport processes in aqueous environments. *PHREEQC* is capable of performing direct (predictive) and inverse (deductive) modelling calculations taking into account complex redox reactions, mixing of solutions, surface complexation, dissolution-precipitation, kinetically controlled reactions and evaporation and dilution with the account for water gained or lost from minerals (Parkhurst and Appelo, 2013). Further explanations about the direct and inverse modelling approaches can be found elsewhere (*e.g.* Zhu and Anderson, 2002). An additional advantage of *PHREEQC* is that it may be used with different thermodynamic databases. The electron balance concept in *PHREEQC* assumes that free electrons do not exist in an aqueous solution and electrons may only enter or leave the system through homogeneous (*i.e.* reactions occur within the solution phase) and heterogeneous (*i.e.* reactions occur between a solid surface and/or gas and the solution phase) redox reactions. This means that an electron which enters the system though one reaction must be removed by another. Thus, the electron-balance equation in *PHREEQC* is written as:

$$\sum_{j=1}^{J} C_{e_j}\alpha_j + \sum_{k=1}^{K} C_{e_k}\alpha_k = 0 \tag{16}$$

For which $C_{e_j}\alpha_j$ is the number of electrons released or consumed in aqueous redox reaction i, and $C_{e_k}\alpha_k$ is the number of electrons released or consumed in the dissolution reaction of a prescribed phase k. The symbol α_j denotes the mole transfers between valence states of each redox-active element and α_k denotes the mole transfers from solids or gases into or out of solution.

3. Redox equilibrium and disequilibrium in natural systems

The characterization of the redox state of natural systems is a basic component of the study of many natural processes together with contamination problems and their remediation approaches. The mobility of many elements is largely dependent on the redox conditions and related processes. For instance, both iron and manganese are much more soluble in their reduced forms (*i.e.* Fe(II) and Mn(II)) compared with their oxidized forms (*i.e.* Fe(III) and Mn(IV)). Therefore, the dissolved concentrations of both elements tend to be quite low under the most frequent pH and redox conditions encountered in natural aqueous systems (*i.e.* high pe and circumneutral pH values, which favour the presence of both of their relatively insoluble oxidized forms; White, 2013). Other elements display the opposite evolution with respect to redox conditions (*e.g.* Cu and Ni). In the presence of a redox potential low enough to allow the transformation of sulfate to sulfide, the total dissolved concentrations of both Cu and Ni tend to decrease due to the precipitation of very insoluble sulfides. For trace elements such as As(III)/(V), Se(IV)/(VI) and Cr(III)/(VI), changes in the redox state determine their mobility and bioavailability and, thus, determine the most appropriate remediation strategies that can be used in a particular system (Sigg, 2000).

The redox state of any aqueous system is determined by the concentrations of oxidized and reduced species. For any given element, its oxidized and reduced species will form redox couples. Each of these couples can be characterized by a redox potential, which can be determined from the activities of the oxidized and reduced species forming the couple by means of the Nernst equation (equation 2).

Aqueous systems can only be considered at overall redox equilibrium if the redox potential of all the present redox couples is the same and the value of that potential (*i.e.* Eh or pe) is the redox potential of the system. Unfortunately, most natural hydrogeochemical systems (if not all of them) are only rarely at redox equilibrium and, therefore, each of their redox couples usually has its own redox potential (see, for instance, the works by Lindberg and Runnels, 1984, and Stefánsson *et al.*, 2005). The main reason for this redox disequilibrium is the variable and slow reaction kinetics of oxidation and reduction reactions. As a general trend, half reactions requiring the transference of just one electron are fast, such as, for instance, the reduction reaction of ferric to ferrous iron. In fact, the redox pair Fe(II)/Fe(III) is one of the few examples for which, if dominant in a system, consistent potentiometric and calculated redox potentials can be obtained (Nordstrom *et al.*, 1979; Macalady *et al.*, 1990; Grenthe *et al.*, 1992). In contrast, half reactions involving the transference of more than one electron tend to be much slower because they involve more extensive changes in molecular configuration and structure, such as, for instance, the oxidation reaction of sulfide to sulfate:

$$HS^- + 2H_2O \leftrightarrow SO_4^{2-} + 9H^+ + 8e^- \quad (17)$$

Redox reactions involving the transference of just one electron are characterized by reaction times generally shorter than a few years and frequently shorter that a few hours. In contrast, redox reactions requiring the transference of more than one electron may take up to several thousands of years to attain equilibrium especially in the absence of biological catalysts (Bruno, 1997).

Another important factor controlling redox kinetics is the coordination environment of the particular element subject to oxidation or reduction. For instance, the reductive dissolution of ferric oxyhydroxides usually requires many bonds to be broken in the mineral structure in order to allow the release of Fe(III), and this may depend on the possibility of forming new bonds between the iron at the mineral surface and adsorbed reductants, metal complexes or organic compounds (Christensen *et al.*, 2000, and references therein).

Finally, an additional and key factor determining the rate of redox processes in natural systems is the role played by the metabolic processes of microorganisms.

Microorganisms are able to catalyse the transfer of electrons from the reduced to the oxidized species, obtaining energy in the transformation. For instance, sulfate-reducing bacteria may use reaction 17 in their respiration, increasing the rate of a process that otherwise could take thousands of years to reach thermodynamic equilibrium. Other processes, such as those involving C, S or N, commonly require microbial catalysis to operate at rates which are fast enough compared to the time-limited water residence times in aquifers (Christensen *et al.*, 2000). It should be noted, however, that microorganisms kinetically facilitate the transfer of electrons, thereby increasing the reaction rate, but reactions that actually occur are determined by their thermodynamic feasibility (Nelson, 2004).

3.1. Estimation of the redox potential of natural systems and its use in equilibrium calculations

Owing, in part, to the limitations imposed by the common redox disequilibrium present in natural waters (explained above), it is frequently very difficult (if not impossible) to obtain a meaningful Eh or pe value which can be used in redox calculations. Moreover, measured Eh values are rarely representative of the redox state of most natural systems due, not only to redox disequilibrium, but also to analytical difficulties in measuring with the Pt-electrode. Such difficulties may be caused by the lack of electroactivity at the Pt surface, by the presence of multi component-derived mixing potentials or even by poisoning of the electrode (Appelo and Postma, 2005).

One of the most classic examples showing the discrepancy between measured Eh values and distribution of redox couples was presented by Lindberg and Runnels (1984). Those authors reported that, for a wide group of groundwaters, the differences between the potentiometrically measured Eh values and the values calculated using analysed concentrations of redox couples exceeded 1000 mV in some cases. Moreover, the values of redox potential calculated using different redox couples were also notably different. In some cases, more reliable potentiometric measurements can be obtained under conditions where the redox pair Fe(II)/Fe(III) is predominant (Nordstrom *et al.*, 1979; Macalady *et al.*, 1990; Grenthe *et al.*, 1992) because of the fast kinetics of the involved redox processes. Apart from the iron redox couple, reversible and thermodynamically meaningful Eh measurements may be possible in waters that contain significant amounts of dissolved Mn or sulfide. This includes many acid mine waters, Fe-rich groundwaters and sulfide-rich sediments. In contrast, meaningful Eh measurements are rarely achievable when the dominant redox-sensitive elements are C, N, O, H and oxidized S, as is usually the case in most surface waters (Langmuir, 1997).

For these reasons, the use of redox couples is considered by many authors as the only way to obtain reliable Eh values to be used in geochemical calculations (Deutsch, 1997; Drever, 1997). The methodology is based on the analytical determination of the dissolved components involved in the different redox pairs present in the target system, in order to calculate the redox potential corresponding to each of the pairs. Subsequently, two main approximations can be adopted: (1) comparison of calculated and potentiometrically measured redox potentials to identify which of the redox pairs is controlling the measured Eh (Langmuir, 1997), or (2) use of calculated redox potentials as representative values for the disequilibrium situation present in the studied system. For both approximations, either homogeneous or heterogeneous redox couples can be used. Heterogeneous redox couples, formed commonly by a solid phase and a dissolved component (*e.g.* $Fe(OH)_3(s)/Fe^{2+}$), are frequently considered 'redox buffers' because they may be able to minimize, by means of dissolution and precipitation processes, changes in the redox potential in response to additions of oxidizing or reducing agents (Drever, 1997).

The assumption of redox equilibrium in speciation calculations implies that a single Eh (or pe) value must be chosen and that all the redox species present in the target problem will be distributed according to that potential. A simple application of this type of calculation is shown in Example 1 for the PHREEQC code (Parkhurst and Appelo, 2013). In this example, the distribution of redox species for a concentration of 2.7 mmol kg^{-1} of Fe of water, at 25°C and pH 3, is done according to a given pe value of 11 (*i.e.* Eh ~ 650 mV):

EXAMPLE 1

```
SOLUTION 1 Redox speciation using measured potential
temp 25; pH 3; pe 11; redox pe; units mmol/kgw; density 1; Fe 2.7
```

In addition to obtaining the distribution of ferric and ferrous iron species equilibrated with the given redox potential value, this calculation allows for the calculation of the saturation state of the hypothetical solution with respect to several iron phases included in the database used. In this case, the results indicate that ferric iron species represent <10% of total iron (not shown) and the following saturation indices (SI) are obtained for iron phases:

	Saturation indices		
Phase	SI	log IAP	log K(298 K,1 atm)
$Fe(OH)_3$(a)	−0.69	4.20	4.89 $Fe(OH)_3$
Goethite	5.20	4.20	−1.00 FeOOH
Hematite	12.42	8.41	−4.01 Fe_2O_3

As an alternative to the assumption of redox equilibrium, some geochemical codes, including *PHREEQC* (Parkhurst and Appelo, 2013), allow the treatment of redox disequilibrium by using multiple redox pairs. To do so, the concentrations of the different components of each relevant redox pair are used to calculate, by means of the Nernst equation, that couple's individual redox potential value. This allows for the quantification of redox couples within the observed overall redox disequilibrium situation. This type of treatment is displayed in Example 2 for the *PHREEQC* code:

EXAMPLE 2

```
SOLUTION 1 Redox potential calculation using redox couple
temp 25; pH 3; pe 4; redox Fe(2)/Fe(3); units mmol/kgw;
density 1; Fe(2) 1.4; Fe(3) 1.3; N(5) 2; N(3) 1; S 2
```

In this example, the redox potential of the system is calculated using the Nernst equation and the concentrations of Fe(II) and Fe(III) provided. Subsequently, the value obtained is used to distribute sulfur species according to redox equilibrium with the Fe(II)/Fe(III) redox couple (*i.e.* using the same redox potential). However, since both the concentrations of nitrate (NO_3^-) and nitrite (NO_2^-) are known, the code will calculate a distinct redox potential for this couple. Some of the main results obtained are as follows:

Redox couples

Redox couple	pe	Eh (volts)
Fe(2)/Fe(3)	11.7959	0.6978
N(3)/N(5)	11.4355	0.6765

Distribution of species

Species/Molality	
Fe(2)	1.400e–003
Fe(3)	1.300e–003
N(3)	1.000e–003
N(5)	2.000e–003
S(–2)	0.000e+000
S(6)	2.000e–003

Saturation indices

Phase	SI	log IAP	log K(298 K, 1 atm)
$Fe(OH)_3$(a)	−0.18	4.72	4.89 $Fe(OH)_3$
FeS(ppt)	−86.94	−90.85	−3.92 FeS
Goethite	5.72	4.72	−1.00 FeOOH
Hematite	13.44	9.43	−4.01 Fe_2O_3
Mackinawite	−86.21	−90.85	−4.65 FeS
Melanterite	−3.93	−6.14	−2.21 $FeSO_4{:}7H_2O$
Pyrite	−136.58	−155.06	−18.48 FeS_2

As can be observed in the results from Example 2, the calculated redox potential using the iron and nitrogen couples are quite similar but not exactly the same. Sulfur speciation is obtained according to the redox potential of the iron couple, resulting in a 100% of the dissolved sulfur being present as sulfate. It is interesting to note that the large difference in the distribution of iron species compared to Example 2 (same total iron concentration but almost 50% of Fe(III) in this example *vs.* <10% Fe(III) in Example 1) is associated with a change of <50 mV in the corresponding redox potential. From another point of view, this implies that seemingly minor differences in the measured redox potential (*e.g.* within ± 50 mV) may lead to significantly different distributions of redox-active elements and also the saturation indices and hence stability of their related mineral phases (*e.g.* compare the saturation indices for ferric phases in Examples 2 and 3).

Even though the approach of multiple redox couples may conceptually be more suitable than the use of a single Eh value, it is often very difficult or even impossible to calculate the redox potential for each possible redox couple in the target system. This is mainly due to analytical problems (*e.g.* the analytical detection limits for some species being too high to allow for accurate and precise measurements and the lack of suitable analytical methods for some species) and to the fact that some redox-active species are solid phases that may be poorly characterized or involved in dissolution/precipitation reactions which are not within equilibrium (Christensen *et al.*, 2000).

3.2. Main redox reactions in natural systems

In most natural systems, atmospheric oxygen and organic matter are the major oxidant and reductant, respectively (Langmuir, 1997). As a general trend, organic matter tends to be unstable in water and is liable to decomposition by different oxidants present in the target system. These oxidants are commonly known as Terminal Electron Acceptors (TEAs) when the process is microbially mediated. The most frequent TEAs in most aquatic redox processes are those related to O, N, Mn, Fe and S and, therefore, the main redox-controlling reactions in natural systems are linked to their presence.

In most groundwater systems, the primary dissolved redox-active species are usually the ions SO_4^{2-}, HS^-, Fe^{2+}, Mn^{2+}, NH_4^+, NO_2^- and NO_3^-, together with the gases CH_4, N_2O and O_2 (Christensen *et al.*, 2000). Apart from these dissolved species, the presence of minerals and precipitates should also be taken into account in order to fully characterize the redox state of a system. Some of the most important redox components of this type are iron and manganese oxides and carbonates, organic matter, sulfide and sulfate minerals and the ions Fe^{2+}, Mn^{2+}, NH_4^+ and SO_4^{2-} associated with exchange sites on solid surfaces. The evaluation of the definite presence of these species and their potential role in redox processes for a given system is one of the most complex and challenging geochemical tasks. An excellent review and discussion of redox conditions in groundwater pollution plumes and summary of sampling and analytical approaches can be found in Christensen *et al.* (2000).

Some of the most typical redox reactions in natural systems, together with their associated ΔG and standard redox potentials (ΔEh), are displayed in Table 1. The energy yield (ΔG_r) for any particular system from the reactions presented in Table 1 can be calculated explicitly from the particular composition variables (*e.g.* concentrations, activities, *etc.*) of the reactants and products involved in the target system (Jakobsen and Postma, 1999; Christensen *et al.*, 2000; Canfield, 2005; Park *et al.*, 2006) as follows:

$$\Delta G_r = \Delta G^o + RT\ln(IAP) = \Delta G^o + 2.303RT\log(IAP) \quad (18)$$

where ΔG^o is the standard free energy, R is the universal gas constant (mol^{-1} K^{-1}), T is temperature (in Kelvin) and IAP the ionic activity product for the reaction.

In Table 1, the oxidized species in any reaction or redox couple are able to oxidize the reduced species of any redox reaction with a higher value of associated ΔG (or lowest redox potential) and, in turn, they themselves are reduced in the process (Langmuir, 1997; Nelson, 2004). For instance, molecular oxygen (O_2) can be reduced to H_2O coupled to the oxidation of Fe(II) into Fe(III) (expressed in Table 1 as the

Table 1. Representative reactions from the main respiratory pathways of organic matter (acetate) mineralization and degradation in nature using both H_2 and acetate as electron donors, together with their associated standard Gibbs free energies and redox potential (as ΔEh) (modified and expanded from Canfield *et al.*, 2005).

Reaction	ΔG^o (kJ per reaction)	ΔEh (V)
Oxic respiration		
$O_2 + 2H_2 \rightarrow 2H_2O$	–456	1.18
$O_2 + \frac{1}{2}C_2H_3O_2^- \rightarrow HCO_3^- + \frac{1}{2}H^+$	–402	1.04
Denitrification		
$\frac{4}{5}H^+ + \frac{4}{5}NO_3^- + 2H_2 \rightarrow \frac{2}{5}N_2 + \frac{12}{5}H_2O$	–460	1.19
$\frac{4}{5}NO_3^- + \frac{3}{5}H^+ + \frac{1}{2}C_2H_3O_2^- \rightarrow \frac{2}{5}N_2 + HCO_3^- + \frac{1}{5}H_2O$	–359	0.93
Mn reduction (pyrolusite)		
$4H^+ + 2MnO_2 + 2H_2 \rightarrow 2Mn^{2+} + 4H_2O$	–440	1.14
$\frac{7}{2}H^+ + 2MnO_2 + \frac{1}{2}C_2H_3O_2^- \rightarrow 2Mn^{2+} + HCO_3^- + 2H_2O$	–385	1.00
Fe reduction (freshly precipitated amorphous FeOOH)		
$8H^+ + 4FeOOH + 2H_2 \rightarrow 4Fe^{2+} + 8H_2O$	–296	0.77
$\frac{15}{2}H^+ + 4FeOOH + \frac{1}{2}C_2H_3O_2^- \rightarrow HCO_3^- + 4Fe^{2+} + 6H_2O$	–241	0.62
Sulfate reduction		
$H^+ + \frac{1}{2}SO_4^{2-} + 2H_2 \rightarrow 2H_2O + \frac{1}{2}H_2S$	–98.8	0.26
$\frac{1}{2}H^+ + \frac{1}{2}SO_4^{2-} + \frac{1}{2}C_2H_3O_2^- \rightarrow \frac{1}{2}H_2S + HCO_3^-$	–43.8	0.11
Methanogenesis		
$\frac{1}{2}H^+ + \frac{1}{2}HCO_3^- + 2H_2 \rightarrow CH_4 + \frac{3}{2}H_2O$	–74.8	0.19
$\frac{1}{2}H_2O + \frac{1}{2}C_2H_3O_2^- \rightarrow CH_4 + \frac{1}{2}HCO_3^-$	–19.9	0.05

Values of ΔG are standardized to a $4e^-$ transfer. All the calculations at 25°C and unit activity for all reactants and products.

precipitated phase FeOOH) but the reverse transformation is not thermodynamically favoured. If two redox couples have the same redox potential for a given system, they would be in redox equilibrium (Banwart *et al.*, 1999). Note that the redox reactions included in Table 1 are thermodynamically favoured but this does not imply that they will inevitably occur or that they should proceed strictly *via* a single process. In fact, most of them tend to proceed by way of much more complex intermediate processes (Christensen *et al.*, 2000). Apart from this, the actual reactants and products involved in the reactions may differ from those considered in Table 1, especially for the organic matter and for the iron and manganese solids. For instance, the degradation of aged organic matter seems to be much slower than is the case with younger organic matter, and this seems to be the key in understanding the differences in redox behaviour between aquifers (Appelo and Postma, 2005).

For practical reasons, redox couples (and reactions) are frequently ordered according to their relative ability to oxidize or reduce other members of the quoted series and this may be graphically represented in so-called redox ladders. Redox ladders are a very useful way to represent graphically whether a given process is thermodynamically favoured or not. Because redox potentials (Eh or pe) depend on both pH and activities of the species involved in the redox pair, redox ladders are always represented for fixed values of these parameters.

One of the most important applications of the redox ladder is its use to infer the theoretical order (from a thermodynamic point of view) that a sequence of redox reactions should follow under a given set of conditions. To illustrate this point, the reactions given in Table 1 were selected to construct the redox ladder shown in Fig. 3. These reactions represent, in simple terms, the most important redox reactions in natural waters including the oxidation of organic matter coupled with the corresponding reduction. An excellent illustration of redox ladders and the conventional rules for constructing them was given by Sposito (2008; pp. 154–156).

If the redox pairs O_2/H_2O and SO_4^{2-}/HS^- are considered, the redox ladder in Fig. 3 suggests that the thermodynamically favoured reaction would be the oxidation of HS^- to SO_4^{2-} and the production of H_2O, *i.e.* the reduction of O_2 to H_2O coupled to the oxidation of HS^- to SO_4^{2-}. This coupling would occur at both pH 5 and pH 7.

Redox ladders can be very useful for the preliminary estimation of the expected evolution of a given system initiated by a change in the redox conditions (Nelson, 2004). For instance, the input of oxygen into the system represented by the redox ladder at pH 7 in Fig. 3 would favour thermodynamically the oxidation of any available organic matter (represented as CH_2O) to CO_2 before any other reaction may proceed, because this couple encompasses the largest difference in the redox potential. Once the organic matter is consumed, Fe^{2+} would tend to be oxidized and precipitate as $Fe(OH)_{3(s)}$ until the available oxygen has been exhausted. From that point on, the Eh of the system should be (thermodynamically speaking) defined by the redox pair $Fe^{2+}/Fe(OH)_{3(s)}$ and the next redox reaction to proceed would be the reduction of NO_3^- to N_2, and so on.

In a closed system with an excess of organic matter, the succession of reactions that would be observed would be conceptually analogous but slightly different. That would

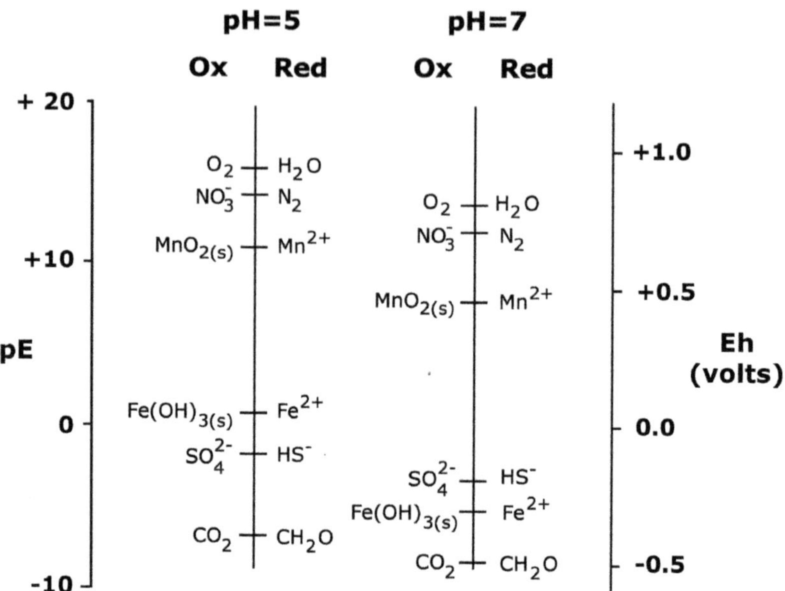

Figure 3. Representation of the main redox couples in natural systems as redox ladders for pH values of 5 and 7. All the activities of reactants and products have been assumed to be equal to 1. Note the change in the redox potential and relative position of redox couples with the pH change (adapted with permission from Scott and Morgan, 1990. Copyright 1990, the American Chemical Society.).

be the case, for instance, for a lake undergoing eutrophication (see *e.g.* Stumm and Morgan, 1996). The whole process can be evaluated easily with the assistance of geochemical modelling as illustrated in the following *PHREEQC* example:

EXAMPLE 3

```
SOLUTION 1 Organic matter titration and redox ladder
temp 25; pH 7; pe 4; redox pe; units mmol/l; density 1; O(0) 0.1; N(5)
0.1; S(6) 0.1; C 0.1

EQUILIBRIUM_PHASES 1
Pyrolusite   0 0.0001
Fe(OH)3(a)   0 0.0001

REACTION 1
CH2O  1
600 micromoles in 100 steps
```

Similar examples can be found in earlier texts (*e.g.* Scott and Morgan, 1990; Appelo and Postma, 2005). In example 3, 600 micromoles of organic matter, represented in its simplest form as CH_2O, are added to a system at pH 7 and 25°C which contains 0.1 mmol L^{-1} of different redox-active elements in their oxidized forms. The oxidized forms of Fe and Mn present in the system are represented by two mineral phases:

amorphous ferrihydrite ($Fe(OH)_3$) and pyrolusite (MnO_2), respectively. The sequence of reactions obtained (or Terminal Electron Acceptor Processes; TEAP), assuming an excess of organic matter, can be deduced from the evolution of the pH, pe and dissolved concentrations of redox-active species obtained (Fig. 4). That sequence corresponds to the successive reduction of O_2, NO_3^-, MnO_2, $Fe(OH)_3$ and SO_4^{2-}, with each step producing a buffering effect in the pH and redox potential of the system (Fig. 4). Once each TEAP is exhausted by oxidation of the excess organic matter, there is a drop in the redox potential of the system until the value corresponding to the next TEAP is reached (Fig. 4). The overall process produces a progressive decrease in the redox potential together with a net pH increase. A similar evolution of processes and parameters is frequently observed in many aquifers, lakes and sediments (Christensen *et al.*, 2000; Chapelle, 2001).

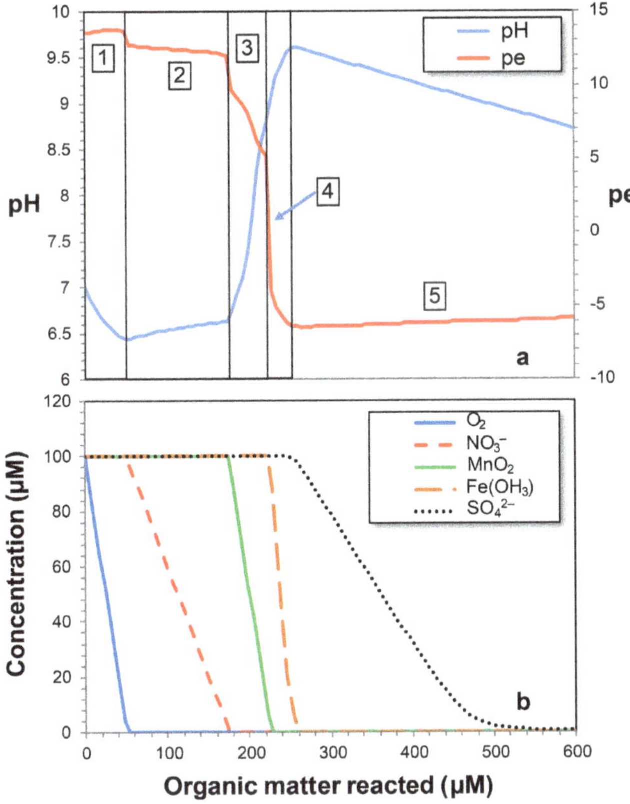

Figure 4. Main geochemical results obtained in a simulation of the degradation of organic matter by different successive terminal electron acceptor processes (TEAP). (a) Evolution of pH and pe during the successive oxidation of organic matter by [1] O_2, [2] NO_3^-, [3] MnO_2, [4] $Fe(OH)_3$ and [5] SO_4^{2-}; and (b) evolution of concentrations of the terminal electron acceptors (TEAs) involved during the process.

3.3. The role of biological processes in redox reactions

Most of the redox couples which involve organic matter in Table 1 are generally catalysed by non-photosynthetic organisms, which tend to decompose organic matter through energy-yielding redox reactions (Stumm and Morgan, 1996). These organisms help to transfer electrons from one species (electron donor) to another (electron acceptor), therefore increasing the rate of redox processes and regulating the redox state in the system. By catalysing redox processes, microorganisms obtain energy that can be used for metabolic processes; it also must be borne in mind, however, that the development of different groups of metabolisms is constrained strictly by the thermodynamically favoured redox reactions pertaining under any given conditions. Thus, the detailed sequence of redox chemical reactions is paralleled by an ecological succession of microorganisms (aerobic heterotrophs, denitrifiers, manganese, iron and sulfate reducers and methane bacteria) and there is a tendency for more energy-yielding mediated reactions to dominate over the rest (Stumm and Morgan, 1996). Again, the redox ladder approach is a simple and effective way of representing this prevalence for the geochemical conditions in the system of interest, even though the actual pathways followed in biotic redox reactions may be considerably more complex than suggested by the coupling of the corresponding half reactions. For instance, aerobic organisms which use oxygen to oxidize organic matter obtain more energy from this process than any other group, because they couple redox reactions with a larger difference of redox potentials (Fig. 3).

Not all the energy released during such processes is effectively available because some must be sacrificed in order to store the gained energy as adenose triphosphate (ATP). Thus, a minimum threshold energy level exists for each TEAP before microbial metabolism is able to proceed. The available energy associated with each respirative pathway can be compared with the respective thresholds of energy requirement for the reaction to proceed in order to identify the thermodynamically favoured reaction (or TEAP) under different chemical conditions. Figure 5 shows the Gibb's free energy change (ΔG_r) for the respirative pathways of organic matter presented in Table 1 depending on the concentrations of electron donors (H_2 and acetate), as presented by Canfield *et al.* (2005).

The thermodynamic feasibility of ferric iron reduction, sulfate reduction and methanogenesis depends heavily on electron donor concentrations (Fig. 5). For instance, at a partial pressure of $H_{2(g)}$ (P_{H_2}) of 10^{-7} atm (equivalent to ~0.08 nM $H_{2(aq)}$), ferric iron reduction (of amorphous FeOOH) would be highly favoured (−75 kJ per 2 moles of the oxidized H_2), whereas the other processes are not energetically feasible (Fig. 5a). If the metabolic activity of iron-reducing bacteria (IRB) results in maintenance of a P_{H_2} value near their threshold energy level, other metabolic groups would be inhibited and IRB would dominate in the habitat. After exhaustion of amorphous iron oxides, P_{H_2} should rise until the next respiration process becomes energetically favourable. Sulfate reduction (by sulfate-reducing bacteria, SRB) would be feasible at a P_{H_2} value of ~10^{-5} atm. Under these conditions, a free-energy yield of ~−20 kJ per 2 moles of H_2 oxidized is obtained (Fig. 5a). By analogy with IRB, this is

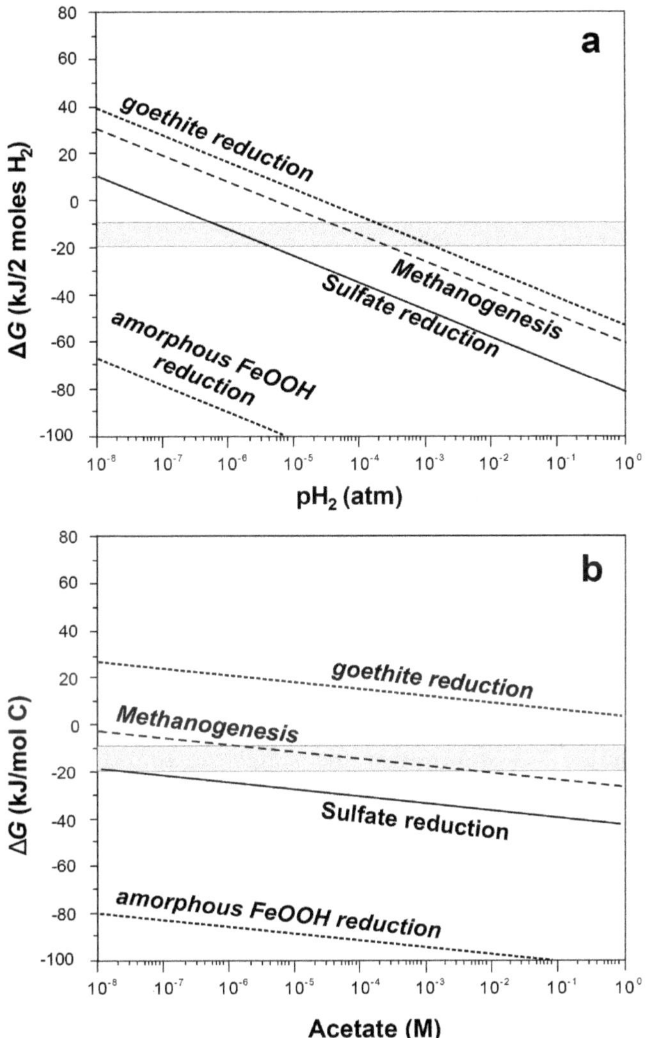

Figure 5. Gibbs free energy variation for different respirative pathways (with different terminal electron acceptor processes at different partial pressure of $H_{2(g)}$) (a) and acetate concentrations (b) as electron donors. Free energy has been calculated for four e^- transfer reactions at pH = 7 and for reasonable environmental concentrations of reactants and products. The oxic respiration, denitrification and Mn reduction are all highly favourable in all examined conditions and they are not included in the figure. The conditions included in the shadowed areas represent the minimum threshold energy level for microbial growth. Adapted from Canfield *et al.* (2005) with the permission of Elsevier..

enough to sustain SRB activity while at the same time excluding methanogens. Again, if SRB activity can maintain a P_{H_2} value near their threshold energy level, they would inhibit the presence of methanogens until all dissolved sulfate is depleted and a P_{H_2}

rises further to allow for methanogenesis. Similar results are obtained if acetate is considered as an electron donor (Fig. 5b).

An additional observation is that the energetics of Fe(III) reduction is highly dependent on the nature of the available solid iron phase. For example, at a P_{H_2} of 10^{-7}atm, reduction of the amorphous phase is favoured whereas reduction of crystalline goethite is not (Canfield *et al.*, 2005).

The situation in which different groups of microorganisms usually metabolize at near their threshold energy levels and thereby exclude the activity of microbial groups with lower energy yields is the basis of the concept of 'competitive exclusion' (Lovley and Goodwin, 1988). As different respirative pathways (or TEAPs) appear to have a characteristic H_2-utilization efficiency, each 'microbial zone' is characterized by a well defined range of hydrogen concentrations. Therefore, an implication of the competitive exclusion approach is that $H_{2(g)}$ contents may be used as a diagnostic value to identify the dominant biologically mediated redox processes operating within the system. In fact, empirical steady-state H_2 concentration ranges for each TEAP have been obtained from field studies: NO_3^- reduction <0.1 nM; Fe(III)-reduction, 0.2–0.8 nM; SO_4^{2-} reduction, 1–4 nM; methanogenesis, 5–30 nM (*e.g.* Christensen *et al.*, 2000; Chapelle, 2001). Note that the sequence of different electron-accepting processes would proceed as predicted by equilibrium thermodynamics in the overall classical redox zonation scheme, which will be explained in detail in section 3.4 below.

The competitive exclusion approach has been used widely to differentiate predominant TEAP in pristine and contaminated anaerobic aquifers (Christensen *et al.*, 2000; Heimann *et al.*, 2010, and references therein). However, an increasing number of studies has shown that the simple relationship between hydrogen concentrations and the corresponding TEAPs is not always observed to hold, and that two or more TEAPs can occur simultaneously in the same zone (*e.g.* Heimann *et al.*, 2010, and references therein). Despite this, hydrogen concentration is still a very valuable parameter in analysing the energetics of the microbial processes (Hoehler *et al.*, 1998; Christensen *et al.*, 2000). Furthermore, the energy yield (ΔG_r) calculation (Equation 18)[1] for any particular system may be used as a test to identify the active TEAPs (all TEAPs characterized by a free energy value lower than the threshold value would be presumed to be active). Ideally, the energy yield will become a judgment of whether a single redox process dominates (*i.e.* spatially segregated redox processes) or, alternatively, several processes can occur concomitantly.

3.4. Redox zonations

In the presence of organic matter as the main reductant, the typical sequence of redox reactions explained above may be dominant in spatially different regions of aquifers or sediments. These regions are often adjacent and are referred to as redox zones (Stumm

[1] A measure of the energy available in an environment can be obtained in another way through the potential difference (ΔEh, the difference between the redox potentials for the electron-accepting and donating reactions: Bethke, 2008) related to the ΔG_r value by the known equation: $\Delta Eh = -(\Delta G_r/nF)$. See Table 1.

and Morgan, 1996; Christensen *et al.*, 2000; Appelo and Postma, 2005). In the cases where a redox zonation is developed, it may be characterized by changes in aqueous hydrogeochemistry, corresponding to the appearance or disappearance of different redox species. For instance, as shown in Fig. 6, if a progressively more reducing redox zonation is developed, some of the dissolved electron acceptors would tend to disappear sequentially (*e.g.* O_2, NO_3^-, SO_4^{2-}) whereas various by-products would tend to appear sequentially in the solution (*e.g.* Fe^{2+}, Mn^{2+}, H_2S, CH_4). In some systems, such changes take place with increasing travel distance from a source (*e.g.* aerobic waters recharging aquifers rich in organic matter) or with increasing depth in lakes or sediments. The size of the redox zones and the distance or depth for which the redox zonation is developed may range from less than a few centimetres to several hundreds of metres, depending on the mineralogical, geochemical, microbiological and hydrogeological features of the target system.

One of the main problems to which the concept of redox zonations has been applied most frequently is the understanding of pollution plumes. The redox state in these plumes ranges from the most reduced conditions, closer to the contaminant source, to progressively more oxidized conditions towards the front and outer parts (Christensen *et al.*, 2000). However, the boundaries between the plume regions dominated by different redox processes are not always well defined. From a thermodynamic point of view, such redox zones should ideally be spatially segregated, but in many cases different redox processes take place in the same region of the target system due to the influence of biological activity and other kinetic factors. An additional drawback impinging the identification of redox zones, based on the presence of some given species, is caused by the previously explained slow rate of many redox transformations, which may allow different redox species to remain in solution and be transported over long distances under thermodynamically unfavourable conditions (Chapelle, 2001).

In spite of these limitations, the concept of redox zonation has been applied frequently for the characterization of redox states in natural systems. An environmentally relevant classification of redox environments was proposed by Berner (1981). This classification system is mainly based on the presence or absence of some key dissolved redox-sensitive components (*e.g.* dissolved oxygen and sulfide) and on the stable mineralogy present under each redox environment (Fig. 6). Berner (1981) differentiated between oxic and anoxic environments and, within the latter group, between sulfidic and non-sulfidic environments. Appelo and Postma (2005) suggested a further subdivision in the post-oxic environment into a nitric and a ferrous zone, depending on the predominance of one or the other. Although very simple, this classification may be used, together with modifications suggested by other authors, as a first approach to some of the most common types of redox environments in natural systems.

3.5. Geochemical modelling of redox transformations

3.5.1. Redox equilibrium vs. kinetic assumption

During the assessment of many problems of element mobility, it is frequently convenient to incorporate redox reactions in geochemical and reactive transport

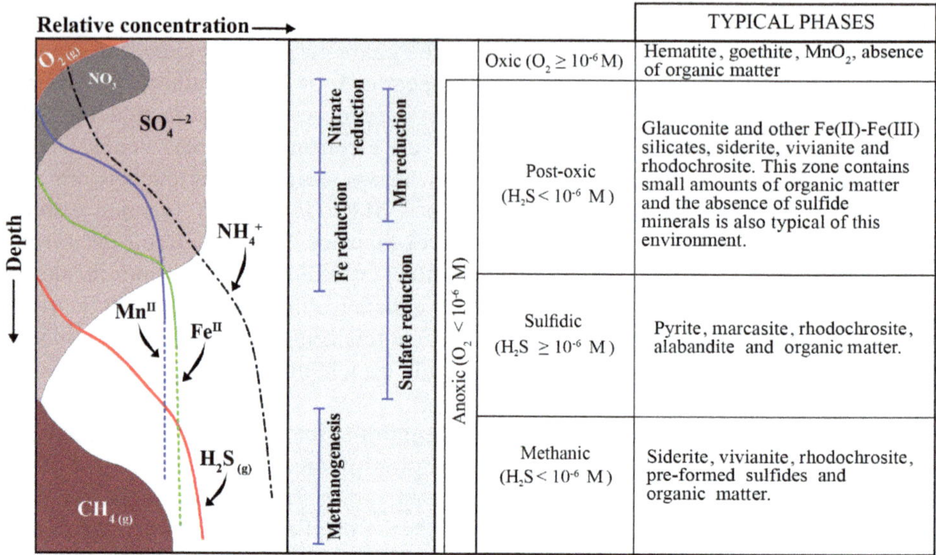

Figure 6. Development of redox zones in groundwater systems modelled using *PHREEQC* with organic matter as the main electron donor. The data on the left show the evanescence of diagnostic electron acceptors such as O_2, NO_3^- and SO_4^{2-} and the sequential evolution of the redox-reaction by-products: Fe(II), Mn(II), H_2S and CH_4. The classification of groundwater redox zones proposed by Berner (1981) was used here as a guide to the expected dominant redox species and the typical associations of solid phases with each zone.

models. This section presents briefly the main approaches for the treatment of redox processes in such models; excellent reviews on the subject can be found elsewhere (*e.g.* Hunter *et al.*, 1998; Schäffer, 2000; Chapelle, 2001; Bethke, 2008).

The most straightforward way to estimate the distribution of redox species would be a simple equilibrium approach using measured or estimated values of redox potentials. This assumption requires that the target redox reactions proceed quickly, *i.e.* that the time to reach thermodynamic equilibrium is shorter than the temporal resolution of the model and that they are fully reversible (Keating and Bahr, 1998; Schäfer, 2000). Unfortunately, neither of these assumptions is usually fulfilled in natural systems, especially if the degradation of organic matter is involved where the rate of fermentation of organic matter controls the overall redox process (see *e.g.* Morris and Stumm, 1967; Lindberg and Rubbells, 1984; Postma and Jakobsen, 1996; Jakobsen and Postma, 1999). Thus, other approaches are frequently necessary to deal with problems involving redox processes. One possible methodology is the treatment of redox processes according to the local equilibrium approach, if the target processes can be considered as close to equilibrium at the spatial or temporal scale required for the problem (Knapp, 1989; Zhu and Anderson, 2002; Bethke, 2008). Even though redox reactions are rarely at true thermodynamic equilibrium, this approach may still be used, *e.g.* if their rate is significantly faster than the rest of target processes (*e.g.* water flow).

However, problems may arise with reactions that are irreversible in natural systems for kinetic reasons, because a model assuming local equilibrium will, by definition, treat the process as reversible and may lead to results inconsistent with field observations (*e.g.* coexistence of thermodynamically unstable redox species; Hunter *et al.*, 1998, and references therein).

A second possible methodology is the use of the kinetic approach, in which rate expressions must be provided for each redox process included, preferably based on previous laboratory or field measurements. All rate expressions must be coupled to each other and to the rest of the processes (*e.g.* advection, diffusion or other equilibrium and non-equilibrium geochemical processes) in order to estimate the distribution of the target redox species in the system (Hunter *et al.*, 1998). Within this group of kinetic expressions, probably the most usual is the first-order rate equation. This expression, in its simplest form, can be written as:

$$\frac{dC_i}{dt} = -kC_i \tag{19}$$

where the variations in the concentration of a given reactant (C_i) with time (t) are given by the product of a constant proportionality factor (k) and the reactant concentration itself. In this approach, the user must specify explicitly and force the sequence of thermodynamically feasible redox reactions. Therefore, one of the main drawbacks of this strategy is the possibility of obtaining numerical results that are thermodynamically inconsistent (Schäfer, 2000), especially if the geochemistry of the target system is highly variable. In spite of this disadvantage, the use of rate expressions can adequately describe many important geomicrobiological redox processes (*e.g.* the oxidation of dissolved and solid-phase organic matter, the oxidation and reduction of insoluble mineral phases, *etc.*; Roden, 2008).

In the case of more complex systems in which large spatial and temporal variations in the rate of redox processes are expected, it is very common to use more complex kinetic expressions which are able to adapt the rate constants to geochemical constraints. This is the case for many microbially mediated redox reactions. The basic kinetic expression traditionally used to describe microbial metabolism is the Monod equation, which is formally similar to the Michaelis-Menten equation and can be generalized as shown by the following equation (Bethke, 2008; Roden, 2008):

$$R = R_{\max}[X]\frac{[S]}{k_m + [S]} \tag{20}$$

where R is the rate of substrate metabolism (*e.g.* in units of mol L^{-1} day^{-1}), R_{max} is the maximum rate (at saturating substrate concentration), $[X]$ is the biomass concentration (*e.g.* in g L^{-1}), $[S]$ is the substrate concentration (*e.g.* in units of mol L^{-1}) and K_m is the half-saturation substrate concentration, at which $R = 0.5R_{max}$.

An intermediate possibility between the two approaches explained above (local equilibrium and kinetics) is the so-called "partial redox disequilibrium approach" proposed by McNab and Narasimhan (1994) and described as "combined approach" in other texts (Schäffer, 2000). In this approach, rate expressions are only assigned to the redox reactions of organic matter degradation and local equilibrium is assumed for the rest of inorganic reactions, including redox speciation. Prior to the assignment of kinetic rate parameters, the thermodynamically viable degradation reactions must be identified and this is done using the redox potential specified for the inorganic species. This treatment implies that redox disequilibrium is considered to be restricted to organic compounds, whereas inorganic redox reactions are fast enough to allow equilibrium to be attained. Thus, theoretical redox potentials are no longer the exclusively controlling variables but still remain as the driving force for redox reactions (Schäfer, 2000).

3.5.2. Some applications of geochemical modelling of redox processes

Case study 1: Role of ferric phases in redox conditions

A simple but informative modelling example of the role played by different iron species and mineral phases in the control of the redox conditions in natural environments is developed below for the *PHREEQC* code (Example 4). In this example, virtually all of the iron initially present in the system is in the form of dissolved Fe(II) but is oxidized by atmospheric oxygen to Fe(III) at a kinetically controlled rate derived from empirical observations by Singer and Stumm (1970). With the increase of the Fe(III) dissolved concentrations, oversaturation with respect to different mineral phases may be attained (goethite, ferrihydrite and schwertmannite in this case), which can trigger their precipitation. The interplay between these processes determines the evolution of the redox conditions in the system. This evolution can be monitored easily by the progressive changes in redox potentials, which need to be calculated from the modelled activities of Fe^{2+} and Fe^{3+}, either with the geochemical code or simply by applying the Nernst equation (Equation 2).

The geochemical modelling of the different variations in this example is straightforward and includes some basic steps that may be applied, with some adaptations, to a wide variety of geochemical modelling problems. Firstly, Fe(II) and Fe(III) must be decoupled to allow the kinetic treatment of the oxidation of the former to the latter. This step is necessary because if it were not taken, the *PHREEQC* code would assume aqueous redox equilibrium (Parkhurst and Appelo, 2013). To proceed with the decoupling, two different and independent iron species are defined, namely Fe_di and Fe_tri, representing Fe(II) and Fe(III), respectively.

EXAMPLE 4

TITLE Kinetically controlled oxidation of Fe(II) to Fe(III) and precipitation of ferric phases

```
SOLUTION_MASTER_SPECIES
  Fe_di     Fe_di+2      0.0      Fe_di    55.847
  Fe_tri    Fe_tri+3     0.0      Fe_tri   55.847
```

In addition, the speciation reactions for all the relevant Fe(II) and Fe(III) species must be rewritten as a function of the newly defined Fe_di and Fe_tri species, respectively. Similar examples of this type of decoupling for Fe and other elements/species can be found in earlier texts (*e.g.* Appelo and Postma, 2005).

```
SOLUTION_SPECIES
 Fe_di+2 = Fe_di+2
 log_k      0.0
 Fe_tri+3 = Fe_tri+3
 log_k      0.0
 #
 # Fe+2 species
 #
 Fe_di+2 + H2O = Fe_diOH+ + H+
          -log_k              -9.5
          -delta_h 13.20      kcal
          -gamma        5.0   0.0
```

The dissolution reactions of the relevant mineral phases should also be rewritten in terms of the decoupled Fe(II) and Fe(III) species already defined. As given in Example 5 below, only the dissolution or precipitation of goethite (FeOOH), ferrihydrite ($Fe(OH)_{3(am)}$) and schwertmannite ($Fe_8O_8(OH)_6SO_4$) will be considered. For goethite and ferrihydrite, the thermodynamic data are taken from the phreeqc.dat database distributed with the *PHREEQC* code (Parkhurst and Appelo, 2013) and for schwertmannite, they correspond roughly to the thermodynamic data proposed by Bigham *et al.* (1996).

EXAMPLE 5

```
PHASES

Goethite_tri
 Fe_triOOH + 3H+ = Fe_tri+3 + 2H2O
 -log_k       -1.0
 -delta_h     -14.48 kcal

Fe_tri(OH)3(a)
 Fe_tri(OH)3 + 3H+ = Fe_tri+3 + 3H2O
 -log_k    4.891

Schwertmannite
 Fe_tri8O8(OH)6(SO4) + 22H+ = 8Fe_tri+3 + SO4-2 + 14H2O
 log_k     18
```

For this example, a simple solution at pH 3 and 25°C, in equilibrium with atmospheric oxygen and containing only iron and sulfate is assumed:

SOLUTION 1

```
 Temp      25
 pH        3
 pe        8
 redox     pe
```

```
  units      mmol/kgw
  density    1
  Fe_di      30
  Fe_tri     0.001
  S(6)       30
  -water     1#kg

EQUILIBRIUM PHASES
  O2(g)       -0.67 10
```

The parameters relative to the kinetic oxidation of Fe(II) to Fe(III) must also be defined. In this case, a total rection time of 1000 days in 1000 steps is simulated using a rate equation similar to that developed in example no. 9 presented in the PHREEQC manual (Parkhurst and Appelo, 2013); this in turn is derived from an expression proposed by Singer and Stumm (1970).

```
KINETICS 1
  Fe_di_ox
  -formula    Fe_di       -1 Fe_tri  1
  -m          1
  -m0         1
  -tol        1e--008
  -steps      86400000 in 1000 steps # seconds
  -step_divide 100
  -runge_kutta 3
  -bad_step_max 500

INCREMENTAL_REACTIONS true
RATES
Fe_di_ox
  -start
  10 Fe_di = TOT("Fe_di")
  20 if (Fe_di<= 0) then goto 200
  30 p_O2 = SR("O2(g)")
  40 moles = (2.91e--9 + 1.33e12 * (ACT("OH-"))^2 * p_O2) * Fe_di * TIME
  200 SAVE moles
  -end
```

Finally, equilibrium with atmospheric oxygen is imposed and the equilibration of the mineral phases considered is permitted only if oversaturation is attained. For the purposes of this example, only one mineral phase at a time will be allowed to precipitate. For instance, if only ferrihydrite precipitation is considered, the following lines should be added:

```
EQUILIBRIUM PHASES
  Fe_tri(OH)3(a) 0 0
    O2(g) -0.67 10
END
```

The main results obtained in the various simulations carried out for the mineral phases included in this example are summarized in Fig. 7. Precipitation of the individual ferric phases results in different evolutions of the main geochemical variables included in the simulations, *i.e.* pe, pH and Fe(III) dissolved concentrations (Fig. 7). In the simulations without mineral precipitation, the main process is the oxidation of ferrous to ferric iron, which leads to a progressive pH and pe increase with time (Fig. 7a), together with an increase in dissolved Fe(III) (Fig. 7b). In the simulations with mineral precipitation, the increasing pH trend is reversed once the precipitation of the target mineral phase begins (Fig. 7a). As expected, the values and evolution of redox potential during the simulated time are also clearly controlled by the precipitated ferric mineral. The pe values in the simulations focused on ferrihydrite and schwertmannite exhibit a trend similar to that where mineral precipitation was suppressed. In the case of goethite precipitation, however, the trend is notably more reducing (Fig. 7a). It follows that the dissolved concentrations of ferric iron are several orders of magnitude lower in the goethite simulations compared to ferrihydrite and schwertmannite (Fig. 7b). These results highlight the key role that the selection of mineral phases may play in the characterization of redox features and processes.

Case study 2: Speciation of selenium in reducing and oxidizing environments

Selenium exists in natural waters and soils mainly as selenium-IV (as selenite; SeO_3^{2-}) and selenium-VI (as selenate; SeO_4^{2-}), organo-selenium (*e.g.* dimethylselenide,

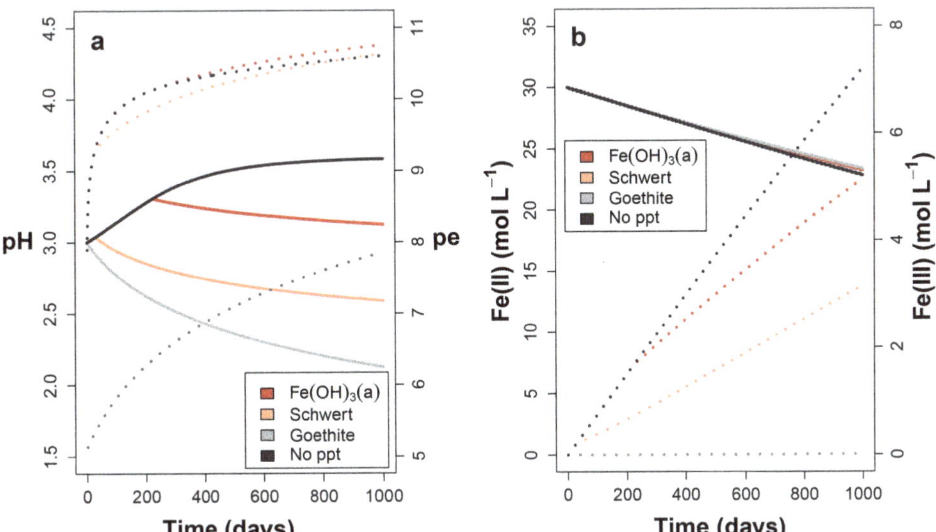

Figure 7. Simulation of the kinetic oxidation of ferrous iron by atmospheric oxygen over 1000 days with the precipitation of different Fe phases (Schwert: schwertmannite). The results from simulations without the precipitation (No ppt) of any mineral phase are also displayed for comparison. (a) Evolution of pH (solid lines) and pe (dotted lines); (b) evolution of the concentrations of ferrous (solid lines) and ferric iron (dotted lines).

$(CH_3)_2Se$) and as a variety of organic and surface complexes (Mayland *et al.*, 1991; Luoma and Rainbow, 2008). Selenate and selenite are the selenium chemically reactive and mobile forms whereas selenides (Se^{-II}) and elemental selenium (Se^0) are insoluble and hence less mobile.

In the following example, the modelling of the oxidation and reduction of Se was carried out in an aqueous environment containing total dissolved concentrations of 100 mM Na, 10 mM Ca, 1 mM Fe, 15 mM SO_4^{2-} and 92 mM Cl^- . A partial pressure of $10^{-2.5}$ bar for $CO_{2(g)}$ was assumed. Changes in redox state were simulated across a pH gradient of 2–12 and pe values from –10 (highly reducing) to +20 (highly oxidizing). To display the dominant Se species at equilibrium as a function of redox potential and pH for a given aqueous concentration we used the *PHREEQC*-based program *PhreePlot* (Kinniburgh and Cooper, 2009) to construct Pourbaix diagrams. Figure 8a shows a simplified Pourbaix diagram for selenium at a total concentration of 10 μM. Speciation calculations show the distribution of Se species and the expected change in Se state when environmental conditions differ. The fully oxidized selenate form is the dominant species in a wide pH range at highly oxic conditions whereas selenite (SeO_3^{2-}) or biselenite ($HSeO_3^-$) are the major Se species under mildly reducing conditions at pH >2.5. These predictions are in agreement with previous studies (*e.g.* Elrashidi *et al.*, 1989; White *et al.*, 1991) where SeO_4^{2-} is commonly found in aerated alkaline and acidic soils and waters. At medium redox range, SeO_3^{2-} is found mostly under alkaline conditions whereas $HSeO_3^-$ is more dominant in acid soils (Elrashidi *et al.*, 1987; Elrashidi *et al.*, 1989). Selenic (H_2SeO_4; pK_{a2} = 1.91) and the weaker selenous (H_2SeO_3; $pK_{a1} = 2.57$, $pK_{a2} = 7.30$) acids are rarely found in environmental systems and exist only under very low pH. In strongly reducing conditions (low pe), thermodynamic calculations suggest that selenium exists predominantly in the Se^{-II} state as elemental Se^0 (Fig. 8a). Reduction of Se^0 leads to the formation of foul-smelling poisonous hydrogen selenide gas (H_2Se) which dissociates in water at pH > 4 to form the anion HSe^- (or Se^{2-}) (Fig. 8b). Microbially mediated reduction of SeO_3^{2-} to Se^0/Se^{-II} (see reaction 29) has been well documented (Lovley, 1993). Reduction of sulfate (*e.g.* reaction 30), nitrate and the precipitation of insoluble compounds with metals such as Fe (reactions 31 and 32) are important processes that maybe associated with selenium redox transformations under varying redox conditions. For example, the model predicts a dual precipitation of elemental selenium (Se^0) and iron sulfide (FeS) at slightly alkaline pH. In natural systems, such processes may occur abiotically but biologically mediated pathways are equally important. Sulfate-reducing bacteria (*e.g. Desulfobacter* and *Desulfovibrio*) use sulfate as a terminal electron acceptor for anaerobic respiration, resulting in the production of sulfide (Nealson and Saffarini, 1994; Vignais and Billoud, 2007). The general transformation reactions between the various species of selenium and some other linked reactions are summarized in reactions 21–32. The review article by Seby *et al.*, (2001) is recommended for readers interested in detailed thermodynamic data for selenium.

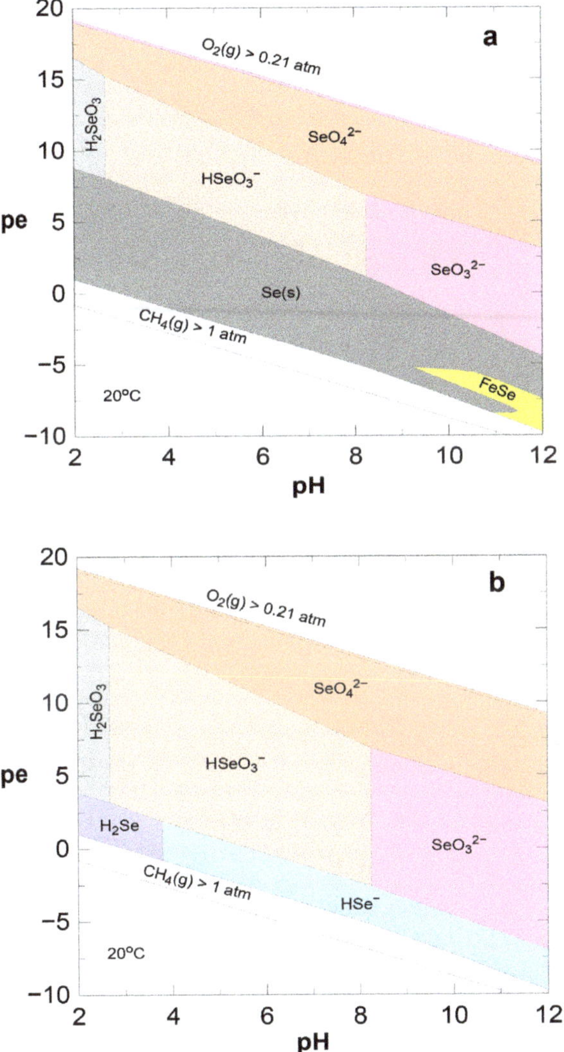

Figure 8. Predicted pH-pe stability field for Se-Fe-S-Ca-H_2O system using total Se of 10 μM at 20°C in the absence of Se surface complexations. (a) Pourbaix diagram showing the dominant Se species; (b) same model calculations as Fig. 8a but after supressing the precipitation of Se^0 and FeSe.

$$HSeO_4^- \leftrightarrow SeO_4^{2-} + H^+ \tag{21}$$

$$H_2SeO_3 \leftrightarrow HSeO_3^- + H^+ \tag{22}$$

$$HSeO_3^- \leftrightarrow SeO_3^{2-} + H^+ \tag{23}$$

$$SeO_4^{2-} + 2H^+ + 2e^- \leftrightarrow SeO_3^{2-} + H_2O \tag{24}$$

$$SeO_4^{2-} + 3H^+ + 2e^- \leftrightarrow HSeO_3^- + H_2O \quad (25)$$

$$H_2Se(g) \leftrightarrow H_2Se(aq) \leftrightarrow HSe^- + H^+ \quad (26)$$

$$HSe^- \leftrightarrow Se^{2-} + H^+ \quad (27)$$

$$HS^- \leftrightarrow S^0 + H^+ + 2e^- \quad (28)$$

$$SeO_3^{2-} + 6H^+ + 4e^- \leftrightarrow Se^0 + 3H_2O \quad (29)$$

$$SO_4^{2-} + 9H^+ + 8e^- \leftrightarrow HS^- + 4H_2O \quad (30)$$

$$FeSe + H^+ \leftrightarrow Fe^{+2} + HSe^- \quad (31)$$

$$FeSe_2 + 2H^+ + 2e^- \leftrightarrow Fe^{+2} + 2HSe^- \quad (32)$$

It is well established that selenium oxyanions form complexes on the surface of soil reactive oxides such as iron and aluminium oxides (Hayes *et al*., 1987; Balistrieri and Chao 1990; Peak, 2006). The high point of zero charge (~8–9.5) of many Fe- and Al-oxides makes them positively charged over a wide range of soil and water pH values. It is important, therefore, to consider surface complexation reactions in geochemical speciation modelling. In the following example, we consider the development of sparingly soluble selenite and selenate complexes on the surface of amorphous iron oxides. The Se-complex formation on the surface of hydrous ferric oxide (Hfo), or ferrihydrite, can be described by the following set of reactions (Dzombak and Morel, 1990):

$$Hfo\_wOH + SeO_4^{2-} + H^+ \leftrightarrow Hfo\_wSeO_4^- + H_2O \qquad \log K = 7.73 \quad (33)$$

$$Hfo\_wOH + SeO_4^{2-} \leftrightarrow Hfo\_wOHSeO_4^{2-} \qquad \log K = 0.80 \quad (34)$$

$$Hfo\_wOH + SeO_3^{2-} + H^+ \leftrightarrow Hfo\_wSeO_3^- + H_2O \qquad \log K = 12.69 \quad (35)$$

$$Hfo\_wOH + SeO_3^{2-} \leftrightarrow Hfo\_wOHSeO_3^{2-} \qquad \log K = 5.17 \quad (36)$$

In an extension of the previous simulation for Se speciation (Fig. 8), we included the above reactions in a *PHREEQC* model while maintaining the same conditions as before. The properties of Hfo including binding-site densities were defined according to Dzombak and Morel (1990). These properties are listed in Table 2. The Se surface complexation was modelled in *PHREEQC* using the Generalized Two-Layer Model (Dzombak and Morel, 1990). The electrostatic double-layer model was used with explicit calculations of the diffuse-layer composition that counter-balances the surface charge.

Model calculations presented as a Pourbaix diagram in Fig. 9 show that SeO_3^{2-} forms strong complexes with Hfo surface whereas SeO_4^{2-} is sorbed only weakly and hence the sub-dominant sorbed Hfo_wSeO_4^- and Hfo_wOHSeO_4^{2-} are not shown on the plot for clarity. It is important to note that under mildly reducing environments and normal soil pH values of 6–8, selenium exists mostly as sorbed SeO_3^{2-} on iron oxides. This modelling exercise demonstrated that changes in soil or water redox chemistry and surface complexation reactions can lead to large changes in Se solubility and mobility in natural systems. For instance, irrigation of alkaline soils, where most SeO_4^{2-} are very

Table 2. Properties of hydrous ferric oxides commonly used in surface complexation reactions (data from Dzombak and Morel, 1990).

Molecular Weight	89 g Hfo/mole Fe
Specific Surface area	600 m^2/g
Stoichiometry	$Fe_2O_3.H_2O$
Pristine point of zero charge	8.1
High-affinity site density	0.005 moles sites/mole Fe
Low-affinity site density	0.2 moles sites/mole Fe

weakly associated with oxides' surfaces, can result in release of SeO_4^{2-} to subsurface environments.

Selenium is an essential plant micro-nutrient which is acquired by plant roots as selenate, selenite or organo-selenium compounds. It is important to note that plants cannot take up colloidal elemental Se or metal selenides (White *et al.*, 2004). In rice paddy soils, redox forms of selenium predominate but they are mostly sorbed on mineral surfaces making it less soluble in soil solution and less available for plant

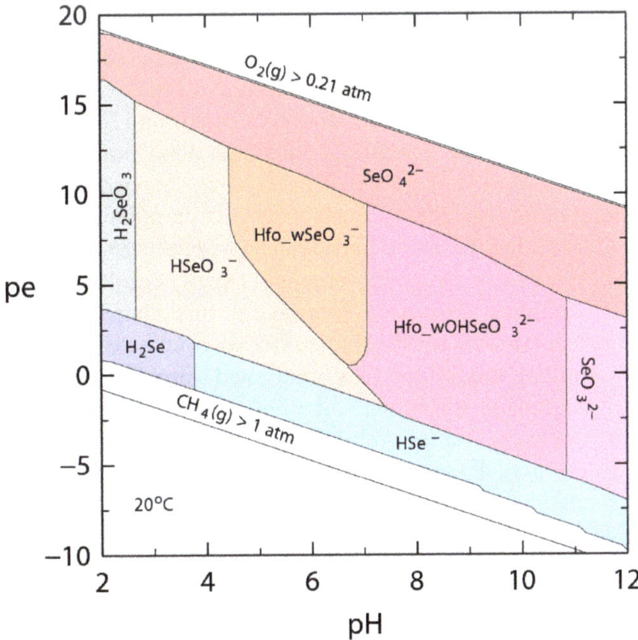

Figure 9. Predicted pH-pe stability field for Se-Fe-S-Ca-H_2O system using total Se of 10 μM at 20°C. The model considers the formation of Se-oxyanion complexes on a hydroxy iron oxide (Hfo) surface.

uptake. Figure 9 shows that increasing pH can increase the solubility of Se. Hence, liming soil would be a suitable strategy to increase the availability of Se to plants by raising pH. This increase in Se bioavailability is probably a consequence of the reduced adsorption capacity of iron oxides and clay minerals for oxyanions such as selenate and selenite. The amount of lime required to reach the required Se concentration in soils can be modelled easily by *PHREEQC* or other geochemical speciation software.

4. General limitations of the geochemical modelling of redox processes and final remarks

Geochemical models are simplified mathematical representations of the physical world. Scientists have made significant attempts to build sophisticated computer codes that are able to simulate natural processes. It is important to understand that natural processes are dynamic and often complex, and that a series of assumptions must be made to simplify numerical calculations. This simplification of reality may lead to a variety of limitations which impinge upon the predictive power and accuracy of the geochemical model.

Currently, different thermodynamic databases (and programs) are used widely to perform geochemical calculations, including redox processes. For instance, the present version of *PHREEQC* is distributed with eight different thermodynamic databases and the user must select one of them in order to perform a specific calculation. However, none of the thermodynamic databases distributed with the modelling codes are comprehensive compilations of all aqueous species and minerals that may be needed. Consequently, differences in the dissolved species included, minerals, equilibrium constants, reaction equations, approaches for activity coefficient calculation, ranges of temperature (or pressure) conditions covered, *etc.*, are found in various databases. Just as for any other geochemical reaction, redox reactions may be affected by these differences and the user must be aware of them when selecting a suitable or, at least, optimum thermodynamic database. Furthermore, treatment of redox processes in highly saline waters (brines) may be particularly problematic as redox reactions have not been parameterized fully for the Pitzer model usually employed for these solutions.

An important limitation discussed throughout this chapter is the fact that the disequilibrium, rather than the equilibrium, of redox couples is the general rule in most natural waters. Another modelling (and analytical) problem is that only a limited number of dissolved ions in water and soil solution are known to provide meaningful redox potential measurements. While potentiometric measurements may be representative of the actual redox potentials in systems controlled by the dissolved contents of Fe, S and to some extent, V and U species, most redox couples do not give meaningful redox potential measurements (Runnells and Lindberg, 1990).

References

Ahmed, I.A.M., Hamilton-Taylor, J., Bieroza, M., Zhang, H. and Davison, W. (2014) Improving and testing geochemical speciation predictions of metal ions in natural waters. *Water Research*, **67**, 276–291.

Amme, M. (2002) Geochemical modelling as a tool for actinide speciation during anoxic leaching processes of nuclear fuel. *Aquatic Geochemistry*, **8**, 177–198.
Appelo, C.A.J. and Postma, D. (2005) *Geochemistry, Groundwater and Pollution*, Second Edition. Taylor and Francis, London.
Auqué, L., Gimeno, M.J., Gómez, J. and Nilsson, A.C. (2008) Potentiometrically measured Eh in groundwaters from the Scandinavian Shield. *Applied Geochemistry*, **23**, 1820–1833.
Baas Becking, L.G.M., Kaplan, I.R. and Moore, D. (1960) Limits of the natural environment in terms of pH and oxidation-reduction potentials. *Journal of Geology*, **68**, 243–284.
Balistrieri, L.S. and Chao, T.T. (1990) Adsorption of selenium by amorphous iron oxyhydroxide and manganese dioxide. *Geochimica et Cosmochimica Acta*, **54**, 739–751.
Ball, J.W. and Nordstrom, D.K. (1991) *User's Manual for WATEQ4F, with Revised Thermodynamic Data Base and Test Cases for Calculating Speciation of Major, Trace, and Redox Elements in Natural Waters*. U.S. Geological Survey, Open-File Report 91-183. Menlo Park, California, USA.
Banwart, S.A., Wikberg, P. and Puigdomenech, I. (1999) Protecting the redox stability of a deep repository: concepts, results and experience from the Äspö hard rock laboratory. *Geological Society of London Special Publication*, **157**, 85–99.
Bard, A.J. and Faulkner, L.R. (2001) *Electrochemical Methods: Fundamentals and Applications*. 2nd edition. Wiley, New York.
Berner, R.A. (1981) A new geochemical classification of sedimentary environments. *Journal of Sedimentary Research*, **51**, 359–365.
Bethke, C.M. (2008) *Geochemical and Biogeochemical Reaction Modeling*. Cambridge University Press, Cambridge, UK.
Bethke, C.M. and Yeakel, S. (2014) *GWB Essentials Guide. The Geochemist's Workbench*. Release 10. Aqueous Solutions LLC. University of Illinois Research Park.
Bigham, J.M., Schwertmann, U., Traina, S.J., Winland, R.L. and Wolf, M. (1996) Schwertmannite and the chemical modeling of iron in acid sulphate waters. *Geochimica et Cosmochimica Acta*, **60**, 2111–2121.
Borch, T., Kretzschmar, R., Kappler, A., Cappellen, P.V., Ginder-Vogel, M., Voegelin, A. and Campbell, K. (2010) Biogeochemical redox processes and their impact on contaminant dynamics. *Environmental Science and Technology*, **44**, 15–23.
Brümmer, G. (1986) Heavy metal species, mobility and availability in soils. Pp. 169–192 in: *The Importance of Chemical Speciation in Environmental Processes* (M. Bernhard, F.E. Brinckman and P.J. Sadler, editors) Dahlem Workshop Reports, **33**. Springer-Verlag, Berlin Heidelberger.
Bruno, J. (1997) Trace element modelling. Pp. 593–621 in: *Modelling in Aquatic Chemistry* (I. Grenthe and I. Puigdomenech, editors). Nuclear Energy Agency, OECD Publications.
Canfield, D.E., Thamdrup, B. and Kristensen, E. (2005) *Aquatic Geomicrobiology*. Elsevier, Amsterdam, pp. 65–94.
Cappellen, P.V. and Ingall, E.D. (1996) Redox stabilization of the atmosphere and oceans by phosphorus-limited marine productivity. *Science*, **271**, 493–496.
Chapelle, F. (2001) *Ground-Water Microbiology and Geochemistry*. Second Edition, John Wiley & Sons , New York.
Chien, C.C., Jr, Medina Jr., A.M., Pinder, G.F., Reible, D.D., Sleep, B. and Zheng, C. (2003) *Contaminated Ground Water and Sediment: Modeling for Management and Remediation*. Lewis Publishers, CRC Press LLC, Boca Raton, Florida, USA.
Choppin, G.R. (2007) Actinide speciation in the environment. *Journal of Radioanalytical and Nuclear Chemistry*, **273**, 695–703.
Christensen, T.H., Bjerg, P.L., Banwart, S.A., Jakobsen, R., Heron, G. and Albrechtsen, H.-J. (2000) Characterization of redox conditions in groundwater contaminant plumes. *Journal of Contaminant Hydrology*, **45**, 165–241.
Cullen, W.R. and Reimer, K.J. (1989) Arsenic speciation in the environment. *Chemical Reviews*, **89**, 713–764
Davis, G.B., Doherty, G. and Ritchie, A.I.M. (1986) A model of oxidation in pyritic mine wastes: part 2: Comparison of numerical and approximate solutions. *Applied Mathematical Modelling*, **10**, 323–329.
Dean, W.E. (1999) The carbon cycle and biogeochemical dynamics in lake sediments. *Journal of*

Paleolimnology, **21**, 375–393.
Deutsch, W.J. (1997) *Groundwater Geochemistry: Fundamentals and Applications to Contamination*. CRC Press, Boca Raton, Florida, USA.
Drever, J.I. (1997) *The Geochemistry of Natural Waters: Surface and Groundwater Environments*. Prentice Hall, Upper Saddle River, New Jersey, USA.
Dzombak, D.A. and Morel F.M.M. (1990) *Surface Complexation Modeling: Hydrous Ferric Oxide*. John Wiley & Sons Inc., New York.
Elrashidi, M.A., Adriano, D.C., Workman, S.M. and Lindsay, W.L. (1987) Chemical equilibria of selenium in soils: a theoretical development. *Soil Science*, **144**, 141–152.
Elrashidi, M.A., Adriano, D.C. and Lindsay, W.L. (1989) Solubility, speciation, and transformations of selenium in soils. Pp. 51–63 in: *Selenium in Agriculture and the Environment* (L.W. Jacobs, editor). SSSA special publication, **23**, Soil Science Society of America, Wisconsin, USA.
Fegley, B. (2013) *Practical Chemical Thermodynamics for Geoscientists*. Academic Press, Kidlington, Oxford, UK.
Förstner, U. (1993) Metal speciation – general concepts and applications. *International Journal of Environmental Analytical Chemistry*, **51**, 5–23.
Ganguly, J. (2009) *Thermodynamics in Earth and Planetary Sciences*. Springer-Verlag, Berlin Heidelberg.
Garrels, R.M. and Christ, C.L. (1965) *Solutions, Minerals, and Equilibria*. Freeman, Cooper, San Francisco, California, USA.
Ginder-Vogel, M. and Fendorf, S. (2007) Biogeochemical uranium redox transformations: potential oxidants of uraninite. Pp. 293–319 in: *Developments in Earth and Environmental Sciences, Adsorption of Metals by Geomedia II: Variables, Mechanisms, and Model Applications* (M.O. Barnett and D.B. Kent, editors). Elsevier, The Netherlands.
Godfrey, L.V. and Falkowski, P.G. (2009) The cycling and redox state of nitrogen in the Archaean ocean. *Nature Geoscience*, **2**, 725–729.
Goldberg, E.D. (1954) Marine geochemistry 1. Chemical scavengers of the sea. *Journal of Geology*, **62**, 249–265.
Grenthe, I., Stumm, W., Laaksoharju, M., Nilsson, A.C. and Wikberg, P. (1992) Redox potentials and redox reactions in deep groundwater systems. *Chemical Geology*, **98**, 131–150.
Gustafsson, J.P. (2009) *Visual MINTEQ ver. 2.61*. Department of Land and Water Resources Engineering, KTH Royal Institute of Technology, SE-100 44.
Hayes, K.F., Roe, A.L., Brown Jr., G.E., Hodgson, K.O., Leckie, J.O. and Parks, G.A. (1987) In situ X-ray absorption study of surface complexes: selenium oxyanions on α-FeOOH. *Science*, **238**, 783–786.
Hayes, J.M. and Waldbauer, J.R. (2006) The carbon cycle and associated redox processes through time. *Philosophical Transactions of the Royal Society B Biological Sciences*, **361**, 931–950.
Heimann, A., Jakobsen, R. and Blodau, C. (2010) Energetic constraints on H_2-dependent terminal electron accepting processes in anoxic environments: a review of observations and model approaches. *Environmental Science and Technology*, **44**, 24–33.
Hoehler, T.M., Alperin, M.J., Albert, D.B. and Martens, C.S. (1998) Thermodynamic control on hydrogen concentrations in anoxic sediments. *Geochimica et Cosmochimica Acta*, **62**, 1745–1756.
Hostettler, J.D. (1984) Electrode electrons, aqueous electrons, and redox potentials in natural waters. *American Journal of Science*, **284**, 734–759.
Hunter, K.S., Wang, Y. and Van Cappellen, P. (1998) Kinetic modeling of microbially-driven redox chemistry of subsurface environments: coupling transport, microbial metabolism and geochemistry. *Journal of Hydrology*, **209**, 53–80.
Jakobsen, R. and Postma, D. (1999) Redox zoning, rates of sulphate reduction and interactions with Fe-reduction and methanogenesis in a shallow sandy aquifer, Rømø, Denmark. *Geochimica et Cosmochimica Acta*, **63**, 137–151.
Karlson, U. and Frankenberger, W.T. (1993) Biological properties of metal alkyl derivatives. Pp. 185–227 in: *Metal Ions in Biological Systems* (H. Sigel and A. Sigel, editors). Vol. **29**. Marcel Dekker, New York.
Keating, E.H. and Bahr, J.M. (1998) Reactive transport modeling of redox geochemistry: Approaches to chemical disequilibrium and reaction rate estimation at a site in northern Wisconsin. *Water Resources*

Research, **34**, 3573–3584.
Keizer, M.G. and Van Riemsdijk, W.H. (1994) *ECOSAT: A Computer Program for the Calculation of Speciation and Transport in Soil–water Systems*. Wagening Agricultural University, Wagening, The Netherlands.
Kinniburgh, D.C. and Cooper, D.M. (2009) *PhreePlot-Creating Graphical Output with PHREEQC*. Manual. http://www.phreeplot.org.
Kinniburgh, D.G., Milne, C.J., Benedetti, M.F., Pinheiro, J.P., Filius, J., Koopal, L.K., and van Riemsdijk, W.H. (1996) Metal ion binding by humic acid: application of the NICA-Donnan model. *Environmental Science & Technology*, **30**, 1687–1698.
Kinniburgh, D.G., van Riemsdijk, W.H., Koopal, L.K., Borkovec, M., Benedetti, M.F. and Avena, M.J. (1999) Ion binding to natural organic matter: competition, heterogeneity, stoichiometry and thermodynamic consistency. *Colloids and Surfaces A: Physicochemical and Engineering Aspects*, **151**, 147–166.
Knapp, R.B. (1989) Spatial and temporal scales of local equilibrium in dynamic fluid-rock systems. *Geochimica et Cosmochimica Acta*, **53**, 1955–1964.
Kulik, D.A., Wagner, T., Dmytrieva, S.V., Kosakowski, G., Hingerl, F.F., Chudnenko, K.V. and Berner, U.R. (2013) GEM-Selektor geochemical modeling package: revised algorithm and GEMS3K numerical kernel for coupled simulation codes. *Computers and Geosciences*, **17**, 1–24.
Langmuir, D. (1997) *Aqueous Environmental Geochemistry*. Prentice Hall, Upper Saddle River, New Jersey, USA.
Lindberg, R.D. and Runnells, D.D. (1984) Ground water redox reactions: an analysis of equilibrium state applied to Eh measurements and geochemical modeling. *Science*, **225**, 925–927.
Li, Y., Yu, S., Strong, J. and Wang, H. (2012) Are the biogeochemical cycles of carbon, nitrogen, sulfur, and phosphorus driven by the "FeIII–FeII redox wheel" in dynamic redox environments? *Journal of Soils and Sediments*, **12**, 683–693.
Lovley, D.R. and Goodwin, S. (1988) Hydrogen concentrations as an indicator of the predominant terminal electron-accepting reactions in aquatic sediments. *Geochimica et Cosmochimica Acta*, **52**, 2993–3003.
Lovley, D. R. (1993) Dissimilatory metal reduction. *Annual Review of Microbiology*, **47**, 263–290.
Luoma, S.N. and Rainbow, P.S. (2008) *Metal Contamination in Aquatic Environments: Science and Lateral Management*. Cambridge University Press, Cambridge, UK.
Macalady, D.L., Langmuir, D., Grundl, T. and Elzerman, A. (1990) Use of model-generated Fe^{3+} ion activities to compute Eh and ferric oxyhydroxide solubilities in anaerobic systems. Pp. 350–367 in: *Chemical Modeling in Aqueous Systems* (E.A. Jenne, editor). American Chemical Society Symposium Series **93**. American Chemical Society, Washington, D.C.
Magalhães, D. de P., Marques, M.R. da C., Baptista, D.F. and Buss, D.F. (2015) Metal bioavailability and toxicity in freshwaters. *Environmental Chemistry Letters*, **13**, 69–87.
Marini, L. (2007) *Geological Sequestration of Carbon Dioxide. Thermodynamics, Kinetics and Reaction Path Modeling*. Developments in Geochemistry, **11**. Elsevier, Amsterdam, 453 pp.
Markich, S.J. (2002) Uranium speciation and bioavailability in aquatic systems: an overview. *The Scientific World Journal*, **2**, 707–729.
Mattigod, S.V. and Zachara, J.M. (1996) Equilibrium modeling in soil chemistry. Pp. 1309–1358 in: *Methods of Soil Analysis. Part 3 Chemical Methods* (D.L. Sparks, editor). SSSA Special Publication, **5**. Soil Science Society of America, Madison, Wisconsin, USA.
Mayland, H.F., Gough, L.P. and Stewart, K.C. (1991) *Selenium Mobility in Soils and its Absorption, Translocation, and Metabolism in Plants*. USGS Circular No. 1064. pp. 57–64.
McNab, W.W. and Narasimhan, T.N. (1994) Modeling reactive transport of organic compounds in groundwater using a partial redox disequilibrium approach. *Water Resources Research*, **30**, 2619–2635.
Meeussen, J.C. (2003) ORCHESTRA: An object-oriented framework for implementing chemical equilibrium models. *Environmental Science & Technology*, **37**, 1175–1182.
Morford, J.L. and Emerson, S. (1999) The geochemistry of redox sensitive trace metals in sediments. *Geochimica et Cosmochimica Acta*, **63**, 1735–1750.
Morris, J.C. and Stumm, W. (1967) Redox equilibria and measurements of potentials in the aquatic environment. Pp. 270–285 in: *Equilibrium Concepts in Natural Water Systems* (W. Stumm, editor). Advances in

Chemistry Series, **67**. American Chemical Society, Washington, D.C.
Nealson, K.H. and Saffarini, D. (1994) Iron and manganese in anaerobic respiration: environmental significance, physiology, and regulation. *Annual Reviews in Microbiology*, **48**, 311–343.
Nelson, E.G. (2004) *Principles of Environmental Geochemistry*. Thomson-Brooks/Cole, Pacific Grove, California, USA.
Nordstrom, D.K., Jenne, E.A. and Ball, J.W. (1979) Redox equilibria of iron in acid mine waters. Pp. 51–79 in: *Chemical Modeling in Aqueous Systems: Speciation, Sorption, Solubility and Kinetics* (E.A. Jenne, editor). ACS Symposium Series, **93**. American Chemical Society, Washington, D.C.
Nordstrom D.K. and Campbell K.M. (2014) Modeling low-temperature geochemical processes. Pp. 27–68 in: *Treatise on Geochemistry, Second Edition* (H.D. Holland and K.K. Turekian, editors). Elsevier, Oxford, UK.
Oelkers, E.H. and Schott, J. (editors) (2009) *Thermodynamics and Kinetics of Water–Rock Interaction*. Reviews in Mineralogy and Geochemistry, **70**, Mineralogical Society of America and Geochemical Society. Chantilly, Virginia, USA.
Parker, D.R., Norvell, W.A. and Chaney, R.L. (1995) *GEOCHEM-PC – A Chemical Speciation Program for IBM and Compatible Personal Computers*. Soil Science Society of America and American Society of Agronomy.
Parkhurst, D.L. and Appelo, C.A.J. (2013) *Description of Input and Examples for PHREEQC Version 3–A Computer Program for Speciation, Batch-Reaction, One-Dimensional Transport, and Inverse Geochemical Calculations*. In: U.S. Geological Survey Techniques and Methods. U.S. Geological Survey, Denver, Colorado, USA.
Parkhurst, D.L., Thorstenson, D.C. and Plummer, L.N. (1980) *PHREEQE- A Computer Program for Geochemical Calculations*. U.S. Geological Survey, Water-Resources Investigations Report 80-96, Denver, Colorado, USA.
Park, J., Sanford, R.A. and Bethke, C.M. (2006) Geochemical and microbiological zonation of the Middendorf aquifer, South Carolina. *Chemical Geology*, **230**, 88–104.
Peak, D. (2006) Adsorption mechanisms of selenium oxyanions at the aluminum oxide/water interface. *Journal of Colloid and Interface Science*, **303**, 337–345.
Postma, D. and Jakobsen, R. (1996) Redox zonation: Equilibrium constraints on the Fe(III)/SO_4-reduction interface. *Geochimica et Cosmochimica Acta*, **60**, 3169–3175.
Prommer, H., Barry, D.A. and Davis, G.B. (2000) Numerical modelling for design and evaluation of groundwater remediation schemes. *Ecological Modelling*, **128**, 181–195.
Robie, R.A. and Hemingway, B.S. (1995) *Thermodynamic Properties of Minerals and Related Substances at 298.15 K and 1 bar (10^5 pascals) Pressure and at Higher Temperatures*. United States Geological Survey, No. B - 2131.
Roden, E.E. (2008) Microbiological controls on geochemical kinetics 1: Fundamentals and case study on microbial Fe(III) oxide reduction. Pp. 335–415 in: *Kinetics of Water–Rock Interaction* (S.L. Brantley, J.D. Kubicki and A.F. White, editors). Springer, New York.
Runnells, D.D. and Lindberg, R.D. (1990) Selenium in aqueous solutions: The impossibility of obtaining a meaningful Eh using a platinum electrode, with implications for modeling of natural waters. *Geology*, **18**, 212–215.
Sahai, N. and Sverjensky, D.A. (1998) GEOSURF: A computer program for modeling adsorption on mineral surfaces from aqueous solution. *Computers and Geoscience*, **24**, 853–873.
Schäfer, W. (2000) Implementation of redox reactions in groundwater models. Pp. 111–119 in: *Redox* (J. Schüring, H.D. Schulz, W.R. Fischer, J. Böttcher and W.H.M. Duijnisveld, editors). Springer, Berlin, Heidelberg.
Schecher, W.D. and McAvoy, D.C. (2001) *MINEQL+: A Chemical Equilibrium Modeling System; Version 4.5 for Windows Workbook*. Environmental Research Software.
Scott M.J. and Morgan, J.J. (1990) Energetics and conservative properties of redox systems. Pp. 368–378 in: *Chemical Modeling of Aqueous Systems II* (D.C. Melchior and R.L. Bassett, editors). ACS Symposium Series, **416**. American Chemical Society, Washington, D.C.
Seby, F., Potin-Gautier, M., Giffaut, E., Borge, G. and Donard, O.F.X. (2001) A critical review of

thermodynamic data for selenium species at 25 C. *Chemical Geology*, **171**, 173–194.
Sigg, L. (2000) Redox potential measurements in natural waters: significance, concepts and problems. Pp. 1–12 in: *Redox: Fundamentals, Processes and Applications* (J. Schüring, H.D. Schulz, W.R. Fischer, J. Böttcher, and W.H.M. Duijnisveld, editors). Springer, Berlin, Heidelberg.
Sillén, L.G., Martell, A.E. and Bjerrum, J. (1964) *Stability Constants of Metal-ion Complexes*, 2nd Edition. Chemical Society, London.
Singer, P.C. and Stumm, W. (1970) Acidic mine drainage: the rate-determining step. *Science*, **167**, 1121–1123.
Sposito, G. (2008) *The Chemistry of Soils*, 2nd Edition. Oxford University Press, Oxford, New York.
Sposito, G. and Coves, J. (1988) *SOILCHEM: A Computer Program for the Calculation of Chemical Speciation in Soils*. Kearney Foundation of Soil Science, University of California.
Stefánsson, A., Arnórsson, S. and Sveinbjörnsdóttir, Á.E. (2005) Redox reactions and potentials in natural waters at disequilibrium. *Chemical Geology*, **221**, 289–311.
Stumm, W. and Morgan, J.J. (1996) *Aquatic Chemistry: Chemical Equilibria and Rates in Natural Waters*, 3rd Edition. John Wiley & Sons, New York.
Tessier, A. and Campbell, P.G.C. (1987) Partitioning of trace metals in sediments: relationships with bioavailability. *Hydrobiologia*, **149**, 43–52.
Tessier, A. and Turner, D.R., editors (1995) *Metal Speciation and Bioavailability in Aquatic Systems*. IUPAC Series on Analytical and. Physical Chemistry of Environmental Systems. John Wiley & Sons, New York.
Templeton, D.M., Ariese, F., Cornelis, R., Danielsson, L.-G., Muntau, H., van Leeuwen, H.P. and Lobinski, R. (2000) Guidelines for terms related to chemical speciation and fractionation of elements. Definitions, structural aspects, and methodological approaches (IUPAC Recommendations 2000). *Pure Applied Chemistry*, **72**, 1453–1470.
Thorstenson, D.C. (1984) *The Concept of Electron Activity and Its Relation to Redox Potentials in Aqueous Geochemical Systems*. U.S. Geological Survey Open-File Report 84-072, Denver, Colorado, USA.
Tipping, E. (1998) Humic ion-binding model VI: an improved description of the interactions of protons and metal ions with humic substances. *Aquatic Geochemistry*, **4**, 3–47.
Tipping, E., Lofts, S. and Sonke, J.E. (2011) Humic Ion-Binding Model VII: a revised parameterisation of cation-binding by humic substances. *Environmental Chemistry*, **8**, 225–235.
Young, S.D., Zhang, H., Tye, A.M., Maxted, A., Thums, C. and Thornton, I. (2005) Characterizing the availability of metals in contaminated soils. I. The solid phase: sequential extraction and isotopic dilution. *Soil Use and Management*, **21(s2)**, 450–458.
van der Lee, J. (1993) *CHESS, Another Speciation and Surface Complexation Computer Code*. CIG-Ecole Mines, Paris.
Vignais, P.M. and Billoud, B. (2007) Occurrence, classification, and biological function of hydrogenases: an overview. *Chemical Reviews*, **107**, 4206–4272.
White, W.M. (2013) *Geochemistry*. Wiley-Blackwell, Rawang, Malaysia
White, A.F., Benson, S.M., Yee, A.W., Wollenberg, H.A. and Flexser, S. (1991) Groundwater contamination at the Kesterson Reservoir, California: 2. Geochemical parameters influencing selenium mobility. *Water Resources Research*, **27**, 1085–1098.
White, P.J., Bowen, H.C., Parmaguru, P., Fritz, M., Spracklen, W.P., Spiby, R.E., Meacham, M.C., Mead, A., Harriman, M., Trueman, L.J., Smith, B.M., Thomas, B. and Broadley, M.R. (2004) Interactions between selenium and sulphur nutrition in *Arabidopsis thaliana*. *Journal of Experimental Botany*, **55**, 1927–1937.
Wolery, T.J. (1992) *EQ3/6, A Software Package for Geochemical Modeling of Aqueous Systems: Package Overview and Installation Guide* (Version 7.0). LLNL, UCRL-MA-110662 PT I.
Zhang, P.-C. and Brady, P.V. (2002) *Geochemistry of Soil Radionuclides*. SSSA special publication no. **59**, Soil Science Society of America, Madison, Wisconsin, USA
Zhu, C. and Anderson, G. (2002) *Environmental Applications of Geochemical Modeling*. Cambridge University Press, Cambridge, UK.

EMU Notes in Mineralogy, Vol. 17 (2017), Chapter 11, 313–356

Breakdown of organic contaminants in soil by manganese oxides: a short review

KAREN L. JOHNSON[1,*], CLARE M. MCCANN[2] and CATHERINE E. CLARKE[3]

[1] *School of Engineering, Durham University, South Road, Durham DH1 3LE, UK, e-mail: karen.johnson@durham.ac.uk*
[2] *School of Civil Engineering and Geoscience, Newcastle University, Newcastle-upon-Tyne NE1 7RU, UK, e-mail: clare.mccann@newcastle.ac.uk*
[3] *Department of Soil Science, Stellenbosch University, P/Bag X1, Matieland 7602, South Africa, e-mail: cdowding@sun.ac.za*

Manganese oxides are natural, ubiquitous and important components of the soil environment. Much scientific research has focused upon their interactions with inorganic contaminants in soil. Yet there is a large body of literature that shows clearly that Mn oxides are powerful oxidants of numerous organic compounds. It is widely suspected that Mn oxides are inherently involved in the soils' natural defence mechanism against such contaminants. With the number of sites being classified as organically contaminated on the increase, and a surge in emerging organic contaminants, exogenous Mn oxide addition may provide a potential remediation option. This chapter reviews the reactions between synthetic and natural Mn oxides with three key classes of organic contaminants: phenols, N-containing compounds and polycyclic aromatic hydrocarbons (PAHs). Historically, the mechanistic understanding of reactions between organics and Mn oxides centred on the oxidative transformation of representative components of natural organic matter (NOM) *via* the well documented process of reductive dissolution. However, the fate of organic contaminant transformations in soils by Mn oxides is dependent upon the processes of sorption, oxidation and polymerization. There is a strong interdependence between Mn oxides, NOM and organic contaminants. NOM is the main sink for organics and Mn oxides can transform phenolic moieties, a prominent component of NOM. Factors affecting the interactions between Mn oxides, NOM and contaminant/contaminant-NOM complexes are discussed. Overall, transformation of organic contaminants by Mn oxides follows a pathway which involves complex formation, electron transfer and radical formation followed either by polymerization to create larger more aromatic molecules or further oxidation resulting in smaller molecules. Manganese oxides, have demonstrated the capacity to react with some of the most recalcitrant organic pollutants, such as PCBs and PAHs, yet the mechanisms by which these reactions occur is still unclear. Mn oxides theoretically show excellent potential as soil amendments. Further research on environmentally relevant systems rather than clean chemical model systems are required in conjunction with field remediation studies.

DOI: 10.1180/EMU-notes.17.9

1. Introduction

1.1. Overview

Manganese (Mn) oxide (a collective term used to include oxide, hydroxide and oxyhydroxide) surfaces are highly reactive hotspots in soils. These overlooked components of soil ecosystems are powerful *in situ* microreactors, acting as sinks and oxidants for inorganic and organic contaminants, respectively. Mn is the tenth most abundant element in Earth's upper continental crust (Gilkes and McKenzie, 1988). With an average abundance of 650 mg/kg Mn in soil (Gilkes and McKenzie, 1988), Mn oxides exert chemical influences far out of proportion to their relative abundance due to their large surface areas, small particle sizes, and the redox chemistry they exhibit. The major focus of this chapter is the redox reactions between Mn oxides and recalcitrant organic contaminants. It should be noted that other minerals present in soil are capable of oxidizing organic molecules (*e.g.* Wang *et al.*, 1983; Wang and Huang, 1991) and reviews on these are available (*e.g.* Huang and Hardie, 2011).

Over recent years, research into the environmental fate of persistent organic pollutants (POPs) such as polycyclic aromatic hydrocarbons (PAHs) and polychlorinated biphenyls (PCBs), has increased due to growing concern over the toxicity of these compounds. Studies involving synthetic Mn oxides and organic contaminants suggest that Mn oxides play a significant role in soil's natural defence mechanism against increasing concentrations of these compounds. This has led to speculation that Mn oxides may have the potential to act as remediation amendments. In this chapter we focus on the reactions between Mn oxides and four main classes of anthropogenic POPs: phenols, polyhalogenated compounds, nitroaromatics and polycyclic aromatic hydrocarbons. Additionally, we review interactions between Mn oxides and natural organic matter (NOM), the main sink for contaminants in soil. Current gaps in our understanding regarding transformation of organic contaminants by Mn oxides in soils are highlighted as well as possible areas for future research.

1.2. Soil contamination

Any unwanted substance introduced by man into the environment is referred to as a 'contaminant' (Megharaj *et al.*, 2011). Detrimental effects or damage by the contaminants lead to 'contamination', a process by which a resource (in this case soil) is rendered unfit for use (Megharaj *et al.*, 2011). The final destinations for contaminants, whether organic or inorganic, natural or anthropogenic, are geosorbents (Luthy *et al.*, 1997), of which soils represent the major component. Soil can therefore be thought of as a global contaminant reservoir which has caused great public concern due to possible adverse impacts upon human health (Menzie *et al.*, 1992; Abrahams, 2002). In Europe, the main sources of soil contamination have been identified as heavy metals and mineral oil (European Environment Agency, 2010).

Attempts to measure the true extent of contamination in soils is difficult due to lack of data. Potentially contaminating activities have occurred at ~3 million sites in Europe, from which >8% (nearly 250,000 sites) are in need of remediation and only 80,000 have been remediated (European Environment Agency, 2010). Furthermore, it is reported that by 2025, the number of sites requiring remediation may increase by >50%. This is due to improvements in data collection and therefore increasing designation of contaminated sites, thus, soil contamination is an ever-growing environmental issue (European Environment Agency, 2010). Contaminated land is an international problem; in the USA the Environmental Protection Agency (EPA) has identified 1,739 locations which have been designated as Superfund sites containing abandoned hazardous wastes (U.S. Environmental Protection Agency, 2014). A search of the US EPA Superfund website (http://www.epa.gov/superfund/sites/, 20th February, 2014) selecting 'all' organic contaminant groups and 'soil' as the contaminant media revealed that 1010 sites matched the criteria. Overall, 58% of all Superfund sites are organically contaminated soils, putting into perspective what a large problem it has become. Contamination in developing regions such as China, India and Africa is less well documented but the situation is likely to be equally, or more severe, due to less stringently enforced environmental regulation. Indeed, worldwide concern is growing with at least 44 nations having surface soil guidance values for carcinogenic PAHs (Jennings, 2012).

For this reason, remediation technology is an ever-expanding field, which is broadly divided into physical, chemical and biological strategies. Remediation often incorporates a combination of methods to offer the most effective approach (Scullion, 2006; Strange *et al.*, 2008). A summary of the main remediation strategies for organic contaminants has been provided in Table 1, collated from the reviews of Hamby (1996), Semple *et al.* (2001) and Scullion (2006). In the European Union (EU), there has been a general move away from the use of traditional techniques of incineration and disposal to landfill towards more innovative technologies, some of which may be applied *in situ* (*i.e.* directly to the soil) or *ex situ* following the excavation of the soil. In general, *in situ* technologies are favoured as they offer minimal disruption, although *ex situ* methods do have the benefit of offering better management and control (Scullion, 2006).

When land is contaminated with recalcitrant organics which are not amenable to biodegradation processes, stabilization/solidification, chemical oxidation, chemical reduction and dehalogenation and incineration can provide potential solutions. The problem with all of these treatments is their sustainability credentials as they all require large inputs of chemicals and energy. The definition of a 'sustainable' treatment (Bardos *et al.*, 2011) is that the benefit of undertaking remediation is greater than its impact, where both benefit and impact are measured in terms of environmental, economic and social indicators.

Since the early 2000s the reuse of wastes as resources has led to many materials, such as composts and mineral byproducts from industry, being considered for use as effective *in situ* soil-amendment remediation technologies. Waste mineral and organic-matter amendment strategies also present potentially financially viable options for treatment of contaminated land which is deemed to be of low economic value.

Table 1. Summary of the main strategies employed in remediation of organically contaminated land.

Classification	Technology
Physical	Removal to landfill
	Capping
	Soil washing
	Physico-chemical washing
	Vapour extraction
	Incineration
	Vitrification
	Stabilization/Solidification
	Thermal desorption
	Vapour stripping
	Infilling with non-contaminated soil
	Air sparging
	Wash/pump and treat
Chemical	Solvent extraction
	Dehalogenation
	In situ flushing
	Surface amendments
	Oxidation
	Reduction
	Hydrolysis
	Dechlorination
Biological	*Microbial activity*
	Windrow turning
	Land farming
	Bioventing
	Bioslurry
	Biopiles
	Bioreactors
	Composting
	Biostimulation
	In situ bioremediation
	Plant activity
	Phytostabilization
	Phytoextraction
	Phytodegredation
	Natural biological activity
	Monitored attenuation

Recently, links have been shown between public health and brownfield sites, although it is not certain whether these are for toxicological or psychological reasons (Bambra *et al.*, 2014, Morrison *et al.*, 2014). Thus, the exploration of more cost-effective and sustainable remediation technologies to regenerate contaminated land is more important than ever.

Research into these more sustainable soil-amendment technologies has been directed at inorganic contaminant remediation (Park *et al.*, 2011; Tica *et al.*, 2011; Bolan *et al.* 2014). Traditionally, the main waste material studied for organic contaminant remediation has been compost (Jones and Healey, 2010; Beesley *et al.*, 2011; Sayara *et al.*, 2011; Marchal *et al.*, 2013). Compost has been used to increase microbial activity and reduce contaminant concentrations by encouraging bioremediation (*e.g.* Semple *et al.*, 2001; Hickman and Reid, 2008; Antizar-Ladislao, 2010; Lashermes *et al.*, 2010; Zhang *et al.*, 2014). Composts can also chemically immobilize organic contaminants, trapping them in hydrophobic domains, making them less prone to leaching and plant uptake (Jones and Healey, 2010). Due to the complexity of interactions between introduced composts and the native soil, these methods are never assured of success and some have even reported either an increase in mobility and/or toxicity (*e.g.* Pugilisi *et al.*, 2009).

There is little in the literature about using Mn oxides for the treatment of organically contaminated soil. We found no published manuscripts on Mn oxides as amendments for organic-contaminated soils in either lysimeter trials or at the field scale. Because Mn oxide is naturally present in the soil, it has the potential to been viewed as less invasive in comparison to other less sustainable methods currently available for recalcitrant organic remediation. Clean waste supplies of Mn oxide have recently been identified (*e.g.* Clarke *et al.*, 2012), making soil amendment with Mn oxide theoretically possible.

1.3. Organic contaminants in soils

The interaction of organic contaminants with soil is complicated by the innate complexity of soil. In general, detailed information and understanding of organic contaminants in soils is lacking but there are a few good reviews on their sources, fate, transport and distribution in soils (Jones and de Voogt, 1999; Meijer *et al.*, 2003; Lohmann *et al.*, 2007; Nam *et al.*, 2008). By comparison, the occurrence, sources, pathways and fate of organic contaminants in aquatic environments such as groundwater are much better studied and understood (*e.g.* Huerta *et al.*, 2012; Jurado *et al.*, 2012; Lapworth *et al.*, 2012; Luo *et al.*, 2014).

It is known, however, that mineral oil is the most commonly found organic contaminant in European soils with 33.7% of contaminated sites reporting it (Valentin *et al.*, 2013). Mineral oil is not included in this review because it is not considered a problem contaminant and there are several suitable techniques for its remediation such as co-composting with organic matter and/or bioremediation (Jones and Healey, 2010). The remaining 66.3% of organically contaminated soils in Europe contain an incredibly diverse array of organic compounds. This is a continually expanding area as (1) new compounds are developed and (2) improvement in analytics means that more compounds are discovered *in situ* in the environment over time.

Anthropogenic organic contaminants broadly fall into two classes: (1) accidental byproducts from burning fossil fuels and plastics which are derived from oil; and (2) chemicals intentionally synthesized specifically for industrial uses (*e.g.* pesticides, herbicides, insecticides, human and veterinary pharmaceuticals, natural and synthetic hormones, personal-care products, surfactants, plasticizers, fire retardants). Many of these synthesized chemicals are not easily degraded by microorganisms or minerals in the natural environment. Due to their hydrophobic and lipophilic natures coupled to their low solubilities, low volatilities and/or resistance to degradation they are known as persistent organic pollutants (POPs) (Andrews *et al.*, 2004). Further environmental issues with POPs come from their toxicity, long half-lives, tendency to bio-accumulate and undergo long-range atmospheric transport. The UNEP Stockholm Convention currently regulates POPs, having published the famous 'dirty dozen' in 2001 (UNEP, 2001), along with the nine new POPS in 2009. REACH (Registation, Evaluation, Authorisation and Restriction of Chemicals) legislation came into force in the EU in 2007 and has gone some way to prevent the production of new chemicals which would be harmful to the environment and/or human health but investigations into existing mixes of different contaminants found in the environment are limited (*e.g.* Wenzel *et al.*, 2002).

Academic research has revealed evidence for toxic (and carcinogenic) effects of many other emerging organic contaminants (EOCs), several of which partition into soils, including antibiotics (*e.g.* triclosan), industrial surfactants (*e.g.* nonylphenols) and flame retardants (*e.g.* bromophenol, BP, or polybrominated diphenyl ether, PBDE). In addition to toxicity and carcinogenicity, another concern has been the endocrine disrupting potential of some chemicals. Many organic contaminants including phenolic chemicals (such as nonylphenols, triclosan, bisphenol A and phthalates) and other brominated chemicals (such as PCBs and PBDEs) mimic human (as well as fish, reptiles and amphibian) hormones and can cause health problems for this reason (Casas *et al.*, 2013). In addition to the growing health concern of antibiotic resistance caused by over use of antiseptic and antibiotic chemicals (*e.g.* triclosan and fluoroquinoline, respectively), remediation of emerging organic contaminants (EOCs) in soils and sludges is an area of active research. Several of these EOCs (including antibiotics and hormones) contain phenolic and anilinic components in their structures and have been shown to be highly susceptible to oxidation by Mn oxides (*e.g.* Zhang and Huang, 2003; Lin *et al.*, 2013; Lu and Gan, 2013).

In this review we consider four classes of recalcitrant organic compounds, phenols, polychlorinated biphenyl compounds, nitrogen-containing aromatic compounds and polycyclic aromatic hydrocarbons. We present their provenance and fate in the soil/water environment and ability to be oxidized/transformed by Mn oxides.

1.3.1. Phenols

There is a lack of data available on background levels of phenol in soils. Phenols are naturally present in surface waters and groundwaters as degradation products of lignin and are thought to have a controlling role in terrestrial ecosystem nutrient cycling (Hattenschwiler and Vitousek, 2000). Anthropogenic inputs of phenolic wastes are important and are associated with the production of phenolic resins and bisphenol-A (an intermediate in the manufacture of nylon), aniline and other organic chemicals. Phenol can be biodegraded rapidly under both aerobic and anaerobic conditions but where it does persist it can be ecotoxic (*e.g.* inhibiting spore germination and hyphal growth of saprotrophic fungi; Hattenshwiler and Vitousek, 2000).

Halogenated phenols are more of a concern as these are more ecotoxic (and can be carcinogenic) and persistent, with 3.6% of contaminated sites found to contain chlorinated phenols (European Environment Agency, 2010). Chlorinated phenols are used as insecticides (*e.g.* 2,4 toluene), herbicides (*e.g.* 2,4,5 toluene) and fungicides (Silvex) and are also used in the synthesis of other important industrial chemicals including PCBs (ATSDR, 2015). Even though chlorophenols are largely non-biodegradable, there have been numerous studies on the ability of microorganisms to transform chlorophenols (Olaniran and Igbinosa, 2011). There is even evidence that microorganisms are evolving to do so and consequently there is great interest in engineering new microorganisms capable of chlorophenol degradation (Olaniran and Igbinosa, 2011). Many studies have shown that chlorophenols are persistent in aquatic environments but far fewer have investigated the fate and behaviour of chlorinated compounds in soil (Lallai and Mura, 2004). Triclosan is a chlorophenol which is an emerging contaminant, used as an antimicrobial pesticide in thousands of everyday products like toothpaste and soap. Triclosan accumulates in the environment as it is not broken down microbiologically and has been linked to antibiotic resistance, a major public health crisis (Levy and Marshall, 2004). For this reason it survives wastewater treatment and persists in sludge that is recycled to land. Like triclosan, bisphenol-A (BPA) is a hormone disruptor that has been shown to alter both male and female sex hormones in animal studies and there is rising concern about this issue as these endocrine-disrupting chemicals are not biologically degraded (Lin *et al.*, 2009).

Brominated and perfluorinated phenols are also of increasing concern due to their persistence, high bioaccumulation potential and toxicity (Hites, 2004); they are present in flame retardants (*e.g.* polybrominated diphenyl ether, PBDE) and are used in construction materials and electronic equipment (de Wit, 2002). PBDE-OHs, the hydroxylated products of PBDEs are of even greater concern due to their increased toxicity and hormone-disrupting potential (Lin *et al.*, 2014).

1.3.2. Polychlorinated biphenyls (PCBs)

Polychlorinated biphenyls (PCBs) consist of at least two fused aromatic rings (biphenyls) with more than one chlorine substituent group and are, perhaps, the most problematic of all chlorinated compounds. The use of PCBs was banned in many countries in the 1970s due to their persistence, toxicity and carcinogenicity. They were used in transformers, capacitors and as hydraulic and heat transfer fluids due to their excellent stability and thermal properties. Due to these properties PCBs still remain in the soil from the 1970s: there is cause for concern for environmental health as their fate in soil is far from understood (Hickey, 1999). PCB-contaminated soils are generally treated by incineration but this process can emit dioxins which are more toxic than the parent compound.

1.3.3. N-containing aromatic compounds

N-containing aromatic compounds have one or more functional N groups attached or within the aromatic ring structure. They are used in explosives and as starting materials for pesticide and dye production. There is no information available on the percentage of sites contaminated with N-containing aromatic compounds and this may be because many are associated with the defence industry. The greatest known one-off industrial leak of N-containing aromatic compounds occurred in China in 2005 when an explosion at a chemical factory resulted in the 100 tonnes of benzene and nitrobenzene seeping to the Songhua River (Fu *et al.*, 2008). However, on an everyday basis, the azo dye industry is probably the most contaminating. Azodyes released into the environment without proper treatment represent 15% of global dye production, *i.e.* 150 tonnes per day (Bandara *et al.*, 1999). A variety of effluent treatment options are available, but many are expensive and only partially effective, thus colour removal from effluents emanating from textile, photographic, paper and food industries remains a pertinent issue in wastewater treatment. Because many man-made N-containing aromatic compounds have been identified as toxic, carcinogenic and persistent against degradation (Purohit and Basu, 2000), this problem also extends to the soil environment.

1.3.4. Polycyclic aromatic hydrocarbons (PAHs)

Polycyclic aromatic hydrocarbons (PAHs) are aromatic hydrocarbons with two or more fused benzene rings (Haritash and Kaushik, 2009). Polycyclic aromatic hydrocarbons are the end-products of hydrocarbon combustion under reducing conditions; thus there are both natural and anthropogenic sources of PAHs. Major anthropogenic sources include fossil-fuel combustion, metallurgical processes and the manufacture of creosote and bitumen (Peng *et al.*, 2008). PAHs are therefore ubiquitously present and are of great concern due to their toxic, mutagenic, teratogenic and carcinogenic nature (Patnaik, 1992; Adonis *et al.*, 2003). Their low water solubility and strong partitioning into natural organic matter (NOM) hinders bioremediation techniques which are more effective when the contaminant is soluble and therefore bioavailable (Zhang *et al.*, 1995). It is estimated that PAHs are present at ~13.3% of contaminated sites (European Environment Agency, 2010). PAHs with greater molecular weights have the lowest water solubility and lowest vapour pressures, and therefore their persistence in the environment is greatest (Haritash and Kaushik, 2009). More than 100 PAH compounds exist in the environment; however, based upon their toxicity and abundance, 16 have been included in the United States Soil Protection Agency's list of priority contaminants as shown in Table 2. None of the remediation solutions listed in Table 1 is entirely suitable for PAH breakdown as these compounds are non-polar and tend to partition into soil organic matter.

Table 2. US EPA 16 priority PAHs and selected physical-chemical properties. Reproduced from Bojes and Pope (2007) with the permission of Elsevier (based on ATSDR, 2005).

PAH	Structure (# of rings)	Molecular weight (g/mole)	Solubility (mg/L)
Naphthalene	2	128.17	31
Acenaphthene	3	154.21	3.8
Acenaphthylene	3	152.20	16.1
Anthracene	3	178.23	0.045
Phenanthrene	3	178.23	1.1
Fluorene	3	166.22	1.9
Fluoranthene	4	202.26	0.26
Benzo(a)anthracene	4	228.29	0.011
Chrysene	4	228.29	0.0015
Pyrene	4	202.26	0.132
Benzo(a)pyrene	5	252.32	0.0038
Benzo(b)fluoranthene	5	252.32	0.0015
Benzo(k)fluoranthene	5	252.32	0.0008
Dibenz(a,h)anthracene	6	278.35	0.0005
Benzo(g,h,i)perylene	6	276.34	0.00026
Indeno[1,2,3-cd]pyrene	6	276.34	0.062

2. Manganese oxides and natural organic matter

2.1. Manganese oxide geochemistry

Mn oxides are one of the most powerful oxidizing agents found on Earth. Mn cations convert the oxygen in H_2O into gaseous oxygen in the atmosphere, soils and water *via* photosynthesis (Dismukes, 1986). Manganese is the second most abundant transition metal in the Earth's crust after Fe and is present at an average of 650 mg/kg in soils (Gilkes and Mckenzie, 1988). The most common form of Mn oxide in both soil and marine sediments is birnessite ($[Na,Ca,K]_xMn_2O_4{\cdot}1.5H_2O$), which is often simplified and denoted as δ-MnO_2 in the geochemical literature (Burns and Burns, 1977). Birnessite precipitates in soils mainly as a result of Mn(II) oxidation promoted by bacteria and fungi, forming highly reactive biominerals (*e.g.* Sparks, 2003; Villalobos *et al.*, 2006) as well as coatings (Post, 1999) nodules and concretions (Gasparatos, 2013).

Because Mn is present, in the Earth's crust, at ~$1/100^{th}$ of the concentration of Fe (Tebo *et al.*, 2004), Mn oxides tend to be overlooked. However, because Mn oxides tend to be present as fine-grained aggregates (Post, 1999), they exert an influence far out of proportion to their total concentration. Secondly, Mn oxides are much stronger oxidants than Fe oxides, taking part in more oxidative reactions than Fe and therefore reduced more rapidly in anoxic environments. Interestingly, Mn (III) oxides have greater redox potential (1.51 V) than that of Mn (IV) oxides (1.23 V) (see Table 3).

The Mn oxidation state in soil is controlled by redox potential and pH (see Fig. 1). In the soil environment manganese is present as Mn(II), Mn(III) and Mn(IV). The occurrence of Mn(II) in the environment is thermodynamically favourable in the absence of oxygen at low pH and can be present in solution or solid form. Mn(III) and Mn(IV) are favoured under aerobic conditions at high pH (Tebo *et al.*, 2004) and until recently it was thought they were both primarily found in solid form. However, Madison *et al.* (2013) have revised the redox sequence for sediments and soils (see Fig. 2) to include Mn(III) (and Fe(III)) complexed in solution with an organic ligand.

This new redox sequence may have significant implications for the role of manganese in both natural organic matter (NOM) and organic contaminant oxidation. As Mn(III) is present in suboxic zones this extends the reach of Mn's ability to transform natural organic matter although its redox potential will be dependent on how strongly it is complexed to an organic ligand. In addition, Mn(III) in solution can act as both an oxidant (*e.g.* Wang *et al.*, 2014) and a reductant (*e.g.* Chen *et al.*, 2013). Importantly, unlike Fe which can be removed in anoxic environments as a sulfide, Mn oxides are effectively recycled at oxic/anoxic interfaces and soluble Mn (as III or II) is released which typically reprecipitates at the oxic boundary. If soluble Mn is present as Mn(III)

Table 3. Standard-state reduction potentials of half reactions that involve important elements in soils (reproduced from McBride, 1994 with the permission of Oxford University Press). Half reactions involving Mn oxides are highlighted in bold.

Reaction	E_0 (volts)
$\mathbf{Mn^{3+} + e^- \leftrightarrow Mn^{2+}}$	**1.51**
$\mathbf{MnOOH_{(s)} + 3H^+ + e^- \leftrightarrow Mn^{2+} + 2H_2O}$	**1.45**
$1/5NO_3^- + 6/5H^+ + e^- \leftrightarrow 1/10N_{2(g)} + 3/5H_2O$	1.25
$\mathbf{1/2MnO_{2(s)} + 2H^+ + e^- \leftrightarrow 1/2Mn^{2+} + H_2O}$	**1.23**
$1/4O_{2(g)} + H^+ + e^- \leftrightarrow 1/2H_2O$	1.23
$Fe(OH)_{3(s)} + 3H^+ + e^- \leftrightarrow Fe^{2+} + 3H_2O$	1.06
$1/2NO_3^- + H^+ + e^- \leftrightarrow 1/2NO_2^- + 1/2H_2O$	0.83
$Fe^{3+} + e^- \leftrightarrow Fe^{2+}$	0.71
$1/2O_{2(g)}^{2-} + H^+ + e^- \leftrightarrow 1/2H_2O_2$	0.68
$1/8SO_4^{2-} + 5/4H^+ + e^- \leftrightarrow 1/8H_2S + 3/5H_2O$	0.30
$1/6N_{2(g)} + 4/3H^+ + e^- \leftrightarrow 1/3NH_4$	0.27
$1/8CO_{2(g)} + H^+ + e^- \leftrightarrow 1/8CH_{4(g)} + 1/4H_2O$	0.17
$H^+ + e^- \leftrightarrow 1/2H_{2(g)}$	0.00

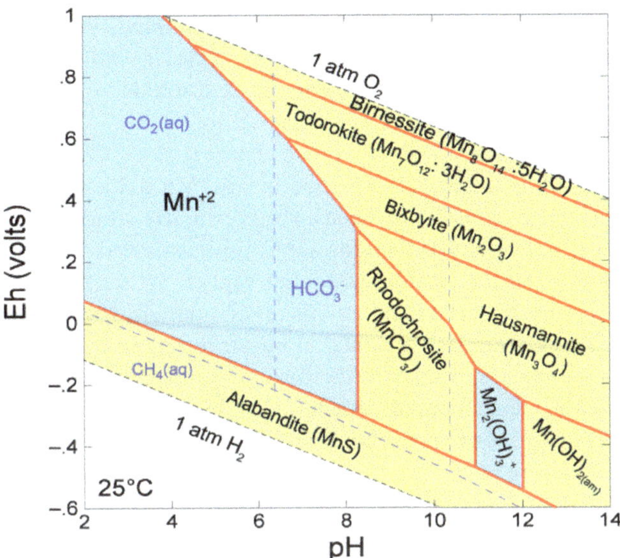

Figure 1. Eh-pH diagram for the Mn-H_2O-CO_2 system at 25°C showing which species predominate under any given condition of Eh or pH.

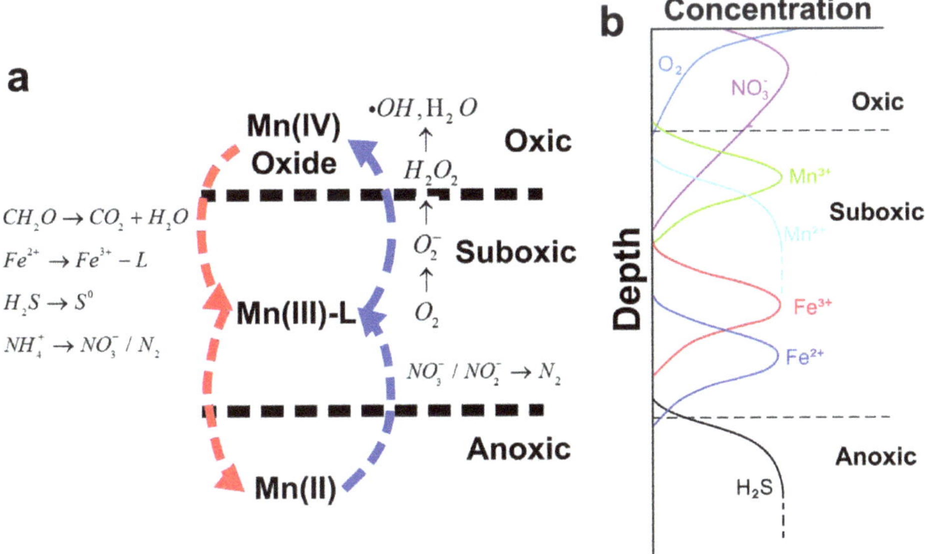

Figure 2. Mn cycle and revised sedimentary porewater redox profile. (a) The Mn cycle with one-electron Mn reactions included explicitly. More details are available in the original paper. (b) An update of the Froelich *et al.* (1979) paradigm for sediment biogeochemistry with Mn(III) and Fe (III) included as major components of the cycles. Reprinted with the permission of the American Association for the Advancement of Science from Madison *et al.* (2013).

it could potentially oxidize NOM and/or organic contaminants in its soluble form. For this reason, Mn oxides may play an important role in both NOM and organic contaminant transformations in both oxic and suboxic soil and marine sediments even though they are present at small concentrations.

Although Mn oxides have a low point of zero charge (pH <2; Chorover and Amistadi, 2001), they are not solely negatively charged in soil. Mn oxide octahedra comprise a manganese cation (either Mn(IV) or occasionally Mn(III) which is surrounded by 6 oxygen anions (O^{2-})). The oxygen anions act as bridges to link the metal cations and thereby form the solid lattice whilst balancing charge. At the mineral edges and at vacancy sites, however, there is a break in this regular pattern with O atoms that are not charge balanced. These are called terminal or non-bridging oxygens (McBride, 1994). These terminal O anions are usually in the form of hydroxyl groups (OH) and can be variably protonated or deprotonated depending on pH. Peacock and Sherman (2007) proposed that there are three types of these terminal O anions, O_I, O_{II} and O_{III} as shown in Fig. 3. 'O_I' oxygens are singly coordinated and are modelled as $-OH^{-1/3}$ or $-OH_2^{+2/3}$. 'O_{II}' oxygens are doubly coordinated and are modelled as either $-O^{-2/3}$ or $-OH^{+1/3}$. The triply coordinated 'O_{III}' oxygens are assumed to be inert.

As the charge on the terminal oxygens varies with pH (Peacock and Sherman, 2007), at low pH, terminal oxygens will tend to be present in the form $-OH_2^{+2/3}$ (O_I sites) and $-OH^{+1/3}$ (O_{II} sites); hence manganese oxides tend to be more reactive with negatively charged organic molecules at low pH.

2.2. Natural Organic Matter (NOM)

2.2.1. Background

Natural organic matter is an extremely complex and diverse group of organic molecules. There is no consensus on the primary binding mechanism holding the individual molecules together. It is known that NOM consists of long aliphatic chains

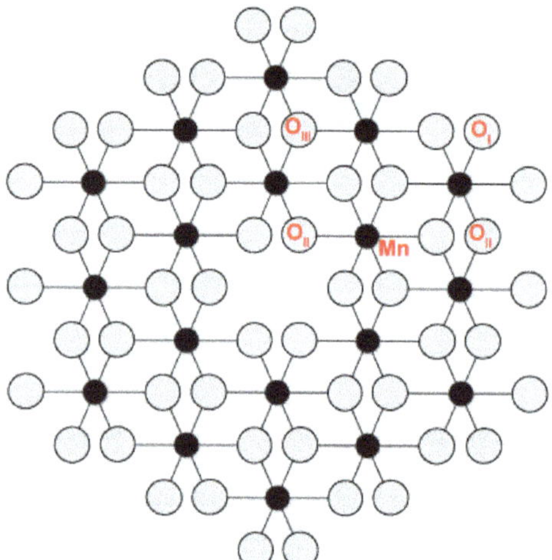

Figure 3. Surface functional groups on hexagonal birnessite showing the Mn octahedra as viewed from above (black circles represent Mn, white circles represent O atoms). Oxygen groups, O_I, O_{II} and O_{III} have a different charge dependent on the number of bonds (I, II or III) with Mn atoms. Reprinted with the permission of Elsevier from Peacock and Sherman (2007).

and aromatic rings of carbon atoms coiled into a three-dimensional macromolecular structure, with phenolic and carboxylic functional groups being seen as the most important reactive groups (Stevenson, 1994). NOM is a key driver and the main accumulator of POPs in the soil environment (Cousins *et al.*, 1999; Agarwal and Bucheli, 2011; Tremolada *et al.*, 2012). At neutral/acidic pH the protonated molecule provides a micelle-like region into which organic contaminants may migrate and can be held by electrostatic forces such as H bonding (Stevenson, 1994). However, this situation may change with an alteration in pH. Increasing alkalinity, for example, has been shown to create negative repulsive charges on SOM leading to an opening of the macromolecular structure and a release of sequestered organic contaminants.

Few studies have looked at the organic matter transformations in conjunction with POP remediation in soils, even though NOM can act as both an oxidant (quinone moieties can be electron acceptors) and a reductant (phenolic moieties can be electron donors). Functional groups in NOM-like phenols can also act as redox mediators (also known as electron shuttles) which can be reversibly oxidized and reduced, thereby enabling them to act as electron carriers in multiple redox reactions. Redox mediators can accelerate reactions by lowering the activation energy of the total reaction (Van der Zee and Cervantes, 2009). In addition, NOM can act as a large sink for cations. For this reason, NOM can significantly affect the reactivity of both contaminants and mineral surfaces and thus affect the environmental fate of contaminants (Van der Zee and Cervantes, 2009). In the following section we summarize interactions between Mn oxides and organic carbon which illustrates that Mn oxides are an important factor influencing interactions between organic contaminants and NOM. This is because Mn oxides are capable of transforming organic contaminants and NOM both separately, and as coupled organic contaminant-NOM complexes, *via* radical formation and oxidative coupling (Stone, 1987a,b).

2.2.2. Dissolved organic carbon (DOC)

Dissolved organic carbon is a complex mixture of the water-soluble fraction of NOM. It is heterogeneous and current theory has it that it has a supramolecular structure where relatively small organic molecules are held together by hydrogen bonds and cation bridging mechanisms (Maoz and Chefetz, 2010). It is important to note that although Ca and Mg have been investigated as potential bridging cations in DOC, Mn (II and III) are potential contenders for cation bridging which have not yet been considered in the literature. DOC is operationally defined as material which passes through 0.2 or 0.45 μm filters (Zsolnay, 2003) and contains both aromatic and aliphatic hydrocarbon molecules. These structures have polar functional groups (carboxyl, hydroxyl, amide and amine) which can be attracted to mineral surfaces. DOC plays an important role in the sorption and transport of organic contaminants (Stevenson, 1994) and can affect the oxidation and hydrolysis of organic contaminants by minerals such as Mn oxides.

2.2.3. Interactions between Mn oxides and DOC

Mn oxides are known to interact strongly with DOC (Tipping and Heaton, 1983; Stone and Morgan, 1984a,b; Lovley, 1991) and are likely to take part in all the mechanisms presented in Fig. 4. Organic synthetic chemists have used Mn oxides for many years to

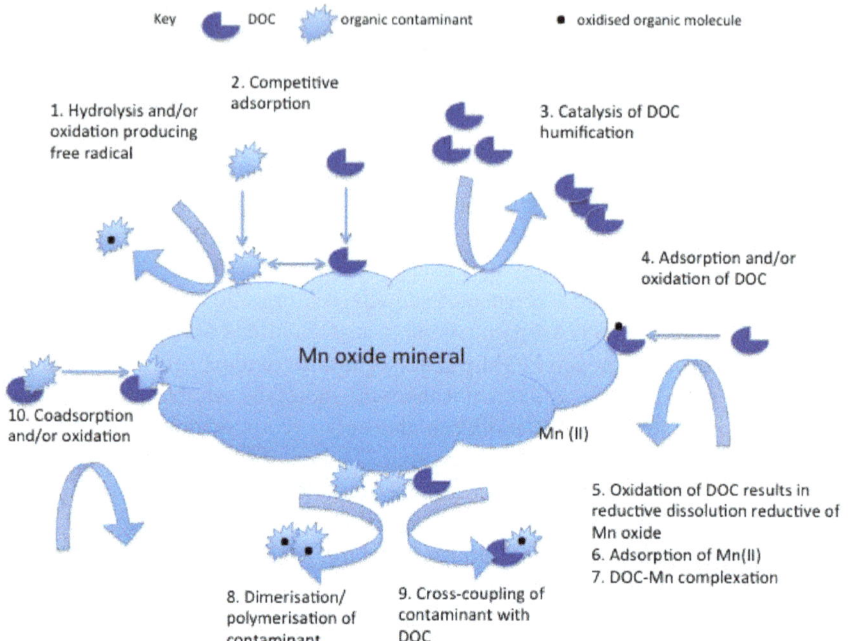

Figure 4. Impact of DOC on oxidation and hydrolysis of contaminants by Mn oxide. Briefly, Process 1 represents direct hydrolysis and/or oxidation of the organic contaminant by the mineral surface. When DOC is present in solution, this can also be adsorbed to the mineral surface and is in competition with organic contaminants for surface sites (Process 2). DOC can also be humified (Process 3) which could also affect contaminant–mineral interactions by changing the nature of DOC. Adsorption and/or oxidation (Process 4) of DOC could result in reductive dissolution of the mineral (Process 5) with concomitant release of Mn(II) or Mn(III) which could either be readsorbed to the mineral surface (Process 6) or complex with the DOC oxidation products (Process 7). Process 8 represents oxidation of the contaminant leading to dimerization or polymerization of organic contaminants with each other. Process 9 represents oxidation of the contaminant leading to cross-coupling with DOC. Process 10 represents co-adsorption of a DOC-organic contaminant complex which could result in oxidation and release of oxidation products. Examples of these processes are given throughout the chapter. Adapted from Polubesova and Chefetz (2014).

oxidize organic molecules, *e.g.* converting CH_4 into higher hydrocarbons (Shen *et al.*, 1993) or alcohols into aldehydes or ketones (Vileno *et al.*, 1999). However, in the laboratory, pH is controlled and Mn oxides oxidize organic contaminants under ideal conditions, at low pH. Under natural conditions, due to their low pH_{zpc} (usually <3), Mn oxides carry a negative charge in soil environments which means that they are not as good at adsorbing DOC as Fe oxides (Chorover, 2001).

The detailed study of interaction between Mn oxides and DOC began in 1984 with Stone and Morgan (1984a,b) publishing their extensive survey of the reduction of Mn oxides by organic compounds, though it must be noted that Tipping and Heaton (1983) had made a significant contribution to the field prior to this. The oxidation of organic compounds and subsequent reduction of Mn oxide by these compounds is termed 'reductive dissolution', and this occurs at the oxide mineral surface when electron transfer takes place between the metal centre (oxidant) and the electron donating part of the organic molecule (reductant). The mechanisms for these processes have been studied in detail (Stone and Morgan, 1984a,b; Stone, 1987a,b) with the following steps determined:

1. diffusion of the reductant molecules to the Mn oxide surface;
2. formation of a surface complex between the reductant and Mn oxide surface;
3. surface chemical reaction *via* inner sphere (or outer sphere) and precursor complex formation, electron transfer or breakdown of successor complex;
4. desorption of the oxidized reductant;
5. movement of Mn(II) from crystal lattice to the adsorbed layer;
6. desorption of Mn(II) followed by autocatalytic oxidation of Mn(II) to Mn(IV);
7. release of reaction products and diffusion away from the Mn oxide surface.

The early studies by Stone and Morgan concentrated on the reactions between Mn oxides and representative components of NOM such as phenols. Phenolic monomers are common in natural organic matter and make up to 10% of DOC in soil (Thorn *et al.*, 1992). Phenolic compounds also take part in both biotic and abiotic humification processes (Gianfreda *et al.*, 2006). Mn oxides are known to transform organic molecules such as phenols into phenoxy radicals (*e.g.* Stone, 1987b) which can then either be mineralized to carbon dioxide and water (*e.g.* Lamaita *et al.*, 2005) or be polymerized into humic substances with a high degree of aliphaticity (*e.g.* Chang Chien *et al.*, 2009). In the presence of humic substances the polymerization process can lead to the incorporation of the phenoxy radicals into the humic material leading to more aromatized humic substances (Li *et al.*, 2012).

A well studied example of an abiotic humification process is the Maillard reaction, a sugar-amino-acid condensation reaction which forms compounds known as melanoidins. The Maillard reacton is catalysed by Mn oxides such as birnessite (Jokic *et al.*, 2001; Chorover, 2004). Birnessite also catalyses both the polyphenol model (Shindo and Huang, 1982) and the integrated polyphenol Maillard pathways (Hardie *et al.*, 2009) of humic polycondensation and so its global role in the stabilization of carbon in sediments may be important. While Mn oxides are known to catalyse humification abiotically (*e.g.* Shindo and Huang, 1982; Hardie *et al.*, 2009) they are also known to break down abiotically high-molecular-weight humic substances in DOC into low-molecular-weight substances (pyruvate, acetone, formaldehyde, acetaldehyde in addition to CO_2 and CO) (Sunda and Keiber, 1994). Whether Mn oxides break down or build up humic substances is not well understood but is thought to be determined by environmental conditions including the presence of oxygen (Chang Chien *et al.*, 2009) and/or humic substances (Li *et al.*, 2012).

Several authors have since studied the influence of DOC on Mn oxide-organic contaminant interactions. Kang *et al.* (2004) evaluated naturally occurring phenols (present in DOC) as mediators for transformation of the fungicide cyprodinil by birnessite. Mediators are reactive chemicals (such as phenols) that facilitate oxidative coupling of contaminants but are not transformed directly by the mineral oxidant. Polubesova and Chefetz (2014) suggested that the influence of DOC on mineral–organic contaminant interactions is dependent on the chemical structure and properties of the contaminant and also the DOC as well as any mediators which may be present in solution (*e.g.* phenolic monomers).

Figure 4 shows that transformations of the DOC and/or contaminants on birnessite surfaces (processes 1–4, 8–10) are accompanied by mineral dissolution (processes 5–7) resulting in release of metal cations which can be complexed with DOC and/or retained at the mineral surface especially by carboxylic moieties present in the DOC (*e.g.* Banerjee and Nesbitt, 1999; 2001), or released into solution as low-molecular-weight molecules (*e.g.* Sunda and Keiber, 1994).

It is therefore important to consider the literature on the role of Mn oxides in transforming DOC (and hence influence composition of NOM) when considering the fate of organic contaminants. This is for two reasons: (1) NOM is a large sink for organic contaminants; and (2) Mn oxides can transform phenolic moieties which are prevalent in DOC/NOM and many organic contaminant streams (Kinney *et al.*, 2006) into reactive phenoxy radical moieties. Further reactions can lead to either cross-coupling reactions between phenoxy radicals within contaminant molecules (as represented by process 8 in Fig. 4) or between phenoxy radicals in contaminant molecules and DOC molecules (as represented by process 9 in Fig. 4). The role of DOC must not be overlooked. Humification of DOC itself as catalysed by birnessite (represented by process 3 in Fig. 4) is not covered in detail in this chapter but the reader is directed to Stevenson (1994) for a comprehensive review of humification and to Huang and Hardie (2011) for more details on the role of Mn oxides in humification reactions.

Mn oxide precipitation and dissolution is controlled by redox cycles which are most extreme in soils which are subjected to wetting and drying sequences (WADS). Climate change is predicted to increase the intensity and frequency of extreme flooding events (Milly *et al.*, 2002). It has been suggested that drying of soil samples results in oxidation of the soil organic matter by Mn oxides thus explaining the increase in soluble Mn and dissolved organic carbon seen in laboratory-stored soils (Dowding *et al.*, 2005). This is based on the premise that drying results in a dramatic decrease in pH on the mineral surface, due to increased hydrolysis of surface cations (Mortland and Raman, 1968), which increases the oxidizing potential of the Mn oxide. This increased acidity may allow reactions to occur which are not thought to be thermodynamically favourable within the environmental pH range (pH 4–9). There is little literature on the reaction of organic contaminants with mineral surfaces under drying conditions. However, the increased sorption and subsequent oxidation of PAHs onto dry Mn oxide surfaces suggests that mineral–organic interactions during wetting and drying cycles could be an interesting avenue of research (Clarke *et al.*, 2012).

In addition, the synthetic Mn oxides which are used almost exclusively in laboratory studies are different from natural Mn oxides. O'Reilly and Hochella (2003) found that natural and synthetic Mn oxides differ in amounts of impurities, crystal form, crystal size and microtopography. Natural Mn oxides frequently contain heterogeneously distributed organic matter coatings both on their surface (O'Reilly and Hochella, 2003) and potentially trapped within the mineral matrix. The organic coatings may theoretically make the Mn oxide surface more hydrophobic, which makes subsequent OC interactions with natural Mn oxides easier. This may also make active hydroxyl sites less accessible to organic contaminants, thereby making the formation of a suitable surface precursor complex for electron transfer potentially more difficult. Clarke *et al.* (2012) found that synthetic manganite (MnOOH) converted more anthracene to anthraquinone than a natural manganite (per m^2 of surface area). The reasons for this were not postulated but may relate to the surface sites being taken by existing organic coatings as these were present in the samples.

Below we concentrate on laboratory studies of the reactions between Mn oxides (both synthetic and natural) and single organic contaminants (with or without humic substances present). We focus on studies involving birnessite as this is the most common mineral in soils (Burns and Burns, 1977). The reader is referred to Remucal and Ginder-Vogel (2014) for an excellent review of a wider selection of Mn oxides reacting with organic contaminants in water. It is important to note, however, that in soils there are not only (multiple) organic contaminants but also NOM and so interactions are likely to be even more complex than discussed below in single-contaminant studies.

3. Interaction mechanisms between Mn oxides and synthetic organic contaminants

3.1. Background

Publications on reactions between laboratory-synthesized birnessite and organic contaminants have increased in recent years (see Table 4). Despite this, there is still a general lack of data on the reactions between naturally occurring Mn oxides found in soil (as opposed to synthetic ones) and contaminants, either in laboratory or field conditions. The role of both dissolved DOC in porewaters and adsorbed DOC on mineral surfaces is another area which has not been explored extensively although there is a good review of the existing data given by Polubesova and Chefetz (2014). The degree of interdependence between Mn oxides, NOM and organic contaminants as highlighted in this chapter, makes transformation mechanisms difficult to study. This may be the reason for the lack of data on Mn oxide and organic-contaminant transformations in soils, with the role that Mn oxides play in soil NOM transformations having been largely overlooked to date. One exception is a study by Brunetti *et al.* (2007) where Mn oxide was used to catalyse the humification of phenolic compounds in olive-mill wastewater and the waste effluent could potentially be used as a liquid soil amendment (see Fig. 4, process 9) to increase soil carbon content.

If organic contaminants are dissolved in porewaters and not associated with NOM, then they may be free to be oxidized by mineral surfaces such as those of birnessite, assuming the mineral surface itself is NOM free. Knowledge of the physical and

Table 4. Organic contaminants oxidized by Mn oxides in laboratory experiments.

Phenols and PCBs			Nitroaromatics			PAHs		
Compound	Author	Year	Compound	Author	Year	Compound	1st Author	Year
chlorophenol	Ulrich and Stone	1989	aliphatic amine	Laha and Luthy	1990			
chlorophenol	Pizzigalo *et al.*	1995						
2,4-D (dichloro-phenoxy acetic acid)	Cheney *et al.*	1996	aniline	Sulzberger *et al.*	1997			
[h]chlorophenol	Park *et al.*	1999	aniline	Klausen *et al.*	1997			
triclosan	Zhang and Huang	2003	azo dye*	Liu and Tang	2000			
EE2	de Rudder *et al.*	2004	azo dye	Ge *et al.*	2003			
PCB	Pizzigalo *et al.*	2004	aromatic amine	Li *et al.*	2003			
glyphosate	Barrett and McBride	2005	atrazine	Shin *et al.*	2000			
hexachlorobenzene	Yuan *et al.*	2006	azo dye	Liu and Qu	2002			
phenol	Jung *et al.*	2008	cyprodinil	Kang *et al.*	2004	[h]phenanthrene	Parikh	2004
phenol	Rao *et al.*	2008	N-oxides	Zhang and Huang	2005			
bisphenol A	Lin *et al.*	2009	azo dye	Wang *et al.*	2006	phenanthrene	Napola	2006
napthol	Shin *et al.*	2009	oxytetracycline	Rubert and Pederson	2006			
chlorinated phenol	Zhao *et al.*	2009	mercaptobenzo-thiazole	Li *et al.*	2008			
mercaptobenzothiazole	Dong *et al.*	2010	sulfadiazine	Liu *et al.*	2009			
Bisphenol F	Lu *et al.*	2011	azo dye	Chakrabarti *et al.*	2009			
bisphenol	Gao *et al.*	2011	azo dye*	Clarke *et al.*	2010			

[h]PCB	Huang *et al.*	2011	azo dye*	Clarke and Johnson	2010			
phenol	Zhang *et al.*	2011	lincosamide	Chen *et al.*	2010			
[h]phenol	Li *et al.*	2012	carbamezapine	He *et al.*	2012			
glyphosphate	Ndjeri *et al.*	2013	morin dye	Polzer *et al.*	2012	pyrene	ChangChien *et al.*	2011
[h]nonylphenol	Lu and Gan	2013	methylene blue	Zhao *et al.*	2013	anthracene*	Clarke *et al.*	2012
acetaminophenol	Xiao *et al.*	2013	diclofenac	Huguet *et al.*	2013	napthalene	Garcia *et al.*	2013
pentachlorophenol	DiLeo *et al.*	2013	azo dye*	Clarke *et al.*	2013			
bisphenol	Lin *et al.*	2013						
brominated phenol	Lin *et al.*	2014				fluorene	Villalobos *et al.*	2014

* natural as well as synthetic manganese oxides
[h] reactions studied with and without humic substances

chemical properties of organic contaminants allows broad predictions to be made about their mobility and fate in the environment, and the likelihood of reaction occurring with *in situ* soil Mn oxides. We have grouped organic contaminants into four categories, phenols, polychlorinated biphenyls, N-containing aromatics and poly-cyclic aromatic hydrocarbons in Table 4. Table 4 states whether or not natural Mn oxides have been used (which may have NOM coatings) and because the presence of NOM (and hence DOC) is also an important determinant of reaction mechanism (*e.g.* Li *et al.*, 2012), as shown in Fig. 4, then the effect of the addition of humic substances on the reactions is also noted.

The processes that control the fate of organic contaminants in soils by Mn oxides can be divided into three general categories:

(1) sorption processes that leave the structure of the molecule intact (see Fig. 4, process 2);
(2) oxidation processes which transform the parent molecule into one or several compounds having a different chemical and physical behaviour (Fig. 4, processes 1, 8–10);
(3) polymerization – transformation of the parent molecules into radicals which can either polymerize into dimers or oligomers (Fig. 4, process 8) or become attached to another organic molecule(s) (*e.g.* DOC) (Fig. 4, process 9).

3.1.1. Sorption

Sorption of the organic molecule to the mineral surface is often a rate-limiting factor in oxidation reactions (Fig. 4, process 1). Most organic compounds adsorb to NOM in soil in preference to mineral surfaces (Stevenson, 1994) due to their hydrophobic nature. Sorption or partitioning of hydrophobic organic compounds (HOCs) to surfaces occurs as a result of weak solute–solvent interactions rather than specific sorbate-sorbent interactions (Means, 1995). Hydrophobic compounds cause a local ordering of water molecules, which decreases the entropy of the system and forces molecules out of solution and onto weakly hydrating or uncharged surfaces. This is known as the 'iceberg' effect (Frank and Evans, 1945). Larger molecules will tend to disturb more water molecules and therefore will be more prone to being forced out of solution than smaller molecules. Thus, water solubility is inversely proportional to molecular size and accordingly inversely proportional to sorption of non-polar compounds to soils. The hydrophobicity of a compound is commonly measured by the octanol/water partitioning coefficient (K_{ow}) which is defined as the ratio of concentration of an organic compound in octanol to its concentration in water after equilibrating with the two solvents. Ulrich and Stone (1989) observed a direct correlation between K_{ow} and the partitioning of chlorophenols on Mn oxide surfaces.

Sorption of HOCs onto mineral surfaces can be influenced by organic or biogenic coatings. The sorption of HOCs onto kaolinite and hematite surfaces has been shown to be proportional to the amount and aromatic nature of the humic coatings (Murphy *et al.*, 1990). Sorption of HOCs on biofilms has only received limited attention (Johnsen and Karlson, 2004; Wicke *et al.*, 2005). Biofilms are bacterial communities of high cell density which provide dynamic geochemical environments at the aqueous/mineral interface. They are held together by extracellular polymeric substances (EPS), which consist largely of proteins and polysaccharides. The EPS are largely responsible for the physical and physicochemical properties of biofilms (Flemming *et al.*, 1999). The emulsifying properties of EPS have been noted and Dohse and Lion (1994) were able to show that adding EPS to a soil resulted in decreasing the partitioning of PAHs onto soil particles and increasing the mobility of PAHs in contamination plumes. Mineral dissolution and precipitation are ongoing processes within the biofilm and biogenic minerals are often integrated within the EPS. Biogenic Mn oxides associated with biofilms have received growing attention due to their metal scavenging and redox cycling capacity (Haack and Warren, 2003; Toner *et al.*, 2005, 2006). The combination of sorption capacity of the bacterial cells, EPS and the Mn oxides has led to the whole system being termed a biosorbent (Toner *et al.*, 2006). Most studies of biofilm sorption have focused on metals (Prasad and Pandey, 2000; Prasad and Pandey, 2001; Quintelas and Tavares, 2001; Toner *et al.*, 2006), but very little information is available on the capacity of biofilms to sorb organic contaminants. It is anticipated that EPS around minerals may form a suitable sorbent for hydrophobic partitioning which may provide sufficient contact of the PAH and the adjacent mineral phase.

Sorption to minerals can occur chemically when there is sufficient electrostatic charge between the mineral surface and the HOC. Inner-sphere sorption occurs where there is direct contact between the organic molecule and the mineral surface and the organic molecule and metal centre are effectively bonded. Outer-sphere sorption occurs when there is a water molecule between the organic molecule and mineral surface. The metal's coordination sphere remains intact and there is no bonding between the metal centre and the organic molecule (McBride, 1994).

Compounds with mono-substituted hydrophilic functional groups (such as OH, Cl, nitro-groups) tend to be more polar and therefore sorption is related to pH. For these more polar organic contaminants, sorption tends to increase as pH decreases. However, sorption also depends on the number and position of the functional groups on the organic compound which not only affects polarity but can also change steric chemistry which is also likely to affect sorption. Studies of sorption of organic compounds to redox active compounds such as Mn oxides are also complicated by the fact that oxidation can take place rapidly and so the surface precursor complex does not survive for long enough time periods to measure sorption. In general, sorption of organic contaminants to Mn oxides takes place at the reactive hydroxyl sites which are shown in Fig. 3 (Peacock and Sherman, 2007).

The charge on these edge hydroxyl sites depends on pH and can be positive at low pH, neutral at the zero point of charge and negative at high pH. The likelihood of an organic molecule being charged depends on how likely it is to undergo protonation (gain of a hydrogen atom) or more likely deprotonation (loss of hydrogen atom to leave a negatively charged species). This reaction depends on the presence of ionizable hydrogen atoms within the molecule and is measured by pKa. The pKa value for phenol is 10 meaning that at pH 10, both phenol and negatively charged phenolate ions are present in equal quantities. At pH 6 the phenolate ion is present at 10^{-4} times lower concentration than that of the phenol molecule. Both the concentration and charge of the organic species and the active hydroxyl sites on the Mn oxide surface determine whether or not a surface complex will form and whether this will be inner or outer sphere (Stone, 1987a).

3.1.2. Oxidation

Thermodynamic data allow a first approximation as to whether an organic compound can be oxidized spontaneously by an oxidant, but due to kinetic constraints and sorption limitations, some thermodynamically spontaneous reactions may not occur at a significant rate (McBride, 1994). Thermodynamics are important, however, in establishing which reactions are theoretically possible. The reduction of Mn(III/IV) oxides, together with their standard redox potentials (E^0), are shown in Table 3. Theoretically, any organic molecule with an E^0 value of <1.5 V should be oxidized by manganite (γ-MnOOH) and any organic with an E^0 value of <1.23 V should be oxidized by MnO_2 (Table 3).

Table 3 lists standard-state redox potentials which are calculated at pH 0 using a concentration of 1M for reactants and products. Such conditions are not realistic for environmental conditions where oxidant and reductant concentrations are significantly lower and the pH range is between 4 and 9. The Nernst equation can be used to translate the E^0 potential of a half reaction into a redox potential (Eh) relevant to environmental conditions. As an example, when environmentally relevant concentrations of H^+ and oxidant and reductant are used, Eh values decrease from 1.5 V to 0.61 V for Mn(IV) and from 1.23 V to 0 V for Mn(III) (Stone, 1991).The reduction potentials of representative organic compounds are also known *e.g.* hydroquinone has an Eh of 0.699 V with an environmentally relevant Eh of 0.196 V, ascorbate has an Eh of 0.40 V with an environmentally relevant Eh of −0.103 V and oxalate has an Eh of −0.18 V with an environmentally relevant Eh of −0.69 V. It is clear from these values that Mn oxides are capable of oxidizing all of these organic molecules (Stone, 1991).

Note that although calculation of Eh values to ascertain the likelihood of a redox reaction occurring is useful, factors such as dissolved organic carbon (DOC) in porewaters must be taken into account. Banerjee and Nesbitt (1999, 2001) found that DOC can reduce Mn oxide and form strong soluble Mn(III)-organic complexes, some of which attach to the mineral surface, thereby blocking reactive hydroxyl sites for reactions with other organic species. The prevalence of aqueous Mn(III) in many natural environments has only recently been established (Madison *et al.*, 2013). Therefore when DOC is present, the role of aqueous Mn(III), which can act as both an oxidant and a reductant, must be taken into account in organic-contaminant transformations.

In general, the reaction kinetics of the oxidation of organic contaminants by Mn oxide is controlled by either the rate of surface precursor complex formation (*i.e.* the rate of sorption) or by the rate of electron transfer within the precursor complex (Zhang *et al.*, 2008). Both of these factors are affected by pH, with oxidation of many polar organic compounds increasing with decreasing pH (Stone, 1987; Ulrich and Stone, 1989; Laha and Luthy, 1990; Zhang and Huang, 2003, 2005b). The increased oxidation of organic compounds at low pH has been attributed to:

- positive charging of any amide groups present in the organic contaminant resulting in electrostatic attraction to the oxide surface (Laha and Luthy, 1990);
- increased sorption of phenolic groups (Stone, 1987b);
- the decrease of the negative charge on Mn oxide due to ZPC<3 (Chorover *et al.*, 2004);
- the increased redox potential of the Mn oxide/Mn(II) couple (see Fig. 1);
- enhanced removal of Mn(II) from the oxide surface, exposing new reactive sites (Klausen *et al.*, 1997).

When Mn oxides oxidize organic molecules the reaction is initiated *via* formation of a complex between an electronegative group on the organic molecule (*e.g.* phenol or amide group) and the manganese metal centre with subsequent electron transfer resulting in the formation of free radicals such as phenoxy radicals (Ulrich and Stone, 1989; Park *et al.*, 1999; Zhang and Huang, 2005a) (see Fig. 4, process 1). Further oxidation of the free phenoxy radicals can result in partial or complete breakdown of the organic molecule and release of CO_2 with or without lower-molecular-weight organic intermediates (*e.g.* Cheney *et al.*, 1996; Nasser *et al.*, 2000).

3.1.3. Polymerization

Without further oxidation by Mn oxides, as phenoxy radicals are so reactive, they can undergo nucleophilic addition to *o*-quinones (*e.g.* Park *et al.*, 1999) or dimerization and/or polymerization into larger more humic-like molecules (*e.g.* Li *et al.*, 2012) (Fig. 4, process 8). In the presence of humic substances or constituents of humic substances such as syringlaldehyde (4-hydroxy-3,5-dimethoxybenzaldehyde) it has been shown that Mn oxides can catalyse the incorporation of the organic contaminant into the humic substance (Fig. 4, process 9), effectively composting the contaminant (Park *et al.*, 1999; Li *et al.*, 2012; Lu and Gan, 2013). This is not always the case: Klausen *et al.* (1997) found that the addition of humic acid to a solution of aqueous Mn oxide and aniline inhibited aniline oxidation and proposed that the humic acid was blocking reactive sites (Fig. 4, process 2).

Note that DOC can also significantly influence the sorption and oxidation of organic contaminants by Mn oxides by competing with contaminants for adsorption sites on mineral surfaces (Fig. 4, process 2) and participating in and/or accelerating mineral dissolution and metal complexation (Fig. 4, processes 4–7).

3.2. Phenols

The oxidation of phenols by Mn oxides to make phenoxy radicals has been well documented (Stone and Morgan, 1984b; Stone, 1987; Ulrich and Stone, 1989; Petrie *et al.*, 2002). Phenols such as catechol, resorcinol, hydroquinone, mono-substituted benzoic acids and methoxy-aromatics can all be oxidized by Mn oxides (Stone, 1987a,b). Ulrich and Stone (1989) proposed that the reaction mechanism between a Mn(III) oxide and a phenol would take place *via* the following steps as shown in Fig. 5.

Mn(IV) oxides (or more specifically Mn(IV)-OH sites) could also undergo the same reaction with Mn(III) being the reduced Mn species. Inner sphere complexes are represented below in Fig. 5 but outer sphere complexes may also be formed.

Many contaminants which contain phenol groups (or hydroxyl groups) can theoretically be oxidized by Mn oxides *via* the formation of phenoxy (or hydroxyl) radicals followed either by formation of quinones or dimerization or polymerization to create larger, more aromatic molecules. The redox reaction is initiated by a one-electron transfer from the phenol to the metal centre to form a phenoxy radical (Waters, 1971). The radical species can then enter into a number of competitive pathways depending on pH and reductant concentration and whether or not other organic molecules such as humic substances are present (*e.g.* Li *et al.*, 2012). The effect of pH on phenolic oxidation rate shows a curved response with pH dependence being greater above pH 6 and constant below 4 (Stone, 1987b). The coupling of two radical species results in dimer formation and dimers are more susceptible to oxidation than monomers so polymerization is often observed to be self-perpetuating (Fig. 4, process 8) (Stone, 1987b). Oxidative coupling of phenolic compounds may be considered more beneficial than further oxidation to quinone species. The coupling of semiquinones results in the formation of stable humic acids, which is essentially a composting process (Huang, 2000). Coupling reactions require lower activation energy compared to electron transfer reactions (Chang and Allan, 1971), thus should be more kinetically favourable. Coupling reactions are also influenced significantly by the presence of oxidants and reductants. In general, when phenoxy radicals are generated, when oxygen is present,

1) Surface complex formation

$$>Mn(III)\text{-}OH + ArOH = >Mn(III)\text{-}OAr + H_2O$$

↓

2) Electron transfer

$$>Mn(III)\text{-}OAr = (>Mn(II), {}^{\cdot}OAr)$$

↓

3) Release of phenoxy radical

$$(>Mn(II), {}^{\cdot}OAr) + H_2O = >Mn(II)\text{-}OH_2 + {}^{\cdot}OAr$$

↓

4) Release of reduced Mn(II)

Figure 5. Reaction mechanism for oxidation of phenol by Mn(III) oxides. Adapted from Ulrich and Stone (1989).

mechanism A (see Fig. 6) takes place resulting in the formation of quinones which can be further completely oxidized to CO_2 and/or to smaller aliphatic compounds (Chang Chien *et al.*, 2009). In the presence of humic substances (reductants), however, mechanism B (Fig. 6) takes place and phenoxy radicals tend to be coupled to existing humic substances (Li *et al.*, 2012).

Stone and Morgan (1984a,b) established that the presence of electron-withdrawing substituents on the phenolic aromatic ring led to a decrease in reactivity between Mn oxides and the substituted phenol, whereas the presence of electron-donating groups (*e.g.* methoxy groups) enhances reactivity. The general order to reactivity of substituted phenol to Mn oxides was seen to be methoxy > methyl > chloro > carboxyl > nitro. In addition to the nature of the additional functional group, the position of the functional group on phenolic compounds also influences the rate of oxidation by Mn oxides (Stone and Morgan, 1984a; Ulrich and Stone, 1989). Phenols with electron-withdrawing functional groups in the *ortho* and parapositions are stronger reductants than their *meta* equivalents. This is thought to be a consequence of the π-electron-donating capacity of *para* and *ortho* isomers, which facilitates resonance stabilization of the phenoxy radical. In addition, *ortho* and *para*

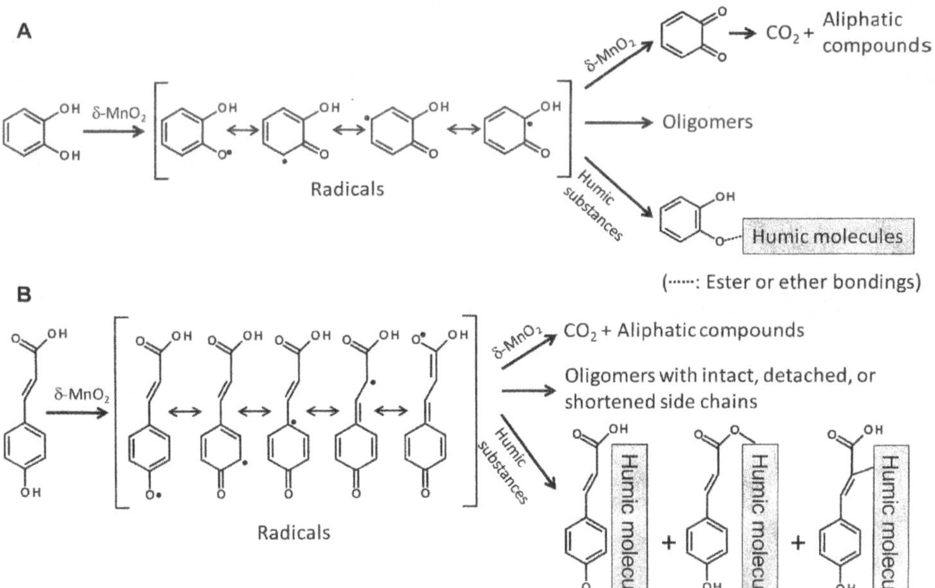

Figure 6. Potential oxidative transformations of two phenolic monomers (A = catechol; B = p-coumaric acid) reacting with MnO_2. The upper reaction for each phenolic monomer with MnO_2 takes place in the absence of humic substances and oxidation of phenolic groups leads to phenolic radicals resulting in the subsequent formation of quinone (also shown as Process 1 in Fig. 4). The lower reaction for each phenolic monomer with MnO_2 takes place in the presence of humic substances and shows oxidation of phenolic groups to phenolic radicals with subsequent cross-coupling with the humic substance (also shown as Process 9 in Fig. 4). Figure reproduced from Li *et al.* (2012) with the permission of Elsevier.

isomers promote the formation of precursor of inner-sphere mineral-organic complexes, which is a function of nucleophilicity of the phenolate ion (Ulrich and Stone, 1989).

Glyphosate is neither a phenol nor a persistent organic contaminant in soil, but it is the most commonly used pesticide worldwide (Barrett and McBride, 2005). Figure 7 shows how the substituent hydroxyl group in the glyphosate molecule can react with Mn oxides. Barrett and McBride (2005) concluded that the C-P bond was cleaved at the Mn oxide surface *via* the mechanism shown in Fig. 7. Further details are available in the original paper but the authors also found evidence for C–N bond cleavage to create the degradation product sarcosine [2-(Methylamino)acetic acid].

3.2.1. Halogenated phenols

Ulrich and Stone (1989) were the first authors to study the reaction between chlorinated phenols and Mn oxides. The reaction mechanism is the same as for phenols presented in Fig. 5. The phenol group on the chlorophenol forms a surface complex with the

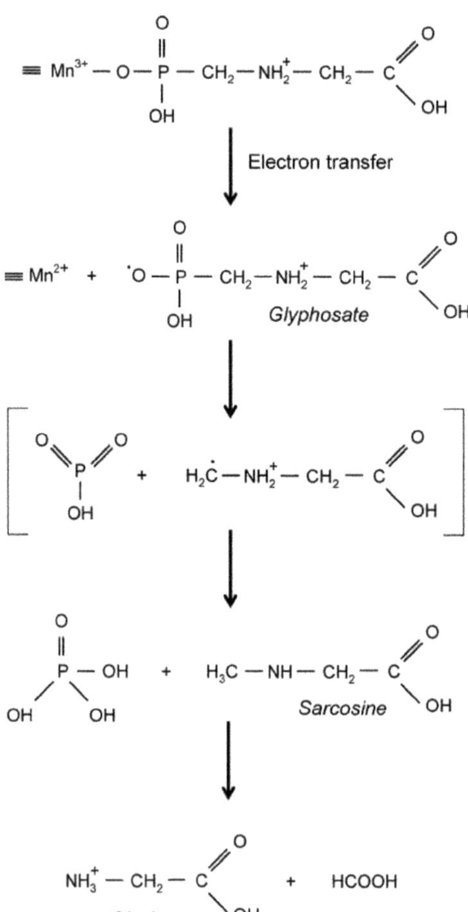

Figure 7. Possible reaction mechanism between glyphosate and Mn oxide. The Mn oxide surface is represented by three bonds to an Mn(III) ion. Reprinted with permission from Barrett and McBride (2005). Copyright 2005, American Chemical Society.

manganese cation and electron transfer creates a phenoxy radical with subsequent reactions including further oxidation (Fig. 4, process 1), dimer formation or polymerization (Fig. 4, process 8). However, the chlorine atom is very electronegative and so this makes a simple mono-substituted chlorophenol less susceptible to oxidation than phenols. Ulrich and Stone (1989) found that both the number and position of chloro-substitutes affects the reactivity of Mn oxide towards chlorophenols which is released.

More complicated chlorophenols with larger substituted groups can be more reactive towards Mn oxides and undergo oxidation of the phenol group to form dimer or polymeric products (Fig. 4, process 8) which are usually more hydrophobic and less mobile and toxic. Triclosan and chlorropheneare emerging contaminants and good examples of chlorophenols with electron-donating groups which increase their reactivity to Mn oxide compared to phenol (Zhang and Huang, 2003). Triclosan and chlorophene also show increased oxidation at low pH (Zhang and Huang, 2003).

Nonetheless, oxidation of some halogenated phenols by Mn oxides is not beneficial. Lin *et al.* (2014) have recently examined the role of Mn oxides in the oxidation of poly-brominated diphenyl ether (PBDE), another emerging contaminant which acts as a flame-retardant added to plastics and foam products to make them difficult to burn. The oxidation of simple bromophenols(BPs) to hydroxylated polybrominated diphenyl ethers (OH-PBDEs) is shown to occur at low pH *via* oxidative coupling of brominated phenoxy radicals (see Fig. 8 and Fig. 4, process 8) and is an example of the product (OH-PBDE) of oxidation by Mn oxides being more toxic than the original chemical (either BP or PBDE) (Dingemans *et al.*, 2008).

3.2.2. Polychlorinated biphenyls

No specific mechanism has been suggested for the reaction between PCBs and Mn oxides. Pizzigallo *et al.* (2004) have reported that MnO_2 alone could partially remove five PCBs from soil. The authors examined mechano-chemical mixing of PCBs with Mn oxides and

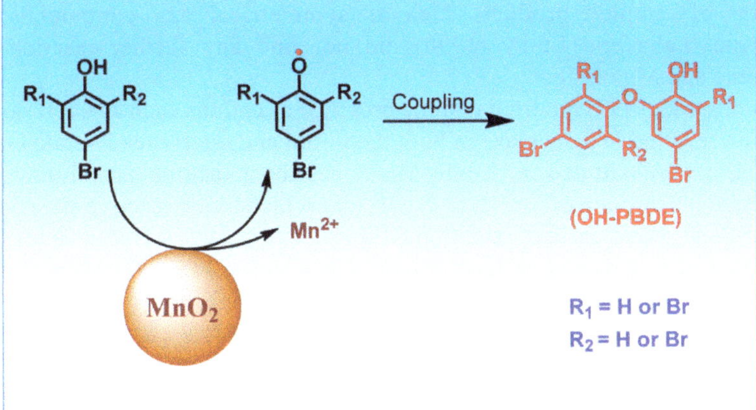

Figure 8. Proposed transformation of polybrominated diphenyl ether (PBDE) to OH-PBDE *via* the oxidative coupling of bromophenoxy radicals. Reprinted with permission from Lin *et al.* (2014). Copyright 2014, American Chemical Society.

found that removal was more successful when there were fewer substituted chlorine groups. For example, complete removal was possible in 10 days for a 2,2′-dichlorobiphenyl; however, after 90 days, removal of only ~30% and 20% for 2′,3,4-trichlorobiphenyl and 3,3′,4,4′-tetrachlorobiphenyl, respectively, was found. The position of the chlorine atoms was also important with greater reactivity of PCBs which had chlorine atoms in the *ortho* position rather than *meta* or *para* positions, probably because of resonance and steric effects (Pizzigallo *et al.*, 2004). However, Huang *et al.*, (2011) found no relation between degree of chlorination in PCBs and reactivity with Mn oxide when microwave radiation was used to enhance the reaction. It is therefore thought that the mechano-chemical mixing allowed radicals, *e.g.* chloro-phenoxyl, to form which then underwent oxidative coupling reactions (Pizzigallo *et al.*, 1995).

3.3. N-containing compounds

Aniline was the first N-containing organic to be studied reacting with birnessite (Laha and Luthy, 1990; Klausen, 1997). Aromatic amines can be removed by aerobic treatment through biodegradation, auto-oxidation and adsorption (Field *et al.*, 1992; van der Zee and Cervantes, 2009) but where they do persist they are of concern because they are carcinogenic. Zhang *et al.* (2008) modelled the reactions between several N-containing compounds (including anilines, fluorquinolones, aromatic N-oxides and tetracyclines) and Mn oxides. Those authors proposed that the hydroxylation of N-containing aromatics by Mn oxides could be summarized by a similar reaction as Ulrich and Stone (1989) proposed between phenols and Mn oxides (Fig. 5) but with the aromatic represented as ArNH instead of ArOH. Electron transfer takes place between the amino group and the Mn centre in the same way as between the phenoxy group and the metal centre.

Laha and Luthy (1990) found that aniline is ultimately oxidized by Mn oxide, *via* two one-electron steps, to symmetrical azobenzene. The reaction between Mn oxide and aniline is initiated by the formation of arylamino radicals followed by subsequent formation of coupled products such as symmetrical azo compounds such as azobenzene. Laha and Luthy (1990) did not find any further reaction between azobenzene and Mn oxide.

Klausen *et al.* (1997) reported that the presence of cations such as Ca(II) and Mn(II) in solution resulted in a decrease in both the initial and quasi-steady-state kinetics for the oxidation of substituted anilines by Mn oxides under anaerobic conditions. This was attributed to the buildup of Mn(II) or Ca(II) ions which block reactive sites (see Fig. 4, process 6) and retard aniline oxidation (Klausen *et al.*, 1997). Humic acids have also been shown to retard oxidation of aniline and Klausen *et al.* (1997) suggested that humic acids were reducing the oxide and generating Mn(II) ions (see Fig. 4, processes 4–7). Banerjee and Nesbitt's (1999, 2001) work on the reaction between humate and Mn oxides suggests that the reaction results in strong Mn(III)-humate and oxalate complexes which block reactive surface sites thereby preventing further reaction. Inhibition of phenol oxidation by phosphate ions (Stone and Morgan, 1984a) also suggests that competition between inorganic ligands for reactive hydroxyl sites may be a contributing factor.

The order to reactivity with substitution of the parent aniline molecule is as follows: methoxy > methyl > chloro > carboxy > nitro (Laha and Luthy, 1990) just as it is for phenols. This makes nitro-substituted anilines the most unreactive of the group. In addition, there is the possibility that carboxyl and nitro groups might be preferentially adsorbed (over the amino group) to reactive hydroxyl sites on the Mn oxide surface thereby preventing the oxidation of the amino group and subsequent radical formation (Wheeler and Gonzalez, 1964).

Zhang and Huang (2005) proposed that the piperazine moieties in fluoroquinoline were the main part of the molecule to complex and be oxidized by Mn oxide (see Fig. 9). N-dealkylation and hydrolysis were suggested pathways as well as possible coupling reactions.

3.3.1. Azo compounds

Azo compounds contain the –N=N- functional group. These compounds are well known for their use as dyes (azo dyes). Azo dyes are recalcitrant by design having to resist light, chemical bleaching, water, sweat and microbial attack, making the treatment of textile effluent a problematic issue. Oxidation has been considered an attractive technology for treating dye waste water (Zhao *et al.*, 2009) but the use of Mn oxides in the oxidative degradation of dyes has received only limited attention.

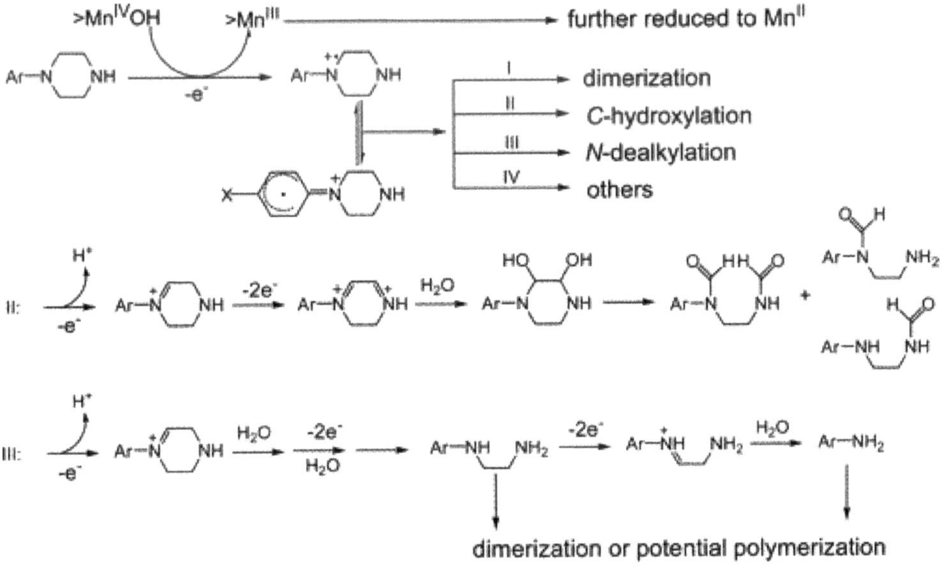

Figure 9. Proposed reaction mechanism for the reaction between Mn oxide and fluoroquinolines (FQ). The redox reaction is initiated by generating a surface complex between the FQ molecule and the Mn oxide surface. Within the surface complex, one electron is transferred from FQ to the surface-bound Mn(IV) (represented by >Mn(IV)) to produce FQ radicals. The radicals can then undergo coupling reactions (Process 8 in Fig. 4) *via* various pathways presented above. More details can be found in the original paper. Reprinted with permission from Zhang and Huang (2005a).

Manganese oxides have shown the capacity to decolorize a number of acid azo dyes either on their own (Liu and Tang, 2000; Ge and Qu, 2003; Clarke *et al.*, 2010) or as catalysts in the electrochemical decolorization processes (Liu and Qu, 2002; Wang *et al.*, 2006). Manganese-mediated oxidation of acid azo dyes are favoured at low pH values. This can be attributed partly to the anionic nature of acid azo dyes, thus sorption of the dye is enhanced at lower pH values as the positive charge on the oxide surfaces increase (Liu and Tang, 2000; Ge and Qu, 2003; Clarke *et al.*, 2010, 2013).

The reaction kinetics and mechanisms of acid azo dye oxidative decolorization reactions by natural Mn oxides present in a mine-waste product have been studied in detail (Clarke and Johnson, 2010; Clarke *et al.*, 2010, 2013).The oxidation of the common dye, acid orange 7 (AO7) (Fig. 10) by Mn oxides is initiated on the hydroxyl group of AO7 and proceeds *via* successive electron transfers from the dye molecule to the oxide surface resulting in the asymmetric cleavage of the azo bond. The terminal reaction products were shown to be 1,2 naphthoquinone, 4-hydroxybenzenesulfonate and coupling products involving 1,2 naphthoquinone and benzenesulfonate radicals (Clarke *et al.*, 2010). Sorption of AO7 to the Mn oxide is rapid and an outer-sphere complex is formed between the dye and the mineral surface. Transfer of the first electron was shown to be the rate-limiting step of the oxidation reaction (Clarke and Johnson, 2010).

The decolorization of the amine-containing dye, acid yellow 36, was shown to occur rapidly when reacted with Mn oxides, but decolorization was achieved without cleavage of the azo bond (Clarke *et al.*, 2013). The reaction pathway involves the formation of a number of colourless intermediate products, some of which hydrolyse in a Mn oxide-independent step. Decolorization of the dye is rapid and is observed before the cleavage of the azo bond, which is a slower process. The terminal oxidation products were observed to be *p*-benzoquinone and 3-hydroxybenzenesulfonate.

The impact of dissolved salts (releasing cations and inorganic ligands) on the oxidation of dyes by Mn oxides is inconclusive. Increased dye decolorization was observed after the addition of $NaNO_3$ to the Mn oxide-containing reaction solution and was attributed to the compaction of the diffuse double layer, which would improve dye-mineral contact (Liu and Tang, 2000). Conversely, Ge and Qu (2003) observed a decrease of decolorization with the addition of NO_3^-, Cl^- and SO_4^{2-} ions, which was ascribed to competition of these anions with the dye molecule for sorption sites on the oxide surface.

3.4. Polycyclic aromatic hydrocarbons

Remediating PAH-contaminated soils and even studying PAH–mineral interactions is problematic due to their low solubility (and therefore availability). In soils, PAHs partition strongly into NOM, which limits PAH–mineral contact and thus interaction. In the absence of NOM or in low organic-matter environments (*e.g.* subsoils, aquifers, landfills), PAHs will partition onto mineral surfaces and this is likely to result in mineral–PAH interactions. Schwarzenbach and Westall (1981) proposed a threshold value of 1 g of organic carbon per kg of sorbent, below which the mineral phase begins

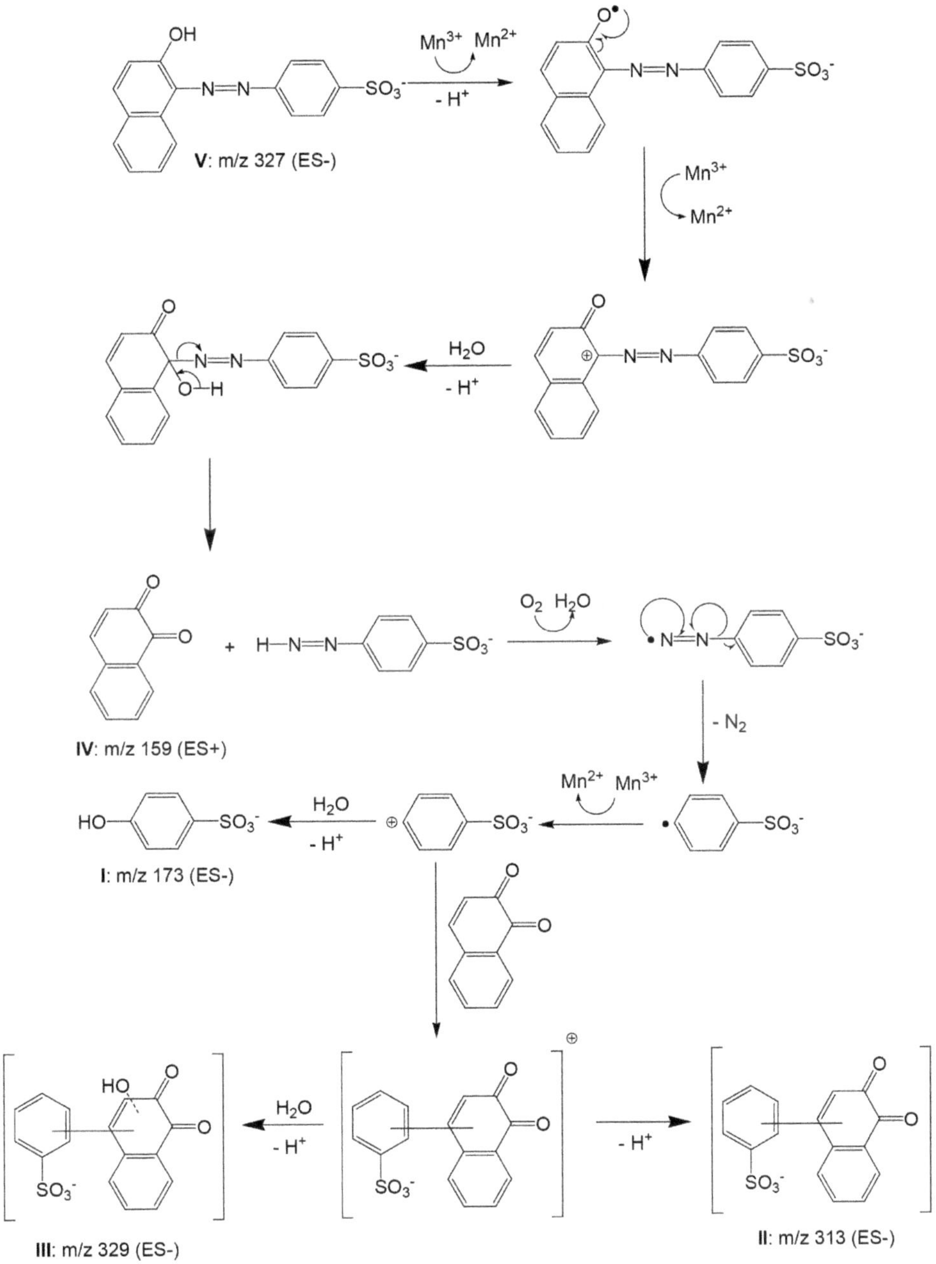

V: m/z 327 (ES-)

IV: m/z 159 (ES+)

I: m/z 173 (ES-)

III: m/z 329 (ES-)

II: m/z 313 (ES-)

Figure 10. Proposed reaction mechanism for the oxidation of AO7 by Mn tailings (m/z values from the negative ion mode analysis) showing interaction between Mn oxide surface (represented here by Mn(III)) and dye molecule initiated at the phenol group resulting in eventual asymmetric cleavage of the azo bond. More details can be found in the original paper. Reprinted with permission from Clarke *et al.* (2010).

to contribute substantially to sorption of hydrophobic organic compounds. Manganese oxides with their high oxidative capacity are the most likely minerals to oxidize PAHs. Despite this there are relatively few studies that have evaluated oxidation of non-polar polycyclic aromatic hydrocarbons (PAHs) by Mn oxides. Maximum oxidation of PAHs is achieved only when water-miscible organic solvents such as acetone are used in reactions to enhance PAH solubility (Eibes *et al.*, 2005). Oxidative breakdown of PAHs has been achieved by enzymatic peroxidases from white rot fungi *via* Mn(III)-organic complexes which serve as easily diffusible, non-specific oxidants (Eibes *et al.*, 2010; Field *et al.*, 1992; Wariishi *et al.*, 1992).

For mineral–PAH interactions, PAHs must first sorb onto the mineral surface prior to any oxidation, but very few studies have been conducted on the sorption of PAHs onto Mn oxide minerals. Villalobos *et al.* (2014) compared oxidation of anthracene, fluorene, phenanthrene and fluoranthene on acid birnessite and δ-MnO_2. Only anthracene and fluorene were oxidized, with both the degradation rate and extent of oxidation being greatest with δ-MnO_2 (Villalobos *et al.*, 2014). Phenanthrene and fluoroanthene did not react with either birnessite mineral, but sorption of both PAHs was observed on the δ-MnO_2. The lower sorption potential of acid birnessite was postulated to relate to the higher surface charge of the acid birnessite (due to increased layer vacancies) which results in a layer of water molecules creating a hydrophilic surface which is repulsive to hydrophobic PAHs. Villalobos *et al.* (2014) also suggested that the physical size of the PAH molecule directly affects the sorption of the molecule onto acid birnessite. These authors propose that small and linear PAHs such as anthracene and fluorene sorb more easily onto acid birnessite than the larger phenathrene and fluoranthene molecules. It is thought that the hydrated cations associated with the vacancy sites in acid birnessite make it difficult for larger PAH molecules to adsorb as PAHs are hydrophobic (Villalobos *et al.*, 2014). The larger PAH molecules are geometrically more likely to encounter hydrophilic site vacancies than the narrower PAH molecules meaning that smaller or more linear PAHs would be more likely to be adsorbed than larger PAHs. Similar steric factors have been highlighted as affecting the sorption of PAHs onto goethite surfaces, with linear molecules having a higher interaction energy than non-linear molecules (Tunega *et al.*, 2009). Villalobos *et al.* (2014) postulated that these steric factors could be the reason why pyrene does not react with birnessite (Jung *et al.*, 2008), despite it having a redox potential similar to anthracene which is oxidized readily by Mn oxides (Clarke *et al.*, 2012; Villalobos *et al.*, 2014). Even mechano-chemical techniques have not facilitated the oxidation of larger PAHs. Napola *et al.* (2006) observed no significant increase in the rate of phenanthrene oxidation when birnessite was added to spiked soil and Joseph-Ezra (2014) found Mn oxides to be ineffective in the mechano-chemical breakdown of pyrene.

Clarke *et al.* (2012) demonstrated the oxidation of anthracene to anthraquinone on a crystalline Mn oxide ore waste consisting predominantly of manganite (MnOOH). Oxidation of anthracene was favoured by decreasing the pH of the solution (Fig. 11). Removal of surface water, through the gentle evaporation of spiked Mn oxide-water

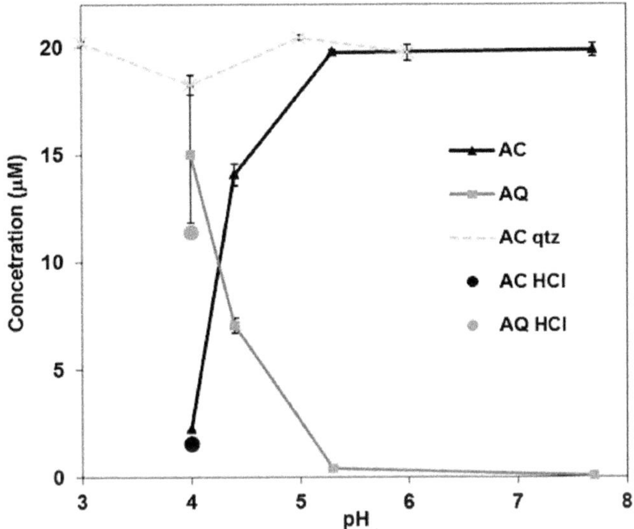

Figure 11. Anthracene (AC) and anthraquinone (AQ) concentrations after reacting anthracene with crystalline Mn oxide ore waste consisting predominantly of manganite (MnOOH) and a quartz control (AC qtz) in a series of pH-adjusted acetate buffers. Grey and black closed circles (AQ and AC HCl, respectively) represent samples reacted at pH 4 in DI using 0.1 M HCl to control pH. Reproduced from Clarke *et al.* (2012) with the permission of Elsevier.

slurries, strongly enhanced the oxidation of anthracene without pH adjustment. Both Clarke *et al.* (2012) and Villalobos *et al.* (2014) demonstrate the inhibiting effect of surface water in PAH–Mn oxide interactions.

Clarke *et al.* (2012) also observed anthracene oxidation under nitrogen purging, suggesting that the Mn oxides were not acting as a catalyst for air oxidation. Conversely, tunnel Mn oxides have been shown to act as catalysts for the air oxidation of 9H-Fluorene to 9-fluoronone (Opembe *et al.*, 2008). Opembe *et al.* (2008) found that tunnel Mn oxides did not directly oxidize 9H-Fluorene but were responsible for the homolytic decomposition of the intermediate hydroperoxides to yield the ketone product. However, the catalytic capacity of the Mn oxide was substantially reduced after the initial reaction.

Garcia *et al.* (2013) have recently suggested that oxidation of naphthalene by a range of Mn oxides (including birnessite) may take place *via* a Mars-van Krevelen mechanism as has been demonstrated for naphthalene oxidation using a cerium oxide catalyst (Solsona *et al.*, 2009) although this is unconfirmed. Briefly, the mechanism for naphthalene oxidation is thought to involve strong adsorption of the naphthalene at an anionic oxygen site on the oxide lattice leading to the formation of an activated complex followed by electron transfer and then by desorption of reaction products (up to and including CO_2) and a reaction of the reduced Mn oxide with oxygen to restore it to its initial state. Opembe *et al.* (2008) also highlighted the role of lattice oxygens in PAH transformations, but the ubiquity of water in natural systems may preclude these types of reactions occurring in soils.

Mn oxides also play an important role in immobilizing the biodegradation products of PAHs. Lee *et al.* (2009) showed that birnessite was essential in the irreversible fixing of the PAH biodegradation products. The mutant bacterial strain Sphingobiumyanoikuyae B8/36 converted phenathrene to cis-phenanthrenedihydrodiol in spiked soils (Lee *et al.*, 2009). In the presence of birnessite, this breakdown product was assimilated irreversibly into the soil organic matter through the oxidative coupling of radical species (Fig. 4, process 9) while in the absence of birnessite the breakdown product accumulated (Lee *et al.*, 2009).

4. Summary

Organically contaminated land is an international problem and is high on the agenda of environmental organizations in much of Europe and North America. The introduction of the EU Landfill Directive in 2006 has resulted in a move away from 'dig and dump' as a remediation strategy in Europe and there is increasing interest across the world in the development of sustainable remediation technologies which take into account environmental, economic and social factors.

Mn oxides can oxidatively transform many persistent organic pollutants and other emerging contaminants. Environmentally, the addition of Mn oxides to soil as a remediation strategy has great potential due to the 'slow release' nature of the oxidizing capacity and ability of Mn oxides to regenerate microbially in soils. However, there is very little research in this area. Economically, cheap sources of Mn oxides may be available including clean mine tailings produced in the South African Mn mining area, as well as Mn oxide-containing water treatment residuals from the clean water treatment industry which are available across the world. Consequently, sources of uncontaminated natural Mn oxides which may be suitable for cost-effective land remediation are attainable at local and international scales. Socially, the idea of using Mn oxides as potential soil amendments is likely to be uncontroversial, as Mn oxides are already considered a beneficial component of soil's defence against contamination. In addition, there is a vast body of literature documenting Mn oxides' ability to immobilize many potentially toxic inorganic elements giving them potential to become 'all-round' remediation soil amendments for so-called 'cocktail' (*i.e.* organic and inorganically contaminated) sites.

Manganese oxides can react with organic contaminants to form complexes between the manganese metal centre and electronegative functional groups (such as phenols) within the organic contaminant. Electron transfer from the manganese metal centre to the organic molecule results in the formation of an organic radical (such as a phenoxy radical). Further oxidation can result in the phenoxy radical being broken down into smaller more aliphatic molecules (up to and including to CO_2) or phenoxy radicals can undergo dimerization/polymerization to become larger more humic-like molecules. Perhaps the most unexploited future research area to come out of this review is the ability of Mn oxides to polymerize organic contaminants onto existing natural organic matter as well as enhancing humification of uncontaminated natural organic matter.

There is much potential for exploring Mn oxide catalysis of co-composting technologies of organic contaminants with uncontaminated organic waste streams. In addition, an important area for research is the role of Mn (III) in oxidizing (and reducing) organic contaminants as its prevalence in many natural environments has only recently been discovered (Madison *et al.*, 2013).

5. Future outlook

As highlighted in the review above, our understanding of various Mn oxide–organic interactions has been enhanced considerably in recent times. The challenge going forward is to investigate these interactions in more complex and heterogeneous systems such as soils. Most studies mentioned in this review consist of 'clean' model systems and limited information is available for the capacity for these reactions under natural pedogenic conditions.

In reality, soils often harbour multiple rather than single organic contaminants and NOM constituents; competition studies are required to fully understand contaminant transformations in multi-element POP and POP-like organic and inorganically contaminated soils. Further research is needed to understand the viability of amending organically contaminated soils with compostable material in conjunction with Mn oxides in order to enhance humification rates and simultaneously lock up organic contaminants in humic substances.

Field trials are essential to determine the feasibility of Mn oxides as soil amendments. There is a complete lack of information on the use of Mn oxides (either natural or synthetic) in historically (and heterogeneously) organically contaminated soils. The only paper we have found reporting the addition of Mn oxide to real soils is by Huang *et al.* (2011) which discussed the use of microwave radiation in combination with added Mn oxides to break down historically PCB-contaminated soils. Field-based research is fundamental to understanding the role Mn oxides inherently play in organic-contaminant transformations in real soils. In addition, if Mn oxides are to be considered as *in situ* amendments the impacts of their addition upon soil biological receptors must be assessed ensuring no negative side effects upon essential ecosystem functions and services. For example, the potential for Mn(II) to build-up in the soil solution of organically contaminated soils needs investigation. In natural soils Mn oxide cycling is mediated by microbial populations. Future research should seek to address biotic, as well as biotic–abiotic, interactions between organic contaminants and Mn oxides. This is pertinent because microbial communities will ultimately control the long-term sustainability of Mn oxide amendments, *via* biological regeneration and precipitation.

Finally, there is need to obtain a better understanding of the role that Mn(III)-organic complexes play in organic-contaminant transformations. The higher redox potential and solubility of these complexes makes them an appealing oxidizing agent for POP degradation.

Acknowledgements

The authors thank Karen Hudson-Edwards, Imad Ahmed and an anonymous reviewer for their helpful comments on this chapter.

References

Abrahams, P.W. (2002) Soils: their implications to human health. *Science of the Total Environment*, **291**, 1–32.

Adonis, M., Martínez, V., Riquelme, R., Ancic, P., González, G., Tapia, R., Castro, M., Lucas, D., Berthou, F. and Gil, L. (2003) Susceptibility and exposure biomarkers in people exposed to PAHs from diesel exhaust. *Toxicology Letters*, **144**, 3–15.

Agarwal, T. and Bucheli, T.D. (2011) Is black carbon a better predictor of polycyclic aromatic hydrocarbon distribution in soils than total organic carbon? *Environmental Pollution*, **159**, 64–70.

Andrews, J.E., Brimblecombe, P., Jickells, T.D., Liss, P.S. and Reid, B.J. (2004) *An Introduction to Environmental Geochemistry, 2nd Edition.* Blackwell Publishing, Padstow, Cornwall, UK.

Antizar-Ladislao, B. (2010) Bioremediation: working with bacteria. *Elements*,**6**, 389–394.

ATSDR (2005) Toxicology profile for polyaromatic hydrocarbons. Agency for Toxic Substances and Disease Registry, Atlanta, Georgia, USA. Available at: http://www.atsdr.cdc.gov/substances/toxsubstance.asp?toxid=25

ATSDR (2015) Toxicological profile for chlorophenols. Available at: http://www.atsdr.cdc.gov/toxprofiles/tp107.pdf

Bambra, C., Robertson, S., Kasim, A., Smith, J., Cairns-Nagi, J.M., Copeland, A., Finlay, N. and Johnson, K.L. (2014) Healthy land? An examination of the area-level association between brownfield land and morbidity and mortality in England. *Environment and Planning A*, **46**, 433–454.

Banerjee, D. and Nesbitt, H.W. (1999) XPS study of reductive dissolution of birnessite by oxalate: rates and mechanistic aspects of dissolution and redox processes. *Geochimica et Cosmochimica Acta*, **63**, 3025–3038.

Banerjee, D. and Nesbitt, H.W. (2001) XPS study of reductive dissolution of birnessite by humate with constraints on reaction mechanism. *Geochimica et Cosmochimica Acta*, **65**, 1703–1714.

Bandara, J., Mielczarski, J.A. and Kiwi, J. (1999) 1. Molecular mechanism of surface recognition. Azo dyes degradation on Fe, Ti, and Al oxides through metal sulfonate complexes. *Langmuir*, **15**, 7670–7679.

Bardos, P., Bone, B., Boyle, R., Ellis, D., Evans, F., Harries, N.D. and Smith, J.W.N. (2011) *Applying Sustainable Development Principles to Contaminated Land Management using the SuRF-UK Framework.* Remediation Spring 2011.

Barrett, K.A. and McBride, M.B. (2005) Oxidative degradation of glyphosate and aminomethyl phosphonate by manganese oxide. *Environmental Science & Technology*, **39**, 9223–9228.

Beesley L., Moreno-Jimenez, E., Gomez-Eyles, J.L., Harris, E., Robinson, B. and Sizmur, T. (2011) A review of biochars' potential role in the remediation, revegetation and restoration of contaminated soils. *Environmental Pollution*, **159**, 3269–3282.

Bojes, H.K. and Pope, P.G. (2007) Characterization of EPA's 16 priority pollutant polycyclic aromatic hydrocarbons (PAHs) in tank bottom solids and associated contaminated soils at oil exploration and production sites in Texas. *Regulatory Toxicology and Pharmacology*, **47**, 288–295.

Bolan, N., Kunhikrishnan, A., Thangarajan, R., Kumpiene, J., Park, J., Makino T., Kirkham M.B., and Scheckel, K. (2014) Remediation of heavy metal(loid)s contaminated soils – To mobilize or to immobilize? *Journal of Hazardous Materials*, **266**,141–166.

Brunetti, G., Senesi, N. and Plaza, C. (2007) Effects of amendment with treated and untreated olive soil mill wastewaters on soil properties, soil humic substances and wheat yield. *Geoderma*, **138**, 144–152.

Burns, R.G and Burns, V.M. (1977) Mineralogy and crystal-chemistry of deep-sea manganese nodules, a polymetallic resource of the 21st century. *Philosophical Transactions of the Royal Society A – Maths, Physics and Engineering Sciences*, **286**, 283–301.

Casas, M., Chevrier, C., Den Hond, E., Fernandez, M., Pierik, F., Philippat, C., Slama, R., Toft, G., Vandentorren, S., Wilhelm, M. and Vrijheid, M. (2013) Exposure to brominated flame retardants, perfluorinated compounds, phthalates and phenols in European birth cohorts: ENRICO evaluation, first human biomonitoring results, and recommendation. *International Journal of Hygiene and Environmental Health* **216**, 230–242.

Chakrabarti, S., Dutta, B. and Apak, R. (2009) Active manganese oxide: a novel adsorbent for treatment of wastewater containing azo dye. *Water, Science and Technology*, **60**, 3017–3024.

Chang, H.M. and Allan, G. (1971) Oxidation. Pp. 433–485 in: *Lignins* (K. Sarkanen and C. Ludwig, editors). Wiley Interscience, New York.

Chang Chien, S.W., Chen, H.L., Wang, M.C. and Seshaiah, K. (2009) Oxidative degradation and associated mineralization of catechol, hydroquinone and resorcinol catalyzed by birnessite. *Chemosphere*, **74**, 1125–1133.

Chang Chien, S.W., Chang, C.H., Chen, S.H., Wang, M.C., Rao, M.M. and Veni, S.S. (2011) Oxidative degradation of pyrene in contaminated soils by delta-MnO_2 with or without sunlight irradiation. *Science of the Total Environment*, **409**, 4078–4086.

Chen, W.R., Ding, Y, Johnston, G.T., Teppen, B.J., Boyd, S.A. and Li, H. (2010) Reaction of lincosamide antibiotics with manganese dioxide in aqueous solution. *Environmental Science & Technology*, **44**, 4486–4492.

Chen, W.R., Lui, C., Boyd, S.A., Teppen, B.J. and Li, H. (2013) Reduction of carbadox mediated by reaction of Mn(III) with oxalic acid. *Environmental Science & Technology*, **47**, 1357–1364.

Cheney, M.A. Sposito, G., McGrath, A.E. and Criddle, R.S. (1996) Abiotic degradation of 2,4-D (dichlorophenoxyacetic acid) on synthetic birnessite: a calorespirometric method. *Colloids and Surfaces A*, 107,131–140.

Chorover, J. and Amistadi, M.K. (2001) Reaction of forest floor organic matter at goethite, birnessite and smectite surfaces. *Geochimica et Cosmochimica Acta*, **65**, 95–109.

Chorover, J., Amistadi, M.K and Chadwick, O.A. (2004) Surface charge evolution of mineral–organic complexes during pedogenesis in Hawaiin basalt. *Geochimica et Cosmochimica Acta*, **68**, 4859–4876.

Clarke, C.E., and Johnson, K.L. (2010) Oxidative breakdown of acid orange 7 by a manganese oxide containing mine waste: insight into sorption, kinetics and reaction dynamics. *Applied Catalysis B: Environmental*, **101**, 13–20.

Clarke, C.E., Kielar, F., Talbot, H.M. and Johnson, K.L. (2010) Oxidative decolorization of acid azo dyes by a Mn oxide containing waste. *Environmental Science and Technology*, **44**, 1116–1122.

Clarke, C.E., Tourney, J. and Johnson, K.L. (2012) Oxidation of anthracene using waste Mn oxide minerals: the importance of wetting and drying sequences. *Journal of Hazardous Materials*, **205-206**, 126–130.

Clarke, C.E., Kielar, F. and Johnson, K.L. (2013) The oxidation of acid azo dye AY 36 by a manganese oxide containing mine waste. *Journal of Hazardous Materials*, **246-247**, 310–318.

Cousins, I.T., Gevao, B. and Jones, K.C. (1999) Measuring and modelling the vertical distribution of semi-volatile organic compounds in soils. I: PCB and PAH soil core data. *Chemosphere*, **39**, 2507–2518.

de Rudder, J., Wiele, T., Dhouge, W., Comhaire, F. and Verstraete, W. (2004) Advanced water treatment with manganese oxide for the removal of 17 alpha-ethynylestradiol (EE2). *Water Research*, **38**, 184–192.

de Wit, C.A. (2002) An overview of the brominated flame retardants in the environment. *Chemosphere*, **46**, 583–614.

Dingemans, M.M.L., de Groot, A., van Kleef, R.G., Bergman, A., van den Berg, M., Vijverberg, H.P. and Westerink, R.H. (2008) Hydroxylation increases the neurotoxic potential of BDE-47 to affect exocytosis and calcium homeostasis in PC12 cells. *Environmental Health Perspectives*, **116**, 637–643.

Dismukes, G.C. (1986) The metal centres of the photosynthetic oxygen-evolving complex. *Photochemistry and Photobiology*, **43**, 99–115.

Dohse, D.M. and Lion, L.W. (1994) Effect of microbial polymers on the sorption and transport of phenanthrene in a low-carbon sand. *Environmental Science & Technology*, **28**, 541–548.

Dong, J., Zhang, L., Liu, H., Liu, C., Gao, Y and Sun, L. (2010) The oxidative degradation of 2-mercaptobenzothiazole. *Fresensius Environmental Bulletin*, **19**, 1615–1622.

Dowding, C.E., M.J. Borda, M.V. Fey and D. L. Sparks. 2005. A new method for gaining insight into the chemistry of drying mineral surfaces using ATR-FTIR. *Journal of Colloid and Interface Science*, **292**, 148–151.

Eibes, G., Lu-Chau, T., Feijoo, G., Moreira, M.T. and Lema, J.M. (2005) Complete degradation of anthracene by manganese peroxidase in organic solvent mixtures. *Enzyme and Microbial Technology*, **37**, 365–372.

Eibes, G., McCann, C., Pedezert, A., Moreira M.T., Feijoo, G. and Lema, J.M. (2010) Study of mass transfer and biocatalyst stability for the enzymatic degradation of anthracene in a two-phase partitioning bioreactor. *Biochemical Engineering Journal*, **51**, 79–85.

European Environment Agency (2010) The European environment state and outlook 2010 – Soil. Available at: http://www.eea.europa.eu/soer/europe/soil?b_start:int=12.

Field, J.A., Dejong, E., Costa, G.F. and Debont, J.A.M. (1992) Biodegradation of polycyclic aromatic-hydrocarbons by new isolates of white rot fungi. *Applied Environmental Microbiology*, **58**, 2219–2226.

Flemming, C.H., Wingender, J., Mortiz, R., Werner, B. and Mayer, C. (1999) Physico-chemical properties of biofilms – A short review. Pp. 1–24 in: *Biofilms in the Aquatic Environment.* (C.W. Keevil, A.G.D. Holt and C. Dow, editors). Royal Society of Chemistry, Cambridge, UK.

Frank, H.S. and Evans, M.W. (1945) Free volume and entropy in condensed systems. 3. Entropy in binary liquid mixtures – partial molal entropy in dilute solutions – structure and thermodynamics in aqueous electrolytes. *Journal of Chemical Physics*, **13**, 507–532.

Froelich, P.N., Klinkhammer, G.P., Bender, M.L., Luedtke, N.A., Heath, G.R., Cullen, D., Dauphin, P., Hammond, D., Hartman, B. and Maynard, V. (1979) Early diagenesis of organic matter in pelagic sediments of the eastern equatorial Atlantic: suboxic diagenesis. *Geochimica et Cosmochimica Acta*, **43**, 1075–1090.

Fu, W., Fu, H., Skøtt, K. and Yang, M. (2008) Modelling the spill in the Songhua River after the explosion in the petrochemical plant in Jilin. *Environmental Science and Pollution Research*, **15**, 178–181.

Gao, N., Hong, J., Yu, Z. and Huang, W. (2011) Transformation of Bisphenol A in the presence of manganese dioxide. *Soil Science*, **176**, 265–272.

Garcia, T., Sellick, D., Varda, F., Vazquez, I., Dejoz, A., Agouram, S., Taylor, S.H. and Solsona, B. (2013) Total oxidation of napthalene using bulk manganese oxide catalysts. *Applied Catalysis A: General*, **450**, 169–177.

Gasparatos, D. (2013) Sequestration of heavy metals from soil with Fe–Mn concretions and nodules. *Environmental Chemistry Letters*, **11**, 1–9.

Ge, J.T. and Qu, J.H. (2003) Degradation of azo dye acid red B on manganese dioxide in the absence and presence of ultrasonic irradiation. *Journal of Hazardous Materials*, **100**, 197–207.

Gianfreda, L., Iamarino, G. Scelza, R. and Rao, M.A. (2006) Oxidative catalysts for the transformation of phenolic pollutants: a brief review. *Biocatalysis and Biotransformation*, **24**, 177–187.

Gilkes, R.J. and McKenzie, R.M. (1988) Geochemistry and mineralogy of manganese in soils. Pp. 23–35 in: *Manganese in Soils and Plants* (R. Graham and Uren, N, editors). Springer, The Netherlands.

Haack, E. A. and Warren, L.A. (2003) Biofilm hydrous manganese oxyhydroxides and metal dynamics in acid rock drainage. *Environmental Science & Technology*, **37**, 4138–4147.

Hamby, D.M. (1996) Site remediation techniques supporting environmental restoration activities – A review. *Science of the Total Environment*, **191**, 203–224.

Hardie, A.G., Dynes, J.J., Kozak, L.M. and Huang, P.M. (2009) The role of glucose in abiotic humification pathways as catalysed by birnessite. *Journal of Molecular Catalysis A: Chemical*, **308**, 114–126.

Haritash, A.K. and Kaushik, C.P. (2009) Biodegradation aspects of polycyclic aromatic hydrocarbons (PAHs): A review. *Journal of Hazardous Materials*, **169**, 1–15.

Hattenschwiler, S. and Vitousek, P.M. (2000) The role of polyphenols in terrestrial ecosystem nutrient cycling. *Trends in Ecology and Evolution*, **15**, 238–243.

He, Y., Xu, T., Zhang, Y., Guo, C., Li, L. and Wang, Y. (2012) Oxidative transformations of carbamazepine by manganese oxides. *Environmental Science Pollution Research*, **19**, 4206–4213.

Hickman, Z.A. and Reid, B.J. (2008) Increased microbial catabolic activity in diesel contaminated soil following addition of earthworms and compost. *Soil Biology and Biochemistry*, **40**, 2970–2976.

Hickey, W.J. (1999) Transformation and fate of polychlorinated biphenyls in soils and sediments. Pp. 213–232 in: *Bioremediation in Contaminated Soils* (D.C. Adriano, J.-M. Bollag, W.T. Frankenberger and W.T. Sims, editors). Agronomy Monograph no. **37**, American Agronomy Society, Madison, Wisconsin, USA.

Hites, R.A. (2004) PBDEs in the environment and in people: a meta-analysis of concentrations. *Environmental Science & Technology*, **38**, 945–956.
Huang, G., Zhao, L., Dong, Y. and Zhang, Q. (2011) Remediation of soils contaminated with polychlorinated biphenyls by microwave-irradiated manganese dioxide. *Journal of Hazardous Materials*, **186**, 128–132.
Huang, P.M. (2000) Abiotic catalysis. Pp. B303–332 in: *Handbook of Soil Science* (M.E. Sumner, editor). CRC Press, Boca Raton, Florida, USA.
Huang, P.M. and Hardie, A.G. (2011) Role of abiotic catalysis in the transformation of organics, metals, metalloids, and other inorganics. Pp. 1–29 in: *Handbook of Soil Sciences, 2nd Edition* (P.M. Huang, Y. Li, and M.E. Sumner, editors). CRC Press, Boca Raton, Florida, USA.
Huerta, B., Rodríguez-Mozaz, S. and Barceló, D. (2012) Pharmaceuticals in biota in the aquatic environment: Analytical methods and environmental implications. *Analytical and Bioanalytical Chemistry*, **404**, 2611–2624.
Huguet, M., Deborde, M., Papot, S. and Gallard, H. (2013) Oxidative decarboxylation of diclofenac by manganese oxide filter bed. *Water Research*, **47**, 5400–5408.
Jennings, A.A. (2012) Worldwide regulatory guidance values for surface soil exposure to carcinogenic or mutagenic polycyclic aromatic hydrocarbons. *Journal of Environmental Management*, **110**, 82–102.
Johnsen, A.R. and Karlson, U. (2004) Evaluation of bacterial strategies to promote the bioavailability of polycyclic aromatic hydrocarbons. *Applied Microbiology and Biotechnology*, **63**, 452–459.
Jokic, A., Frenkel, A.I., Vairavamurthy, M.A. and Huang, P.M. (2001) Birnessite catalysis of the Maillard reaction: Its significance in natural humification. *Geophysical Research Letters*, **28**, 3899–3902.
Jones, D.L. and Healey, J.R. (2010) Organic amendments for remediation: putting waste to good use. *Elements*, **6**, 369–374.
Jones, K.C. and de Voogt, P. (1999) Persistent organic pollutants (POPs): state of the science. *Environmental Pollution*, **100**, 209–221.
Joseph-Ezra, H., Nasser, A., Ben-Ari, J. and Mingelgrin, U. (2014) Mechanochemically enhanced degradation of pyrene and phenanthrene loaded on magnetite. *Environmental Science & Technology*, **48**, 5876–5882.
Jung, J.-W., Lee, S., Ryu, H., Nam, K. and Kang, K.-H. (2008) Enhanced reactivity of hydroxylated polycyclic aromatic hydrocarbons to birnessite in soil: Reaction kinetics and non-extractable residue formation. *Environmental Toxicology and Chemistry*, **27**, 1031–1038.
Jurado, A., Vàzquez-Suñé, E., Carrera, J., López de Alda, M., Pujades, E. and Barceló, D. (2012) Emerging organic contaminants in groundwater in Spain: A review of sources, recent occurrence and fate in a European context. *Science of the Total Environment*, **440**, 82–94.
Kang, K.H., Dec, J., Park, H. and Bollag, J.-M. (2004) Effect of phenolic mediators and humic acid on cyprodinil transformation in the presence of birnessite. *Water Research*, **38**, 2737–2745.
Kinney, C.A., Furlong, E.T., Zaugg, S.D., Burkhardt, M.R., Werner, S.L., Cahill, J.D. and Jorgensen, G.R. (2006) Survey of organic wastewater contaminants in biosolids destined for land application. *Environmental Science & Technology*, **40**, 7207–7215.
Klausen, J., Haderlein, S.B. and Schwarzenbach, R.P. (1997) Oxidation of substituted anilines by aqueous MnO_2: Effect of co-solutes on initial and quasi steady-state kinetics. *Environmental Science & Technology*, **31**, 2642–2649.
Laha, S. and Luthy, R.G. (1990) Oxidation of aniline and other primary aromatic-amines by manganese-dioxide. *Environmental Science & Technology*, **24**, 363–373.
Lallai, M. and Mura, G. (2004) Biodegradation of 2-chlorophenol in forest soil: effect of inoculation with aerobic sewage sludge. *Environmental Toxicology and Chemistry*, **23**, 325–330.
Lamaita L., Peluso, M.A., Sambeth, J.E. and Thomas, H.J. (2005) Synthesis and characterisation of manganese oxides employed in VOCs abatement. *Applied Catalysis B: Environmental*, **61**, 113–119.
Lapworth, D.J., Baran, N., Stuart, M.E. and Ward, R.S. (2012) Emerging organic contaminants in groundwater: A review of sources, fate and occurrence. *Environmental Pollution*, **163**, 287–303.
Lashermes, G., Houot, S. and Barriuso, E. (2010) Sorption and mineralisation of organic pollutants during different stages of composting. *Chemosphere*, **79**, 455–462.
Lee, S., Ryu, H. and Nam, K. (2009) Phenanthrene metabolites bound to soil organic matter by birnessite following partial biodegradation. *Environmental Toxicology and Chemistry*, **28**, 946–952.

Levy, S.B. and Marshall, B. (2004) Antibacterial resistance worldwide: causes, challenges and responses. *Nature Medicine*, **10**, S122–S129.

Li, C., Zhang, B., Ertunc, T., Schaeffer, A. and Ji, R. (2012) Birnessite-induced binding of phenolic monomers to soil humic substances and nature of the bound residue. *Environmental Science & Technology*, **46**, 8843–8850.

Li, F., Liu, C., Liang, C., Li, X. and Zhang, L. (2008) The oxidative degradation of 2-mercaptobenzothiazole at the interface of beta-Mn dioxide and water. *Journal of Hazardous Materials*, **154**, 1098–1105.

Li, H., Lee, L.S., Schulze, D.G. and Guest, C.A. (2003) Role of soil manganese in the oxidation of aromatic amines. *Environmental Science & Technology*, **37**, 2686–2693.

Lin, K., Liu, W. and Gan, J. (2009) Oxidative removal of bisphenol A by manganese dioxide: Efficacy, products and pathways. *Environmental Science & Technology*, **43**, 3860–3864

Lin, K., Peng, Y., Huang, X. and Ding, J. (2013) Transformation of bisphenol A by manganese-oxide coated sand. *Environmental Science and Pollution Research*, **20**, 1461–1467.

Lin, K., Yan, C. and Gan, J. (2014) Production of hydroxylated polybrominated diphenyl ethers (OH-PBDEs) from bromophenols by manganese dioxide. *Environmental Science & Technology*, **48**, 263–267.

Liu, C., Zhang, L., Li, F., Wang, Y., Gao, Y., Li, X., Cao, W., Feng, C. Dong, J. and Sun, L. (2009) Dependence of sulfadiazine oxidative degradation on physicochemical properties of manganese dioxides. *Industrial &and Engineering Chemistry Research*, **48**, 10408–10413.

Liu, H.J. and Qu, J.H. (2002) Decolorization of reactive bright red K2G dye: electrochemical process catalyzed by manganese mineral. *Water Science and Technology*, **46**, 133–138.

Liu, R.X. and Tang, H.X. (2000) Oxidative decolorization of direct light red F3B dye at natural manganese mineral surface. *Water Research*, **34**, 4029–4035.

Lohmann, R., Breivik, K., Dachs, J. and Muir, D. (2007) Global fate of POPs: Current and future research directions. *Environmental Pollution*, **150**, 150–165.

Lovley, D. (1991) Dissimilatory Fe(III) and Mn(IV) reduction. *Microbiological Reviews*, **55**, 259–287.

Lu, Z. and Gan, J. (2013) Oxidation of nonylphenol and octylphenol by manganese dioxide: kinetics and pathways. *Environmental Pollution*, **180**, 214–220.

Lu, Z., Lin, K. and Gan, J. (2011) Oxidation of bisphenol F (BPF) by manganese dioxide. *Environmental Pollution*, **159**, 2546–2551.

Luo, Y., Guo, W., Ngo, H.H., Nghiem, L.D., Hai, F.I., Zhang, J., Liang, S. and Wang, X.C. (2014) A review on the occurrence of micropollutants in the aquatic environment and their fate and removal during wastewater treatment. *Science of the Total Environment*, **473-474**, 619–641.

Luthy, R.G., Aiken, G.R., Brusseau, M.L., Cunningham, S.D., Gschwend, P.M., Pignatello, J.J., Reinhard, M., Traina, S.J., Weber Jr, W.J. and Westall, J.C. (1997) Sequestration of hydrophobic organic contaminants by geosorbents. *Environmental Science & Technology*, **31**, 3341–3347.

Madison, A.S., Tebo, B.M., Mucci, A., Sundby, B. and Luther, G.W. (2013) Abundant porewater Mn(III) as a major component of the sedimentary redox system. *Science*, **341**, 875–878.

Maoz, A. and Chefetz, B. (2010) Sorption of pharmaceuticals carbamazepine and naproxen to dissolved organic matter: Role of structural fractions. *Water Research*, **44**, 981–989.

Marchal, G., Smith, K.E.C., Rein, A., Winding, A., Trapp, S. and Karlson, U.G. (2013) Comparing the desorption and biodegradation of low concentrations for phenanthrene sorbed to activated carbon, biochar and compost. *Chemosphere*, **90**, 1767–1778.

McBride, M.B. (1994) *Environmental Chemistry of Soils*. Oxford University Press, New York.

Means, J.C. (1995) Influence of salinity upon sediment–water partitioning of aromatic hydrocarbons. *Marine Chemistry*, **51**, 3–16.

Megharaj, M., Ramakrishnan, B., Venkateswarlu, K., Sethunathan, N. and Naidu, R. (2011) Bioremediation approaches for organic pollutants: A critical perspective. *Environment International*, **37**, 1362–1375.

Meijer, S.N., Ockenden, W.A., Sweetman, A., Breivik, K., Grimalt, J.O. and Jones, K.C. (2003) Global distribution and budget of PCBs and HCB in background surface soils: Implications for sources and environmental processes. *Environmental Science & Technology*, **37**, 667–672.

Menzie, C.A., Potocki, B.B. and Santodonato, J. (1992) Ambient concentrations and exposure to carcinogenic PAHs in the environment. *Environmental Science & Technology*, **26**, 1278–1284.

Milly, P.C.D., Wetherald, R.T., Dunne, K.A. and Delworth, T.L. (2002) Increasing risk of great floods in a changing climate. *Nature*, **415**, 514–517.

Morrison, S., Fordyce, F.M. and Scott, M. (2014) An initial assessment of spatial relationships between respiratory cases, soil metal content, air quality and deprivation indicators in Glasgow, Scotland, UK: relevance to the environmental justice agenda. *Environmental Geochemistry and Health*, **36**, 319–332.

Mortland, M.M. and Raman, K.V. (1968) Surface acidity of smectites in relation to hydration, exchangeable cation and structure. *Clays and Clay Minerals*, **16**, 393–398.

Murphy, E.M., Zachara, J.M. and Smith, S.C. (1990) Influence of mineral-bound humic substances on the sorption of hydrophobic organic-compounds. *Environmental Science & Technology*, **24**, 1507–1516.

Nam, J.J., Thomas, G.O., Jaward, F.M., Steinnes, E., Gustafsson, O. and Jones, K.C. (2008) PAHs in background soils from Western Europe: Influence of atmospheric deposition and soil organic matter. *Chemosphere*, **70**, 1596–1602.

Napola, A., Pizzigallo, M.D.R., Di Leo, P., Spagnuolo, M. and Ruggiero, P. (2006) Mechanochemical approach to remove phenanthrene from a contaminated soil. *Chemosphere*, **65**, 1583–1590.

Nasser, A., Sposito, G. and Cheney, M.A. (2000) Mechanochemical degradation of 2,4-D adsorbed on synthetic birnessite. *Colloids and Surfaces A: Physicochemical and Engineering Aspects*, **163**, 117–123.

Ndjeri, M., Pensel, A., Peulon, S., Haldys, V., Desmazieres, B. and Chausse, A. (2013) Degradation of glyphosate and AMPA (amino methylphosphonic acid) solutions by thin films of birnessite electrodeposited: A new design of material for remediation processes. *Colloids and Surface Analysis*, **435**, 154–169.

O'Reilly, S.E and Hochella, M.F. (2003) Lead sorption efficiencies of natural and synthetic Mn and Fe oxides. *Geochimica et Cosmochimica Acta*, **67**, 4471–4487.

Olaniran, A. and Igbinosa, E.O. (2011) Chlorophenols and other related derivatives of environmental concern: Properties, distribution and microbial degradation processes. *Chemosphere*, **83**, 1297–1306.

Opembe, N.N., Son, Y.-C., Sriskandakumar, T. and Suib, S.L. (2008) Kinetics and mechanism of 9H-fluorene oxidation catalyzed by manganese oxide octahedral molecular sieves. *ChemSusChem*, **1**, 182–185.

Parikh, S.J., Chorover, J. and Burgos, W.D. (2004) Interaction of phenanthrene and its primary metabolite (1-hydroxy-2-napthoic acid) with estuarine sediments and humic fractions. *Journal of Contaminant Hydrology*, **72**, 1–22.

Park, J.W., Dec, J., Jim, J.-E. and Bollag, J.M. (1999) Transformation of chlorophenols enhanced by the addition of syringaldehyde. *Environmental Science & Technology*, **33**, 2028–2034.

Park, J.H., Lamb, D., Paneerselvam, P., Choppala, G., Bolan, N. and Chung, J-W. (2011) Role of organic amendments on enhanced bioremediation of heavy metal(loid) contaminated soils. *Journal of Hazardous Materials*, **185**, 549–574.

Patnaik, P. (1992) *Health hazard, which includes toxic, corrosive, carcinogenic, and teratogenic properties, exposure limits. A Comprehensive Guide to the Hazardous Properties of Chemical Substances*. John Wiley & Sons, USA.

Peacock, C.L. and Sherman, D.M. (2007) Sorption of Ni by birnessite: Equilibrium controls on Ni in seawater. *Chemical Geology*, **238**, 94–106.

Peng, R.H., Xiong, A.S., Xue, Y., Fu, X.Y., Gao, F., Zhao, W., Tian, Y.S. and Yao, Q.H. (2008) Microbial biodegradation of polyaromatic hydrocarbons. *FEMS Microbiology Reviews*, **32**, 927–955.

Petrie, R.A., Grossl, P.R. and Sims, R.C. (2002) Oxidation of pentachlorophenol in manganese oxide suspensions under controlled Eh and pH environments. *Environmental Science & Technology*, **36**, 3744–3748.

Pizzigallo, M.D.R., Ruggiero, P., Crecchio, C. and Mininni, G. (1995) Manganese and iron oxides as reactants for oxidation of chlorophenols. *Soil Science Society of America Journal*, **59**, 444–452.

Pizzigallo, M.D.R., Napola, M., Spagnuolo, M. and Ruggiero, P. (2004) Mechanochemical removal or organo-chlorinated compounds by inorganic components of soil.*Chemosphere*, **55**, 1485–1492.

Polubesova, T. and Chefetz, B. (2014) DOM-affected transformation of contaminants on mineral surfaces: A review. *Critical Reviews in Environmental Science and Technology*, **44**, 223–254.

Polzer, F., Wunder, S., Lu, Y. and Ballauf, M. (2012) Oxidation of an organic dye catalysed by MnOx nanoparticles. *Journal of Catalysis*, **289**, 80–87.

Post, J.E. (1999) Mn oxide minerals: crystal structures, economic and environmental significance. *Proceedings of the National Academy of Sciences of the United States of America*, **96**, 3447–3454.
Prasad, B.B. and Pandey, U.C. (2000) Separation and preconcentration of copper and cadmium ions from multielemental solutions using Nostoc muscorum-based biosorbents. *World Journal of Microbiology and Biotechnology*, **16**, 819–827.
Prasad, B.B. and Pandey, U. C. (2001) Separation and preconcentration of lead from aquatic environment using Microcystis-based biosorbent. *Environmental Technology*, **22**, 771–780.
Pugilisi, E., Vernile, P., Bari, G., Spagnuolo, M., Trevisan, M., de Lillo, E. and Ruggiero, P. (2009) Bioaccessibility, bioavailability and ecotoxicity of pentachlorophenol in compost amended soils. *Chemosphere*, **77**, 80–86.
Purohit, V. and Basu, A.K. (2000) Mutagenicity of nitroaromatic compounds. *Chemical Research in Toxicology*, **13**, 673–692.
Quintelas, C. and Tavares, T. (2001) Removal of chromium(VI) and cadmium(II) from aqueous solution by a bacterial biofilm supported on granular activated carbon. *Biotechnology Letters*, **23**, 1349–1353.
Rao, M.A., Iamarino, G., Scelza, R., Russo, F. and Gianfreda, L. (2008) Oxidative transformation of aqueous phenolic mixtures by birnessite-mediated catalysis. *Science of the Total Environment*, **407**, 438–446.
Remucal, C.K. and Ginder-Vogel, M. (2014) A critical review of the reactivity of manganese oxides with organic contaminants. *Environmental Science Processes & Impact*, **16**, 1247.
Rubert, K.F. and Pederson, J.A. (2006) Kinetics of oxytetracycline reaction with a hydrous manganese oxide. *Environmental Science & Technology*, **40**, 7216–7223.
Sayara, T., Borràs, E., Caminal, G., Sarrà, M. and Sánchez, A. (2011) Bioremediation of PAHs-contaminated soil through composting: Influence of bioaugmentation and biostimulation on contaminant biodegradation. *International Biodeterioration and Biodegradation*, **65**, 859–865.
Schwarzenbach, R.P. and Westall, J. (1981) Transport of nonpolar organic compounds from surface water to groundwater. Laboratory sorption studies. *Environmental Science & Technology*, **15**, 1360–1367.
Scullion, J. (2006) Remediating polluted soils. *Naturwissenschaften*, **93**, 51–65.
Semple, K.T., Reid, B.J. and Fermor, T.R. (2001) Impact of composting strategies on the treatment of soils contaminated with organic pollutants. *Environmental Pollution*, 1**12**, 269–283.
Shen, Y.F., Zerger, R.P., Deguzma, R.N., Suib, S.L., McCurdy, L., Potter, D.I. and Oyoung, C.L. (1993) Manganese oxide octahedral molecular sieves - preparation, characterisation and applications. *Science*, **260**, 511–515.
Shin, H., Lim, D., Lee, D. and Kang, K. (2009) Reaction kinetics and transformation products of 1-napthol by Mn oxide-mediated oxidative coupling reactions. *Journal of Hazardous Materials*, **165**, 540–547.
Shin, J.Y., Buzgo, C.M. and Cheney, M.A. (2000) Mechanochemical degradation of atrazine adsorbed on four synthetic manganese oxides. *Colloids and Surfaces A: Physicochemical and Engineering Aspects*, **172**, 113–123.
Shindo, H. and Huang, P.M. (1982) Role of Mn(IV) oxide in abiotic formation of humic substances in the environment. *Nature*, **298**, 363–365.
Solsona, B., García, T., Murillo, R., Mastral, A., Ntainjua Ndifor, E., Hetrick, C., Amiridis, M. and Taylor, S. (2009) Ceria and Gold/Ceria catalysts for the abatement of poly-cyclic aromatic hydrocarbons: An in-situ DRIFTS study. *Topics in Catalysis*, **52**, 492–500.
Sparks, D.L. (2003) *Environmental Soil Chemistry, 2nd edition*. Academic Press, San Diego, California, USA.
Stevenson, F.J. (1994) *Humus Chemistry: Genesis, Composition, Reactions*. John Wiley and Sons, New York.
Stone, A.T. (1987a) Adsorption of organic reductants and subsequent electron transfer on metal oxide surfaces. Pp. 446–461 in: *Geochemical Processes at Mineral Surfaces* (A.D. James and F.H. Kim, editors). American Chemical Society.
Stone, A.T. (1987b) Reductive dissolution of manganese(III/IV) oxides by substituted phenols. *Environmental Science & Technology*, **21**, 979–988.
Stone, A.T. (1991) Oxidation and hydrolysis of ionizable organic pollutants at hydrous metal oxide surfaces. *Rates of Soil Chemical Processes*. Soil Science Society of America Special Publication no. **27**, pp. 231–254.
Stone, A.T. and Morgan, J.J. (1984a) Reduction and dissolution of manganese(III) and manganese(IV) oxides by

organics. 1. Reaction with hydroquinone. *Environmental Science & Technology*, **18**, 450–456.
Stone, A.T. and Morgan, J.J. (1984b) Reduction and dissolution of manganese(III) and manganese(IV) oxides by organics: 2. Survey of the reactivity of organics. *Environmental Science & Technology*, **18**, 617–624.
Strange, J., Langdon, N. and Harris, M. (2008) *Contaminated Land Investigation, Assessment and Remediation, 2nd edition*. Thomas Telford, London.
Sulzberger, B., Canonica, S, Egli, T., Giger, W., Klausen, J. and von Gunten, U. (1997) Oxidative transformations of contaminants in natural and in technical systems. *Chimia*, **51**, 900–907.
Sunda, W.G. and Keiber, D.J. (1994) Oxidation of humic substances by manganese oxides yields low molecular weight substrates. *Nature*, **367**, 62–64.
Tebo, B.M., Bargar, J.R., Clement, B.G., Dick, G.J., Murray, K.J., Parker, D., Verity, R. and Webb, S.M. (2004) Biogenic manganese oxides: Properties and mechanisms of formation. *The Annual Review of Earth and Planetary Sciences*, **32**, 287–328.
Thorn, K., Arterburn, J. and Mikita, M. (1992) 15N and 13C NMR investigation of hydroxylamine-derivatised humic substances. *Environmental Science & Technology*, **26**, 107–116.
Tica, D., Udovic, M. and Lestan, D. (2011) Immobilization of potentially toxic metals using different soil amendments. *Chemosphere*, **85**, 577–583.
Tipping, E. and Heaton, M.J. (1983) The adsorption of aquatic humic substances by 2 oxides of manganese. *Geochimica et Cosmochimica Acta*, **47**, 1393–1397.
Toner, B., Fakra, S., Villalobos, M., Warwick, T. and Sposito, G. (2005) Spatially resolved characterization of biogenic manganese oxide production within a bacterial biofilm. *Applied and Environmental Microbiology*, **71**, 1300–1310.
Toner, B., Manceau, A., Webb, S. and Sposito, G. (2006) Zinc sorption to biogenic hexagonal-birnessite particles within a hydrated bacterial biofilm. *Geochimica et Cosmochimica Acta*, **70**, 27–43.
Tremolada, P., Guazzoni, N., Smillovich, L., Moia, F. and Comolli, R. (2012) The effect of the organic matter composition on POP accumulation in soil. *Water, Air, and Soil Pollution*, **223**, 4539–4556.
Tunega, D., Gerzabek, M.H., Haberhauer, G., Totsche, K.U. and Lischka, H. (2009) Model study on sorption of polycyclic aromatic hydrocarbons to goethite. *Journal of Colloid and Interface Science*, **330**, 244–249.
Ulrich, H.J. and Stone, A.T. (1989) Oxidation of chlorophenols adsorbed to manganese oxide surfaces. *Environmental Science & Technology*, **23**, 421–428.
UNEP (2001) *Final Act of the Plenipotentiaries on the Stockholm Convention on Persistent Organic Pollutants*. United Nations Environment Program Chemicals, Geneva, Switzerland. http://www.pops.int/documents/meetings/dipcon/25june2001/conf4_finalact/en/FINALACT-English.pdf
U.S. Environmental Protection Agency (2014) Search Superfund Site Information. Available at: http://cumulis.epa.gov/supercpad/cursites/ srchrslt.cfm?start=1&CFID=12297806&CFTOKEN=34599551&jsessionid=e0301c2aaa7a41d5242212454670273402 53.
Valentin, L., Nousianinen, A. and Mikkonnen, A. (2013) Introduction to organic contaminants in soil: concepts and risks. Pp. 1–30 in: *Emerging Organic Contaminants in Sludges: Analysis, Fate and Biological Treatment* (T. Vicent, G. Caminal, E. Eljarrat and D. Barceló, editors). Springer-Verlag, Berlin, Heidelberg, Germany.
Van der Zee, F.R. and Cervantes, F.J. (2009) Impact and application of electron shuttles on the redox (bio)transformation of contaminants: A review. *Biotechnology Advances*, **27**, 256–277.
Vileno, E., Zhou, H., Zhang, Q.H., Suib, S.L., Corbin, D.R. and Koch, T.A. (1999) Synthetic todorokite produced by microwave heating: An active oxidation catalyst. *Journal of Catalysis*, **187**, 285–297.
Villalobos, M., Lanson, B., Manceau, A., Toner, B. and Sposito, G. (2006) Structural model for the biogenic Mn oxide produced by *Pseudomonas putida*. *American Mineralogist*, **91**, 489–502.
Villalobos, M., Carrillo-Cárdenas, M., Gibson, R., López-Santiago, N.R. and Morales, J.A. (2014) The influence of particle size and structure on the sorption and oxidation behaviour of birnessite: II. Adsorption and oxidation of four polycyclic aromatic hydrocarbons. *Environmental Chemistry*, **11**, 279–288.
Wang, A.M., Qu, J.H., Liu, H.J. and Lei, P.J. (2006) Dyes wastewater treatment by reduction oxidation process in an electrochemical reactor packed with natural manganese mineral. *Journal of Environmental Sciences – China*, **18**, 17–22.
Wang, M.C. and Huang, P.M. (1991) Nontronite catalysis in polycondensation of pyrogallol and glycine and the

associated reactions. *Soil Science Society America Journal*, **55**, 1156–1161.

Wang, T.S.C., Wang, M.C. and Huang, P.M. (1983) Catalytic synthesis of humic substances by using aluminas as catalysts. *Soil Science*, **136**, 226–246.

Wang, Z.M., Xiong, W., Tebo, B.M. and Giammar, D.E. (2014) Oxidative UO_2 dissolution induced by soluble Mn(III). *Environmental Science & Technology*, **48**, 289–298.

Wariishi, H., Valli, K. and Gold, M.H. (1992) Manganese(II) oxidation by manganese peroxidase from the basidiomycete phanerochaete-chrysosporium – kinetic mechanism and role of chelators. *Journal of Biological Chemistry*, **267**, 23688–23695.

Waters, W.A. (1971) Mechanism of one-electron oxidation of phenols – fresh interpretation of oxidative coupling reactions of plant phenols. *Journal of the Chemical Society B-Physical Organic*, **10**, 2026–2029.

Wenzel, K.D., Manz, M., Hubert, A. and Schuurmann, G. (2002) Fate of POPs (DDX, HCHs, PCBs) in upper soil layers of pine forests. *Science of The Total Environment*, **286**, 143–54.

Wheeler, O.H. and Gonzalez, D. (1964) Oxidation of primary aromatic amines with manganese dioxide. *Tetrahedron*, **20**, 189–193.

Wicke, D., Reemstma, T. and Böckelmann, U. (2005) Laboratory investigations on the exchange of PAH between aqueous phase and biofilms or soil bacteria. In *the 9th International FZK/TNO Conference on Soil–water Systems* (Bordeaux) *Abstracts.*

Xiao, H., Song, H., Xie, H., Huang, W., Tan, J. and Wu, J. (2013) Transformation of acetaminophen using manganese dioxide-mediated processes: Reaction rates and pathways. *Journal of Hazardous Materials*, **250–251**, 138–146.

Yuan, S., Tian, M. and Lu, X. (2006) Microwave remediation of soil contaminated with hexachlorobenzene. *Journal of Hazardous Materials*, **137**, 878–885.

Zhang, H.C. and Huang, C.H. (2003) Oxidative transformation of triclosan and chlorophene by manganese oxides. *Environmental Science & Technology*, **37**, 2421–2430.

Zhang, H.C. and Huang, C.H. (2005a) Reactivity and transformation of antibacterial N-oxides in the presence of manganese oxide. *Environmental Science & Technology*, **39**, 593–601.

Zhang, H.C. and Huang, C.H. (2005b) Oxidative transformation of fluoroquinolone antibacterial agents and structurally related amines by manganese oxide. *Environmental Science & Technology*, **39**, 4474–4483.

Zhang, W., Bouwer, E., Wilson, L. and Durant.N. (1995) Biotransformation of aromatic hydrocarbons in subsurface biofilms. *Water Science and Technology*, **31**, 1–14.

Zhang, H.C., Chen, W.R. and Huang, C.H. (2008) Kinetic modelling of oxidation of antibacterial agents by manganese oxide. *Environmental Science & Technology*, **42**, 5548–5554.

Zhang, Q., Cheng, X., Zheng, C., Feng, X., Qiu, G., Tan, W. and Liu, F. (2011) Roles of manganese oxides in degradation of phenol under UV Vis irradiation: adsorption, oxidation and photocatalysis. *Journal of Environmental Science*, **23**, 1904–1910.

Zhang, Y., Lashermes, G., Houst, S., Zhu, Y.G., Barriuso, E. and Garnier, P. (2014) COP-compost: a software to study the degradation of organic pollutants in composts. *Environmental Science and Pollution Research*, **21**, 2761–2776.

Zhao, L., Yu, Z., Peng, P., Huang, W. and Dong, Y. (2009) Oxidative transformation of tetrachlorophenols and trichlorophenols by manganese dioxide. *Environmental Toxicology and Chemistry*, **28**, 1120–1129.

Zhao, G., Li, J., Ren, X., Hu, J., Hu, W. and Wang, X. (2013) Highly active MnO_2 nanosheet synthesis from graphene oxide templates and their application inefficient oxidative degradation of methylene blue. *Royal Society of Chemistry Advances*, **3**, 12909–12914.

Zsolnay, A. (2003) Dissolved organic matter: artefacts, definitions and functions. *Geoderma*, **113**, 187–209.

EMU Notes in Mineralogy, Vol. 17 (2017), Chapter 12, 357–378

Role of redox-reactive minerals in the reuse and remediation of mine wastes

KAREN A. HUDSON-EDWARDS and DAVID KOSSOFF

Department of Earth and Planetary Sciences, Birkbeck, University of London, Malet St., London WC1E 7HX, UK e-mail: k.hudson-edwards@bbk.ac.uk; jkoss02@mail.bbk.ac.uk

Mining, oil and gas and other extractive industries are vital and irreplaceable constituents of the modern global economy. The overall demand for the products of these industries rises inexorably with economic growth and developing prosperity. However, these industries produce vast quantities of potentially harmful waste. Many of the elemental, and hence mineralogical, components of the waste stream are redox active. This review focuses on redox-reactive minerals sourced from this waste stream in reuse and remediation schemes. Copper-, manganese- and iron-bearing mine wastes are used as pigments, in fertilizers, sorbents of toxic compounds in water treatment systems, and for the production of SO_2 and H_2SO_4. Some solid mine wastes can be remediated by phytostabilization, where plants are used to induce the precipitation of secondary redox-reactive minerals that sequester contaminants. Liquid wastes are remediated using a variety of abiotic and biotically-assisted schemes such as anoxic limestone drainages and permeable reactive barriers. These schemes use phases such as zero-valent iron and Fe-oxyhydroxides, and produce mineralogical by-products such as sulfides, green rust and oxyhydroxides. Further research is needed to optimize the reuse and remediation schemes in mine wastes and to develop new and innovative systems employing redox active minerals.

1. Introduction

Today's society requires metals, metalloids and other mineral products to maintain industrial and household activities. The extraction of these products generates considerable quantities of mine wastes because the valuable parts generally constitute a small proportion of the ores from which they originate. Lottermoser (2010) estimated that for every tonne of ore consumed, the same amount of solid wastes would be generated, giving 20,000–25,000 Mt of solid wastes produced annually on a global scale. This figure is predicted to grow over the next 100 years due to increasing demand for mineral resources, coupled with lower ore grades (Gordon *et al.*, 2006). Mining activities produce solid, gaseous and liquid mine wastes (Hudson-Edwards *et al.*, 2011). Solid mine wastes include flue ashes and dusts, slags, tailings and waste rock (Fig. 1). Solid mine wastes cause environmental problems because their toxic components can be leached into surface and ground waters and taken up by plants or

DOI: 10.1180/EMU-notes.17.10

Figure 1. Examples of mine wastes. (a) Lead- and zinc-rich waste rock and neutral mine drainage (NMD) at the former Cymrheidol Mine, Wales. (b) Tailings impoundment, Rabbit Lake, Saskatchewan, Canada; (c) Acid mine drainage (AMD) from a former adit, Cymrheidol Mine, Wales; (d) Flow-textured slag from Hegeler Zn-smelting facility, Illinois (USA), reprinted from Piatak and Seal (2010), with permission from Elsevier; (e) Metalliferous dendrites in a silicate matrix of spherical, smelter-derived flue ash particles, reprinted from Shukurov *et al.* (2014), with permission from Elsevier.

humans (*e.g.* from dust inhalation). Waters can be contaminated by the dissolution of mine-waste minerals, forming basic (pH 8–12, BMD), circumneutral (pH 6–8, NMD) or acidic (pH –3–5, AMD) mine drainage (Nordstrom and Alpers, 1999; Nordstrom, 2011). AMD is the most common and most detrimental type of fluid mine waste.

The principle of 'zero waste' is a desirable environmental objective which would significantly reduce the footprint of the mining and extractive industries. This objective maps onto principles 2 and 3 of the 12 principles of green engineering, namely: "it is better to prevent waste than to treat or clean up waste after it is formed", and "separation and purification operations should be designed to minimize energy consumption and materials use" (Anastas and Zimmerman, 2003). There are, however, two principal obstacles impinging on this otherwise desirable objective. Firstly, there is the economic issue; *i.e.* whether the proposed reuse will be economically viable. The second issue, particularly with metal-ferruginous tailings, is the problems posed by the introduction of contaminant and potentially toxic elements into the wider environment. To some extent, however, the economic question, in the light of the zero-waste objective, may be a false dichotomy. This is particularly the case if rigorous environmental auditing is applied to the entirety of the waste material (Jenkins and Yakovleva, 2006). For example, what may be said to be an uneconomic process in the context of an isolated profit and loss centre, such as a mining company, may well become an economic or desirable project in the context of wider society. It is likely that the on-going activities of mining enterprises will increasingly require what has been described as an 'operational social licence' (Warhurst, 2001; Azapagic, 2004). An important constituent of a mining company's social licence should be the minimization of the volume of waste. This could be achieved partially by the utilization of the waste material for other purposes (*i.e.* reuse), or by rendering the material safe according to set environmental guidelines (*i.e.* remediation) (Lottermoser, 2011).

Redox-reactive minerals such as Fe- and Mn-oxyhdyroxides and oxyhydroxysulfates and metal sulfides are both used and produced during the reuse and remediation of mine wastes. This chapter gives an overview of the minerals and processes involved in these important activities. For further information, the reader is referred to the articles discussed in this chapter, especially the overviews given by Blowes *et al.* (2003), Fortin and Beveridge (1997), Johnson and Hallberg (2005) Lottermoser (2010, 2011) and Rankin (2011, chapters 6, 11 and 12).

2. Reuse of mine wastes containing redox-reactive minerals

2.1. Definition of, and potential problems associated with, reuse of mine wastes

This section overviews the reuse of Cu-, Mn- and, particularly, Fe-bearing mine waste. The 'reuse' of tailings described here encompasses both existing and newly proposed and innovative uses for the material, thereby distinguishing them from 'remining' (*i.e.* re-processing old mine wastes using conventional mining technologies), 'reprocessing' (re-processing old mine wastes using unconventional technology) and 'remediation' (see definition in the previous section; Lottermoser, 2011).

Boulding (1966) proposed two end-member global economies. One end member is defined as a linear economy, where the world receives a fresh flow of resources, and, as a consequence, disposes of the resultant waste. In contrast, the opposite end member is a proposed circular global economy, which is underpinned by the fundamental observation that the Earth is effectively closed to the addition of further matter. The former model is unsustainable in the long term and, in some instances, in the medium term too (*e.g.* the discussion on Cu which follows). Every effort, therefore, should be made to shift the balance of the global economy towards the latter and away from the former end member. Therefore, the reuse, reprocessing and remining, rather than just the safe storage or remediation of mining waste, should be viewed as a means towards this end. Two examples that serve to convey the scale of these processes are the comprehensive environmental audits of Fe (Wang *et al.*, 2007) and Cu cycling (Graedel *et al.*, 2004) which have been undertaken on regional and worldwide scales. In the case of Cu, it is perhaps pertinent to note that over the course of the 20th century in North America, 40 Mt were collected and recycled from post-consumer waste, 56 Mt accumulated in landfills (or were lost through dissipation) and 29 Mt of Cu waste was produced in the form of tailings and slag and stored in waste reservoirs (Spatari *et al.*, 2005). The authors further estimate that the Cu reserves of North America are limited to 113 Mt, of which ~50% can be feasibly extracted. Thus, Cu is very much a limited resource, emphasizing the importance of utilizing the significant portion of this metal currently, and previously, placed within waste reservoirs.

The potential problems posed by the introduction of toxic contaminants into the environment constitute a significant impediment to the effective reuse of mined material. For example, in sulfide tailings, Fe is usually the dominant chalcophilic cation (Keith and Vaughan, 2000) but Fe cannot at present be extracted economically from redox-reactive sulfide minerals. These minerals include pyrrhotite ($Fe_{1-x}S$, where x ranges from 0 to 0.2), marcasite (FeS_2) and pyrite (FeS_2). Arsenic is often found associated with arsenopyrite (FeAsS) which is the most abundant arsenic-bearing mineral (Smedley and Kinniburgh, 2002), and with pyrite, which is then described as arsenian pyrite (*e.g.* Kossoff *et al.*, 2012). Hence, any proposal to reuse Fe-bearing waste should be examined in the light of possible high As levels. In mine wastes, the highly toxic metalloid antimony (Sb; USEPA, 2014) is also found associated with Fe-bearing minerals, particularly those which have undergone, even moderate, oxidation (Ritchie *et al.*, 2013). Similarly Cu is often associated with Zn (*e.g.* in sphalerite, ZnS and chalcopyrite, $CuFeS_2$), while Zn is commonly associated with potentially toxic Cd (*e.g.* in sphalerite or wurtzite, ZnS) and, often, Pb (*e.g.* in galena, PbS; Blowes *et al.*, 1995). Therefore, any proposal to reuse Fe- and particularly pyrite rich-tailings should be considered only after assaying Sb and, in particular, As. Similarly, any proposal to reuse Cu in tailings should be cross-checked against Zn and, particularly Cd and Pb concentrations. Moreover, the bioavailability and transport of toxic elements is constrained by their partitioning into minerals and by the solubility of these minerals (*e.g.* Brown and Callas, 2011). Thus, any scheme to reuse mine waste ideally requires a thorough elemental audit and mineralogical characterization of the material in question.

2.2. Industrial reuse of mine waste

In terms of bulk utilization, the most important current reuse application of mine tailings is in civil engineering construction (*e.g.* Hammond, 1988; Yellishetty *et al.*, 2008). The mining industry itself has long used tailings for at-surface storage built on-site. For example, the dams of tailing impoundments are themselves often constructed from this material (Younger and Wolkersdorfer, 2004; Bussière *et al.*, 2007). Tailings with an added binder, usually cement, may also be used to backfill old workings (Sivakugan *et al.*, 2006). Tailings can also be utilized in the manufacture of specific building materials such as bricks (*e.g.* Fang *et al.*, 2011; Zhao *et al.*, 2009) and cementitious raw materials (*e.g.* Li *et al.*, 2010). The following discussion will concentrate on those applications involving redox or, alternatively described as, electron-transfer processes. The discussion will describe proposed tailings applications as pigments, fertilizers, absorbents in the water-treatment industry, semi-conductors utilized in photovoltaic panel manufacture and, finally, as a source of sulfur dioxide (SO_2) and sulfuric acid (H_2SO_4). Note that the latter two applications largely describe applications specifically for redox-reactive pyrite, rather than tailings in general.

The electrons in transition metals occupy partially filled *d* orbital sub-shells. Electrons in particular orbitals can absorb light at individually unique wavelengths and, as a consequence, move to defined higher energy levels. Hence, the transition metals and their compounds may be employed as pigments. The transition metals Cr, Mn, Fe, Co and Cu have variable valence states, which depend on the oxidation potential of the environment they occupy. Furthermore, as the oxidation state varies, so does the wavelength of the visible light absorbed. For example, Cr^{3+} compounds are green, while Cr^{6+} compounds are yellow. Although the reuse of tailings for pigment manufacture is as yet applied on a small scale there are some examples of successful pilot projects that are described in the literature (Dengxin *et al.*, 2008). For reasons of safety, Fe oxides are becoming the pigments of choice due to their non-toxicity compared with that of the heavy metal-based pigments (*e.g.* Rosner *et al.*, 2005). Minerals such as magnetite (Fe_3O_4), hematite (Fe_2O_3) and goethite (FeOOH) are commonly used, conferring red, brown and yellow colours, respectively. These phases, at a workable pigment particle size of <10 μm, have been separated from steel-waste sludge at a Mittal steel plant in the Republic of South Africa (Legodi and De Waal, 2007). Therefore, it is feasible that such a process could be successfully employed for the reuse of Fe-rich mine tailings. At Lowber in southwestern Pennsylvania, a comparatively large project involved the reuse of Fe oxy/hydroxide sludge (principally, in this case, goethite) as a pigment. This was sourced from the discharge emanating from a long-abandoned coal mine. One thousand tonnes of raw material was supplied to a pigment manufacturer (Hoover Color Corp), indicating the large global scale of this potential resource (Hedin, 2003). On a smaller scale, a successful pilot study performed by Dengxin *et al.* (2008) found that Fe pigment may be obtained from cyanide tailings using ammonia to purify the reduced iron sulfate and with urea ($CO(NH_2)_2$) as precipitant.

Acid mine drainage emanating from historic coal mining run-off and spoil at a headwaters tributary of Seaton Creek, Pennsylvania (Ohio River Basin) is being treated passively, with metallic Mn being recovered. The Mn collected has been utilized as a pottery glaze, albeit on an as yet uncommercial scale (Denholm *et al.*, 2008). Manganese is also used in the manufacture of steel, batteries, chemicals, fertilizers and animal feeds, and it is possible that some of these markets may also be serviced by the reused Mn. Extenders are commonly used in paint formulation both to confer structural stability and for economic reasons. Saxena and Dhimole (2006) investigated the possibilities of reusing Cu-rich tailings in this manner. They found that the tailings, after initial sieving and grinding, acted as both a first class and economical paint extender.

Based upon its properties as a semiconductor and the resultant availability of electrons in the conduction band, pyrite has been proposed as an abundant and cheap material for the manufacture of photovoltaic cells (Wadia *et al.*, 2009). It has been estimated that just 10% of the pyrite disposed of annually as mining waste in six US states would be sufficient to supply the entirety of the US primary power demand (Puthussery *et al.*, 2010). Unlike water-treatment tailings reuse, however, the research has so far been conducted only on laboratory-synthesized pyrite. Moreover, the viability of pyrite's utility as a semi-conductor has been questioned (Hadlington, 2012).

A more established use of pyrite-bearing tailings, and indeed the ore itself, is for the manufacture of sulfuric acid. In this process the pyrite is roasted in order that it may be oxidized quickly. The sulfide anion is oxidized to SO_2 while the Fe is also oxidized, from a di- to a tri-valent state (*e.g.* equation 3, from Runkel and Sturm, 2009).

$$2FeS_{2(s)} + 11/2O_{2(g)} = Fe_2O_{3(s)} + 4SO_{2(g)} \qquad \Delta H = -1666 \text{ kJ mol}^{-1} \qquad (1)$$

Historically, the SO_2 produced in this process found large-scale use in the paper industry as a bleaching agent, *e.g.* in Sweden (Canepa *et al.*, 2009), and as a first step in the manufacture of H_2SO_4. Largely for economic reasons pyrite roasting has been discontinued in many countries in favour of sulfur extraction, as a useful by-product, from the metal smelting, oil and natural gas industries (Rappold and Lackner, 2010). Sites with a history of pyrite roasting often exhibit serious ongoing adverse environmental effects (Pérez-López *et al.*, 2009; Oliveira *et al.*, 2012). In China, however, pyrite roasting continues to be a large-scale sulfur-manufacturing process (Yang *et al.*, 2009).

2.3. Mine wastes as sources of plant nutrients

Mine tailings may be exploited as the source of essential plant nutrients as the demand for agricultural production is increasing in tandem with increasing global population. A significant proportion of plant enzymes have metal constituents and, in the case of the redox-active metals, their role as electron carriers is fundamental to their efficient functioning (Andreini *et al.*, 2008). In plants, just as with animals, Fe is the most frequently utilized redox-active metal. This is almost certainly a function of its high availability, as it is the fourth most abundant element within the Earth's crust. Iron deficiency can be a significant agricultural problem which may be addressed by the application of Fe-rich fertilizers (Wiersma, 2005; Caliskan *et al.*, 2008). In Portugal, pyrite from the Neves-Corvo Cu mine has been trialled as a possible soil improver for the calcareous alkaline, high-pH coastal soils (Castelo-Branco *et al.*, 1999). The oxidation of bivalent Fe in pyrite to trivalent ferric hydroxide provides protons that serve to reduce the soil pH (Equation 2) and to provide micronutrients which may be contained within the primary pyrite structure.

$$4FeS_{2(s)} + 15O_{2(g)} + 14H_2O_{(l)} = 4Fe(OH)_{3(s)} + 16H^+_{(aq)} + 8SO_4^{2-}{}_{(aq)} \qquad (2)$$

Although Zn and Cu and particularly, Pb, were detected in the treated soils, their levels remained below the regulatory maxima. Indeed the authors concluded that: "..... the use of pyrite, obtainable as a waste product from the ore milling industry, may indeed be a useful amendment and effective fertilizer for agricultural purposes" (Castelo-Branco *et al.*, 1999, p. 366). A similar conclusion was reached on the application of gold-mine waste, containing pyrite as a dominant phase, to sunflower crops (Divya, 2007 and references therein). Moreover, in China, further research has been conducted into the use of Fe-rich tailings (after initial magnetic separation) as a fertilizer. The fractionated tailings conferred a better physical condition on the soil and supplied essential micronutrients. Indeed, a study reported that rice and soy crop yields improved by ~12–15% after application of the tailings (Zhang *et al.*, 2006). Copper, to a lesser extent than Fe, acts as an electron carrier in plants and, just as is the case with Fe, low background concentrations of this essential metal can impede plant growth (Yamasaki *et al.*, 2008). After pre-treatment with organic manures and mineral acids, Cu-rich tailings have been trialled with some success for fertilizer use (Tamgale, 2005 and references therein). Manganese too is an essential plant micronutrient and, although it is a redox-active metal, it is reported as having structural, rather than electron-carrying functions (*e.g.* Hebbern *et al.*, 2009). Manganese-rich tailings have also been trialled successfully in China as a fertilizer improving the yield and quality of capsicum plants (Zhou *et al.*, 2009). To add a cautionary note, for some years the product Ironite© has been marketed in the USA as a tailings-derived fertilizer. It has been established that the product is not only a source of the micronutrients needed, but also a potential source of both As and Pb contamination (Dubey and Townsend, 2004; Williams *et al.*, 2006). Ironite is now banned in Canada and is the subject of several litigation actions in the USA (United States Environmental Protection Agency, 2012).

2.4. Water purification

Tailings have been the subject of some research as a possible contaminant absorbent filter in the water purification industry; in effect it is proposed that one waste stream may be utilized to moderate another. For example, hexavalent Cr is considerably more mobile and toxic than its trivalent analogue (Shah *et al.*, 2009). Hence, the reduction of Cr^{6+} and its subsequent removal by precipitation is a desirable objective. To that end pyrite-bearing tailings may be employed, and these in turn can be oxidized

$$3FeS_{2(s)} + 15HCrO_{4}^{-}{}_{(aq)} + 57H^{+}_{(aq)} = 3Fe^{3+}_{(aq)} + 15Cr^{3+}_{(aq)} + 6SO_{4}^{2-}{}_{(aq)} + 36H_2O_{(l)} \quad (3)$$

Therefore, this is a potentially significant reuse technology (Zouboulis *et al.*, 1995; Lin and Huang, 2008). Similarly, Mn-rich tailings may potentially also find application as a water-treatment agent for the effective removal of trivalent As from solution. In this case the Mn is reduced and the As is oxidized (Equation 4; Driehaus *et al.*, 1995)

$$H_3AsO_{3(aq)} + MnO_{2(s)} = HAsO_{4}^{2-}{}_{(aq)} + Mn^{2+}_{(aq)} + H_2O_{(l)} \qquad E = 0.67V \quad (4)$$

Note that although tailings reuse has been proposed for water-purification applications, the research outlined thus far has been performed on laboratory-synthesized phases. There are, however, two historical precedents, on a reasonably large scale, for just such a use. Ocherous coal mine waste was used in the Buxton (Debyshire, UK) treatment works as a flocculating agent during the 19th century (Banks *et al.*, 1997). Moreover, in Falun (Sweden) waste water from a copper mine was employed as a flocculating agent, by mixing it with municipal sewage. Indeed, illustrative of the scale of this process, in 1995, some 109,000 m^3 of mine water was blended with 5,900,000 m^3 of sewage (G. Strålin, Falu kommune, pers. comm.; quoted by Banks *et al.*, 1997).

3. Phytostabilization of mine waste

Mine tailings and mine-contaminated soils can be remediated by phytostabilization, *i.e.* the use of plants to transform the species of metals from toxic to non-toxic forms. The plant species used are metal-tolerant which can exclude metals and metalloids from their above-ground biomass so this part of the plant can be used for grazing and human activities (Craw *et al.*, 2007). The elements are either taken up in the plant roots or precipitated in secondary minerals around the roots, so as to reduce mobility and bioavailability (*e.g.* Stoltz and Greger, 2002; Grandlic *et al.*, 2008; Santibáez *et al.*, 2008; Lottermoser, 2010). For example, Kumpiene *et al.* (2012) showed that arsenic was removed by adsorption onto poorly crystalline Fe oxyhdyroxide in compost-, zero-valent iron- and coal fly ash-amended gold mining spoils. Cotter-Howells *et al.* (1999) demonstrated elegantly that a pyromorphite-like ($Pb_5(PO_4)_3(Cl)$) phase formed on the outermost layer of the roots of *Agrostis capillaris* L., a metal-tolerant grass, on Parys Mountain (Wales) mine spoils (Fig. 2). This suggests that other redox-active metal and metalloid species might be taken up and sequestered in P-bearing precipitates.

Barren land known as 'kill zones' that are created by overland flow of AMD, and which are acidic, and contain small amounts of organic matter, have little water-holding capacity and have high metal concentrations (Mendez *et al.*, 2007; Rojas *et al.*, 2014), can be remediated using phytoremediation techniques. For example, vegetative cover was re-established on Fe(III) oxyhydroxide-rich soils created from oxidation and

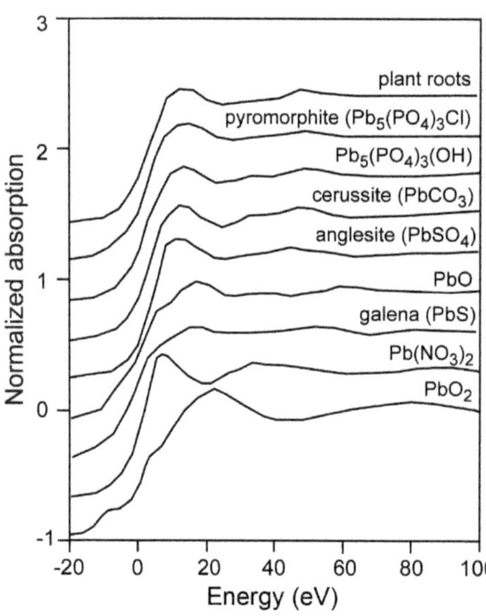

Figure 2. XANES spectra from freeze-dried plant roots from Pb-mining impacted wastes at Parys Mountain, and a range of model (pure) lead compounds. The plant root XANES spectra are similar to those of pyromorphite and Pb-hydroxylapatite ($Pb_5(PO_4)_3(OH)$). The insolubility of such phosphate minerals over a wide range of pH means that the Pb is potentially sequestered for the long term. Redox-active elements could also be taken up and held in similar phosphate minerals. Modified version reproduced with the permission of the Mineralogical Society of Great Britain & Ireland from a paper by Cotter-Howells *et al.* (1999).

hydrolysis reactions during AMD overland flow in Pennsylvania (Lupton *et al.*, 2013). Other mine tailings-affected soils can also be phytoremediated. Knight *et al.* (1997) showed that the Cd- and Zn-hyperaccumulator *Thlaspi caerulescens* could be suitable for remediation of mine spoils with relatively low concentrations of Cd and Zn. Many hyperaccumulators, however, grow too slowly and do not produce enough biomass to make them adequate phytoremediators. Another issue with phytoremediation is that the potentially toxic and hazardous biomass needs to be properly disposed of and managed (Ghosh and Singh, 2005).

4. Remediation of mine waste waters

This part of the chapter describes techniques used to remediate mine waste waters. Most of the schemes described are for AMD, as this is the most common mine waste water and the most expensive to treat (Nordstrom, 2011). A variety of active and passive treatment techniques are used to remediate mine drainage, and AMD in particular. Figure 3 summarizes several methods for mine-waste water treatment, and the redox-reactive minerals (and materials). These are discussed in more detail below, following the classification outlined by Johnson and Hallberg (2005) (Fig. 3).

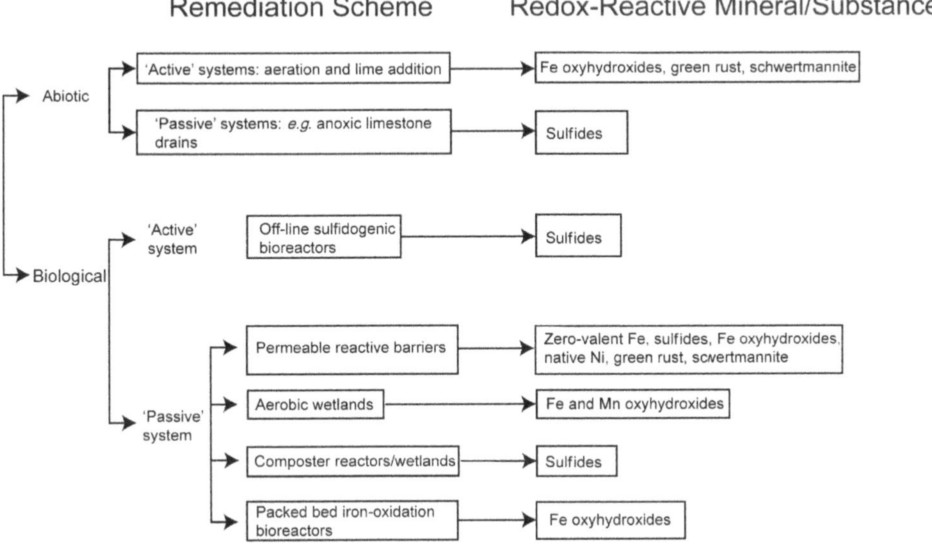

Figure 3. Remediation strategies for mine-drainage waters. The term 'oxyhydroxides' is used to represent oxides, hydroxides and oxyhydroxides. Reprinted (adapted) from Johnson and Hallberg (2005) with permission from Elsevier.

4.1. Abiotic systems: lime addition and limestone drains

For AMD, remediation schemes are designed to raise the pH and ideally, remove potentially toxic metallic and metalloid elements from the waters so that they can be released to streams. Note that raising the pH of waters which contain significant quantities of As and Sb may render these contaminants more mobile. These metalloids are often present as anions in solution and therefore compete with the hydroxyl and other anion groups introduced for binding sites (Olías *et al.*, 2004). Remediation sites are often coupled with oxidation-settling ponds or wetlands (see below; Cravotta and Trahan, 1999). Raising the AMD pH is achieved by adding chemicals, most commonly quicklime (CaO) to form Ca hydroxide in the presence of water (*e.g.* Kalin *et al.*, 2006; Caraballo *et al.*, 2011a), which then releases hydroxide ions, as follows:

$$CaO_{(s)} + H_2O_{(l)} = Ca(OH)_{2\,(s)} \tag{5}$$

$$Ca(OH)_{2\,(s)} = Ca^{2+}_{(aq)} + 2OH^-_{(aq)} \tag{6}$$

$$OH^-_{(aq)} + H^+_{(aq)} = H_2O_{(l)} \tag{7}$$

The other common method of raising the pH of AMD waters is to pass them over limestone ($CaCO_3$) (Hedin *et al.*, 1994; Cravotta and Trahan, 1999; Hammarstrom *et al.*, 2003; Caraballo *et al.*, 2011b; Macias *et al.*, 2012):

$$CaCO_{3\,(s)} + 2H^+_{(aq)} = Ca^{2+}_{(aq)} + H_2CO_{3\,(l)} \tag{8}$$

$$CaCO_{3\,(s)} + H_2CO_{3\,(l)} = Ca^{2+}_{(aq)} + 2HCO^-_{3\,(aq)} \tag{9}$$

$$CaCO_{3\,(s)} + H_2O_{(l)} = Ca^{2+}_{(aq)} + HCO^-_{3\,(aq)} + OH^-_{(aq)} \tag{10}$$

These systems are operated as open (oxic) limestone channels or diversion wells (Arnold, 1991; Ziemkiewicz *et al.*, 1997) or anoxic limestone drains (Turner and McCoy, 1990; Hedin *et al.*, 1994; Whitehead *et al.*, 2005). The latter are used to prevent clogging and armouring of the limestone by secondary phases (Hedin *et al.*, 1994); the effluents produced in the anoxic drain are subsequently exposed to the atmosphere to produce Fe hydroxides (see below). The excess protons in the AMD are neutralized by hydroxide ions or by the limestone (reactions 8 and 9). The hydroxide ions also combine with metal cations that also form hydroxides at specific pH ranges. The most common of these are redox-sensitive Fe hydroxides:

$$4Fe^{2+}_{(aq)} + O_{2\,(g)} + 4H^+_{(aq)} = 4Fe^{3+}_{(aq)} + 2H_2O_{(l)} \tag{11}$$

$$Fe^{3+}_{(aq)} + 3OH^-_{(aq)} = Fe(OH)_{3\,(s)} \tag{12}$$

$$Fe^{3+}_{(aq)} + 3H_2O_{(l)} = Fe(OH)_{3\,(s)} + 3H^+_{(aq)} \tag{13}$$

$$Fe^{3+}_{(aq)} + 3HCO^-_{3\,(aq)} = Fe(OH)_{3\,(s)} + 3CO_{2\,(g)} \tag{14}$$

The Fe^{3+} is produced by the oxidation of Fe^{2+} either inorganically or by Fe-oxidizing bacteria such as *Acidothiobacillus ferrioxidans* (Baker and Banfield, 2003). The $Fe(OH)_3$ produced is relatively metastable, but over time can recrystallize to more stable and more highly crystalline phases such as goethite or hematite (Blowes *et al.*, 2003). The $Fe(OH)_3$

phases also sorb many potentially toxic metallic and metalloid elements (*e.g.* As, Cu, Co, Mn, Ni, Pb, Zn) from the AMD waters (Cravotta and Trahan, 1999).

Other redox-reactive phases are formed by lime and limestone treatment of AMD. These include green rust ($((Fe_4^{2+}Fe_2^{3+}(OH)_{12}))^{2+}(A.2H_2O)$ where A represents SO_3^{2-}, SO_4^{2-}, $2Cl^-$, $2Br^-$ or CO_3^{2-}; Drissi *et al.*, 1995), schwertmannite ($Fe_8O_8(OH)_6SO_4.nH_2O$) and goethite (Lee *et al.*, 2002; Caraballo *et al.*, 2011a). In addition, because the limestone drains are often coupled with other drains filled with active material such as MgO, other minerals can be formed, including gypsum ($CaSO_4.2H_2O$), Zn carbonates and hydroxides, hydrozincite ($Zn_5(CO_3)_2(OH)_6$) and loseyite ($(Mn,Zn)_7(CO_3)_2(OH)_{10}$; Pérez-López *et al.*, 2011; Macias *et al.*, 2012).

Laboratory investigations are being carried out to investigate whether other redox-reactive materials could be used in active systems to remediate mine drainage waters. Bearcock *et al.* (2011) carried out open batch experiments to evaluate the effectiveness of green rust for treating such waters. The green rust was able to rapidly (within 1 h) remove Fe, Al, Cu and Zn, even after being dried and mostly oxidized (Fig. 4). In the Northern Pennines Orefield, pelletized hydrous ferric oxide wastes from coal-mine water treatments have been tested to see if they remove Zn from NMD (Mayes *et al.*, 2009). Thirty-two percent of the Zn was removed over a 10-month period *via* the processes of sorption onto the hydrous ferric oxide and precipitation with secondary calcite, formed as a result of the dissolution of portlandite in the cement and calcium recarbonation.

Several studies have been carried out to determine the mineralogy of redox-reactive phases produced by the weathering of mine tailings in impoundments. For example, the formation of scorodite ($FeAsO_4.2H_2O$) in hardpans (cemented layers) by precipitation and cementation of secondary weathering-mine tailings products can reduce Fe(III) and As(V) concentrations in pore waters, and, as long as conditions remain comparatively oxic, it also acts as a long-term store due to its relatively low solubility (DeSisto *et al.*, 2011). With progressive aging and acidification, mine tailings from the Klondyke State Superfund Site (Arizona, USA), the amounts of Zn sorbed to poorly crystalline Fe and

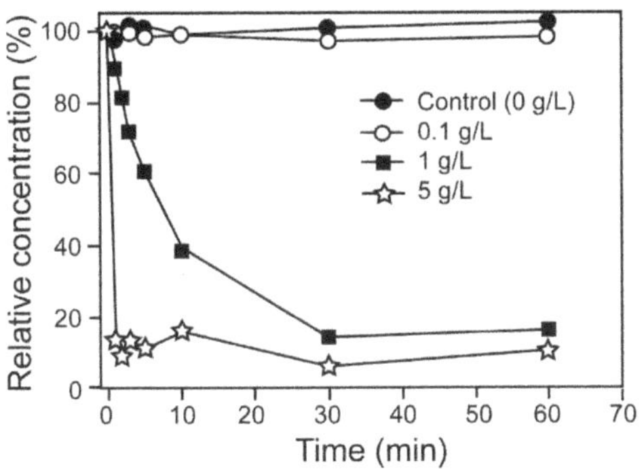

Figure 4. Results of stirred open batch experiments to test the effectiveness of dried green rust on removal of metallic elements from AMD waters in an abiotic active system. The plot shows that green rust removes significant amounts of Fe from AMD water from Parys Mountain, Wales, especially with each increase in solid material (shown as mass per litre of the AMD). The green rust is thought to remove the Fe due to Fe oxyhydroxide precipitation, sorption on active sites on the green rust, and surface-catalysed oxidation of ferrous iron. Reprinted (adapted) from Bearcock *et al.* (2011), with permission from Elsevier.

Mn oxyhydroxides decrease, while those of Zn in sulfate minerals and crystalline Fe oxides increase (Hayes *et al.*, 2011). The remediation potential of these redox-reactive minerals for the long-term storage of contaminants has not been investigated fully, but may be significant.

4.2. Biological systems

4.2.1. Off-line sulfidic bioreactors

Off-line sulfidogenic bioreactors, such as compost bioreactors, wetlands and Permeable Reactive Barriers (PRBs), promote the bacterial production of hydrogen sulfide to produce alkalinity and insoluble metal sulfides (Johnson and Hallberg, 2005). The bioreactors are specifically engineered to optimize H_2S production and protect the microbes from direct contact with AMD. Boonstra *et al.* (1999) and Johnson (2000) suggest that the performance of such bioreactors is more predictable and controlled than passive treatments, they allow selective recovery of desired heavy metals and they can reduce dramatically SO_4^{2-} concentrations in processed waters. They use carbon sources for the electron donors such as ethanol (Bekmezci *et al.*, 2011), lactate (Oyekola *et al.*, 2010) and compost (Chang *et al.*, 2000).

Two sulfidogenic bioreactor technologies include the Biosulfide and the Thiopaq processes (Johnson and Hallberg, 2005). In the Biosulfide system, H_2S produced from the bioreactor is fed into a precipitation tank containing the AMD, and metal-bearing sulfides are then produced. In the Thiopaq system, the AMD and carbon source are fed into the bioreactor, with sulfate reduction and metal-bearing sulfides then forming. Xingyu *et al.* (2013) describe another type of sulfidogenic bioreactor used to treat AMD from the Zijinshang Copper Mine in China. Sulfide (S^{2-}) ions were produced in an up-flow anaerobic sludge bed, two precipitation tanks were used for copper and iron precipitation, and activated sludge was used as the carbon source. The copper and iron removal efficiencies from AMD with concentrations of 100–120 mg/L Cu and 170–200 mg/L Fe were 60.95% and 97.8%, respectively.

4.2.2. Permeable reactive barriers

Permeable reactive barriers are used to remediate mine drainage waters and to prevent them from contaminating aquifers underlying mine wastes. PRBs are excavated trenches situated down-gradient from the mine wastes that are filled with a variety of materials designed to react with the mine drainage waters and render them safe for discharge into aquifers and streams (Blowes *et al.*, 2000; Fig. 5). The most commonly used redox-reactive material is zero-valent iron (Fe^0), which is a product of scrap iron and steel recycling (Blowes *et al.*, 2000; Wilkin and McNeil, 2003; Jambor and Raudsepp, 2005) and is often nano-sized (Bartzas and Komnitsas, 2010). Zero-valent iron is a highly reactive material that can remove a number of redox-active species, including As(III), As(V), Cr(IIII), Cr(VI), Cu(II), Mo(IV), Mo(VI), Se(-II), Se(IV), Se(VI), U(IV), U(VI), V(II), V(III), V(IV) and V(V) (Gu *et al.*, 1999; Ahn *et al.*, 2003; Bartzas and Komnitsas, 2010; Scott *et al.*, 2011).

Organic materials such as compost and wood wastes are also used in many PRBs to support the actions of sulfate-reducing bacteria (SRB). SRB use SO_4^{2-} as an electron

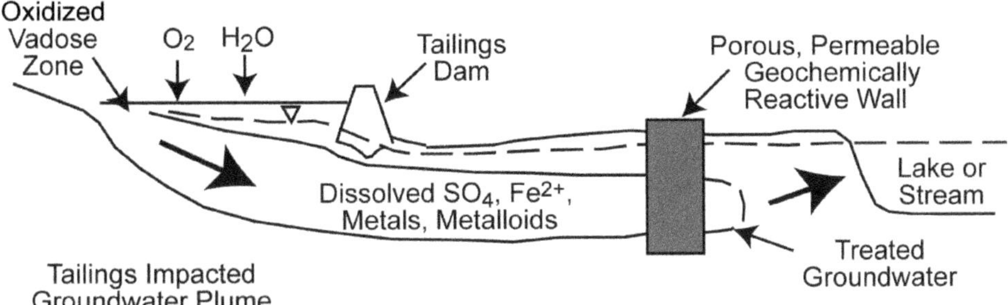

Figure 5. Schematic diagram of a permeable reactive barrier at a mine tailings impoundment. Reprinted (adapted) with permission from Waybrant *et al.* (1998). Copyright 1998, American Chemical Society.

acceptor, and this oxidizes the organic material and generates H_2S, according to the following reaction (Blowes *et al.*, 2003; Oyekola *et al.*, 2010):

$$(CH_2O)_x(NH_3)_y(H_3PO_4)_{z\,(s)} + xSO^{2-}_{4\,(aq)} = 2xHCO^{-}_{3\,(aq)} + xH_2S_{(g)} + 2yNH_{3\,(l)} + 2zH_3PO_{4\,(l)} \quad (15)$$

where $(CH_2O)_x(NH_3)_y(H_3PO_4)_z$ represents organic matter. The H_2S generated in Equation 15 reacts with dissolved metals to form metal sulfides (Blowes *et al.*, 2003):

$$Me^{2+}_{(aq)} + S^{2-}_{(aq)} \rightarrow MeS_{(s)} \quad (16)$$

A variety of other redox-reactive secondary minerals can be formed in PRBs due to the presence of other metals and the corrosion of the Fe^0. These include Fe oxide, hydroxides and oxyhydroxides (*e.g.* maghemite (γ-Fe_2O_3) magnetite (Fe_3O_4) and goethite), pyrite, greigite (Fe_3S_4), native nickel (Ni), covellite (CuS), chalcopyrite, bornite (Cu_5FeS_4), green rust and schwertmannite ($Fe_8O_8OH_6SO_4$) (Jambor and Raudsepp, 2005; Bartzas *et al.*, 2006; Bartzas and Komnitsas, 2010; Caraballo *et al.*, 2010). The Fe oxides, hydroxides and oxyhydroxides, in particular, are important secondary phases because they sorb redox-active species (Scherer *et al.*, 1998). They tend to form directly on the surfaces of the Fe^0, and act as a passive film, semi-conductor or coordinating surface (Scherer *et al.*, 1998). However, all of these secondary minerals can clog the PRB and affect its long-term efficiency. This was demonstrated by Gu *et al.* (1999) who showed, using laboratory column experiments, that HCO_3^- and SO_4^{2-} ions corroded Fe^0, forming $FeCO_3$ and H_2S, the latter of which stimulated SRB populations and increased SO_4^{2-} reduction. Several clogging secondary minerals were observed in the columns, including lepidocrocite (γ-FeOOH), akagnaeite (β-FeOOH), mackinawite ($(Fe,Ni)_{1+x}S$, where x = 0 to 0.11), magnetite, maghemite, goethite, siderite ($FeCO_3$) and amorphous ferrous sulfide.

PRBs are constructed with single or composite cells, depending on the contaminants to be removed and materials used to sequester them. At the Fry Canyon mine site in Utah, for example, a three-reactor PRB comprising hydrous ferric oxide to adsorb uranium, zero-valent iron to reduce and precipitate uranium, and phosphorus-rich bone char to induce formation of insoluble uranium-bearing phosphates, was installed

(Blowes *et al.*, 2000, Fuller *et al.*, 2002; Naftz *et al.*, 2002). This composite PRB was almost completely effective at removing uranium from the mine waste waters. The PRB system constructed to remediate AMD arising from the Aznalcóllar tailings-dam failure in SW Spain in 1998 had three compartments (Gibert *et al.*, 2011). These contained: limestone, to increase alkalinity and neutralize pH; organic material, to induce reduction of SO_4^{2-} by SRB and precipitation of metal sulfides; and zero-valent iron, to continue to generate reducing conditions and generate H_2S for the SRB, and to form Fe oxyhydroxides that would sorb or co-precipitate other contaminants.

4.2.3. Aerobic wetlands

Constructed wetland systems are designed to simulate natural wetlands, in that the latter are significant sinks for contaminant trace elements (Dinges, 1982; Munger *et al.*, 1995). Contaminated mine waste waters are purified by passing through wetlands *via* several processes, including sedimentation, adsorption onto the wetland substrate, uptake by plants and microbial oxidation or reduction and sequestering of the new species in minerals (*e.g.* Matagi *et al.*, 1998; Sheroan and Sheoran, 2006). Aerobic wetlands are normally used to treat NMD waters (Johnson and Hallberg, 2005). Iron and manganese oxides, oxyhydroxides and hydroxides are formed by hydrolysis and/or oxidation according to reactions such as (1), (2) and (12)–(14), above, and the following:

$$4Mn^{2+}_{(aq)} + O_{2\,(g)} + 6H_2O_{(l)} = 4MnOOH_{(s)} + 8H^+_{(aq)} \qquad (17)$$

The oxidation of Fe and Mn can be aided by neutrophilic Fe-oxidizing and Mn-oxidizing bacteria, including *Gallionella ferruginea*, which resides at the aerobic/anaerobic boundary, and *Leptothrix spp.*, which breaks down organically complexed iron (Johnson and Hallberg, 2005). Depending on the pH, the Fe and Mn oxides, oxyhydroxides and hydroxides can co-precipitate or sorb other metallic and metalloid contaminants, including Cu, Cd, Ni and Zn (Stumm and Morgan, 1981). The rates of removal of these contaminants are variable, but can be high, between 75 and 98% (Noller *et al.*, 1994).

4.2.4. Anaerobic composter reactors and wetlands

Composter reactors and other constructed wetlands rely on anaerobic conditions to remediate AMD waters. The processes involved are the same as those in reactions 15 and 16, where the actions of SRB organisms produce metal sulfides that sequester the contaminants. Although the aim of such reactors and wetlands is to produce sulfides, other phases such as metal carbonates and hydroxides can be formed (Neculita *et al.*, 2007). Furthermore, the role of Fe(III)-reducing bacteria in altering solution chemistry and precipitating secondary minerals that sequester contaminants has not been investigated fully (Kalin *et al.*, 2006). Sulfate reduction can only occur when Fe(III) is reduced fully (Stumm and Morgan, 1981), so Fe(III)-reducing bacteria should play a significant role in anaerobic bioreactors and wetlands. Kalin *et al.* (2006) also pointed out that the type of organic wastes used is of key importance to the functioning of these systems, with highly refined organic substances such as ethanol and methanol being the most effective per unit addition (Jones and Gusek, 2004; Tsukamoto and Miller, 2004; Kalin *et al.*, 2006), even though they are the most costly.

The removal of manganese is often problematic in mine-drainage remediation schemes such as constructed wetlands. This is because manganese does not easily form sulfides and because high pH conditions (>8) for the abiotic oxidation of Mn(II) to Mn(IV) (which in turn forms insoluble oxyhydroxide compounds) are required (Hallberg and Johnson, 2005). This was demonstrated at the Wheal Jane pilot passive treatment plant (PPTP) which used several pre-treatment cells, including oxic and anoxic limestone drains, to remove several metals, including manganese. Manganese removal was poor in the cells with pH <5, but better in higher pH (>7) cells. Hallberg and Johnson (2005) added ferromanganese nodules and manganese-oxidizing bacteria to the PPTP, and demonstrated improved Mn removal. This improved scheme was also shown to have a smaller footprint than the original PPTP, suggesting it had great promise for removing soluble Mn from mine waste waters.

4.2.5. Packed bed iron-oxidation bioreactors

Iron-oxidizing bioreactors rely on the activities of iron-oxidizing autotrophic archaea and bacteria such as *Acidithiobacillus ferrooxidans* (*e.g.* Mousavi *et al.*, 2007) to oxidize the Fe(II) of AMD and produce Fe(III) oxyhydroxides (see reactions 11–14). Other Fe(II)-oxidizing prokaryotes that have different affinities for Fe(II) and flourish best at specific Fe(II) concentration, temperature and pH ranges, are probably involved in this process and thus could be exploited for this technology, but these have not been characterized in the same detail as *At. ferrooxidans* (Hallberg and Johnson, 2001).

5. Conclusions

The global quantity of mine wastes is enormous and presents a significant problem in terms of where it may be stored or disposed of. This issue becomes ever more pressing as the global demand for land both for urban and agricultural use increases as a function of increasing population. Furthermore, the waste often contains significant concentrations of potentially toxic contaminants. Therefore, it is incumbent on the extractive industries, both as a whole and as the constituent individual companies, to address these issues if these organizations are to have an operating 'social licence'. The reuse of materials is problematic, principally because of the issue of potentially toxic contaminants. However, if society is to move from a linear towards a circular economy, further research is necessary in order that alternative uses for this, the largest of the anthropocentric waste streams, might be found. Many of the elemental, and hence mineralogical, components of the waste stream are redox active. This review addresses this property in particular giving some prospective and current relevant applications.

Mine wastes can potentially be the source of metal and metalloid contamination to the wider environment. This contamination is normally, but not universally, associated with acidification (AMD). Using chemical means to increase the pH of the effluent emanating from a waste depositary reduces the acidity, but it is not of itself a redox-based procedure. It does, however, allow for the precipitation of solid phases, most significantly redox-reactive ferric or oxidized Fe minerals (*e.g.* goethite and hematite) which will go on to act as major contaminant sinks.

Redox-reactive minerals play a very important role in the stabilization of contaminants within an impoundment. Every effort should be made to optimize conditions (*e.g.* Eh, pH and hydration state) to facilitate this stabilization. Metal-tolerant plants are used to accumulate specific metals of concern, although care must be taken that these plants do not enter the food chain.

The treatment of metal-mine drainage by the use of sulfate-reducing bacteria is a clear way forward and, if implemented on a sufficiently large scale, may potentially lead to the economic recovery of targeted metals. Permeable reactive barriers use either inorganic, organic or, most often, a combination of approaches to 'clean' a contaminated groundwater flow, while allowing for basin recharge. Similarly, both aerobic and anaerobic wetland treatments rely on micro-organism consortia to remove contaminant metals and metalloids from mine drainage. Packed bed iron-oxidation bioreactors also utilize this combination approach. Hence, it is only by the application of thorough, ongoing and multi-disciplinary research programmes that the potentially serious adverse environmental effects of the extractive industries can hope to be contained satisfactorily.

References

Ahn, J.S., Chon, C.M., Moon, H.S. and Kim, K.W. (2003) Arsenic removal using steel manufacturing byproducts as permeable reactive materials in mine tailing containment systems. *Water Research*, **37**, 2478–2488.

Anastas, P.T. and Zimmerman, J.B. (2003) Design through the 12 principles of green engineering. *Environmental Science & Technology*, **37**, 94A–101A.

Andreini, C., Bertini, I., Cavallaro, G., Holliday, G. L. and Thornton, J.M. (2008) Metal ions in biological catalysis: from enzyme databases to general principles. *Journal of Biological Inorganic Chemistry*, **13**, 1205–1218.

Arnold, D.E. (1991) Diversion wells – a low-cost approach to treatment of acid mine drainage. Pp. 39–50 in: *Proceedings of the 12th Annual West Virginia Surface Mine Drainage Task Force Symposium*. Charleston, West Virginia, Mining and Reclamation Association.

Azapagic, A. (2004) Developing a framework for sustainable development indicators for the mining and minerals industry. *Journal of Cleaner Production*, **12**, 639–662.

Baker, B.J. and Banfield, J.F. (2003) Microbial communities in acid mine drainage. *FEMS Microbiology Ecology*, **44**, 139–152.

Banks, D., Younger, P.L., Arnesen, R.T., Iversen, E.R. and Banks, S.B. (1997) Mine-water chemistry: the good, the bad and the ugly. *Environmental Geology*, **32**, 157–174.

Bartzas, G. and Komnitsas, K. (2010) Solid phase studies and geochemical modelling of low-cost permeable reactive barriers. *Journal of Hazardous Materials*, **183**, 301–308.

Bartzas, B., Komnitsas, K. and Paspaliaris, I. (2006) Laboratory evaluation of Fe0 barriers to treat acidic leachates. *Minerals Engineering*, **19**, 505–514.

Bekmezci, O.K., Ucar, D., Kaksonen, A.H. and Sahinkaya, E. (2011) Sulfidogenic bioreatment of synthetic acid mine drainage and sulfide oxidation in anaerobic baffled reactor. *Journal of Hazardous Materials*, **189**, 670–676.

Bearcock, J.M., Perkins, W.T. and Pearce, N.J.G. (2011) Laboratory studies using naturally occurring "green rust" to aid metal mine water remediation. *Journal of Hazardous Materials*, **190**, 466–473.

Blowes, D.W., Al, T., Lortie, L., Gould, W.D. and Jambor, J.L. (1995) Microbiological, chemical, and mineralogical characterization of the Kidd Creek mine tailings impoundment, Timmins area, Ontario. *Geomicrobiology Journal*, **13**, 13–31.

Blowes, D.W., Ptacek, C.J., Benner, S.G., McRae, C.W.T. and Puls R.W. (2000) Treatment of dissolved metals and nutrients using permeable reactive barriers. *Journal of Contaminant Hydrology*, **45**, 120–135.

Blowes, D.M., Ptacek, C.J., Jambor, J.L. and Weisener, C.G. (2003) The geochemistry of acid mine drainage. Pp. 149–204 in: *Treatise on Geochemistry*, Vol. **9** (H.D. Holland and K.K. Turekian, editors). Elsevier-Pergamon, Oxford, UK.

Boonstra, J., van Lier, R., Janssen, G., Kijkman, H. and Buisman, C.J.N. (1999) Biological treatment of acid mine drainage. Pp. 559–567 in: *Biohydrometallurgy and the Environment Toward the Mining of the 21st Century, vol. 9B* (R. Amils and A. Ballester, editors). Elsevier, Amsterdam.

Boulding, K.E. (1966) The economics of the coming spaceship earth. Pp. 3–14 in: *Environmental Quality in a Growing Economy* (H. Jarrett, editor). Resources for the Future/Johns Hopkins University Press. Baltimore, Maryland, USA.

Brown, G.E. and Calas, G. (2011) 'Environmental mineralogy'. Understanding element behavior in ecosystems. *Comptes Rendus Geoscience*, **343**, 90–112.

Bussiere, B.B.B. (2007) Colloquium 2004: Hydrogeotechnical properties of hard rock tailings from metal mines and emerging geoenvironmental disposal approaches. *Canadian Geotechnical Journal*, **44**, 1019–1052.

Caliskan, S., Ozkaya, I., Caliskan, M. and Arslan, M. (2008) The effects of nitrogen and iron fertilization on growth, yield and fertilizer use efficiency of soybean in a Mediterranean-type soil. *Field Crops Research*, **108**, 126–132.

Canepa, P., Kossoff, D., Hudson-Edwards, K., Dubbin, W. and Alfredsson, M. (2009) Hematite contaminated by heavy metals. *Geochimica et Cosmochimica Acta*, **73**, A189.

Caraballo, M.A., Santofimia, E. and Jarvis, A.P. (2010) Metal retention, mineralogy, and design considerations of a mature permeable reactive barrier (PRB) for acidic mine water drainage in Northumberland, U.K. *American Mineralogist*, **95**, 1642–1649.

Caraballo, M.A., Macias, F., Nieto, J.M., Castillo, J., Quispe, D. and Ayora, C. (2011a) Hydrochemical performance and mineralogical evolution of a dispersed alkaline substrate (DAS) remediating the highly polluted acid mine drainage in the full-scale passive treatment of Mina Esperanza (SW Spain). *American Mineralogist*, **96**, 1270–1277.

Caraballo, M.A., Macias, F., Rotting, T.S., Nieto, J.M. and Ayora, C. (2011b) Long term remediation of highly polluted acid mine drainage: a sustainable approach to restore the environmental quality of the Odiel river basin. *Environmental Pollution*, **159**, 3613–3619.

Castelo-Branco, M., Santos, J., Moreira, O., Oliveira, A., Pereira Pires, F., Magalhaes, I., Dias, S., Fernandes, L., Gama, J. and Vieira e Silva, J. (1999) Potential use of pyrite as an amendment for calcareous soil. *Journal of Geochemical Exploration*, **66**, 363–367.

Chang, I.S., Shin, P.K. and Kim, B.H. (2000) Biological treatment of acid mine drainage under sulphate-reducing conditions with solid waste materials as substrate. *Water Research*, 34, 1269–1277.

Cotter-Howells, J.D., Champness, P.E. and Charnock, J.M. (1999) Mineralogy of Pb-P grains in the roots of *Agrostis capillaris* L. by ATEM and EXAFS. *Mineralogical Magazine*, **63**, 777–789.

Cravotta, C.A. and Trahan, M.K. (1999) Limestone drains to increase pH and remove dissolved metals from acidic mine drainage. *Applied Geochemistry*, **14**, 581–606.

Craw, D., Rufaut, C., Haffert, L. and Paterson, L. (2007) Plant colonization and arsenic uptake on high arsenic mine wastes, New Zealand. *Water, Air and Soil Pollution*, **179**, 351–364.

Dengxin, L., Guolong, G., Fanling, M. and Chong, J. (2008) Preparation of nano-iron oxide red pigment powders by use of cyanided tailings. *Journal of Hazardous Materials*, **155**, 369–377.

Denholm, C., Danehy, T., Busler, S., Dolence, R. and Dunn, M. (2008) Sustainable passive treatment of mine drainage: demonstration of manganese resource recovery. *Proceedings of the American Society of Mining and Reclamation*, Richmond, Virginia, USA, pp. 285–297.

Dengxin, L., Guolong, G., Fanling, M. and Chong, J. (2008) Preparation of nano-iron oxide red pigment powders by use of cyanided tailings. *Journal of Hazardous Materials*, **155**, 369–377.

DeSisto, S.L., Jamieson, H.E. and Parsons, M.B. (2011) Influence of hardpan layers on arsenic mobility in historical gold mine tailings. *Applied Geochemistry*, **26**, 2004–2018.

Dinges, R. (1982) *Natural Systems for Water Pollution Control*. Can Nostrand Reinhold Co, New York, 252 p.

Divya, S. (2007) *Use of gold ore tailings (GOT) as a source of micronutrients to sunflower (Helianthus annuus L.) in a vertisol.* MSc. thesis: University of Agricultural Sciences, Dharwad, India.

Drissi, S.H., Refait, M., Abdelmoula, M. and Genin, J.M.R. (1995) The preparation and thermodynamic properties of Fe(II)-Fe(III) hydroxide-carbonate (green rust 1); Pourbaix diagram of iron in carbonate-containing aqueous media. *Corrosion Science*, **37**, 2025–2041.

Driehaus, W., Seith, R. and Jekel, M. (1995) Oxidation of arsenate (III) with manganese oxides in water treatment. *Water Research*, **29**, 297–305.

Dubey, B. and Townsend, T. (2004) Arsenic and lead leaching from the waste derived fertilizer ironite. *Environmental Science & Technology*, **38**, 5400–5404.

Fang, Y., Gu, Y., Kang, Q., Wen, Q. and Dai, P. (2011) Utilization of copper tailing for autoclaved sand-lime brick. *Construction and Building Materials*, **25**, 867–872.

Fortin, D. and Beveridge, T.J. (1997) Microbial sulfate reduction within sulfide mine tailings. Formation of diagenetic Fe sulfides. *Geomicrobiology Journal*, **14**, 1–21.

Fuller, C.C., Piana, M.J., Bargar, J.R., Davis, J.A. and Kohler, M. (2002) Evaluation of apatite materials for use in permeable reactive barriers for the remediation of uranium-contaminated groundwater. Pp. 255–280 in: *Handbook of Groundwater Remediation Using Permeable Reactive Barriers – Applications to Radionuclides, Trace Metals, and Nutrients* (D.L. Naftz, S.J. Morrison, J.A. Davis and C.C. Fuller, editors). Academic Press, San Diego, California, USA.

Ghosh, M. and Singh, S.P. (2005) A review on phytoremediaiton of heavy metals and utilization of its by products. *Asian Journal on Energy and Environment*, **6**, 214–231.

Gibert, O., Röttling, T., Cortina, J., de Pablo, J., Ayora, C., Carrera, J. and Bolzicco, J. (2011) In-situ remediaiton of acid mine drainage using a permeable reactive barrier in Aznalcóllar (Sw Spain). *Journal of Hazardous Materials*, **191**, 287–295.

Gordon, R.B., Bertram, M. and Graedel, T.E. (2006) Metal stocks and sustainability. *Proceedings of the National Academy of Science*, **103**, 1209–1214.

Graedel, T., Bertram, M., Kapur, A., Reck, B. and Spatari, S. (2004) Exploratory data analysis of the multilevel anthropogenic copper cycle. *Environmental Science & Technology*, **38**, 1253–1261.

Grandlic, C.J., Mendez, M.O., Chorover, J., Machado, B. and Maier, R.M. (2008) Plant growth-promoting bacteria for phytostabilization of mine tailings. *Environmental Science & Technology*, **42**, 2079–2084.

Gu, B., Phelps, T.F., Liang, L., Dickey, M.J., Kinsall, B.L. and Jacobs, G.K. (1999) Biogeochemical dynamics in zero-valent iron columns: implications for permeable reactive barriers. *Environmental Science & Technology*, **33**, 2170–2177.

Hadlington, S. (2012) Setback for fool's gold photovoltaics. *RSC Chemistry World*, http://www.rsc.org/chemistryworld/2012/08/setback-fools-gold-photovoltaics.

Hallberg, K.B. and Johnson, D.B. (2001) Biodiversity of acidophilic microorganisms. *Advances in Applied Microbiology*, **49**, 37–84.

Hallberg, K.B. and Johnson, D.B. (2005) Biological manganese removal from acid mine drainage in constructed wetlands and prototype bioreactors. *Science of the Total Environment*, **338**, 115–124.

Hammarstrom, J.M., Sibrell, P.L. and Belkin, H.E. (2003) Characterization of limestone reacted with acid mine drainage in a pulsed limestone bed treatment system at the Friendship Hill National Historical Site, Pennsylvania, USA. *Applied Geochemistry*, **18**, 1705–1721.

Hammond, A. (1988) Mining and quarrying wastes: a critical review. *Engineering Geology*, **25**, 17–31.

Hayes, S.M., O'Day, P.A., Webb, S.M., Maier, R.M. and Chorover, J. (2011) Changes in zinc speciation with mine tailings acidification in a semiarid weathering environment. *Environmental Science & Technology*, **45**, 7166–7172.

Hebbern, C.A., Laursen, K.H., Ladegaard, A.H., Schmidt, S.B., Pedas, P., Bruhn, D., Schjoerring, J.K., Wulfsohn, D. and Husted, S. (2009) Latent manganese deficiency increases transpiration in barley (*Hordeum vulgare*). *Physiologia Plantarum*, **135**, 307–316.

Hedin, R.S. (2003) Recovery of marketable iron oxide from mine drainage in the USA. *Land Contamination and Reclamation*, **11**, 93–98.

Hedin, R.S., Watzlaf, G.R. and Nairn, R.W. (1994) Passive treatment of acid-mine drainage with limestone. *Journal of Environmental Quality*, **23**, 1338–1345.

Hudson-Edwards, K.A., Jamieson, H.E. and Lottermoser, B.G. (2011) Mine wastes: past, present, future.

Elements, **7**, 375–380.

Jambor, J.L. and Raudsepp, M. (2005) Mineralogy of permeable reactive barriers for the attenuation of subsurface contaminants. *The Canadian Mineralogist*, **43**, 2117–2140.

Jenkins, H. and Yakovleva, N. (2006) Corporate social responsibility in the mining industry: Exploring trends in social and environmental disclosure. *Journal of Cleaner Production*, **14**, 271–284.

Johnson, D.B. (2000) Biological removal of sulfurous compounds from inorganic wastewaters. Pp. 175–206 in: *Environmental Technologies to Treat Sulfur Pollution: Principles and Engineering* (P.N.L. Lens and L.W. Hulshoff Pol, editors). International Association of Water Quality, London.

Johnson, D.B. and Hallberg, K.B. (2005) Acid mine drainage remediation options: a review. *Science of the Total Environment*, **338**, 3–14.

Jones, J. and Gusek, J. (2004) Scaling up design challenges for large scale sulphate reducing bioreactors. In: *Proceedings of the 2004 Ontario MEND Workshop, Sludge Management and Treatment of Weak Acid or Neutral pH Drainage*. May 26–27, 2004, Sudbury, Ontario, Canada, p. E7.

Kalin, M., Fyson, A. and Wheeler, W.N. (2006) The chemistry of conventional and alternative treatment systems for the neutralization of acid mine drainage. *Science of the Total Environment*, **366**, 395–408.

Keith, C. and Vaughan, D.J. (2000) Mechanisms and rates of sulphide oxidation in relation to the problems of acid rock (mine) drainage. Pp. 117–139 in: *Environmental Mineralogy: Microbial Interactions, Anthropogenic Influences, Contaminated Land and Waste Management* (J. Cotter-Howells, L. Campbell, E. Valsami-Jones and M. Batchelder, editors). Mineralogical Society of Great Britain & Ireland Book Series, **9**, London.

Knight, B., Zhao, F.J., McGrath, S.P. and Shen, Z.G. (1997) Zinc and cadmium uptake by the phyeraccumulator Thlaspi caerulescnes in contaminated soils and its effects on the concentration and chemical speciation of metals in soil solution. *Plant and Soil*, **197**, 71–78.

Kossoff, D., Hudson-Edwards, K.A., Dubbin, W.E., Alfredsson, M. and Geraki, T. (2012) Cycling of As, P, Pb and Sb during weathering of mine tailings: implications for fluvial environments. *Mineralogical Magazine*, **76**, 1209–1228.

Kumpiene, J., Fitts, J.P. and Mench, M. (2012) Arsenic fractionation in mine spoils 10 years after aided phytostabilization. *Environmental Pollution*, **166**, 82–88.

Lee, G., Bigham, J.M. and Faure, G. (2002) Removal of trace metals by coprecipitation with Fe, Al and Mn from natural waters contaminated with acid mine drainage in the Ducktown Mining District, Tennessee. *Applied Geochemistry*, **17**, 569–581.

Legodi, M. and De Waal, D. (2007) The preparation of magnetite, goethite, hematite and maghemite of pigment quality from mill scale iron waste. *Dyes and Pigments*, **74**, 161–168.

Li, C., Sun, H., Yi, Z. and Li, L. (2010) Innovative methodology for comprehensive utilization of iron ore tailings: Part 2: The residues after iron recovery from iron ore tailings to prepare cementitious material. *Journal of Hazardous Materials*, **174**, 78–83.

Lin, Y.T. and Huang, C.P. (2008) Reduction of chromium (VI) by pyrite in dilute aqueous solutions. *Separation and Purification Technology*, **63**, 191–199.

Lottermoser, B.G. (2010) *Minewastes: Characterization, Treatment and Environmental Impacts*, 3rd edition. Springer, Berlin, 400 pp.

Lottermoser, B.G. (2011) Recycling, reuse and rehabilitation of mine wastes. *Elements*, **7**, 405–410.

Lupton, M.K., Rojas, C., Drohan, P. and Bruns, M.A. (2013) Vegetation and soil development in compost-amended iron oxide precipitates at a 50-year-old acid mine drainage barrens. *Restoration Ecology*, **21**, 320–328.

Macias, F., Caraballo, M.A., Rotting, T.S., Perez-Lopez, R., Nieto, J.M. and Ayora, C. (2012) From highly polluted Zn-rich acid mine drainage to non-metallic waters: implementation of a multi-step alkaline passive treatment system to remediate metal pollution. *Science of the Total Environment*, **433**, 323–330.

Matagi, S.V., Swai, D. and Mugabe, R. (1998) A review of heavy metal removal mechanisms in wetlands. *African Journal for Tropical Hydrobiology and Fisheries*, **8**, 23–35.

Mayes, W.M., Potter, H.A.B. and Jarvis, A.P. (2009) Novel approach to zinc remove from circum-neutral mine wastes using pelletised recovered hydrous ferric oxide. *Journal of Hazardous Materials*, **162**, 512–520.

Mendez, M.O., Glenn, E.P. and Maier, R.M. (2007) Phytostabilization potential of quailbush for mine tailings.

Journal of Environmental Quality, **36**, 245–253.

Mousavi, S.M., Yaghmaei, S. and Jafari, A. (2007) Influence of process variables on biooxidation of ferrous sulfate by an indigenous Acidithiobacillus ferrooxidans. Part II: Bioreactor experiments. *Fuel*, **86**, 993–999.

Munger, A.S., Shutes, R.B.E., Revitt, D.M. and House, M.A. (1995) An assessment of metal removal from highway runoff by a natural wetland. *Water Science and Technology*, **32**, 169–175.

Naftz, D.L., Fuller, C.C., Davis, J.A., Morrison, S.J., Feltcorn, E.M., Freethey, G.W., Rowland, R.C., Wilkowske, C. and Piana, M. (2002) Field demonstration of three permeable reactive barriers to control uranium contamination in groundwater, Fry Canyon, Utah. Pp. 401–434 in: *Handbook of Groundwater Remediation Using Permeable Reactive Barriers – Applications to Radioculides, Trace Metals, and Nutrients* (D.L. Naftz, S.J. Morrison, J.A. Davis and C.C. Fuller, editors). Academic Press, San Diego, California, USA.

Neculita, C.M., Zagury, G.J. and Bussiere, B. (2007) Passive treatment of acid mine drainage in bioreactors using sulfate-reducing bacteria: critical review and research needs. *Journal of Environmental Quality*, **36**, 1–16.

Noller, B.N., Woods, P.H. and Ross, B.J. (1994) Case studies of wetland filtration of mine waste water in constructed and naturally occurring systems in northern Australia. *Water Science and Technology*, **29**, 257–266.

Nordstrom, D.K. (2011) Mine waters: acidic to circumneutral. *Elements*, **7**, 393–398.

Nordstrom, D.K. and Alpers, C.N. (1999) Geochemistry of acid mine waters. Pp. 133–160 in: *The Environmental Geochemistry of Mineral Deposits* (G.S. Plumlee and M.J. Logsdon, editors). Reviews in Economic Geology, **6A**.

Oyekola, O.O., van Hille, R.P. and Harrison, S.T. (2010) Kinetic analysis of biological sulphate reduction using lactate as carbon source and electron donor: Effect of sulphate concentration. *Chemical Engineering Science* **65**, 4771–4781.

Olías, M., Nieto, J.M., Sarmiento, A.M., Cerón, J.C. and Cánovas, C.R. (2004) Seasonal water quality variations in a river affected by acid mine drainage: the Odiel River (South West Spain). *Science of The Total Environment*, **333**, 267–281.

Oliveira, M.L.S., Ward, C.R., Izquierdo, M., Sampaio, C.H., de Brum, I.A.S., Kautzmann, R.M., Sabedot, S., Querol, X. and Silva, L.F.O. (2012) Chemical composition and minerals in pyrite ash of an abandoned sulphuric acid production plant. *Science of the Total Environment*, **430**, 34–47.

Oyekola, O.O., van Hille, R.P. and Harrison, S.T. (2010) Kinetic analysis of biological sulphate reduction using lactate as carbon source and electron donor: Effect of sulphate concentration. *Chemical Engineering Science*, **65**, 4771–4781.

Pérez-López, R., Sáez, R., Álvarez-Valero, A.M., Nieto, J.M. and Pace, G. (2009) Combination of sequential chemical extraction and modelling of dam-break wave propagation to aid assessment of risk related to the possible collapse of a roasted sulphide tailings dam. *Science of the Total Environment*, **407**, 5761–5771.

Pérez-López, R., Macías, F.., Caraballo, M.A., Nieto, J.M., Román-Ross, G. and Tucoulou, R. (2011) Mineralogy and geochemistry of Zn-rich mine-drainage precipitates from an MgO passive treatment system by synchrotron-based X-ray analysis. *Environmental Science & Technology*, **45**, 7826–7833.

Piatak, N.M. and Seal, R.R. II (2010) Mineralogy and the release of trace elements from slag from the Hegeler zinc smelter, Illinois (USA). *Applied Geochemistry*, **25**, 302–320.

Puthussery, J., Seefeld, S., Berry, N., Gibbs, M. and Law, M. (2010) Colloidal iron pyrite (FeS_2) nanocrystal inks for thin-film photovoltaics. *Journal of the American Chemical Society*, **133**, 716–719.

Rankin, W.J. (2011) *Minerals, Metals and Sustainability: Meeting Future Material Needs*. CRC Press, Boca Raton, Florida, USA, 440 pp.

Rappold, T. and Lackner, K. (2010) Large scale disposal of waste sulfur: From sulfide fuels to sulfate sequestration. *Energy*, **35**, 1368–1380.

Ritchie, V.J., Ilgen, A.G., Mueller, S.H., Trainor, T.P. and Goldfarb, R.J. (2013) Mobility and chemical fate of antimony and arsenic in historic mining environments of the Kantishna Hills District, Denali National Park and Preserve, Alaska. *Chemical Geology*, **335**, 172–188.

Rojas, C., Martínez, C.E. and Bruns, M.A. (2014) Fe biogeochemistry in reclaimed acid mine drainage

precipitates – Implications for phytoremediation. *Environmental Pollution*, **184**, 231–237.
Rosner, D., Markowitz, G. and Lanphear, B. (2005) J. Lockhart Gibson and the discovery of the impact of lead pigments on children's health: a review of a century of knowledge. *Public Health Reports*, **120**, 296–300.
Runkel, M. and Sturm, P. (2009) Pyrite roasting, an alternative to sulphur burning, www.saimm.co.za/Journal/v109n08p491.pdf, accessed 15/11/2012.
Santibáez, C., Verdugo, C. and Ginocchio, R. (2008) Phytostabilization of copper mine tailings with biosolids: implications for metal uptake and productivity of *Lolium perenne*. *Science of the Total Environment*, **395**, 1–10.
Saxena, M. and Dhimole, L.K. (2006) Utilization and value addition of copper tailing as an extender for development of paints. *Journal of Hazardous Materials*, **129**, 50–57.
Scherer M.M., Balko B.A. and Tratnyek P.G. (1998) The role of oxides in reduction reactions at the metal–water interface. Pp. 301–322 in: *Mineral–Water Interfacial Reactions: Kinetics and Mechanisms* (D.L. Sparks and T.J. Grundl, editors). ACS Symposium Series No. 715, American Chemical Society, Washington, D.C.
Scott, R.B., Popescu, I.C., Crane, R.A and Noubactep, C. (2011) Nano-scale metallic iron for the treatment of solutions containing multiple inorganic contaminants. *Journal of Hazardous Materials*, **186**, 280–287.
Shah, B., Shah, A. and Singh, R. (2009) Sorption isotherms and kinetics of chromium uptake from wastewater using natural sorbent material. *International Journal of Environmental Science & Technology*, **6**, 77–90.
Sheoran, A.S. and Sheoran, V. (2006) Heavy metal removal mechanism of acid mine drainage in wetlands: a critical review. *Minerals Engineering*, **19**, 105–116.
Shukurov, N., Kodirov, O., Peitzsch, M., Kersten, M., Pen-Mouratov, S. and Steinberger, Y. (2014) Coupling geochemical, mineralogical and microbiological approaches to assess the health of contaminated soil around the Almalyk mining and smelter complex, Uzbekistan. *Science of the Total Environment*, **476-477**, 447–459.
Sivakugan, N., Rankine, R., Rankine, K. and Rankine, K.S. (2006) Geotechnical considerations in mine backfilling in Australia. *Journal of Cleaner Production*, **14**, 1168–1175.
Smedley, P.L. and Kinniburgh, D.G. (2002) A review of the source, behaviour and distribution of arsenic in natural waters. *Applied Geochemistry*, **17**, 517–568.
Spatari, S., Bertram, M., Gordon, R.B., Henderson, K. and Graedel, T.E. (2005) Twentieth century copper stocks and flows in North America: a dynamic analysis. *Ecological Economics*, **54**, 37–51.
Stoltz, E. and Greger, M. (2002) Accumulation properties of As, Cd, Cu, Pb and Zn by four wetland plant species growing on submerged mine tailings. *Environmental and Experimental Botany*, **47**, 271–280.
Stumm, W. and Morgan, J. (1981) *Aquatic Chemistry, second edition*. John Wiley and Sons, New York, 780 pp.
Tamgale, S.D. (2005) *Improving efficiency of copper ore tailings (COT) – a source of micronutrients*. PhD. Thesis, University of Agricultural Sciences, Dharwad, India.
Tsukamoto, T., and Miller, G. (2004) Lateral flow, flushable bioreactors for the treatment of acid mine drainage. In: *Proceedings of the 2004 Ontario MEND Workshop, Sludge Management and Treatment of Weak Acid or Neutral pH Drainage*. May 26–27, 2004, Sudbury, Ontario, Canada, p. E6.
Turner, D. and McCoy, D. (1990) Anoxic alkaline drain treatment system, a low cost acid mine drainage treatment alternative. Pp. 73–75 in: *Symposium of Surface Mining Hydrology, Sedimentology and Reclamation* (D.H. Graves and R.W. DeVore, editors). University of Kentucky, Lexington, Kentucky, USA.
United States Environmental Protection Agency (2012) Release of heavy metals from Ironite. *Land Risk Management Research*, https://archive.epa.gov/nrmrl/archive-lrpcd/web/html/ironite.html. Accessed 08/11/2012.
United States Environmental Protection Agency (2014) Priority pollutants. https://www.epa.gov/sites/production/files/2015-09/documents/priority-pollutant-list-epa.pdf
Wadia, C., Alivisatos, A.P. and Kammen, D.M. (2009) Materials availability expands the opportunity for large-scale photovoltaics deployment. *Environmental Science & Technology*, **43**, 2072–2077.
Wang, T., Muller, D.B. and Graedel, T. (2007) Forging the anthropogenic iron cycle. *Environmental Science & Technology*, **41**, 5120–5129.
Warhurst, A. (2001) Corporate citizenship and corporate social investment. *Journal of Corporate Citizenship*, **1**, 57–73.

Waybrant, K.R., Blowes, D.W. and Ptacek, C.J. (1998) Selection of reactive mixtures for use in permeable reactive walls for treatment of mine drainage. *Environmental Science & Technology*, **32**, 1972–1979.

Whitehead, P.G., Cosby, B.J. and Prior, H. (2005) The Wheal Jane wetlands model for bioremediation of acid mine drainage. *Science of the Total Environment*, **338**, 125–135.

Wiersma, J.V. (2005) High rates of Fe-EDDHA and seed iron concentration suggest partial solutions to iron deficiency in soybean. *Agronomy Journal*, **97**, 924–934.

Wilkin, R.T. and McNeil, M.S. (2003) Laboratory evaluation of zero-valent iron to treat water impacted by acid mine drainage. *Chemosphere*, **53**, 715–725.

Williams, A.G.B., Scheckel, K.G., Tolaymat,T. and Impellitteri, C.A. (2006) Mineralogy and characterization of arsenic, iron, and lead in a mine waste-derived fertilizer. *Environmental Science & Technology*, **40**, 4874–4879.

Xingyu, L., Zou, G., Wang, S., Zou, L., Wen, J., Ruan, R. and Wang, D. (2013) A novel low pH sulfidogenic bioreactor using activated sludge as carbon source to treat acid mine drainage (AMD) and recovery metal sulfides: pilot scale study. *Minerals Engineering*, 48, 51–55.

Yamasaki, H., Pilon, M. and Shikanai, T. (2008) How do plants respond to copper deficiency? *Plant Signaling and Behavior*, **3**, 231–232.

Yang, C., Chen, Y., Peng, P., Li, C., Chang, X. and Wu, Y. (2009) Trace element transformations and partitioning during the roasting of pyrite ores in the sulfuric acid industry. *Journal of Hazardous Materials*, **167**, 835–845.

Yellishetty, M., Karpe, V., Reddy, E.H., Subhash, K.N. and Ranjith, P.G. (2008) Reuse of iron ore mineral wastes in civil engineering constructions: A case study. *Resources, Conservation and Recycling*, **52**, 1283–1289.

Younger, P.L. and Wolkersdorfer, C. (2004) Mining impacts on the fresh water environment: technical and managerial guidelines for catchment scale management. *Mine Water and the Environment*, **23**, 2–80.

Zhang, S., Xue, X., Liu, X., Duan, P., Yang, H., Jiang, T., Wang, D. and Liu, R. (2006) Current situation and comprehensive utilization of iron ore tailing resources. *Journal of Mining Science*, **42**, 403–408.

Zhao, F.-Q., Zhao, J. and Liu, H.-J. (2009) Autoclaved brick from low-silicon tailings. *Construction and Building Materials*, **23**, 538–541.

Zhou, Z., Xu, L., Xie, J. and Liu, C. (2009) Effect of manganese tailings on capsicum growth. *Chinese Journal of Geochemistry*, **28**, 427–431.

Ziemkiewicz, P.F., Skousen, J.G., Brant, D.L., Sterner, P.L. and Lovett, R.J. (1997) Acid mine drainage treatment with armored limestone in open limestone channels. *Journal of Environmental Quality*, **26**, 1017–1024.

Zouboulis, A., Kydros, K. and Matis, K. (1995) Removal of hexavalent chromium anions from solutions by pyrite fines. *Water Research*, **29**, 1755–1760.

EMU Notes in Mineralogy, Vol. 17 (2017), Chapter 13, 379–403

Functionality and applications of redox-active porous metal–organic framework structures

THOMAS DEVIC* and CHRISTIAN SERRE*

Institut Lavoisier Versailles, UMR 8180 CNRS – Université de Versailles St Quentin en Yvelines, 45 avenue des Etats-Unis, 78035 Versailles, France,
E-mail: thomas.devic@uvsq.fr

While Metal–Organic Frameworks (MOFs) are now considered as promising materials for various applications related to their tunable porosity (gas storage and capture, separation of fluids, controlled drug release, *etc.*), their potential redox-activity has rarely been exploited. This chapter gives an overview of this rapidly growing field from its origins to 2013. Aspects related to the nature of the redox-active moieties in play, which can be either organic or inorganic, as well as the synthesis procedures available to produce electroactive MOFs, are discussed first. In a second step, redox-driven insertion of neutral or charged species within MOFs and associated characterization techniques are presented. Finally, applications of redox-active MOFs, which range from enhanced gas capture to Li-ion battery electrodes are overviewed briefly.

1. Introduction

Porous Coordination Polymers (PCPs), also known as Metal–Organic Frameworks (MOFs) have emerged in recent decades as a promising class of new crystalline porous materials. These micro- or meso-porous hybrid solids are built-up from inorganic entities (isolated cations, molecular clusters, chains, layers, *etc.*) connected through polytopic ligands (typically poly-carboxylates, phosphonates and azolates) to define cavities of various sizes and shapes. These materials can have a vast variety of chemical composition, both on the inorganic and inorganic parts. The inorganic cation is either an alkali, alkaline earth, transition metal, main group element, lanthanide or actinide, while the organic ligand is built up from a core surrounded by a variable number of complexing groups, with both the nature of the core and the complexing groups being adjustable. This leads to an unprecedented structural diversity, with materials presenting tunable pore size, pores shape and surface functionality. In terms of

* Current addresses:
Thomas Devic: Institut des Matériaux Jean Rouxel, UMR 6502 CNRS – Université de Nantes, 2 rue de la Houssinière, 44322 Nantes cedex 03, France. E-mail: Thomas.devic@cnrs-imn.fr
Christian Serre: Institut des Matériaux Poreux de Paris, FRE 2000 CNRS Ecole Normale Supérieure, Ecole Supérieure de Physique et de Chimie Industrielles de Paris, 75005 Paris, France

DOI: 10.1180/EMU-notes.17.11

porosity, these solids can exhibit a 'permanent' micro- (pore diameter Ø = 3−20 Å) or meso- (Ø = 20−60 Å) porosity with surface area and pore volume sometimes exceeding 6000 $m^2\ g^{-1}$ and 2 $cm^3\ g^{-1}$ respectively, or for some of them a 'dynamic' porosity, with their pore size and shape adapting to its content while the solid remains crystalline (Kitagawa *et al.*, 2004; Fletcher *et al.*, 2005; Férey and Serre, 2009). The rich functionality and designability of MOFs compounds make them promising new materials in various fields of application, mostly related to their porosity (fluid capture, separation and storage, controlled drug release) but also dependent on their chemical composition (catalysis, magnetism, conductivity, optics, bioactivity). Such aspects have been covered by exhaustive reviews; the interested reader is referred to special issues in *Chem. Rev. (MOFs special issue, 2012)* and *Chem. Soc. Rev.* (Long and Yaghi, 2009) devoted solely to MOFs, and covering both the synthesis and structural aspects of such materials, as well as all the applications cited above.

Nevertheless, benefits associated with the potential redox activity of these porous frameworks have been considered only in recent years. One of the main interests in these systems relies on the variety of redox centres available: purely inorganic (cation), organic (core of the ligand) or hybrid (complex with non-innocent ligands leading to delocalized electronic states). On the other hand, in order to exploit these centres, one needs to combine noticeable porosity (permanent or dynamic), redox activity, stability (especially during the redox process) and sometimes reasonably good charge transport properties.

This chapter gives an overview of this emerging field, covering the following items: (1) the nature of the redox-active moieties in play; (2) their specific synthesis procedures (especially electrochemically assisted synthesis); and (3) the exploitation of the redox activity, and the related applications in various fields, highlighting when possible the interplay between redox activity, porosity and other physicochemical properties.

2. Nature of the redox-active species

As noted above, MOFs can be made from a great variety of inorganic cations. These metal centres and their environment play a major role in the physicochemical properties of the MOFs. These cations can be distinguished by the number of oxidation states available in typical synthesis and working conditions of MOFs (mainly aerobic conditions). As an example, some metal centres have a single, positively charged stable oxidation state, such as Al^{III}, Zn^{II}, Cd^{II}, Sc^{III} and Zr^{IV}. For the resulting solids, attempts to reduce post-synthetically the metal centres would lead typically to the degradation of the material, either through the formation of metal(0) or the irreversible reduction of the ligand. On the other hand, transition metals can lead to more than one accessible (positive) oxidation state: *e.g.* the 3d cations $Ti^{IV/III}$, $Mn^{IV/III/II}$, $Co^{III/II}$, $Ni^{III/II}$, as well as $V^{V/IV/III}$, $Fe^{III/II}$ and $Cu^{II/I}$. The latter examples are the most studied, and MOFs in which these cations are oxidized/reduced to a significant extent (typically from 10 to 100%), sometimes in a reversible manner, have been reported. Examples of $Fe^{III/II}$

porous coordination polymers are shown in Fig. 1. In these systems, although the nature of the cation is of paramount importance, the structure of the solid, and especially the nature of the inorganic sub-units, also plays a major role in the resulting properties (see below).

Regarding the organic part, carboxylates are probably the most widely used complexing groups, although alternative groups, such as phosphonates, azolates and pyridine derivatives are also encountered frequently. Most of the archetypical MOFs (*e.g.* MOF-5, HKUST-1, MIL-53, MIL-100) are based on simple, polycarboxylate ligands built up from an aromatic, electrochemically inert, core, such as 1,3,5-benzenetricarboxylate (BTC) or 1,4-benzenedicarboxylate (BDC). It is, nevertheless, relatively easy to construct electroactive ligands, simply by grafting complexing groups on a known redox active core: examples of both oxidizable and reducible ligands, based on either purely organic (*e.g.* triarylamine, tetrathiafulvalene, naphthalenetetracarboxydiimide, viologen (Zhang *et al.*, 2001; Sun *et al.*, 2006), tetracyanoquinodimethane (Abrahams *et al.*, 2010)) or hybrid (*e.g.* ferrocene (Fc) (Hirai *et al.*, 2012), $Ru(bipy)_3$ (Kent *et al.*, 2010), Ru(salen) (Falkowski *et al.*, 2011)) cores suitable for the preparation of redox active MOFs are shown in Fig. 2. In these cases, the redox activity of the core is only weakly affected by the complexation (see *e.g.* the case of TCNQ-based CPs (Li *et al.*, 2013)). Alternatively, the binding group can be involved in the redox processes, such as in the case of quinone/phenol (Abrahams *et*

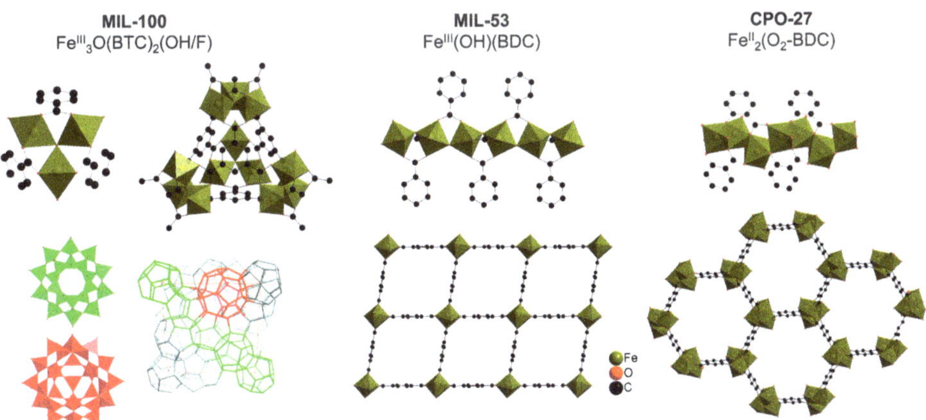

Figure 1. Examples of redox-active Fe-based MOFs: crystal structure of Fe^{III} or Fe^{II}-based porous coordination polymers. From left to right: MIL-100 (Yoon *et al.*, 2010), MIL-53 (Férey *et al.*, 2007) and CPO-27 (Bloch *et al.*, 2011), where MIL and CPO stand for Materials Institute Lavoisier and Coordination Polymers from Oslo, respectively. BTC = 1,3,5-benzenetricarboxylate; BDC = 1,4-benzenedicarboxylate; O_2-BDC = 2,5-dioxido-1,4-benzenedicarboxylate. Top: view of the inorganic subunits (iron trimer for MIL-100, chains for MIL-53 and CPO-27), bottom: view of the structure along the axis of the pores (MIL-53 and CPO-27). For MIL-100, the structure consists of hybrid supertetrahedra comprising the aforementioned trimers and BTC ligands (top), which are further connected to give rise to a zeotypic structure exhibiting two types of mesoporous cages (bottom).

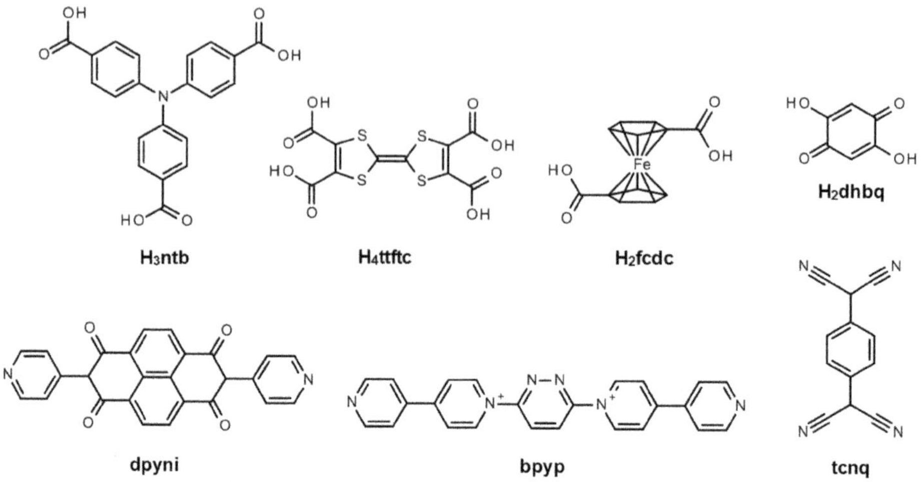

Figure 2. Examples of redox-active ligands used for the preparation of MOFs: they can be oxidizable (H_3ntb, H_4ttftc, H_2fcdc), reducible (dpyni, bpyp, tcnq) or both (H_2dhbq). ntb = 4,4′,4″-nitrilotrisbenzoate (Cheon and Paik Suh, 2009); ttftc = tetrathiafulvalenetetracarboxylate (Nguyen *et al.*, 2010); fcdc; 1,1′-ferrocenedicarboxylate; dhbq = 2,5-dioxidobenzoquinone (Abrahams *et al.*, 2012); dpyni = N,N'-di-(4-pyridyl)-1,4,5,8-naphthalenetetracarboxydiimide (Mulfort and Hupp, 2007); bpyp = 3,6-bis(4′-pyridyl-1′-pyridinio)pyridazine (Zhang *et al.*, 2001); tcnq = 7,7,8,8-tetracyano-p-quinodimethane (Miyasaka *et al.*, 2010).

al., 2012) (see H_2dhbq in Fig. 2) or dithiolene derivatives (Kambe *et al.*, 2013). Here, the final redox response will rely on the interplay between the inorganic cation and the ligand.

3. Synthesis issues and electrochemical characterization

3.1. Conventional synthesis

Although alternative methods exist, MOFs are typically synthesized upon heating, either at ambient or autogeneous pressure (solvothermal conditions). Typically, a metallic salt and a prescribed organic ligand (often in its acidic form when anionic) are mixed in a polar solvent (water, alcohol, amide). As an example, the formation of MIL-53(Fe) follows the reaction:

$$H_2BDC + Fe^{III}Cl_3 + H_2O \rightarrow Fe^{III}(OH)(BDC) + 3HCl$$

In a few cases, inorganic or organic additives (hydrochloric, hydrofluoric, nitric or acetic acid, sodium or potassium hydroxide, amine, *etc.*) can be added (Stock and Biswas, 2012). The use of redox-active ligands may require special care; these are often more fragile than the inactive ones as they could degrade under "harsh" solvothermal conditions such as high temperature or very acidic pH. This is especially true when one tries to embed them in a MOF framework in their radical forms. Indeed, examples of direct synthesis of MOFs comprising ligands either oxidized or reduced in their radical forms are rare (Maspoch *et al.*, 2003; Yong *et al.*, 2011), one rare example involving ttftc is discussed in section 3.2. In most cases, such radical-containing MOFs are generated by post-synthesis treatment under milder conditions.

In conventional syntheses, the formation of the coordination polymers implies only acid-base reactions. Alternatively, processes involving redox reactions can be envisaged: metal(0) can be used as a reactant, which can then be oxidized chemically *in situ* to produce the desired M^{n+} cations and further the hybrid solid. Indeed, few Fe^{III}- and Cr^{III}-based MOFs have been prepared from Fe^0 (Horcajada *et al.*, 2010; Nouar *et al.*, 2012) and Cr^0 (Férey *et al.*, 2004). One of the main interests in this technique is the modulation of the kinetics of the reaction. Oxidation and dissolution of metal(0) are the kinetically limiting steps of the MOFs formation, slowing down (compared to the use of metal salts) the whole growth process, but leading to better crystallinity of the final MOF. Although this process is often not fully controlled, and the nature of the oxidative species not always clearly identified, it can also be exploited to prepare mixed-metal MOFs.

3.2. Electrochemically assisted synthesis

A more rational strategy involves the electrochemical oxidation of the metal(0), allowing, in principle, careful control of the oxidation state and amount of the resulting cation. This strategy was first developed by BASF Corp to prepare the Cu^{II} trimesate HKUST-1 (currently sold as Basolite C-300) (Mueller *et al.*, 2006). In this procedure, a Cu^0 plate was electrolysed (galvanostatic conditions) in a solution of trimesic acid in methanol. The resulting solid was found to contain fewer impurities (probably nitrate anions) than that obtained from Cu^{II} nitrate by conventional synthesis. Martinez Joaristi *et al.* (2012) extended this approach to other MOFs of interest (and to other metals), such as Zn^{II} imidazolate ZIF-8, Al^{III} trimesate MIL-100 and Al^{III} terephthalate MIL-53 (Martinez Joaristi *et al.*, 2012). All these syntheses were carried out under mild conditions with temperature below the boiling point, but this approach was recently extended to solvothermal conditions (*i.e.* in a closed vessel with temperature above the boiling point), allowing the preparation of other MOFs such as MIL-100(Fe) (Campagnol *et al.*, 2013). Compared to conventional solvothermal conditions, electrochemically assisted synthesis usually allows us to speed up the reaction, and in some cases offer a better control of the particle size. This technique can also be used to grow MOFs coating on metallic surfaces: by partial oxidation of a Cu^0 electrode; Ameloot *et al.* were able to electrochemically prepare thick (>1 μm) dense films of HKUST-1 on a Cu plate (Ameloot *et al.*, 2009). Using printed circuit boards, they then extended this approach to the preparation of patterned surfaces, MOFs particles growing exclusively on areas covered by copper.

This electrochemically assisted method can also be used with redox-active ligands. The conventional hydrothermal reaction of H_4ttftc (see Fig. 2) with KCl was shown to produce two polymorphs formulated as $K_2(H_2ttftc)$ in which the ttf core is in its reduced form (ttf^0). Under electrochemical oxidative conditions, another compound of formula

$K(H_2ttftc)$ was isolated, the ttf core being in its oxidized cation radical form $ttf^{+\cdot}$ (Nguyen *et al.*, 2010).

As an alternative to the 'direct synthesis' route, redox-active moieties can be grafted, post-synthesis, on inactive porous frameworks, such as a Zn^{II}- or Al^{III}-based MOFs. This approach requires the presence of reactive groups (either organic or inorganic) at the surface of the pores, which can react to form covalent bonds with the redox species. It was mainly devoted to the introduction of redox-active coordination complexes within porous MOFs, such as ferrocene in MIL-53(Al) (Meilikhov *et al.*, 2009), V^{IV} (Ingleson *et al.*, 2008) and Mn^{II} (Bhattacharjee *et al.*, 2011) complexes in the Zn aminoterephthalate IRMOF-3, or Fe acetylacetonate in the Zn polyacarloxylate UMCM-1 (Tanabe *et al.*, 2010). These grafting reactions (single or multiple steps) usually require mild synthesis conditions, allowing the introduction of redox-active groups that can not be inserted by 'direct synthesis' methods, either because they would degrade or compete with other reactants for the formation of coordination bonds.

3.3. Electrochemical characterization

Numerous MOFs comprising potentially redox active moieties (mostly inorganic) have been reported. In order to assess their redox activity, few electrochemical characterizations of MOFs by solid state cyclic voltammetry (CV) have been reported (Domenech *et al.*, 2007; Zhang *et al.*, 2010; Abrahams *et al.*, 2012; Hirai *et al.*, 2012). In most cases, the solids were deposited on glassy carbon electrodes (frequently by drop-casting), or mixed with a carbon paste in order to achieve decent electronic conductivity suitable for CV measurements. Zn-, Cu- (Domenech *et al.*, 2007) and Fe- (Babu *et al.*, 2010) based MOFs have been studied. In all cases, it has been shown that the complete reduction of the metallic cation to M^0 is irreversible, and leads to the dissolution of the materials. Alternatively, for Fe^{III}-based MOFs, the reduction to Fe^{II} was found to be partially reversible, thanks to the similar coordination environment of both cations (Férey *et al.*, 2007; Fateeva *et al.*, 2010).

Combined spectro-electrochemical techniques (CV and UV-vis spectroscopy) have been proposed to study the redox activity of MOFs and shed some light on the nature of the reactive part of the framework (Usov *et al.*, 2012). This technique is particularly useful when the redox process leads to drastic changes in the light-absorption properties such as the organic redox centre of the dpyni type (see Fig. 2). Nevertheless, even if the redox reaction appears reversible by CV measurements, further characterization (IR spectroscopy, XRD analysis) is needed to assess the integrity of the MOF framework.

4. Redox-driven insertion

Post-synthesis insertion of guests within pores driven by redox processes is both a way to assess the redox activity and to use it. Depending on the chemical systems in play, it can involve either the oxidation or the reduction of the framework, insertion of either anionic, cationic or neutral species, both in molecular or polymeric form. Various examples are summarized in Fig. 3 and are discussed below.

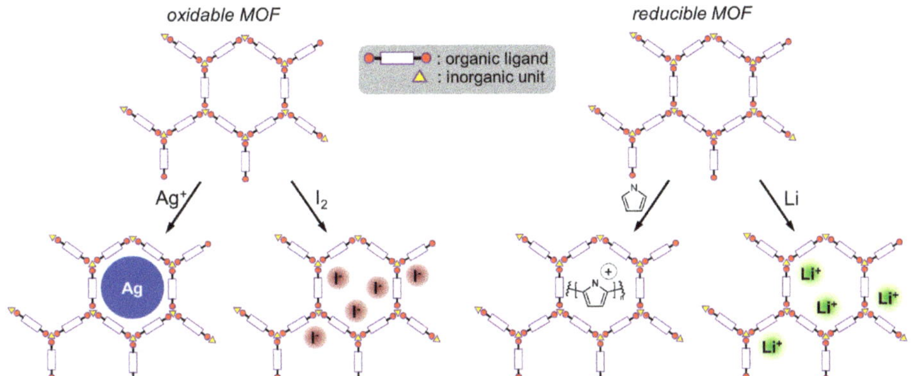

Figure 3. Redox-driven post-synthesis insertions of various guests within porous MOFs. From left to right: silver nanoparticles, iodides, oxidized polypyrrole and lithium ions.

4.1. Insertion of cationic species through the reduction of the framework

One of the most studied reactions is the insertion of alkali ions, especially lithium, but also sodium and potassium, into neutral, reducible matrices. Various examples involving the reaction of MOFs with alkaline derivatives, such as alkali alone (Mulfort and Hupp, 2007), alkyl lithium, alkali naphthalenide (Mulfort and Hupp, 2008) have been reported, mainly by Hupp and coworkers. They all involve the reduction of the organic part of the framework, such as naphthalenetetracarboxydiimide (dpyni) (Mulfort and Hupp, 2007) or a tetrazine derivative (Mulfort *et al.*, 2009). Similar to many insertion reactions, proving directly and beyond doubt that the ions are within the pores and not outside at the surface of the crystallites is still challenging. Nevertheless, 'indirect' clues could be given by X-ray diffraction (XRD), infrared (IR) analysis, gas sorption experiments, which in these cases all agreed with the insertion of the ions. Using this chemical method, limited amounts of ions are often introduced within the framework (5–30% of the theoretical value). The main difficulty lies in the nature of the reducing agents; when added in excess, these strong reducing agents may reduce species other than the expected ones and lead to the collapse of the framework.

Alternatively, lithium ions can be introduced electrochemically. Using two-electrode Swagelok-type cells (one made of a mixture of MOF and carbon black, the second one of a lithium foil), lithium ions were introduced in Fe^{III} carboxylate-based solids such as MIL-53(Fe) (Férey *et al.*, 2007) and MIL-68 (Fateeva *et al.*, 2010). Up to 60% of Fe^{III} was reduced to Fe^{II} in a reversible manner. The same strategy was applied to other solids based on oxidizable ligands, such as $K_2(H_2ttftc)$ or MIL-132/133 (Nguyen *et al.*, 2010). In the latter case, the process involving the de-insertion (oxidation)/insertion (reduction) of the alkali ions led to the amorphization of the material. It was nevertheless possible to reversibly oxidize/reduce 100% of the redox active centres.

More sophisticated species can also be introduced through the reduction of the framework. For example, using the V^{IV} MOF MIL-47, the structure of which is similar to that of MIL-53 but less flexible (see Fig. 1), Meilikhov *et al.* (2010) were able to introduce cobaltocenium complexes through the reduction of 50% of V^{IV} to V^{III}. Interestingly, using XRD, these species were located precisely within the pores proving unambiguously the insertion reaction.

Finally, a similar strategy was used by Yanai *et al.* (2008) to grow polypyrrole nanoslits. Pyrrole was first inserted between the layer of an Fe^{III}-based 2-D coordination polymer. The polymerization of pyrrole was initiated by addition of water, and involved both the departure of anion (sulfonate) from the interlayer space and the reduction of Fe^{III} to Fe^{II}. Similar reactions performed on the redox inert Co^{III} analogue were unsuccessful, highlighting the key role of the redox process during the polymerization reaction.

4.2. Insertion of neutral species into pores through the oxidation of the framework

The examples below deal with the synthesis of metal(0) nanoparticles (NPs). From a practical point of view, a metallic salt is inserted into the pores of the porous solid which is further reduced by the framework to produce metal(0) NPs, ideally with the shape and size controlled by those of the pores. Electron microscopy is then the technique of choice to characterize the resulting composites. This method is an alternative to the two-step synthesis (first inclusion, then reduction) developed for inert MOF matrices.

The redox-assisted method was first proposed by Hong *et al.* (2001) for the growth of ultrafine Ag^0 nanowires through the reduction of a molecular porous organic solid (here a hydroquinone derivative). This method was then applied to MOFs. Using redox-reactive Ni^{II} cyclam-based solids, the group of M.P. Suh prepared various Ag (Moon *et al.*, 2005), Au (Suh *et al.*, 2006) and Pd (Moon and Suh, 2010) NPs by this technique. A similar approach was used to prepare Pd NPs from a nitrilotrisbenzoate-based MOF (ntb, see Fig. 2) (Cheon and Paik Suh, 2009).

Although this method appears appealing, control of the reduction step, and hence of the size of the resulting nanoparticles is not straightforward. Even if the MOF may appear to remain intact from XRPD analysis, the aggregation of the metal atoms through the formation of strong metal–metal bonds often affords local bond breaking within the framework. As a consequence, the final nanoparticles are generally larger than the pores.

4.3. Insertion of anionic species through the oxidation of the framework

Halides can be inserted easily in oxidizable MOFs. In the case of iodine (the most studied halide), the reaction can be performed by suspending the solid in a solution of I_2, or even simply upon exposure of the solid to I_2 vapour.

Suh *et al.* (2006) first reported the insertion of iodide with the porous MOF BOF-1 (Choi and Suh, 2004). This is later built up from Ni^{II}(cyclam) ligands, and the insertion reaction was shown to be associated with the oxidation of up to 66% Ni^{II} to Ni^{III}, in agreement with electron paramagnetic resonance analyses. Reduction of iodine is known to afford various polyatomic anionic derivatives, which can be tricky to identify. In the present case, the redox reaction was performed from a single crystal to single crystal, allowing the unambiguous identification of the resulting anion, here the linear polyiodide I_3^-. Using a similar approach, but on an Fe^{II}-based material, Horike *et al.* (2013) incorporated V-shaped I_5^- anions in the pores. Augusti *et al.* (2009) considered the reactivity of the Hofmann clathrate $Fe(pz)[Pt^{II}(CN)_4]$ (py = pyrazine). Upon exposure to I_2, half of the square planar Pt^{II} ions were oxidized to Pt^{IV} while maintaining the crystal structure. The resulting I^- anions were found to be bound to the Pt^{IV} centres, leading to an octahedral geometry after oxidation. Similar results were obtained using Br_2 and Cl_2 as oxidizing agents.

5. Applications

In this section, properties or potential applications specifically associated with the redox activity will be described. They can appear after a first redox-driven redox insertion reaction, or be derived from the redox reaction itself. Note that the electrochemical applications of MOFs have been reviewed recently (Morozan and Jaouen, 2012); here, the focus is on those where the redox activity plays a significant role.

5.1. Enhanced gas storage or capture

The idea here is to tune the gas sorption properties of a given porous MOF through the modulation of the host–guest interactions. More precisely, in many cases attempts are made to increase these interactions in order to enhance the storage capacity at a given pressure or temperature range, or to increase the separation factor between two gases (one interacting strongly with the framework, the other, not).

Whatever the polarity of gas or vapour adsorbed, for a given pore size or geometry, polar walls usually lead to stronger host–guest interactions than non-polar ones. As a consequence, the introduction of charged species within the pores appears to offer a promising way of enhancing such interactions. With this in mind, the solids reduced by lithium or other alkalis (see Section 4.3) were considered for enhanced H_2 storage and CO_2 capture.

In the case of redox-active MOFs, not only the cations (similarly to what is done with cationic zeolites or MOFs) but also the reduced organic anions can act as strong adsorption sites. It should nevertheless be mentioned that for gas storage in rigid MOFs, such a strategy should be beneficial at low pressure or high temperature (enhancement of the interaction leads to greater adsorption capacity at low pore coverage), but detrimental at high pressure, due to the steric hindrance of the species inserted (Fig. 4).

Regarding H_2 storage, although high sorption capacity (up to 10 wt.%) can be achieved at low temperature (77 K) in few MOFs, this capacity drops significantly (<1%) at ambient temperature, hampering their practical use. Increasing the H_2–framework interaction is thus mandatory. Many theoretical studies have pointed

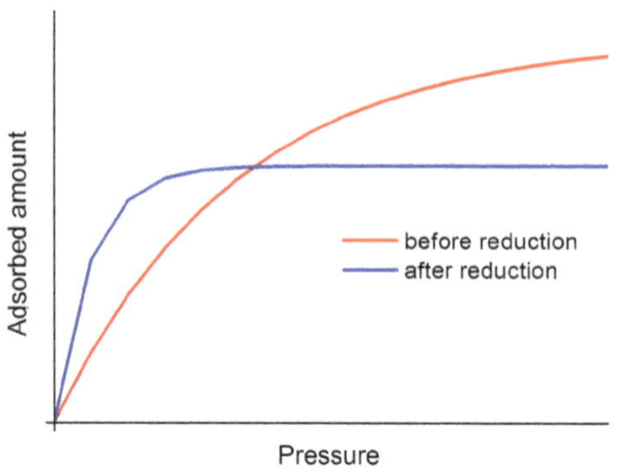

Figure 4. Idealized gas-sorption isotherm in a rigid MOF before (red line featuring exponential growth curve) and after (blue line, with logistic-like curve) reduction, illustrating the gain and loss of sorption capacity at low and high pressure, respectively.

out the significant benefits which should arise from redox insertion of alkali (Han and Goddard, 2007; Dalach *et al.*, 2008; Mavrandonakis *et al.*, 2008; Ghoufi *et al.*, 2012; Stergiannakos *et al.*, 2012; Han *et al.*, 2013). Nevertheless, experimental studies are less numerous and limited to two types of reducible interwoven structures (Mulfort and Hupp, 2007, 2008; Mulfort *et al.*, 2009). Upon slight alkali doping, H_2 capacities were found to increase. Nevertheless, the heats of adsorption were found to remain almost unchanged, indicating the absence of strong additional host-H_2 interactions upon alkali insertion. In these specific cases, the enhanced sorption capacity is probably associated with a structural change (shift of the interwoven networks), leading to an increase in the pore volume. Experimental data on other MOFs, preferably rigid, are still needed to validate this approach.

The same strategy was proposed for the capture of CO_2 with theoretical studies predicting an increase of the CO_2 *vs.* CH_4 selectivity upon alkali doping (Xu *et al.*, 2010; Wu *et al.*, 2010; Babarao and Jiang, 2011; Mu *et al.*, 2011). From an experimental standpoint, such a phenomenon was observed in an interwoven framework but attributed once again to a change of pore volume, leading to a decrease of the CH_4 adsorption (and thus increase of the selectivity) (Bae *et al.*, 2011). More recently, on similar systems, Leong *et al.* (2013) noticed an increase in the enthalpy of adsorption of CO_2 upon sodium doping, which was attributed to strong interactions between the CO_2 molecules and both alkali cations and organic anions. Here also, there is still a clear lack of experimental data on a broad range of systems to validate the effect of the reduction on CO_2 capture.

Mixed Pd^0 MOF systems obtained by redox processes (see section 4.2) were also considered for the enhanced adsorption of H_2 (Cheon and Paik Suh, 2009). For a porous Zn^{II} nitrilotrisbenzoate (ntb) MOF, incorporation of 3 wt.% of Pd^0 was shown to increase the H_2 sorption capacity, especially at room temperature (from 0.13 to 0.3 wt.% at 95 bar, 298 K), but with a slight decrease in the isosteric heat of adsorption. The benefit here was attributed to a spillover effect, *i.e.* the dissociation of H_2 molecules on the surface of the noble metal nanoparticles. Note, however, that the more chemically fragile structure of MOFs, compared to carbons or zeolites, leads often, upon inclusion of metal(0) NPs, to a strong reduction of the pore volume and surface area due to a partial degradation of the framework during the insertion or reduction process.

The redox activity of the framework itself, without any further insertion, can be used to modulate the separation properties. The mesoporous iron trimesate MIL-100(Fe) (see Fig. 1) is an example (Yoon *et al.*, 2010). In its raw form, it contains only $Fe^{III}O_6$ octahedra assembled in oxocentred trimeric units. Each cation bears a terminal ligand, which can be neutral (H_2O, 2/3) or anionic (OH^- or F^-, 1/3), leading to neutral moieties (Fig. 5, top). Activation at 150°C leads to the departure of coordinated water molecules and the appearance of coordinatively accessible pentacoordinated Lewis acid Fe^{III} sites. In this form, the solid was shown to separate propene a little from propane (Fig. 5, bottom), the separation arising from the increased interaction between the Fe^{III} sites and the C=C double bond. When the same solid is heated (in the absence of oxygen) at 250°C under vacuum, bound anions (F^-, OH^-) can also be partially eliminated, this phenomenon being associated with the reduction of Fe^{III} to Fe^{II}, as confirmed by Mössbauer spectroscopy. In this form, MIL-100(Fe) separates propane from propene more efficiently (Fig. 5, bottom), with a separation factor increasing from 5 to almost 30. This effect was attributed to a stronger interaction of propene with Fe^{II} than Fe^{III} sites, due to an electron back-donation effect of the reduced cation (Yoon *et al.*, 2010).

Coordinatively accessible Fe^{II} sites can also be used for the separation of O_2/N_2 mixtures. Such separation is usually based on kinetics effects, or preferential adsorption of N_2 over O_2, *e.g.* with cationic zeolites. Bloch *et al.* (2011) have shown that the CPO-27(Fe^{II}) solid (see Fig. 1) adsorbs O_2 preferentially over N_2, in a reversible manner at low temperature (T < 226 K). This effect was attributed to the

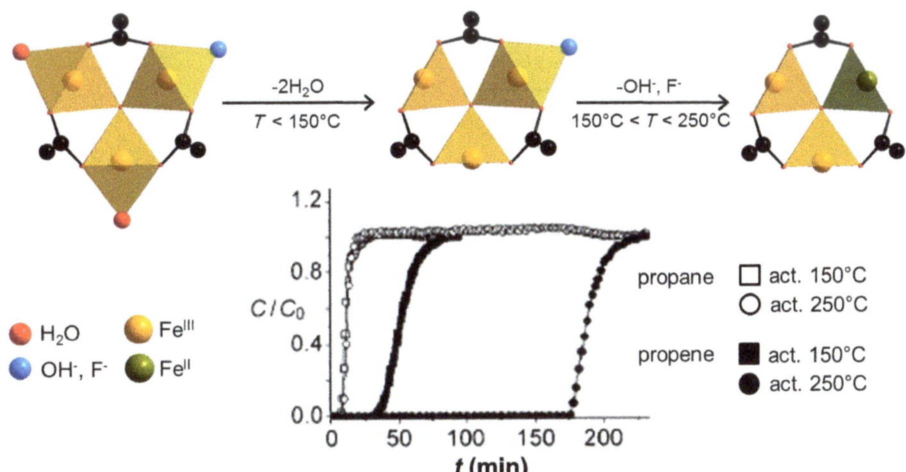

Figure 5. Top: illustration of the thermal reduction of Fe^{III} to Fe^{II} in the inorganic trimeric unit consisting of MIL-100(Fe); bottom: breakthrough curves on an equimolar binary mixture (P = 0.25 kPa) of propane and propene in MIL-100(Fe) after activation at 150°C (squares) and 250°C (circles) (lower part after Yoon *et al.*, 2010).

selective binding of O_2 to the Fe^{II} sites *via* electron transfer interactions. At higher temperature, as evidenced by Mössbauer spectroscopy, IR and neutron diffraction experiments, this binding becomes irreversible, and is associated with the partial oxidation of Fe^{II} to Fe^{III} and the formation of peroxide anions. The instability of the present material precludes any practical use, but other more stable Fe^{II}-based solids might lead to the same properties under experimental conditions closer to real separation processes.

5.2. MOFs as electrodes for Li-ion batteries

MOFs have been proposed as an alternative to conventional electrode materials such as layered oxides, phosphates, sulfates, *etc.* in Li-ion batteries.

The basic principle lies on the insertion/de-insertion of Li within the pores and subsequent reduction/oxidation of the material, leading to electronic motion, with the porosity favouring the ionic mobility (Fig. 6, top). On the contrary, the insulating nature of most MOFs (see below) appears as a strong limitation for such applications. As a consequence, mixing with black carbon (5–30 wt.%) to achieve significant electronic conductivity is necessary.

The first report of such an application deals with the use of the Zn^{II} 1,3,5-benzenetrisbenzoate MOF, labelled as MOF-177 (Li *et al.*, 2006). This solid exhibits a very high degree of porosity, but no reversible redox activity. As a consequence, the

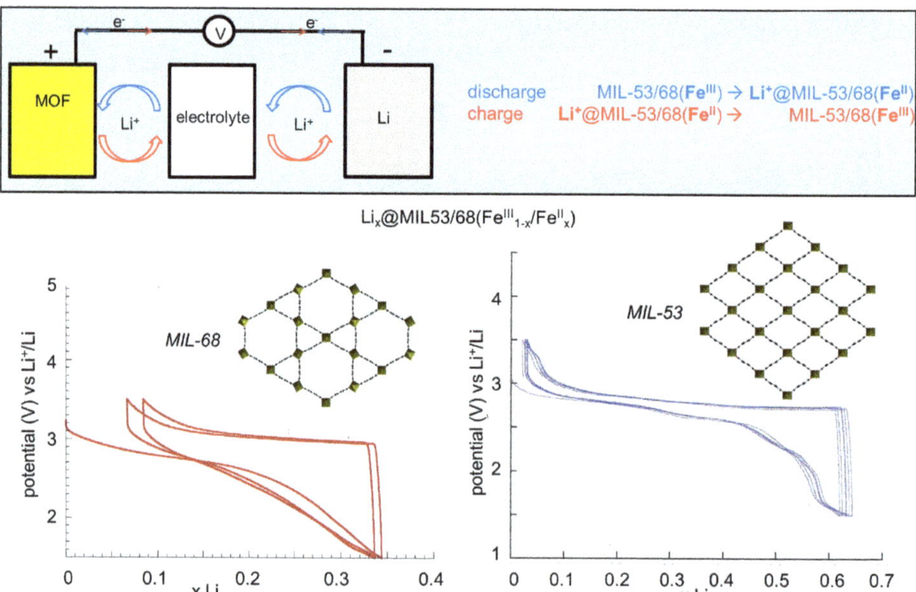

Figure 6. Schematic diagram of a Li ion battery using an Fe-based MOF as a positive electrode (upper) and the corresponding charge-discharge curve for the MIL-68 and MIL-53 materials (lower) at C/10 (1 eq Li in 10 h) and C/40 (1 eq Li in 40 h), respectively (after Férey *et al.*, 2007; Fateeva *et al.*, 2010).

reduction of the material involves the irreversible formation of Zn^0 and reduction of entrapped solvent molecules (N,N'-diethylformamide). Similar results were obtained for Mn^{II}-based MOFs (Liu *et al.*, 2013). Using Zn^{II} formate ($Zn(HCO_2)_2$), Saravanan *et al.* (2010) were, however, able to achieve large reversible capacities (up to 500 mAh g^{-1}), but once again associated with the complete degradation of the material to Li formate and Zn^0, which were then the redox-active species. MOFs can thus be considered here as 'sacrificial solids' to produce metal or Li oxide, which can then be used further as electrode materials (Morozan and Jaouen, 2012).

On the other hand, solids based on the reducible cation Fe^{III} were found to maintain their structure upon cycling. For the flexible Fe^{III} terephthalate MIL-53(Fe) (Férey *et al.*, 2007), up to 0.6 Li/Fe were reversibly inserted into the structure, this insertion being associated with the reduction of Fe^{III} to Fe^{II} at ~3 V *vs.* Li^+/Li, as indicated by Mössbauer spectroscopy and EXAFS studies (Fig. 6, bottom) (Combarieu *et al.*, 2009). When considering its rigid polymorph MIL-68(Fe), which consists of identical inorganic chains (see Fig. 1) but presents smaller triangular and larger hexagonal pores (see Fig. 6), only ~0.35 Li/Fe were reversibly introduced (Fateeva *et al.*, 2010). This indicates that the limitation of capacity is not related to pore-filling issues, but is rather an intrinsic property of the inorganic sub-unit. Indeed, theoretical calculations indicate that the introduction of 0.5 Li/Fe leads to a stabilized, localized mixed-valence Fe^{III}/Fe^{II} state, while further reduction leads to an irreversible structural transformation (Combelles *et al.*, 2010). In order to increase the capacity (75 mAh g^{-1} for MIL-53(Fe)), organic redox-active species (quinone) were introduced in the pores. As expected, the initial capacity increases, but reduces rapidly upon cycling, as a consequence of the departure of the organic species from the pores (Combarieu *et al.*, 2009).

Alternatives lie in the use of redox-active organic linkers. A few simple aromatic hydrocarbonated Li dicarboxylates (terephthalate, muconate) were shown to be reversibly reducible, with rather large capacity (up to 200 mAh g^{-1}), but at low potential (typically <~1 V *vs.* Li^+/Li) (Armand *et al.*, 2009; Walker *et al.*, 2011; Burkhardt *et al.*, 2013). Using more reducible organic cores (aromatic ketone (Xiang *et al.*, 2008; Chen *et al.*, 2009; Walker *et al.*, 2010; Goriparti *et al.*, 2013), dimiide (Renault *et al.*, 2011) or ttf (Nguyen *et al.*, 2010)), it was possible to reach higher potential (up to 3.5 V *vs.* Li^+/Li), but with generally lower capacity. The approach was extended recently to Na ion batteries (Park *et al.*, 2012). In these systems, contrary to the cases of MIL-53(Fe) and MIL-68(Fe), alkali Li insertion/de-insertion processes are associated with structure changes, the ions being constituent parts of the framework and not simply inserted within cavities.

Although MOFs can act as positive electrodes for alkali ion batteries, they can often not compete with other materials regarding capacities. Improvements can certainly be found upon a more appropriate choice of combination of both organic and inorganic redox-active species within the solids (Hmadeh *et al.*, 2012).

5.3. MOFs as supercapacitors

As large porosity in the electrode is one of the key points needed to achieve large capacitance in supercapacitors, MOFs have been proposed as potential candidates (Morozan and Jaouen, 2012; Gong *et al.*, 2013). Han *et al.* (2013) studied the behaviour of Co^{II} dicarboxylates (such as the Co^{II} terephthalate MOF-71), for example. These solids were shown to present large capacitances (up to 200 F g^{-1}) and good cyclabilites (up to 1000 cycles) (Lee *et al.*, 2012, 2013). Nevertheless, no clear evidence of the mechanism in play was given. More recently, Miles *et al.* (2013) showed that this activity was not associated with the MOF itself, but with the Co^{II} hydroxide formed upon hydrolysis of the MOF. Although the MOF solid here again seems to be a precursor of the truly active material, its initial structure may affect the morphology of the resulting hydroxide and further influence the final activity.

5.4. Electronic conductivity

Conducting coordination polymers are not particularly rare, and have been reviewed recently (Givaja *et al.*, 2011; Hendon *et al.*, 2012), as were the requirements for device preparation (Allendorf *et al.*, 2011). Nevertheless, most of these solids, the properties of which are generally associated with a mixed-valence state (either organic, *e.g.* with ttf (Nguyen *et al.*, 2010), dithiolene or tcnq (Avendano *et al.*, 2011; Ballesteros-Rivas *et al.*, 2011; Miyasaka, 2013; Zhang *et al.*, 2013) ligand derivatives, or inorganic), are dense phases without any noticeable porosity. Examples combining both porosity and 'true' electronic conductivity are still very scarce and will be discussed below (Takaishi *et al.*, 2009; Kobayashi *et al.*, 2010; Leclerc *et al.*, 2011; Gándara *et al.*, 2012; Hmadeh *et al.*, 2012; Narayan *et al.*, 2012; Sun *et al.*, 2013). Indeed, achievement of a significant electronic conductivity within a solid requires both a conducting pathway (related to orbital overlap), and, if possible, a mixed-valence system, hosted either by the organic or inorganic part. While the 3-D connectivity in MOF structures should allow the transport through delocalized valence or conduction bands, many porous solids are built up from carboxylate or phosphonate binding groups, which prevent any efficient overlap (Hendon *et al.*, 2012; Yang *et al.*, 2012a,b). As a consequence, such MOFs are mostly insulators ($\sigma < 10^{-8}–10^{-10}$ S cm^{-1}). Note that few carboxylate-type MOFs have been listed as 'semi-conductors' (such as MOF-5, UiO-66, MIL-125), based simply on optical measurements and not on conduction measurements. Although charges might be photogenerated (see Section 5.6), the electronic structure of such solids might be better described from molecular orbital than truly delocalized bands (when calculated, the latter appear flat (Hendon *et al.*, 2012; Yang *et al.*, 2012a,b); as a consequence, they can not be mentioned reasonably as semi-conductors (Nasalevich *et al.*, 2013)).

The combination of porosity and electronic conductivity was first reported in the 3-D MOF formulated $Cu^{II}(Ni^{II}(pdt))$, (pdt = pyrazine 2,3-dithiolate). This solid exhibits a significant porosity ($S_{BET} \approx 400$ m^2 g^{-1}), and consists of dithiolene moieties, known to lead readily to delocalized electronic states. Upon partial oxidation with iodine, the conductivity was shown to increase from ~10^{-8} to 10^{-4} S cm^{-1}, due to the partial oxidation of the dithiolene complexes, and thus to the charge transport through a hybrid organic-inorganic pathway (Takaishi *et al.*, 2009; Kobayashi *et al.*, 2010). More recently, other examples involving either oxygenated (benzenetriscatecholate) (Hmadeh *et al.*, 2012) or sulfurated (2,5-dithioterephthalate) (Sun *et al.*, 2013) ligands have been reported. In those cases, using either transient current measurement or I-V curves, reasonably high charge mobility (~10^{-2} cm^2 V^{-1} s^{-1}) and conductivity (~10^{-1} S cm^{-1}) were observed, respectively.

Inorganic conductive pathways can also be achieved, *e.g.* through M-*X*-M (*X* = halide) (Otsubo *et al.*, 2011) or M-O-M chains (Leclerc *et al.*, 2011). If one considers the MIL-47 solid (see Section 4.1), it can be isolated both in its oxidized V^{IV} (formula $V^{IV}O(bdc)$) and its reduced V^{III} form (formula $V^{III}(OH)(bdc)$). Complex impedance spectroscopy showed that both compounds are insulators with similar conductivity (Fig. 7). Under controlled oxidative conditions, it is possible, nevertheless, to isolate an intermediate mixed-valence V^{III}/V^{IV} form (formula $V_x^{III}V_{1-x}^{IV}(OH_x)(bdc)$), the conductivity of which increases by a factor 10^3-10^7. Such enhancement was attributed to the partial oxidation of the cation leading to (poor) semi-conducting behaviour (Leclerc *et al.*, 2011).

Finally, conductivity can also be achieved through a pure organic pathway. Narayan *et al.* (2012) reported recently a Zn^{II}-based MOF consisting of an extended ttf tetracarboxylate derivative, which combines both significant porosity (S_{BET} > 600 m^2 g^{-1}) and high charge mobility (~10^{-1} cm^2 V^{-1} s^{-1}). This high mobility was attributed to an electronic delocalization involving the overlap of the extended π orbital of the organic ligands, arising from the specific packing of the ttf cores.

5.5. Catalysis

MOFs have been tested widely, mostly for acid, but also base or enantioselective catalysis, and these aspects have been reviewed (Farrusseng *et al.*, 2009; Corma *et al.*,

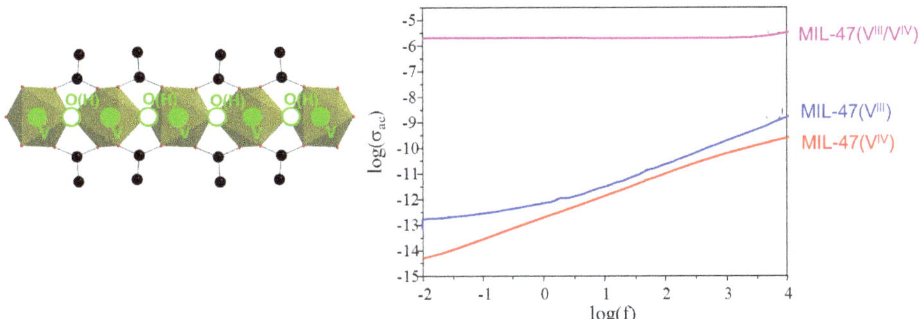

Figure 7. Left: View of the inorganic chain in MIL-47; right: real part of the conductivity obtained from complex impedance spectroscopy for MIL-47 in various states: fully reduced (MIL-47(V^{III}) or $V^{III}(OH)(BDC)$), fully oxidized (MIL-47(V^{IV}) or $V^{IV}(O)(BDC)$) and mixed valence (MIL-47(V^{III}/V^{IV}) or $V^{III}{}_x V_{1-x}^{IV}(OH_x)(BDC)$, x ~ 0.5) (after Leclerc *et al.*, 2011).

2010; Yoon *et al.*, 2012). In all cases, the idea is to exploit the confinement of the catalytic reaction within the pores to improve the catalytic performances (high TON, high selectivity, easy recovery, *etc.*) compared to either homogenous catalysts or dense heterogeneous ones. We selected a few examples where the catalytic activity is associated with redox processes (photocatalysis will be treated briefly in Section 5.6). Catalytic activity can, in these cases, be associated either with the inorganic nodes, or with the ligands. In the latter cases, the ligands are mostly complexes known to be active in homogenous catalysis (*e.g.* Schiff bases, porphyrins) bearing complexing groups at the periphery to act as linkers in the framework.

MFU-4, a porous Co^{II}-based bis-triazolate presenting coordinatively unsaturated metal sites (CUS) was reported by Denysenko *et al.*, (2012). These authors have shown that the solid can be reversibly oxidized to Co^{III} with molecular oxygen (which remains bound to the cation), and that this oxidized form is active for the oxidation of CO into CO_2. The Fe^{III} trimesate MIL-100(Fe) (see above) was found to be highly efficient for the Friedel-Craft benzylation of benzylchloride to diphenylmethane, while the redox inert Cr^{III} analogue is almost inactive (Horcajada *et al.*, 2007). The high activity of the Fe^{III}-based solid was attributed to its redox activity. As a consequence, most Fe-based MOFs presenting CUS (MIL-100, Fe(BTC), CPO-27,...) can be considered as potential heterogeneous catalysts (Dhakshinamoorthy *et al.*, 2010, 2011). V^{IV} MOFs of the MIL-47 type (see above) were considered by Leus *et al.* (2010, 2012) for the oxidation of cyclohexene by terbutylperoxide. Significant catalytic activities were found, without any obvious degradation of the framework. While the solids do not present any CUS when prepared, Leus *et al.* proposed a possible catalytic pathway, involving both the oxidation of V^{IV} to V^{V} and the reversible rupture of V-carboxylate bonds.

In the previous examples, the geometry of the catalytic site is the result of the structural arrangement, and thus cannot be fully 'designed'. Alternatively, using the knowledge extracted from homogeneous catalysis studies, it is possible to incorporate directly within the framework 'optimized' single-site catalytic centres, typically as the core of the ligands. Porphyrin-based MOFs are case examples (Zou and Wu, 2012). As Mn^{III}-based porphyrins are known to be active in homogeneous oxidation catalysis, few groups have reported the preparation of MOFs, either porous (Farha *et al.*, 2011) or not (Suslick *et al.*, 2005), involving a Mn^{III} porphyrin derivative as a linker, or simply occluded within the pores (Alkordi *et al.*, 2008). As expected, these solids have been found to be catalytically active both for the oxidation of alkanes and the epoxidation of alkenes.

Falkowski *et al.* (2011) reported a porous and chiral Zn^{II}-based MOF involving a Ru^{III} NN′-bis(salicylidene)ethylenediamine derivative as a ligand. Upon reaction with strong reducing agents such as $LiBEt_3H$ or $NaB(OMe)_3H$, the solid was reduced to an Ru^{II} form, remarkably in a single-crystal-to-single-crystal manner. Whereas the oxidized Ru^{III} form was found to be inactive, the reduced Ru^{II} form catalyses the asymmetric cyclopropanation of substituted olefins. The same group incorporated (bipy)(Cp^*)IrCl complexes (bipy= 2,2′-bipyridine, Cp* = pentamethylcyclopenta-dienyl) as ligands in a Zr-based dicarboxylate of the UiO-67 type (Wang *et al.*, 2011,

2012a). After oxidation of the complex with Ce^{IV}, the solid was found to be active for the oxidation of water. Thorough spectroscopic studies allowed unambiguous identification of the active catalytic site, with the Ir complex arising from the oxidation of the Cp^{*-} ring.

MOFs were also tested as electrocatalysts, focusing mostly on Cu-based systems. These solids were tested for oxidation reactions, such as transformation of methanol to dimethylcarbonate (Jia *et al.*, 2013) or oxidation of ethanol (Ishimoto *et al.*, 2013), and for reduction reactions, such as the reduction of CO_2 to oxalic acid (Senthil Kumar *et al.*, 2012) or the four-electron reduction of O_2 (Mao *et al.*, 2012). Nevertheless, the MOF itself is often degraded during the electrocatalytic process, the active species being the degradation product rather than the MOF. This renders the structure/ properties rationalization scheme inadequate.

5.6. Photo-induced redox activity

Light harvesting and photo-related properties (especially photocatalysis) of MOFs have been reviewed (Gomes Silva *et al.*, 2010; Wang *et al.*, 2012b; Horiuchi *et al.*, 2013). The introduction of adequate organic and inorganic moieties within the MOF allows charges (electrons and holes) to be generated upon light excitation, which can then be used either to activate substrates within the pore (photocatalysis) or migrate over long distances to be harvested *via* electron transfer reactions (Wang *et al.*, 2012b). In the latter case, as mentioned above, the insulating nature of most MOFs precludes efficient charge transport properties, contrary to what is found in dense phases (TiO_2, ZnO).

The photo-induced charge transfer could involve either the framework only or both the framework and the guest. For example, triazine carboxylate-based MOFs were shown to present photo-induced charge transfer, with the carboxylate group acting as the donor and the triazine as the acceptor (Fu *et al.*, 2011, 2012). Using a Zn dicarboxylate MOF with a methylviologen cation embedded in the pores, Fu *et al.* were also able to generate a 'purely organic' charge transfer, with the carboxylate acting here also as a donor, and the entrapped methylviologen as the acceptor. In this particular case, photochromic behaviour, associated with the oxidation of the methylviologen, was observed (Zeng *et al.*, 2012). In the case of porphyrin-based MOFs, such photo-induced charge transfers were used to perform photocatalysis, *e.g.* the reduction of H^+ to H_2 (Zn^{II} porphyrin) (Fateeva *et al.*, 2012) or oxidation of phenols and sulfides (Sn^{IV} porphyrin) (Xie *et al.*, 2011).

Inorganic nodes can also be involved in the charge-transfer process: the first example was reported by Alvaro *et al.* (2007) who observed a ligand-to-cluster charge transfer (LCCT) in the Zn terephthalate MOF-5 upon UV irradiation, with the holes generated and electrons being used to oxidize or reduce sacrificial agents, or to degrade phenol. Few prominent examples deal with the photoactive cation Ti^{IV}. Upon UV irradiation, the Ti^{IV} terephthalate MIL-125, the solely known porous crystalline Ti-based MOF, exhibits a strong colour change (from white to dark blue) (Dan-Hardi *et al.*, 2009). As confirmed by electron paramagnetic resonance studies, this phenomenon is associated with the reduction of Ti^{IV} to Ti^{III}, with entrapped benzylalcohol molecules acting as sacrificial donors. In the same vein, the amino-funtionalized analogue MIL-125-NH_2, which absorbs slightly more photons in the visible range (de Miguel *et al.*, 2012), was used to photocatalytically degrade CO_2 to formate (Fu, Sun *et al.*, 2012) and produce water from aqueous solution of triethanolamine (Horiuchi *et al.*, 2012). Similarly, the amino-functionalized Zr terephthalate UiO-66-NH_2 was shown to act as a good photocatalyst, both for the oxidation of cyclic alkanes and alcohols (Long *et al.*, 2012) and the reduction of Cr^{VI} to Cr^{III} in aqueous solutions (Shen *et al.*, 2013). In the latter case, the good chemical stability of Zr^{IV}-based MOFs is a clear advantage/prerequisite for such application.

6. Conclusion

While MOFs already appear promising in applications related to fluid capture and storage, the exploitation of their redox activity is still in its infancy. Indeed, literature dealing with such a phenomenon is far less extensive than those related to purely porosity-driven applications. By contrast, very diverse aspects of redox activity have been explored, ranging from halide capture, enhanced hydrogen storage and photocatalysis to charge transport. In each case, a few studies have established the proof of concept, but further efforts are needed to reach high performances. For other domains of application, such as biomedicine, redox activity has not yet been exploited. From a synthesis standpoint, the development of new materials is thus of utmost importance in order to improve the critical weak points of such materials, and achieve, for example: (1) high stabilities upon full oxidation or reduction of the active centre; or (2) good charge- (electron and hole) transport properties.

Acknowledgements

The authors acknowledge past and present coworkers, both in Versailles and elsewhere, and more specifically those involved in the chemistry (T.L.A. Nguyen, A. Fateeva, P. Horcajada and especially G. Férey), characterization (G. Clet, A. Vimont, M. Daturi and J.-M. Grenèche) and applications (R. Demir-Cakan, M. Morcrette, J.-M. Tarascon, S. Devautour-Vinot and J.-S. Chang) of redox-active MOFs.

References

Abrahams, B.F., Elliott, R.W., Hudson, T.A. and Robson, R. (2010) A new class of easily generated TCNQ2-based coordination polymers. *Crystal Growth and Design*, **10**, 2860–2862.

Abrahams, B.F., Bond, A.M., Le, T.H., McCormick, L.J., Nafady, A., Robson, R. and Vo, N. (2012) Voltammetric reduction and re-oxidation of solid coordination polymers of dihydroxybenzoquinone. *Chemical Communications*, **48**, 11422–11424.

Agusti, G., Ohtani, R., Yoneda, K., Gaspar, A.B., Ohba, M., Sanchez-Royo, J.F., Munoz, M.C., Kitagawa, S. and Real, J.A. (2009) Oxidative addition of halogens on open metal sites in a microporous spin-crossover coordination polymer. *Angewandte Chemie International Edition*, **48**, 8944–8947.

Alkordi, M.H., Liu, Y., Larsen, R.W., Eubank, J.F. and Eddaoudi, M. (2008) Zeolite-like metal–organic frameworks as platforms for applications: On metalloporphyrin-based catalysts. *Journal of the American*

Chemical Society, **130**, 12639–12641.
Allendorf, M.D., Schwartzberg, A., Stavila, V. and Talin, A.A. (2011) A roadmap to implementing metal–organic frameworks in electronic devices: Challenges and critical directions. *Chemistry – A European Journal*, **17**, 11372–11388.
Alvaro, M., Carbonell, E., Ferrer, B., Llabrés i Xamena, F.X. and Garcia, H. (2007) Semiconductor behavior of a metal–organic framework (MOF). *Chemistry – A European Journal*, **13**, 5106–5112.
Ameloot, R., Stappers, L., Fransaer, J., Alaerts, L., Sels, B.F. and De Vos, D.E. (2009) Patterned growth of metal–organic framework coatings by electrochemical synthesis. *Chemistry of Materials*, **21**, 2580–2582.
Armand, M., Grugeon, S., Vezin, H., Laruelle, S., Ribière, P., Poizot, P. and Tarascon, J.-M. (2009) Conjugated dicarboxylate anodes for Li-ion batteries. *Nature Materials*, **8**, 120–125.
Avendano, C., Zhang, Z., Ota, A., Zhao, H. and Dunbar, K.R. (2011) Dramatically different conductivity properties of metal–organic framework polymorphs of Tl(TCNQ): An unexpected room-temperature crystal-to-crystal phase transition. *Angewandte Chemie International Edition*, **50**, 6543–6547.
Babarao, R. and Jiang, J.W. (2011) Cation characterization and CO_2 capture in Li+-exchanged metal–organic frameworks: From first-principles modeling to molecular simulation. *Industrial and Engineering Chemistry Research*, **50**, 62–68.
Babu, K.F., Kulandainathan, M.A., Katsounaros, I., Rassaei, L., Burrows, A.D., Raithby, P.R. and Marken, F. (2010) Electrocatalytic activity of BasoliteTM F300 metal-organic-framework structures. *Electrochemistry Communications*, **12**, 632–635.
Bae, Y.-S., Hauser, B.G., Farha, O.K., Hupp, J.T. and Snurr, R.Q. (2011) Enhancement of CO_2/CH_4 selectivity in metal-organic frameworks containing lithium cations. *Microporous and Mesoporous Materials*, **141**, 231–235.
Ballesteros-Rivas, M., Ota, A., Reinheimer, E., Prosvirin, A., Valdés-Martinez, J. and Dunbar, K.R. (2011) Highly conducting coordination polymers based on infinite M(4,4′-bpy) chains flanked by regular stacks of non-integer TCNQ radicals. *Angewandte Chemie International Edition*, **50**, 9703–9707.
Bhattacharjee, S., Yang, D.-A. and Ahn, W.-S. (2011) A new heterogeneous catalyst for epoxidation of alkenes via one-step post-functionalization of IRMOF-3 with a manganese(ii) acetylacetonate complex. *Chemical Communications*, **47**, 3637–3639.
Bloch, E.D., Murray, L.J., Queen, W.L., Chavan, S., Maximoff, S.N., Bigi, J.P., Krishna, R., Peterson, V.K., Grandjean, F., Long, G.J., Smit, B., Bordiga, S., Brown, C.M. and Long, J.R. (2011) Selective binding of O_2 over N_2 in a redox active metal–organic framework with open iron(II) coordination sites. *Journal of the American Chemical Society*, **133**, 14814–14822.
Burkhardt, S.E., Bois, J., Tarascon, J.-M., Hennig, R.G. and Abruna, H.D. (2013) Li-carboxylate anode structure–property relationships from molecular modeling. *Chemistry of Materials*, **25**, 132–141.
Campagnol, N., Van Assche, T., Boudewijns, T., Denayer, J., Binnemans, K., De Vos, D. and Fransaer, J. (2013) High pressure, high temperature electrochemical synthesis of metal-organic frameworks: Films of MIL-100 (Fe) and HKUST-1 in different morphologies. *Journal of Materials Chemistry A*, **1**, 5827–5830.
Chen, H., Armand, M., Courty, M., Jiang, M., Grey, C.P., Dolhem, F., Tarascon, J.-M. and Poizot, P. (2009) Lithium salt of tetrahydroxybenzoquinone: Toward the development of a sustainable Li-ion battery. *Journal of the American Chemical Society*, **131**, 8984–8988.
Cheon, Y.E. and Paik Suh, M. (2009) Enhanced hydrogen storage by palladium nanoparticles fabricated in a redox-active metal–organic framework. *Angewandte Chemie International Edition*, **48**, 2899–2903.
Choi, H.J. and Suh, M.P. (2004) Dynamic and redox active pillared bilayer open framework: Single-crystal-to-single-crystal transformations upon guest removal, guest exchange, and framework oxidation. *Journal of the American Chemical Society*, **126**, 15844–15851.
Combarieu, G.D., Hamelet, S., Millange, F., Morcrette, M., Tarascon, J.-M., Férey, G. and Walton, R.I. (2009) In situ Fe XAFS of reversible lithium insertion in a flexible metal organic framework material. *Electrochemistry Communications*, **11**, 1881–1884.
Combelles, C., Ben Yahia, M., Pedesseau, L. and Doublet, M.-L. (2010) Design of electrode materials for lithium-ion batteries: The example of metal–organic frameworks. *Journal of Physical Chemistry C*, **114**, 9518–9527.
Corma, A., Garcia, H. and Llabres i Xamena, F.X. (2010) Engineering metal organic frameworks for

heterogeneous catalysis. *Chemical Reviews*, **110**, 4606–4655.

Dalach, P., Frost, H., Snurr, R.Q. and Ellis, D.E. (2008) Enhanced hydrogen uptake and the electronic structure of lithium-doped metal–organic frameworks. *Journal of Physical Chemistry C*, **112**, 9278–9284.

Dan-Hardi, M., Serre, C., Frot, T.O., Rozes, L., Maurin, G., Sanchez, C.M. and Férey, G.R. (2009) A new photoactive crystalline highly porous titanium(IV) dicarboxylate. *Journal of the American Chemical Society*, **131**, 10857–10859.

de Miguel, M., Ragon, F., Devic, T., Serre, C., Horcajada, P. and Garcia, H. (2012) Evidence of photoinduced charge separation in the metal–organic framework MIL-125(Ti)-NH2. *ChemPhysChem*, **13**, 3651–3654.

Denysenko, D., Werner, T., Grzywa, M., Puls, A., Hagen, V., Eickerling, G., Jelic, J., Reuter, K. and Volkmer, D. (2012) Reversible gas-phase redox processes catalyzed by Co-exchanged MFU-4l(arge). *Chemical Communications*, **48**, 1236–1238.

Dhakshinamoorthy, A., Alvaro, M. and Garcia, H. (2010) Aerobic oxidation of thiols to disulfides using iron metal-organic frameworks as solid redox catalysts. *Chemical Communications*, **46**, 6476–6478.

Dhakshinamoorthy, A., Alvaro, M., Hwang, Y.K., Seo, Y.-K., Corma, A. and Garcia, H. (2011) Intracrystalline diffusion in metal organic framework during heterogeneous catalysis: Influence of particle size on the activity of MIL-100 (Fe) for oxidation reactions. *Dalton Transactions*, **40**, 10719–10724.

Domenech, A., Garcia, H., Domenech-Carbo, M.T. and Llabres-i-Xamena, F.X. (2007) Electrochemistry of metal-organic frameworks: A description from the voltammetry of microparticles approach. *Journal of Physical Chemistry C*, **111**, 13701–13711.

Falkowski, J.M., Wang, C., Liu, S. and Lin, W. (2011) Actuation of asymmetric cyclopropanation catalysts: Reversible single-crystal to single-crystal reduction of metal–organic frameworks. *Angewandte Chemie International Edition*, **50**, 8674–8678.

Farha, O.K., Shultz, A.M., Sarjeant, A.A., Nguyen, S.T. and Hupp, J.T. (2011) Active-site-accessible, porphyrinic metal–organic framework materials. *Journal of the American Chemical Society*, **133**, 5652–5655.

Farrusseng, D., Aguado, S. and Pinel, C. (2009) Metal–organic frameworks: Opportunities for catalysis. *Angewandte Chemie International Edition*, **48**, 7502–7513.

Fateeva, A., Horcajada, P., Devic, T., Serre, C., Marrot, J., Grenèche, J.M., Morcrette, M., Tarascon, J.M., Maurin, G. and Férey, G. (2010) Synthesis, structured characterization, and redox properties of the porous MIL–68(Fe) solid. *European Journal of Inorganic Chemistry*, 3789–3794.

Fateeva, A., Chater, P.A., Ireland, C.P., Tahir, A.A., Khimyak, Y.Z., Wiper, P.V., Darwent, J.R. and Rosseinsky, M.J. (2012) A water-stable porphyrin-based metal–organic framework active for visible-light photocatalysis. *Angewandte Chemie International Edition*, **51**, 7440–7444.

Férey, G., Serre, C., Mellot-Draznieks, C., Millange, F., Surblé, S., Dutour, J. and Margiolaki, I. (2004) A hybrid solid with giant pores prepared by a combination of targeted chemistry, simulation, and powder diffraction. *Angewandte Chemie International Edition*, **43**, 6296–6301.

Férey, G., Millange, F., Morcrette, M., Serre, C., Doublet, M.-L., Grenèche, J.-M. and Tarascon, J.-M. (2007) Mixed-valence Li/Fe-based metal–organic frameworks with both reversible redox and sorption properties. *Angewandte Chemie International Edition*, **46**, 3259–3263.

Férey, G. and Serre, C. (2009) Large breathing effects in three-dimensional porous hybrid matter: Facts, analyses, rules and consequences. *Chemical Society Reviews*, **38**, 1380–1399.

Fletcher, A.J., Thomas, K.M. and Rosseinsky, M.J. (2005) Flexibility in metal-organic framework materials: Impact on sorption properties. *Journal of Solid State Chemistry*, **178**, 2491–2510.

Fu, Y., Sun, D., Chen, Y., Huang, R., Ding, Z., Fu, X. and Li, Z. (2012) An amine-functionalized titanium metal–organic framework photocatalyst with visible-light-induced activity for CO_2 reduction. *Angewandte Chemie International Edition*, **51**, 3364–3367.

Fu, Z., Chen, Y., Zhang, J. and Liao, S. (2011) Correlation between the photoactive character and the structures of two novel metal organic frameworks. *Journal of Materials Chemistry*, **21**, 7895–7897.

Fu, Z., Zhang, J., Zeng, Y., Tan, Y., Liao, S., Chen, H. and Dai, J. (2012) Synthesis and structure of a mixed crystal containing tris(4-pyridiniumyl)-1,3,5-triazine and benzenetetracarboxylate ions: Constructing a new photochromic molecular system via self-assembly. *CrystEngComm*, **14**, 786–788.

Gándara, F., Uribe-Romo, F.J., Britt, D.K., Furukawa, H., Lei, L., Cheng, R., Duan, X., O'Keeffe, M. and Yaghi,

O.M. (2012) Porous, conductive metal-triazolates and their structural elucidation by the charge-flipping method. *Chemistry – A European Journal*, **18**, 10595–10601.
Ghoufi, A., Deschamps, J. and Maurin, G. (2012) Theoretical hydrogen cryostorage in doped MIL-101(Cr) metal organic frameworks. *Journal of Physical Chemistry C*, **116**, 10504–10509.
Givaja, G., Amo-Ochoa, P., Gomez-Garcia, C.J. and Zamora, F. (2011) Electrical conductive coordination polymers. *Chemical Society Reviews*, **41**, 115–147.
Gomes Silva, C., Corma, A. and Garcia, H. (2010) Metal–organic frameworks as semiconductors. *Journal of Materials Chemistry*, **20**, 3141–3156.
Gong, Y., Li, J., Jiang, P.-G., Li, Q.-F. and Lin, J.-H. (2013) Novel metal(II) coordination polymers based on N,N′-bis-(4-pyridyl)phthalamide as supercapacitor electrode materials in an aqueous electrolyte. *Dalton Transactions*, **42**, 1603–1611.
Goriparti, S., Harish, M.N.K. and Sampath, S. (2013) Ellagic acid – a novel organic electrode material for high capacity lithium ion batteries. *Chemical Communications*, **49**, 7234–7236.
Han, S.S. and Goddard, W.A. (2007) Lithium-doped metal–organic frameworks for reversible H_2 storage at ambient temperature. *Journal of the American Chemical Society*, **129**, 8422–8423.
Han, S.S., Jung, D.H., Choi, S.-H. and Heo, J. (2013) Lithium-functionalized metal–organic frameworks that show >10 wt.% H_2 uptake at ambient temperature. *ChemPhysChem* **14**, 2698–2703.
Hendon, C.H., Tiana, D. and Walsh, A. (2012) Conductive metal-organic frameworks and networks: fact or fantasy? *Physical Chemistry Chemical Physics*, **14**, 13120–13132.
Hirai, K., Uehara, H., Kitagawa, S. and Furukawa, S. (2012) Redox reaction in two-dimensional porous coordination polymers based on ferrocenedicarboxylates. *Dalton Transactions*, **41**, 3924–3927.
Hmadeh, M., Lu, Z., Liu, Z., Gandara, F., Furukawa, H., Wan, S., Augustyn, V., Chang, R., Liao, L., Zhou, F., Perre, E., Ozolins, V., Suenaga, K., Duan, X., Dunn, B., Yamamoto, Y., Terasaki, O. and Yaghi, O.M. (2012) New porous crystals of extended metal-catecholates. *Chemistry of Materials*, **24**, 3511–3513.
Hong, B.H., Bae, S.C., Lee, C.-W., Jeong, S. and Kim, K.S. (2001) Ultrathin single-crystalline silver nanowire arrays formed in an ambient solution phase. *Science*, **294**, 348–351.
Horcajada, P., Chalati, T., Serre, C., Gillet, B., Sebrie, C., Baati, T., Eubank, J.F., Heurtaux, D., Clayette, P., Kreuz, C., Chang, J.-S., Hwang, Y.K., Marsaud, V., Bories, P.-N., Cynober, L., Gil, S., Férey, G., Couvreur, P. and Gref, R. (2010) Porous metal–organic-framework nanoscale carriers as a potential platform for drug delivery and imaging. *Nature Materials*, **9**, 172–178.
Horcajada, P., Surblé, S., Serre, C., Hong, D.-Y., Seo, Y.-K., Chang, J.-S., Grenèche, J.-M., Margiolaki, I. and Férey, G. (2007) Synthesis and catalytic properties of MIL-100(Fe), an iron(III) carboxylate with large pores. *Chemical Communications*, 2820–2822.
Horike, S., Sugimoto, M., Kongpatpanich, K., Hijikata, Y., Inukai, M., Umeyama, D., Kitao, S., Seto, M. and Kitagawa, S. (2013) Fe^{2+}-based layered porous coordination polymers and soft encapsulation of guests *via* redox activity. *Journal of Materials Chemistry A*, **1**, 3675–3679.
Horiuchi, Y., Toyao, T., Saito, M., Mochizuki, K., Iwata, M., Higashimura, H., Anpo, M. and Matsuoka, M. (2012) Visible-light-promoted photocatalytic hydrogen production by using an amino-functionalized Ti(IV) metal–organic framework. *Journal of Physical Chemistry C*, **116**, 20848–20853.
Horiuchi, Y., Toyao, T., Takeuchi, M., Matsuoka, M. and Anpo, M. (2013) Recent advances in visible-light-responsive photocatalysts for hydrogen production and solar energy conversion – from semiconducting TiO_2 to MOF/PCP photocatalysts. *Physical Chemistry Chemical Physics*, **15**, 13243–13253.
Ingleson, M.J., Barrio, J.P., Guilbaud, J.-B., Khimyak, Y.Z. and Rosseinsky, M.J. (2008) Framework functionalisation triggers metal complex binding. *Chemical Communications*, 2680–2682.
Ishimoto, T., Ogura, T., Koyama, M., Yang, L., Kinoshita, S., Yamada, T., Tokunaga, M. and Kitagawa, H. (2013) A key mechanism of ethanol electrooxidation reaction in a noble-metal-free metal organic framework. *Journal of Physical Chemistry C*, **117**, 10607–10614.
Jia, G., Gao, Y., Zhang, W., Wang, H., Cao, Z., Li, C. and Liu, J. (2013) Metal-organic frameworks as heterogeneous catalysts for electrocatalytic oxidative carbonylation of methanol to dimethyl carbonate. *Electrochemistry Communications*, **34**, 211–214.
Kambe, T., Sakamoto, R., Hoshiko, K., Takada, K., Miyachi, M., Ryu, J.-H., Sasaki, S., Kim, J., Nakazato, K., Takata, M. and Nishihara, H. (2013) Pi-conjugated nickel Bis(dithiolene) complex nanosheet. *Journal of the*

American Chemical Society, **135**, 2462–2465.

Kent, C.A., Mehl, B.P., Ma, L., Papanikolas, J.M., Meyer, T.J. and Lin, W. (2010) Energy transfer dynamics in metal organic frameworks. *Journal of the American Chemical Society*, **132**, 12767–12769.

Kitagawa, S., Kitaura, R. and Noro, S.-I. (2004) Functional porous coordination polymers. *Angewandte Chemie International Edition*, **43**, 2334–2375.

Kobayashi, Y., Jacobs, B., Allendorf, M.D. and Long, J.R. (2010) Conductivity, doping, and redox chemistry of a microporous dithiolene-based metal organic framework. *Chemistry of Materials*, **22**, 4120–4122.

Leclerc, H., Devic, T., Devautour-Vinot, S., Bazin, P., Audebrand, N., Férey, G., Daturi, M., Vimont, A. and Clet, G. (2011) Influence of the oxidation state of the metal center on the flexibility and adsorption properties of a porous metal organic framework: MIL-47(V). *Journal of Physical Chemistry C*, **115**, 19828–19840.

Lee, D.Y., Shinde, D.V., Kim, E.-K., Lee, W., Oh, I.-W., Shrestha, N.K., Lee, J.K. and Han, S.-H. (2013) Supercapacitive property of metal-organic-frameworks with different pore dimensions and morphology. *Microporous and Mesoporous Materials*, **171**, 53–57.

Lee, D.Y., Yoon, S.J., Shrestha, N.K., Lee, S.-H., Ahn, H. and Han, S.-H. (2012) Unusual energy storage and charge retention in Co-based metal organic-frameworks. *Microporous and Mesoporous Materials*, **153**, 163–165.

Leong, C.F., Faust, T.B., Turner, P., Usov, P.M., Kepert, C.J., Babarao, R., Thornton, A.W. and D'Alessandro, D.M. (2013) Enhancing selective CO_2 adsorption via chemical reduction of a redox-active metal-organic framework. *Dalton Transactions*, **42**, 9831–9839.

Leus, K., Muylaert, I., Vandichel, M., Marin, G.B., Waroquier, M., Van Speybroeck, V. and Van Der Voort, P. (2010) The remarkable catalytic activity of the saturated metal organic framework V-MIL-47 in the cyclohexene oxidation. *Chemical Communications*, **46**, 5085–5087.

Leus, K., Vandichel, M., Liu, Y.-Y., Muylaert, I., Musschoot, J., Pyl, S., Vrielinck, H., Callens, F., Marin, G.B., Detavernier, C., Wiper, P.V., Khimyak, Y.Z., Waroquier, M., Van Speybroeck, V. and Van Der Voort, P. (2012) The coordinatively saturated vanadium MIL-47 as a low leaching heterogeneous catalyst in the oxidation of cyclohexene. *Journal of Catalysis*, **285**, 196–207.

Li, Q., Yan, P., Hou, G., Wang, Y. and Li, G. (2013) Three alkaline-earth metal complexes with 3D networks constructed from a 7,7,8,8-tetracyanoquinodimethane ligand: synthesis, structure and electrochemical properties. *Dalton Transactions*, **42**, 7810–7815.

Li, X., Cheng, F., Zhang, S. and Chen, J. (2006) Shape-controlled synthesis and lithium-storage study of metal-organic frameworks Zn4O(1,3,5-benzenetribenzoate)2. *Journal of Power Sources*, **160**, 542–547.

Liu, Q., Yu, L., Wang, Y., Ji, Y., Horvat, J., Cheng, M.-L., Jia, X. and Wang, G. (2013) Manganese-based layered coordination polymer: Synthesis, structural characterization, magnetic property, and electrochemical performance in lithium-ion batteries. *Inorganic Chemistry*, **52**, 2817–2822.

Long, J.R. and Yaghi, O.M. (2009) Editorial: The pervasive chemistry of metal–organic frameworks. *Chemical Society Reviews*, **38**, 1213–1214.

Long, J., Wang, S., Ding, Z., Wang, S., Zhou, Y., Huang, L. and Wang, X. (2012) Amine-functionalized zirconium metal-organic framework as efficient visible-light photocatalyst for aerobic organic transformations. *Chemical Communications*, **48**, 11656–11658.

Mao, J., Yang, L., Yu, P., Wei, X. and Mao, L. (2012) Electrocatalytic four-electron reduction of oxygen with copper (II)-based metal-organic frameworks. *Electrochemistry Communications*, **19**, 29–31.

Martinez Joaristi, A., Juan-Alcaniz, J., Serra-Crespo, P., Kapteijn, F. and Gascon, J. (2012) Electrochemical synthesis of some archetypical Zn^{2+}, Cu^{2+}, and Al^{3+} metal organic frameworks. *Crystal Growth and Design*, **12**, 3489–3498.

Maspoch, D., Ruiz-Molina, D., Wurst, K., Domingo, N., Cavallini, M., Biscarini, F., Tejada, J., Rovira, C. and Veciana, J. (2003) A nanoporous molecular magnet with reversible solvent-induced mechanical and magnetic properties. *Nature Materials*, **2**, 190–195.

Mavrandonakis, A., Tylianakis, E., Stubos, A.K. and Froudakis, G.E. (2008) Why Li doping in MOFs enhances H_2 storage capacity? A multi-scale theoretical study. *Journal of Physical Chemistry C*, **112**, 7290–7294.

Meilikhov, M., Yusenko, K. and Fischer, R.A. (2009) Turning MIL-53(Al) redox-active by functionalization of the bridging OH-Group with 1,1′-Ferrocenediyl-Dimethylsilane. *Journal of the American Chemical Society*,

131, 9644–9645.
Meilikhov, M., Yusenko, K., Torrisi, A., Jee, B., Mellot-Draznieks, C., Pöppl, A. and Fischer, R.A. (2010) Reduction of a metal–organic framework by an organometallic complex: Magnetic properties and structure of the inclusion compound $[(\eta^5-C_5H_5)2Co]_{0.5}$@MIL-47(V). *Angewandte Chemie International Edition*, **49**, 6212–6215.
Miles, D.O., Jiang, D., Burrows, A.D., Halls, J.E. and Marken, F. (2013) Conformal transformation of [Co(bdc)(DMF)] (Co-MOF-71, bdc = 1,4-benzenedicarboxylate, DMF = N,N-dimethylformamide) into porous electrochemically active cobalt hydroxide. *Electrochemistry Communications*, **27**, 9–13.
Miyasaka, H. (2013) Control of charge transfer in donor/acceptor metal–organic frameworks. *Accounts of Chemical Research*, **46**, 248–257.
Miyasaka, H., Motokawa, N., Matsunaga, S., Yamashita, M., Sugimoto, K., Mori, T., Toyota, N. and Dunbar, K.R. (2010) Control of charge transfer in a series of Ru2(II,II)/TCNQ two-dimensional networks by tuning the electron affinity of TCNQ units: A route to synergistic magnetic/conducting materials. *Journal of the American Chemical Society*, **132**, 1532–1544.
MOFs special issue (2012) Metal–Organic Frameworks. *Chemical Reviews*, **112**, 673–1268.
Moon, H.R. and Suh, M.P. (2010) Flexible and redox-active coordination polymer: Control of the network structure by pendant arms of a macrocyclic complex. *European Journal of Inorganic Chemistry*, **2010**, 3795–3803.
Moon, H.R., Kim, J.H. and Suh, M.P. (2005) Redox-active porous metal–organic framework producing silver nanoparticles from AgI ions at room temperature. *Angewandte Chemie International Edition*, **44**, 1261–1265.
Morozan, A. and Jaouen, F. (2012) Metal organic frameworks for electrochemical applications. *Energy and Environmental Science*, **5**, 9269–9290.
Mu, W., Liu, D. and Zhong, C. (2011) A computational study of the effect of doping metals on CO_2/CH_4 separation in metal-organic frameworks. *Microporous and Mesoporous Materials*, **143**, 66–72.
Mueller, U., Schubert, M., Teich, F., Puetter, H., Schierle-Arndt, K. and Pastré, J. (2006) Metal–organic frameworks – prospective industrial applications. *Journal of Materials Chemistry*, **16**, 626–636.
Mulfort, K.L. and Hupp, J.T. (2007) Chemical reduction of metal–organic framework materials as a method to enhance gas uptake and binding. *Journal of the American Chemical Society*, **129**, 9604–9605.
Mulfort, K.L. and Hupp, J.T. (2008) Alkali metal cation effects on hydrogen uptake and binding in metal-organic frameworks. *Inorganic Chemistry*, **47**, 7936–7938.
Mulfort, K.L., Wilson, T.M., Wasielewski, M.R. and Hupp, J.T. (2009) Framework reduction and alkali-metal doping of a triply catenating metal–organic framework enhances and then diminishes H_2 uptake. *Langmuir*, **25**, 503–508.
Narayan, T.C., Miyakai, T., Seki, S. and Dinca, M. (2012) High charge mobility in a tetrathiafulvalene-based microporous metal–organic framework. *Journal of the American Chemical Society*, **134**, 12932–12935.
Nasalevich, M.A., Goesten, M.G., Savenije, T.J., Kapteijn, F. and Gascon, J. (2013) Enhancing optical absorption of metal-organic frameworks for improved visible light photocatalysis. *Chemical Communications*, **49**, 10575–10577.
Nguyen, T.L.A., Demir-Cakan, R., Devic, T., Morcrette, M., Ahnfeldt, T., Auban-Senzier, P., Stock, N., Goncalves, A.-M., Filinchuk, Y., Tarascon, J.-M. and Férey, G. (2010) 3-D coordination polymers based on the tetrathiafulvalenetetracarboxylate (TTF-TC) derivative: Synthesis, characterization, and oxidation issues. *Inorganic Chemistry*, **49**, 7135–7143.
Nouar, F., Devic, T., Chevreau, H., Guillou, N., Gibson, E., Clet, G., Daturi, M., Vimont, A., Greneche, J.M., Breeze, M.I., Walton, R.I., Llewellyn, P.L. and Serre, C. (2012) Tuning the breathing behaviour of MIL-53 by cation mixing. *Chemical Communications*, **48**, 10237–10239.
Otsubo, K., Wakabayashi, Y., Ohara, J., Yamamoto, S., Matsuzaki, H., Okamoto, H., Nitta, K., Uruga, T. and Kitagawa, H. (2011) Bottom-up realization of a porous metal-organic nanotubular assembly. *Nature Materials*, **10**, 291–295.
Park, Y., Shin, D.-S., Woo, S.H., Choi, N.S., Shin, K.H., Oh, S.M., Lee, K.T. and Hong, S.Y. (2012) Sodium terephthalate as an organic anode material for sodium ion batteries. *Advanced Materials*, **24**, 3562–3567.
Renault, S., Geng, J., Dolhem, F. and Poizot, P. (2011) Evaluation of polyketones with N-cyclic structure as

electrode material for electrochemical energy storage: case of pyromellitic diimide dilithium salt. *Chemical Communications*, **47**, 2414–2416.

Saravanan, K., Nagarathinam, M., Balaya, P. and Vittal, J.J. (2010) Lithium storage in a metal organic framework with diamondoid topology – a case study on metal formates. *Journal of Materials Chemistry*, **20**, 8329–8335.

Senthil Kumar, R., Senthil Kumar, S. and Anbu Kulandainathan, M. (2012) Highly selective electrochemical reduction of carbon dioxide using Cu based metal organic framework as an electrocatalyst. *Electrochemistry Communications*, **25**, 70–73.

Shen, L., Liang, S., Wu, W., Liang, R. and Wu, L. (2013) Multifunctional NH_2-mediated zirconium metal–organic framework as an efficient visible-light-driven photocatalyst for selective oxidation of alcohols and reduction of aqueous Cr(VI). *Dalton Transactions*, **42**, 13649–13657.

Stergiannakos, T., Tylianakis, E., Klontzas, E., Trikalitis, P.N. and Froudakis, G.E. (2012) Hydrogen storage in novel Li-doped corrole metal–organic frameworks. *Journal of Physical Chemistry C*, **116**, 8359–8363.

Stock, N. and Biswas, S. (2012) Synthesis of metal–organic frameworks (MOFs): Routes to various MOF topologies, morphologies, and composites. *Chemical Reviews*, **112**, 933–969.

Suh, M.P., Moon, H.R., Lee, E.Y. and Jang, S.Y. (2006) A redox-active two-dimensional coordination polymer: Preparation of silver and gold nanoparticles and crystal dynamics on guest removal. *Journal of the American Chemical Society*, **128**, 4710–4718.

Sun, L., Miyakai, T., Seki, S. and Dinca, M. (2013) Mn2(2,5-disulfhydrylbenzene-1,4-dicarboxylate): A microporous metal–organic framework with infinite $(-Mn-S-)_\infty$ chains and high intrinsic charge mobility. *Journal of the American Chemical Society*, **135**, 8185–8188.

Sun, Y.-Q., Zhang, J. and Yang, G.-Y. (2006) Molecular self-assemblies of a p-conjugated redox-active bipyridinium cation with magnetic dimetallic oxalate-bridged trimeric clusters. *Dalton Transactions*, 1685–1690.

Suslick, K.S., Bhyrappa, P., Chou, J.-H., Kosal, M.E., Nakagaki, S., Smithenry, D.W. and Wilson, S.R. (2005) Microporous porphyrin solids. *Accounts of Chemical Research*, **38**, 283–291.

Takaishi, S., Hosoda, M., Kajiwara, T., Miyasaka, H., Yamashita, M., Nakanishi, Y., Kitagawa, Y., Yamaguchi, K., Kobayashi, A. and Kitagawa, H. (2009) Electroconductive porous coordination polymer $Cu[Cu(pdt)_2]$ composed of donor and acceptor building units. *Inorganic Chemistry*, **48**, 9048–9050.

Tanabe, K.K., Allen, C.A. and Cohen, S.M. (2010) Photochemical activation of a metal–organic framework to reveal functionality. *Angewandte Chemie International Edition*, **49**, 9730–9733.

Usov, P.M., Fabian, C. and D'Alessandro, D.M. (2012) Rapid determination of the optical and redox properties of a metal-organic framework via in situ solid state spectroelectrochemistry. *Chemical Communications*, **48**, 3945–3947.

Walker, W., Grugeon, S., Mentre, O., Laruelle, S.P, Tarascon, J.-M. and Wudl, F. (2010) Ethoxycarbonyl-based organic electrode for Li-batteries. *Journal of the American Chemical Society*, **132**, 6517–6523.

Walker, W., Grugeon, S., Vezin, H., Laruelle, S., Armand, M., Wudl, F. and Tarascon, J.-M. (2011) Electrochemical characterization of lithium 4,4′-tolane-dicarboxylate for use as a negative electrode in Li-ion batteries. *Journal of Materials Chemistry*, **21**, 1615–1620.

Wang, C., Xie, Z., deKrafft, K.E. and Lin, W. (2011) Doping metal–organic frameworks for water oxidation, carbon dioxide reduction, and organic photocatalysis. *Journal of the American Chemical Society*, **133**, 13445–13454.

Wang, C., Wang, J.-L. and Lin, W. (2012a) Elucidating molecular iridium water oxidation catalysts using metal–organic frameworks: A comprehensive structural, catalytic, spectroscopic, and kinetic study. *Journal of the American Chemical Society*, **134**, 19895–19908.

Wang, J.-L., Wang, C. and Lin, W. (2012b) Metal–organic frameworks for light harvesting and photocatalysis. *ACS Catalysis*, **2**, 2630–2640.

Wu, D., Xu, Q., Liu, D. and Zhong, C. (2010) Exceptional CO_2 capture capability and molecular-level segregation in a Li-modified metal–organic framework. *Journal of Physical Chemistry C*, **114**, 16611–16617.

Xiang, J., Chang, C., Li, M., Wu, S., Yuan, L. and Sun, J. (2008) A novel coordination polymer as positive electrode material for lithium ion battery. *Crystal Growth and Design*, **8**, 280–282.

Xie, M.-H., Yang, X.-L., Zou, C. and Wu, C.-D. (2011) A Sn^{IV} porphyrin-based metal–organic framework for the selective photo-oxygenation of phenol and sulfides. *Inorganic Chemistry*, **50**, 5318–5320.

Xu, Q., Liu, D., Yang, Q., Zhong, C. and Mi, J. (2010) Li-modified metal–organic frameworks for CO_2/CH_4 separation: a route to achieving high adsorption selectivity. *Journal of Materials Chemistry*, **20**, 706–714.

Yanai, N., Uemura, T., Ohba, M., Kadowaki, Y., Maesato, M., Takenaka, M., Nishitsuji, S., Hasegawa, H. and Kitagawa, S. (2008) Fabrication of two-dimensional polymer arrays: Template synthesis of polypyrrole between redox-active coordination nanoslits. *Angewandte Chemie International Edition*, **47**, 9883–9886.

Yang, L.-M., Ravindran, P., Vajeeston, P. and Tilset, M. (2012a) Formation of an intermediate band in isoreticular metal-organic framework-993 (IRMOF-993) and metal-substituted analogues M-IRMOF-993. *Journal of Materials Chemistry*, **22**, 16324–16335.

Yang, L.-M., Ravindran, P., Vajeeston, P. and Tilset, M. (2012b) Properties of IRMOF-14 and its analogues M-IRMOF-14 (M = Cd, alkaline earth metals): electronic structure, structural stability, chemical bonding, and optical properties. *Physical Chemistry Chemical Physics*, **14**, 4713–4723.

Yong, G., Li, Y., She, W. and Zhang, Y. (2011) Isostructural metal–anion radical coordination polymers with tunable phosphorescent colors (deep blue, blue, yellow, and white) induced by terminal anions and metal cations. *Chemistry – A European Journal*, **17**, 12495–12501.

Yoon, J., Seo, Y.K., Hwang, Y., Chang, J.S., Leclerc, H., Wuttke, S., Bazin, P., Vimont, A., Daturi, M., Bloch, E., Llewellyn, P., Serre, C., Horcajada, P., Grenèche, J.M., Rodrigues, A. and Férey, G. (2010) Controlled reducibility of a metal–organic framework with coordinatively unsaturated sites for preferential gas sorption. *Angewandte Chemie International Edition*, **49**, 5949–5952.

Yoon, M., Srirambalaji, R. and Kim, K. (2012) Homochiral metal organic frameworks for asymmetric heterogeneous catalysis. *Chemistry Reviews*, **112**, 1196–1231.

Zeng, Y., Fu, Z., Chen, H., Liu, C., Liao, S. and Dai, J. (2012) Photo- and thermally induced coloration of a crystalline MOF accompanying electron transfer and long-lived charge separation in a stable host-guest system. *Chemical Communications*, **48**, 8114–8116.

Zhang, J.-W., Yan, P.-F., Li, G.-M., Liu, B.-Q. and Chen, P. (2010) Systematic study on electrochemical properties of a series of TCNQ lanthanide complexes. *Journal of Organometallic Chemistry*, **695**, 1493–1498.

Zhang, J., Matsushita, M.M., Kong, X.X., Abe, J. and Iyoda, T. (2001) Photoresponsive coordination assembly with a versatile logs – stacking channel structure based on redox-active ligand and cupric ion. *Journal of the American Chemical Society*, **123**, 12105–12106.

Zhang, Z., Zhao, H., Kojima, H., Mori, T. and Dunbar, K.R. (2013) Conducting organic frameworks based on a main-group metal and organocyanide radicals. *Chemistry – A European Journal*, **19**, 3348–3357.

Zou, C. and Wu, C.-D. (2012) Functional porphyrinic metal-organic frameworks: Crystal engineering and applications. *Dalton Transactions*, **41**, 3879–3888.

EMU Notes in Mineralogy, Vol. 17 (2017), Chapter 14, 405–442

Removal of arsenic, nitrate, persistent organic pollutants and pathogenic microbes from water using redox-reactive minerals

ANNA A. BOGUSH, JONG KYU KIM and LUIZA C. CAMPOS

Civil, Environmental and Geomatic Engineering Department, University College London, Chadwick Building, Gower Street, London WC1E 6BT, UK, e-mail: a.bogush@ucl.ac.uk (annakhol@gmail.com), j.k.kim@ucl.ac.uk, l.campos@ucl.ac.uk

Water pollution is a major global problem at present, involving inorganic and organic contaminants such as heavy metals, pesticides and pharmaceutical compounds, as well as pathogens. Therefore, the development of water-management strategies and appropriate water-treatment technologies has been the subject of intense research for decades. This chapter reviews the potential use of redox-reactive minerals as reactive media in environmental cleanup technologies especially in the removal of arsenic, nitrate, persistent organic pollutants and pathogenic microbes. The properties and applications of five classes of redox-reactive materials are summarized: zero-valent metals, mixed-valence iron oxyhydroxides, Fe-bearing clays, mixed-valence manganese oxides and titanium and zinc oxides. Examples and case studies that demonstrate the significant role of these reactive materials in water-treatment processes are also provided.

1. Introduction

Rapid industrial development, urban growth and domestic and agricultural activities are the main contributors to the quality deterioration of water bodies (*i.e.* surface and ground waters) and soils. Pollution of aquatic and terrestrial environments is the contamination of streams, lakes, groundwater, rivers and soils by substances harmful to the entire ecosystem. The major soil and water pollutants are chemical, biological or physical materials. These pollutants are the major threat to human health, especially in densely populated countries such as China and India.

Inorganic pollutants include heavy metals (*e.g.* Cd, Pb, Cr, Cu, V and Hg) and metalloids (*e.g.* As, Se and Sb) and these are regarded collectively as the most serious environmental contaminants globally due to their high toxicity and links to many forms of cancer (Riley *et al.*, 1992; Smedley and Kinniburgh, 2002; Mishra *et al.*, 2010; Tchounwou *et al.*, 2012). The treatment of heavy metals and metalloids is essential due to their persistence in the environment (Dangerous Substances Directive, 1976; Bradl, 2005). Organic pollutants include petroleum and its derivatives, polycyclic aromatic

DOI: 10.1180/EMU-notes.17.12

hydrocarbons (PAHs), and chlorinated hydrocarbons. These contaminants affect the aquatic ecosystem by depletion of dissolved oxygen (LENNTECH (a)). They are also known to have direct health effects including cancer, damage to the nervous system and reproductive disorders (Rao and Rao, 1997; Mosharraf Hossain *et al.*, 2012). Biological pollution is another major concern caused by invasive biological species including viruses, bacteriophage and bacteria. Millions of people die of illnesses contracted through infectious microorganisms, and most cases are caused by microbiological contamination of water supplies (Inamori and Fujimota, 2007). The major sources of biological contamination are domestic and agricultural activities.

Various water-treatment processes utilize chemical, physical and/or biological solutions to remove pollutants from water (Cheremisinoff, 2002; Bradl, 2005). Biological slow sand filtration was the sole method used to treat water until the advent of rapid gravity filters at the end of 19th century, followed by the introduction of chlorination and chemical coagulation (Bowles *et al.*, 1983). The appearance of organic and inorganic compounds and emergent pathogens in water led to the development of many other processes. Water-treatment processes currently available include, among others, adsorption, membrane filtration, oxidation, ion exchange and photocatalysis. Oxidants employed in water treatment include chlorine, ozone and hydroxyl radicals ($^{\bullet}OH$). However, some of these oxidants have been found to produce by-products as a result of either reactions with compounds in water or as a natural decay of the product itself (Legube *et al.*, 1989; McGuire *et al.*, 1990). Some of these by-products are caused by the application, in particular, of chlorine, including halogenated organics such as thrihalomethanes, which are said to be carcinogenic. As result, there is a need for the development and application of chemicals that are efficient but non-toxic or harmful to human health.

There is increasing interest in employing reduction-oxidation (redox) reactions using redox-reactive minerals in environmental remediation technologies due to their cost effectiveness and potential for the development of sustainable remediation technologies. Redox reactions transform the valence states of toxic contaminates (*e.g.* dissolved Cr^{VI}) to less reactive and less toxic forms (*e.g.* less soluble Cr^{III}) (Borch *et al.*, 2010). Redox-reactive materials have been evaluated to break down chlorinated solvents (*e.g.* trichloroethane, trichloroethylene, dichloroethene, tetrachloroethylene, vinyl chloride) in groundwater of permeable reactive zones (Sivavec *et al.*, 1997; USEPA, 2008c; Karn *et al.*, 2009; Bardos *et al.*, 2011) and the removal of redox-active elements such as U and Cr (see Fig. 1 from Ahmed *et al.*, 2010b; Dickinson and Scott, 2010; Crane *et al.*, 2011), and Cr (Schrick *et al.*, 2004; Cao and Zhang, 2006).

Amongst the redox-reactive minerals used in water treatment are iron-bearing phases (magnetite, green rust, zero-valent iron (ZVI), *etc.*), iron-bearing clay phases, sulfides, manganese oxide (MnO_2) and titanium oxide (TiO_2). Besides their redox properties, these materials have unique surface characteristics that lead to exceptional sorption and ion exchange properties. These redox-reactive media can be very useful in water-cleanup processes and environmental protection technologies. This chapter provides a review with examples of promising water-treatment processes involving redox-

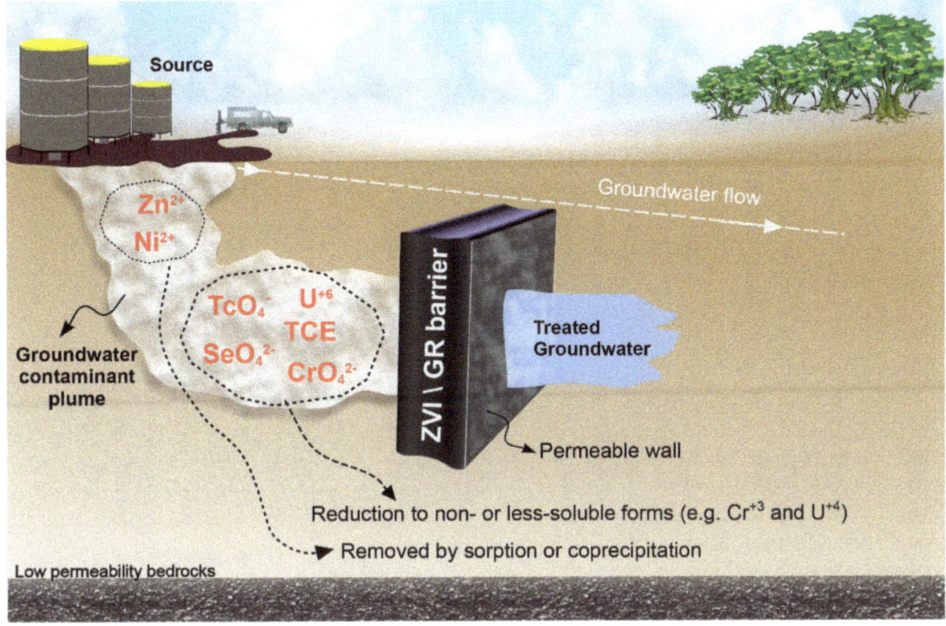

Figure 1. Permeable reactive barrier system: Zero-valent iron/ green rust (ZVI/GR) to remove inorganic and organic contaminants from groundwater. (Reproduced with the permission of the Environmental Chemistry Group of the Royal Society of Chemistry from Ahmed *et al.*, 2010)

reactive minerals for the removal of arsenic, nitrate, persistent organic pollutants and pathogenic microbes.

1.1 Properties of some redox-reactive minerals

In this section, we will discuss the properties of five groups of redox-reactive minerals: zero-valent metals, mixed-valence iron oxyhydroxides, Fe-bearing clays, mixed-valence manganese oxides and titanium and zinc oxides. The main properties of these redox-reactive minerals are summarized in Table 1.

Zero-valent metals (*e.g.* Fe^0, Zn^0, Ni^0, *etc.*) are used as strong reducing agents in water treatment. For example, the corrosion of zero-valent iron (ZVI) in water under the aerobic or anaerobic condition can be described by the following reactions:

$$2Fe^0_{(s)} + O_{2(g)} + 2H_2O_{(l)} \rightarrow 2Fe^{2+}_{(aq)} + 4OH^-_{(aq)} \text{ (aerobic, fast process)} \quad (1)$$

$$Fe^0_{(s)} + 2H_2O_{(l)} \rightarrow Fe^{2+}_{(aq)} + H_{2(g)} + 2OH^-_{(aq)} \text{ (anaerobic, slow process)} \quad (2)$$

In the first reaction, zero-valent iron reduces the oxygen present in water, leading to a decrease in redox potential (often < -0.5 V) and an increase in pH. In equation 2, ZVI reduces water to hydrogen and hydroxide, and an increase in pH occurs. Products of ZVI corrosion such as Fe^{2+} and H_2 are also reducing agents (Tratnyek *et al.*, 2003). However, the condition, media and nature of the contaminants can influence

Table 1. Properties of the redox-reactive minerals.

Redox-reactive mineral		Redox property	Redox processes	Parameters	Application
Zero-valent metals (ZVM)	Zero-valent iron (ZVI)	$Fe^{2+}_{(aq)} + 2e^{-} \rightarrow Fe_{(s)}$, $E = -0.44$ V; $\Delta G = +42.5$ kJ/electron; formation of oxidizing intermediates (O_2^{-}, H_2O_2, $^{\bullet}OH$, Fe(IV)) under oxic conditions	Direct reduction at the metal surface; reduction by ferrous iron and hydrogen (with catalyst); oxidation by oxidizing intermediates	mZVI (>0.1 nm) with SSA $\leqslant$ 1 m^2/g; nZVI (20–100 nm) with SSA $\approx$ 10–40 m^2/g; Fe(0) content 20–90%; bcc Fe(0) and Fe_3O_4 were identified in nZVI	Reductive and oxidative treatment of polluted waters containing inorganic (Cr, As, U, Th, Cl, N) and organic (chlorinated hydrocarbons, pesticides, viruses, pathogenic microbes) contaminants
	Zero-valent zinc (ZVZ)	$Zn^{2+}_{(aq)} + 2e^{-} \rightarrow Zn_{(s)}$, $E = -0.76$V, $\Delta G = +73.6$ kJ/electron	Reduction at the metal surface	–	Catalyst, reductive treatment of polluted waters
Mixed Fe(II)-Fe(III) minerals	Magnetite ($Fe^{2+}Fe_2^{3+}O_4$)	E varies by almost 1 V	Reduction and adsorption at the mineral surface	Spinel group; $pH_{PZC} \approx 6$; SSA varies from ~2 to 257 m^2/g	Reductive and adsorptive treatment of polluted waters from inorganic contaminants
	Green rusts ($[Fe^{II}_{6-x}Fe^{III}_{x}OH_{12}]_x^{+}$) with interlayers of anions (Cl^{-}, SO_4^{2-} and CO_3^{2-}) and water molecules	E varies, because of the variable composition (Fe^{2+}/Fe^{3+} ratio)	Reduction and adsorption at the surface of green rust	Structure of a pyroaurite type; PZC of $GRSO_4$ = 8.3±0.1 and PZC of $GRCO_3$ = 8.35±0.05; GR particle size $\approx$ 50–200 nm	Reductive and adsorptive treatment of polluted waters from inorganic contaminants

Fe-bearing clay minerals	Fe^{3+}/Fe^{2+} clay minerals (*e.g.* smectite – Fe content < 30 wt.%)	*E* varies considerably, because of the Fe oxidation state, the clay mineral structure, the total Fe content, and the ordering of structural cations	Redaction and adsorption processes at the active site of the Fe-bearing clay mineral	Reducing agents: bacteria, dithionite, hydrazine, hydrogen gas, hydrogen sulfide; Structural Fe^{2+} content changed from 10% to 90%	Reductive treatment of polluted waters containing inorganic and organic contaminants
Manganese oxides	Pyrolusite	E varies	oxidation and absorption	single-chain structure	Oxidative treatment of polluted waters containing inorganic and organic contaminants
	Birnessite	E varies	surface-mediated oxidation and absorption	layer structures	Oxidative treatment of polluted waters containing inorganic and organic contaminants
ZnO	Hexagonal wurtzite, cubic zincblende	$ZnO+h\nu \rightarrow e^{-}+h^{+}$, E_g=3.3 eV; $e^{-}+O_2 \rightarrow O_2^{-}$; $H_2O+h^{+} \rightarrow {}^{\bullet}OH+H^{+}$; $^{\bullet}OH$+orgaric pollutants→Intermediates→CO_2+H_2O	Photocatalysis and oxidation processes	Wurtzite crystal structure; SSA of ZnO ≈ 2.5–12 m^2/g	Photocatalytic oxidative degradation of organic pollutants (herbicides, insecticides and pesticides)
TiO_2	Anatase, brookite and rutile	$TiO_2+h\nu \rightarrow e^{-}+h^{+}$, E_g=3.2 eV; $H_2O+h^{+} \rightarrow {}^{\bullet}OH+H^{+}$; $O_2+e^{-} \rightarrow O_2^{-}$; $O_2^{-}+H^{+} \rightarrow HO_2^{\bullet}$; $h^{+}+O_2^{-} \rightarrow 2O^{\bullet}$	Photocatalysis and oxidation processes	Tetragonal crystal; TiO_2 particle size ≈ 6–104 nm with SSA ≈ 14–254 m^2/g	Photocatalytic oxidative degradation of organic pollutants (herbicides, insecticides and pesticides)

E – Standard reduction potential and ΔG – Gibbs free energy changes at 25C and pH 7.

significantly the mechanism of reactions of Fe^0, with contaminants involving multiple competing pathways (see Fig. 2 from Matheson and Tratnyek, 1994; Dickinson and Scott, 2010). Under oxic conditions, Fe(II) oxidation leads to the formation of intermediates (O_2^-, H_2O_2, $^{\bullet}OH$ and Fe(IV)), which are able to oxidize pollutants such as As (Leupin and Hug, 2005).

Matheson and Tratnyek (1994) described a model of possible pathways for the reduction of chlorinated solvents by ZVI. The model proposed three reaction pathways: (1) direct electron transfer from ZVI to the adsorbed halocarbon occurs, resulting in dechlorination and production of Fe^{2+}; (2) Fe^{2+} dechlorinates the halocarbon, producing Fe^{3+}; (3) H_2 from the anaerobic reduction of water can dechlorinate halocarbon if a catalyst is present. Understanding the relative importance of these reaction pathways is essential for the development and performance assessment of ZVI-based remediation technologies.

Micro- and nanoscale zero-valent iron (mZVI>0.1 μm and nZVI<0.1 μm) has been investigated widely in water-treatment applications (Sayles *et al.*, 1997; Li *et al.*, 2006; Fu *et al.*, 2014). ZVI is used to treat the following types of contaminants (Yan *et al.*, 2013): organic pollutants (chlorinated solvents, pesticides, azo-dyes, flame retardants, and antibiotics); inorganic contaminants (nitrate, arsenic, hexavalent chromium, heavy metals and radionuclides). The characteristics of nZVI are summarized in the comprehensive reviews by Nurmi *et al.* (2005) and Yan *et al.* (2013). The key advantage of nZVI is its large surface area which provides more reactive sites with

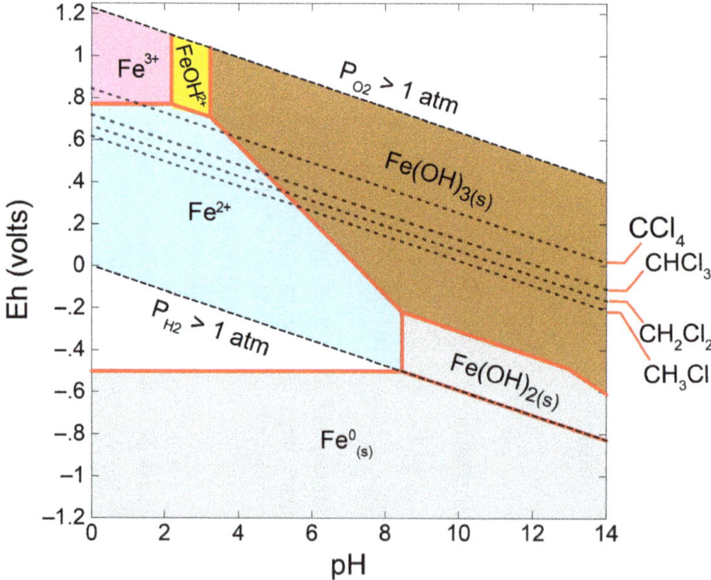

Figure 2. Pourbaix diagram for the Fe^0–H_2O system constructed with the suppression of hematite, magnetite, goethite and FeO. Modified, with the permission of the Royal Society of Chemistry, using *Geochemist's Workbench*, from Matheson and Tratnyek (1994).

more rapid degradation of contaminants compared to mZVI (USEPA, 2008b). For example, the average specific surface area (SSA) of the nZVI is in the range of 10–40 m^2/g and SSA of conventional microscale iron powder is typically $\leqslant 1$ m^2/g (Cao and Zhang, 2006). The disadvantages of nZVI are: difficulty in preparing stable and well dispersed nZVI; limited life of nZVI particles; the agglomeration or clumping of particles to each other or to the soil surface (USEPA, 2008a; Yan *et al.*, 2013). Agglomeration may be caused by groundwater conditions, surface properties of the particles, the age of the materials or shipping conditions (USEPA, 2008a). Using surface coating of nZVI with polyelectrolytes or non-ionic surfactant can improve the particle stability (Yan *et al.*, 2013).

Mixed Fe(II)-Fe(III) minerals (*e.g.* magnetite (Fe_3O_4) and green rusts) contain the ferrous (Fe^{2+})–ferric (Fe^{3+}) iron redox couple, and can control several environmental processes (reduction of several chemicals, redox mediating reactions, biological nutrient cycling, *etc.*). Thus, these minerals play important roles in the reductive treatment of polluted waters containing different contaminants. Magnetite is a common, naturally occurring mineral and a member of the spinel group. Magnetite can form during biotic and abiotic reduction of Fe(III) oxides and as a result of the oxidation of Fe(II) and Fe(0) (Anthony *et al.*, 1997; Cornell and Schwertmann, 2003; Gorski, 2009; Gorski *et al.*, 2010). At room temperature, Fe_3O_4 has a spinel face-centred cubic unit cell with the chemical formula often written as $Fe^{3+}_A[Fe^{2+}Fe^{3+}]_BO_4$, where A and B represent the tetrahedral and octahedral sites, respectively (Santos-Carballal *et al.*, 2014). Magnetite stoichiometry ($x = Fe^{2+}/Fe^{3+}$) can range from 0 to 0.5, where 0.5 corresponds to the most reduced form (stoichiometric magnetite), and 0 is the completely oxidized form (maghemite: γ-Fe_2O_3) (Gorski *et al.*, 2010). The specific surface areas for synthetic magnetite, synthetic nano-scale mesoporous magnetite, mechanically activated magnetite and three commercial magnetite chemicals (two ALFA chemicals with 97% and 99.999% purity and one Sigma Aldrich (<50 nm; $\geqslant$98% purity)) are 95.2-95.3, 247-257, 0.5-6.1, 5.59, 2.76 and 52.5 m^2/g (Sun *et al.*, 1998; Illés and Tombácz, 2004; Balâž *et al.*, 2010; Crane *et al.*, 2011). Gorski *et al.* (2010) investigated the factors controlling rates of contaminant (particularly for nitrobenzene) reduction by magnetite (Fe_3O_4) which are still poorly understood. They showed that rates of nitrobenzene ($ArNO_2$) reduction became almost five orders of magnitude faster as the particle stoichiometry increased from 0.31 to 0.50. They also proposed that both redox and Fe^{2+} diffusion processes plays important role in contaminant reduction by magnetite. The pH value of the point of zero charge (PZC) of magnetite in the absence of multivalent cations is about 6 (Sun *et al.*, 1998; Pang *et al.*, 2007; Ficai *et al.*, 2012). The presence of excess cations such as Fe^{2+} or Fe^{3+} in magnetite can greatly influence the zeta potential (Sun *et al.*, 1998; Marmier *et al.*, 1999; Illés and Tombácz, 2004; Pang *et al.*, 2007). Magnetite particles are positively and negatively charged at the dispersion pH values below and above the PZC, respectively (Sun *et al.*, 1998; Pang *et al.*, 2007). Sun *et al.* (1998) explained that at pH $< pH_{PZC}$ the protonation may be significant, at pH $\approx pH_{PZC}$ coagulation occurs, and at pH $> pH_{PZC}$ deprotonation and surface metal ion hydrolysis dominate.

Green rusts (GRs) are layered double hydroxide compounds and are intermediate phases in the formation of Fe oxides and oxyhydroxides such as goethite, lepidocrocite and magnetite (Schwertmann and Fechter, 1994). GRs are products of abiotic and microbially induced corrosion of iron and steel (Génin *et al.*, 1998; Kumar *et al.*, 1999; O'Loughlin *et al.*, 2003). The general formula of green rust is $[Fe^{II}_{(1-x)}Fe^{III}_{x}(OH)_2]^{x+}$ $\cdot[(x/n)A^{n-})\cdot mH_2O]^{x-1.2}$ where x = Fe^{III}/Fe_{total}, A^{n-} denotes interlayer anions with charge n, and m is moles of water molecules (Ahmed *et al.*, 2010a,b). Green rusts have a structure of a pyroaurite type consisting of positively charged trioctahedral iron hydroxide layers of variable composition ($[Fe^{II}_{6-x}Fe^{III}{}_{x}OH_{12}]^{x+}$) with interlayers of anions such as carbonate, sulfate and chloride and water molecules (Bernal *et al.*, 1959; Brindley and Bish, 1976; Hansen *et al.*, 1994; Schwertmann and Fechter, 1994). Fougerite ($[Fe^{2+}_4Fe^{3+}_2(OH)_{12}][CO_3]\cdot 3H_2O$) is the natural analogue of green rust (Genin *et al.*, 2001). There is an analytical challenge to handling, sampling and characterizing green rusts due to its quick transformation into ferric phases (*e.g.* goethite and magnetite) under oxidizing conditions (Ahmed *et al.*, 2010b). Ahmed *et al.* (2010a,b) developed a chemostat reactor combined with *in situ* time-resolved X-ray scattering measurements in order to characterize accurately GR and understand its formation and further transformation. Synthesized GR particle sizes vary from 50 to 200 nm (Ruby *et al.*, 2003, 2006; Bocher *et al.*, 2004; Guilbaud *et al.*, 2013). Guilbaud *et al.* (2013) determined that the point of zero charge (PZC) values are 8.3±0.1 and 8.35±0.05 for GR containing SO_4^{2-} ($GRSO_4$) and GR containing CO_3^{2-} ($GRCO_3$), respectively.

Fe-bearing clay minerals, such as rectorite (RAr-1 (illite:smectite = 50:50), 4.9 wt.% Fe, SSA – 144 m^2/g); illite (IMt-1, 12.3 wt.% Fe, SSA – 5 m^2/g); nontronite (NAu-2, 23.4 wt.% Fe, SSA – 271 m^2/g)) are important redox-reactive phases in subsurface soils and sediments, and play an important role in contaminant reduction (White and Peterson, 1996; Ernstsen *et al.*, 1998; Hofstetter *et al.*, 2003, 2006; Schwarzenbach *et al.*, 2010; Neumann *et al.*, 2011; Liu *et al.*, 2012). Gorski *et al.* (2012a) quantified the electron-accepting and -donating capacities (Q_{EAC} and Q_{EDC}) at applied potentials (Eh) of –0.60 V and +0.61 V, respectively, for four natural Fe-bearing smectites (SWa-1, 12.6 wt.% Fe), Na-rich montmorillonite (SWy-2, 2.3 wt.% Fe), and two nontronites (NAu-1, 21.2 wt.% Fe and NAu-2, 19.2 wt.% Fe). They showed that for SWa-1 and SWy-2 sample, all the structural Fe was redox-active over the tested Eh range but for NAu-1 and NAu-2, a significant fraction (~18%) of the structural Fe was redox-inactive. Iron may be distributed randomly or clustered in the octahedral sheet. Khaled and Stuki (1991) showed that the total layer charge of unaltered (oxidized) Fe-bearing smectites (SWa-1) measured by K^+, Ca^{2+}, Zn^{2+} or Cu^{2+} varies from 66.3 to 77.3 $cmol_c/kg$. Structural iron (*e.g.* Fe^{3+}) in clay minerals can be reduced to Fe^{2+} by different reducing agents (*e.g.* hydrazine, hydrogen gas, hydrogen sulfide) but the two most commonly used agents are dithionite and metal-reducing bacteria (Stucki *et al.*, 1984, 1987; Neumann *et al.*, 2011). Bacteria including *Shewanella*, *Geobacter*, *Pseudomonas* and *Bacillus* may reduce structural iron in clay (Stucki and Getty, 1986; Stucki *et al.*, 1987; Kostka *et al.*, 1999; Stucki, 2005). Fe^{2+} released can reduce organic and inorganic pollutants. Neumann *et al.*, (2011) showed that the redox properties of structural Fe are affected by its bonding environment in the clay's lattice, the total Fe content, the ordering of structural cations and the Fe oxidation state. Gorski *et al.* (2012b) investigated the redox properties of structural Fe in an iron-bearing clay mineral (ferruginous smectite, SWa-1), including the distribution of apparent reduction potentials of Fe^{3+}/Fe^{2+} pairs, the reversibility and pH dependence of electron transfer to and from structural Fe. Large changes in the standard reduction potential were observed for ferruginous smectite (−0.28 V to −0.54 V) as the structural Fe^{2+} content changed from 10% to 90% (Gorski *et al.*, 2012b). Lear and Stucki (1989) showed that specific surface area (SSA) of swelling SWa-1 (SSA $\approx$ 720 m^2/g) decreased with increasing Fe^{2+} in the octahedral sheet and about 20% of the layers collapsed completely in Fe-bearing smectite. Reduction of Fe^{3+} to Fe^{2+} in the dioctahedral structure of clay minerals is reflected in an increase in the negative surface charge and in cation fixation (Stucki, 2005). Kostka *et al.* (1999) added that SSA decreased by 170 m^2/g and the cation exchange capacity increased from 0.81 to 1.05 mEq/g in SWa-1 upon reduction of structural Fe by bacteria. In addition, Khaled and Stucki, (1991) showed that the total layer charge of chemically reduced Fe-bearing smectites (SWa-1) measured by K^+, Ca^{2+}, Zn^{2+} or Cu^{2+} varies from 77.9 to 99.7 $cmol_c/kg$.

Manganese-bearing minerals such as manganese oxides play an important role in removal of inorganic and organic pollutants (Moore *et al.*, 1990; Tournassat *et al.*, 2002; Tebo *et al.*, 2004; Mohan and Pittman, 2007; Con *et al.*, 2013). Manganese can form oxides with main oxidation states of +2 and +3 (*e.g.* MnO_2, Mn_2O_3, Mn_3O_4). Therefore, Mn oxides are capable of acting either as reducing agents ($Mn^{2+} - e^- = Mn^{3+} - e^- = Mn^{4+}$) or oxidizing agents ($Mn^{4+} + e^- = Mn^{3+} + e^- = Mn^{2+}$) (Fierro, 2006). Manganese oxides are used for many different applications in water treatment, soil and sediment remediation, metal removal and recovery, and they are also used as catalysts, sorbents and electrical conductors (Tebo *et al.*, 2004). Pyrolusite is manganese dioxide (MnO_2, single chain structure) formed under oxidizing conditions in manganese-bearing hydrothermal deposits, bogs, lakes and oceanic systems (Anthony *et al.*, 1997). Birnessite, with an empirical formula of $(Na,Ca)_{0.5}(Mn^{4+},Mn^{3+})_2O_4 \cdot 1.5H_2O$, is a major manganese-bearing mineral occurring in soil, aquifers and oceanic and aquatic systems (Anthony *et al.*, 1997; Tournassat *et al.*, 2002). Lanson *et al.* (2000) determined the structural formula of hexagonal birnessite to be $H^{+}_{0.33}Mn^{3+}_{0.111}Mn^{2+}_{0.055}(Mn^{4+}_{0.722}Mn^{3+}_{0.111}\square_{0.167})O_2 \cdot (H_2O)_{0.50}$, where $\square$ is a Mn vacancy in the octahedral layer. These vacancies, in turn, can be reactive sorption sites for pollutants (Tournassat *et al.*, 2002). The presence of Mn(III), Mn(II) and vacant sites within Mn oxide minerals cause negative surface charges, which can be compensated by cation sorption.

Titanium dioxide (TiO_2: anatase, brookite and rutile) is a widely used semiconductor photocatalyst (Fujishima *et al.*, 2000; Hashimoto *et al.*, 2005). Titanium dioxide (TiO_2) is chemically stable in acidic and basic conditions and costs relatively little (Castellote and Bengtsson, 2011). When aqueous TiO_2 suspensions are irradiated with light energy greater than the band-gap energy (E_g), TiO_2 is activated and can combine with water or dissolved oxygen (or both) to form highly reactive species, including $O_2^{\bullet-}$, $^{\bullet}OH$, $HO_2^{\bullet}$ and $O^{\bullet}$, which can oxidize and degrade a range of contaminants (Table 1; USEPA,

2008a; Hashimoto *et al.*, 2005). The band-gap energy of TiO_2 (anatase) is 3.2 eV, which corresponds to photons with a wavelength of 388 nm (Castellote and Bengtsson, 2011). Rutile has a 3.0 eV band-gap energy. In general, anatase gives better reaction results than rutile for hydrogen production in photocatalysis. There are several reasons why anatase is more efficient in photocatalysis. It is probably because of the higher reduction potential of photogenerated electrons in anatase than in rutile, *i.e.* the bottom of the conduction band of anatase is located at a point 0.1 V more negative than that of rutile (Hashimoto *et al.*, 2005), and the recombination of electron-hole pairs is slower in anatase than in rutile. Suttiponparnit *et al.* (2011) showed that nanoscale titanium oxide particles have different specific surface areas: (a) TiO_2 (P25) nanoparticles (27 nm) with SSA = 57.4 m^2/g; (b) anatase TiO_2 nanoparticles of 6, 16, 26, 38, 53 and 104 nm with SSA of 253.9, 102.1, 61.5, 41.2, 29.7 and 15 m^2/g, respectively; (c) rutile TiO_2 particle of 102 nm with SSA of 13.8 m^2/g. Nanoscale TiO_2 has been shown to mineralize a variety of herbicides, insecticides and pesticides *via* photocatalysis and can convert other contaminants to less toxic compounds (Konstantinou and Albanis, 2003).

Zinc oxide, a II-VI compound semiconductor, has been used as a pigment in rubber, sun-blocking ointments and paint, as well as a food supplement. Recently its application in water treatment has been acknowledged increasingly as a suitable alternative for TiO_2 due to its comparable band-gap energy (3.37 eV) and lower cost of production (Yu and Yu 2008; Sapkota *et al.*, 2011; Doria *et al.*, 2013; Hoseinzadeha *et al.*, 2014). ZnO has a large excitation binding energy (60 meV) at room temperature (Yu and Yu, 2008). ZnO nanoparticles with excitation wavelengths of 300 nm and 350 nm exhibited a strong and wide photoluminescence signal in the range 400 to 550 nm, with the energy of the excitation light being greater than the band-gap energy, having two obvious photoluminescence peaks at ~420 and 480 nm, respectively (Liqiang *et al.*, 2006). These photoluminescence signals are attributed to excitonic photoluminescence, which results mainly from surface-oxygen vacancies and defects of ZnO nanoparticles (Liqiang *et al.*, 2003).

2. Removal of arsenic

Arsenic in drinking-water is a hazard to human health, potentially causing arsenicosis, a chronic disease with a significant latency period for non-cancer (skin lesions, hyperkeratosis, melanosis, blackfoot disease, diabetes and neurological dysfunction) and cancer effects including those of bladder, skin and lung cancers (National Research Council, 1999; WHO, 2001, 2003a; Roberts *et al.*, 2004; Hopenhayn, 2006). Arsenic enters drinking water supplies from natural deposits or from agricultural, mining or industrial practices. Arsenic is present at various concentrations (<10–5300 μg/L) in the groundwater of several countries, including Bangladesh, India, Mongolia, Thailand, China, Vietnam, Nepal, Myanmar, Cambodia, Pakistan, Mexico, Argentina, Chile, Hungary, Romania, Germany, USA and Canada (Kinniburgh and Smedley, 2001; Smedley and Kinniburgh, 2002; Welch and Stollenwerk, 2003; Sun,

2004; Vaughan, 2006; Hopenhayn, 2006; Edmunds *et al.*, 2015). The WHO guideline for arsenic in drinking water is 10 μg/L.

Arsenic can occur in the environment in different oxidation states (−3, 0, +3 and +5), and in natural waters is mostly found in inorganic form as trivalent arsenite [As(III)] or pentavalent arsenate [As(V)]. Redox potential and pH are the most important factors controlling arsenic speciation (Masscheleyn *et al.*, 1991). Arsenic(V) is generally the main As species in oxygen-rich surface waters and As(III) usually predominates in groundwater (Smedley and Kinniburgh, 2002). The common pentavalent (+5) species in natural waters are AsO_4^{3-}, $HAsO_4^{2-}$ and H_2AsO_4, and trivalent (+3) species are $As(OH)_3$, $As(OH)_4^-$, $AsO_2OH_2^-$ and AsO_3^{3-} (Mohan and Pittman, 2007). Toxicity depends upon the chemical species and, in general, As(III) is 20 to 60 times more toxic than As(V) (Korte and Fernando, 1991). The toxicity of arsenic for aquatic organisms decreases in the order: arsenite (As^{3+}) > arsenate (As^{5+}) > monomethylarsonic acid (CH_5AsO_3) > dimethylarsinic acid ($C_2H_7AsO_2$) > arsenobetaine ($C_5H_{11}AsO_2$) (Hindmarsh and McCurdy, 1986; Shiomi, 1994; CCME, 1999).

The most common treatment processes for arsenic removal from water include chemical precipitation, oxidation, biological oxidation, co-precipitation, adsorption, ion exchange, filtration and membrane technology (reverse osmosis, electrodialysis and nanofiltration) (NSF, 2001; Bissen and Frimmel, 2003; Holm and Wilson, 2006; Mondal *et al.*, 2006; Mohan and Pittman 2007; Jain and Singh, 2012). Table 2 shows the efficiency of the main treatment process for arsenic removal. Redox-reactive minerals play a significant role in arsenic removal and can be used in several water-treatment technologies. Sometimes a combination of processes can be used, *e.g.* oxidation by filtration through zero-valent ions (Leupin and Hug, 2005).

2.1. Chemical precipitation

Chemical precipitation is an effective process used for arsenic removal and is usually applied in water treatment plants by coagulation-flocculation (Fig. 3). Four precipitation processes using redox compounds can be used: iron coagulation, iron/manganese precipitation and lime softening (Table 3).

Table 2. Main treatment processes for arsenic removal from water.

Technology	Efficiency removal	Reference
Chemical precipitation	95%	Vu *et al.* (2003)
Adsorption on different media	up to 99%	Mohan and Pittman (2007)
Chemical oxidation and filtration	80–95%	Jain and Singh (2012)
Reverse osmosis	99%	NSF (2001)
Ion exchange	>90%	Mondal *et al.* (2006)
Aeration	20–25%	Holm and Wilson (2006)

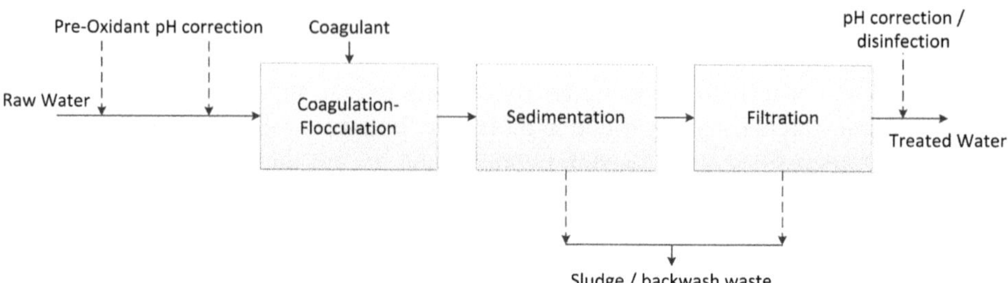

Figure 3. Schematic representation of a chemical precipitation process for water treatment. Dashed lines indicate optional processes.

Arsenic in drinking water can be removed effectively with alum if a pre-oxidant such as chlorine, potassium permanganate, ozone or hydrogen peroxide is used (Table 4). For precipitation with iron metals, ferric chloride or ferric sulfate can be used. Iron-based coagulants are usually more effective at removing As(V) than aluminium salts. This is because iron hydroxides are more stable than aluminium hydroxides in the pH range 5.5−8.5 (Jain and Singh 2012). Lakshmanan *et al.* (2008) found that the efficiency of removal of As(V) was significantly greater than As(III) when iron, titanium and zirconium coagulants (in decreasing order of efficiency) were used at pH values of 6.5, 7.5 and 8.5. Therefore, depending on the raw water characteristics, pH adjustment prior to coagulant addition may be required to achieve the optimum coagulation/precipitation. A final pH correction after disinfection is usually required to ensure acceptable pH levels in the distribution system.

When raw water (*e.g.* groundwater) contains both iron and arsenic, the best method is to oxidize Fe(II) by an aeration/oxidation process (Fig. 4) and the resulting Fe(III)

Table 3. Chemical precipitation processes for arsenic removal (modified from Mondal *et al.*, 2006).

Process	Efficiency of removal (%)	Advantages	Disadvantages
Precipitation with iron	60−90	Proven and reliable	Use of chemical; high-As contaminated sludge; dose of oxidizing chemicals influences the removal efficiency.
Precipitation with Fe/Mn	40−90	Proven and reliable	Higher and lower pH reduces efficiency; use of chemical; high-As contaminated sludge; dose of oxidizing chemicals influence the removal efficiency.
Lime softening	80−90	Proven and reliable; reduces corrosion	Sulfate ions influence efficiency; secondary treatment is required; use of chemicals.

Table 4. Main oxidants used in water treatment.

Oxidant	Advantages	Disadvantages
Chlorine	Well established technology Low cost	Highly corrosive and toxic. Reacts with some organic compounds to create disinfection by-product.
Ozone	No harmful by-products Generated onsite so few problems with transport/handling Increase dissolved oxygen (DO) content in water	Complex technology Highly reactive and corrosive. Provides no residual in drinking water. Relatively high implementation cost.
Potassium permanganate	Controls taste and odour Oxidizes iron and manganese Control of DBP formation Oxidizes As(III) very quickly	Higher relative cost. No primary disinfection capability. Possibly pink colour in water. Difficulty in handling.
Hydrogen peroxide	Does not produce residuals	Reacts with various compounds.

precipitates the arsenic. Leupin and Hug (2005) found that during the oxidation of Fe(II) by dissolved oxygen, As(III) was partially oxidized and As(V) adsorbed on the hydrous ferric oxides (HFO) that formed. To improve arsenic removal, in this case, it is necessary to use an oxidant to oxidize both Fe(II) and As(III). Arsenic(V) then adsorbs onto the iron hydroxide precipitates that are ultimately filtered out of solution.

Roberts *et al.* (2004) investigated arsenic removal with Fe(II) and Fe(III) by passive treatment. They showed that application of Fe(II) was more advantageous than the Fe(III) applied usually because oxidation of Fe(II) by dissolved oxygen caused partial oxidation of As(III) and formation of Fe(III) (hydr)oxides with greater sorption capacities. Roberts *et al.* (2004) showed that five hourly additions of 5 mg of Fe(II)/L or

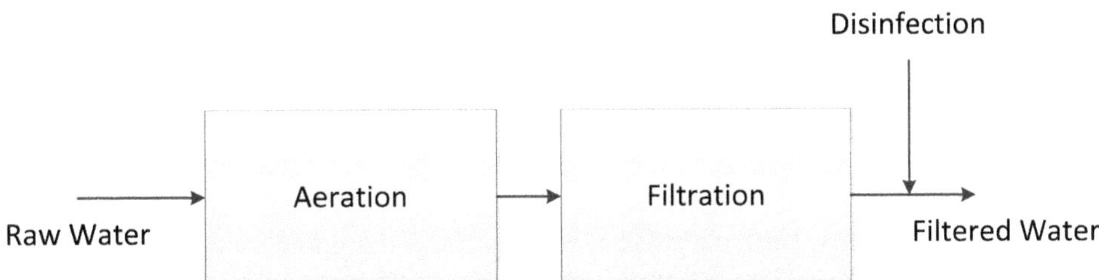

Figure 4. Schematic representation of aeration process for water treatment.

eight additions of 2.5 mg/L every 30 min to water containing 500 mg/L As(III), 3 mg/L P and 30 mg/L Si would suffice to attain the 50 ppb limit without an added oxidant. The removal efficiency of arsenic depends heavily on the initial iron and arsenic concentrations; where the Fe:As mass ratio should be at least 20:1 (Jain and Singh, 2012). Arsenic removal also depends on the anion composition of arsenic-contaminated waters. Groundwater in Bangladesh, for instance (Kinniburgh and Smedley, 2001), has high concentrations of silicate and phosphate which can influence negatively the rate of arsenic removal (Meng *et al.*, 2000, 2002).

2.2 Oxidation and adsorption

Oxidation and adsorption technologies have been used widely to treat As-polluted groundwater, reducing arsenic concentrations to <10 μg/L. Amongst the redox-based adsorbents for the removal of arsenic are iron-based sorbents (*e.g.* granular ferric hydroxide, zero-valent iron (ZVI), iron coated sand, *etc.*), red mud, activated carbon, biochar, lignite and peat (Sun *et al.*, 2006; Mohan and Pittman, 2007; Jain and Singh, 2012). Natural hematite, goethite and magnetite, for example, were also found to be suitable adsorbents to remove both As(III) and As(V) from solutions, with hematite having the highest sorption capacity, especially at acidic pH (Gimenez *et al.*, 2007). Raven *et al.* (1998) investigated arsenic adsorption on ferrihydrite. They concluded that the adsorption reaction was relatively fast and almost complete within the first few hours. They also showed that adsorption maxima arsenite and arsenate were ~0.6 and 0.25 mol As/mol Fe, respectively. Magnetite is effective in arsenic remediation (Yean *et al.*, 2005; Yavuz *et al.*, 2010). Yavuz *et al.* (2010) showed that percentage of As removal increases with decrease of magnetite particle size: (1) the magnetite particle size (MPS) was 300 nm and the As percentage removal (AsPR) was ~25–29%; (2) MPS = 20 nm and AsPR = ~91–97%; and (3) MPS = 12 nm and AsPR = ~98–99%. Joshi and Chaudhuri (1996) showed that iron oxide-coated sand might be a good material for use in small systems or home-treatment units for removing As(III) and As(V) from groundwater. Iron oxide-coated sand has more pores and a large surface area (Lo *et al.*, 1997). The system temperature has a significant influence on iron-phase formation, *i.e.* at low temperature (60°C) only an amorphous iron phase is produced; at 150°C goethite and hematite are produced; and at 300–500°C only hematite forms (Lo *et al.*, 1997). Guo and Chen (2005) used bead cellulose loaded with iron oxyhydroxides (BCF) for the removal of arsenate and arsenite from aqueous systems. They determined that the adsorption capacity for arsenite and arsenate was 99.6 and 33.2 mg/g BCF at pH 7.0 with Fe content of 220 g/L. They also showed that arsenate removal is better at acidic pH but arsenite – in a wide pH range of 5–11.

Using ZVI to remove arsenic from polluted waters is quite promising, but it is still difficult to predict the mechanism and efficiency of this process due to variations in aqueous parameters (pH, water composition, oxic or anoxic conditions). For example, under oxic conditions; Fe(0) corrodes, forming various amorphous and crystalline iron-bearing phases (hematite, maghemite, lepidocrocite, magnetite, green rust) which can adsorb As(III) and As(V) (Su and Puls, 2001a, 2003; Manning *et al.*, 2002; Bang *et al.*,

2005; Leupin and Hug, 2005; Lien and Wilkin, 2005). By contrast, under anoxic conditions arsenic is adsorbed on the surface of iron in an unspecified form (Leupin and Hug, 2005). Leupin and Hug (2005) investigated the fate of arsenic and iron in a filter containing a mixture of iron filings and sand with synthetic groundwater similar to groundwater in Bangladesh. They proposed the following reactions that can occur in this system under oxic conditions:

$$Fe(0) + 2H_2O + 1/2O_2 \rightarrow Fe(II) + H_2O + 2OH^- \quad (3)$$

$$Fe(II) + 1/4O_2 + H_2O \rightarrow Fe(III) + 1/2H_2O + OH^- \quad (4)$$

$$Fep(III) + 3H_2O \rightarrow Fe(OH)_3 + 3H^+ \quad (5)$$

$$Asp(III) + intermediates\ (^{\cdot}OH;\ Fe(IV)) \rightarrow As(IV) \quad (6)$$

$$As(IV) + O_2 \rightarrow As(V) + {}^{\cdot}O_2^- \quad (7)$$

It was found that Fe(II) existed partly as a dissolved compound and partly as an adsorbed complex on the surfaces of iron filings and sand. Iron(II) oxidation led to the formation of intermediate complexes (O_2^-, H_2O_2, OH and Fe(IV)), some of which are able to oxidize As(III) (Leupin and Hug, 2005). Those authors concluded that the hydrous ferric oxides formed in this system were excellent substrates for the adsorption of As(V). They found that 15–18 mg of Fe(III) are needed to remove 90% of 500 mg/L As(V) (with 3 mg/L of P and 20 mg/L of Si). Subsequently, the USEPA (2008a) showed that nanoscale zero-valent iron (nZVI) may be more effective than hydrous ferric oxide because nZVI has a high reactive surface area and can degrade contaminants more rapidly and completely than macroscale ZVI under similar environmental conditions (USEPA, 2008a). Du *et al.* (2013) developed a bifunctional resin-supported nanosized ZVI (N-S-ZVI) composite by combining the oxidation properties of $nZVI/O_2$ with adsorption features of iron oxides and anion-exchange resin N-S (as a supporter, a dispersant of nZVI and an adsorbent of anions). The maximum adsorption capacities of N-S-ZVI for As(III) and As(V) were 121 and 125 mg/g, respectively (Du *et al.*, 2013).

Manganese minerals such as manganese oxides (pyrolusite and birnessite) also play an important role in arsenic removal. The main reactions occurring in this process are: (1) reductive dissolution of MnO_2 surface; (2) oxidation of As(III) by MnO_2; and (3) adsorption of As(V) on the MnO_2 surface (Moore *et al.*, 1990; Tournassat *et al.*, 2002; Mohan and Pittman, 2007). Nesbitt *et al.* (1998) showed that the oxidation of As(III) results in the release of As(V) and Mn(II) according to the reactions:

$$2MnO_2 + H_3AsO_3 + H_2O \rightarrow 2MnOOH^* + H_3AsO_4 \quad (8)$$

$$2MnOOH^* + H_3AsO_3 + 4H+ \rightarrow 2Mn^{2+} + H_3AsO_4 + 3H_2O \quad (9)$$

where MnOOH* is a Mn(III) intermediate.

Con *et al.* (2013) investigated the properties of nano-dimensional MnO_2 (30–50 nm) prepared by redox reaction between $MnSO_4$ and $KMnO_4$ in a water–ethanol solution. The separate MnO_2 nano particles were coated on already-existing MnO_2 (pyrolusite) grains to create adsorption material for treatment of arsenic in a water environment

(Con *et al.*, 2013). The maximum adsorption capacity (84 mg/g) of arsenic on nano-MnO_2 coated on pyrolusite was about ten times greater than the maximum adsorption (8.67 mg/g) of MnO_2 produced by electrolysis oxidation coating on the same substrate (Con *et al.*, 2013).

Chakravarty *et al.* (2002) studied a low-cost ferruginous manganese ore (FMO), which consisted mainly of pyrolusite (MnO_2) and goethite (FeOOH), for the removal of arsenic from groundwater. Those authors showed that FMO can remove both As(III) and As(V) in the pH range 2.0–8.0 without any pre-treatment. They used FMO for the purification of six groundwater samples containing arsenic concentrations in the range 0.04–0.18 ppm and concluded that arsenic removals are almost 100% in all cases. The cost of the FMO is ~50–56 US$ per metric tonne (Chakravarty *et al.*, 2002).

'Greensand' is a green iron-rich clay-like mineral (*e.g.* glauconite) that has ion-exchange properties. The greensand is treated with $KMnO_4$ until the sand grains are coated with a layer of manganese oxides of various Mn valence states. Arsenic-removal processes by greensand comprise oxidation, ion exchange and adsorption. Arsenic complexes displace species from the manganese oxide (presumably OH^- and H_2O) to become bound to the greensand surface (in effect, an exchange of ions occurs). The manganese oxide surface oxidizes As(III) to As(V), which is then adsorbed onto the surface. As a result of the electron transfer and adsorption of As(V), reduced manganese (Mn(II)) is released from the surface. Subramanian *et al.* (1997) investigated the effectiveness of manganese greensand filtration (MGSF) in batch and column experiments. Batch experiments showed that the removal efficiency of arsenic by MGSF up to the interim maximum acceptable concentration (IMAC = 25 μg/L) was 60% for 24 h equilibration. In a column experiment, the removal efficiency was 41%. However, in the presence of Fe(II), especially at an Fe/As concentration ratio of 20, manganese greensand gave an overall efficiency of 83% and a throughput volume of 1440 L (Subramanian *et al.*, 1997).

Red mud, a by-product of the alumina production process, is also a promising material for the remediation of arsenic-contaminated waters (Altundogan *et al.*, 2000, 2002; Mohan and Pittman, 2007). The capacities were 4.3 μmol/g at a pH of 9.5 for As(III) and 5.1 μmol/g at a pH of 3.2 for As(V) (Altundogan *et al.*, 2002). Untreated red mud can contain high concentrations of hazardous elements including As, Pb, Hg, Cd and Cr (Ruyters *et al.*, 2011; Wang and Liu, 2012) and, therefore, requires purification before use. Seawater-neutralized red mud (Bauxol) has a higher sorption capacity with respect to arsenic (14.4 μmol/g) than unmodified red muds (Genc-Fuhrman *et al.*, 2003, 2004a,b, 2005). Furthermore, Bauxol activated by acid and heat treatment increased arsenic adsorption (Genc-Fuhrman *et al.*, 2004a,b).

In general, arsenic adsorption depends on the arsenic oxidation state, pH and property of the adsorbent surface. Usually the adsorption media is packed into a column. During the adsorption process contaminated water enters at the top of the column and flows downward through the bed and leaves through the underdrain system. The media may become clogged with suspended solids present in the feed water, resulting in increased head loss across the bed. When the maximum head loss is achieved, the media must be backwashed to remove clogging. In order to minimize an increase in head loss, it is recommended that the influent water to the column has less than 5 mg/L of suspended solids (Reynolds and Richards, 1996), otherwise a pre-filter will be necessary (Fig. 5). When the adsorbent is exhausted, the column media must be regenerated or disposed of, and replaced with new media.

2.3. *In situ* treatment technologies

Conventional engineering technologies for the treatment of groundwater, such as pump and treat, are expensive. Therefore, *in situ* technologies, like Permeable Reactive Barriers (effective passive remediation technology combining subsurface fluid-flow management with chemical and biological treatment), have received increased attention in recent years (USEPA, 1998; Blowes *et al.*, 2000; Wilkin *et al.*, 2009). For example, Blowes and Ptacek (1994) proposed the use of iron as a reactive barrier material and subsequently this technology has been developed and improved (USEPA, 1998). Lackovic *et al.* (2000) investigated ZVI as a main material for permeable reactive barriers (PRBs) to remove arsenic from groundwater. A mechanism of arsenic removal by ZVI using a combination of abiotic surface precipitation and adsorption was proposed. Those authors also demonstrated that both As(III) and As(V) could be removed effectively from aqueous solution under anoxic conditions without the use of a preoxidation step.

Zero-valent iron, used for As removal, can be corroded under the influence of dissolved oxygen (DO), forming various iron-bearing phases such as ferric oxide, oxyhydroxide and hydroxide. Several studies have shown that arsenic precipitated, co-precipitated and adsorbed onto ZVI corrosion products, including iron hydroxides, oxyhydroxides and mixed-valence Fe(II)–Fe(III) green rusts (Farrell *et al.*, 2001; Manning *et al.*, 2002; Su and Puls, 2003; Bang *et al.*, 2005; Leupin and Hug, 2005; Lien and Wilkin, 2005). Beaulieu and Ramirez (2013) applied a commercially available product containing ZVI, organic carbon substrate and sulfate to an aquifer contaminated by dissolved arsenic and iron. The PRB was capable of removing 97% of the arsenic, achieving 5 μg/L of total arsenic for ~2 y.

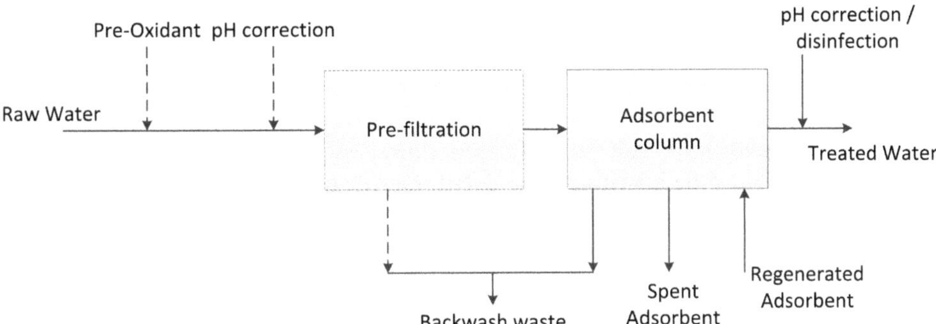

Figure 5. Schematic representation of adsorption process for water treatment. Dashed lines indicate optional processes.

It was shown that sodium-sulfate-type groundwater is better suited for arsenic treatment by ZVI than calcium-bicarbonate-type groundwater (Beak and Wilkin, 2009). This is because carbonate minerals (calcite, aragonite, *etc.*) can form in PRBs due to pH increase during the corrosion of the metallic iron (Beak and Wilkin, 2009). Precipitation of carbonate minerals can cause decreases in porosity, permeability and ZVI reactivity because iron surfaces become coated, blocking pores and cementing materials (USEPA, 2003; Beak and Wilkin, 2009). Deeper in the aquifer, sulfate from groundwater can be reduced to sulfides (usually iron sulfides) which provide additional surfaces for arsenic removal. Beak and Wilkin (2009) investigated the PRB core samples using XANES analysis and concluded that As is probably present in three forms in the PRB (*i.e.* As(V) and As(III) sorbed to Fe (oxy)hydroxides and As(III) sorbed to Fe sulfide phases. Rao *et al.* (2009) investigated the effects of humic acid on arsenic(V) removal by ZVI from groundwater. They concluded that the removal rate of arsenic by Fe(0) was inhibited in the presence of humic acid due to the formation of soluble Fe-humates in groundwater. Humic acid can also block adsorption sites (Gu *et al.*, 1994). The binding capacity of humic acid for dissolved Fe is estimated to be ~0.75 mg Fe/mg of humic acid (Rao *et al.*, 2009). Inorganic anions in groundwater, such as bicarbonate, sulfate, nitrate, silicate, phosphate and chromate, can also inhibit arsenic removal due to competition for adsorption sites on corrosion products (Su and Puls, 2001a,b, 2003; Tyrovola *et al.*, 2006). Therefore, the effect of inorganic and organic compounds on the removal of arsenic from different groundwaters should be taken into account in the design of ZVI PRBs.

Other materials such as iron-bearing minerals (*e.g.* goethite, magnetite, akaganeite, ferrihydrite, *etc.*), granular ferric hydroxide, granular ferric oxide, furnace slag, hydrous ferric oxide loaded into activated carbon or furnace slag, and composite materials (iron oxides with calcium oxides and limestone; compost with zero-valent iron, *etc.*) have also been proposed for permeable reactive barriers (USEPA, 1998; Sasaki *et al.*, 2008; Gibert *et al.*, 2010).

Zero-valent iron-based permeable reactive barrier technologies are very suitable, effective and reliable for arsenic remediation and potentially cost-effective options (they require minimal operation and maintenance expenditure), but there is still a need for more investigation of the chemistry, geochemistry, hydrogeology and microbiology involved for long-term field applications. Direct injection of nZVI into contaminated aquifers could be less costly than traditional pump-and-treat methods and using PRBs (USEPA, 2008a). To use this direct injection technology, the geological and hydrological conditions, geochemical properties of aquifer, concentration and type of contaminant must be investigated in detail in order to optimize the effectiveness of the nZVI (USEPA, 2008a). Gavaskar *et al.* (2005) indicated that the price for nZVI varied from $20 to $77 per pound depending on the quantity purchased. They explained that the price of nZVI has decreased in more recent years because of a decrease in the price of raw materials and increase in manufacturing capacity and number of suppliers (Gavaskar *et al.*, 2005). Information on the cost of nZVI direct injection is limited. The USEPA (2008a), for example, provided total remediation costs for three sites which varied from \$255,500 to \$4,000,000. USEPA (2008a) and Gavaskar *et al.* (2005) estimated the total nZVI injection implementation cost ranged from \$250,000 to \$1,400,000. The total nZVI injection implementation cost is quite variable and depends on many factors including: site type, environmental conditions and the geology of areas to be treated, type of contaminants, concentrations and speciation of contaminants, volume of contaminated water, installation of a monitoring well, sampling, extent of the plume, nZVI injection, post-injection sampling, reporting and any challenges that may have occurred during remediation (USEPA, 2008a; Gavaskar *et al.*, 2005).

3. Removal of nitrate

Nitrate (NO_3^-) is an essential nutrient for animals and plants (nitrogen is part of DNA and proteins) but in large concentrations may cause some health problems. Nitrate is a tasteless, colourless and odourless compound that cannot be detected unless the water is analysed chemically (Self and Waskom, 2008). The maximum recommended contaminant concentration in drinking water, according to the US Environmental Protection Agency and the World Health Organization, is: (1) 45 ppm as nitrate (NO_3^-) and 10 ppm as nitrogen in nitrate (NO_3-N) (EPA, 2009); and (2) 50 ppm as nitrate and 11 ppm as nitrogen in nitrate (WHO, 2011).

Nitrates occur naturally in soil, rocks and surface and ground waters but become potentially more hazardous because they are applied as agricultural fertilizers. The common sources of nitrite include animal waste, manure, septic systems, industrial processes, municipal wastewater and sludge (Self and Waskom, 2008). Nitrates are very soluble and dissolve easily in natural waters, causing widespread contamination. Elevated concentrations of nitrate in surface waters promote excessive growth of plankton (mainly algae) and plants causing eutrophication, as well as death of aquatic living organisms (insects, invertebrates, fish, *etc.*) due to anoxia. High concentrations of nitrate in drinking water may cause chronic diseases, birth defects, gastric problems due to the formations of nitrosamines, as well as methemoglobinemia (blue baby syndrome) which can result in brain damage and death (Camargo and Alonso, 2006; Self and Waskom, 2008).

Drinking water treatments such as conventional purification (*e.g.* coagulation, flocculation, sedimentation and filtration) are not suitable for nitrate removal. The common technologies used to remove nitrate from drinking water are chemical denitrification, distillation, ion exchange, reverse osmosis and electrodialysis (Follett and Hatfield, 2001). These methods have some limitations and difficulties when applied in small communities (Huang *et al.*, 1998; Jeong *et al.*, 2012).

3.1. Application of iron-bearing minerals

Redox-reactive minerals such as Fe(II)-bearing minerals can, potentially, remove nitrate from drinking water. It was proposed that Fe(II) can reduce nitrate to nitrite, which converts to N_2 and/or might react with dissolved organic matter (Postma *et al.*, 1991; Davidson *et al.*, 2003; Burgin and Hamilton, 2007). For example, green rusts

A pilot-scale PRB at a lead-smelting facility operated by ASARCO, East Helena, Montana, USA.

The ASARCO East Helena lead smelter is located near East Helena in Montana (USA). The plant operated from 1888 to 2001. Wilkin *et al.* (2009) showed that groundwater underneath the site is contaminated with arsenic, selenium, lead, cadmium and zinc. Plumes of arsenic and selenium have migrated offsite whereas the occurrence of other dissolved metals is restricted within site boundaries. In June 2005, a 9.1 m long (perpendicular to groundwater flow), 13.7 m deep, and 1.8 to 2.4 m wide (in the direction of groundwater flow) pilot-scale permeable reactive barrier was installed at that area. The reactive barrier was designed to treat groundwater contaminated with moderately high concentrations of both As(III) and As(V), as well as other elements. The reactive barrier was installed over three days using bio-polymer slurry methods and modified excavating equipment for deep trenching (Wilkin *et al.*, 2009). The reactive medium was composed entirely of granular iron which was selected based on long-term laboratory column experiments. The trench was backfilled with a 7.6 m-thick layer of granular iron (from 6.1 to 13.7 m below ground surface) and a 6.1 m-thick layer of sand (from 0 to 6.1 m below ground surface). The top of the granular iron zone was located >2 m above the maximum groundwater level observed during site characterization studies. The base of the granular iron zone is located ~1 m above the confining ash tuff deposit. The PRB contains ~174 t of granular iron with an estimated initial porosity of ~50% (Wilkin *et al.*, 2009).

A monitoring network consisting of 40 multilevel wells was installed in 2005 within and around the PRB. Wilkin *et al.* (2009) collected and analysed groundwater samples within 1, 4, 12, 15 and 25 months of operation. Arsenic concentrations were >25 ppm in wells located hydraulically upgradient of the PRB; within the PRB, arsenic concentrations were reduced to 0.01 ppm after 2 y of monitoring.

Wilkin *et al.* (2009) concluded that ZVI can be used effectively to treat groundwater contaminated with arsenic, given appropriate groundwater geochemistry and hydrology.

Source : Wilkin *et al.* (2009)

(GRs) are excellent reductants for nitrate (Hansen *et al.*, 1996; Hansen and Koch, 1998). Hansen *et al.* (1996) suggested that sulfate green rust $[Fe_4^{II}Fe_2^{III}(OH)_{12}SO_4 \cdot yH_2O]$ should be considered as a possible important reductant for the reduction of nitrate to ammonium in subsoils, sediments or aquifers where microbially mediated reduction rates are small. Hansen *et al.* (1994) estimated the free energy of sulfate GRs formation, and indicated that GRs were stable in strongly reducible and non-acidic condition at equilibrium state. They also demonstrated that sulfate green rust reacts with nitrate producing ammonia and magnetite according to the following reaction (Hansen *et al.*, 1996):

$$[Fe_4^{II}Fe_2^{III}(OH)_{12}]SO_4(s) + 1/4NO_3^- + 3/2OH^- \leftrightarrow SO_4^{2-} + 1/4NH_4^+ + 2Fe_2O_4(s) + 6\cdot 1/4H_2O \quad (10)$$

Hansen and Koch (1998) determined that active nitrate reducing sites were located at the green rust surface and that the accessibility of nitrate to these sites controls the reaction rate.

Redox-reactive iron-bearing clay phases (Stucki *et al.*, 1984) can also play a significant role in nitrate reduction according to Day (2010) who determined a clear relationship between the amount of Fe(II) present in the clay structure and the amount of nitrate removed from a dilute solution. The reactive sites were located on the edge surfaces of the clay layers. The reduced form of Na^+-SWa-1 smectite took up a small amount of nitrate due to coulombic repulsion between the negatively charged smectite surface and nitrate anions. This process can inhibit significantly nitrate reduction by Fe(II)-bearing smectite (Day, 2010). In order to solve this problem, Su *et al.* (2012) proposed the modification of smectite with polydiallyldimethylammonium chloride (poly-DADMAC) which resulted in the creation of a positively charged surface on the smectite where nitrate could be adsorbed easily. Then nitrate can be reduced to nitrite by the structural Fe(II) in the redox-modified smectite (Su *et al.*, 2012). Those authors demonstrated extensive nitrate reduction by the positively charged, cationic polymer-modified, reduced-Fe ferruginous smectite and concluded that these novel redox-reactivated organoclays have important adsorption and reduction properties. However, there are still many questions concerning the mechanisms and kinetics of nitrate reduction by redox-reactive iron-bearing clay.

Micro- and nano-scale zero-valent iron can also reduce nitrate (Young *et al.*, 1964; Huang *et al.*, 1998; Till *et al.*, 1998; Alowitz and Scherer, 2002; Westerhoff, 2003) through the following equation:

$$4Fe^0 + NO_3^- + 7H_2O \leftrightarrow 4Fe^{2+} + NH_4^+ + 10OH^- \quad (11)$$

Alkalinity is produced in this process and, therefore, pH is an important parameter which can influence the reaction kinetics (Huang *et al.*, 1998; Huang and Zhang; 2002; Choe *et al.*, 2004; Miehr *et al.*, 2004). Choe *et al.* (2004) investigated nitrate reduction by ZVI under anaerobic and various pH conditions. The pH rise was neutralized by the addition of acids (HCl and CH_3COOH), and NO_3^- reduction occurred continuously and completely with the appearance of green rusts at pH 6.5. Thus, the formation of green rust leads to nitrate reduction in this process.

3.2. Electrochemical methods

Electrochemical methods for nitrate removal show high potential as effective and low cost technologies which do not require the addition of chemicals before or after treatment, and which produce small amounts of sludge (Koparal and Ogutveren, 2002). The latter authors investigated the feasibility of the nitrate removal from water by an electrocoagulation method with a bipolar, packed-bed reactor filled with iron Raschig rings. The following reactions occurred in this system:

$$2H_2O + 2e \leftrightarrow H_2 + 2OH^- \text{ at the cathode} \quad (12)$$

$$Fe^0 \leftrightarrow Fe^{2+} + 2e \text{ at the anode} \quad (13)$$

$$Fe^{2+} + 2OH^- \leftrightarrow Fe(OH)_2 \leftrightarrow Fe(OH)_3 \text{ in the solution} \quad (14)$$

Koparal and Ogutveren (2002) found that the concentration of nitrate decreased from 300 mg/L to 50 mg/L in ~10 min with an energy consumption of about 0.7×10^{-4} kWh/g at 80 V. They also showed that electrocoagulation is more cost effective than electroreduction (energy consumption of 1×10^{-3} kWh/g).

Jeong *et al.* (2012) investigated the performance of a ZVI packed-bed bipolar electrolytic cell for nitrate removal (Fig. 6). They used ZVI not only as conductive particles but also as electron donors. In their experiment pH increased slightly due to the consumption of hydrogen ions during nitrate reduction, with iron corrosion as a product, according to the following equations:

$$NO_3^- + Fe^0 + 2H^+ \rightarrow Fe^{2+} + NO_2^- + H_2O \quad (15)$$

$$NO_3^- + 4Fe^0 + 10H^+ \rightarrow 4Fe^{2+} + NH_4^+ + 3H_2O \quad (16)$$

$$2NO_3^- \; 5Fe^0 + 12H^+ \rightarrow 5Fe^{2+} + N_{2(g)} + 6H_2O \quad (17)$$

$$NO_2^- + 3Fe^0 + 8H^+ \rightarrow 3Fe^{2+} + NH_4^+ + 2H_2O \quad (18)$$

$$2NO_2^- + 3Fe^0 + 8H^+ \rightarrow 3Fe^{2+} + N_{2(g)} + 4H_2O \quad (19)$$

$$6NO_3^- + 10Fe^0 + H_2O \rightarrow 5Fe_2O_{3(s)} + 6OH^- + 3N_{2(g)} \quad (20)$$

$$NO_3^- + 2.82Fe^0 + 0.75Fe^{2+} + 2.25H_2O \rightarrow NH_4^+ + 1.19Fe_3O_{4(s)} + 0.5OH^- \quad (21)$$

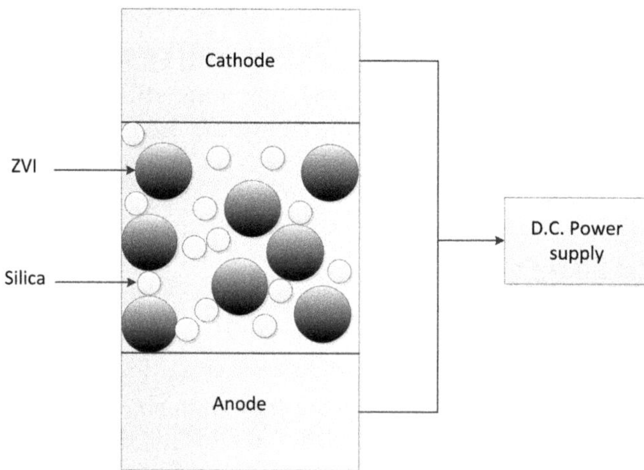

Figure 6. Schematic representation of a ZVI packed-bed bipolar electrolytic cell.

As a result of these redox reactions, ZVI removed >99% of the nitrate and the influx nitrate was converted to ammonia (20% to a maximum of 60%) and nitrite (always less than 0.5 mg/L as N in the effluent). The mechanism of nitrate removal was nitrate reduction in the lower part of the reactor under acidic conditions, followed by adsorption in the upper part of the reactor under alkaline conditions (Jeong *et al.*, 2012).

4. Removal of persistent organic pollutants

Some organic pollutants are toxic and cannot be degraded by natural environmental degradation. These organic pollutants, termed 'persistent organic pollutants' (POPs), refer to a variety of chemical compounds which contain carbon and are resistant to environmental degradation by chemical, biological and photolysis processes (Ritter *et al.*, 2007). POPs have low water contents, high fat solubilities and low vapour pressure. It has been proven that some POPs cause death or illness including certain cancers (*e.g.* breast, prostate) and endometriosis, neurobehavioural disorders, and disruption of the endocrine system (Ritter *et al.*, 2007). As a result of major threats to human health, the United Nations Environment Programme Governing Council (UNEPGC) has short-listed an initial twelve POPs substances for elimination, including organochlorine pesticides, cancer-causing polychlorinated biphenyls (PCBs) and the super-toxic dioxins and furans (Ritter *et al.*, 2007). Tonnes of POPs are being developed and manufactured every day and used in various products, including pesticides, industrial chemicals, food additives, pharmaceutical and personal care products (PPCPs), endocrine disruptor, surfactants and fuel additives (Pal *et al.*, 2010).

4.1. Conventional treatment technologies for POPs

Conventional treatment technologies for POPs in water consist of coagulation/flocculation, biofiltration, filtration and disinfection. The efficiency of POPs removal using these conventional technologies may be a joint function of the pollutant's structure and the treatment process applied.

Coagulation/flocculation, which removes only hydrophobic compounds associated with particle or colloidal material with a large organic carbon content, is not regarded as an efficient way of removing most POPs because most of these compounds are polar and hydrophilic (Yu *et al.*, 2008). Activated sludge, which uses air and a biological floc composed of bacteria and protozoa, is commonly used in municipal sewage and industrial wastewater-treatment plants (Jones *et al.*, 2002). It is possible, therefore, that POPs are easily removed by this treatment process.

Adsorption processes using either powdered activated carbon (PAC) or granular activated carbon (GAC) could play an important role in the removal of POPs (Yu *et al.*, 2008). Hydrophobic interactions are the principal mechanism in activated carbon adsorption of organic compounds (Yoon *et al.*, 2003). Therefore, POPs with higher octanol/water partition coefficients (K_{ow}) could be removed by activated carbon.

Powdered activated carbon with microfiltration (PAC-MF) systems and PAC with ultrafiltration (PAC-UF) systems, which combine PAC adsorption with low-pressure-driven membrane filtration, have shown great potential to achieve the removal of POPs from water. They have been considered as an alternative process for the remove of low molar mass compounds, including nano-sized POPs such as phenol, clofibric acid (a pharmaceutical product), methaldehyde (pesticide), which could not be removed by the membrane alone. Figure 7 shows the PAC with membrane hybrid process, in which the membrane module is arranged in connection with the adsorption reactor and operated in cross-flow mode (Delgado *et al.*, 2012). However, the PAC with membrane filtration system carries a high capital cost and energy consumption compared with conventional activated carbon processes.

Some conventional technologies have been adapted with redox processes for degradation of POPs in water and wastewater treatment. These include the ZVI with absorbent process and semiconductor (photocatalyst) with membrane filtration process (Zhang *et al.*, 2005; Luo *et al.*, 2007; Madaeni and Ghaemi, 2007; Zhu *et al.*, 2009; Doria *et al.*, 2013; Khraisheh *et al.*, 2014). As redox processes are the main topic of this book, they will be discussed in more detail in the next section.

4.2. Reduction and oxidation process for POPs

Redox processes are used in a wide variety of POPs remediation schemes for water. Generally, ZVI has been applied widely as a reductive material for the dechlorination of chlorinated organic compounds such as carbon tetrachloride (CT), trichloroethylene (TCE) and dichloroethylene (PCE) in ground and surface waters. Gillham and O'Hannesin (1994), using the results of CT reduction by ZVI, concluded that the reductive dechlorination of CT occurred at the metal surface. Matheson and Tratnyek, (1994) and Muftikian *et al.* (1995) proposed that the dechlorination process is a consequence of direct oxidative corrosion of the iron by TCE according to the following equations:

$$Fe^0 \rightarrow Fe^{2+} + 2e \tag{22}$$

$$C_2HCl_3 + 3H^+ + 6e^- \rightarrow C_2H_4 + 3Cl^- \text{ (on an iron phase surface)} \tag{23}$$

Semiconductor materials (such as TiO_2 anatase phase, TiO_2 rutile phase, ZnO, CdS, ZnS, WO_3, WSe_2, $SrTiO_3$, Si, GaAs, CdTe, SnO_2, Fe_2O_3, *etc.*), which can be regarded as photocatalysts, have also been applied to remove or degrade POPs in water. These materials can generate oxidation reactions from photo-generated holes, and reduction reactions from photo-generated electrons. Among those photocatalysts, TiO_2 has been used widely in water-treatment applications. Titanium dioxide (TiO_2) is the most industrially suitable photocatalyst because it has the most efficient photoactivity, the greatest stability and the lowest cost (Hashimoto *et al.*, 2005). Figure 8 shows redox

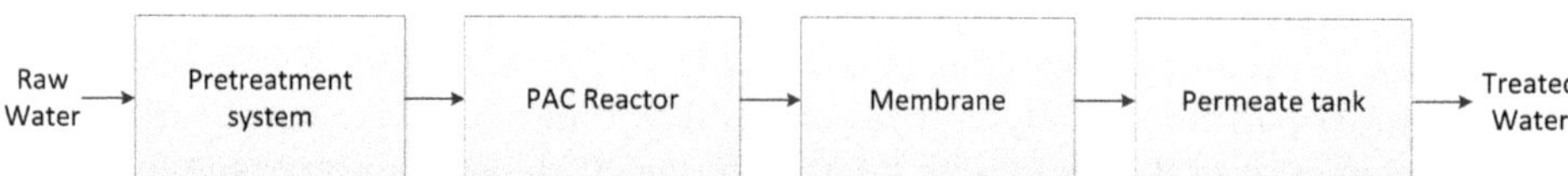

Figure 7. Schematic representation of the PAC with a membrane hybrid system.

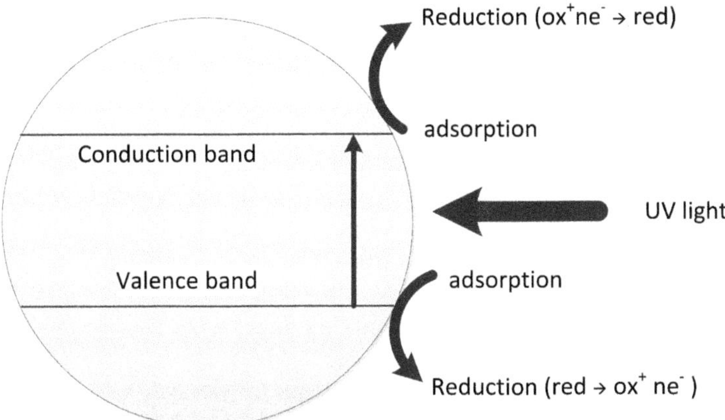

Figure 8. Redox reactions involved in the photocatalysis process.

reactions in the photocatalysis process. Khraisheh *et al.* (2014) researched the removal of pharmaceutical and personal care products, which are classified as emerging and persistent pollutants, from waters using novel TiO_2 with a coconut-shell powder composite. Three pharmaceutical and personal-care-products pollutants were removed with an efficiency of 99% from water. Doria *et al.* (2013) investigated the removal of metaldehyde (pesticide) using ZnO/Laponite composite using batch tests, and achieved greater removal efficiencies with combined redox and adsorption reactions than with an adsorption reaction only.

The photocatalyst membrane hybrid process has been investigated extensively because it has great potential for the development of efficient treatment systems compared with photocatalytic treatment or filtration system only. Membrane processes are well known in the chemical, medical, water and wastewater treatment and many other industries. Madaeni and Ghaemi (2007) investigated the effects of coating membrane surface with TiO_2 particles and UV radiation in creating self-cleaning to avoid the problem of fouling the membrane surface. Ma *et al.* (2009) invented a novel Ag-TiO_2/hydroxylapatite/Al_2O_3 bioceramic composite membrane which has anti-fouling properties and highly efficient bacterial adsorption properties. Ag-TiO_2 in this novel membrane acted as a powerful photocatalyst to attack bacteria in water.

Redox-reactive nanoparticles (NPs) have also been applied extensively to remove POPs in water. The nano-scale zero-valent iron (nZVI) technology, in which the advantages of the reductive potential of ZVI can be maintained, and the large surface area to volume can be exploited, gives rapid and high degradation efficiency for organic pollutants. Moreover, stable NPs with various chemical additives (*e.g.* polyelectrolytes, polymers, and surfactants) or NPs immobilized on inorganic or organic substrates have also been reported. Zhu *et al.* (2011) investigated the dechlorination of recalcitrant PCBs, which are one of the POPs, and a group of aromatic compounds notorious for their toxicity and persistence in the environment (UNEP, 2010). Nano-scale Ni/Fe particles were applied and compared to NZVI and nano-Ni^0. The

Chlorinated-solvent source-zone remediation *via* ZVI-Clay soil mixing at Marine Corps Base Camp Lejeune, North Carolina, USA.

ZVI-Clay was applied *in situ* with soil mixing to treat a chlorinated solvent, TeCA, TCE, and degradation products, consisting primarily of cDCE, tDCE, and VC at Site 89, Camp Lejeune, North Carolina. The dimensions of the treated zone include a surface footprint of 3010 m^2, a volume of 22,900 m^3 of soil was treated to an average depth of 7.6 m by mixing with 2% ZVI and 3% bentonite. Soil-mixing equipment (Fig. 9) included a crane, rotary table, and mixing auger were used in this field.

Figure 9. Soil mixing equipment (left) and mixing was completed using 3 m diameter augers (right) – Photos' credit: *Professor Charles Shackelford* (CSU).

In several of the soil- and groundwater-monitoring sites, the total concentration of CVOC decreased more than 99% within 1 y of completion of the mixing of ZVI-Clay with soil.

Source : Olson *et al.* (2012)

dechlorination products were more prevalent in the presence of nano-scale Ni/Fe than other redox-reactive particles, and biphenys, cyclohexyl-benzene and 1-alkyl-benzenes were detected as the main by-products. The conclusion is that the reactive nano-scale Ni/Fe can retain PCBs in subcritical water for rapid catalytic hydrodechlorination. Zhu *et al.* (2010) reported that nano-scale Cu/Fe bimetals have been used in the dechlorination of hexachlorobenzene (HBC), a representative polychlorinated additive in agricultural antiseptics. They found that HCB reduction was significantly increased by nano-scale Cu/Fe. HCB removal was nearly complete within 48 h of reaction with nano-scale Cu/Fe.

5. Removal of bacteria and other microorganisms

Bacteria and microorganisms that are not removed through filtration are usually inactivated/killed during disinfection, which is the last stage in conventional water treatment. Redox-reactive compounds commonly used for water disinfection include silver and/or copper, and these have been used for centuries due to their bactericidal and algaecidal properties. The advantage of silver and/or copper is that they do not produce by-products such as trihalomethanes during disinfection while maintaining an effective residual. Most modern silver-copper systems use electrolytic ions generated to control their concentration. The electrolytic ion generators consist of positive and negative electrodes made of the silver and/or copper contained in a vessel. To release the metal ions into the water, a DC power source is used to provide current with a potential of a few volts. The concentration of metal ions released in the water depends on the current and water flow passing the electrodes. Silver-copper systems have been used successfully in hospitals for *Legionella* control (Box 3) and ionization units range from US$40,000 to 80,000 with maintenance costs varying from US$1500 to 4000 for electrode replacement (Lin *et al.*, 2001). Silver ion systems have also been applied successfully to disinfect rainwater for human consumption in rural communities in Mexico (Adler *et al.*, 2011).

Silver-copper ionization system for Legionella's control in hospitals

A long-term study was carried out in 16 hospitals in Pennsylvania, USA, to determine the efficacy of silver-copper ionization as a disinfection method in controlling Legionella in water systems. Surveys were conducted during 1995 and 2000 to determine the hospitals' experience with the maintenance of the system, contamination of water with Legionella, and occurrence of hospital-acquired Legionnaires' Disease. Each hospital had an average of 435 beds. All 16 hospitals had cases of Legionnaires' Disease prior to installing the silver-copper disinfection system. Prior to this, most of the hospitals had other disinfection systems, including superheat and flush, ultraviolet light, and hyperchlorination. After installation, in 1995, 50% of the hospitals reported 0% positivity, and 43% still reported 0% in 2000, with respect to cases of Legionnaires' Disease. After 1995, no new cases of hospital-acquired Legionnaires' Disease were reported in any of these hospitals.

Source: Stout and Yu (2003)

TiO_2 and ZnO are redox-reactive compounds that have also been used for microbial disinfection of water by photocatalytic reaction. A recent study of the degradation of Methylene Blue found that TiO_2 nanoparticles generated more reactive oxygen species (ROS) and led to increased loss of cell viability than the ZnO nanoparticles for three of the four species of bacteria examined (*S. aureus*, *B. subtilis*, *E. coli* and *P. aeruginosa*)

(Barnes *et al.*, 2013). The ZnO particles produced less ROS than the TiO_2 nanoparticles under ultraviolet light, and they were shown to be toxic to two of the bacterial species even under dark conditions. *P. aeruginosa* cells were resistant to all types of treatment indicating a potential limitation to the application of these nanoparticles for water disinfection with these redox-reactive compounds. Another study investigating inactivation of *E. coli* by TiO_2/Cu nanosurfaces (*i.e.* DC-magnetron sputtered thin films) in the dark and under low-intensity actinic light found TiO_2/Cu sputtered layers to be sensitive to actinic light (Baghriche et *al.*, 2012). This work demonstrated the spectral characteristics of Cu/CuO indicating that Cu does not substitute for Ti^{4+} in the crystal lattice. The hybrid composite TiO_2/Cu sample produced fast bacterial inactivation times (<5 min) under diffuse actinic light (4 mW/cm^2). A direct relation between the film optical absorption obtained by diffuse reflectance spectroscopy and bacterial inactivation kinetics by the TiO_2/Cu samples was also demonstrated. Baghriche et *al.* (2012) suggested that the bacterial inactivation mechanism by TiO_2/Cu occurred *via* interfacial charge transfer involving charge transfer between TiO_2 and Cu.

Metallic gold is another redox-reactive compound that has been used successfully in microbial disinfection. Biodegradable Au nanocomposite hydrogels have been developed recently using acrylamide and wheat protein isolate (Jayaramudu *et al.*, 2013). The gold nanoparticle composite hydrogels have been shown to be a potential candidate for antibacterial applications.

5.1. Effects of redox compounds on health and DBP production

Silver is not particularly toxic to human beings, and although large doses of this metal used for certain medical treatments have been found to cause discoloration of the skin, hair and nails (argirosis), no problems have been noted with the low concentrations needed to disinfect water. Treatment of drinking water with silver produces no by-products and the World Health Organization's (WHO) guidelines for drinking water quality indicate levels of silver of up to 0.1 mg/L can be tolerated without risk to health (WHO, 2003b). There is also insufficient evidence for negative health effects of silver ions (Lin *et al.*, 2002) and copper-silver ionization (LENNTECH (b)) and their by-products.

Titanium dioxide is the most widely used photocatalyst because of its efficient photoactivity, high stability, low cost and negative toxicity for humans and animals. A study by Richardson *et al.* (1996) observed a single type of organic DBP (tentatively identified as 3-methyl-2,4-hexanedione) in ultra-filtered raw water treated with TiO_2/UV alone.

6. Conclusions

Redox-reactive minerals (green rusts, iron-bearing clay phases, ZVI, manganese oxides, TiO_2, ZnO, Ag, *etc.*) play a significant role in water-treatment processes. In this chapter we have demonstrated that proven treatment technologies using redox minerals are effective at reducing the levels of arsenic, nitrate, organic pollutants and pathogenic microbes to concentrations which are less than the recommended maximum contaminant concentrations in surface and ground water. Selection of an appropriate cost-effective treatment technology for pollutant removal from drinking water requires a detailed investigation of the quality and composition of the raw water as well as consideration of simplicity, sustainability, feasibility, reliability and costs of chosen technologies.

Knowledge is still limited on the fate, transport and toxicological effects of some redox-reactive minerals (particularly nano-scale materials such as nZVI) in the environment. For example, agglomeration of nanoscale redox-reactive minerals (nRAM) in surface and underground waters and their interaction with other compounds present in the system can impact the reactivity and mobility (transport) of nRAM. Studies related to surface modification (*e.g.* coating) of nanoparticles or addition of compounds which can inhibit nRAM agglomeration and improve water-treatment technologies with increasing mobility and activity of nRAM within aquifers are required. Further comprehensive investigations are needed to understand fully reaction mechanisms and redox-reactive mineral fates in contaminated waters, as well as to determine cost-effective and environmentally friendly waste-water management. Investigation of the kinetic and thermodynamic controls on the redox reactivity of RAM (particularly complex heterogeneous redox-reactive minerals) is very important for the future.

Acknowledgements

We are grateful for comments, questions and careful and helpful reviews by Karen A. Hudson-Edwards, Imad A.M. Ahmed, Kevin Murphy and the anonymous reviewers.

References

Adler, I., Hudson-Edwards, K.A. and Campos, L.C. (2011) Converting rain into drinking water: Quality issues and technological advances. *Water Science and Technology: Water Supply*, **11**, 659–667.

Ahmed, I.A.M., Benning, L.G., Kakonyi, G., Sumoondur, A.D., Terrill, N.J. and Shaw, S. (2010a) Formation of green rust sulphate: A combined in situ time-resolved X-ray scattering and electrochemical study. *Langmuir*, **26**, 6593–6603.

Ahmed, I.A.M., Shaw, S., Kakonyi, G. and Benning, L.G. (2010b) In situ studies of green rust formation using synchrotron-based X-ray scattering. Helping to develop a new range of environmental materials. *ECG Bulletin*, 3–5.

Alowitz, M.J. and Scherer, M.M. (2002) Kinetics of nitrate, nitrite, and Cr(VI) reduction by iron metal. *Environmental Science & Technology*, **36**, 299–306.

Altundogan, H.S., Altundogan, S., Tumen, F. and Bildik, M. (2000) Arsenic removal from aqueous solutions by adsorption on red mud. *Waste Management*, **20**, 761–767.

Altundogan, H.S., Altundogan, S., Tumen, F. and Bildik, M. (2002) Arsenic adsorption from aqueous solutions by activated red mud. *Waste Management*, **22**, 357–363.

Anthony, J.W., Bideaux, R.A., Bladh, K.W. and Nichols, M.C. (editors) (1997) *Handbook of Mineralogy*, Volume III, Mineralogical Society of America, Chantilly, Virginia, USA. http://www.handbookof mineralogy.org/.

Baghriche, O., Rtimi, S., Pulgarin, C., Sanjines, R. and Kiwi, J. (2012) Innovative TiO_2/Cu nanosurfaces inactivating bacteria in the minute range under low-intensity actinic light. *ACS Applied Materials & Interfaces*, **4**, 5234–5240.

Baláž, P., Timko, M., Kovác, J., Bujnáková, Z., Durišin, J., Myndyk, M. and Šepelák, V. (2010) Magnetic properties and sorption activity of mechanically activated magnetite Fe_3O_4. *Acta Phisica Polonica A*, **118**, 1005–1007.

Bang, S., Korfiatis, G.P. and Meng, X. (2005) Removal of arsenic from water by zero-valent iron. *Journal of Hazardous Materials*, **121**, 61–67.

Bardos, P., Bone, B., Elliott, D., Harton, N., Henstock, J. and Nathanail, P. (2011) *A Risk/Benefit Approach to the Application of Iron Nanoparticles for the Remediation of Contaminated Sites in the Environment.* Defra Research Project Final Report.

Barnes, R.J., Molina, R., Xu, J., Dobson, P.J. and Thompson, I.P. (2013) Comparison of TiO_2 and ZnO nanoparticles for photocatalytic degradation of methylene blue and the correlated inactivation of gram-positive and gram-negative bacteria. *Journal of Nanoparticle Research*, **15**, 1432–1443.

Beak, D.G. and Wilkin, R.T. (2009) Performance of a zerovalent iron reactive barrier for the treatment of arsenic in groundwater: Part 2. Geochemical modeling and solid phase studies. *Journal of Contaminant Hydrology*, **106**, 15–28.

Beaulieu, B. and Ramirez, R.E. (2013) Arsenic remediation field study using a sulfate reduction and zero-valent iron PRB. *Groundwater Monitoring & Remediation*, **33**, 85–94.

Bernal, J.O., Dasgupta, D.R. and Mackay, A.L. (1959) The oxides and hydroxides of iron and their structural interrelationships. *Clay Minerals Bulletin*, **4**, 15–30.

Bissen, M. and Frimmel, F.H. (2003) Arsenic – a review – Part II: Oxidation of arsenic and its removal in water treatment. *Acta Hydrochimica et Hydrobiologica*, **31**, 97–107.

Blowes, D.W. and Ptacek, C.J. (1994) *System for Treating Contaminated Groundwater.* US Patent number US5362394.

Blowes, D.W., Ptacek, C.J., Benner, S.G., McRae, C.W.T., Bennett, T.A. and Puls, R.W. (2000) Treatment of inorganic contaminants using permeable reactive barriers. *Journal of Contaminant Hydrology*, **45**, 123–137.

Borch, T., Rretzschmar, R., Kappler, A., van Cappellen, Ph., Ginder-Vogel, M., Voegelin, A. and Campbell, K. (2010) Biogeochemical redox processes and their impact on contaminant dynamics. *Environmental Science & Technology*, **44**, 15–23.

Bocher, F., Gehin, A., Ruby, C., Ghanbaja, J., Abdelmoula, M. and Genin, J.M.R. (2004) Coprecipitation of Fe(II–III) hydroxycarbonate green rust stabilised by phosphate adsorption. *Solid State Sciences*, **6**, 117–124.

Bowles, B.A., Drew, W.M. and Hirth, G. (1983) Application of the slow sand filter process to the treatment of small town water supplies. *Annual Convention Australian Water and Wastewater Association*, **28**, 1–28

Bradl, H.B. (2005) *Heavy Metals in the Environment: Origin, Interaction and Remediation.* Elsevier Academic Press, Amsterdam.

Brindley, G.W. and Bish, D.L. (1976) Green rust: a pyroaurite type structure. *Nature*, **262**, 353.

Burgin, A.J. and Hamilton, S.K. (2007) Have we over-emphasized the role of denitrification in aquatic ecosystems? A review of nitrate removal pathways. *Frontiers in Ecology and the Environment*, **5**, 89–96.

Camargo, J.A. and Alonso, Á. (2006) Ecological and toxicological effects of inorganic nitrogen pollution in aquatic ecosystems: A global assessment. *Environment International*, **32**, 831–849.

Cao, J. and Zhang, W. (2006) Stabilization of chromium ore processing residue (COPR) with nanoscale iron particles. *Journal of Hazardous Materials*, **B132**, 213–219.

Castellote, M. and Bengtsson, N. (2011) Principles of TiO_2 Photocatalysis. Pp. 5–10 in: *Application of Titanium Dioxide Photocatalysis to Construction Materials* (Y. Ohama and D. Van Gemert, editors). RILEM State-of-the-Art Reports 5.

CCME (1999) Canadian sediment quality guidelines for the protection of aquatic life: Arsenic. In: *Canadian Environmental Quality Guidelines.* Canadian Council of Ministers of the Environment, Winnipeg.

Chakravarty, S., Dureja, V., Bhattacharyya, G., Maity, S. and Bhattacharjee, S. (2002) Removal of arsenic from groundwater using low cost ferruginous manganese ore. *Water Research*, **36**, 625–632.

Cheremisinoff, N.P. (2002) *Handbook of Water and Wastewater Treatment Technologies.* Butterworth-

Heinemann, USA.

Choe, S., Liljestrand, H.M. and Khim, J. (2004) Nitrate reduction by zero-valent iron under different pH regimes. *Applied Geochemistry*, **19**, 335–342.

Con, T.H., Thao, Ph., Dai, T.X. and Loan, D.K. (2013) Application of nano dimensional MnO_2 for high effective sorption of arsenic and fluoride in drinking water. *Environmental Sciences*, **1**, 69–77.

Cornell, R.M. and Schwertmann, U. (2003) *The Iron Oxides: Structure, Properties, Reactions, Occurrence, and Uses*. VCH, New York.

Crane, R.A., Dickinson, M., Popescu, I.C. and Scott, T.B. (2011) Magnetite and zero-valent iron nanoparticles for the remediation of uranium contaminated environmental water. *Water Research*, **45**, 2931–2942.

Dangerous Substances Directive (1976) Council Directive 76/464/EEC on pollution caused by certain dangerous substances discharged into the aquatic environment of the Community (OJ L 129, 18.05.1976, p. 23).

Davidson, E.A., Chorover, J. and Dail, D.B. (2003) A mechanism of abiotic immobilization of nitrate in forest ecosystems: The ferrous wheel hypothesis. *Global Change Biology*, **9**, 228–236.

Day, Z.B. (2010) *Abiotic Nitrate Reduction by Redox Activated Iron-bearing Smectites*. Master thesis, University of Illinois, Urbana, 57 pp.

Delgado, L.F., Charles, F., Glucina, K. and Morlay, C. (2012) The removal of endocrine disrupting compounds, pharmaceutically activated compounds and cyanobacterial toxins during drinking water preparation using activated carbon – a review. *Science of the Total Environment*, **435-436**, 509–525.

Dickinson, M. and Scott, T.B. (2010) The Application of zero-valentiron nanoparticles for the remediation of a uranium-contaminated waste effluent. *Journal of Hazardous Materials*, **178**, 171–179.

Doria, F.C., Borges, A., Kim, J.K., Nathan, A., Joo, J.C. and Campos, L.C. (2013) Removal of metaldehyde through photocatalytic reactions using nano-sized zinc oxide composites. *Water Air and Soil Pollution*, **22**, 1434–1443.

Du, Q., Zhang, Sh., Pan, B., Lu, L., Zhang, W. and Zhang, Q. (2013) Bifunctional resin-ZVI composites for effective removal of arsenate through simultaneous adsorption and oxidation. *Water Research*, **47**, 6064–6074.

Edmunds, W.M., Ahmed, K.M. and Whitehead, P.G. (2015) A review of arsenic and its impacts in groundwater of the Ganges-Brahmaputra-Meghna delta, Bangladesh. *Environmental Science: Processes and Impacts*, **17**, 1032–1046.

EPA (2009) *National Primary Drinking Water Regulations*, 816-F-09-0004.

Ernstsen, V., Gates, W.P. and Stucki, J.W. (1998) Microbial reduction of structural iron in clays – A renewable source of reduction capacity. *Journal of Environmental Quality*, **27**, 761–766.

Farrell, J., Wang, J., O'Day, P. and Conklin, M. (2001) Electrochemical and spectroscopic study of arsenate removal from water using zero-valent iron media. *Environmental Science & Technology*, **35**, 2026–2032.

Ficai, D., Andronescu, E., Ficai, A., Voicu, G., Vasile, B., Ionita, B. and Guran, C. (2012) Synthesis and characterization of mesoporous magnetite based nanoparticles. *Current Nanoscience*, **8**, 875–879.

Fierro, J.L.G. (Editor) (2006) *Metal Oxides: Chemistry and Application*. Taylor & Francis Group, CRC Press, Boca Raton, Florida, USA.

Follett, R.F. and Hatfield, J.L. (editors) (2001) *Nitrogen in the Environment: Sources, Problems, and Management*. Elsevier Science Publishers, Amsterdam.

Fu, F., Dionysiou, D.D. and Liu, H. (2014) The use of zero-valent iron for groundwater remediation and wastewater treatment: A review. *Journal of Hazardous Materials*, **267**, 194–205.

Fujishima, A., Rao, T.N. and Tryck, D.A. (2000) Titanium dioxide photocatalysis. *Journal of Photochemistry and Photobiology C*, **1**, 1–21.

Gavaskar, A., Tatar, L. and Condit, W. (2005) *Cost and Performance Report: Nanoscale Zero-Valent Iron Technologies for Source Remediation*. NAVFAC, California, 54 pp.

Genc-Fuhrman, H., Tjell, J.C., McConchie, D. and Schuiling, O. (2003) Adsorption of arsenate from water using neutralized red mud. *Journal of Colloid and Interface Science*, **264**, 327–334.

Genc-Fuhrman, H., Tjell, J.C. and McConchie, D. (2004a) Adsorption of arsenic from water using activated neutralized red mud. *Environmental Science & Technology*, **38**, 2428–2434.

Genc-Fuhrman, H., Tjell, J.C. and McConchie, D. (2004b) Increasing the arsenate adsorption capacity of neutralized red mud (Bauxsol). *Journal of Colloid and Interface Science*, **271**, 313–320.

Genc-Fuhrman, H., Bregnhoj, H. and McConchie, D. (2005) Arsenate removal from water using sand–red mud columns. *Water Research*, **39**, 2944–2954.

Génin, J.M. R., Refait, Ph., Olowe, A.A., Abdelmoula, M., Fall, I. and Drissi, S.H. (1998) Identification of green rust compounds in the aqueous corrosion processes of steels; the case of microbially induced corrosion and use of 78 K CEMS. *Hyperfine Interactions*, **112**, 47–51.

Génin, J.R., Refait, P., Bourrie, G., Abdelmoula, M. and Trolard, F. (2001) Structure and stability of the Fe(II)–Fe(III) green rust "fougerite" mineral and its potential for reducing pollutants in soil solutions. *Applied Geochemistry*, **16**, 559–570.

Gibert, O., Pablo, J., Cortina, J-L. and Ayora, C. (2010) In situ removal of arsenic from groundwater by using permeable reactive barriers of organic matter/limestone/zero-valent iron mixtures. *Environmental Geochemistry and Health*, **32**, 373–378.

Gillham, R.W. and O'Hannesin, S.F. (1994) Enhanced degradation of halogenated aliphatics by zero-valent iron, *Ground Water*, **29**, 958–967.

Gimenez, J., Maȑtnez, M., Pablo, J., Rovira, M. and Duroc, L. (2007) Arsenic sorption onto natural hematite, magnetite, and goethite. *Journal of Hazardous Materials*, **141**, 575–580.

Gorski, C.A. (2009) *Redox Behavior of Magnetite in the Environment: Moving Towards a Semiconductor Model*. PhD thesis, University of Iowa.

Gorski, C.A., Aeschbacher, M., Soltermann, D., Voegelin, A., Baeyens, B., Fernandes, M.M., Hofstetter, T.B. and Sander, M. (2012a) Redox properties of structural Fe in clay minerals. 1. Electrochemical quantification of electron-donating and -accepting capacities of smectites. *Environmental Science & Technology*, **46**, 9360–9368.

Gorski, C., Kluepfel, L., Voegelin, A., Sander, M. and Hofstetter, T.B. (2012b) Redox properties of structural Fe in clay minerals: 2. Electrochemical and spectroscopic characterization of electron transfer irreversibility in ferruginous smectite, SWa-1. *Environmental Science & Technology*, **46**, 9369–9377.

Gorski., C.A., Nurmi, J.T., Tratnyek, P.G., Hofstetter, T.B. and Scherer, M.L. (2010) Redox behavior of magnetite: implications for contaminant reduction. *Environmental Science & Technology*, **44**, 55–60.

Gu, B., Schmitt, J., Chen, Z., Liang, L. and McCarthy, J.F. (1994) Adsorption and desorption of natural organic matter on iron oxide: Mechanisms and models. *Environmental Science & Technology*, **28**, 38–46.

Guilbaud, R., White, M.L. and Poulton, S.W. (2013) Surface charge and growth of sulphate and carbonate green rust in aqueous media. *Geochimica et Cosmochimica Acta*, **108**, 141–153.

Guo, X. and Chen, F. (2005) Removal of arsenic by bead cellulose loaded with iron oxyhydroxide from groundwater. *Environmental Science & Technology*, **39**, 6808–6818.

Hansen, H.C.B. and Koch, C.B. (1998) Reduction of nitrate to ammonium by sulphate green rust: Activation energy and reaction mechanism. *Clay Minerals*, **33**, 87–101.

Hansen, H.C.B., Borggaard, O.K. and Sorensen, J. (1994) Evaluation of the free energy of formation of Fe(II)-Fe(III) hydroxide-sulphate (green rust) and its reduction of nitrite. *Geochimica et Cosmochimica Acta*, **58**, 2599–2608.

Hansen, H.C.B., Koch, C.B., Nanckekrogh, H., Borggaard, O.K. and Sorensen, J. (1996) Abiotic nitrate reduction to ammonium: Key role of green rust. *Environmental Science & Technology*, **30**, 2053–2056.

Hashimoto, K., Irie, H. and Fujishima, A. (2005) TiO_2 photocatalysis: a historical overview and future prospects. *Japanese Journal of Applied Physics*, **44**, 8269–8285.

Hindmarsh, J.T. and McCurdy, R.F. (1986) Clinical and environmental aspects of arsenic toxicity. *CRC Critical Reviews in Clinical Laboratory Sciences*, **23**, 315–347.

Hofstetter, T.B., Neumann, A. and Schwarzenbach, R.P. (2006) Reduction of nitroaromatic compounds by Fe(II) species associated with iron rich smectites. *Environmental Science & Technology*, **40**, 235–242.

Hofstetter, T.B., Schwarzenbach, R.P. and Haderlein, S.B. (2003) Reactivity of Fe(II) species associated with clay minerals. *Environmental Science & Technology*, **37**, 519–528.

Holm, T.R. and Wilson, S.D. (2006) *Chemical Oxidation for Arsenic Removal*. Illinois State Water Survey, 53 pp.

Hopenhayn, C. (2006) Arsenic in drinking water: Impact on human health. *Elements*, **2**, 103–107.

Hoseinzadeh, E., Alikhani, M.Y., Samarghandi, M.R. and Shirzad-Siboni, M. (2014) Antimicrobial potential of synthesized zinc oxide nanoparticles against gram positive and gram negative bacteria. *Desalination and Water Treatment*, **52**, 4969–4976.

Huang, C.P., Wang, H.W. and Chiu, P.C. (1998) Nitrate reduction by metallic iron. *Water Research*, **32**, 2257–2264.

Huang, Y.H. and Zhang, T.C. (2002) Kinetics of nitrate reduction by iron at near neutral pH. *Journal of Environmental Engineering ASCE*, **128**, 604–611.

Illés, E. and Tombácz, E. (2004) The role of variable surface charge and surface complexation in the adsorption of humic acid on magnetite. *Colloids and Surfaces A: Physicochemical and Engineering Aspects*, **230**, 99–109.

Inamori, Y. and Fujimota, N. (2007) Microbial/biological contamination of water. In: *Water Quality and Standards* (S. Kubota and Y. Tsuchiya, editors). *Encyclopedia of Life Support Systems* (EOLSS), Developed under the Auspices of the UNESCO, Eolss Publishers, Oxford, UK. (http://www.eolss.net)

Jayaramudu, T., Raghavendra, G.M., Varaprasad, K., Sadiku, R. and Raju, K.M. (2013) Development of novel biodegradable Au nanocomposite hydrogels based on wheat: For inactivation of bacteria. *Carbohydrate Polymers*, **92**, 2193–2200.

Joshi, A. and Chaudhuri M. (1996) Removal of arsenic from ground water by iron oxide-coated sand. *Journal of Environmental Engineering*, **122**, 769–771.

Jain, C.K. and Singh, R.D. (2012) Technological options for the removal of arsenic with special reference to South East Asia. *Journal of Environmental Management*, **107**, 1–18.

Jeong, J., Kim, H., Kim, J. and Park, J. (2012) Electrochemical removal of nitrate using ZVI packed bed bipolar electrolytic cell. *Chemosphere*, **89**, 172–178.

Jones, O.A.H., Voulvoulis, N. and Lester, J.N. (2002) Aquatic environmental assessment of the top 25 English prescription pharmaceuticals. *Water Research*, **36**, 5013–5022.

Karn, B, Kuiken, T. and Otto, M. (2009) Nanotechnology and *in situ* remediation: A review of the benefits and potential risks. *Environmental Health Perspectives*, **117**, 1823–1831.

Khaled, E.M. and Stucki, J.W. (1991) Iron oxidation effects on cation fixation in smectites. *Soil Science Society of America Journal*, **55**, 550–554.

Khraisheh, M., Kim, J., Campos, L., Al-Muhtaseb, A.H., Al-Hawari, A., Al Ghouti, M. and Walker, G.M. (2014) Removal of pharmaceutical and personal care products (PPCPs) pollutants from water by novel TiO_2-coconutshell Powder (TCNSP) composite. *Journal of Industrial and Engineering Chemistry*, **20**, 979–987.

Kinniburgh, D.G. and Smedley, P.L. (2001) *Arsenic Contamination of Groundwater in Bangladesh*. British Geologic Survey Report WC/00/19 (www.bgs.ac.uk/arsenic/Bangladesh).

Konstantinou, I.K. and Albanis, T.A. (2003) Photocatalytic transformations of pesticides in aqueous titanium dioxide suspensions using artificial and solar light: intermediates and degradation pathways. *Applied Catalysis B: Environmental*, **42**, 319–335.

Koparal, A.S. and Outveren, U.B. (2002) Removal of nitrate from water by electroreduction and electrocoagulation. *Journal of Hazardous Materials*, **89**, 83–94.

Korte, N.E. and Fernando, Q. (1991) A review of arsenic(III) in groundwater. *Critical Reviews in Environmental Control*, **21**, 1–39.

Kostka, J.E., Wu, J., Nealson, K.H. and Stucki, J.W. (1999) The impact of structural Fe(III) reduction by bacteria on the surface chemistry of smectite clay minerals. *Geochimica et Cosmochimica Acta*, **63**, 3705–3713.

Kumar, A.V.R., Singh, R. and Nigam, R.K. (1999) Mössbauer spectroscopy of corrosion products of mild steel due to microbiologically influenced corrosion. *Journal of Radioanalytical and Nuclear Chemistry*, **242**, 131–137.

Lackovic, J.A., Nikolaidis, N.P. and Dobbs, G.M. (2000) Inorganic arsenic removal by zero-valent iron. *Environmental Engineering Science*, **17**, 29–39.

Lakshmanan, D., Clifford, D.A. and Samanta, G. (2008) *Arsenic Coagulation with Iron, Aluminum, Titanium, and Zirconium Salts*. WERC and AWWA Research Foundation and U.S. Department of Energy, USA.

Lanson, B., Drits, V.A., Silvester, E.J. and Manceau, A. (2000) Structure of H-exchanged hexagonal birnessite and its mechanism of formation from Na-rich monoclinic buserite at low pH. *American Mineralogist*, **85**, 826–838.

Lear, P.R. and Stucki, J.W. (1989) Effects of iron oxidation state on the specific surface area of nontronite. *Clays and Clay Minerals*, **37**, 547–552.

Legube, B., Croue, J.P., De Latt, J. and Dore, M. (1989) Ozonation of an extracted aquatic fulvic acid: theoretical and practical aspects. *Ozone: Science and Engineering*, **11**, 69–91.

LENNTECH (a) Organic compounds in freshwater. (http://www.lenntech.com/aquatic/organic-pollution.htm)

LENNTECH (b) Disinfectants Copper-silver ionization. (http://www.lenntech.com/processes/disinfection/chemical/disinfectants-copper-silver-ionization.htm)

Leupin, O.X. and Hug, S.J. (2005) Oxidation and removal of arsenic(III) from aerated groundwater by filtration through sand and zero-valent iron. *Water Research*, **39**, 1729–1740.

Li, X., Elliott, D.W. and Zhang, W. (2006) Zero-valent iron nanoparticles for abatement of environmental pollutions: Materials and engineering aspects. *Critical Reviews in Solid State and Materials Sciences*, **31**, 111–122.

Liu, D., Dong, H., Bishop, M.E., Zhang, J., Wang, H., Xie, S., Wang, S., Huang, L. and Eberl, D.D. (2012) Microbial reduction of structural iron in interstratified illite-smectite minerals by a sulfate-reducing bacterium. *Geobiology*, **10**, 150–162.

Lien, H.L. and Wilkin, R.T. (2005) High-level arsenite removal from groundwater by zero-valent iron. *Chemosphere*, **59**, 377–386.

Lin, Y.E., Stout, J.E. and Yu, V.L. (2001) Control of *Legionella*. Pp. 505–512 in: *Disinfection, Sterilization, and Preservation* (S.S. Block, editor), Lippincott Williams & Wilkins.

Lin, Y.E., Vidic, R.D., Stout, J.E. and Yu, V.L. (2002) Negative effect of high pH on biocidal efficacy of copper and silver ions in controlling *Legionella pneumophila. Applied and Environmental Microbiology*, **68**, 2711–2715.

Liqiang, J., Yichun, Q., Baiqi, W., Shudan, L., Baojiang, J., Libin, Y., Wei, F., Honggang, F. and Jiazhong, S. (2006) Review of photoluminescence performance of nano-sized semiconductor materials and its relationships with photocatalytic activity. *Solar Energy Materials & Solar Cells*, **90**, 1773–1780.

Liqiang, J., Xiaojun, S., Jing, S., Weimin, C., Zili, X., Yaoguo, D. and Honggang, F. (2003) Review of surface photovoltage spectra of nanosized semiconductor and its applications in heterogeneous photocatalysis. *Solar Energy Materials & Solar Cells*, **79**, 133–151.

Lo, S.L., Jeng, H.T. and Lai, C.H. (1997) Characteristics and adsorption properties of iron-coated sand. *Water Science & Technology*, **35**, 63–70.

Luo, Y., Liu, J., Xia, X., Li, X., Fang, T., Li, S., Ren, Q., Li, J. and Jia, Z. (2007) Fabrication and characterization of TiO_2/short MWNTs with enhanced photocatalytic activity. *Materials Letters*, **61**, 2467–2472.

Ma, N., Fan, X., Quan, X. and Zhang, Y. (2009) Ag-TiO_2/HAP/Al_2O_3 bioceramic composite membrane: Fabrication, characterization and bactericidal activity. *Journal of Membrane Science*, **336**, 109–117.

Madaeni, S.S. and Ghaemi, N. (2007) Characterization of self-cleaning RO membranes coated with TiO_2 particles under UV irradiation. *Journal of Membrane Science*, **303**, 221–233.

Manning, B.A., Hunt, M.L., Amrhein, C. and Yarmoff, J.A. (2002) Arsenic (III) and arsenic (V) reactions with zerovalent iron corrosion products. *Environmental Science & Technology*, **36**, 5455–5461.

Marmier, N., Delisée, A. and Fromage, F. (1999) Surface complexation modeling of Yb(III), Ni(II), and Cs(I) sorption on magnetite. *Journal of Colloid and Interface Science*, **201**, 54–60.

Masscheleyn, P.H., Delaune, R.D and Patrick Jr., W.P. (1991) Effect of redox potential and pH on arsenic speciation and solubility in a contaminated soil. *Environmental Science & Technology*, **25**, 1414–1419.

Matheson, L.J. and Tratnyek, P.G. (1994) Reductive dehalogenation of chlorinated methanes by iron metal. *Environmental Science & Technology*, **28**, 2045–2053.

McGuire, M.J. Ferguson, D.W. and Gramith, J.T. (1990) Overview of ozone technology for organics control and disinfection. Conference proceedings, AWWA Seminar on Practical Experiences with Ozone for Organics Control and Disinfection, Cincinnati, Ohio, USA.

Meng, X.G., Bang, S.B. and Korfiatis, G.P. (2000) Effect of silicate, sulfate and carbonate on arsenic removal by ferric hydroxide. *Water Research*, **34**, 1255–1261.

Meng, X.G., Korfiatis, G.P., Bang, S.B. and Bang, K.W. (2002) Combined effects of anions on arsenic removal by iron hydroxides. *Toxicology Letters*, **133**, 103–111.

Miehr, R., Tratnyek, M.M., Bandstra, J.Z., Scherer, M.M., Alowitz, M.J. and Bylaska, E.J. (2004) Diversity of contaminant reduction reactions by zerovalent iron: Role of the reductate. *Environmental Science &*

Technology, **38**, 139–147.

Mishra, S., Dwivedi, S.P. and Singh, R.B. (2010) A review on epigenetic effect of heavy metal carcinogens on human health. *The Open Nutraceuticals Journal*, **3**, 188–193.

Mohan, D. and Pittman, C.U. Jr. (2007) Arsenic removal from water/wastewater using adsorbents – a critical review. *Journal of Hazardous Materials*, **142**, 1–53.

Mondal, P., Majumder, C.B. and Mohanty, B. (2006) Laboratory based approaches for arsenic remediation from contaminated water: recent developments. *Journal of Hazardous Materials*, **B137**, 464–479.

Moore, J.N., Walker, J.R. and Hayes, T.H. (1990) Reaction scheme for the oxidation of As(III) to arsenic(V) by birnessite. *Clays and Clay Minerals*, **38**, 549–555.

Mosharraf Hossain, M., Nazmul Islam, K.M. and Rahman, I.M.M. (2012) An Overview of the Persistent Organic Pollutants in the Freshwater System. Pp. 455–470 in: *Ecological Water Quality – Water Treatment and Reuse* (K. Voudouris, editor). InTech, Available from: http://www.intechopen.com/books/ecological-water-quality-water-treatment-and-reuse/an-overview-of-the-persistent-organic-pollutants-in-the-fresh-water-system

Muftikian, R., Fernando, Q. and Korte, N. (1995) A method for the rapid dechlorination of low-molecular-weight chlorinated hydrocarbons in water. *Water Research*, **29**, 2434–2439.

National Research Council (1999) *Arsenic in Drinking Water*. National Academy Press, Washington, D.C.

Nesbitt, H.W., Canning, G.W. and Bancroft, G.M. (1998) XPS study of reductive dissolution of 7Å-birnessite by H_3AsO_3 with constraints on reaction mechanism. *Geochimica et Cosmochimica Acta*, **62**, 2097–2110.

Neumann, A., Sander, M. and Hofstetter, T.B. (2011) Redox properties of structural Fe in smectite clay minerals. Pp. 361–379 in: *Aquatic Redox Chemistry* (P.G. Tratnyek, T.J. Grundl, S.B. Haderlein, editors). ACS Symposium Series, **1071**. Washington, DC.

NSF International (2001) *Environmental Technology Verification Report: Removal of Arsenic in Drinking Water. Hydranautics ESPA2-4040 Reverse Osmosis Membrane Element Module*. NSF 01/20/EPADW395, March 2001.

Nurmi, J.T., Tratnyek, P.G., Sarathy, V., Baer, D.R., Amonette, J.E., Pecher, K., Wang, C., Linehan, J.C., Matson, D.W., Lee Penn, R. and Driessen, M.D. (2005) Characterization and properties of metallic iron nanoparticles: spectroscopy, electrochemistry, and kinetics. *Environmental Science & Technology*, **39**, 1221–1230.

O'Loughlin, E.J., Kelly, S.D., Cook, R.E., Csencsits, R. and Kemner, K.M. (2003) Reduction of uranium(VI) by mixed iron(II)/iron(III) hydroxide (green rust): formation of UO_2 nanoparticles. *Environmental Science & Technology*, **37**, 721–727.

Olson, M.R., Sale, T.C., Shackelford, C.D., Bozzini, C. and Skeean, J. (2012) Chlorinated solvent source-zone remediation via ZVI-Clay soil mixing: 1-Year results. *Ground Water Monitoring & Remediation*, doi: 10.1111/j1745-6592.2012.01391.x

Pal, A., Gin, K.Y.H., Lin, A.Y.C. and Reinhard, M. (2010) Impacts of emerging organic contaminants on freshwater resources: Review of recent occurrences, sources, fate and effects. *Science of the Total Environment*, **8**, 6062–6069.

Pang, S.C., Chin, S.F. and Anderson, M.A. (2007) Redox equilibria of iron oxides in aqueous-based magnetite dispersions: Effect of pH and redox potential. *Journal of Colloid and Interface Science*, **311**, 94–101.

Postma, D., Boesen, C., Kristiansen, H. and Larsen, F. (1991) Nitrate reduction in an unconfined sandy aquifer – water chemistry, reduction processes, and geochemical modeling. *Water Resources Research*, **27**, 2027–2045.

Rao, M.P. and Rao, P.S.C. (1997) Organic pollutants in groundwater: 1. Health effects. *Soil Science Fact Sheet SL-54*, Institute of Food and Agricultural Science, University of Florida, Gainesville, Florida, USA.

Rao, P., Mak, M.S.H., Liu, T., Lai, K.C.K. and Lo, I.M.C. (2009) Effects of humic acid on arsenic(V) removal by zero-valent iron from groundwater with special references to corrosion products analyses. *Chemosphere*, **75**, 156–162.

Raven, K.P., Jain, A. and Loeppert, R.H. (1998) Arsenite and arsenate adsorption on ferrihydrite: kinetics, equilibrium, and adsorption envelopes. *Environmental Science & Technology*, **32**, 344–349.

Reynolds, T.D. and Richards, P.A. (1996) *Unit Operations and Processes in Environmental Engineering, 2nd ed*. PWS Publishing Co., Boston, London, 798 pp.

Richardson, S.D., Thruston, Jr, A.D. and Collette, T.W. (1996) Identification of TiO_2/UV Disinfection Byproducts in Drinking Water. *Environmental Science & Technology*, **30**, 3327–3334.

Riley, R.G., Zachara, J.M. and Wobber, F.J. (1992) *Chemical contaminants on DOE lands and selection of contaminant mixtures for subsurface science research.* U.S. Department of Energy, Washington, D.C.

Ritter, L., Solomom, K.R., Forget, J., Stemeroff, M. and O'Leary, C. (2007) *Persistant Organic Pollutant.* United Nations Environment Programme. Retrieved 2007-09-16.

Roberts, L.C., Hug, S.J., Ruettimann, T., Billah, M., Khan, A.W. and Rahman, M.T. (2004) Arsenic removal with iron(II) and iron(III) waters with high silicate and phosphate concentrations. *Environmental Science & Technology*, **38**, 307–315.

Ruby, C., Gehin, A., Abdelmoula, M., Genin, J.M.R. and Jolivet, J.P. (2003) Coprecipitation of Fe(II) and Fe(III) cations in sulphated aqueous medium and formation of hydroxysulphate green rust. *Solid State Sciences*, **5**, 1055–1062.

Ruby, C., Åssa, R., Gehin, A., Cortot, J., Abdelmoula, M. and Genin, J.M. (2006) Green rusts synthesis by coprecipitation of FeII– FeIII ions and mass-balance diagram. *Comptes Rendus Geoscience*, **338**, 420–432.

Ruyters, S., Mertens, J., Vassilieva, E., Dehandschutte, R.B., Poffijn, A. and Molders, E. (2011) The red mud accident in Ajka (Hungary): Plant toxicity and trace metal bioavailability in red mud contaminated soil. *Environmental Science & Technology*, **45**, 1616–1622.

Santos-Carballal, D., Roldan, A., Grau-Crespo, R. and de Leeuw, N.H. (2014) ADFT study of the structures, stabilitys and redox behaviour of the major surfaces of magnetite Fe_3O_4. *Physical Chemistry Chemical Physics*, **16**, 21082–21095.

Sapkota, A., Anceno, A.J., Baruah, S., Shipin, O.V. and Dutta, J. (2011) Zinc oxide nanorod mediated visible light photoinactivation of model microbes in water. *Nanotechnology*, **22**, 215703.

Sasaki, K., Nukina, S., Wilopo, W. and Hirajima, T. (2008) Removal of arsenate in acid mine drainage by a permeable reactive barrier bearing granulated blast furnace slag: Column study. *Materials Transactions*, **49**, 835–844.

Sayles, G., You, G., Wang, M. and Kupferle, M.J. (1997) DDT, DDD, and DDE dechlorination by zero-valent iron. *Environmental Science & Technology*, **31**, 3448–3454.

Schrick, B., Hydutsky, B.W., Blough, J.L. and Mallouk, T.E. (2004) Delivery vehicles for zerovalent metal nanoparticles in soil and groundwater. *Chemistry of Materials*, **16**, 2187–2193.

Schwarzenbach, R.P., Egli, T., Hofstetter, T.B., von Gunten, U. and Wehrli, B. (2010) Global water pollution and human health. *Annual Review of Environmental Research*, **35**, 109–136.

Schwertmann, U. and Fechter, H. (1994) The formation of green rust and its transformation to lepidocrocite. *Clay Minerals*, **29**, 87–92.

Self, J.R. and Waskom, R.M. (2008) *Nitrates in Drinking Water.* Colorado State University, Fact Sheet N0. 517.

Shiomi, K. (1994) Arsenic in marine organisms: Chemical forms and toxicological aspects. Pp. 261–282 in: *Arsenic in the Environment: Part II, Human Health and Ecosystem Effects* (J.O. Nriagu, editor). John Wiley & Sons, Inc., New York.

Sivavec, T.M., Mackenzie, P.D., Horney, D.P. and Baghel, S.S. (1997) Redox-active media for permeable reactive barriers. Pp. 753–759 in *International Containment Technology Conference and Exhibition*, St. Petersburg, FL, USDOE.

Smedley, P.L. and Kinniburgh, D.G. (2002) A review of the source, behaviour and distribution of arsenic in natural waters. *Applied Geochemistry*, **17**, 517–568.

Stout, J.E. and Yu, V.L. (2003) Experiences of the first 16 hospitals using copper–silver ionization for Legionella control: Implications for the evaluation of other disinfection modalities. *Infection Control Hospital and Epidemiology*, **24**, 563–568.

Stucki, J.W. (2005) Iron redox processes in smectites. Pp. 395–406 in: *Handbook of Clay Science* (F. Bergaya, B.K.G. Theng and G. Lagaly, editors). Elsevier, Amsterdam.

Stucki, J.W. and Getty, P.J. (1986) Microbial reduction of iron in nontronite. *Agronomy Abstracts*, 279.

Stucki, J.W., Golden, D.C. and Roth, C.B. (1984) Preparation and handling of dithionite-reduced smectite suspensions. *Clays and Clay Minerals*, **32**, 191–197.

Stucki, J.W., Komadel, P. and Wilkinson, H.T. (1987) Microbial reduction of structural iron(III) in smectites. *Soil Science Society of America Journal*, **51**, 1663–1665.

Su, C. and Puls, R.W. (2001a) Arsenate and arsenite removal by zerovalent iron: kinetics, redox transformation, and implications for in situ groundwater remediation. *Environmental Science & Technology*, **35**, 1487–1492.

Su, C. and Puls, R.W. (2001b) Arsenate and arsenite removal by zerovalent iron: effects of phosphate, silicate, carbonate, borate, sulfate, chromate, molybdate, and nitrate, relative to chloride. *Environmental Science & Technology*, **35**, 4562–4568.

Su, C. and Puls, R.W. (2003) In situ remediation of arsenic in simulated groundwater using zerovalent iron: laboratory column tests on combined effects of phosphate and silicate. *Environmental Science & Technology*, **37**, 2582–2587.

Su, K., Radian, A., Mishael, Y., Yang, L. and Stucki, J.W. (2012) Nitrate reduction by redox-activated polydiallyldimethylammonium-exchanged ferruginous smectite. *Clays and Clay Minerals*, **60**, 464-472.

Subramanian, K.S., Viraraghavan, T., Phommavong, T. and Tanjore, S. (1997) Manganese greensand for removal of arsenic in drinking water. *Water Quality Research Journal Canada*, **32**, 551–561.

Sun, G. (2004) Arsenic contamination and arsenicosis in China. *Toxicology and Applied Pharmacology*, **198**, 268–271.

Sun, Z.X., Su, F.W., Forsling, W. and Samskog, P.O. (1998) Surface characteristics of magnetite in aqueous suspension. *Journal of Colloid and Interface Science*, **197**, 151–159.

Sun, H., Wang, L., Zhang, R., Sui, J. and Xu, G. (2006) Treatment of groundwater polluted by arsenic compounds by zero valent iron. *Journal of Hazardous Materials*, **B129**, 297–303.

Suttiponparnit, K., Jiang, J., Sahu, M., Suvachittanont, S., Charinpanitkul, T. and Biswas, P. (2011) Role of surface area, primary particle size, and crystal phase on titanium dioxide nanoparticle dispersion properties. *Nanoscale Research Letters*, **6**, 27–35.

Tebo, B.M., Bargar, J.R., Clement, B.G., Dick, G.J., Murray, K.J., Parker, D., Verity, R. and Webb, S.M. (2004) Biogenic manganese oxides: properties and mechanisms of formation. *Annual Review of Earth and Planetary Sciences*, **32**, 287–328.

Tchounwou, P.B., Yedjou, C.G., Patlolla, A.K. and Sutton, D.J. (2012) Heavy metals toxicity and the environment. *Molecular, Clinical and Environmental Toxicology – Experientia Supplementum*, **101**, 133–164.

Tratnyek, P.G., Scherer, M.M., Johnson, T.J. and Matheson, L.J. (2003) Permeable reactive barriers of iron and other zerovalent metals. Pp. 371–421 in: *Chemical Degradation Methods for Wastes and Pollutants; Environmental and Industrial Applications. Environmental Science and Pollution Control* (M.A. Tarr, editor). Marcel Dekker, New York.

Till, B.A., Weathers, L.I. and Alverez, P.J.J. (1998) Fe(0)-supported autotrophic denitrification. *Environmental Science & Technology*, **32**, 634–639.

Tournassat, C., Charlet, L., Bosbach, D. and Manceau, A. (2002) Arsenic(III) oxidation by birnessite and precipitation of manganese(II) arsenate. *Environmental Science & Technology*, **36**, 493–500.

Tyrovola, K., Nikolaidis, N.P., Veranis, N., Kallithrakas-Kontos, N. and Koulouridakis, P.E. (2006) Arsenic removal from geothermal waters with zero-valent iron – effect of temperature, phosphate and nitrate. *Water Research*, **40**, 2375–2386.

UNEP (2010) *Stockholm Convention on Persistent Organic Pollutants*. Available from: http://chm.pops.int/Convention/tabid/54/language/en US/Default.aspx#convtext, 2010.

USEPA (1998) *Permeable Reactive Barrier Technologies for Contaminant Remediation*. EPA 600/R-98/125, DC 20460: U.S. Environmental Protection Agency, Washington.

USEPA (2003) *Capstone Report on the Application, Monitoring, and Performance of Permeable Reactive Barriers for Ground-water Remediation; Volume 1, Performance Evaluations at Two Sites*. USEPA National Risk Management Research Laboratory, EPA/600/R-03/045a, Cincinnati, Ohio, USA.

USEPA (2008a) *Nanotechnology for Site Remediation Fact Sheet. Solid Waste and Emergency Response: Fact Sheet*. National Service Center for Environmental Publications.

USEPA (2008b) *Office of Superfund, Remediation and Technology Innovation. Nanotechnology: Practical Considerations for Use in Groundwater Remediation*. National Association of Remedial Project Managers Annual Training Conference. Portland, Oregon, USA.

USEPA (2008c) *Selected Sites Using or Testing Nanoparticles for Remediation*. https://clu-in.org/download/

remed/nano-site-list.pdf

Vaughan, D.J. (2006) Arsenic. *Elements*, **2**, 71–75.

Vu, K.B., Kaminski, M.D. and Nue, L. (2003) Review of arsenic removal technologies for contaminated groundwaters. Report ANL-CMT-03/2, United States Department of Energy, USA. www.ipd.anl.gov/anlpubs/2003/05/46522.pdf

Wang, P. and Liu, D. (2012) Physical and chemical properties of sintering red mud and Bayer red mud and the implications for beneficial utilization. *Materials*, **5**, 1800–1810.

Welch, A.H. and Stollenwerk, K.G. (2003) *Arsenic in Ground Water: Geochemistry and Occurrence.* Kluwer Academic Publishers, New York, Boston, Dordrecht, London, Moscow.

Westerhoff, P. (2003) Reduction of nitrate, bromate and chlorate by zero valent iron (Fe^0). *Journal of Environmental Engineering*, **129**, 10–16.

White, A. and Peterson, M. (1996) Reduction of aqueous transition metal species on the surfaces of Fe(II)-containing oxides. *Geochimica et Cosmochimica Acta*, **60**, 3799–3814.

Wilkin, R.T., Acree, S.D., Ross, R.R., Beak, D.G. and Lee, T.R. (2009) Performance of a zerovalent iron reactive barrier for the treatment of arsenic in groundwater: Part 1. Hydrogeochemical studies. *Journal of Contaminant Hydrology*, **106**, 1–14.

WHO (2001) *Arsenic and arsenic compounds. Environmental Health Criteria 224.* International Programme on Chemical Safety, Geneva, Switzerland.

WHO (2003a) *Arsenic, Drinking-water and Health Risk Substitution in Arsenic Mitigation: A Discussion Paper.* SDE/WSH/03.06. Prepared by G. Howard, Loughborough University, UK.

WHO (2003b) *Silver in Drinking-water.* Background document for development of WHO Guidelines for Drinking-water Quality.

WHO (2011) *Guidelines for Drinking-water Quality: Nitrate and nitrite in drinking-water*, 4th edition. World Health Organization, Geneva, Switzerland.

Yan, W., Lien, H., Koel, B.E. and Zhang, W. (2013) Iron nanoparticles for environmental clean-up: recent developments and future outlook. *Environmental Science: Processes and Impacts*, **15**, 63–77.

Yavuz, C.T., Mayo, J.T., Suchecki, C., Wang, J., Ellsworth, A.Z., D'Couto, H., Quevedo, E., Prakash, A., Gonzalez, L., Nguyen, C., Kelty, C. and Colvin, V.L. (2010) Pollution magnet: nano-magnetite for arsenic removal from drinking water. *Environmental Geochemistry and Health*, **32**, 327–334.

Yean, S., Cong, L., Yavuz, C.T., Mayo, J.T., Yu, W.W., Colvin, V.L. and Tomson, M.B. (2005) Effect of magnetite particle size on adsorption and desorption of arsenite and arsenate. *Journal of Materials Research*, **20**, 3255–3264.

Yoon, Y., Westerhoff, P., Snyder, S.A. and Esparza, M. (2003) HPLC-fluorescence detection and adsorption of bisphenol A, 17β-estradiol, and 17α-ethynylestradiol on powdered activated carbon. *Water Research*, **37**, 3530–3537.

Young, G.K., Bungay, H.R., Brown, L.M. and Parsons, W.A. (1964) Chemical reduction of nitrate in water. *Journal of Water Pollution Control Federation*, **36**, 395–398.

Yu, J. and Yu, D. (2008) Hydrothermal synthesis and photocatalytic activity of zinc oxide hollow spheres. *Environmental Science & Technology*, **42**, 4902–4907.

Yu, Z.R., Peldszus, S. and Huck, P.M. (2008) Adsorption characteristics of selected pharmaceuticals and an endocrine disrupting compound – naproxen, carbamazepine and nonylphenol – on activated carbon. *Water Research*, **42**, 2873–2882.

Zhang, X., Zhou, M. and Lei, L. (2005) Preparation of an Ag-TiO_2 photocatalyst coated on activated carbon by MOCVD. *Materials Chemistry and Physics*, **91**, 73–79.

Zhu, H., Jia,Y., Wu, X. and Wang, H. (2009) Removal of arsenic from water by supported nano zero-valent iron on activated carbon. *Journal of Hazardous Materials*, **172**, 1591–1596.

Zhu, N., Luan, H., Yuan, S., Chen, J., Wu, X. and Wang, L. (2010) Effective dechlorination of HCB by nanoscale Cu/Fe particles. *Journal of Hazard Materials*, **176**, 1101–1105.

Zhu, N., Yi-Li and Zhang, F. (2011) Catalytic dechlorination of polychlorinated biphenyls in subcritical water by Ni/Fe nanoparticles. *Chemical Engineering Journal*, **171**, 919–925.

Subject and Author index

Numbers refer to the page where a definition or explanation of, and/or a *figure* or a **table** for, a given subject is found. Author names are given with the number of the first page of the paper in which they are involved.

A

Abiotic systems, lime addition and limestone drains, 366
Acero, P., 273
Activation barriers in mineral redox reactions, 67
Adsorption mechanisms of As(III) on 6 nm maghemite, 9
Aeration process for water treatment, *417*
Ag-Cu ionization system for control of Legionella in hospitals, 431
Ahmed, I.A.M., 1, 273
Anaerobic compost reactors, 370
Anionic species, inserted into pores, 386
Anthracene, reacted with magnetite, *345*
Anthraquinone, reacted with magnetite, *345*
Archaeological artefacts, Mössbauer and Raman spectroscopy, 47
Arsenic removal from water, 414, **415**
Arsenic(V), reduction by Fe, 210
Arsenopyrite oxidation, 108
Auqué, L.F., 273

B

Bacteria, Fe(III) reduction, 200, formation of manganese oxide minerals by, 173, removal, redox-reactive compounds used for water disinfection, 431
Bacteriogenic Mn oxides, 184
Band gap, *10*
Banded iron formations (BIF), 130, trace element enrichment and isotope records, *153*
Bioconversion of Fe(III), 197
Biogenic Mn oxides, *187*
Biogeochemical, evolution, iron minerals as archives of Earth's, 121, modelling of redox processes in low-*T* natural systems, 273, redox processes of sulfide minerals, 95
Biological processes, formation of sulfide minerals through, 99, in redox reactions, 292
Biomedical applications, magnetic nanoparticles, 217
Biominerals, applications of magnetic, 207
Bioreduction, 25
Black shales, trace-element concentrations, 156
Bogush, A.A., 405
Bromophenoxy radicals, oxidative coupling of, *339*
Buffer material, waste package and buffer material, interface between, 238

C

Campos, L.C., 405
Catalysis of redox reactions, *23*
Catalysts, MOFs as, 393
Chalcopyrite oxidation, 109
Chanéac, C., 5
Chemical, information, iron minerals as archives of, 122, kinetics analysis, of mineral redox reactions, 65, precipitation process for water treatment, *416*
Chlorinated-solvent source-zone remediation, 430
Chromium(VI), reductive transformation by Fe(II)-bearing phases, 208
Chromium, 144
Clarke, C.E., 313
Clean technologies, application in, 113, redox-reactive minerals in, 1
Coker, V.S., 95, 197
Compost bioreactors, 368
Conduction band, *10*
Contaminant-magnetite interactions, 208
Corrosion behaviour of Fe-bearing waste canisters, 236
Crystal structure, influence of particle size of iron oxides on, 8
Cultural heritage, characterization of iron-bearing minerals, 33

D

Degradation of organic matter, simulation, *291*
Devic, T., 379
Dissolution of sulfide minerals, 104
DOC, impact of, on oxidation and hydrolysis of contaminants by Mn oxide, *326*

DOI: 10.1180/EMU-notes.17.index

E

Earth's biogeochemical evolution, 121
Eh-pH, diagram for the Mn-H_2-CO_2 system, *323*, stability diagrams, 278
Electrochemical, characterization of MOFs, 382, methods for removing nitrate from drinking water, 425, methods, time-resolved, 83
Electrodes for Li-ion batteries, MOFs as, 390
Electronic conductivity, coordination polymers, 392
Electronic structure of a solid, *10*
Elements in soils, reduction potentials of half-reactions of, **322**
Enzymatic and non-enzymatic oxidation, 184
Equilibrium, calculations, 284, redox potentials, 273
EXAFS, study of bacteriogenic Mn oxides, 185

F

Fe phases, radionuclide interaction with, *234*
Fe(II) adsorption, redox transformations induced by, 18
Fe(II) oxidation, 201
Fe(III) reduction, 198
Fe-based MOF, *381*
Fe-bearing, minerals, in candidate far-field lithologies, 242, minerals in GDF host rocks, **244**, minerals in the removal of nitrate from drinking water, 421, waste canisters, corrosion behaviour of, 236
Felmy, A.R., 229
Ferromanganese crusts, 131, 155
Fractionation theory, stable isotope, 134, *138*
Free energy diagrams for thermally activated chemical reaction processes, 68
Free energy relationships, 73
Free enthalpy of transformation of iron oxides, *7*
FTIR, study of bacteriogenic Mn oxides, 185

G

Galena oxidation, 107
Gas sorption isotherm, *388*
Gas storage or capture (enhanced), 387
Gázquez, F., 33
Geochemical modelling of redox transformations, 295
Geochemical reaction kinetics, 58
Geological disposal facility (GDF), **231**
Gibbs free energy variation, for different respirative pathways, *293*
Gilbert, B., 55
Glyphosate, possible reaction mechanism between Mn oxide and, *338*
Goethite, *13*, nanorods, dissolution of, *17*
Green rust removal, *367*
Greigite, oxidation of Fe, 204
Groundwater systems, redox zones in, *296*

H

Health effects of silver ionization, 432
Hematite, *13*, electronic structure, *21*
Heme peroxidases, animal, Mn(II) oxidation mediated by Ca^{2+}-binding, 182
Historical and cultural heritage, 46
Hollandite, Mn oxide with tunnel structure, *185*
Hudson-Edwards, K.A., 1, 357
Hydrolysis of contaminants by Mn oxide, *326*
Hydrous ferric oxides, properties of, **305**
Hydroxides, free energies of transformation of, *57*

I

In situ treatment technologies, 421
Industrial reuse of mine wastes, 360
Interfacial, electron transfer, 73, electron transfer process, *20*
Iodides, inserted into MOFs, *385*
Iron isotope composition of natural materials, *141*
Iron minerals, as archives of Earth's redox and biogeochemical evolution, 121, cycling, pathways of, *142*
Iron oxide structure, influence on redox activity, 5
Iron oxides, free energies of transformation of, *57*
Iron oxyhydroxides, 124
Iron redox chemistry, impact on nuclear waste disposal, 229
Iron sulfide formation, 102, 126
Iron-based records of Earth's redox and biogeochemical evolution, 145

J

Jaroso Ravine, Mars analogue site, 41
Johnson, K.L., 313
Jolivet, J.-P., 5
Jones, M., 173

K

Kim, J.K., 405
Kinetic assumption, *vs.* redox equilibria, 295
Kinetic oxidation of ferrous iron, *301*
Klingelhöfer, G., 33
Konhauser, K., 121
Kossoff, D., 357

L

Lalonde, S.V., 121
Lee, S.-W., 173
Lithium ions inserted into MOFs, *385*
Lloyd, J.R., 197

M

Mackinawite, uptake by, 111
Maghemite, oxidation of magnetite to, *16*
Magnetic minerals, tailoring, 206
Magnetic nanoparticles, bioconversion of Fe(III) to, 197
Magnetite, 13, adsorption onto, size-dependent reduction, 24, nanoparticles, 79, oxidation of, *16*, oxidation of Fe, 204
Manganese oxide minerals, formation of, by bacteria, 173, **176**
McCann, C.M., 313
MCO (multi-copper oxidase), 178
Mechanistic descriptions of mineral redox reactions, 61
Medina, J., 33
Mercury(II), reduction by Fe, 210
Metal reduction, microbial, 101
Metal–organic framework (MOF) structures, redox-active, 379
Microbes, pathogenic, removal from water, 405
Microbial activity, effect of, on Fe-bearing phases in the near field, 241
Microbial metal reduction, 101
Microbial sulfate reduction, 100
MIMOS II, Mössbauer instrument, 38
Mine wastes, redox-reactive minerals, in the reuse and remediation of, 357
Mineral redox reactions, logarithmic timescale, 59
Mineral-reaction intermediates, chemical dynamics methods for, 80
Mn cycle, *323*
Mn oxide, possible reaction mechanism between glyphosate and, *338*
Mn oxides and natural organic matter, 321
Mn tailings, oxidation of AO7 by, *343*
Mnx protein complex, Mn(II) binding and oxidation by, *179*
Mo isotope compositions of diverse Precambrian formations, 145
Mo isotope compositions of marine sediments , *143*
Mo isotopes in shales, 158
Molecular modelling, radionuclide interaction with redox-active Fe components, 258
Molybdenum, the most abundant transition metal in seawater, 128
Morphologies of iron oxides, *13*
Mössbauer spectra of maghemite, *19*
Mössbauer spectroscopy, characterization of iron-bearing minerals in Mars, 33
Multi-copper oxidase (MCO), 178
Multi-electron redox reactions, 76

N

Nanoparticles, magnetic, bioconversion of Fe(III) to, 197
Natural organic matter, soil contaminant, 325
Natural systems, redox-reactive minerals in, 1
N-containing, aromatic compounds, soil contaminants, 319, compounds, reacting with birnessite, 340, species, log-transformed activities of, **278**
Near field, redox-active components in the, 236
Neptunium(V), reduction by Fe, 214
Neutral species, inserted into pores, 386
Nitrate removal from water, 423
Non-traditional stable isotope systems, 140
Nuclear waste disposal, impact of iron redox chemistry on, 229

O

Opalinus clay, **247**
Opportunity, Mars Rover, 34
Organic compounds, reduction by Fe, 215
Organic contaminants, breakdown of, in soil, by Mn oxides, 313
Organic pollutants, persistent, removal from water, 427
Organically contaminated land, **316**
Oxidants used in water treatment, **417**
Oxidases, Mn(II), 177
Oxidation and adsorption, in the treatment of polluted water, 418, for persistent organic pollutants, 428
Oxidation of contaminants by Mn oxide, *326*, **330**
Oxidation of magnetite, to maghemite, *16*
Oxidized, iron nanoparticles, *78*, polypyrrole inserted into MOFs, *385*
Oxyhydroxides, free energies of transformation of, *57*

P

Particle reactivity, influence of particle size of iron oxides on, 6
Partitioning coefficients of phosphates, *152*
Pathogenic microbes, removal from water, 405
Pattrick, R.A.D., 229

Peacock, C., 121
Pearce, C.I., 229
Permeable reactive barrier system, 368, *369*, *407*
Phenolic monomers, oxidative transformation of, *337*
Phenols, as soil contaminants, 318, oxidation of by Mn oxides, 335
Photocatalysis and redox reactions, 429
Photo-induced redox activity, 395
Photostabilization of mine waste, 363
pH-pe diagram showing approximate positions of some natural environments, *280*
Phylonetic tree of domain bacteria, *175*
Plant nutrients, mine wastes as a source of, 362
Plutonium(V) and (VI), reduction by Fe, 214
Polychlorinated biphenyls (PCBs), soil contaminants, 319
Polycyclic aromatic hydrocarbons (PAH), soils contaminated with, 320, **321**, 342
Primary and secondary mineral phases, 203
Proton adsorption, reactivity towards, 15
Pump probe method for studying reactions, 80
Pyrite, oxidation, 105, uptake by, 110

R

Radiation damage, effect of, on Fe-bearing phases in the near field, 241
Radical species, formation of, 76
Radionuclide interaction with redox-active Fe components, 251
Radionuclides relevant to geological disposal of high-activity wastes, half-lives of, **233**
Raman spectrometer, ExoMars 2018 mission, 39
Raman spectroscopy, characterization of iron-bearing minerals in Mars, 33
Ramsdellite, Mn oxide with tunnel structure, *185*
Rates of multistep reactions, 75
Reaction balance, *19*
Reactions from the main respiratory pathways of organic matter, **288**
Recommended reading, 3
Redox, catalysis, 22, conditions in a GDF, post-closure, 234, cycling, 198, evolution, iron minerals as archives of, 121, iron mineralogy as a function of, 122, potentials in natural systems, 278, processes, biogeochemical modelling of, in low-*T* natural systems, 273, reactions, timescales of, 55
Redox-active, Fe components, radionuclide interaction with, 251, ligands, *382*, minerals, 1, minerals, in the reuse and remediation of mine-wastes, 357, minerals, removal of As, organic pollutants and pathogenic microbes from water, 405, minerals, properties of, 407, **408**, porous metal–organic framework structures, 379,
Reduction potentials for iron redox phases, **60**
Reductive dissolution of 6-line ferrihydrite nanoparticles, 70
Remediation, of mine wastes, 357, mechanisms facilitated by biogenic magnetite, *208*
Río Tinto, Mars analogue site, 41
Rock paintings in the field, Mössbauer spectroscopy, 48
Romano, C., 173
Rosso, K.M., 229
Rull, F., 33

S

Schematic adsorption mechanisms of As(III) on 6 nm maghemite, *9*
Sedimentary porewater redox profile, *323*
Sedimentary sulfide minerals, iron isotope compositions of, 145
Serre, C., 379
Silicate glass alteration enhanced by iron, *238*
Silver nanoparticles inserted into MOFs, *385*
Soil contamination, 313
Spain, Altamira Cave, rock paintings, 48
Stable isotope, fractionation, of trace elements, 133, nomenclature and conventions, 134
Structural defects, influence of particle size of iron oxides on, 6
Structures and compositions, of sulfide minerals, 95, **97**, *98*
STXM, study of bacteriogenic Mn oxides, 185
Sulfidic bioreactors, 368
Sulfur cycle, under biomining conditions, *114*
Supercapacitors, MOFs as, 392
Synthesis, of MOFs, 382
Synthetic analogues, radionuclide interaction with redox-active Fe components, 251

T

Tebo, B.M., 173
Technetium(VII), reduction by Fe, 213
Thermodynamic, basis for mineral redox reactions, 60, modelling, radionuclide interaction with redox-active Fe components, 258, stability, influence of particle size of iron oxides on, 6
Timescales, of redox reactions, 55
Todorokite, Mn oxide with tunnel structure, *185*
Toxic metals bound to sulfide, 109
Trace elements, uptake of, by iron minerals, 124

U

Uranium(VI), reduction by Fe, 211

V

Valence band, *10*
Vaughan, D.J., 95
Venegas, G., 33

W

Waste (nuclear) package, *232*
Waste package and buffer material, interface between, 238
Water purification, tailings as a possible contaminant absorbent filter, 363
Water treatment, oxidants used in, **417**
Watts, M.P., 197
Waychunas, G.A., 55

X

XANES spectra from freeze-dried plant roots, *364*
XANES, study of bacteriogenic Mn oxides, 185
XAS, study of bacteriogenic Mn oxides, 185

www.ingramcontent.com/pod-product-compliance
Ingram Content Group UK Ltd.
Pitfield, Milton Keynes, MK11 3LW, UK
UKHW050923290726
14058UKWH00011B/685